GLOBAL PERSPECTIVES ON RIVER CONSERVATION

Science, Policy and Practice

Palmiet River, flowing through the Kogelberg Biosphere Reserve, south-western Cape, South Africa, 2 km above its estuary. The river flows through the smallest (the size of Wales in the UK) but most diverse plant Kingdom in the World – the Fynbos Biome. Its conservation is vital to this core community of fire-adapted plants, but threats include over-abstraction of water for Cape Town, several regulating dams, an interbasin transfer, alien invasive plants, and pesticide pollution from fruit-farming activities.

GLOBAL PERSPECTIVES ON RIVER CONSERVATION

Science, Policy and Practice

Edited by

P.J. Boon, B.R. Davies and G.E. Petts

John Wiley & Sons, Ltd

Chichester • New York • Weinheim • Brisbane • Singapore • Toronto

Copyright © 2000 by John Wiley & Sons Ltd,
Baffins Lane, Chichester,
West Sussex PO19 1UD, England

National   01243 779777
International (+44) 1243 779777
e-mail (for orders and customer service enquiries): cs-books@wiley.co.uk
Visit our Home Page on http://www.wiley.co.uk or http://www.wiley.com

Other Wiley Editorial Offices

John Wiley & Sons, Inc., 605 Third Avenue,
New York, NY 10158-0012, USA

WILEY-VCH Verlag GmbH, Pappelallee 3,
D-69469 Weinheim, Germany

Jacaranda Wiley Ltd, 33 Park Road, Milton,
Queensland 4064, Australia

John Wiley & Sons (Asia) Pte Ltd, 2 Clementi Loop #02-01,
Jin Xing Distripark, Singapore 129809

John Wiley & Sons (Canada) Ltd, 22 Worcester Road,
Rexdale, Ontario M9W 1L1, Canada

Library of Congress Cataloging-in-Publication Data

Global perspectives on river conservation : science, policy, and practice / edited by P.J.
Boon, B.R. Davies, and G.E. Petts.
        p.   cm.
    Includes bibliographical references.
    ISBN 0-471-96062-4
    1. Stream conservation.   I. Boon, P. J.   II. Davies, B. R. (Bryan Robert)   III. Petts,
Geoffrey E.
QH75.G59 2000
333.91'6216—dc21   00-029014

British Library Cataloguing in Publication Data

A catalogue record for this book is available from the British Library

ISBN 0-471-96062-4

Typeset in 9/11pt Times by Mayhew Typesetting, Rhayader, Powys
Printed and bound in Great Britain by Bookcraft (Bath) Ltd, Midsomer Norton
This book is printed on acid-free paper responsibly manufactured from sustainable
forestry, in which at least two trees are planted for each one used for paper production.

CONTENTS

LIST OF CONTRIBUTORS

First named/correspondence authors are indicated by an asterisk.

J.D. Allan School of Natural Resources and Environment, University of Michigan, Ann Arbor, MI 48109-1115, USA

*K.E. Baer Upper Chattahoochee Riverkeeper, Atlanta, GA 303118, USA

A.C. Benke Aquatic Biology Program, Department of Biological Sciences, Box 870206, University of Alabama, Tuscaloosa, AL 35487-0206, USA

*J.P. Benstead Institute of Ecology, University of Georgia, Athens, GA 30602, USA

J. Bøgestrand National Environmental Research Institute, Vejlsøvej 25, DK-8600 Silkeborg, Denmark

P.J. Boon Scottish Natural Heritage, 2 Anderson Place, Edinburgh, EH6 5NP, UK

B. Bose School of Water Resources Engineering, Jadavpur University, Calcutta – 700032, India

*A.J. Boulton Division of Ecosystem Management, University of New England, Armidale, NSW 2350, Australia

I. Campbell Department of Biological Sciences, Monash University, Clayton 3168, Victoria, Australia

S. Choowaew Faculty of Environmental and Resource Studies, Mahidol University, Salaya, Nakornpathom 73170, Thailand

K.J. Collier National Institute of Water and Atmospheric Research, Hamilton, New Zealand

*B.R. Davies Freshwater Research Unit, University of Cape Town, Rondebosch, 7701, South Africa

*D. Dudgeon Department of Ecology and Biodiversity, The University of Hong Kong, Pokfulam Road, Hong Kong

H.T. El-Zanfaly National Research Centre, Water Pollution Control Department, Dokki, Cairo, Egypt

J. Gagneur Laboratoire d'Hydrobiologie, Université Paul Sabatier, 31062 Toulouse Cedex, France

*J.L. Gardiner Philip Williams & Associates Ltd, Pacific North-west Office, 322 NW 5th Avenue, Portland, OR 97209, USA

*B. Gopal School of Environmental Sciences, Jawaharlal Nehru University, New Delhi, 110067, India

A.B. Goswami School of Water Resources Engineering, Jadavpur University, 89/2 Maharani Indira Devi Road, Pallisree, Calcutta – 700060, India

D.M. Harper Department of Biology, University of Leicester, Leicester, LE1 7RH, UK

S.-C. Ho School of Biological Sciences, Universiti Sains Malaysia, 11800 Penang, Malaysia

*B.P. Hooper Department of Geographical Sciences and Planning, The University of Queensland, Brisbane 4072, Queensland, Australia

*T.M. Iversen   National Environmental Research Institute, Vejlsøvej 25, DK-8600 Silkeborg, Denmark

*J.R. Karr   University of Washington, Box 355020, 222A Fishery Sciences, Seattle, WA 98195-5020, USA

*P.A. Khaiter   Rostov State Economic Academy, 69 Bolshaya Sadovaya St., Rostov-on-Don, 344104, Russia

*L. Li   Institute of Geography, Chinese Academy of Sciences, Building 917, Da Fun Road, Beijing, 100100, China

C. Liu   Institute of Geography, Chinese Academy of Sciences, Building 917, Da Fun Road, Beijing, 100100, China

P.V. Loiselle   Aquarium for Wildlife Conservation, Surf Avenue and West 8th Street, Brooklyn, NY 11224, USA

B.L. Madsen   Danish EPA, Strandgade 29, DK-1401 Copenhagen K, Denmark

M. Meador   United States Geological Survey, Water Resources Division, Raleigh, NC, USA

R.D. Margerum   Integrated Resource Management Research Pty Ltd, 126 Hawken Drive, St Lucia 4067, Australia

H. Mou   Institute of Geography, Chinese Academy of Sciences, Building 917, Da Fun Road, Beijing, 100100, China

A.M. Nikanorov   Hydrochemical Institute, 198 Stachki Avenue, Rostov-on-Don, 344104, Russia

M. Núñez-Ferrera   MeRida Incorporated, Apartado 276-1250, Escazu, Costa Rica

*J.H. O'Keeffe   Institute for Water Research, Rhodes University, Grahamstown 6140, South Africa

J. Oldfield   School of Geography and Environmental Sciences, University of Birmingham, Edgbaston, Birmingham, B15 2TT, UK

P. Paaby-Hansen   MeRida Incorporated, Apartado 276-1250, Escazu, Costa Rica

*N. Pacini   Via B.P. D'Arezzo 96, 04020, Itri, Italy

*C.G. Palmer   Institute for Water Research, Rhodes University, Grahamstown 6140, South Africa

B. Peckham (Deceased)   Formerly of Department of Law, Rhodes University, Grahamstown 6140, South Africa

N.C. Perala-Gardiner   Philip Williams & Associates Ltd, Pacific North-west Office, 322 NW 5th Avenue, Portland, OR 97209, USA

*G.E. Petts   School of Geography and Environmental Sciences, University of Birmingham, Edgbaston, Birmingham, B15 2TT, UK

K. Prach   Institute of Botany, Czech Academy of Sciences, Dukelska 145, CZ-379 82, Trebon, Czech Republic

*C.M. Pringle   Institute of Ecology, University of Georgia, Athens, GA 30602, USA

J. Quinn   National Institute of Water and Atmospheric Research, Hamilton, New Zealand

N. Raminosoa   Université d'Antananarivo, Service de Zoologie, BP 901 101, Antananarivo, Madagascar

K.J. Riseng   Department of Biology, University of Michigan, Ann Arbor, MI 48109, USA

F.N. Scatena   International Institute of Tropical Forestry, Rio Piedras, Puerto Rico

*N.J. Schofield   Land and Water Resources R&D Corporation, Canberra, ACT, Australia

F. Sheldon   Co-operative Research Centre for Freshwater Ecology, University of Canberra, ACT, Australia

*K.B. Showers   African Studies Center, Boston University, 270 Bay State Road, Boston, MA 02215, USA

C.D. Snaddon   Freshwater Research Unit, University of Cape Town, Rondebosch, 7701, South Africa

F. Soltau   Department of Public Law, University of Cape Town, Rondebosch, 7701, South Africa

R. Sparks   Illinois Water Resources Center, University of Illinois, 278 Environmental and Agricultural Sciences Building, MC-635, 1101 West Peabody Drive, Urbana, IL 61801, USA

*E.H. Stanley   Center for Limnology, University of Wisconsin, 680 N. Park St, Madison, WI 53706-1492, USA

M.L.J. Stiassny   Department of Herpetology and Ichthyology, American Museum of Natural History, 79th Street at Central Park West, New York, NY 10024, USA

M.C. Thoms   Co-operative Research Centre for Freshwater Ecology, University of Canberra, ACT, Australia

M. Uys   Institute for Water Research, Rhodes University, Grahamstown 6140, South Africa

A. Vadineanu   UNESCO-Cousteau Chair of Ecology and Environmental Management, University of Bucharest, Bucharest, Romania

*M.J. Wishart   Freshwater Research Unit, University of Cape Town, Rondebosch, 7701, South Africa

M.G. Yereschukova   Hydrochemical Institute, 198 Stachki Avenue, Rostov-on-Don, 344104, Russia

INTRODUCTION

RIVER CONSERVATION: A GLOBAL IMPERATIVE

B.R. Davies, P.J. Boon and G.E. Petts

"... I was looking at a river bed. And the story it told of a river that flowed made me sad to think it was dead."

From the song 'Horse with no name' by Dewey Bunnell. America

The background

This book has been long in the making. Intended as a sequel to *River Conservation and Management* (Boon *et al.*, 1992), which sprang from a conference held at York University in 1990, the first discussions for this volume began as long ago as 1994. Its production has not been an easy undertaking, with many problems of selection (authors/coverage/topics), conceptualization, identification of appropriate markets and style and, of course, the editing and writing processes (responsibilities, time frames, correspondence with authors). On the way we have had to overcome many barriers of language, perspectives and size. The result, we hope, is a co-operative statement on humanity's 'love–hate' relationship with that most diverse and globally vital collection of ecosystems that we generally refer to as 'rivers'.

Perhaps, as we enter the new Millennium it is appropriate to take stock, not just of the problems of 'Y2K', but also of the effects of 'Y6G': that point at which our population reached 6 billion (using computer terminology: 6-giga). In fact, according to recent statements by the United Nations and World Health Organisation, the Y6G mark was passed as we prepared this introduction in October 1999.

Ironically, a population of such magnitude will,

more than likely, cause far less general concern than Y2K itself, although its arrival is far more significant in ecological terms. The stresses on global resources are already legion, but that most precious of resources – fresh, potable water – provided in the main by the rivers of the planet, is in our opinion the most crucial of all. In this context, the pages of this book contain a myriad facts and figures that lead one to the inevitable conclusion that all is not well; far from well.

Carving the landscape over aeons, much more even than the seas and oceans of the world, rivers are the single most important entities influencing the topography and geomorphology of the planet. Rivers have also been an integral part of human development throughout history. As Davies and Walker (1986) point out, earliest recorded civilizations sprang up along watercourses: they provided irrigation, drinking water, a means of transport for fledgling commercial and economic development, as well as food in the form of sustainable fisheries. They have also formed convenient political boundaries, created obstacles to expansion, and provided defence to communities threatened by militaristic neighbours. Many cultures revere rivers and have woven a series of mythologies around them, giving them personae, moods and human traits. Yet the same human communities that have relied so much on rivers have an ambivalent approach to them. The dual emotions of fear of their dynamics coupled with greed for resources have led to massive exploitation and abuse:

- over-consumption of water and biota
- manipulation of natural droughts and floods
- organic and industrial pollution on varying but often very large scales

- manipulation of flow regimes for water supply and redistribution
- channelization and containment in the name of flood control
- mining river beds and banks for alluvial minerals, fill and aggregate

Indeed, it is likely that rivers have suffered the single most intense onslaught of all the world's ecosystems over the past 50 years of human history. Rapidly escalating human populations require power, food and water, all provided in the main by impoundment and flow regulation by dams. According to McCully (1996) some 50 000 'large' and 'major' dams have been constructed world-wide, while the total number of dams on the planet (excluding small, farm storages which more often than not fall outside local and regional controls) hovers around the 800 000 mark. (The term 'large dam' coined by the International Commission on Large Dams – ICOLD – refers to structures 'with a wall equal to or higher than 15m from base to crest'. A dam qualifies as 'major' if it has one or more of the following attributes: a wall >150 m from base to crest; a storage capacity ≥ 25 km^3; a power output ≥ 1000 MW; or a wall volume of ≥ 15 m$^3 \times 10^6$.)

At present, approximately five times the global annual river flow to the oceans now resides behind dam walls. This extraordinary industry and energy to control the flowing waters of the world has happened in the equivalent of a 'blink of an eyelid' in terms of geological time. As McCully (1996) has detailed, it has led to a multitude of ecological, socio-political and socio-economic problems stretching far beyond the rivers themselves, deep into oceanic and climatic processes, as well as reaching into the very fabric of human societies. Some of the effects include:

- the forced removal of some 65 million people to make way for storage reservoirs
- the loss of economic fisheries – inshore coastal, estuarine and riverine – through loss of sediment and nutrient inputs (accumulating in reservoirs) and blockage of migration routes
- floodplain destruction – e.g. flood curtailment and loss of the flood pulse (*sensu* Junk *et al.*, 1989)
- massive impacts on riverine, estuarine (and probably also, coastal and oceanic) biodiversity
- geomorphological degradation of river channels
- salinization of soils (at present some 1 million more hectares are lost to salinization than are placed under irrigation each year: McCully, 1996)

- large-scale evaporative losses of water from reservoirs in arid and semi-arid regions (e.g. Lake Nasser (Aswan High Dam) evaporates some 11 km^3 yr^{-1})
- production of greenhouse gases (carbon dioxide and methane) by reservoirs
- alteration to and loss of habitat on a vast scale
- a range of complex and interactive disruptions of river, floodplain, estuarine and coastal food chains
- the spread of water-borne diseases, and the creation of new habitat for disease vectors
- political tensions and conflict between nation states that share river basins (e.g. Turkey/Iraq; Iran/Iraq; Namibia/Botswana; Israel/Palestine; Israel/Syria)
- leaching of toxic materials from soils (e.g. mercury in northern Canada; selenium in the western United States)
- reservoir-induced seismicity
- loss of fertile farmlands through inundation
- fragmentation of fluvial and terrestrial landscapes and habitat, and a loss of biodiversity

All this through one human activity in and around rivers – the construction of dams. Other activities, such as the use of rivers as conduits for human and industrial wastes, would generate a similar list of problems for the conservation of river ecosystems, habitats and species.

We remain to be convinced that the purported benefits of such water-supply and redistribution schemes, and hydropower/flood-control projects have been properly costed, particularly in terms of ecosystems, habitats, communities and species, and the services they provide human societies. Indeed, McCully (1996) asserts that if correctly costed to include their overall environmental effects, very few schemes to supply water for potable supply, or hydropower production, come out as realistically economically viable. Moreover, the majority of recently constructed dams are an ageing cohort: we have no experience of how dam walls will behave with time, no rigorous protocols exist for their proper maintenance, and many dams are completely neglected, or infrequently inspected.

Indeed, given these problems, there is a growing, world-wide, ground-swell of opposition to large and major dams (e.g. in India, Brazil, Guatemala, Thailand). Peoples directly affected by large-scale manipulation of rivers are saying '. . . so far and no further' and a series of NGOs has developed around river conservation concerns – e.g. the International

Rivers Network based in Berkeley, California, and its news publication *World Rivers Review*. Interestingly, the first moves to decommission regulating dams on a handful of rivers, particularly those in conservation-sensitive areas in the USA (and recently in France), is predicted to increase momentum in the coming century (e.g. Arnould, 1997; Cantrell, 1997; McCully, 1997; Wegner, 1997).

At a rough guess, well over half of the human population of the world has no access at present to water-borne sanitation and/or secure and safe water supplies. Indeed, water-borne diseases such as malaria (250 million cases and 2 million deaths each year in Africa alone), enteroviruses, dracunculiasis, elephantiasis, river blindness, cholera, schistosomiasis, typhoid and, particularly, diarrhoeal diseases, including giardiasis and amoebic dysentery, are the greatest daily killers of humans, and for many hundreds of millions of people the daily collection of water is both time and energy consuming and hazardous. Unless alternative ways of distributing water, properly costed payment for the resource, and educational programmes to curb over-consumption are implemented, it is difficult to envisage a reduction in the rate of dam building, pollution, and flow regulation, particularly in the developing world. As such, the threats to rivers as renewable resources, and as functional ecosystems, will continue to escalate with obvious consequences for ecosystem, habitat, and species conservation. This is the *raison d'être* for this book.

Before describing the overall structure and content of the book, perhaps a few words are needed to clarify terminology. In their ground-breaking book on river regulation, Ward and Stanford (1979) used the title *The Ecology of Regulated Streams* on the understanding that the term 'stream' covered any fluvial system – flowing aquatic systems on any scale from the 'English trout brook', or 'stream-let', to the vast Amazon basin. Throughout our book we are using the term river to mean the same, from the vast to the tiny, and from the perennial to the most extreme intermittent system that may flow only once in a few decades, and then, perhaps only for a few hours. Secondly, we are aware of different terms relating to rivers and their catchments. Throughout this book we have tended to use the terms 'basin', 'catchment' and 'watershed' inter-changeably. Where authors (predominantly North American) have used the word 'watershed', it is taken to mean 'catchment' or 'basin' and vice versa. Thus, watershed management in the United States of America is the equivalent of catchment management in southern Africa, Australia, or the UK.

Structure of the book

In developing this book we have attempted to examine carefully the potential market for the work. This is a book prepared not simply for academic teachers and researchers, and for post-graduate and contract research workers in the fields of water resources management and conservation, although clearly these sectors are part of the target audience. We have also addressed issues that will be of great interest and concern for urban and rural planners, policy makers and politicians, and for workers in non-governmental organizations concerned with sustainable resource management, conservation and community development.

The book is divided into two sections. Part I comprises 12 chapters designed to give the user a broad overview of river conservation worldwide by dividing the globe into a number of large and geographically (or politically) distinct regions. This allows an investigation of many of the planet's river systems and their regional problems, and some of the challenges, successes and failures for river conservation. We have compiled the contributions in the form of a circumnavigation of the globe.

Starting in the Western Hemisphere with the United States and Canada, Karr *et al.* (Chapter 1) explore the north-temperate, sub-arctic, and arid-zone rivers of a region where, after the considerable damage caused by rural and urban development, industrialization, and river manipulation, some attention is being directed towards the development of a conservation ethos and a strong conservation-oriented research base. To the south, some of the complexities of the tropical and sub-tropical river systems of Latin American and the Caribbean nations – a region that generates the greatest river on the planet, the Amazon – are examined in Chapter 2 (Pringle *et al.*). Here, the first contrasts in the approaches to, and the many problems of, river conservation – north:south; developed:developing worlds – begin to surface.

Attention is then directed towards the countries of the European Community (EC) and Scandinavia (Chapter 3, Iversen *et al.*) where some of the problems of the industrialized world – pollution, urban growth, industrialized agriculture and dense human populations – have been concentrated over the past two to three centuries, but also where many important river conservation techniques have developed recently. Further eastwards, Khaiter *et al.* (Chapter 4) review the situation in Central and Eastern Europe, including the European components of the Russian Federation of

States. It is clear from this contribution that river conservation in this part of the world faces enormous challenges.

Switching to Africa, four chapters deal with river conservation issues in disparate climatic and geo-political zones. Wishart *et al.* (Chapter 5) address the complexities of river conservation in the arid to semi-arid zones of North Africa and the Middle East, an area of ancient civilizations, diverse cultures and religious practices, and many human tensions and contrasts. Chapter 6 by Pacini and Harper moves to Central and Tropical Africa where vast populations of urban and rural poor daily face an array of life-threatening water-borne diseases, as well as a daily struggle to procure water. Here emerge the major contrasts between the developed and developing worlds. Similar problems face many of the countries of the Southern African Development Community (SADC), as described by Davies and Wishart (Chapter 7), with the additional problem for conservation of the 'forgotten' civil wars that have raged throughout the region for decades (e.g. Moçambique) and that still rage as we write (e.g. Angola). Chapter 8 takes us to 'offshore' Africa, in the form of Madagascar, that most extraordinary and bio-diverse of large islands of the planet. Here, Benstead *et al.* review the almost insurmountable problems facing river conservation through over-population, deforestation on a massive scale, and soil erosion. The problems of this island state are vast and if its unique flora and fauna are to be retained for future generations, then serious external efforts will have to be forthcoming *now*.

The Indian Sub-continent, including Sri Lanka, Bangladesh, Pakistan and the regions of the Himalayan Arc, are dealt with in Chapter 9 (Gopal *et al.*). In common with most African nations, and despite the reverential, religious and cultural approaches to the rivers of this region, river conservation as a science, in policy development, and in practice, is virtually non-existent. Here, all the ills of the First World – rapid urbanization and industrialization, and associated pollution and river degradation – have been thrust onto the most densely populated region of the planet. The scale of river degradation and pollution is similar to that facing Madagascar and parts of Africa, with virtually no controls either in hand, or planned. This situation is further highlighted in Chapter 10 on China, and Central and Eastern Asia (Li *et al.*), and in Chapter 11 on south-east Asia (Dudgeon *et al.*), both of which emphasize the constraints and conflicts that prevail in the Far East. Finally, the geographical

overview of Part I moves to continental Australia and New Zealand (Chapter 12, Schofield *et al.*) where industrialized and agriculturally developed communities are turning some of their attention to the development of conservation tools and technologies for the assessment of river 'health'. The contrasts between this final chapter and the preceding ones (5–11) are startling to say the least.

Part II of the book, entitled '*Constraints and Opportunities – Problems and Solutions*', is itself divided into four sections:

- an introductory overview (Chapter 13)
- *Geographical Settings* (Chapters 14–16)
- *Key Constraints* (Chapters 17–19); and lastly
- *Conservation in Practice* (Chapters 20–24)

In Chapter 13, Gardiner and Perala-Gardiner examine the philosophies of nature conservation, ecosystem use and sustainability, in relation to ecosystem health and human population growth, highlighting the serious problems and shortcomings that prevail world-wide. Under the section on geographical settings three major conservation issues are examined. The first (Chapter 14, Wishart *et al.*) contrasts the disparities in approaches to conservation of river ecosystems that are evident in developed world (so-called 'First World') and developing world ('Third World') human communities. In many respects, Pringle's review of the approaches to the conservation of contrasting river types – tropical versus temperate systems (Chapter 15) – mirrors the views expressed in Chapter 14 for most tropical systems happen to occur in regions that are socio-economically 'Third World' (see Wishart *et al.*, Chapter 14 for definition), whilst the converse in true for a large proportion of temperate systems. Baer and Pringle (Chapter 16) then take the reader through the problems associated with rivers that run through urban environments, examining the attitudes to urban rivers and the tools for their conservation.

Under the sub-heading, *Key Constraints,* we examine three issues. In Chapter 17, Stanley and Boulton detail the constraints placed on conservation that are forced by river size, whilst Chapter 18 explores the same in relation to rivers with highly variable flow regimes (Boulton *et al.*). In Chapter 19 (Davies *et al.*), the problems caused by breaching river basin integrity and, in particular, the genetic and evolutionary/species conservation implications of interbasin transfers of water (IBTs) alert the reader to some of the complexities of world-wide water manipulation projects.

Finally, under *Conservation in Practice*, five chapters look at aspects of conservation methodologies and approaches that are at present available for use in river conservation. O'Keeffe and Uys (Chapter 20) review river classification tools and their relevance to conservation issues and to river health, while Chapter 21 (Showers) discusses the vital role of community participation (including NGO activities) in river conservation, as well as the development of the ground-swell in community objection to large-scale water supply, irrigation and hydroelectric development projects on rivers world-wide. In Chapter 22, Palmer *et al.* review river and catchment legislation in a number of disparate countries and communities, whilst Chapter 23 (Petts *et al.*) details the history and development of river restoration philosophy and techniques in developed economies with special reference to the UK, USA and Australia. Lastly, American and Australian experiences of integrated catchment management (ICM) (Hooper and Margerum) illustrate examples of the holistic approaches to river conservation that will be needed in the new Millennium if we are in any way to succeed in conserving our river systems and their biota.

Differences in perception and priorities

We have gleaned two interesting experiences while putting this book together. Both are closely intertwined and serve to illustrate what can only be perceived at present to be an ever-widening gulf in global river conservation. The first concerns the understanding of the word 'conservation' itself, while the second concerns the differing priorities given to conservation by countries that are at different stages of economic development.

It was only as the first geographical-review manuscripts began to arrive that, to our consternation, we realized that our concept of 'river conservation' differed markedly from those of several invited authors, and that the difference depended upon whether or not these authors were of developed- or developing-world origin (see Wishart *et al.*, this volume Part II; also Wishart and Davies, 1998).

In the case of developed-world contributions, chapters were crafted around our stated objectives: to examine the science, policy and practice for each region's ecosystems and biota – to detail conservation programmes, techniques and legislation aimed at the sustainable utilization of rivers as water resources and the protection of their habitats, and their indigenous and endemic flora and fauna. However, four of the regional reviews, originating from countries in the developing world, had a different slant. Here the term 'conservation' was used in an entirely different context, and focused on the conservation of water *quality* for human potable supply, water *quantity* for human and livestock use, and river channels for flood control.

In the first category, authors concentrated on water pollution and its control. In the second, we were often presented with very large data sets on water-storage volumes, numbers of reservoirs, and water consumption statistics across a variety of user sectors. In the third, data on river lengths with dykes, levees, and canalization were presented. In three cases there was no mention of biological or ecosystem conservation as we understood it. This observation is not meant in any way to be hypercritical of our colleagues – clearly different forces are at work in different parts of the world; that is what interested us.

Tied to the first problem was the second component that leapt out of the pages: namely, differences in priorities. The rivers described in these chapters were viewed as entities for exploitation by human populations, rather than as ecosystems in their own right, to be used in a sustainable manner. The strong message from this dichotomy is that conservation as it is beginning to be practised in many developed economies is simply not on the agenda in many developing economies. The reason for this is probably quite complex.

First of all, we must not lose sight of the fact that conservation, and an awareness of the issues surrounding conservation, is hardly a well-developed art *in any corner of the globe*. The industrialized nations have, for centuries, systematically pillaged not only their own resources, but also those of other countries, particularly in the case of so many past colonial powers. It is only in relatively recent times that the idea of *sustainable development* (a term that is probably an oxymoron and better described as *sustainable utilization* or *sustainable management*), has begun to take root. It might, therefore, be tempting to interpret our observations of developing countries from the moral high-ground of developed-world models as 'the only way to go'. This is neither a correct observation, nor is it a solution.

From the data presented to us as editors, and in the light of our own varied backgrounds and experiences of the 'two' worlds, poverty, education and time seem to be the three factors driving the differences in approach and perception. Most developing nations suffer the problem of mass poverty, and lack of education and training. On the other hand, the developed 'G7' economies have relatively little by way of urban and rural poor, in contrast to an abundance of resources (an over-abundance and over-consumption in almost all instances). In such contrasting worlds it is inevitable that there will be both a clash of interests and a difference of perception.

Although clearly at odds with our original concept for the book, the differences in perception and priorities presented by our colleagues were an honest reflection of their approaches. However, this left us with the dual problems of imbalance and mismatch. Our solution has been difficult and time consuming but, we feel, has had satisfactory results. In each case, and respecting the standpoint of each author, we arranged either for help from new authors in the region, or for a rewrite, whilst retaining elements of the original contribution (where possible) as well as original authorship.

Yet the very difference of approach to river conservation (and presumably to the conservation of many other natural resources) should be of concern to all who read this book. Developing communities do not necessarily need many of the technologies that are being developed by communities with 'more time' (and more guilt) to reflect on conservation philosophies and techniques. They need trained personnel; they need appropriate, 'low-tech technologies'; they need funds for research; they need literature and information at *affordable prices*; but most of all, they need time and understanding. Hopefully, both exist. If not, humanity faces a dual escalating crisis: a catastrophic loss of global biodiversity on a scale equal to, or greater than, known past ('natural') extinction events on our planet, and the continued, accelerating degradation of river systems to the point where they cannot provide us with that most fundamental of substances for all life – clean, potable water. The choice is ours – sound, scientifically based stewardship, or accelerating and unsustainable exploitation?

Acknowledgements

B.R. Davies wishes to acknowledge financial support from both the University of Cape Town and the South African Water Research Commission for travel to the United Kingdom and Eire during 1998 which allowed meetings between the three editors to take place at a crucial phase in the production of the book. The continued support and encouragement of Dr Kathryn Jagoe-Davies throughout the past five years is acknowledged with love and great appreciation.

References

Arnould, M. 1997. Loire dams to be dismantled for salmon. *World Rivers Review* 12: 11.

Boon, P.J., Calow, P. and Petts, G.E. (eds) 1992. *River Conservation and Management*. John Wiley, Chichester.

Cantrell, S. 1997. US dam removals documented. *World Rivers Review* 12: 9.

Davies B.R. and Walker, K.F. 1986. Introduction to the Ecology of River Systems. In: Davies, B.R. and Walker, K.F. (eds), *The Ecology of River Systems. Monographiae Biologicae*, Dr W Junk Publishers, Dordrecht, 60: 1–8.

Junk, W.J., Bayley, P.B. and Sparks, R. 1989. The flood pulse concept in river-floodplain systems. In: Dodge, D.P. (ed.), *Proceedings of the International Large River Symposium, Canadian Special Publications in Aquatic Science* 106: 110–127.

McCully, P. 1996. *Silenced Rivers: The Ecology and Politics of Large Dams*. Zed Books, London and New Jersey.

McCully, P. 1997. Taking down bad dams. *World Rivers Review* 12: 8.

Ward, J.V. and Stanford, J.A. (eds) 1979. *The Ecology of Regulated Streams*. Plenum Press, New York.

Wegner, D. 1997. An effort to restore the Colorado River and Glen Canyon gathers steam. *World Rivers Review* 12: 10–11.

Wishart M.J. and Davies, B.R. 1998. The increasing divide between First and Third Worlds: science, collaboration and conservation of Third World aquatic ecosystems. *Freshwater Biology* 39: 557–567.

PART I
Geographical Overview

1

River conservation in the United States and Canada

J.R. Karr, J.D. Allan and A.C. Benke

The regional context

Rivers in Canada and the United States, like rivers world-wide, are shaped by their landscapes as much as landscapes are shaped by rivers. Both are defined by regional geology, topography, rainfall, temperature and living organisms. As a result of the complex interactions of climate, running water and land, rivers have been changing for millennia, but the rapid growth over the past 200 years of human populations and their technologies has been a new force for change, altering US and Canadian rivers radically from what they were 300 years ago.

THE MAIN RIVERS

The 24.2 million km^2 of the United States and Canada (16% of the world's total land area) extends from 25 to 70°N latitude. The continent is marked by major landscape features, such as the western and eastern mountain ranges (the Rocky Mountains and the Coast Range in the West, and the Appalachians in the East) and the Laurentian Great Lakes on the border between the two countries. This variation in physical environment has given rise to at least six major terrestrial biomes: tundra, coniferous forest, temperate deciduous forest, grassland, desert and chaparral. Consequently, the major rivers vary tremendously in drainage area, run-off volume, channel form and biotic attributes (Figure 1.1, Table 1.1). The watersheds of two major rivers (Colorado and Rio Grande) include Mexico, an area beyond the scope of this chapter. (Biogeographically,

Mexico is part of North America but the focus in this chapter arbitrarily excludes Mexico.)

Total river run-off averages 8200 km^3 yr^{-1} or 17.6% of the world total (Shiklomanov, 1993; World Resources Institute: WRI, 1994). Approximately two-thirds comes from the surface, one-third from groundwater. Large rivers such as the Mississippi, Colorado and Columbia run through several biomes (Table 1.1). The continent's western border illustrates the extreme range of physical features and river characteristics: annual precipitation varies from less than 10 cm in the arid south-western US to 400–500 cm along the north-western coastal mountains (Fisher, 1995; Mackay, 1995; Oswood et al., 1995; Patrick, 1995). Precipitation in the central temperate grassland ranges from 20 cm in the Intermountain West to 100 cm in the East (Brown and Matthews, 1995; Patrick, 1995). In the temperate deciduous forests of the eastern United States and south-eastern Canada, precipitation varies from 70 cm in Wisconsin and Ontario to more than 200 cm in the southern Appalachian Mountains (Mackay, 1995; Patrick, 1995; Webster et al., 1995). Precipitation may exceed 100 cm in the Canadian Rocky Mountains but declines to 20–70 cm in coniferous forest and tundra biomes of Alaska and northern Canada (Mackay, 1995; Oswood et al., 1995).

Consider the contrast in climate-caused run-off between two western rivers. The extensive Colorado River basin (635 000 km^2) in the arid Southwest (Figure 1.1) has a mean annual discharge of only 640 m^3 s^{-1} (Meybeck, 1988); precipitation is low and evaporation is high throughout. Present-day heavy withdrawals of water in the United States reduce the Colorado's

Global Perspectives on River Conservation: Science, Policy and Practice.
Edited by P.J. Boon, B.R. Davies and G.E. Petts. © 2000 John Wiley & Sons Ltd.

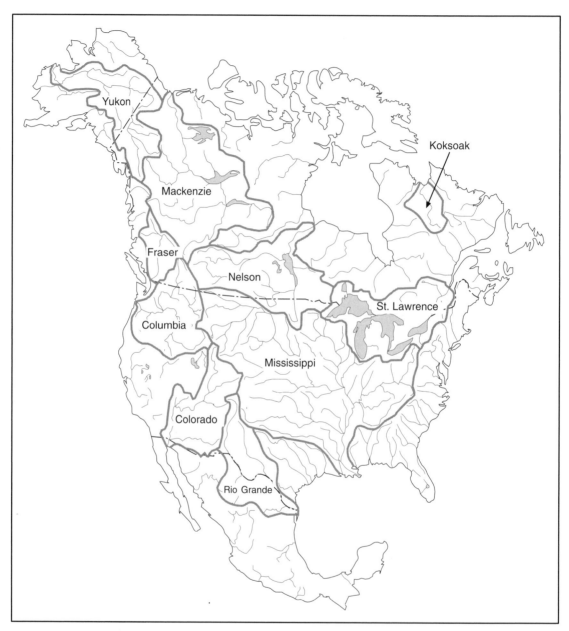

Figure 1.1   *Major river basins in the United States and Canada. See Table 1.1 for basin statistics and characteristics*

discharge to a trickle where it crosses into Mexico. In contrast, the only slightly larger Columbia River basin (670 000 km^2) sustains a discharge of almost 8000 m^3 s^{-1} because precipitation is higher and evaporation is lower.

The Mississippi River basin, the largest river basin in North America, encompasses nearly 14% of the continent – 42% of the area of the conterminous 48 US states (Table 1.1, Figure 1.1). The Colorado, Rio Grande (both partly in Mexico) and Columbia (also partly in Canada) are the other large river basins in the conterminous 48 states. Because of precipitation, several smaller basins (e.g. Mobile, Sacramento–San Joaquin) discharge more water than either the Colorado

Table 1.1 *Characteristics of the major river basins in the United States and Canada, most with a mean annual discharge exceeding 2000 m³ s⁻¹. Colorado River and Rio Grande are included because of their length and their large basins. Degree of fragmentation is based on the criteria of Dynesius and Nilsson (1994): unaffected, moderately affected, and strongly affected. Biomes are grassland = GL, temperate deciduous forest = TDF, desert = D, coniferous forest = CF, arctic tundra = TR, alpine tundra = TL*

River basin	Mean annual discharge (m³ s⁻¹)	Drainage area (km² × 10³)	River length[a] (km)	Degree of fragmentation	Major biomes
Mississippi	18 400[b]	3 267[b]	5 970	Strongly	GL, TDF, CF, TL
St Lawrence	10 700[b]	1 025[b]	3 320	Strongly	TDF, CF
Mackenzie	9 600[b]	1 800[b]	4 240	Moderately	CF, TR, TL
Columbia	7 960[b]	670[b]	2 240	Strongly	CF, D, TL
Yukon	6 200[b]	770[b]	3 180	Unaffected	TR, TL, CF
Fraser	3 750[c]	225[c]	1 375[d]	Moderately	CF, TL
Nelson	3 500[b]	1 150[b]	2 570	Strongly	CF, TL
Koksoak	2 420[e]	133	858[f]	Strongly	CF
Colorado	640[b]	635[b]	2 330	Strongly	D, GL, CF, TL
Rio Grande	100[b]	670[b]	3 030	Not rated	D, GL, CF

[a] Showers (1989), Gleick (1993)
[b] Meybeck (1988)
[c] Czaya (1981)
[d] Northcote and Larkin (1989)
[e] Dynesius and Nilsson (1994)
[f] Service Hydrologie et Cartographie, Ministère de l'Environnement et de la Faune, Gouvernement du Quebec

or Rio Grande. In Canada, the St Lawrence River (shared with the US) and the Mackenzie River have the second and third largest discharges in North America; along with the Nelson basin, they dominate much of Canada. In Alaska, the Yukon River is by far the largest in area and discharge, representing the continent's fifth highest discharge.

The great diversity of climate, geology, land-forms and biome types in the United States and Canada supports an extremely rich freshwater biota. At least 1061 species and sub-species of native freshwater fish occur in North America (790 full species; Page and Burr, 1991; Warren and Burr, 1994; Campbell, 1997), perhaps the world's richest temperate freshwater fish fauna. This diversity is not evenly distributed, however: the greatest concentration of species is found in the east, particularly in the south-eastern United States (Figure 1.2).

Biodiversity of invertebrates is also very high within the US and Canada, comprising 338 native crayfish taxa (decapod crustaceans in the families Astacidae and Cambaridae: Taylor *et al.*, 1996), 297 freshwater mussel taxa (281 species and 16 sub-species of the families Margaritiferidae and Unionidae: J.D. Williams *et al.*, 1993), and 342 freshwater snail taxa (Lydeard and Mayden, 1995). We are not aware of any catalogues of other freshwater invertebrates, such as aquatic insects, but more than 500 species of aquatic insects can sometimes be found in a single stream (e.g. Morse *et al.*,

1983). A complete list of North American aquatic insect genera can be found in Merritt and Cummins (1996).

DEVELOPMENT PRESSURES AND HUMAN IMPACTS

Rivers in the 48 conterminous US states and in all but the most northern and western regions of Canada have been heavily influenced by withdrawal of water for irrigation or domestic and industrial consumption; construction of dams for power generation, navigation and flood control; channelization to enhance drainage and navigation; discharge of domestic, industrial and agricultural pollutants; overharvest by sport and commercial fishing; intentional and accidental introduction of alien species; grazing by domestic livestock; and removal of natural vegetation to permit crop planting, logging or urbanization and the resulting hydrologic alterations and sedimentation from soil erosion. These activities degrade, fragment or destroy rivers. Humans will continue to occupy North America with inevitable effects on rivers; we can and should strive to limit negative effects and to repair damage done in the past.

Approximately 11% of freshwater run-off in North America (United States and Canada only) is withdrawn for human use (WRI, 1990); the United States uses more water than any other country in the world (546 km³

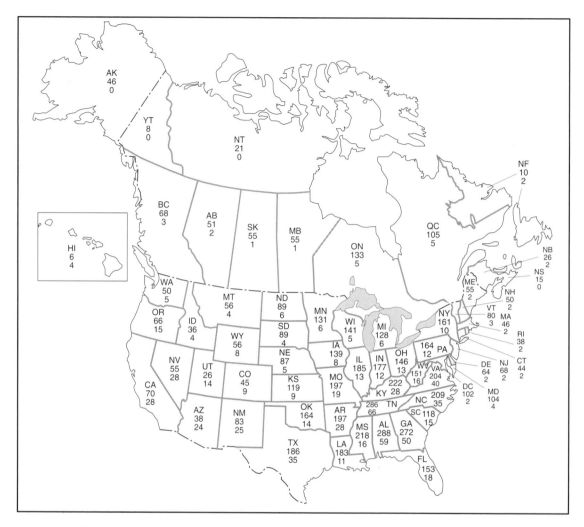

Figure 1.2   *Endangered riverine fishes of North America. Shown for each state, province, and territory are the number of full, riverine fish species native to that place (top) and the number of those species at risk (bottom). The highest proportion of endangered fish species inhabit the arid Southwest; a high number of endangered taxa are found in the species-rich Southeast. Source: Master* et al. *(1998), Natural Heritage Central Databases*

withdrawn per year: Shiklomanov, 1993). Per capita consumption of water in both the United States and Canada is also among the world's highest (1868 m^3 and 1688 m^3 per year, respectively: WRI, 1994). Water use in the United States is divided into three broad categories: domestic, 13%; industrial, 45%; and agricultural, 42% (WRI, 1994). In Canada, the percentages shift towards industry (18%, 70% and 12%, respectively). In contrast, the global average is heavily weighted towards agriculture (8%, 23%, 69%, respectively).

Although river networks are sometimes interrupted by natural barriers like waterfalls, an important

characteristic of natural rivers is the linear connection among network segments. During the past century, humans have fragmented nearly all North American rivers with dams (Dynesius and Nilsson, 1994), especially outside the tundra and coniferous forests of Canada and Alaska (see Table 1.1). (Fragmentation is measured as an integration of frequency of dams in the main channel and tributaries, and the degree of flow regulation (Dynesius and Nilsson, 1994).) Within the conterminous 48 states, for example, the Pascagoula River in Mississippi is the only large river system not fragmented by dams that flows directly to the sea and

Table 1.2 *Five attributes of water resources altered by the cumulative effects of human activity, with examples of degradation in watersheds of the Pacific Northwest (modified from Karr, 1995)*

Attribute	Components	Degradation in Northwest watersheds
Water quality	Temperature; turbidity; dissolved oxygen; acidity; alkalinity; organic and inorganic chemicals; heavy metals; toxic substances	Increased temperature Oxygen depletion Chemical contaminants
Habitat structure	Substrate type; water depth and flow velocity; spatial and temporal complexity of physical habitat	Sedimentation and loss of spawning gravel Obstructions interfering with movement of adult and juvenile salmonids Lack of coarse woody debris Destruction of overhanging banks and riparian vegetation Lack of deep pools Altered abundance and distribution of constrained and unconstrained channel reaches
Flow regime	Water volume; flow timing	Reduced low flows and increased high flows limit survival of salmon and other aquatic organisms at various phases in their life cycles
Food (energy) source	Type, amount, and size of organic particles entering stream; seasonal pattern of energy availability	Altered supply of organic material from riparian corridor Reduced or unavailable nutrients from the carcasses of adult salmon and lampreys after spawning
Biotic interactions	Competition; predation; disease; parasitism; mutualism	Increased predation on young by native and alien species Overharvest by sport and commercial fishers Genetic swamping by less fit hatchery fish Alien diseases and parasites often associated with hatchery or aquaculture operations

exceeds $350\,\mathrm{m^3\,s^{-1}}$. Most rivers whose flows exceed this level are fragmented by dams many times. Only a few East Coast rivers are moderately fragmented. In Canada, the St Lawrence and Nelson Rivers are heavily fragmented, and the Mackenzie moderately fragmented. The few large unaffected river systems in northern Canada and Alaska account for 65% of the unaffected basins in North America and Eurasia, including the largest (Yukon River with a discharge of $6370\,\mathrm{m^3\,s^{-1}}$).

Few medium-sized rivers remain unfragmented. Out of more than 5×10^6 river km in the conterminous 48 states, only about 42 rivers longer than 200 km remain relatively natural and undammed (Benke, 1990). Only two large interior rivers whose discharge exceeds $350\,\mathrm{m^3\,s^{-1}}$ are unfragmented in the United States – the Yellowstone River (Mississippi basin) and the Salmon River (Columbia basin). Interior river basins in Canada are also becoming more fragmented, but we are unaware of a comprehensive survey other than that of the Canadian Heritage River System (see 'Application of Legislation').

Of the negative effects human activities have on rivers, the discharge of chemical contaminants is probably the most widely recognized. It is also the primary focus of most regulation aimed at protecting water resources, even though chemical pollution is often not the most important factor responsible for a river's degradation (Table 1.2). The nature and effects of pollutants across the United States and Canada have not been assessed with any planned, statistically rigorous, sampling design or strategy, or using any consistent indicators. Nonetheless, the most recent biennial report of the US Environmental Protection Agency (USEPA, 1994a) showed that 36% of surveyed kilometres were impaired, that agriculture was the leading cause of impairment, and that bacteria and sediments were the leading pollutants (Figure 1.3). Effluent from municipal sewage treatment was lower than in previous years but still the second most common source of river pollution. The relative influence of pollution sources changes regionally, even within local river basins. Sediment delivery from sheet and rill erosion on farm fields, for example, is especially high in the Midwest, along river floodplains in many regions, in the dry grasslands of eastern Oregon and Washington, and in the interior of California (Figure 1.4).

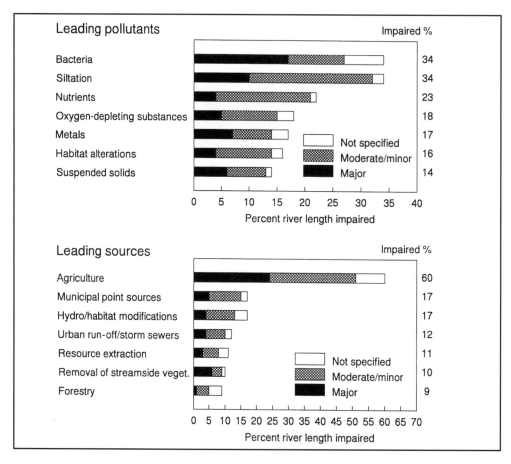

Figure 1.3 *Impaired river lengths in the United States (pollutants and sources) from a small sample that may or may not be representative of all US rivers. Source: USEPA (1994a)*

Water quantity is also problematic in much of temperate North America. Although surface waters are abundant in some regions, other areas lack enough water to supply present human demands (Figure 1.5). Water supply issues dominate in the Southwest, where the amount of water is limited (6% of US supply but 31% of use). In contrast, water quality issues dominate in the East, where water supplies generally exceed demand (37% of supply and 8% of use: *Population-Environment Balance*, 1990). Where precipitation cannot meet demand in an average year, people turn to interbasin transfers or to mining groundwater. Conflicts over water use can be bitter, especially when attempts are made to protect the needs of aquatic organisms. Diversion of water from the Owens Valley of central California to permit Los Angeles to grow out of a desert is perhaps the most famous example of water conflict (Reisner, 1993): pipelines were

dynamited, shots were fired, and armed police were needed to guard the 359 km aqueduct completed in the 1920s.

The extensive alterations of rivers resulting from fragmentation, chemical pollution and water withdrawals have caused widespread degradation of river ecosystems and their biota. Degradation is well documented in individual North American rivers in terms of reduced species richness, the spread of alien taxa, alterations of biogeochemical cycles and declining indices of biological integrity. One expression of the state of North American rivers is the number or percentage of endangered aquatic species. Among terrestrial vertebrates, the Nature Conservancy reports that 17% of mammals and 14% of birds are extinct or endangered. Among aquatic or semi-aquatic groups, however, the percentages are much higher: amphibians, 40%; fish, 37%; crayfish, 51%; and unionid mussels,

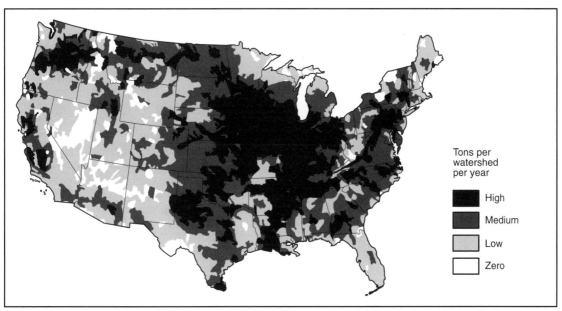

Figure 1.4 *Sediment delivered to rivers and streams in the conterminous 48 states from sheet and rill erosion. Source: National Resources Conservation Service (NRCS) (1996)*

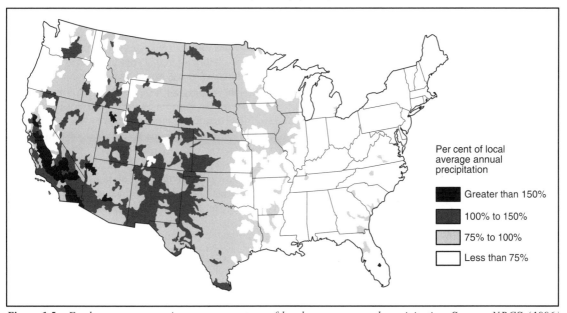

Figure 1.5 *Freshwater consumption as a percentage of local average annual precipitation. Source: NRCS (1996)*

67% (Master, 1998). These biological changes and the factors responsible for them have been described at scales from river basins to the continent (Karr *et al.*, 1985; Benke, 1990; Hughes and Noss, 1992; Allan and Flecker, 1993; Warren and Burr, 1994; Naiman *et al.*, 1995a; Bilby and Naiman, 1998).

The plight of river species is well illustrated by fish. About one-third of North American freshwater fish (364) are estimated to be endangered, threatened or of special concern (Williams *et al.*, 1989). The problem is especially acute in the south-western United States where the percentage of endangered fish is very high (see Figure 1.2). On the other hand, the richer fauna of the south-eastern United States has the greatest number of species at risk.

Local extinction threats are but one regional manifestation of the decline in the biota of North American rivers. For rivers that have been assessed, from 40 to 60% of resident fish species are locally extinct or have declined over the last 150 years (Karr, 1995). Commercial fisheries are down by at least 80% since 1900 and often by 98 to 100% in North American rivers. Atlantic salmon were common before the start of the 20th century in the Penobscot River in Maine but disappeared by mid-century (Palmer, 1994). Chinook salmon in California's Sacramento River fed the influx of human migrants during the Gold Rush (1840s); by the Civil War, 15 years later, they were gone, and fishers moved north to exploit the Columbia River's annual influx of 14–16 million anadromous salmon. Now Columbia River stocks are down to a few hundred thousand fish a year.

Even where commercial and sport catches of fish and shellfish are permitted, it can no longer be assumed that those fish or shellfish are safe to eat. According to USEPA, fish consumption advisories were imposed on 5% of the river kilometres in the US in 1996 (www. epa.gov/ost/fish/). The fish in those rivers were contaminated with toxic chemicals (primarily mercury, PCBs, chlordane and DDT) at levels threatening to the health of anyone who ate them, especially sensitive groups (e.g. pregnant women, children) or those that eat more fish than the average Caucasian male (e.g. native Americans). Furthermore, the number of fish advisories is rising. The 2193 advisories reported for US water bodies in 1996 represent an increase of 72% over 1993. Most (87%) of the 2617 advisories reported in 1996 from Canada (primarily mercury, PCBs, dioxins, furans, toxaphene and mirex) came from the provinces of Ontario and Quebec (www.epa.gov/ost/fishadvice/index.html).

Extinction and the threat of extinction, as well as major declines in geographic range, are as widespread for invertebrates as they are for fish. Of the 338 native crayfish taxa of the United States and Canada, roughly half merit conservation action (Taylor *et al.*, 1996). The Nature Conservancy estimates that 36% of the crayfish fauna of North America is extinct or endangered, and an additional 26% are rare or vulnerable (Master, 1990). Limited natural range, habitat alteration and the spread of non-native species are primary causes (Taylor *et al.*, 1996). The rusty crayfish *Orconectes rusticus* is a noted example of a species extending its range through 'bait-bucket' introductions and

subsequent colonization and also of aggressive displacement of native species (Butler and Stein, 1985).

Of the 297 native freshwater mussels of the United States and Canada, 213 taxa (72%) are considered endangered, threatened or of special concern. Habitat loss from dams, channel modification and introduction of non-native molluscs are viewed as the principal threats (J.D. Williams *et al.*, 1993). The Nature Conservancy described 55% of North America's mussels as extinct or endangered (Master, 1990). Freshwater mussels are arguably the most endangered aquatic biota in North America, and North America contains a uniquely high diversity of freshwater molluscs, concentrated especially in the south-eastern United States (Neves, 1997). Since 1933, for example, 20% of molluscs in the Tennessee River system have been lost, and 46% of the remaining molluscs are endangered or seriously depleted throughout their ranges (Williams *et al.*, 1989). In 1910, more than 2600 commercial mussel fishermen operated on the Illinois River in Illinois; virtually none remains today (Karr, 1995).

Each of these changes, from water withdrawal to decline in commercial fish and shellfish and rising fish consumption advisories, reflects a specific consequence of human interactions with rivers and their landscapes. North American rivers, especially their biotas, have been irreversibly altered. From changes in the genetic and life-history diversity of aquatic organisms to the extinction of species to disruption of the complex chemical, physical and biological processes that sustain riverine ecosystems, the cumulative effects of human actions have compromised the integrity of North American rivers. Growing recognition of the consequences of these issues is changing societal views of rivers, but finding solutions is still difficult: too many people and agencies regard water simply as a liquid for human use and rivers as forces to be controlled.

CATCHMENT VERSUS LOCAL ISSUES

The effects of clearing a riparian corridor, of logging on fragile or steep land, or of building a shopping centre seem limited to a local scale, whereas dams and levees have obvious catchment-level and interbasin effects. In reality, however, virtually all human actions interact with one another in a chain of events that cascades across landscapes. Rivers are affected at all scales – local, catchment, interbasin, regional, national and continental – because of the connectivity of the river network and, in broader context, the water cycle. But the connections between specific human actions and changes in rivers can be difficult to track, especially when diverse human actions are involved. Even when those connections are carefully documented, management challenges frequently arise because watersheds do not coincide with political boundaries, whether city, county, state, provincial or national. The resolution of such conflicts can either be devastating to the rivers involved or lead to wise management.

Widespread recognition of an impending regional water crisis sometimes stimulates planning and compromise. A good example is the Delaware River, an area with relatively high precipitation and one of the largest population concentrations in the United States. Crossing parts of New York, New Jersey and Pennsylvania, the Delaware is among the 74 largest rivers that flow to the sea in North America (Dynesius and Nilsson, 1994). Water quality in the Delaware is relatively high from its headwaters until Trenton, New Jersey, where industrial and municipal effluents rapidly degrade water quality. Despite intensive pressure for hydropower development and water supply over most of a century (Albert, 1987), the upper Delaware has high biological diversity (Sweeney, 1993) and is renowned for diverse fishing opportunities (trout, smallmouth bass, walleye and shad).

The States of New Jersey, New York and Pennsylvania and the city of New York began meeting and discussing the fate of the Delaware River in the early 20th century. From the 1920s until the early 1960s, several dams for hydropower and water supply (particularly for New York City) were built on tributaries, causing moderate fragmentation (Dynesius and Nilsson, 1994). In 1961, the interstate Delaware River Basin Commission (DRBC) was created to manage the basin's affairs. Shortly thereafter, the US Army Corps of Engineers completed a 50-year water resources plan that called for 19 new dam projects for water supply, flood control and recreation (Albert, 1987). By far the largest of the proposed dams was a mainstem dam at Tocks Island, only 10 km upstream from the famous Delaware Water Gap.

The Tocks Island Dam was highly controversial from the beginning, opposed by both landowners and conservation groups (Palmer, 1986; Albert, 1987). Although land acquisition began in 1967, federal funding was slow, and a succession of environmental impact studies raised doubts about the project, particularly the likelihood of eutrophication. With growing public opposition, DRBC recommended de-authorization in 1975. By that time, much of the land was already owned by the federal government. The victory for conservationists became complete when a large section of the main stem was designated a national wild and scenic river in 1978 and more than 4000 ha of land that would have been flooded were added to national park lands. The history of the Delaware River shows the value of an interstate commission in managing an entire river basin. It also demonstrates the influence that citizens (before the conservation movement became popular; see Showers, this volume) and environmental studies can have on decision-making when local and catchment issues are integrated.

Yet this success story is the exception rather than the rule. Difficulties in watershed management frequently arise when the watershed encompasses multiple political units. Even the catchments of small rivers are governed by a complex web of overlapping and fragmented jurisdictional units (see Hooper and Margerum, this volume). The River Raisin in south-eastern Michigan, a catchment of 2700 km^2, includes five counties, 41 townships, six cities and 10 villages, which all exert some measure of local planning control (Allan *et al.*, 1997). In addition, at least three state-level and six federal-level government agencies have some authority over water quality or land management. The resolution of such overlapping and sometimes contradictory responsibilities requires much cooperation and an appreciation that catchment-based or other regional management is essential (Doppelt *et al.*, 1993; Hooper and Margerum, this volume).

MAIN AREAS AT RISK

Defining main areas at risk in the United States and Canada can be quite subjective because different regions have different problems, each of critical importance to that region. The criteria for defining risk may thus differ substantially in different regions. For example, should definition of risk be based on how much water remains in a river after human use, how much chemical pollution is present, how extensively hydrological behaviour has been altered, how severely target species are threatened, how much total biological diversity has been altered or biological integrity lowered, or how impaired ecological 'function' is (e.g. in terms of food webs, productivity, or future evolutionary potential)? Alternatively, perhaps risk should be seen from the perspective of likely future human activities and their consequences rather than viewing risk as the cumulative effects of past actions. Good arguments can be made for using all these criteria, but taken singly each of these criteria describes only one aspect of ecological risk in a river. Management approaches that look at one of these criteria and not others may not discover or retard damage until substantial degradation has come about.

In terms of the degree of endangerment of selected animal groups, two regions of North America might arguably be considered main areas at risk: the south-western and south-eastern United States. Water supply is the dominant issue in the Southwest, particularly withdrawal of water from the Colorado River. A 1922 compact between the US and Mexico assumed the river had an average annual flow of 18 million acre-feet (maf; 1 acre-foot = 1233 m^3); present estimates put available supply closer to 15 maf in 1990. Most water allocated from the Colorado is transferred outside the basin's boundaries, to the cities of San Diego and Los Angeles (Morrison *et al.*, 1996). Today, only about 10% of the Colorado's annual flow reaches Mexico, and in a typical year even this remnant is consumed before reaching the Gulf of California (National Research Council: NRC, 1996a).

Thus, it is no surprise that the highest proportion of fish species threatened with extinction occurs in the US Southwest (e.g. 63% of species at risk in Arizona; Figure 1.2). The 17 western states – including 235 counties (22% of the 17-state total) with irrigated agriculture that relies on water from rivers with listed species – contain 50 species listed in the Endangered Species Act (ESA) that are harmed by agriculture. The number of ESA-listed species per county increases as the area of irrigated agriculture dependent on surface water increases within a county (Moore *et al.*, 1996). Programmes to protect endangered fish thus come into direct conflict with western water allocation.

Unlike the Southwest, the Southeast has plentiful water supplies, but its rivers have been greatly modified by dam construction, other river modifications, and urban development. These activities, imposed on a region whose aquatic biodiversity is naturally high, have led to the greatest number of threatened species on the continent. The Mobile River basin, centred on Alabama, exemplifies the main problem in this region. With a mean annual discharge of 1900 m^3 s^{-1}, it is the largest system wholly in the US east of the Mississippi River basin and one of the three most fragmented large river systems in North America (Dynesius and Nilsson, 1994). Development in the basin continues with little concern for biodiversity. Yet recent analyses show that Alabama, drained primarily by the Mobile River system, represents an aquatic biodiversity hot spot (Ward *et al.*, 1992; Lydeard and Mayden, 1995; Stolzenberg, 1997).

Alabama supports 38% of the native freshwater fish species found in North America, 43% of the gill-breathing snails, 60% of the mussels and 52% of the freshwater turtles. Many of these forms are endemic. The Cahaba River, an unimpounded tributary, contains at least 130 species of fish, far more than are found in most states or provinces (see Figure 1.2). Extensive alteration of these rich aquatic environments has produced a biodiversity crisis. Ten percent of the fish species, 65% of the snails, 69% of the mussels and 43% of the turtles are considered either extinct, endangered, threatened or of concern (Lydeard and Mayden, 1995). Unfortunately, most state leaders are oblivious to the importance of river conservation and historically have opposed federal influence of any kind. Thus, few rivers in Alabama receive federal protection, and no state river protection programme is in place. Recent growth in river conservation groups, however, may help enlighten state leaders.

Rivers in other regions are similarly at risk. In the Pacific Northwest, hydropower dams, poor forestry practices, overgrazing and urban development continue to devastate salmon and other populations. Overharvest by sport and commercial fishers and massive releases of hatchery fish worsen the salmon problem. Dams on rivers, such as the highly fragmented Connecticut River, in the Northeast, degrade river ecosystems and anadromous fish populations. Large and small water development projects and deforestation in central and eastern Canada have destroyed the natural features of many rivers. In the Midwest, dams and drainage projects have reduced many streams and rivers to mere drainage ditches amidst amber waves of grain. Clearly, no region of North America, not even north-western Canada and Alaska, is currently untouched; no region is immune to additional human 'progress'. Commercial fisheries in US rivers are no longer at risk because, for all practical purposes, they are gone. Free-flowing rivers for recreation are rare. Only persistent conservationists, informed citizens and enlightened politicians and government agencies can prevent continued degradation.

Information for conservation

SCIENTIFIC DATA

River conservation strategies must be firmly based on scientific knowledge to achieve conservation goals, but for most of the 20th century the science behind river conservation in North America seldom escaped from narrow academic disciplinary boxes. Fortunately, such narrowness is diminishing, as teams of scientists, particularly at universities and federal laboratories, provide information directly applicable to river conservation.

Scientific research has demonstrated the connectivity of water and associated biological resources. Hydrological connectivity in space and time, for example, has at least four dimensions: upstream to downstream, or longitudinal; between channel and groundwater, or vertical; between channel and riparian, to floodplains and uplands, or horizontal; and in time (Ward, 1989). Water, plants, animals, nutrients, debris and the by-products of human society move along and among these continuities, exerting a major influence on the river and its inhabitants. Transport downstream and the migration of fish upstream carry nutrients, energy, toxic chemicals and organisms (including alien species) throughout the river network. Riparian corridors serve as buffer zones and a transition between terrestrial and aquatic environments. Channels are shaped by complex interactions of flow, geological substrates and source materials, including organic components like leaves or woody debris. Scientific understanding of all these issues has expanded in the last two decades, and continued research throughout North America is adding to it daily. Renewed efforts in research (Naiman *et al.*, 1995a,b) and in university education (NRC, 1996c) are needed to advance knowledge still further. The bulk of accumulating scientific knowledge resides in professional publications, but many outstanding popular magazines and books have made the state of North America's rivers more obvious to conservationists (Hume, 1992; Doppelt *et al.*, 1993; Palmer, 1994; Wilcove and Bean, 1994; Karr and Chu, 1999). A direct outcome of scientific research is the revolution in biological monitoring and assessment (see next section).

Despite the scientific advances, gaps remain. Knowledge of medium to large rivers, for example, is rudimentary in comparison with what is known about small streams (Patrick, 1996). There are no nationwide biological survey programmes in the United States or Canada, although a few states have made major advances (e.g. Ohio, Maine, Florida, Illinois). The lack of nationally consistent programmes or evaluation protocols designed for river conservation limits the ability to make sound decisions and thus to use resources wisely, especially scarce federal resources. Evaluation protocols should connect degradation to specific environmental threats. Instead they focus on permits issued or fines levied. Too often, selection

criteria for conservation priorities (past and present) focus on scenic or recreational values rather than on scientific data (see 'Application of Legislation').

Yet even if technical and scientific knowledge were complete, the success of conservation programmes would be constrained by a legal and regulatory framework mired in outdated philosophies and policies that ignore, even contradict, the science (Boon, 1992; Karr, 1995).

CLASSIFICATION AND EVALUATION SCHEMES

Classification

Significant advances in North American river ecology stem from efforts to organize or classify spatial patterns. Defining hierarchical relationships of tributaries within drainage networks (Horton, 1945; Strahler, 1957), longitudinal zonation for fishes (Burton and Odum, 1945; Huet, 1954) and macroinvertebrates (Illies and Botosaneanu, 1963), and the River Continuum Concept (Vannote *et al.*, 1980) are all based on spatial classification. Stream classification has a long history and has generated a series of comprehensive reviews (Macan, 1961; Illies and Botosaneanu, 1963; Hawkes, 1975; Waason, 1989; Hudson *et al.*, 1992; Naiman *et al.*, 1992). Physical, chemical and biological data, alone or together, have been used as the basis of classification. A significant challenge, unmet in many of these schemes, is transcendence of regions. Features that worked well in one location have rarely been transferable to areas of different zoogeographical, geological or climatic history. Recent efforts to classify river systems have emphasized geomorphological and landscape features (Naiman *et al.*, 1992); however, whether or not this emphasis will be sufficient to satisfy management needs is uncertain. Tension persists between efforts to develop general schemes applicable anywhere in North America and regional schemes tailored to local circumstances.

Hierarchical classification offers a solution to the challenge of constructing a classification that is effective across spatial scales from continents to local catchments. Hierarchy theory (O'Neill *et al.*, 1986) holds that smaller units develop or act within constraints set by the larger units in which they are nested. For example, the geoclimatic features of a catchment constrain the types of valley segments, and valley segment types constrain the stream reaches. The relevance of a hierarchical approach to river classification is evident in the wide utility of stream-order classification (Horton, 1945), which forms the template for the river continuum (but see Hughes and Omernik (1981) for a caution about the coarseness of stream-order in classification). The physical structure of stream channels appears to map well into a spatial hierarchy of microhabitats within macrohabitats, macrohabitats within stream reaches, stream reaches within valley segments, and segments nested within catchments (Frissell *et al.*, 1986; Hawkins *et al.*, 1993). Habitat classification protocols based on this conceptual approach are widely used by stream assessment teams working for US state and federal agencies.

Using large-scale patterns in climate, soils and vegetation, ecologists and geographers have classified terrestrial ecosystems into ecoregions (Abell *et al.*, 2000; United States: Bailey *et al.*, 1994; Canada: Ecoregions Working Group, 1989). Ecoregions are large-scale landscape units that include relatively homogeneous ecosystems and are distinguishable from other ecoregions (Omernik and Bailey, 1997). They nonetheless contain a mosaic of sites, which, at a finer scale, can be further subdivided into more detailed land units and effectively delineated by mappable characteristics of climate, geology, soils and vegetation. The process of delineating such ecological units, termed ecoregion analysis, and relating them within a hierarchical framework is increasingly viewed as a crucial step toward ecosystem management. Ecoregions are useful to river classification as descriptors of landscapes within and among river basins (Omernik, 1987, 1995; Omernik and Gallant, 1990), but because the catchments of many medium-sized or larger rivers will span more than one ecoregion, ecoregion and catchment boundaries differ. Thus, neither provides a singular truth (Omernik and Bailey, 1997) and both may be useful for management. Ecoregions group environmental resources and ecosystems into fairly homogeneous spatial units, while catchments define contributions to the quality and quantity of water at a particular point. Within a smaller region, the specific protocols used by US management agencies (e.g. Meador *et al.*, 1993a; Rankin, 1995; Barbour *et al.*, 1999) commonly rely upon the segment–reach–habitat hierarchy described well by Frissell *et al.* (1986).

Evaluation

Evaluation of river condition in North America historically centred on two rather different questions: Is the river of sufficiently high quality to warrant conservation efforts to preserve it? Are water quantity and water quality sufficient to serve human needs; more generally, is water quality degraded? Paradoxically, the search for answers rarely looked at or measured biological condition. Decisions about conservation potential, such as wild and scenic river designation (see 'Statutory Framework'), placed emphasis on very general criteria: scenic, recreational (wildness and free flowing), or cultural heritage values (Benke, 1990). Assessment of water quality emphasized physical or chemical characteristics of water, the fluid. Monitoring programmes emphasized water quantity at gauging stations on medium to large streams or pollution by chemical contaminants; smaller streams were neglected (they were not 'fishable' or 'swimmable' to use the language of the Clean Water Act), and so were important biological changes in rivers.

Because of advances in scientific understanding and greater public appreciation for the value of rivers, evaluation approaches are evolving rapidly, and biological condition is emerging as a central component of local and regional assessments of conservation merit. On the West Coast, King County, Washington, designated $14 million to protect rivers. A scientific committee was convened to define criteria for selecting priority rivers. Four major concepts were included, most with multiple criteria: watershed condition, riparian condition, risk factors and biological condition. Watershed condition was assessed on the basis of the proportion of watershed developed, proportion still forested, extent of ongoing restoration efforts, and proportion of protected watershed land. The primary criterion for riparian condition was the proportion of stream length with a 90 m buffer of natural vegetation. Risk factors included the proportion of the basin inside the recently defined urban growth boundary and the proportion within a designated timber production zone. Five major biological conditions were considered: salmonid richness index (ratio of species present to species expected); salmonid abundance (fish per kilometre); riparian richness index (ratio of riparian plant species present to species expected); stream kilometres accessible to anadromous fish; and richness of other aquatic taxa (such as invertebrates and amphibians). These criteria integrate landscape and stream condition, including the likelihood of future protection, and a broad evaluation of biological condition.

On the East Coast, Angermeier and colleagues (Angermeier *et al.*, 1993; Winston and Angermeier, 1995; Angermeier and Winston, 1997) used fish to assess the conservation value of stream communities. Three components of resource value were assessed in these studies: value of rare biota, ecological value, and fishery value. The researchers developed an index of centres of density (ICD) based on fish, that represents the average importance of an area for biodiversity conservation, an approach that was superior to simple taxon richness. Their index was robust over time, indicating that multiple samples were not necessary to define biodiversity 'hot spots'. Success in conservation efforts would require that the regional landscape's ecological integrity be maintained. The conservation criteria in these studies were similar to standards actually adopted in New Zealand (Collier and McColl, 1992).

Water quality programmes are also shifting to more biological approaches and endpoints. Measures of chemical water quality as a presumed indicator of biological conditions, and biotic indices that measured only the biological effects of organic effluent (Hilsenhoff, 1988), are being replaced by more integrative multimetric indices (Karr, 1981; Karr *et al.*, 1986; Ohio Environmental Protection Agency, 1988; Plafkin *et al.*, 1989; Davis and Simon, 1995; Barbour *et al.*, 1999; Karr and Chu, 1999). Biological evaluation is more scientifically robust for assessing degree of degradation and for identifying the best sites for conservation and protection.

The revolution in biological monitoring and assessment is a recent and direct outcome of scientific research showing that concentrations of chemical contaminants are poor surrogates for the health of aquatic biota. In the United States, this research has led to national efforts to strengthen the biological content of water quality monitoring programmes (USEPA, 1990a). No state used comprehensive biological evaluations in 1981. Three states had them in place, and five were developing them, in 1989. By 1995, 42 states used multimetric biological assessments (Davis *et al.*, 1996); an additional six were developing multimetric biological programmes. States vary in the taxa selected to assess the conditions of rivers: benthic invertebrates (44), fish (29) and algae (4). Twenty-six states sample more than one taxonomic group (Davis *et al.*, 1996). The number of studies and formal protocols is rising rapidly, through active research programmes throughout North America (Table 1.3).

This biological approach makes economic sense as well (Keeler and McLemore, 1996). Chemical surrogates for environmental quality are useful for examining the costs of controlling pollution but are less helpful for understanding the damage caused by

Table 1.3 *River monitoring methods in use or under development in North America. Davis* et al. *(1996) summarize use of multimetric indices in state programmes*

Category	Representative examples
General	Karr and Dudley, 1981; Plafkin *et al.*, 1989; Karr, 1991; Rosenberg and Resh, 1993; Gurtz and Muir, 1994; Barbour *et al.*, 1995; Davis and Simon, 1995; Yoder, 1995; Yoder and Rankin, 1995a,b; Stribling *et al.*, 1996; Karr, 1998; Karr and Chu, 1999
Fish	Karr, 1981; Karr *et al.*, 1986; Hughes and Gammon, 1987; Miller *et al.*, 1988; Lyons, 1992; Meador *et al.*, 1993b; Fore *et al.*, 1994; Lyons *et al.*, 1996; Simon, 1999
Benthic invertebrates	Hilsenhoff, 1988; Ohio EPA, 1988; Plafkin *et al.*, 1989; Barbour *et al.*, 1992; Cuffney *et al.*, 1993a; Lenat, 1993; Resh and Jackson, 1993; Kerans and Karr, 1994; DeShon, 1995; Barbour *et al.*, 1996; Fore *et al.*, 1996
Algae	Bahls, 1993; Porter *et al.*, 1993; Pan *et al.*, 1996
Physical habitat	Ohio EPA, 1988; Meador *et al.*, 1993a; Rankin, 1995
Hydrology	Ward, 1989; Richter *et al.*, 1996
Quality assurance	Klemm *et al.*, 1990, 1993; Cuffney *et al.*, 1993b

pollution. More direct monitoring using biological indicators is 'valuable in both a benefit–cost framework and in implementing an exogenously determined safety standard' (Keeler and McLemore, 1996).

Two national United States efforts to assess freshwater status and trends have been initiated in the last decade. The Environmental Monitoring and Assessment Program (EMAP) was established by the EPA to monitor and assess the status of, and trends in, national ecological resources, including streams (www.epa.gov/emap/). Similarly, the National Water Quality Assessment (NAWQA) Program of the US Geological Survey is collecting extensive data to describe status and trends in the quality of the nation's ground- and surface-water resources (http://water.usgs.gov/nawqa/). In addition, citizen monitoring teams in many regions track the condition of local streams and rivers (see 'Public Awareness and Education' and 'Role of Voluntary Organizations'). Similar monitoring programmes should be routine in and near river conservation areas to identify early signs of degradation before it is too late to reverse the trends.

Biological indices integrate all sources of degradation – chemical, physical and biological –

because the biota are influenced by all three categories of factors. Biological monitoring can improve water resource protection by recognizing degradation caused by non-chemical stressors; it can better target limited funds to improve water resources, it can identify the remaining best places that warrant conservation efforts, and it can determine whether conservation efforts are protecting living systems. Moreover, the condition of living systems offers a compelling and easily understood framework for thinking about and promoting conservation and river health (Karr, 1999; Karr and Chu, 1999).

Although great strides have been made in incorporating biological information into stream monitoring, management and conservation, no widely adopted biological assessment procedure exists for selecting which streams and rivers should have the highest priority for federal or state and provincial protection programmes.

PUBLIC AWARENESS AND EDUCATION

Public awareness

Polls show that Americans increasingly identify themselves as environmentalists. Recently 1032 US and 1011 Canadian citizens were asked, 'How concerned are you personally about environmental problems?' (Dunlap *et al.*, 1993). Eighty-nine percent of respondents in Canada and 85% of respondents in the United States labelled themselves as concerned 'a great deal' or 'a fair amount'. Americans share a strong core of environmental values (Kempton *et al.*, 1995); survey results even suggest that the public has shifted from moderate positions toward pro-environment ones.

Citizens often identify water pollution as the most important environmental issue (e.g. in the Pacific Northwest: Harris and Associates, 1995). Safe, clean water ranked number one among critical national issues for coastal county and city managers across the United States (NOAA press release, May 1997; www. noaa.gov/public-affairs). Fifty-eight percent ranked clean water as important as, or more important than, health care.

A public awareness survey by Darcy, Masius, Benton, and Bowles for American Rivers (unpublished April 1994) indicated near-universal concern (94% of 733 Americans of age 18 and older) about drinking water contaminated with industrial waste; sewage headed the list of river pollutants. Yet individuals described their own knowledge of specific water issues as low. Focus-group discussions (Lauer,

Lally, Victoria, Inc for American Rivers, 1997) with four environmental and river groups revealed a widespread perception that US rivers are in bad shape; industrial pollution and urban run-off were considered the biggest threats, and wildlife protection was the most important, with safe drinking water second, as reasons for protecting rivers. Even among these relatively knowledgeable individuals, threats such as dams, mining and logging were far down the list of identified threats.

In another survey of 51 individuals having grassroots experience in river restoration (Lauer, Lally, Victoria, Inc for American Rivers, 1997), nearly half of those interviewed had more than 10 years' experience in river conservation. Raising public awareness and building public support for river conservation were top priorities for these practitioners, along with concern for water quality and riparian protection. These individuals also commented on government's need to streamline bureaucracy and to provide help in forming partnerships among organizations and agencies.

In short, US and Canadian citizens are aware of, and concerned about, environmental degradation. Water quality and rivers often top their list of concerns. Much of this concern is for human health and also for broad-level protection of nature; surveys reveal little appreciation for such threats to rivers as non-native species (spreading from other regions of North America, as well as from other countries), land-use patterns, or river regulation. Awareness of threats to the integrity of living systems, and to the goods and services they provide, is still developing (Costanza *et al.*, 1997; Daily, 1997; Pimentel *et al.*, 1997). Education is still needed to increase the public's awareness of river issues and to expand the knowledge of people who are already involved in river protection.

Education

Numerous river education activities can be found in schools or river-watch programmes sponsored by non-governmental organizations (NGOs) and by local, state and national governments. Government agencies put out fact-sheets, pamphlets and videos. NGOs and government agencies also publish newsletters and mount public campaigns on radio, television, and the Internet. Large NGOs invariably have a communications staff whose primary role is to obtain press and television coverage of environmental issues. Formal education also exists, of course, largely within secondary schools but also in colleges and universities

(NRC, 1996c). When innovative river programmes are implemented, however, they usually depend on the motivation of a few individuals and are not well established at national or state and provincial levels. Student participation in comprehensive, scientific river conservation programmes is most likely where individual teachers and schools join forces with state-managed river-watch programmes and NGOs.

In North America, volunteer monitoring of water quality and river health contributes significantly to awareness and education through 'hands-on' participation by citizens of all ages (USEPA, 1990b, 1994b; Firehock and West, 1995; Murdoch *et al.*, 1996). Programmes have evolved from cleaning up unsightly litter to chemical monitoring and, most recently, to biological monitoring, usually of macroinvertebrates (Firehock and West, 1995). A survey published in 1994 found 517 freshwater monitoring programmes in the United States, in 45 of the 50 US states (up to 30 programmes in some states), involving nearly 350 000 volunteers. The most common use of monitoring data is to educate participants and their communities about local problems and local decision-making. Volunteer monitors also provide data on water quality trends and actively try to influence environmental legislation and enforcement (Firehock and West, 1995). As of 1994, 53 programmes in 27 states provided data to their state's biennial '305(b) report' under the Clean Water Act (see Figure 1.3) to EPA and Congress (USEPA, 1994a).

The growth of volunteer monitoring has been accompanied by increasingly sophisticated methods, manuals and training. Pioneering efforts by NGOs, such as the Izaak Walton League of America, Project Green, Adopt-a-Stream Foundation (Murdoch *et al.*, 1996), Trout Unlimited, and by local, state and national government agencies, have resulted in well-crafted monitoring protocols with quality assurance/ quality control (QA/QC) plans that provide reliable information on the physical, chemical and biological condition of rivers (Firehock and West, 1995).

Monitoring done by school children appears to be one of the fastest-growing elements of volunteer monitoring (USEPA, 1994b). Colorado Waterwatch is a partnership between the State Division of Wildlife and teachers and students in some 260 schools. Students monitor more than 500 stations 24 times a year, providing the primary data on river conditions for the state of Colorado. The programme's two goals are given equal importance: to provide usable data and to provide an opportunity for students and teachers to learn the value and the ecology of their rivers.

Although some excellent environmental education programmes are being implemented in North American secondary schools (Lieberman, 1995), such programmes are hardly the norm, despite environmental education requirements in about half the states. For many states, curricula and trained teachers are not available for every school (Holtz, 1996). Much environmental education is directed at students in elementary and middle school, with a primary emphasis on developing awareness and interest in scientific issues (Project WILD and Project Learning Tree). Unfortunately, in high school, where students are better prepared to understand the more scientifically challenging issues affecting rivers, environmental education is less prevalent. The Rivers Curriculum Project (www.siue. edu/OSME/river/), which now includes more than 300 schools in 35 states and Canada, focuses on the Mississippi and Illinois Rivers (R.A. Williams *et al.*, 1993). Useful Internet links to river curricula can be found at Web sites maintained by The Izaak Walton League, Project Green (www.earthforce.org/green), the National Consortium for Environmental Education and Training (nceet.snre.umich.edu/nceet.html), and the state of Kentucky's Waterwatch Program (www.state.ky.us/nrepc/water/waterres.html).

Finally, a panel of aquatic scientists concluded that strengthening university programmes is essential if society is to manage the risks involved in river degradation (NRC, 1996c). A vital ingredient in those programmes would be better connections within the educational system among academics, water resource managers and the general public.

In sum, diverse and innovative educational programmes aimed at river conservation are thriving, particularly those of NGOs, often in partnership with governments. Environmental education in secondary schools is sporadic but appears to be growing; it presents an enormous opportunity as new curricula develop and schools become partners with NGOs and government in volunteer monitoring. University programmes are mostly still fragmented.

The framework for conservation

THE SOCIO-POLITICAL AND RELIGIOUS– CULTURAL BACKGROUND

Two waves of human migration have altered North America. The first came more than 10 000 years ago when palaeo-Indians arrived from Asia. These peoples were wanderers – hunters, fishers, and gatherers – who thrived on the abundant resources. Slowly, these earliest Americans developed more sedentary habits and cultures finely tuned to the places and the seasons. The early Americans hardly lived on the land without changing regional biological systems. To the contrary, they altered vegetation and hunted, even decimated, native mammals (Martin and Klein, 1984); some societies, notably irrigation-dependent civilizations in the American Southwest, even collapsed (Reisner, 1993). Yet humans adapted to their changing surroundings and sustained themselves for millennia, in most cases without the local ecological collapse historians have documented in other places (e.g. Easter Island: Ponting, 1991; Catton, 1993). Human societies prospered throughout North America, and most of the regional wealth of forests, rivers and the sea remained.

Then, 500 years ago, European colonists arrived and unwittingly altered for ever the interactions among landscapes, rivers and human populations. Manifest Destiny – the assumption that westward expansion was a God-given right – pushed early waves of European migration across the continent. A new school of meteorology arose whose adherents, many of them government scientists, believed that 'rains follow the plow' (Worster, 1985; Reisner, 1993). Of course, they were wrong.

The consequences of human activity across the North American landscape is evident in its rivers (see 'Development Pressures and Human Impacts'). River degradation plainly shows that stewardship of regional resources has fallen short. In economic terms, the continent's rich natural capital has been squandered, replaced by human and manufactured capital, making many wealthy while others live in poverty. Private property and individual rights continue to assert themselves, especially in the West. Ownership of property, especially land, is seen by many as the authority to do what they want on that land, even public lands, regardless of impacts beyond property lines. For example, the American Heritage Rivers Initiative, proposed by the Clinton administration in 1997, is strongly opposed by the religious right and private-property activists.

At the same time, many citizens now worry that the goods and services supplied by North American landscapes in the past cannot be sustained into the 21st century. Environmental concern, often stemming from visible and unacceptable changes in neighbourhoods and regions, is tempering forces that call for continued high levels of natural resource exploitation.

The growth and dissemination of scientific information is at least partly responsible for changing attitudes. People today care about streams because they recognize the forecast that declining river health predicts about landscape health. The success of local and regional human communities for much of the past 500 years in North America, and for the tribes before that, came in part from the bounty provided by rivers. The decline in rivers, in many respects the lifelines of a landscape, says much about the effect of human activities and about the quality of life in those landscapes.

Citizen responses described in the previous section demonstrate that more people recognize these issues and the shortsighted nature of the philosophies that have guided Western society's conquering of North America and its rivers.

STATUTORY FRAMEWORK

Because water is so important to society, a large body of law has developed to deal with water and river issues in the United States (Goldfarb, 1988; Rodgers, 1994). Laws that deal explicitly with water fall into three broad categories: (1) water quality and quantity; (2) rivers as corridors; and (3) watersheds and watershed ecology. In addition, the Endangered Species Act applies to aquatic systems. Issues associated with these topics are handled at three political levels: federal, state and local.

Water quality and quantity

US federal water law has been evolving since the Rivers and Harbors Act (also called the Refuse Act) was passed in 1899 to protect navigable waterways from domestic effluent and oil spills. Since 1899, a law focused on water quality and promoting clean water has been variously called the Water Pollution Control Act, the Water Quality Act, and now the Clean Water Act. Last reauthorized in 1987, the Clean Water Act is the centrepiece of present US water law. Since 1972, the Clean Water Act's stated objective has been 'to restore and maintain the physical, chemical, and biological integrity of the nation's waters'. The US EPA is responsible for enforcing clean water legislation in collaboration with the states. The EPA establishes minimum water quality standards, and the states formulate policies, regulate activities affecting water quality on the basis of standards and criteria at least as rigorous as those defined by the EPA, and offer incentives – often with EPA assistance – designed to accomplish the broad federal goal.

Unlike national legislation defining water quality goals, laws regulating access to surface water vary regionally. In most of the East, where water is historically abundant, water use is governed by the doctrine of riparian rights. Individuals who own land adjacent to flowing water have the right to use that water as long as some is left for downstream landowners. In the West, which is largely arid and semi-arid, the prior appropriation doctrine defines who has access to surface water (Wilkinson, 1992). Prior appropriation became a part of western US water policy in the mid-19th century when miners, needing water to get minerals out of rock, took control of available water on a first come, first served basis. Water rights established out of these early local customs could be bought and sold, but to retain control over a right, the appropriator had to put the water to 'beneficial use', usually activities such as mining, agriculture, or hydropower. In-stream uses – such as keeping fish and wildlife alive, recreation, or aesthetics – did not qualify for prior appropriation.

Groundwater was regarded as a separate category of water and allocated under the principle that landowners control water beneath their land (Wilkinson, 1992).

Rivers as corridors

Rivers and their associated riparian and floodplain zones attract the interest of disparate societal segments seeking an advantage in using those areas. For decades, the dominant philosophy has been to control rivers with dams or other structures to facilitate navigation, control floods, or drain surrounding wetlands. As a result, 60 to 80% of riparian corridors across the continent have been destroyed (Swift, 1984), largely to provide land for agricultural, urban and industrial development.

As US river corridors were degraded, local, state and national initiatives developed to protect what remained, mostly for recreation. Canoeing, rafting, kayaking, and other sports attracted many people to high-quality rivers, and active lobbies arose to protect the last remaining corridors. The National Wild and Scenic Rivers Act, passed in 1968, sought to keep free-flowing certain US rivers with outstanding natural, cultural, or recreational features. To qualify for wild and scenic status, a river (or segment generally 40 km or longer) must be free-flowing and the river must possess outstanding natural or cultural values. Both

the river and its corridor must be relatively undeveloped. Rivers may qualify within 'wild,' 'scenic' or 'recreational' sub-categories. Selection does not require consideration of scientific data, for example, on the status of the regional biota. Initially conceived and implemented to halt the proliferation of dams, the National Wild and Scenic Rivers Act expressly claims unappropriated water amounts needed to fulfil the purposes of the Act (Doppelt *et al.*, 1993). Unfortunately, the late priority date of those rights limits federal claims on water, and thus limits success in protecting the biota of those rivers, despite calls to make biodiversity protection a new focus of the Wild and Scenic Rivers Act (Raffensperger and Tarlock, 1993).

The Canadian Heritage Rivers System (CHRS) was established in 1984 as a cooperative programme of federal, provincial and territorial governments. Like the US National Wild and Scenic Rivers Act, its objective is to give national recognition to Canada's outstanding rivers and to ensure the long-term management and conservation of their natural, cultural, historical and recreational values. Selection criteria focus on the river's natural and cultural heritage, its recreational opportunities, and the feasibility of managing the river. As in the United States, selecting Canadian rivers for conservation does not consider scientific data.

Watersheds and watershed ecology

Until recently, few agencies or organizations considered watersheds as primary units for management attention, and virtually no laws concentrated on watersheds. Only in 1991 did USEPA form an Office of Wetlands, Oceans, and Watersheds, amalgamating several areas of interest. The agency's 1991 report, *Watershed Events* (USEPA, 1991), broadened an historically narrow view of environmental protection. The report defined an innovative watershed protection approach with five major components: (1) address watershed protection in a holistic fashion, including ecological and human health issues; (2) implement watershed-specific plans, including all stressors from all sources; (3) consider all chemical, physical, and biological effects; (4) coordinate programmes to include all interested parties; and (5) assess progress and develop and improve tools and programmes.

Numerous federal and state initiatives across the nation have adopted the watershed perspective. Problems these initiatives aim to deal with include protection of old-growth forest (Forest Ecosystem Management Team: FEMAT, 1993) and declining anadromous fish in the Pacific Northwest (Spence *et al.*, 1996), degradation in the extensive wetlands and rivers of South Florida (Harwell, 1997), pollution of Chesapeake Bay and flooding in the Mississippi River (Interagency Floodplain Management Review Committee: IFMRC, 1994). Although watersheds are not central to any current legal doctrine, they are central to a growing number of regulatory approaches. Watersheds are also the focus of NGO and scientific initiatives to protect rivers and their landscapes, such as the Pacific Rivers Council's *Entering the Watershed: A New Approach to Save America's River Ecosystems* (Doppelt *et al.*, 1993). Others too have joined the call for ecosystem and watershed management and restoration (Kim and Weaver, 1994; Rapport *et al.*, 1995; Samson and Knopf, 1996; Meffe and Carroll, 1997; Williams *et al.*, 1997). Paradoxically, dissatisfaction with the Endangered Species Act has fuelled some of these advances.

The US Endangered Species Act

The purpose of the Endangered Species Act is to provide a means for conserving the ecosystems that endangered and threatened species depend on and to provide a programme for protecting such endangered and threatened species (NRC, 1995). Passed in 1973, the Act has been called 'the most far-reaching wildlife statute ever adopted by any nation' (Reffalt, 1991), because it extends legal protection to the existence of non-human species.

The Act can be a strong and effective tool for the conservation of aquatic species, but despite its power, the law has drawbacks (Moyle and Yoshiyama, 1994): (1) by the time a population qualifies for listing, it is already severely diminished and prospects of recovery may be dim; (2) use of the law to invoke strong conservation measures makes the Act a focal point for major controversies among affected parties; (3) protecting a listed species takes precedence over protecting an unlisted species, even though unlisted species might be in severe decline; (4) 'quick-fix' measures such as hatcheries and captive rearing often are favoured over longer-term and more difficult programmes such as habitat and ecosystem conservation; and (5) more species deserve listing than can be designated with present funding and staffing and more species qualify for listing than can be managed.

ROLE OF VOLUNTARY ORGANIZATIONS

Voluntary organizations working for river and river corridor conservation are widespread and effective in the United States and growing in importance in Canada. These NGOs range from large, multinational groups such as the Nature Conservancy and Sierra Club, for which rivers are but one part of a broad environmental agenda, to national and regional NGOs focusing specifically on rivers, to local watershed councils operating at the community level. All are not-for-profit organizations that rely mainly on membership fees and grants from private foundations to pay their operating expenses. The staff of larger organizations consists of highly trained professionals, including lawyers, policy specialists, scientists and others skilled in membership campaigns, fund raising and organizational management.

The number of NGOs working directly or indirectly on river conservation in North America is huge and any compilation is quickly outdated. For the US, the *1996–1997 River and Watershed Conservation Directory* (River Network, 1996) includes approximately 3000 organizations and agencies in the United States whose missions directly involve river or watershed conservation. This directory does not include all local chapters of national NGOs or all local programmes because of the spontaneous nature of small, grassroots organizations and because some large groups work on river issues episodically or may accomplish river protection only incidentally (see Showers, this volume).

Large environmental NGOs that work on river conservation include the Nature Conservancy (828 000 members), Sierra Club (550 000 members), World Wildlife Fund of the USA (1.2 million members: Abell *et al.*, 2000), National Wildlife Federation (4 million members), Environmental Defense Fund (200 000 members) and Natural Resources Defense Council (170 000 members). Some NGOs are more concerned with clean drinking water than with aquatic ecosystem health. Still, the Nature Conservancy helps maintain the Natural Heritage database of endangered species (e.g. Figure 1.2); is developing a freshwater classification system to catalogue US aquatic biodiversity (Nature Conservancy, 1997); maintains a network of reserves, including reserves centred on rivers (see '*In Situ* Conservation'); and aids citizen groups in local watershed protection (see 'Restoration and Rehabilitation'). Many of the approximately 400 local chapters of the Sierra Club are active in specific river issues and the national organization concentrates on issues affecting rivers within federal lands.

Several NGOs specifically address river conservation; some maintain Web sites with detailed information on current issues. Anglers are the core members of both the Izaak Walton League (www.iwla.org) and Trout Unlimited (www.tu.org) and both traditionally deal with issues of concern to sportfishers. A wide variety of river enthusiasts with recreational, aesthetic and conservation priorities contribute to groups such as American Rivers (www.amrivers.org), River Network (www.teleport. com/~rivernet/), River Watch Network (River Watch Network, 154 State Street, Montpelier, VT 05602, USA) and the Pacific Rivers Council (www. pacrivers.org). River Network serves as a clearing-house of information and coordinator of activities for the thousands of local watershed groups whose concern is often motivated by direct threats to rivers in their neighbourhoods. The EPA also provides access to information: Office of Wetlands, Oceans and Watersheds (www.epa.gov/owow) and the newsletter *Volunteer Monitor* (owow/monitoring/volunteer/vm_index.html).

Owing to the sheer diversity of river protection NGOs and the need to consolidate for effective advocacy, coalitions of environmental groups are increasing in number. The Hydropower Reform Coalition, for example, focuses on river conservation and restoration through improved operation of hydropower dams, particularly in the US Northeast. Its 31 members include recreational river users, such as fly-fishers and whitewater boaters and conservation organizations, such as preservation trusts and other 'friends-of-the-river' groups. The Save Our Salmon Coalition grew up around rivers and watersheds of the Pacific Northwest, where the extirpation and near-extirpation of hundreds of salmon stocks (Nehlsen *et al.*, 1991) is common knowledge and a widespread concern. Its 44 member groups include Northwest fishers, conservationists, scientists, business people and private citizens. The Coalition to Restore Urban Waters, formed in 1992, includes more than 175 mostly grassroots groups working to restore urban waters.

Professional scientific societies also have non-profit status. They traditionally serve their members by publishing scholarly journals and organizing annual meetings. Excellent technical volumes have also appeared that focus on river conservation (Smith and Hellmund, 1993). Increasingly, however, scientific societies are calling on sub-committees of their members to provide scientific judgement on

environmental issues and prepare 'white papers' on particular topics. For example, *Fisheries*, a monthly publication of the American Fisheries Society, frequently publishes position papers (e.g. Winter and Hughes, 1997) or articles on the conservation status of freshwater biota (e.g. Taylor *et al.*, 1996). The North American Benthological Society (NABS) has a standing committee on conservation, which, among other activities, distributes educational slide sets on rivers and riparian conservation. Papers on freshwater conservation are published in the society's journal (Pringle and Aumen, 1993). In 1997, the NABS Conservation Committee joined the US Natural Resources Conservation Service to establish a graduate student conservation fellowship programme. An *ad hoc* committee of the Ecological Society of America was formed to comment on the management plan for Pacific Northwest forests, in which river corridor protection was a central element. Scientists and scientific societies are increasingly speaking out on river conservation in North America.

Conservation in practice

APPLICATION OF LEGISLATION

The success of laws, even the best laws, is determined by society and its public policy desires, expressed in the way those laws are implemented. Laws can provide a logical framework to protect and conserve rivers, but the laws and implementing regulations must evolve as societal values and attitudes change and as knowledge about the effects of human activities increases. Too much current water law ignores the connectedness of water – the river network – while the regulations trivialize the importance of river conservation. The absorption of wastes and other onslaughts from human society are consistently given higher regulatory priority.

The current statutory framework comprises disparate doctrines peculiar to certain regions (e.g. riparian and prior appropriation doctrines), conflicting jurisdictions and inappropriate dichotomies that defy common sense as well as modern scientific knowledge. National, regional, state and local agencies manage water quality separately from water quantity and allocation, point-source pollution apart from non-point pollution and fish-bearing from non-fish-bearing streams. Bureaucracies vested in the *status quo* are often reluctant to adopt new approaches based on scientific advances, and the scales of policy and action rarely match the scales of rivers and river resources as science understands them today.

The US water quality programme illustrates these complex problems. Despite massive expenditure to protect water quality over the past 25 years, evidence of the expenditure's effectiveness is equivocal at best; no one really knows if other approaches might have yielded better water quality. Under the Clean Water Act, taxpayers and the private sector have spent more than $540 billion on controlling water pollution since 1972, mostly on end-of-pipe techniques to control point sources of pollution, such as municipal and industrial waste discharges (Knopman and Smith, 1993).

An analysis of water quality trends for about 30 variables regularly sampled at more than 300 locations on major US rivers from 1974 to 1981 documents major changes in stream water quality resulting from point-source controls (Smith *et al.*, 1987). Faecal coliform and faecal streptococcal bacteria decreased widely thanks mainly to improved wastewater treatment. Point-source controls seem to have reduced phosphate loads in the Great Lakes and Upper Mississippi regions, but not in other regions. Nitrates and phosphates showed different trends for several reasons: (1) atmospheric nitrogen deposition is an important factor, especially in the eastern half of the United States; (2) phosphorus adsorbs to sediments and can thus be stored for considerable time in river channels; and (3) point-source controls have targeted phosphates because phosphorus is considered more limiting to eutrophication. Nitrate loads to the Great Lakes have increased, but phosphate loads have declined.

Point-source pollution control has thus had some successes: effluent from municipal sewage treatment is now ranked second as a source of river pollution, but control of non-point pollution is rudimentary to non-existent (Smith *et al.*, 1987). Nitrate, chloride, arsenic and cadmium concentrations have risen widely, probably because of increased use of nitrogen fertilizers and salt on highways. Suspended sediment, an important (Figure 1.4) and understudied stream pollutant (Waters, 1995), has both increased and decreased locally, perhaps because of high concentrations in agricultural run-off and countervailing effects of reservoirs, which trap sediment. Agriculture has now surpassed effluent as the leading cause of impairment; bacteria and sediments are the leading pollutants (Figure 1.3).

Such ranking, however, may reflect bureaucratic guesswork more than measured condition in US rivers.

It is nearly impossible to judge how effective spending programmes have been, despite reporting requirements imposed by the US Clean Water Act (Knopman and Smith, 1993; see also Adler *et al.*, 1993).

State reports of impaired river length are limited in value for three important reasons. First, in 1994 only about 11% (630 000 out of 5.6×10^6 km of streams and rivers) were assessed by looking at the biota ('aquatic life use attainment') (Davis *et al.*, 1996). The percentage of river length looked at biologically varies among the states – for example, Alabama (1.1%), Michigan (9.2%) and Washington (0%). Second, even when states have 'aquatic life' goals in their water quality standards, most measure water quality by chemical criteria, assuming that these criteria are reliable indicators of biological condition, which they are not. By chemical standards, 25% of surveyed river kilometres was considered impaired; in contrast, 50% of river kilometres was judged impaired when biological condition was measured directly (Davis *et al.*, 1996). Chemical standards, the core of most state programmes, chronically underestimate the extent of river degradation (Karr, 1991). Third, the states' habit of choosing sampling sites without a plan or statistical sampling design has resulted in a biased sample population.

Changing modern legal doctrine to match the connectedness of water and river resources is vital for river conservation. In an important 1994 decision, the US Supreme Court affirmed the intimate connection between water quality and water quantity, for the first time demonstrating a federal interest in making a legal connection between the two (Karr, 1995; Ransel, 1995). In this case, the Court upheld the Washington Department of Ecology's argument that minimum in-stream flows had to be established to protect high water quality for anadromous fish migration because doing so was necessary to comply with the state's water quality standards, mandated by the Clean Water Act. In contrast, the Washington State Supreme Court had earlier backed away from language in the state's own Water Resources Act of 1971, which expressly recognized the 'natural interrelationships of surface and ground waters'. This decision limits the Department of Ecology's ability to decide conflicts over water rights by essentially denying the connection between ground and surface waters (Johnson and Paschal, 1995). The state Supreme Court gave the prior appropriation doctrine precedence over the public trust doctrine, which requires the protection and perpetuation of natural resources as a judicial imperative. River degradation resulting from these

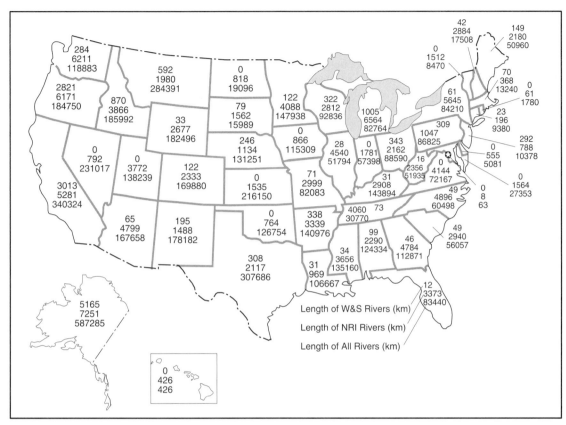

Figure 1.6 *State-by-state comparisons of river lengths (km) showing degree of federal protection. Wild and Scenic (W&S) rivers are those that receive federally legislated protection. Lengths of W&S rivers (top numbers) were obtained from a list provided by the Division of Park Planning and Special Studies of the National Park Service (NPS). The Nationwide Rivers Inventory (NRI) is a register of rivers that potentially qualify as W&S rivers, but they are not legislatively protected by this designation (some of these rivers receive state legislative protection). Lengths of NRI rivers (middle number) were obtained from the Rivers, Trails and Conservation Assistance Program of the National Park Service (www.nps.gov/rtca). In most cases, W&S river lengths are not included in river lengths listed for NRI. Total river lengths in each state (bottom number) were obtained from individual states' Clean Water Act 305(b) reports to the US Environmental Protection Agency (USEPA, 1994a)*

contradictory and narrow approaches has left few high-quality rivers to conserve.

Finally, despite its great potential, the National Wild and Scenic Rivers Act has been underused for conservation. The distribution of designated rivers varies substantially among states and regions; many of them are concentrated in the Pacific Northwest and Alaska (Figure 1.6). Nationally, only 154 river segments and about 17 400 km are protected, less than 0.3% of the 5.6 × 10⁶ river km in the US. Mine drainage affects more than that amount of river and dams drown 60 times that length (Raffensperger and Tarlock, 1993; Cassidy, 1994). Even protected areas can be threatened, however, because the Act typically

protects only segments and a total corridor width of only 0.8 km – as if upstream development and surrounding land use has no effect on the protected segment. In only a few cases are expanded corridor widths or entire watersheds protected, such as the Buffalo National River (Arkansas) and the Ozark National Scenic Riverway (Missouri) (www.nps.gov/parks.html). Specific information on individual rivers may be obtained through a National Park Service website (www.nps.gov/rivers/).

Present selection of rivers for wild and scenic status is based upon their inclusion in the Nationwide Rivers Inventory, a comprehensive survey completed in 1982 (Benke, 1990). This inventory is a register of river segments with potential to qualify as national wild, scenic and recreational rivers (for a current listing, see www.ncrc.nps.gov/rtca/nri/). Assessment criteria for the original inventory were general and qualitative (Benke, 1990): Was the river free-flowing and 40 km or longer? Were the channel and riparian corridor relatively undeveloped? Did the river have outstanding natural and cultural values? Current assessment procedures still lack comprehensive biological criteria. Although the Nationwide Rivers Inventory itself provides no legislated protection, a presidential directive requires each federal agency to avoid adverse effects on rivers identified in the inventory.

Total river lengths listed under both the Nationwide Rivers Inventory and Wild and Scenic Rivers amounts to 153 500 km, which means that only 2.7% out of the total river km in the US (5.6×10^6 km) are considered of sufficient quality to warrant protection. Of those high quality river kilometres, only 11% are protected by Wild and Scenic River status.

The Canadian Heritage Rivers System in 1998 included 30 nominated rivers, of which 19 had been designated (Figure 1.7; www.chrs.ca/); in other words, a management plan has been written describing necessary actions to protect the rivers' outstanding features. Together, designated and nominated rivers comprised a total length of 8244 km. They range from northern rivers nominated by aboriginal groups seeking to protect their heritage to rivers such as the Grand in Ontario, where more than 60 communities participate in a management plan to protect the river and its catchment. Rivers flowing mainly through Crown Lands are often of higher quality and easier to administer than those flowing mainly through private lands.

United States and Canadian river conservation laws differ in a number of respects. The US National Wild and Scenic Rivers programme and the National Park programme are federal initiatives, but 37 states also have legislation to protect outstanding state rivers. US federal legislation is strong, whereas federal control is weak in Canada. Unlike the National Wild and Scenic Rivers programme, the Canadian Heritage Rivers System has no standing legislation. Provinces retain jurisdiction over their natural resources.

IN SITU CONSERVATION

Too often, protection measures are put in place only after a species or site is seriously endangered and has received some notoriety, such as official listing under the US Endangered Species Act (ESA) or National Wild and Scenic Rivers Act. Protection and restoration are thus done on an emergency basis, at a time when species or places may well not recover or even be protected against current actions. Fearing that tens or even hundreds of species or populations will receive the ESA's special protections, a variety of interests have begun some cooperative planning, making 'watershed conservation' the buzzword of the late 1990s. Efforts to avoid the perceived 'hammer of the ESA' include attempts to manage all the salmon of the Columbia River basin (NRC, 1996b; Independent Scientific Group – ISG, 1996) and the 24 federally listed endangered and threatened fish species of the lower Colorado River (plus one proposed, four candidate and 67 species 'at risk' of being listed; US Fish and Wildlife Service: USFWS, 1995). Listings of coho salmon in Oregon and chinook salmon of Puget Sound, Washington, have prompted similar flurries of activity.

The extent of river protection resulting from reserves, specially designated rivers and river corridors, or parks and other protected lands has received little study, but the evidence from the southwestern United States, an area with many federally listed species, is disheartening (Williams, 1991). For native fish, official reserves and refuges (defined as areas managed for one or a few species, rather than for the entire biota) were rare before 1970. The Nature Conservancy established one reserve in 1966 and three small federal reserves have been established since 1970. Federally designated wilderness areas, as well as national monument and national park lands, harbour endangered western fish; some are vital for the protection of threatened species. But none of these reserves was created principally for aquatic conservation. Even large reserves that appear superficially pristine can be impoverished in native biodiversity. A survey of Cataract Canyon of the Colorado River within Canyonlands National Park, Utah, for example, discovered that only seven of 23 collected species were native; fully 83% of the individuals collected were non-native (Valdez and Williams, 1986).

Reserves should protect entire, naturally functioning native communities and the ecosystem in which they are embedded (Moyle and Sato, 1991; Meyer, 1997) – a criterion that potentially excludes many public lands that are managed for multiple uses including resource extraction and recreation. Ideally, reserves should permit the full range of natural variation and

Figure 1.7 *The Canadian Heritage Rivers System (CHRS), established in 1984, includes 30 rivers across Canada, of which 19 have been formally designated. Rivers are considered designated (black) when a management plan has been written to describe the necessary actions to protect the outstanding features of nominated (grey) rivers. Designation of Heritage Rivers is ongoing; consult http://www.chrs.ca/ for updates*

disturbance, including floods and drought, fires and other events that are part of the natural dynamics of regional landscapes (Poff *et al.*, 1997). Reserves should maintain the full complement of native species (although species dominance will fluctuate). In many instances, achieving these goals will pose challenging, perhaps impossible, minimum size requirements for reserves. For naturally small systems such as springs,

or remote areas such as northern Canada, such requirements might be met.

Large rivers in extensively developed areas (e.g. the middle Mississippi River in the upper US Midwest) present special problems; individual reserves are still needed, but they should exist within the context of a regional plan (Moyle and Sato, 1991). The concept of 'beads on a string' and differing levels of protection come into play in these situations. Not all reserves in such areas could provide the highest level of protection. Moyle and Sato propose a system of six classes from 1 (best) to 6 (worst). Class 1 reserves would protect a catchment from ridgeline to ridgeline with a nearly complete native biota. Such status might be possible in mountainous or other less-developed areas or on public land and protected private lands such as those owned by the Nature Conservancy. Class 2 reserves would permit increased human presence – for example, light recreation or walking only – but otherwise provide many of the attributes of class 1 reserves. The potential for restoration to class 1 status is high for class 2 reserves. Many state and federal US lands contain class 2 rivers, but various human activities, including timber harvest, grazing, recreation and associated road-building, stand in the way of class 1 status. A key goal for aquatic conservation should be restoration and maintenance of as many rivers as possible as classes 1 and 2. Class 2 reserves may be the most important because they are numerous, large and already being enjoyed and supported by the public as parks and public lands (Moyle and Sato, 1991).

A strategy for protecting native biota and ecosystems through an aquatic reserve network begins with a spatial analysis: identification of regions for which a reserve network is desirable, possibly using databases for endangered fish; identification of waters with a large percentage of native species; and development of a priority list of places to acquire and manage on the basis of conservation principles (Moyle and Sato, 1991). US Fish and Wildlife Service scientists have begun such a process for terrestrial landscapes and species, termed 'gap analysis' (Scott *et al.*, 1987) and some early efforts are under way to develop an aquatic gap programme (Meixler *et al.*, 1996; see also www.dnr.cornell.edu/hydro2/aquagap.htm).

Gap analysis of aquatic systems would be useful for highlighting the unique and unmet challenges of aquatic conservation. Springs, small streams and headwaters can probably be protected in the same way as a special terrestrial feature. Yet the interconnectedness of river ecosystems and the mobility of aquatic species can complicate efforts to determine the value of individual protected areas. Large rivers, which typically harbour the greatest biodiversity, are unlikely to be protected through governments acquiring and setting aside entire catchments. Hence land management, rather than land acquisition, will form the backbone of aquatic

conservation in most situations (Meixler *et al.*, 1996). Because human actions have already modified so many rivers in North America, ecological restoration must also play a role in designating a network of aquatic reserves.

A plan to protect aquatic biodiversity in California illustrates a multi-tiered approach in a state where dense human populations, high water demand and high native biodiversity are on a collision course (Moyle and Yoshiyama, 1994). Of 115 native fish taxa, nine are extinct, 16 are formally listed as threatened or endangered, 26 qualify for listing and 26 more are likely to qualify soon for listing – nearly 60% of the total fish fauna. In the conservation plan proposed by Moyle and Yoshiyama (1994), only tier 1, which focuses on immediate designation of species facing regional extirpation, is actively used. Rarely attempted in California or elsewhere are more integrative management plans for co-occurring clusters of declining species (tier 2), designation of aquatic diversity management areas to provide systematic regional protection (tier 3), designation of key watersheds for special management and restoration (tier 4), or coordination of activities in cooperative multi-regional programmes (tier 5).

EX SITU CONSERVATION

Some conservationists might view '*ex situ* conservation' as an oxymoron or, at best, a last-ditch strategy. The term includes germplasm banking, DNA collections, maintenance of frozen viable cells and live collections maintained in zoos, aquaria and other facilities for captive propagation (Meffe and Carroll, 1997). *Ex situ* conservation can play a role in the conservation of North American riverine taxa, though such efforts for these taxa are still rare.

Propagation of fish in hatcheries for commercial and recreational fishing has a long history. The success of these efforts is decidedly mixed. Hatcheries are typically used to supplement natural production of harvested species; the goal is usually to mitigate other human activities that have decreased natural production (NRC, 1996b). However, in the Pacific Northwest, for example, the steep decline of many valuable salmon stocks, despite a rising output of hatchery-reared juveniles, has shown that hatcheries alone cannot compensate for habitat loss, overfishing and dams or other barriers. Moreover, for morphological, physiological and behavioural reasons, hatchery fish perform poorly in comparison with wild fish (White *et al.*, 1995). Hatchery fish also have direct adverse effects on wild fish, competing with them for food, even preying on them. Hatchery operations also degrade water quality, reduce flows and divert funds from activities such as habitat protection. In the Pacific Northwest hatchery operations often consume 40% of state fishery budgets (White *et al.*, 1995).

In hatcheries, as in all *ex situ* conservation activities, genetics is fundamental. Healthy species usually contain much genetic diversity, both within local breeding populations (demes) and among them (NRC, 1996b). Loss of diversity through local extirpation, fragmentation of once-connected populations and careless selection of breeding stock for hatcheries all represent serious threats to long-term species survival (Busack and Currens, 1995; NRC, 1996b).

Captive breeding is often criticized on the grounds that human intervention might harm further, that it weakens the incentive to conserve habitat and that loss of genetic variability and changes in behaviour under captive conditions produce individuals ill-prepared to survive in the wild (Johnson and Jensen, 1991; see also Benstead *et al.*, this volume). Disease outbreaks are often worse in hatcheries and may be carried to wild populations. Still, hatcheries and the accumulated knowledge of scientists and hatchery operators represent an extensive, existing infrastructure for the propagation of threatened and endangered species *if* hatchery operations can resolve the accompanying problems. When captive breeding poses the only realistic alternative, it should be carried out, because it offers an opportunity to protect the genetic diversity of endangered populations and propagate enough individuals for reintroduction (see also Benstead *et al.*, this volume). Captive breeding is part of the conservation strategy for some endangered river-dwelling fish and is being developed for some mussels; it may also aid conservation of threatened crayfish.

Redirecting hatchery efforts towards culturing endangered fish species is a recent development (see Schramm and Piper, 1995) whose potential is high to aid recovery, providing that such culture takes place within holistic programmes that include habitat protection and restoration (White *et al.*, 1995). At present, 33 US fish hatcheries (out of a total of 74) work with 28 federally listed fish species and 11 state-listed species in accordance with recovery plans; their goal is to prevent further declines and listings.

Dexter National Fish Hatchery in New Mexico has been a leading site for propagating warm-water fishes of the American Southwest; since 1974, 24 species and

sub-species, mostly inhabitants of rivers or desert springs, have been held at this facility. The Dexter Hatchery pioneered rearing methods, crucial for endangered taxa, such as once-through rather than recirculated water to minimize disease, strict measures to prevent mixing of potentially interbreeding stocks that may interbreed, and extra attention to preventing escape, which might establish non-native populations (Johnson and Jensen, 1991).

Some riverine species present special problems in recovery because of their large body size and long life-span (e.g. Acipenseridae, some Cyprinidae, Catostomidae, Ictaluridae). Augmentation plans for the razorback sucker *Xyrauchen texanus* in the upper Colorado River basin (Modde *et al.*, 1995; Mueller, 1995) show how *ex situ* conservation efforts can aid recovery of such an endangered species. Razorback recovery incorporates genetic data for remnant stocks (Dowling *et al.*, 1996a,b), a breeding strategy to maintain diversity and flows to inundate bottomlands and better manage nursery sites. Larvae are first harvested from natural spawning, reared in dedicated backwaters and hatcheries away from non-native predators until they are large enough to avoid predation and, finally, repatriated. Future work will create habitats that exclude non-native predators, where the razorbacks and other native fish can form self-sustaining populations.

Short-lived, smaller species benefit most from zoos, aquaria and private aquarists. Some live-bearers and killifishes and their allies (families Poeciliidae, Cyprinodontidae and others) are common aquarium species, but others (e.g. Percidae) are not and may require special conditions for propagation. At least 25 of 42 western North American species of the former groups are endangered or extinct in the wild, though most are readily bred and maintained in captivity. Ten or more of the endangered forms are or were artificially maintained as part of recovery efforts (Minckley *et al.*, 1991); a number of these only persist under artificial conditions (e.g. Marsh and Sada, 1993). Again, genetic variation is a major concern because of fragmentation of natural and artificial stocks, small population sizes and other factors.

Although as many as 162 crayfish taxa deserve some form of protection or recovery measure, only four species are now recognized as endangered under the ESA; 47 species receive US state-level protection (Taylor *et al.*, 1996). This fact is particularly disturbing because no major technical obstacles prevent hatchery cultivation of many crayfish species. Although commercial aquaculture focuses on a half-dozen species, all crayfish species that have been studied can be cultured with little difficulty (Huner, 1994) and culture techniques could probably be developed for many or most of the endangered species. Yet we know of no efforts to use captive propagation in crayfish recovery; indeed, few recovery programmes are in place for any crayfish.

Cummings *et al.* (1997) provide an excellent overview of current activities in the conservation and management of freshwater mussels. Relocation of unionid mussels from one site to another is not uncommon; a literature search by Cope and Waller (1995) identified 37 discrete projects that relocated nearly 90 000 individual mussels, often to comply with the Endangered Species Act. Mean mortality was near 50%, but only 16% of the populations were monitored for five years or more. Captive propagation is essentially non-existent. At present, only one ESA-listed endangered species (the tan riffleshell, *Epioblasma florentina walkeri*) is being propagated as part of a recovery programme (R. Neves, Virginia Technical University, Blacksburg, VA, personal communication).

The complex reproductive biology and life cycle of mussels – particularly the need for the larvae of most species to infest a fish host briefly before they can become juveniles – challenge captive propagation. It will be difficult to locate gravid adults of rare, wild populations and to identify the appropriate fish host, known for only about 30% of mussel species. Nevertheless, freshwater mussels' relatively simple behaviour, high fecundity, great longevity and poikilothermic physiology make them otherwise suitable for captive propagation (Neves, 1997). Complete life-cycle propagation is probably achievable, but, in the short term, reliance on wild populations for brood stock is likely to be most successful (R. Neves, personal communication).

RESTORATION AND REHABILITATION

Widespread degradation of river ecosystems makes it urgent to protect the remaining best places. Just as important is the restoration and rehabilitation of already degraded systems (Williams *et al.*, 1997), something proposed several years ago by the US National Research Council (NRC, 1992). In the NRC committee's view, 'Without an active and ambitious restoration program in the United States, our swelling population and its increasing stresses on aquatic ecosystems will certainly reduce the quality of human life for present and future generations.' The committee recommended a national target of 700 000 km of river–

riparian ecosystem restoration to be accomplished within the next 20 years.

The NRC also documented numerous past restoration efforts, whose outcomes and practices gave little reason for optimism. Many efforts failed to use available information and post-project evaluations, when they existed at all, were inadequate. Yet post-project evaluations are essential if river restoration is to succeed (Kondolf, 1995; Williams *et al.*, 1997). Such evaluations should include clear objectives, baseline data, good study design, commitments to the long term and willingness to acknowledge failures. The NRC committee noted several shortcomings of many restoration programmes: (1) programmes were often steeped in too powerful analytical systems; (2) programmes lacked adequate biological insights or imaginative questioning and hypothesizing; and (3) the spatial and temporal scope of programmes was far too small and too fragmented to accomplish their long-term goals.

Thus, little progress has been made toward the NRC's target restoration goal, but some restoration efforts are moving forward and the potential for implementing others is high (Williams *et al.*, 1997; Harwell, 1997).

Hydropower dam relicensing

The Federal Energy Regulatory Commission (FERC), an independent federal commission in the US Department of Energy, licenses private and public dams built for hydropower generation. Licences typically last 30 to 50 years. They stipulate how the dam is operated, including minimum flows, fish passage and, in some cases, how lands are managed. Dams constructed decades ago were subject to few requirements for species and ecosystems. Many dams licensed in the 1950s or earlier are (or will soon be) under FERC's review for relicensing. In 1993, 160 licences expired on 262 dams in 105 rivers. Only about half these relicensings had been completed by 1996; another 550 dams are due for relicensing in the next 15 years. A number of NGOs, as well as the Hydropower Reform Coalition, view the relicensing process as an excellent opportunity to influence how dam operations affect many rivers for the next 30 to 50 years (Bowman, 1996).

Opportunities to influence the relicensing process include commenting on the environmental review required by FERC, commenting at public hearings and filing a motion for intervention, which allows interested individuals, agencies, or groups to become a formal party to the proceedings. Only formal intervention confers the official standing needed to seek judicial review of a FERC decision. Interested parties may then negotiate licence terms to include conservation and recreation concerns. Such negotiations may actually expedite the long and complicated relicensing process because the agreements may allow the dam operator to avoid expensive environmental impact studies. They also help establish a framework for long-term cooperative agreements instead of development-versus-conservation battles.

Relicensing opens the way for any number of improvements to dam operations (Bowman, 1996). Mitigation for recreation might include improved access, such as parking lots and boat launches, flow releases scheduled to peak usage by whitewater boating enthusiasts, and aesthetic improvements to dam and power-generating structures. Examples of mitigation for conservation often involve modifying the flow regime to enhance aquatic life, returning flows to bypassed reaches and maintaining water levels for riparian species. Upstream and downstream fish passage was inadequately provided for at the first licensing of many dams and now poses major difficulties. New construction, changed timing of flow release, hatcheries, transporting young fish downstream in barges, and monetary compensation have all been thrown at this problem. For example, Consumers Power, which operates 11 hydroelectric projects on the Manistee, Muskegon and Au Sable Rivers in Michigan pays $575 000 annually into a habitat restoration fund. Once fish protection devices are installed, this contribution will be reduced, depending on how effective the protection system is.

In 1997 the Hydropower Reform Coalition won a significant victory, persuading FERC to order the removal of Edwards Dam on the Kennebec River, Maine, in order to restore fish populations. This case was the first in which FERC ruled against dam operators, who had fought to keep the dam.

Urban rivers

Most urban rivers face a common set of problems: extensive areas made impervious by transportation networks and building rooftops, untreated and treated wastes from municipal or industrial sources, channel modifications for flood control and shipping, contaminated stormwater run-off and more. Urbanization increases the size and frequency of floods because development increases impervious surface area and reduces infiltration (Arnold and Gibbons, 1996). Increases in peak discharges and duration of high flows make urban waterways more susceptible to stream-channel incision (Booth, 1990). Cities often occupy the floodplains of lowland rivers, transforming them into built structures protected from inundation by various devices. Gravel bars, side channels and woody debris – important components of habitat complexity – are lost in rivers channelized for navigation and rapid drainage of flood waters. Numerous studies report that macroinvertebrate and fish assemblages are harmed by urban development (Jones and Clark, 1987; Lenat and Crawford, 1994; Weaver and Garman, 1994; Karr, 1998).

Restoring these rivers poses special challenges (Sparks, 1995; Schueler, 1996) and some cities are taking steps to meet them. A number of North American cities post signs warning residents not to dump wastes into urban storm drains that discharge into local rivers. A restoration project for the Rouge River in Detroit, Michigan, centres on constructing underground storage reservoirs to hold temporarily and then gradually release rainwater from major storms, lessening sudden floods from storm run-off.

Waterways in the Chicago metropolitan area have become an example of successful restoration of an urban river. The Chicago and Calumet river systems originate in what is now farms and suburbs surrounding the metropolitan area. For more than a century, the watersheds have been used for agriculture, fishing, flood control, industrial water supply, power generation, recreation, transportation, waste assimilation and wildlife habitat. The growth of metropolitan Chicago and the disposal of untreated wastes from homes, slaughterhouses, tanneries, distilleries and other industries caused serious pollution. In 1885 an intense summer storm carried raw sewage to the intake in Lake Michigan for the city's drinking water supply. Some 100 000 Chicago-area residents died of diseases contracted from the contaminated water. The state of Illinois in 1898 created the Chicago Sanitary District and a century of engineering and construction projects began. Channels and canals were built, permanently reversing the flows of the Chicago and Calumet rivers away from Lake Michigan and diverting contaminated water into the Mississippi drainage (inadvertently creating a human-made link between the St Lawrence (Great Lakes) and the Mississippi, facilitating interbasin dispersal of non-native species: Karr *et al.*, 1985).

An extensive network of sewers and wastewater treatment plants now collects and treats wastewater from 125 municipalities with a combined population of 5.5 million. Steady improvements in system design and capacity, particularly a reduction in combined sewer overflows, have measurably reduced chemical contamination, such as ammonia and suspended solids, and increased dissolved oxygen. As a result, fish species in the Chicago and Calumet river systems have increased substantially (Polls *et al.*, 1994). Still, as in many old North American cities, most Chicago-area sewage drains and storm drains are combined, which means that wastewater treatment plants must release untreated sewage into rivers during severe storms because they cannot handle the large volumes of water. Huge underground tunnels have been constructed to store overflows for release later. Expenditure on Chicago waterways has been enormous: since the Sanitary District was created in 1889, some $16.1 billion (in 1996 dollars) have been invested in capital projects to control water pollution.

Parallel efforts by Chicago-area volunteers and NGOs have also contributed to recovery. Since 1979, Friends of the Chicago River has worked to restore the river to ecological health and also to improve public access, recreation and aesthetics. Urban planning along the river is gradually modifying the riverfront, making it attractive and accessible. Re-establishing the river as a key feature of Chicago's landscape has helped draw support from the business community. New parks and trails are nearing completion and Friends' activities are expanding upstream into the watershed. The Friends' work is most effective where feasibility, citizen interest and leadership coincide.

Kissimmee River restoration

An ambitious, long-term plan to restore the Kissimmee River in central Florida is a cornerstone of a larger effort to restore the Florida Everglades. The Kissimmee basin, the upper portion of a hydrological system of 23 000 km^2 that flows through the Everglades (Berger, 1992), consists of a chain of lakes and the Kissimmee River. The river originally meandered 166 km from Lake Kissimmee to Lake Okeechobee within a marshy floodplain that was 1.5–3 km wide (Koebel, 1995). Mean annual discharge of the original river was about 60 m^3 s^{-1} and unusual in that 94% of the floodplain was inundated 50% of the time (Koebel, 1995). The river ecosystem included a mosaic of wetland habitats (Toth *et al.*, 1995), supporting at least 35 fish species, 16 wading bird species and 16 waterfowl species (Perrin *et al.*, 1982; Toth, 1993; Dahm *et al.*, 1995).

Because of a growing human population in the Kissimmee basin and extensive flooding in the late 1940s, the US Army Corps of Engineers designed a flood control plan for the state of Florida. Channelization of the Kissimmee began in 1962 and was completed in 1971, for a cost of about $30 million. A 90 km channel, 9 m deep and 100 m wide, was dug from Lake Kissimmee to Lake Okeechobee, cutting off much of the previously meandering river and draining much of the floodplain (14 000 ha) (Toth, 1993; Koebel, 1995). Channelization destroyed river and marshland habitats (Toth *et al.*, 1995) and devastated invertebrates (Harris *et al.*, 1995), fish (Trexler, 1995), waterbirds (Weller, 1995) and ecological integrity (Karr, 1990; Dahm *et al.*, 1995).

Conservation groups and the US Fish and Wildlife Service had strongly objected to the Kissimmee project before, during and after channelization. A move to restore the Kissimmee started gaining political support soon after channelization was completed. By the late 1970s, the South Florida Water Management District (SFWMD) and the Corps of Engineers had begun a series of studies to evaluate restoration alternatives (Koebel, 1995). Field research demonstrated that re-establishing hydrological connections would lead to a rapid recovery of wetland vegetation and wildlife in the floodplain; resumption of flows in previously stagnated channels flushed decaying organic matter and restored many river channel characteristics. Several restoration alternatives were evaluated and a plan was selected that included re-establishment of more natural river flows by altering water release schedules in the upper chain of lakes and backfilling the dredged channel, and re-establishing 70 km of contiguous river channel and 11 000 ha of floodplain wetland (Koebel, 1995). A test fill to assess construction methods has been completed and major backfilling and restoration which began in 1998 will take 15 years to complete. The SFWMD has established a restoration evaluation programme that will continue over the entire construction period.

The Kissimmee River restoration is the largest river restoration project ever attempted and will be closely watched by river management agencies and aquatic scientists worldwide. Although the goal is to restore much of the original river–floodplain ecosystem, several physical structures will remain to help reduce the effects of catastrophic floods to surrounding areas. The Kissimmee experience makes a strong statement about the financial cost of environmental ignorance. Restoring the Kissimmee River alone has been estimated to cost more than $400 million and it is just part of restoring the entire Kissimmee–Okeechobee–Everglades system, which will eventually cost more than $1 billion (Berger, 1992).

FUTURE THREATS AND THEIR MITIGATION

The human appropriation of renewable fresh water seems destined to expand as the population grows and migrates from place to place, particularly in arid and semi-arid regions where demand already exceeds supply (Postel *et al.*, 1996; Figure 1.5). Although new large dams are unlikely to be constructed in North America, the mining of aquifers, grand engineering schemes for interbasin water transfers and intense pressure on many small streams and rivers are likely to persist.

Even if no additional threats to rivers arise, existing threats will continue indefinitely. The threats range from extinction to local and regional loss of biological integrity (Karr *et al.*, 1985; Allan and Flecker, 1993) to the homogenization of regional, national and continental river biotas through the spread of alien species (Moyle *et al.*, 1986). Habitat destruction and fragmentation from urban and suburban sprawl combine with agricultural run-off and chemical pollution to degrade significant lengths of North American rivers. Major water diversions and dam operations on large rivers and channelization of small streams continue to threaten waters throughout North America. Livestock grazing damages rivers, especially in the dry West. The broad influence of climate change, an important new threat, can only be guessed at (Firth and Fisher, 1992).

These trends are ominous but not hopeless. With public awareness growing and scientific understanding advancing rapidly, we have the opportunity to protect North America's remaining best rivers and to restore the rest. If we want viable rivers in the 21st century, we have no other choice.

Acknowledgements

Many individuals and organizations helped us locate useful information or critically reviewed sections of the manuscript. We thank American Rivers for making available public awareness surveys, Miriam Steiner and Larry Master of the Nature Conservancy for providing data on endangered fish species and Harry Allan of the US Geological Survey and Beth Porter and Charlie Stockman of the National Park Service for data on the Nationwide Rivers Inventory and wild and scenic rivers. We thank Shirley Ann Off of the Canadian Heritage Rivers System and Ginger Murphy of the US Natural Resources Conservation Service for their help in acquiring figures. Thanks to Tim Donohue, Karen Firehock, Gerry Lieberman, Bob Williams and Michaela Zint for assistance with public awareness and education. Margaret Bowman, Beth Norcross and Tom Princen provided insight into the role of voluntary organizations and Mary Lammert did likewise with the topic of classification. Irwin Polls and Laurene von Klan were invaluable sources on restoration of Chicago's waterways. Thanks also to Mark Bain for assistance with *in situ* conservation; to Jay Huner, Buddy Jensen, W.L. Minckley, Tim Modde and Chris Taylor for their help with *ex situ* conservation; to Charley Dewberry and Bob Hughes for their reviews of the entire manuscript; and to Ellen W. Chu for substantive editing. Many more individuals graciously responded to requests for information. Special thanks to Michael Wade in the Geographic Information System Laboratory of the Department of Biological Sciences at the University of Alabama and Cathy Schwartz in the School of Fisheries at the University of Washington for producing several maps.

References

Abell, R.A., Olson, D.M., Dinerstein, E., Hurley, P.T., Diggs, J.T., Eichbaum, W., Walters, S., Wettengel, W., Allnutt, T., Loucks, C.J. and Hedao, H. 2000. *Freshwater Ecoregions of North America: A Conservation Assessment.* Island Press, Washington, DC.

Adler, R.W., Landman, J.C. and Cameron, D.M. 1993. *The Clean Water Act 20 Years Later.* Island Press, Washington, DC.

Albert, R.C. 1987. *Damming the Delaware.* The Pennsylvania State University Press, University Park, Pennsylvania.

Allan, J.D. and Flecker, A.S. 1993. Biodiversity conservation in running waters. *BioScience* 43:32–43.

Allan, J.D., Erickson, D.L. and Fay, J. 1997. The influence of catchment land use on stream integrity across multiple spatial scales. *Freshwater Biology* 37:149–161.

Angermeier, P.L. and Winston, M.R. 1997. Assessing conservation value of stream communities: a comparison of approaches based on centres of density and species richness. *Freshwater Biology* 37:699–710.

Angermeier, P.L., Neves, R.J. and Kauffman, J.W. 1993. Protocol to rank value of biotic resources in Virginia streams. *Rivers* 4:20–29.

Arnold, C.L. and Gibbons, C.J. 1996. Impervious surface coverage: the emergence of a key environmental indicator. *Journal of the American Planning Association* 62:243–258.

Bahls, L.L. 1993. *Periphyton Bioassessment Methods for Montana Streams.* Water Quality Bureau, Department of Health and Environmental Sciences, Helena, Montana.

Bailey, R.G., Avers, P.E., King, T. and McNab, W.H. (eds) 1994. *Ecoregions and Subregions of the United States.* 1:7 map 500 000; coloured with supplementary table of map unit descriptions, compiled and edited by W.H. McNab and R.G. Bailey. US Department of Agriculture First Service, Washington, DC.

Barbour, M.T., Plafkin, J.L., Bradley, B.P., Graves, C.G. and Wisseman, R.W. 1992. Evaluation of EPA's rapid bioassessment benthic metrics: metric redundancy and variability among reference site streams. *Environmental Toxicology and Chemistry* 11:437–449.

Barbour, M.T., Stribling, J.B. and Karr, J.R. 1995. Multimetric approach for establishing biocriteria and measuring biological condition. In: Davis, W.S. and Simon, T.P. (eds) *Biological Assessment and Criteria: Tools for Water Resource Planning and Decision Making.* Lewis Publishers, Boca Raton, Florida, 63–77.

Barbour, M.T., Gerritsen, J., Griffith, G.E., Frydenborg, R., McCarron, E. and White, J.S. 1996. A framework for biological criteria for Florida streams using benthic macroinvertebrates. *Journal of the North American Benthological Society* 15:185–211.

Barbour, M.T., Gerritsen, J., Snyder, B.D. and Stribling, J.B. 1999. *Rapid Bioassessment Protocols for Use in Streams and Wadeable Rivers: Periphyton, Benthic Macroinvertebrates, and Fish.* EPA 841-B-99-002. Office of Water, US Environmental Protection Agency, Washington, DC.

Benke, A.C. 1990. A perspective on America's vanishing streams. *Journal of the North American Benthological Society* 9:77–88.

Berger, J.J. 1992. The Kissimmee riverine–floodplain system. In: National Research Council. *Restoration of Aquatic Ecosystems: Science, Technology and Public Policy.* National Academy Press, Washington, DC, 477–496.

Bilby, R.E. and Naiman, R.J. (eds) 1998. *River Ecology and Management: Lessons from the Pacific Coastal Ecoregion.* Springer, New York.

Boon, P.J. 1992. Essential elements in the case for river conservation. In: Boon, P.J., Calow, P. and Petts, G.E. (eds) *Rivers Conservation and Management.* John Wiley, Chichester, 11–33.

Booth, D.B. 1990. Stream-channel incision following drainage-basin urbanization. *Water Resources Bulletin* 26:407–417.

Bowman, M. 1996. *River Renewal – Restoring Rivers Through Hydropower Dam Relicensing.* Rivers, Trails and Conservation Assistance Program, US National Park Service, Washington, DC.

Brown, A.V. and Matthews, W.J. 1995. Stream ecosystems of the central United States. In: Cushing, C.E., Cummins, K.W. and Minshall, G.W. (eds) *Ecosystems of the World 22. River and Stream Ecosystems.* Elsevier, Amsterdam, 89–116.

Burton, G.W. and Odum, E.P. 1945. The distribution of stream fish in the vicinity of Mountain Lake, Virginia. *Ecology* 26:182–193.

Busack, C.A. and Currens, K.P. 1995. Genetic risks and hazards in hatchery operations: Fundamental concepts and issues. *American Fisheries Society Symposium* 15:71–80.

Butler, M.J. and Stein, R.A. 1985. An analysis of mechanisms governing species displacement in crayfish. *Oecologia* 66:168–177.

Campbell, R.R. 1997. Rare and endangered fishes and marine mammals of Canada: COSEWIC Fish and Marine Mammal Subcommittee Status Reports: XI. *Canadian Field-Naturalist* 111:249–257.

Cassidy, T.J. 1994. 25 years of the Wild and Scenic Rivers Act. *Fisheries (Bethesda, MD)* 19(3):24–25.

Catton, W.R. 1993. Carrying capacities and the death of a culture: a tale of two autopsies. *Sociological Inquiry* 63:202–223.

Collier, K.J. and McColl, R.H.S. 1992. Assessing the natural value of New Zealand rivers. In: Boon, P.J., Calow, P. and Petts, G.E. (eds) *River Conservation and Management.* John Wiley, Chichester, 195–211.

Cope, W.G. and Waller, D.L. 1995. Evaluation of freshwater mussel relocation as a conservation and management strategy. *Regulated Rivers: Research & Management* 11:147–155.

Costanza, R., d'Arge, R., de Groot, R., Farber, S., Grasso, M., Hannon, B., Limburg, K., Haeem, S., O'Neill, R.V., Parvelo, J., Raskin, R.G., Sutton, P. and van den Belt, M. 1997. The value of the world's ecosystem services and natural capital. *Nature* 387:253–260.

Cuffney, T.F., Gurtz, M.E. and Meador, M.R. 1993a. Methods for collecting benthic invertebrate samples as part of the national water-quality assessment program. *US Geological Survey Open-File Report* 93-406, Raleigh, North Carolina.

Cuffney, T.F., Gurtz, M.E. and Meador, M.R. 1993b. Guidelines for the processing and quality assurance of benthic invertebrate samples collected as part of the national water-quality assessment program. *US Geological Survey Open-File Report* 93-407, Raleigh, North Carolina.

Cummings, K.S., Buchanan, A.C., Mayer, C.A. and Naimo, T.J. (eds) 1997. Conservation and management of freshwater mussels II: initiatives for the future. In: *Proceedings of a UMRCC Symposium,* 16–18 October 1995, St Louis, Missouri. Upper Mississippi River Conservation Committee, Rock Island, Illinois.

Czaya, E. 1981. *Rivers of the World.* Van Nostrand Reinhold, New York.

Dahm, C.N., Cummins, K.W., Valett, H.M. and Coleman, R.L. 1995. An ecosystem view of the restoration of the Kissimmee River. *Restoration Ecology* 3:225–238.

Daily, G.C. (ed.) 1997. *Nature's Services: Societal Dependence on Natural Ecosystems.* Island Press, Washington, DC.

Davis, W.S. and Simon, T.P. (eds) 1995. *Biological Assessment and Criteria: Tools for Water Resource Planning and Decision Making.* Lewis Publishers, Boca Raton, Florida.

Davis, W.S., Snyder, B.D., Stribling, J.B. and Stoughton, C. 1996. *Summary of State Biological Assessment Programs for Streams and Wadeable Rivers*. EPA 230-R-96-007. US Environmental Protection Agency, Office of Policy, Planning, and Evaluation, Washington, DC.

DeShon, J.E. 1995. Development and application of the Invertebrate Community Index (ICI). In: Davis, W.S. and Simon, T.P. (eds) *Biological Assessment and Criteria: Tools for Water Resource Planning and Decision Making*. Lewis Publishers, Boca Raton, Florida, 217–243.

Doppelt, B., Scurlock, M., Frissell, C. and Karr, J.R. 1993. *Entering the Watershed: A New Approach to Save America's River Ecosystems*. Island Press, Washington, DC.

Dowling, T.E., Minckley, W.L., Marsh, P.C. and Goldstein, E.S. 1996a. Mitochondrial DNA variability in the endangered razorback sucker (*Xyrauchen texanus*): Analysis of hatchery stocks and implications for captive propagation. *Conservation Biology* 10:120–127.

Dowling, T.E., Minckley, W.L. and Marsh, P.C. 1996b. Mitochondrial DNA diversity within and among populations of razorback sucker (*Xyrauchen texanus*), as determined by restriction endonuclease analysis. *Copeia* 1996:542–550.

Dunlap, R.E., Gallup, G.H. and Gallup, A.M. 1993. Of global concern: results of the Health of the Planet Survey. *Environment* 35:7–15, 33–39.

Dynesius, M. and Nilsson, C. 1994. Fragmentation and flow regulation of river systems in the northern third of the world. *Science* 266:753–762.

Ecoregions Working Group 1989. *Ecoclimatic Regions of Canada, First Approximation*. Ecological Land Classification Series No. 23. Ottawa, Environment Canada. Separate map at 1:7 500 000.

Forest Ecosystem Management Team (FEMAT) 1993. *Forest Ecosystem Assessment: An Ecological, Economic, and Social Assessment*. USDA Forest Service, Portland, Oregon.

Firehock, K. and West, J. 1995. A brief history of volunteer biological water monitoring using macroinvertebrates. *Journal of the North American Benthological Society* 14:197–202.

Firth, P. and Fisher, S.G. (eds) 1992. *Global Climate Change and Freshwater Ecosystems*. Springer-Verlag, New York.

Fisher, S.G. 1995. Stream ecosystems of the western United States. In: Cushing, C.E., Cummins, K.W. and Minshall, G.W. (eds) *Ecosystems of the World 22. River and Stream Ecosystems*. Elsevier, Amsterdam, 61–86.

Fore, L.S., Karr, J.R. and Conquest, L.L. 1994. Statistical properties of an index of biotic integrity used to evaluate water resources. *Canadian Journal of Fisheries and Aquatic Sciences* 51:1077–1087.

Fore, L.S., Karr, J.R. and Wisseman, R.W. 1996. Assessing invertebrate responses to human activities: evaluating alternative approaches. *Journal of the North American Benthological Society* 15:212–231.

Frissell, C.A., Liss, W.S., Warren, C.E. and Hurley, M.D. 1986. A hierarchical framework for stream habitat classification: viewing streams in a watershed context. *Environmental Management* 10:199–214.

Gleick, P.H. 1993. *Water in Crisis: A Guide to the World's Fresh Water Resources*. Oxford University Press, New York.

Goldfarb, W. 1988. *Water Law*, 2nd edition. Lewis Publishers, Chelsea, Michigan.

Gurtz, M.E. and Muir, T.A. 1994. Report of the interagency biological methods workshop. *US Geological Survey Open-File Report* 94-490, Raleigh, North Carolina.

Harris, L. and Associates 1995. A survey on environmental issues in the Northwest. Prepared by Louis Harris and Associates for Gannett News Service and four Gannett newspapers in Washington, Oregon, and Idaho, as partially reported in the *Bellingham* (Washington) *Herald*, 23 April 1995: A-1.

Harris, S.C., Martin, T.C. and Cummins, K.W. 1995. A model for aquatic invertebrate response to Kissimmee River restoration. *Restoration Ecology* 3:181–194.

Harwell, M.A. 1997. Ecosystem management of South Florida. *BioScience* 47:499–512.

Hawkes, H.A. 1975. River zonation and classification. In: Whitton, B.A. (ed.) *River Ecology*. Blackwell, London, 312–374.

Hawkins, C.J., Kerschner, J.L., Bisson, P.A., Bryant, M.D., Decker, L.M., Gregory, S.V., McCullough, D.A., Overton, C.K., Reeves, G.H., Steedman, R.J. and Young, M.K. 1993. A hierarchical approach to classifying stream habitat features. *Fisheries (Bethesda, MD)* 18(6):3–11.

Hilsenhoff, W.L. 1988. Rapid field assessment of organic pollution with a family-level biotic index. *Journal of the North American Benthological Society* 7:65–68.

Holtz, R.E. 1996. Environmental education: A state survey. *Journal of Environmental Education* 27:617–619.

Horton, R.E. 1945. Erosional development of streams and their drainage basins; hydrophysical approach to quantitative morphology. *Geological Society of America Bulletin* 56:275–370.

Hudson, P.L., Griffiths, R.W. and Wheaton, T.J. 1992. Review of habitat classification schemes appropriate to streams, rivers, and connecting channels in the Great Lakes Drainage Basin. In: Busch, W.D.N. and Sly, P.G. (eds) *The Development of an Aquatic Habitat Classification System for Lakes*. CRC Press, Ann Arbor, Michigan, 74–106.

Huet, M. 1954. Biologie, profils en long et en travers des eaux courantes. *Bulletin Français de Pisciculture* 175:41–53.

Hughes, R.M. and Gammon, J.R. 1987. Longitudinal changes in fish assemblages and water quality in the Willamette River, Oregon. *Transactions of the American Fisheries Society* 116:196–209.

Hughes, R.M. and Noss, R.F. 1992. Biological diversity and biological integrity: current concerns for lakes and streams. *Fisheries (Bethesda, MD)* 17(3):11–19.

Hughes, R.M. and Omernik, J.M. 1981. Use and misuse of the terms watershed and stream order. In: Krumholz, L.A.

(ed.) *Proceedings of the Warmwater Streams Symposium.* American Fisheries Society, Bethesda, Maryland, 320–326.

Hume, M. 1992. *The Run of the River: Portraits of Eleven British Columbia Rivers.* University of British Columbia Press, Vancouver, British Columbia.

Huner, J.V. 1994. *Freshwater Crayfish Aquaculture in North America, Europe and Australia: Families Atacidae, Cambaridae, and Parastacidae.* Food Products Press, New York.

Interagency Floodplain Management Review Committee (IFMRC) 1994. *A Blueprint for Change: Sharing the Challenge: Floodplain Management into the 21st Century.* Administration Floodplain Management Task Force, Washington, DC.

Illies, J. and Botosaneanu, L. 1963. Problèmes et méthodes de la classification et de la zonation écologique des eaux courantes, considerées surtout du point de vue faunistique. *Mitteilungen der Internationalen Vereinigung für theoretische und angewandte Limnologie* 12:1–57.

Independent Scientific Group (ISG) 1996. *Return to the River – Restoration of Salmonid Fishes in the Columbia River Ecosystem.* Northwest Power Planning Council, Portland, Oregon.

Johnson, J.E. and Jensen, B.L. 1991. Hatcheries for endangered freshwater fishes. In: Minckley, W.L. and Deacon, J.E. (eds) *Battle Against Extinction: Native Fish Management in the American West.* The University of Arizona Press, Tucson, 199–217.

Johnson, R.W. and Paschal, R. 1995. The limits of prior appropriation. *Illahee* 11:40–50.

Jones, R.C. and Clark, C.C. 1987. Impact of watershed urbanization on stream insect communities. *Water Resources Bulletin* 23:1047–1055.

Karr, J.R. 1981. Assessment of biotic integrity using fish communities. *Fisheries (Bethesda, MD)* 6(6):21–27.

Karr, J.R. 1990. Kissimmee River: restoration of degraded resources. In: Loftin, M.K., Toth, L.A. and Obeysekara, J.T.B. (eds) *Kissimmee River Restoration: Alternative Plan Evaluation and Preliminary Design Report.* South Florida Water Management District, West Palm Beach, Florida, G2–G19.

Karr, J.R. 1991. Biological integrity: a long-neglected aspect of water resource management. *Ecological Applications* 1:66–84.

Karr, J.R. 1995. Clean water is not enough. *Illahee* 11:51–59.

Karr, J.R. 1998. Rivers as sentinels: using the biology of rivers to guide landscape management. In: Naiman, R.J. and Bilby, R.E. (eds) *River Ecology and Management: Lessons from the Pacific Coastal Ecoregion.* Springer, New York, 502–528.

Karr, J.R. 1999. Defining and measuring river health. *Freshwater Biology* 41:221–234.

Karr, J.R. and Chu, E.W. 1999. *Restoring Life in Running Waters: Better Biological Monitoring.* Island Press, Washington, DC.

Karr, J.R. and Dudley, D.R. 1981. Ecological perspective on water quality goals. *Environmental Management* 5:55–68.

Karr, J.R., Toth, L.A. and Dudley, D.R. 1985. Fish communities of midwestern rivers: a history of degradation. *BioScience* 35:90–95.

Karr, J.R., Fausch, K.D., Angermeier, P.L., Yant, P.R. and Schlosser, I.J. 1986. Assessment of biological integrity in running water: a method and its rationale. *Illinois Natural History Survey Special Publication* Number 5, Champaign, Illinois.

Keeler, A.G. and McLemore, D. 1996. The value of incorporating bioindicators in economic approaches to water pollution control. *Ecological Economics* 19:237–245.

Kempton, W., Boster, J.S. and Hartley, J.A. 1995. *Environmental Values in American Culture.* MIT Press, Cambridge, Massachusetts.

Kerans, B.L. and Karr, J.R. 1994. Development and testing of a benthic index of biotic integrity (B-IBI) for rivers of the Tennessee Valley Authority. *Ecological Applications* 4:776–785.

Kim, K.C. and Weaver, R.D. (eds) 1994. *Biodiversity and Landscapes: A Paradox of Humanity.* Cambridge University Press, New York.

Klemm, D.J., Lewis, P.A., Fulk, F. and Lazorchak, J.M. 1990. *Macroinvertebrate Field and Laboratory Methods for Evaluating the Biological Integrity of Surface Waters.* EPA-600-4-90-030. US Environmental Protection Agency, Environmental Monitoring and Support Laboratory, Cincinnati, Ohio.

Klemm, D.J., Stober, Q.J. and Lazorchak, J.M. 1993. *Fish Field and Laboratory Methods for Evaluating the Biological Integrity of Surface Waters.* EPA/600/R-92/111. US Environmental Protection Agency, Environmental Monitoring and Support Laboratory, Cincinnati, Ohio.

Knopman, D.S. and Smith, R.A. 1993. Twenty years of the Clean Water Act. *Environment* 35:16–20, 34–41.

Koebel, J.W. 1995. An historical perspective on the Kissimmee River restoration project. *Restoration Ecology* 3:149–159.

Kondolf, G.M. 1995. Five elements for effective evaluation of stream restoration. *Restoration Ecology* 3:133–136.

Lenat, D.R. 1993. A biotic index for the southeastern United States: derivation and list of tolerance values, with criteria for assigning water-quality ratings. *Journal of the North American Benthological Society* 12:279–290.

Lenat, D.R. and Crawford, J.K. 1994. Effects of land use on water quality and aquatic biota of three North Carolina Piedmont streams. *Hydrobiologia* 294:185–199.

Lieberman, G.A. 1995. *Pieces of a Puzzle: An Overview of the Status of Environmental Education in the United States.* A report prepared for the Pew Charitable Trusts, Philadelphia.

Lydeard, C. and Mayden, R.L. 1995. A diverse and endangered aquatic ecosystem of the southeast United States. *Conservation Biology* 9:800–805.

Lyons, J. 1992. *Using the Index of Biotic Integrity (IBI) to Measure Environmental Quality in Warmwater Streams of Wisconsin.* General Technical Report NC-149. North Central Forest Experiment Station, US Department of Agriculture, St Paul, Minnesota.

Lyons, J., Wang, L. and Simonson, T.D. 1996. Development and validation of an index of biotic integrity for coldwater streams in Wisconsin. *North American Journal of Fisheries Management* 16:241–256.

Macan, T.T. 1961. A review of running water studies. *Verhandlungen der Internationalen Vereinigung für theoretische und angewandte Limnologie* 14:587–662.

Mackay, R.J. 1995. River and stream ecosystems of Canada. In: Cushing, C.E., Cummins, K.W. and Minshall, G.W. (eds) *Ecosystems of the World 22. River and Stream Ecosystems.* Elsevier, Amsterdam, 33–60.

Marsh, P.C. and Sada, D. 1993. *Desert Pupfish* (Cyprinodon macularius) *Recovery Plan.* US Fish and Wildlife Service, Albuquerque, New Mexico.

Martin, P.S. and Klein, R.G. (eds) 1984. *Quaternary Extinctions: A Prehistoric Revolution.* University of Arizona Press, Tucson, Arizona.

Master, L. 1998. The imperiled status of North American aquatic animals. *Biodiversity Network News* 3:1–2, 7–8.

Master, L.L., Flack, S.R. and Stein, B.A. (eds) 1998. *Rivers of Life: Critical Watersheds for Protecting Freshwater Biodiversity.* The Nature Conservancy, Arlington, Virginia.

Meador, M.R., Hupp, C.J., Cuffney, R.F. and Gurtz, M.E. 1993a. Methods for characterizing stream habitat as part of the national water-quality assessment program. *US Geological Survey Open-File Report* 93-408, Raleigh, North Carolina.

Meador, M.R., Cuffney, R.F. and Gurtz, M.E. 1993b. Methods for sampling fish communities as part of the national water-quality assessment program. *US Geological Survey Open-File Report* 93-104, Raleigh, North Carolina.

Meffe, G.K. and Carroll, C.R. 1997. *Principles of Conservation Biology*, 2nd edition. Sinauer Associates Inc, Sunderland, Massachusetts.

Meixler, M.S., Bain, M.B. and Galbreath, G.H. 1996. Aquatic gap analysis: tools for watershed scale assessment of fluvial habitat and biodiversity. In: Le Clerc, M., Capra, H., Valentin, S., Boudreaut and Côté, Y. (eds) *Proceedings of the Second IAHR Symposium on Habitat Hydraulics, Ecohydraulics 2000.* Institute National de la Recherche Scientifique–Eau, Ste -Foy, Quebec, Canada, A665–A670.

Merritt, R.W. and Cummins, K.W. 1996. *An Introduction to the Aquatic Insects of North America*, 3rd edition. Kendall/Hunt Publishing Company, Dubuque, Iowa.

Meybeck, M. 1988. How to establish and use world budgets of riverine materials. In: Lerman, A. and Meybeck, M. (eds) *Physical and Chemical Weathering in Geochemical Cycles.* Kluwer Academic Publishers, Reidel Press, Dordrecht, 247–272.

Meyer, J.L. 1997. Conserving ecosystem function. In: Pickett, S.T.A., Ostfeld, R.S., Shachak, M. and Likens, G.E. (eds) *The Ecological Basis of Conservation: Heterogeneity, Ecosystems, and Biodiversity.* Chapman & Hall, New York, 136–145.

Miller, D.L., Leonard, P.M., Hughes, R.M., Karr, J.R., Moyle, P.B., Schrader, L.H., Thompson, B.A., Daniels, R.A., Fausch, K.D., Fitzhugh, G.A., Gammon, J.R., Halliwell, D.B., Angermeier, P.L. and Orth, D.J. 1988. Regional applications of an index of biotic integrity for use in water resource management. *Fisheries (Bethesda, MD)* 13(5):12–20.

Minckley, W.L., Marsh, P.C., Brooks, J.E., Johnson, J.E. and Jensen, B.L. 1991. Management toward recovery of the Razorback Sucker. In: Minckley, W.L. and Deacon, J.E. (eds) *Battle Against Extinction. Native Fish Management in the American West.* University of Arizona Press, Tucson, Arizona, 303–358.

Modde, T., Scholz, A.T., Williamson, J.H., Haines, G.B., Burdick, B.D. and Pfeifer F.K. 1995. An augmentation plan for the razorback sucker in the Upper Colorado River Basin. *American Fisheries Society Symposium* 15:102–111.

Moore, M.R., Mulville, A. and Weinberg, M. 1996. Water allocation in the American West; Endangered fish versus irrigated agriculture. *Natural Resources Journal* 36:319–357.

Morrison, J.I., Postel, S.L. and Gleick, P.H. 1996. *The Sustainable Use of Water in the Lower Colorado River Basin.* Pacific Institute for Studies in Development, Environment, and Security, Oakland, California.

Morse, J.C., Chapin, J.W., Herlong, D.D. and Harvey, R.S. 1983. Aquatic insects of upper Three Runs Creek, Savanna River Plant, South Carolina. Part II: Diptera. *Journal of the Georgia Entomological Society* 18:303–316.

Moyle, P.B. and Sato, G.M. 1991. On the design of preserves to protect native fishes. In: Minckley, W.L. and Deacon, J.E. (eds) *Battle Against Extinction: Native Fish Management in the American West.* University of Arizona Press, Tucson, Arizona, 155–170.

Moyle, P.B. and Yoshiyama, R.M. 1994. Protection of aquatic biodiversity in California: A five-tiered approach. *Fisheries* 19:6–18.

Moyle, P.B., Li, H.W. and Barton, B.A. 1986. The Frankenstein effect: Impact of introduced fishes on native fishes in North America. In: Stroud, R.H. (ed.) *Fish Culture in Fisheries Management.* American Fisheries Society, Bethesda, Maryland, 415–426.

Mueller, G. 1995. A program for maintaining the razorback sucker in Lake Mojave. *American Fisheries Society Symposium* 15:127–138.

Murdoch, T., Cheo, T.M. and O'Laughlin, K. 1996. *The Streamkeeper's Field Guide: Watershed Inventory and Stream Monitoring Methods.* Adopt-A-Stream Foundation, Everett, Washington.

Naiman, R.J., Lonzarich, D.G., Beechie, T.J. and Ralph, S.C. 1992. General principles of classification and the assessment of conservation potential in rivers. In: Boon, P.J., Calow, P. and Petts, G.E. (eds) *River Conservation and Management.* John Wiley, Chichester, 93–123.

Naiman, R.J., Magnuson, J.J., McKnight, D.M. and Stanford, J.A. (eds) 1995a. *The Freshwater Imperative: A Research Agenda.* Island Press, Washington, DC.

Naiman, R.J., Magnuson, J.J., McKnight, D.M., Stanford,

J.A. and Karr, J.R. 1995b. Freshwater ecosystems and their management: a national initiative. *Science* 270:584–585.

National Research Council (NRC) 1992. *Restoration of Aquatic Ecosystems: Science, Technology, and Public Policy*. National Academy Press, Washington, DC.

National Research Council (NRC) 1995. *Science and the Endangered Species Act*. National Academy Press, Washington, DC.

National Research Council (NRC) 1996a. *River Resource Management in the Grand Canyon*. National Academy Press, Washington, DC.

National Research Council (NRC) 1996b. *Upstream. Salmon and Society in the Pacific Northwest*. National Academy Press, Washington, DC.

National Research Council (NRC) 1996c. *Freshwater Ecosystems: Revitalizing Educational Programs in Limnology*. National Academy Press, Washington, DC.

Natural Resources Conservation Service (NRCS) 1996. *America's Private Land: A Geography of Hope*. Natural Resources Conservation Service, United States Department of Agriculture, Washington, DC.

Nature Conservancy 1997. A classification framework for freshwater communities: *Proceedings of the Nature Conservancy's Aquatic Community Classification Workshop*, New Haven, Missouri, 9–11 April 1996. Arlington, Virginia.

Nehlsen, W., Williams, J.E. and Lichatowich, J.A. 1991. Pacific salmon at the crossroads: stocks at risk from California, Oregon, Idaho, and Washington. *Fisheries (Bethesda, MD)* 16(2):4–21.

Neves, R.J. 1997. A national strategy for the conservation of native freshwater mussels. In: Cummings, K.S., Buchanan, A.C., Mayer, C.A. and Naimo, T.J. (eds) *Conservation and Management of Freshwater Mussels II: Initiatives for the Future*. Proceedings of a UMRCC symposium, 16–18 October 1995, St Louis, Missouri. Upper Mississippi River Conservation Committee, Rock Island, Illinois 1–11.

Northcote, T.G. and Larkin, P.A. 1989. The Fraser River: a major salmonine production system. *Canadian Special Publication in Fisheries and Aquatic Sciences* 106:172–204.

Ohio Environmental Protection Agency 1988. *Biological Criteria for the Protection of Aquatic Life*, Volumes I–III. Ohio Environmental Protection Agency, Columbus, Ohio.

Omernik, J.M. 1987. Ecoregions of the conterminous United States. *Annals of the Association of American Geographers* 77:118–125.

Omernik, J.M. 1995. Ecoregions: a spatial framework for environmental management. In: Davis, W.S. and Simon, T.P. (eds) *Biological Assessment and Criteria: Tools for Water Resource Planning and Decision Making*. Lewis Publishers, Boca Raton, Florida, 49–62.

Omernik, J.M. and Bailey, R.G. 1997. Distinguishing between watersheds and ecoregions. *Journal of the American Water Resources Association* 33:935–949.

Omernik, J.M. and Gallant, G.L. 1990. Defining regions for evaluating environmental resources. In: Lund, H.G. and

Preto, G. (coordinators) *Global Natural Resource Monitoring and Assessments: Preparing for the 21st Century*. American Society of Photogrammetry and Remote Sensing, Bethesda, Maryland, 936–947.

O'Neill, R.V., DeAngelis, D.L., Waide, J.B. and Allen, T.F.H. 1986. *A Hierarchical Concept of Ecosystems*. Monographs in Population Biology 23, Princeton University Press, Princeton, New Jersey.

Oswood, M.W., Irons, J.G. and Milner, A.M. 1995. River and stream ecosystems of Alaska. In: Cushing, C.E., Cummins, K.W. and Minshall, G.W. (eds) *Ecosystems of the World 22. River and Stream Ecosystems*. Elsevier, Amsterdam, 9–32.

Page, L.M. and Burr, B.M. 1991. *Freshwater Fishes*. Peterson Field Guide Series No. 42. Houghton-Mifflin, Boston, Massachusetts.

Palmer, T. 1986. *Endangered Rivers and the Conservation Movement*. University of California Press, Berkeley, California.

Palmer, T. 1994. *Lifelines: The Case for River Conservation*. Island Press, Washington, DC.

Pan, Y., Stevenson, R.J., Hill, B., Herlihy, A. and Collins, G. 1996. Using diatoms as indicators of ecological conditions in lotic systems: a regional assessment. *Journal of the North American Benthological Society* 15:481–495.

Patrick, R. 1995. *Rivers of the United States*, Volume II, *Chemical and Physical Characteristics*. John Wiley, New York.

Patrick, R. 1996. *Rivers of the United States*, Volume III: *The Eastern and Southeastern States*. John Wiley, New York.

Perrin, L.S., Allen, M.J., Rowse, L.A., Montalbano, F., Foote, K.J. and Olinde, M.W. 1982. *A Report on Fish and Wildlife Studies in the Kissimmee River Basin and Recommendations for Restoration*. Florida Game and Fresh Water Fish Commission, Okeechobee, Florida.

Pimentel, D., Wilson, C., McCullom, C., Huang, H., Dwen, P., Flack, J., Tran, Q., Saltman, T. and Cliff, B. 1997. Economic and environmental benefits of biodiversity. *BioScience* 47:747–757.

Plafkin, J.L., Barbour, M.T., Porter, K.D., Gross, S.K. and Hughes, R.M. 1989. *Rapid Bioassessment Protocols for Streams and Rivers: Benthic Macroinvertebrates and Fish*. EPA/444/4-89-01. US Environmental Protection Agency, Assessment and Watershed Protection Division, Washington, DC.

Poff, N.L., Allan, J.D., Bain, M.B., Karr, J.R., Prestagaard, K.L., Richter, B.D. and Sparks, R.P. 1997. The natural flow regime: a paradigm for river conservation and restoration. *BioScience* 47:769–784.

Polls, I., Lanyon, R., Sedita, S.J., Zenz, D.R. and Lue-Hing, C. 1994. *Fact or Fiction: Has the Water Quality Improved in Chicago Area Waterways?* Report No. 94-23, Metropolitan Water Reclamation District of Greater Chicago.

Ponting, C. 1991. *A Green History of the World: The Environment and the Collapse of Great Civilizations*. St Martin's Press, New York.

Population-Environment Balance 1990. A comparison of the

regional distribution of water supply and demand with population growth. Page 6 in Balance data number 27, Washington, DC.

Porter, S.D., Cuffney, R.F., Gurtz, M.E. and Meador, M.R. 1993. Methods for collecting algal samples as part of the national water-quality assessment program. *US Geological Survey Open-File Report* 93-409, Raleigh, North Carolina.

Postel, S.L., Daily, G.C. and Ehrlich, P.R. 1996. Human appropriation of renewable fresh water. *Science* 271:785–788.

Pringle, C.M. and Aumen, N.G. 1993. Current issues in freshwater conservation: introduction to a symposium. *Journal of the North American Benthological Society* 12:174–176.

Raffensperger, C. and Tarlock, A.D. 1993. The Wild and Scenic Rivers Act at 25: the need for a new focus. *Rivers* 4:81–90.

Rankin, E.T. 1995. Habitat indices in water resource quality assessments. In: Davis, W.S. and Simon, T.P. (eds) *Biological Assessment and Criteria: Tools for Water Resource Planning and Decision Making*. Lewis Publishers, Boca Raton, Florida, 181–208.

Ransel, K.P. 1995. The sleeping giant awakes: PUD No. 1 of Jefferson County v. Washington Department of Ecology. *Environmental Law* 25:255–283.

Rapport, D.J., Gaudet, C.L. and Calow, P. (eds) 1995. *Evaluating and Monitoring the Health of Large-Scale Ecosystems*. NATO ASI Series I: Global Environmental Change, Vol. 28. Springer-Verlag, Berlin.

Reffalt, W. 1991. The endangered species lists: chronicles of extinction? In: Kohm, K.A. (ed.) *Balancing on the Brink of Extinction: The Endangered Species Act and Lessons for the Future*. Island Press, Washington, DC, 77–85.

Reisner, M. 1993. *Cadillac Desert: The American West and its Disappearing Water*, revised edition. Penguin Books, New York.

Resh, V.H. and Jackson, J.K. 1993. Rapid assessment approaches to biomonitoring using benthic macroinvertebrates. In: Rosenberg, D.L. and Resh, V.H. (eds) *Freshwater Biomonitoring and Benthic Invertebrates*. Chapman & Hall, New York, 195–233.

Richter, B.D., Baumgartner, J.V., Powell, J. and Braun, D.P. 1996. A method for assessing hydrologic alteration within ecosystems. *Conservation Biology* 10:1163–1174.

River Network 1996. *1996–1997 River and Watershed Conservation Directory*. To The Point Publications, Portland, Oregon.

Rodgers, W.H. 1994. *Environmental Law*, 2nd edition. West Publishing Co, St Paul, Minnesota.

Rosenberg, D.L. and Resh, V.H. (eds) 1993. *Freshwater Biomonitoring and Benthic Invertebrates*. Chapman & Hall, New York.

Samson, F.B. and Knopf, F.L. (eds) 1996. *Ecosystem Management: Selected Readings*. Springer-Verlag, New York.

Schramm, H.L. and Piper, R.G. (eds) 1995. *Uses and Effects of Cultured Fishes in Aquatic Ecosystems*. American Fisheries Society Symposium 15, Bethesda, Maryland.

Schueler, T.R. 1996. Crafting better urban watershed protection plans. *Watershed Protection Techniques* 2:369–372.

Scott, J.M., Csuti, B., Jacobi, J.J. and Estes, J.E. 1987. Species richness: A geographic approach to protecting future biological diversity. *BioScience* 37:782–788.

Shiklomanov, I.A. 1993. World fresh water resources. In: Gleick, P.H. (ed.) *Water in Crisis: A Guide to the World's Fresh Water Resources*. Oxford University Press, New York, 13–24.

Showers, V. 1989. *World Facts and Figures*, 3rd edition. John Wiley, New York.

Simon, T.P. 1999. *Assessing the Sustainability and Biological Integrity of Water Resources Using Fish Assemblages*. CRC Press, Boca Raton, Florida.

Smith, D.S. and Hellmund, P.C. (eds) 1993. *Ecology of Greenways: Design and Function of Linear Conservation Areas*. University of Minnesota Press, Minneapolis, Minnesota.

Smith, R.A., Alexander, R.B. and Wolman, M.G. 1987. Water-quality trends in the Nation's rivers. *Science* 235:1607–1615.

Sparks, R.E. 1995. Need for ecosystem management of large rivers and their floodplains. *BioScience* 45:168–182.

Spence, B.C., Lomnicky, G.A., Hughes, R.M. and Novitzki, R.P. 1996. *An Ecosystem Approach to Salmonid Conservation*. TR-4501-96-6057. ManTech Environmental Research Services Corp., Corvallis, Oregon. (Available from National Marine Fisheries Service, Portland, Oregon.)

Stolzenberg, W. 1997. Sweet home Alabama. *Nature Conservancy* 47:8–9.

Strahler, A.N. 1957. Quantitative analysis of watershed geomorphology. *American Geophysical Union Transactions* 38:913–920.

Stribling, J.B., Snyder, B.D. and Davis, W.D. 1996. *Biological Assessment Methods, Biocriteria, and Biological Indicators. Bibliography of Selected Technical, Policy, and Regulatory Literature*. EPA 230-B-96-001. US Environmental Protection Agency, Office of Policy, Planning, and Evaluation, Washington, DC.

Sweeney, B.W. 1993. Effects of streamside vegetation on macroinvertebrate communities of White Clay Creek in eastern North America. *Proceedings of the Academy of Natural Sciences of Philadelphia* 144:291–340.

Swift, B.L. 1984. Status of riparian ecosystems in the United States. *Water Resources Bulletin* 20:233–238.

Taylor, C.A., Warren, M.L., Fitzpatrick, J.F., Hobbs, H.H., Jezerinac, R.J., Pflieger, W.L. and Robison, H.W. 1996. Conservation status of crayfishes of the United States and Canada. *Fisheries (Bethesda, MD)* 21(4):25–38.

Toth, L.A. 1993. The ecological basis of the Kissimmee River restoration plan. *Florida Scientist* 56:25–51.

Toth, L.A., Arrington, D.A., Brady, M.A. and Muszick, D.A. 1995. Conceptual evaluation of factors potentially affecting restoration of habitat structure within the channelized Kissimmee River ecosystem. *Restoration Ecology* 3:160–180.

Trexler, J.C. 1995. Restoration of the Kissimmee River: a conceptual model of past and present fish communities and its consequences for evaluating restoration success. *Restoration Ecology* 3:195–210.

US Environmental Protection Agency (USEPA) 1990a. *Biological Criteria: National Program Guidance for Surface Waters*. EPA 440-5-90-004. US Environmental Protection Agency, Office of Water Regulations and Standards, Washington, DC.

US Environmental Protection Agency (USEPA) 1990b. *Volunteer Water Monitoring: A Guide for State Managers*. EPA 440/4-90-010. US Environmental Protection Agency, Office of Water, Washington, DC.

US Environmental Protection Agency (USEPA) 1991. *Watershed Events, the Watershed Protection Approach: An Overview*. EPA 503/9-92-002. US Environmental Protection Agency, Office of Wetlands, Oceans, and Watersheds, Washington, DC.

US Environmental Protection Agency (USEPA) 1994a. *The Quality of Our Nation's Water: 1994*. EPA 841-S-94-002, US Environmental Protection Agency, Office of Water, Washington DC.

US Environmental Protection Agency (USEPA) 1994b. *National Directory of Volunteer Environmental Monitoring Programs*. EPA 841-B-94-001. US Environmental Protection Agency, Office of Water, Washington, DC. (www.epa.gov.owow/volunteer/vm_index.htm)

US Fish and Wildlife Service (USFWS) 1995. *Lower Colorado Ecoregion Plan*. US Fish and Wildlife Service, Denver, Colorado.

Valdez, R.A. and Williams, R.D. 1986. *Cataract Canyon Fish Study*. Final Report for US Bureau of Reclamation Contract 6-CS-40-03980. US Bureau of Reclamation, Salt Lake City, Utah.

Vannote, R.L., Minshall, G.W., Cummins, K.W., Sedell, J.R. and Cushing, C.E. 1980. The river continuum concept. *Canadian Journal of Fisheries and Aquatic Sciences* 37:130–137.

Waason, J.G. 1989. Eléments pour une typologie fonctionelle des eaux courantes: 1. Revue critique de quelques approches existantes. *Bulletin d'Ecologie* 20:109–127.

Ward, A.K., Ward, G.M. and Harris, S.C. 1992. Water quality and biological communities of the Mobile River drainage, eastern Gulf of Mexico region. In: Becker, C.D. and Neitzel, D.A. (eds) *Water Quality in North American River Systems*. Battelle Press, Columbus, Ohio, 278–304.

Ward, J.V. 1989. The four-dimensional nature of lotic ecosystems. *Journal of the North American Benthological Society* 8:2–8.

Warren, M.L. and Burr, B.M. 1994. Status of freshwater fishes of the United States: Overview of an imperiled fauna. *Fisheries* 19:6–18.

Waters, T.F. 1995. *Sediments in Streams: Sources, Biological Effects, and Control*. Monograph 7, American Fisheries Society, Bethesda, Maryland.

Weaver, L.A. and Garman, G.C. 1994. Urbanization of a watershed and historical changes in a stream fish assemblage. *Transactions of the American Fisheries Society* 123:162–172.

Webster, J.R., Wallace, J.B. and Benfield, E.F. 1995. Organic processes in streams of the eastern United States. In: Cushing, C.E., Cummins, K.W. and Minshall, G.W. (eds) *Ecosystems of the World 22. River and Stream Ecosystems*. Elsevier, Amsterdam, 117–187.

Weller, M.W. 1995. Use of two waterbird guilds as evaluation tools for the Kissimmee River restoration. *Restoration Ecology* 3:211–224.

White, R.J., Karr, J.R. and Nehlsen, W. 1995. Better roles for stocking in aquatic resource management. *American Fisheries Society Symposium* 15:527–547.

Wilcove, D.S. and Bean, M.J. (eds) 1994. *The Big Kill: Declining Biodiversity in America's Lakes and Rivers*. Environmental Defense Fund, Washington, DC.

Wilkinson, C.F. 1992. *Crossing the Next Meridian: Land, Water, and the Future of the West*. Island Press, Washington, DC.

Williams, J.D., Warren, M.L., Cummins, K.S., Harris, J.L. and Neves, R.J. 1993. Conservation status of freshwater mussels of the United States and Canada. *Fisheries (Bethesda, MD)* 18(9):6–22.

Williams, J.E. 1991. Preserves and refuges for native western fishes: history and management. In: Minckley, W.L. and Deacon, J.E. (eds) *Battle Against Extinction: Native Fish Management in the American West*. University of Arizona Press, Tucson, 171–190.

Williams, J.E., Johnson, J.E., Hendrickson, D.A., Contreras-Balderas, S., Williams, J.D., Navarro-Mendoza, M., McAllister, D.E. and Deacon, J.E. 1989. Fishes of North America, endangered threatened, or of special concern: 1989. *Fisheries (Bethesda, MD)* 14(6):2–20.

Williams, J.E., Wood, C.A. and Dombeck, M.P. (eds) 1997. *Watershed Restoration: Principles and Practices*. American Fisheries Society, Washington, DC.

Williams, R.A., Bidlak, C. and Winnett, D. 1993. At the water's edge, students study their rivers. *Environmental Leadership* 51:80–83.

Winston, M.R. and Angermeier, P.L. 1995. Assessing conservation value using centers of population density. *Conservation Biology* 9:1518–1527.

Winter, B.D. and Hughes, R.M. 1997. American Fisheries Society Biodiversity Position Statement. *Fisheries (Bethesda, MD)* 22(1):22–29.

World Resources Institute (WRI) 1990. *World Resources 1990–91*. Oxford University Press, New York.

World Resources Institute (WRI) 1994. *World Resources 1994–95*. Oxford University Press, New York.

Worster, D. 1985. *Rivers of Empire: Water, Aridity, and the Growth of the American West*. Pantheon, New York.

Yoder, C.O. 1995. Policy issues and management applications for biological criteria. In: Davis, W.S. and Simon, T.P. (eds) *Biological Assessment and Criteria: Tools for Water Resource Planning and Decision Making.* Lewis Publishers, Boca Raton, Florida, 327–344.

Yoder, C.O. and Rankin, E.T. 1995a. Biological criteria development and implementation in Ohio. In: Davis, W.S. and Simon, T.P. (eds) *Biological Assessment and Criteria: Tools for Water Resource Planning and Decision Making.* Lewis Publishers, Boca Raton, Florida, 109–144.

Yoder, C.O. and Rankin, E.T. 1995b. Biological response signatures and the area of degradation value: new tools for interpreting multimetric data. In: Davis, W.S. and Simon, T.P. (eds) *Biological Assessment and Criteria: Tools for Water Resource Planning and Decision Making.* Lewis Publishers, Boca Raton, Florida, 263–286.

2

River conservation in Latin America and the Caribbean

C.M. Pringle, F.N. Scatena, P. Paaby-Hansen and M. Núñez-Ferrera

Introduction

El agua
es la luz
con raíz en la tierra.

Water
is the light
with roots on earth

Beberla
es echarse a caminar
como un río.

To drink it
is to journey
like a river

Raul Banuelos, 1994

The developing nature of Latin America combined with the vastness of its aquatic resources, highly diverse aquatic fauna and flora, rapidly increasing human population, and associated environmental problems, make it a high priority for international riverine conservation efforts. Brazil, Colombia, Ecuador and Peru have been identified by international conservationists as among the 12 'megadiversity' countries which together harbour 70% of the world's biodiversity (Mittermeier, 1989), thus requiring special attention (McNeely *et al.*, 1990). Specific areas in Latin America that have been classified as 'hot spots' of biodiversity and endemism include the Colombian Chocó of western Ecuador, the uplands of western Amazonia, the Atlantic forest region of eastern Brazil, and central Chile (Myers, 1988).

These international conservation priorities were based on high levels of biodiversity and endemism in terrestrial ecosystems. As in many other developing countries (e.g. Benstead *et al.*, this volume) freshwater riverine systems have received comparatively little attention. Comparison of recent regional conservation assessments of freshwater (Olson *et al.*, 1998) and terrestrial (Dinerstein *et al.*, 1995) biodiversity in Latin America and the Caribbean indicate that freshwater biodiversity is more seriously threatened in terms of the geographic extent and severity of threats.

The rich aquatic biodiversity of rivers and streams in Latin America (World Conservation Monitoring Centre, 1992) promises to become an increasingly important criterion for guiding conservation investments. As the largest tropical watershed ($5\,711\,000$ km^2) in the world, the Amazon has well over 2000 fish species, of which 90% are endemic. New species are being discovered every year: a recent study of deep-water habitats of the Amazon revealed more than 240 new fish species (Yoon, 1997).

Over the last few decades, pollution of riverine systems has emerged as a major problem throughout the region (United Nations Economic Commission for Latin America and the Caribbean – UN ECLAC, 1990a,b,c). Among the most important factors contributing to this rapid increase in pollution is the corresponding rapid increase in population growth without the development of waste treatment facilities, land-use planning and pollution control. Latin America contains an estimated 450 million people and the population is growing at an annual rate of 2.1%. It is expected to approach almost 800 million by the year 2020 (World Resources Institute, 1990). In addition, the combined external debt of Latin American countries totalled US$387 billion in 1987, which represents between 10 and 93% of the gross national product of individual nations (World Resources Institute, 1990). Such socio-economic

Global Perspectives on River Conservation: Science, Policy and Practice.
Edited by P.J. Boon, B.R. Davies and G.E. Petts. © 2000 John Wiley & Sons Ltd.

factors have restricted most Latin American governments from responding to environmental concerns.

The maintenance of water quality in Latin America and the Caribbean is being driven primarily by human health concerns (e.g. Pan American Health Organization – PAHO, 1992). About 80% of tropical diseases are apparently water-related and can be attributed to poor or non-existent sewage treatment and lack of safe drinking water (Gladwell and Bonell, 1990). Large segments of the population do not have access to potable water supplies.

This chapter provides a general overview of the status of river systems in Latin America and the Caribbean. While it should be noted that marine coastal communities and coral reefs throughout Latin America and the Caribbean are threatened by sediment, nutrient and pesticide loads from rivers (Glynn *et al.*, 1984; Seeliger *et al.*, 1988; Goenaga, 1991; Connell and Hawker, 1992; Guzman and Jimenez, 1992), these issues are not addressed here. We do emphasize the conservation implications of hydroelectric development, given the magnitude of plans for such development in Latin America.

The regional context

PHYSICAL SETTING AND THE MAIN RIVERS

Knowledge of the spatial distribution of water resources and the human use of these resources is fundamental to understanding water management, conservation and sustainable development. Some of this baseline information is provided by UN ECLAC (1990a,b). Also, the hydrology and water resources of humid tropical regions of Latin America and the Caribbean are reviewed by Griesinger and Gladwell (1993).

Latin America and the Caribbean have relatively abundant water resources. The mean annual rainfall of 1500 mm is 50% higher than the world average. The region also contributes almost one-third of the total worldwide land drainage entering the oceans. However, Latin America also contains some of the most arid areas (e.g. the Atacama Desert in northern Chile).

The main hydrographic divisions of Latin America are illustrated in Figure 2.1 and the area and distribution of human activity in each hydrographic division is listed in Table 2.1. Throughout the mainland, a key physiographic factor affecting the

distribution of water resources is the mountain chain that runs all the way from Mexico to Chile. Using biogeography and dynamic linkages of ecosystems, Olson *et al.* (1998) identified 117 freshwater ecoregions in Latin America and the Caribbean (Figure 2.2). Ecoregions were categorized under major habitat types and included: (1) 12 ecoregions which were large rivers and major tributaries; (2) four large river deltas; (3) 10 montane streams and rivers; (4) 49 wet region rivers and streams; (5) 16 xeric rivers and streams; (6) 11 xeric endorheic (closed-basin) habitats; (6) 17 flooded grasslands and savannas; (7) three cold streams, bogs, swamps and mires (montane or low latitude); and, (8) seven large lakes.

Overall, the freshwater biodiversity of the neotropics is the richest on earth with highly distinctive habitat types, communities and species (World Conservation Monitoring Centre, 1992). Moreover, the geological history and hydrographic features of various neotropical regions (i.e. Central America, the Caribbean islands and South America) are important factors in determining patterns of freshwater biodiversity and are equally critical in discussions of riverine conservation. For example, the fish diversity of South America is extremely high, with an estimated 3000 fish species in the Amazon area and around 2000 species in the Orinoco River (Lowe-McConnell, 1987). In contrast, the San Juan drainage in the geologically younger Central America has a total of 138 fish species (Bussing, 1994), while streams draining Puerto Rico in the Caribbean have very low numbers of fish species (fewer than 10), typical of oceanic islands.

Central America

Land masses in Central America (and the Caribbean) are typically characterized by a high ocean-to-coast ratio. The relatively recent connection between North and South America, together with the presence of riverine connections with both the Atlantic and Pacific Oceans, has strongly determined patterns in the faunal diversity of streams draining the region. Much of Central America has relatively young soils due to recent volcanic activity. Furthermore, ongoing geothermal activity along the volcanic spine of Central America (and extending into South America) creates spatial patterns in stream solute levels which influence the ecology of streams draining the area (Pringle *et al.*, 1993).

Central America has a tropical–equatorial climate, where annual rainfall exceeds 1500 mm throughout

Figure 2.1 *Main hydrographic systems within Latin America and the Caribbean (from UN ECLAC, 1990a)*

most of the region. Many areas have precipitation levels between 2000 and 3000 mm yr^{-1}. Caribbean coastal regions, which are on the windward side of prevailing north-easterly winds, are the areas of highest precipitation with annual rainfall often exceeding 3000 mm. On the Pacific side, rainfall is often 50% less, yet the climate is still very humid.

Most of the rivers that discharge into the Caribbean and Pacific are short. However, there are three systems which are 300–400 km long: the Motagua River (Guatemala), the Lempa River (El Salvador) and the

Coco or Segovia River (between Honduras and Nicaragua). The largest catchment area, however, is drained by the San Juan River (37 km long with 41 600 km^2 in drainage area) on the border between Nicaragua and Costa Rica. Most rivers in Central America have their headwaters located in volcanic mountains with cool clear waters which quickly change into warm and turbid waters in the lowlands. Both Costa Rica and Panama are very narrow mountainous areas resulting in multiple, parallel, riverine systems with relatively small drainage areas.

Table 2.1 *Latin America and the Caribbean: distribution of human activity by major hydrographic system*

Hydrographic system	Area (km²)	Population			Density 1980	Annual mean percentage increase (1960–1980)	GDP (millions of 1980 US dollars)		% GDP		
		1960	1970	1980			1980	1985	Agric.	Indus.	Serv.
Central America and the Caribbean											
California	471 473	2 820 032	4 132 946	5 739 565	12.17	3.62	15 100.8	16 620.4	16.2	20.9	63.0
Caribbean	646 213	16 840 767	22 384 393	28 497 374	44.10	2.66	32 857.4	36 332.2	22.5	24.8	52.6
Caribbean Islands	230 789	20 328 660	24 430 255	30 044 411	130.18	1.97	55 517.0	63 565.5	12.7	33.7	53.6
Gulf of Mexico	474 552	19 724 062	28 113 599	39 783 373	83.83	3.57	109 288.7	120 005.2	8.5	35.4	56.1
North Pacific	328 406	8 121 876	10 618 230	13 820 162	42.08	2.69	26 095.6	28 673.9	19.6	21.8	58.6
Northern Endorheic, Mexico	140 840	699 272	919 139	1 143 122	8.12	2.49	2 807.6	3 090.2	19.5	27.6	52.8
Rio Bravo	214 096	1 621 683	2 243 483	3 112 607	14.54	3.31	10 672.9	11 747.0	8.5	44.0	47.5
Southern Endorheic, Mexico	236 637	2 257 220	2 829 615	3 727 082	15.75	2.54	8 493.3	9 348.0	35.4	23.9	40.7
Yucatan	141 523	832 437	1 098 061	1 710 271	12.08	3.67	3 518.4	3 872.4	17.0	26.6	56.4
South America											
Amazonas	6 157 253	10 414 471	13 684 801	18 416 972	2.99	2.89	19 242.7	19 924.1	22.9	22.1	55.0
Central Chile	116 002	5 466 021	6 809 934	8 324 396	71.76	2.13	18 245.3	18 022.3	7.6	36.8	55.6
Central Venezuela	142 419	4 356 758	6 346 751	8 661 137	60.81	3.50	30 562.5	28 189.1	7.0	29.8	63.2
Guayanas	468 235	895 431	1 149 533	1 186 631	2.53	1.42	1 702.0	1 655.0	14.8	35.3	49.9
Endorheic, Argentina	706 869	3 923 597	4 405 305	5 333 450	7.55	1.55	11 556.8	10 127.7	19.4	27.8	52.8
Maracaibo	101 688	2 450 814	3 296 334	4 114 182	40.46	2.62	12 062.3	11 265.3	8.2	29.3	62.5
North-east Brazil	881 361	12 415 643	15 647 407	19 175 848	21.76	2.20	13 284.2	14 466.3	18.2	16.0	65.8
Orinoco	1 116 599	3 490 445	4 475 929	5 646 580	5.06	2.43	12 278.3	12 069.5	19.0	27.1	53.9
Pacific: Arid Climate	590 419	6 630 031	9 575 465	12 795 372	21.67	3.34	19 070.3	18 977.5	10.7	41.5	47.8
Pacific: Tropical Climate	348 495	12 840 420	17 273 856	21 506 679	61.71	2.61	26 547.9	28 283.2	28.8	22.9	48.3
Pampa	621 207	2 347 099	2 887 705	3 666 148	5.90	2.25	9 896.1	8 672.4	14.9	45.8	39.3
Patagonia	487 645	201 851	285 909	402 784	0.83	3.51	1 691.3	1 482.1	14.5	49.6	35.9
Plate	3 878 926	42 987 903	56 231 1529	70 061 929	18.06	2.47	204 115.6	207 430.0	11.3	37.1	51.6
San Francisco	617 778	10 758 761	13 051 100	15 607 537	25.26	1.88	20 445.7	22 265.0	17.6	21.7	60.7
South Atlantic	795 875	16 634 026	20 799 476	30 500 661	38.32	3.08	73 943.9	80 471.3	9.5	26.9	63.6
South Pacific	343 471	1 405 130	1 564 822	1 736 796	5.06	1.07	2 696.6	2 664.7	27.0	18.0	55.0
Titicaca	112 501	753 130	912 488	1 089 602	9.69	1.86	671.3	619.0	17.7	34.6	47.7
Total Latin America	20 371 272	211 217 540	275 168 065	355 804 671	17.47	2.64	742 366.5	779 839.3			

Sources: GDP 1980 and 1985: ECLAC, Statistics Division, national accounts (computer printout), 1988
GDP Caribbean Islands and Guayanas: United Nations, *National Accounts Statistics: Analysis of Main Aggregates*, 1982 (ST/ESA/STAT/SER.X/2), New York, 1985, Sales No. E.85.XVII.4
Population and Area: National censuses and *United Nations Statistical Yearbook*
Structure of GDP: ECLAC and national reports
GDP by sector: Economic Commission for Latin America (ECLAC), *Distribución regional del producto interno bruto sectorial en los países de América Latina,* Cuadernos Estadísticos de la CEPAL series, No. 6, (E/CEPAL/G.1115), Santiago, Chile, 1981

Notes: 1. Year for the structure of GDP by country: Argentina 1968/Brazil 1980/Chile 1980/Colombia 1975/Ecuador 1965/Panama 1968/Peru 1980/Mexico 1980/ Uruguay 1961
2. The regional distribution of GDP for Bolivia, Costa Rica, Guatemala, Honduras, Nicaragua and Venezuela has been calculated in proportion to the distribution of population in 1980 by administrative region, and the base of national GDP
From UN ECLAC (1990a).

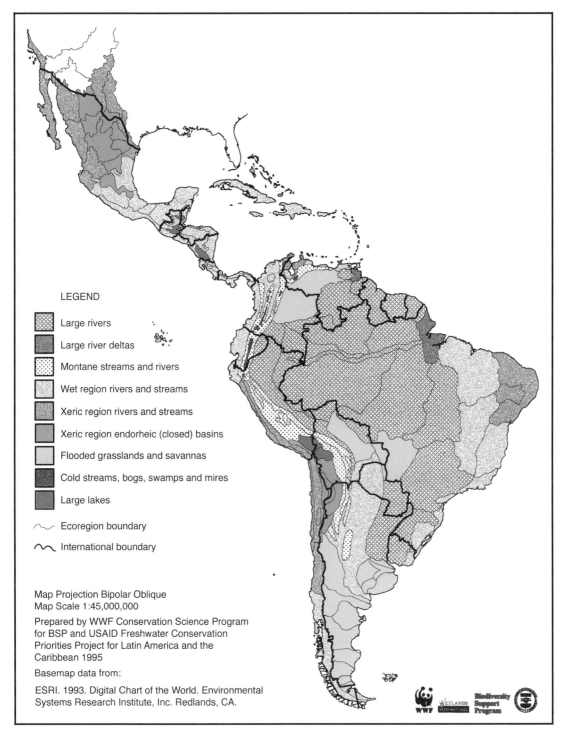

LEGEND

Large rivers

Large river deltas

Montane streams and rivers

Wet region rivers and streams

Xeric region rivers and streams

Xeric region endorheic (closed) basins

Flooded grasslands and savannas

Cold streams, bogs, swamps and mires

Large lakes

Ecoregion boundary

International boundary

Map Projection Bipolar Oblique
Map Scale 1:45,000,000

Prepared by WWF Conservation Science Program
for BSP and USAID Freshwater Conservation
Priorities Project for Latin America and the
Caribbean 1995

Basemap data from:

ESRI. 1993. Digital Chart of the World. Environmental
Systems Research Institute, Inc. Redlands, CA.

Figure 2.2   *Major habitat types of freshwater ecoregions in Latin America and the Caribbean as determined at a special workshop on regional priorities for the conservation of freshwater biodiversity of Latin America and the Caribbean held in Santa Cruz, Bolivia, in 1995 (after Olson* et al., *1998)*

The Caribbean

The Caribbean region exhibits insular conditions with varying freshwater resources ranging from small arid islands (e.g. Antigua, Curaçao) to large islands with abundant freshwater resources (Trinidad and most of the Greater Antilles). As noted above for Central America, because of the north-easterly trade winds, east-facing mountain slopes receive the greatest amount of precipitation and are drained by perennial streams. On the drier sides of the islands, streams are ephemeral or intermittent, only flowing during large storm events.

Most catchments in the Caribbean islands are relatively small (i.e. less than 250 km^2). Many of the smaller islands are noted for their general lack of perennial surface water, while larger islands (e.g. Cuba, Hispaniola, Puerto Rico, Jamaica, Guadeloupe, Martinique, and Trinidad and Tobago) have permanent streams. Larger islands generally have mountainous cores that are surrounded by fertile, lowland coastal plains. Perennial streams on the larger mountainous islands typically have steep, boulder-lined headwaters that flow across narrow coastal plains before entering estuarine or coastal waters. Headwater areas are usually forested with dispersed agriculture and residential land use, while lower courses flow through human-dominated landscapes characterized by agriculture, urban and industrial land use.

South America

South America has some of the highest rainfall and run-off of any continent. As a result of the north–south orientation of the Andean mountains, rivers are generally much shorter, with higher gradient on the Pacific side and longer and larger on the Atlantic side (which contains some 84% of the land area). Notable exceptions to this are located in Brazil and northern Bolivia and Paraguay where the São Francisco, Paraná/Plata and Paraguay Basins have their headwaters in the central plateau of Brazil (Griesinger and Gladwell, 1993). Rivers of South America are primarily rain-fed and have mean annual river discharges varying within broad limits. The geographic distribution of rainfall is extremely uneven, ranging from a low of 1 mm yr^{-1} in Arica, Chile, to almost 8000 mm in Quibdo, Colombia.

The Amazon is the largest river basin in the world. Within the Amazon and Orinoco catchments, a wide diversity of freshwater habitats occurs, including very large rivers (whitewater, blackwater and clearwater rivers), floating meadows, várzea or seasonally flooded forests, swamp forests, cataracts, mangroves, igapo (white sand flooded forests), small rivers and streams and oxbow lakes. Although the Amazon and Orinoco River ecosystems are spatially dominant and widely recognized elements of neotropical freshwater biodiversity, Latin America contains a diverse array of other freshwater habitats and communities, including vast seasonally flooded areas which occur in the Llanos, Pantanal, Chaco and Beni savannas. Cold montane streams and waterfalls occur in the Andes, Amazonian piedmont and other mountainous regions. Other habitats include fogdrip pools and streams of the Pacific deserts of Peru and Chile and closed-basin (endorheic) springs, pools and streams in the Chihuahuan Desert. Many regions support highly distinctive freshwater biotas with large numbers of endemics (World Conservation Monitoring Centre, 1992; Olson *et al.*, 1998).

DEVELOPMENT PRESSURES AND HUMAN IMPACTS

Differences in the availability of water, coupled with variations in human population densities and activities, produce strongly contrasting patterns of use and transformation of water resources within the region (UN ECLAC, 1990a). Many areas of Latin America are still considered to be relatively undisturbed (e.g. almost half is classified as forest and woodland – although much of this is secondary forest). However, the explosion of large urban areas, combined with rapid and extensive land-use change, is threatening water quality and quantity in many areas. Human use of water resources is diverse and highly concentrated along coastal areas; as a result, environmental pollution and water quality issues have typically first emerged in major coastal cities (UN ECLAC, 1990c). A primary factor affecting larger river systems is stream regulation for hydroelectric power.

Very few scientific data exist on effects of development on the 'non-human' ecology of river ecosystems in Latin America (Pringle and Scatena, 1999a). However, reports of the sheer magnitude of river deterioration in many areas often include accounts of biotic degradation (e.g. in riverine fisheries) that are attributable to multiple development impacts (e.g. Fuentes and Quiros, 1988; Quiros, 1993; Goulding *et al.*, 1996).

This section provides a general overview of development pressures on river ecosystems focusing primarily on deforestation; agriculture, pesticides and irrigation; urban and industrial development; and hydropower. A more comprehensive description of development pressures affecting streams and rivers of Latin America is provided by the UN ECLAC (1990c). Freshwater resource development in Latin America is also reviewed by Pringle and Scatena (1999a,b).

Deforestation

Deforestation rates in Latin America and the Caribbean are extremely high. Until the mid-1980s, Costa Rica's deforestation rate was one of the highest in the world, peaking at 40 470 ha yr^{-1}. The rate is now about 8000 ha yr^{-1}, since much of the remaining primary forest is within national park boundaries. Deforestation in the Amazon Basin has become a matter of international alarm. Forest burning in the Amazon region increased by 28% between 1996 and 1997, according to satellite data. Also, 1994 deforestation figures (the most recent available) show a 34% increase since 1991 (Schemo, 1997). Brazil is now losing more rain forest each year than any country in the world; in addition to the 15 000 km^2 that satellite images show are being lost each year, another 10 000 km^2 are thinned through logging beneath the forest canopy (Schemo, 1997).

Logging in lowland areas has often been localized along rivers due to the abundance of timber in floodplain forests, low relative costs of wood extraction and transport and good access to markets (Barros and Uhl, 1995). Logging operations typically move inland once floodplain forests have been cleared; riverine systems denuded of riparian and floodplain zones become extremely vulnerable to the effects of logging within the catchment owing to the loss of the forest buffering capacity.

Effects of logging include: the destruction and/or fragmentation of critical riverine habitat, including upstream floodplains and estuarine mangroves (see Pringle, this volume); alteration of stream sediment loads, turbidity and solute chemistry; changes in hydrology and fluvial geomorphology; shifts in incident light and temperature regimes; and loss of allochthonous energy sources (e.g. Goulding *et al.*, 1996) and structural components such as wood (e.g. Maser and Sedell, 1994) which maintain habitat heterogeneity in the stream channel. Physical and chemical alterations have ecological consequences including loss of in-stream and riparian zone biodiversity, decreases in the abundance of aquatic organisms, shifts in the ratio of photosynthesis to respiration, shifts in the relative importance of different functional feeding groups along stream continua and facilitation of invasion by exotic species. Effects of logging on tropical rivers is reviewed in detail elsewhere (Pringle and Benstead, 2000).

Removal of Amazonian vegetation on a large scale could bring about changes in the region's hydrological cycle and climate large enough that tropical forests may not be able to re-establish themselves (Shukla and Sellers, 1990). Other researchers have claimed that the cumulative effect of deforestation in the Amazon is already having a noticeable impact on the hydrology of the Amazon River itself (Gentry and Lopez-Parodi, 1980): it was suggested that increased flooding was caused by increased sediment loads (derived from deforestation) and subsequent aggradation of the river-bed. However, Richey *et al.* (1989) concluded that there has been no statistically significant change in river discharge over an 83-year-long period. Short-term increases in discharge, such as those found by Gentry and Lopez-Parodi (1980) from 1962 to 1978, fell within the range of long-term oscillations in the hydrograph. It is clear that caution must be exercised in the extrapolation of short-term studies to longer time-frames. Given the restrictive time-scale of much scientific inquiry (i.e. human life-span), such large-scale effects of deforestation will be difficult to identify with certainty in the near future.

Agriculture, pesticides and irrigation

Unlike many other areas of the world, the agricultural frontier in Latin America continues to expand, with many hydrographic regions remaining predominantly in agriculture. Major cash crops in the region include coffee, maize, wheat, soybean, sugar cane, cotton and fruit. Cattle, goats, sheep and pigs are also important in most countries. Fertilizers, pesticides and herbicides are a growing part of agricultural water pollution in Latin America – as cultivation systems have become more intensive, often with 'zero' pest levels required for export of crops such as fruit (van Emden and Peakall, 1996). The problem is often aggravated by local abuses in fertilizer and pesticide application (UN ECLAC, 1990c). The consumption of fertilizers in Latin America and the Caribbean increased by 97% between 1973 and 1985 (UN ECLAC, 1990c) and pesticide imports increased by almost 50% between 1971 and 1973 and between 1983 and 1985 (Postel, 1987). Pesticides currently correspond to about 10% of the world market, with Brazil accounting for about half of the pesticide use in Latin America (van Emden and Peakall, 1996). The problem is aggravated by the fact that pesticides and other chemicals are often employed that are restricted or banned in countries with more stringent environmental legislation. In addition, there is often no centralized authority to regulate pesticide trade, use and application. Brazil ranks among the top five countries in the world in terms of pesticide use, including products such as aldrin, eldrin, ethilic parathion, heptachlor and lindane – which have been banned or restricted in a number of European countries as well as the United States (Hurtado, 1987). Although organic chlorinated compounds such as DDT and aldrin still play a key role in water pollution in the region, apparently phosphates and carbamates are being used increasingly (UN ECLAC, 1990c).

There are few data on the effects of agricultural pesticides and fertilizers on the biota of aquatic ecosystems (but see Food and Agriculture Organization of the United Nations – FAO, 1993). This is due, in part, to the fact that analytical determination of pesticides requires sophisticated methods and equipment as well as highly trained personnel. Most countries in Latin America do not have this infrastructure and, when it does exist, pesticide residues may be measured but biological effects are typically not assessed. A study of 12 river basins in the State of Paraná, Brazil, indicated that 91% of the samples obtained contained agrotoxic residues (Andreoli, 1993). A study of soil and water quality in the Irrigation District of Saldana, Colombia (15 000 ha), found DDT in soils and stream-bottom sediments with values in the order of 74 ppb (Gomez-Sanchez, 1993). The number of recorded human pesticide-related poisonings in areas of Latin America is particularly illuminating: 1800 pesticide poisonings occur per 600 000 persons annually in Central America compared with only 1 per 600 000 persons a year in the United States (UN ECLAC, 1990c).

Irrigation is a major pressure on freshwater resources throughout Latin America and the Caribbean (Graham and Edwards, 1984); it has historically been the main purpose of dam construction in the region since the first large dam was built in 1750 on the Saucillo River in Mexico. Over the past two decades, the area of cultivated land under irrigation in the region has increased by 75% and in some countries, particularly in Central America and the Caribbean, it has exceeded these figures (UN ECLAC, 1990c). While Mexico has the greatest amount of irrigated land, over the last two decades the largest increases have occurred in parts of

central and southern Brazil, Central America and Cuba (UN ECLAC, 1990a). Irrigation is a particularly important form of water use on larger Caribbean islands (Cuba, Jamaica, Dominican Republic and Puerto Rico) which produce sugar cane, rice and in some areas bananas. In a recent workshop on regional needs for water quality monitoring, water resource managers indicated that water pollution (resulting from increasing agricultural and industrial development) is rapidly reaching a level where it will have adverse impacts on natural ecosystems, socio-economic conditions and health of the region (World Meteorological Organization, 1993).

As in other parts of the world, the re-use of drainage water for irrigation is accelerating the process of soil salinization: desertification claims 2250 km^2 of farmland in Mexico each year (Grainger, 1990). An estimated 34% of Peru's coastal area is affected by salinization and drainage problems (Alva *et al.*, 1976). In Chile, there is a growing problem with high saltwater content and concentration of boron in arid and semi-arid areas of the country. Salinization is a particular problem in some valleys of northern Chile (Pena-Torrealba, 1993).

The growing exploitation of groundwater for irrigation and other purposes has led to saltwater intrusions into coastal aquifers in many regions including the Caribbean, Argentina, Brazil, El Salvador and Mexico (UN ECLAC, 1990c). Saltwater intrusion is particularly problematic in rivers of tropical countries where there is a large difference between maximum and minimum flows. As an example, saltwater intrusion in the Guayas River of Ecuador is threatening the river's use as a water supply for the city of Guayaquil during periods of low flow (UN ECLAC, 1990c).

While groundwater contamination by fertilizers and toxic agrochemicals does not at present represent the acute problem that it does in more developed regions, it can be expected to grow. An increasing problem in some South American countries (e.g. Chile) is the increase in groundwater nitrate levels where sewage is used for irrigation (Pena-Torrealba, 1993).

Urban and industrial development

The explosion of large urban areas is severely affecting water quality and quantity throughout Latin America (Griesinger and Gladwell, 1993; Pringle and Scatena, 1999a). This increasing urbanization is typically unaccompanied by sewage treatment facilities and most domestic sewage is discharged untreated into the nearest water body. Estimates suggest that less than 2% of total urban sewage flows in Latin America receive treatment (UN ECLAC, 1990c). According to Nash (1993), sewer systems are accessible to only 41% of the urban population of Latin America and more than 90% of collected waste water is discharged completely untreated into rivers. Polluted drinking water supplies from untreated municipal sewage are responsible for increasing human health problems from water-related diseases such as epidemic cholera and intestinal infections (PAHO, 1990a; Witt and Reiff, 1991; Reiff, 1993).

Historically, the most intense urban pollution has been concentrated in a few major basins; three of the 26 major hydrographic divisions contain almost 40% of the human population and account for 52% of the total gross domestic product of the region (UN ECLAC, 1990a): the Gulf of Mexico; the South Atlantic; and Plate basins (Figure 2.1, Table 2.1).

The Tiete River, which flows through São Paulo, a metropolitan area of more than 20 million people, provides an example of the severity of pollution in many urban areas. The Tiete flows 1120 km from the Atlantic Coast mountain range to the Paraná River which then empties into the Atlantic Ocean at Buenos Aires. During the dry season, an estimated 60% of the flow of the Tiete consists of untreated residential wastes from the city of São Paulo. Industry adds another 4.5 t of chemicals and heavy metals each day. The clean-up of the Tiete (which would involve hooking up 70% of the residences to sewage treatment plants and control of industrial effluents) has received international attention ($450 million has already been disbursed by the Inter-American Development Bank), but apparently the clean-up action has ground to a halt because of corruption and inefficiency (Switkes, 1995).

Most Caribbean islands are undergoing land-use changes that are characterized by a decrease in agricultural land use and an increase in urban and industrial development (Graham and Edwards, 1984; Birdsey and Weaver, 1987). On many islands, sugar plantations and other lowland agricultural developments are being abandoned and converted into urban and industrial land uses. Generally, the rapid shift in population from rural to urban areas has been faster than the construction of wastewater disposal facilities. Notable exceptions are the island of Hispaniola and some smaller isolated islands that have large agrarian-based economies. On most islands, domestic and industrial waste treatment is minimal before being dumped into nearby rivers.

Demands on water resources for disposal and transport of industrial wastes are rapidly increasing. For example, the pulp/paper and iron/steel industries rank among the most important sources of water pollution in the region and they have been growing twice as fast as the economy of Latin American countries as a whole (Postel, 1984; Abramovitz, 1996). In many countries all but the most toxic industrial effluents are discharged directly into the nearest water body. In Ecuador, industrial effluents are generally not treated (UN ECLAC, 1990c). The Maipo River Basin in Chile represents a relatively high degree of waste treatment for the region, yet only 26% of industrial effluent receives treatment (UN ECLAC, 1990c).

The more insidious effects of industrial waste dumping in landfills on groundwater and surface water in Latin America are less well known because of the time-lag effect, yet it can be expected to be a major problem in the decades to come. In Puerto Rico, for example, the industrialization of the 1950s was accompanied by uncontrolled dumping of industrial wastes into the nearest landfill. As a result, Puerto Rico has a heritage of toxic dumps including nine superfund sites (Hunter and Arbona, 1995). Six out of every 10 hazardous toxic sites in Puerto Rico lie over vulnerable limestone aquifers which can act as pollution conduits. The high density of dump sites (138 sites in 8700 km^2) poses a threat to both surface water and groundwater (Hunter and Arbona, 1995).

Hydropower

With increasing industrialization, the regulation of rivers for hydroelectric power is emerging as a major factor affecting water quality and quantity in large rivers throughout Latin America. The rate of dam building has been high since 1950 and the water held in reservoirs has increased by more than 20-fold since 1945 (UN ECLAC, 1985). The international banking system provided relatively easy loans to many governments in Latin America in the 1970s and early 1980s, resulting in the launching of megaprojects such as massive hydroelectric schemes (Inter-American Development Bank, 1984). For example, 79 dams have either been built or are planned for Brazilian Amazonia alone (Figure 2.3; Seva, 1990). Brazil's state-owned power monopoly, Eletronorte, proposes to meet more than half of Brazil's future electricity needs using hydroelectricity from the Amazon region despite the drawbacks of the flat topography, wide floodplains and abundant trees and wildlife (Best and da Silva, 1989).

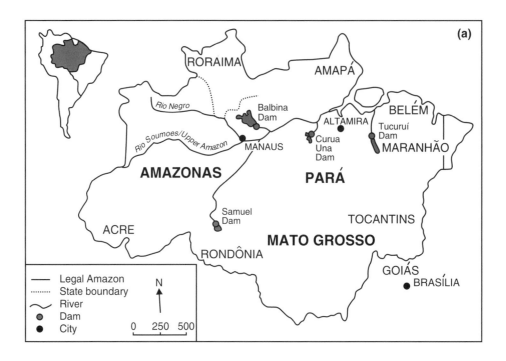

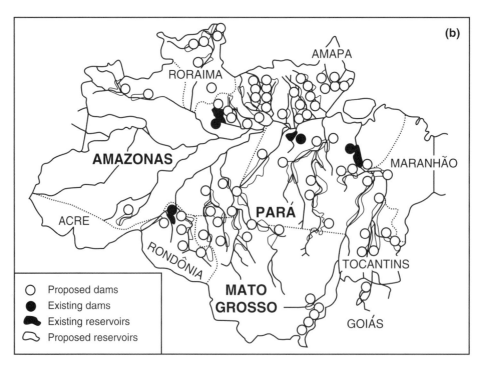

Figure 2.3   *(a) Brazil's Amazon region (shaded in inset) with four existing large dams. (b) Location of proposed and existing hydroelectric projects in the Brazilian Amazon (from Fearnside, 1995)*

While Brazil's financial difficulties have resulted in the postponement of dam-building plans, the overall scale of the plans remains unchanged (Fearnside, 1995) and is consequently an important consideration for future conservation and management in the area. The remainder of this section addresses the conservation implications of hydroelectric development in Latin America with respect to overall decreases in aquatic biodiversity. Effects of hydroelectric development on emissions of greenhouse gases in tropical reservoirs are discussed elsewhere (Fearnside, 1995; Pringle, this volume).

Hydroelectric dams are considered as potentially the most dangerous human activity to Amazonian fisheries in the near future (Bayley and Petrere, 1989; Goulding *et al.*, 1996). The fragmentation of river systems by hydroelectric dams can affect landscape-level patterns of aquatic biodiversity by impeding the passage of migratory tropical aquatic fauna, many of which are economically important food fish. For example, characin fish are prevented from migrating biannually upstream and downstream (Goulding *et al.*, 1996). It is believed that dams will negatively affect catfish populations by interrupting long-distance migration patterns – including the downstream movements of eggs and young (provided that they do not spawn in the upper tributaries) and upstream migrations which annually restore catfish stocks up river (Barthem *et al.*, 1991; Barthem and Goulding, 1997).

Because of their feeding habits, dolphins are also very vulnerable to hydrological modifications which fragment the stream longitudinally and also disrupt lateral riverine floodplain connections (Reeves and Leatherwood, 1994). For example, Amazon River dolphins (*Inia geoffrensis*) feed on up to 50 species of fish throughout the year (Klinowska, 1991). Their diets vary seasonally and are synchronized with the annual flood cycle (Perrin *et al.*, 1989). In the Amazon, the annual wet season flood cycle carries pink Amazon River dolphins into the floodplain (Schmidt-Lynch, 1994). When hydrological modifications (e.g. water abstraction for irrigation, deforestation, channelization) cause water levels in the rivers to drop, Amazon River dolphins are often stranded in drying pools. For example, 40 river dolphins were observed stranded in a 30 m^2 area in 1985 (see Carpino, 1994). Dams also act to fragment dolphin populations into small and genetically isolated sub-populations by preventing them from interbreeding (Perrin *et al.*, 1989). Dams can also affect prey by cutting off the spawning routes of migratory fish, affecting sedimentation patterns, reducing levels of dissolved

oxygen (Vidal, 1993) and increasing levels of hydrogen sulphide (Perrin *et al.*, 1989).

In Central America, major hydroelectric projects are also fraught with multiple problems and it is questionable whether they will survive beyond their first years of operation. The El Cajon (Honduras) and Chixoy (Guatemala) projects have been hindered by silt deposition in reservoirs which occurred much faster than project planners had predicted. Loans from the Inter-American Development Bank (IDB) to create these projects totalled nearly $2 billion; however, the price tag did not include outlays for sustaining the project in perpetuity (e.g. upkeep, dredging). Unfortunately, river basin management plans were not required by IDB as part of the funding criteria. Deforestation in the catchments has led to massive erosion and sedimentation, destroying hydroelectric potential and fisheries (International Rivers Network, 1988a).

As early as 1982, the US Agency for International Development (US AID) wrote in its environmental profile of Honduras, 'Why risk $650 million on a hydroelectric project when the watershed is being degraded by activities of deforestation, shifting agriculture, erosion and sedimentation? It is an act of negligence on the part of the funding agencies to ignore the real problems of watershed management in the El Cajon watershed' (International Rivers Network, 1988b). The Chixoy Dam in Guatemala provides another example of an economic and environmental disaster. The total cost of the dam was $1.2 billion (521% higher than forecast in 1974 (McCully, 1996)). Erosion and massive siltation in the reservoir have taken their toll along with geological problems resulting from the high level of seismic activity in the area. The World Bank's Project Completion Report states that 'with hindsight (Chixoy Dam) has proved to be an unwise and economic disaster' (McCully, 1996).

Other development pressures

Other development pressures include mining, petroleum extraction, exotic species, the lack of sufficient landfill sites and direct dumping of solid wastes into stream channels (particularly in the Caribbean), dredging for navigation, overfishing and the aquarium trade.

Pollution from mining affects water bodies and coastal areas in nearly all South American countries and it is a particularly acute problem in the Andean countries (e.g. Chile and Peru (UN ECLAC, 1990c)). Pollution from gold mining (mercury) and the petroleum industry is becoming much more widespread and severe, as relatively unregulated exploitation spreads into more intact regions (Olson *et al.*, 1998). In most mining operations, only a small portion of the mining effluent receives any treatment and this treatment is often only partial. As an example, the Rimac River of Peru is contaminated by arsenic, cyanide, lead, chrome and selenium from mining activities and is considered to rank among the most polluted rivers of South America. This is of immediate concern because the river is the water source for 60% of the population of Lima (UN ECLAC, 1990c).

Gold mining can be particularly harmful to aquatic systems because of mercury contamination. In the upper Madeira region in the state of Rondônia in Brazil as much as 100 t of mercury may have been discharged into waters of the Río Madeira between 1974 and 1985. Brazil is responsible for between 2 and 11% of the mercury annually discharged into the global environment and apparently Amazonian gold-miners are responsible for most of Brazil's mercury pollution (Goulding *et al.*, 1996).

While mining does not appear to represent as serious a threat to the water resources of Central America and the Caribbean as it does in South America, the negative impacts of mining wastes have the potential to be much more pronounced on small islands, given their limited land and water resources (UN ECLAC, 1990c). In Jamaica and other bauxite-producing countries, the effluent of the bauxite–alumina industry is a major pollutant (Reid, 1978). Sand and gravel mining is a particularly serious factor contributing to deteriorating water quality in the Caribbean. Due to the high humidity, high frequency of hurricanes and lack of timber, most houses in the Caribbean are constructed with cement rather than wood, resulting in an ever-increasing demand for sand and limestone as human populations increase. Coastal sand mines have led to severe coastal erosion and 'drowned' estuaries in many parts of the Caribbean. Many islands have drawn up legislation to restrict gravel mining. For example, sand and gravel mining at the mouths of rivers has been restricted in Puerto Rico since the early 1970s.

Historically, water pollution related to petroleum production was significant in only a few Latin American and Caribbean countries (e.g. Trinidad and Tobago). However, more recent petroleum extraction and oil production is threatening freshwater systems in Venezuela; Argentina; southern Chile; the foothills of the Andes in Bolivia, Ecuador and Peru; the central Amazon Basin; the South Atlantic and the Gulf of Mexico (UN ECLAC, 1990c). Oil spills related to

petroleum industry activities have caused serious freshwater deterioration in the Napo region of Ecuador and are an increasing cause of environmental concern (e.g. Olson *et al.*, 1998). Apparently, the Coatzacoalcos and Tonala Rivers in Mexico have the most extreme levels of hydrocarbon pollution recorded globally (Rose, 1987).

The introduction of exotic aquatic species is a particular problem on 'ocean-locked' Caribbean islands that have a low diversity of aquatic taxa, with some endemic species. This makes them particularly vulnerable to displacement by exotic species. The freshwater fish fauna of Puerto Rico is largely composed of exotic species from the south-eastern United States, Africa and South America (Erdman, 1984; Bunkley-Williams *et al.*, 1994). *Lepomis incisor* and *Ameiurus melaas melas* were introduced in about 1914 and were well established in reservoirs by 1934. *Gambusia holbrookii* was introduced in 1923 for mosquito control and was fairly numerous by 1934 (Erdman, 1984).

During the early part of the 20th century, many fish were purposely introduced into streams of Latin America to 'improve' sport fishing opportunities, for mosquito control, or to provide local sources of protein. Fish aquaculture using exotic species, such as African tilapia and Asian carp, is common in the Amazon Basin. Farming of native fish is relatively recent (Goulding *et al.*, 1996), reflecting the lack of literature on these species. There have been no reports of farmed exotics invading 'natural' riverine ecosystems, but it is possible that some populations have already become established (Goulding *et al.*, 1996). Well-managed fish farming using native rather than exotic fish species is less destructive of the floodplain than livestock ranching and should be encouraged from a conservation perspective (Araujo-Lima and Goulding, 1997). For example, the tambaqui (*Colossoma macropomum*) has been bred successfully in captivity by using hormonal injections and has recently been chosen as a 'flagship' fish with which to launch a project aimed at making fish culture information (i.e. for native species) available and accessible in the Amazon Basin (Goulding *et al.*, 1996; Araujo-Lima and Goulding, 1997).

CATCHMENT VERSUS LOCAL, REGIONAL, NATIONAL AND INTERNATIONAL ISSUES

Streams are continuous systems with dynamic links between headwaters and estuaries; local effects can cumulatively disrupt this linkage. Local development projects (or 'sub-catchment' issues) must therefore be considered in terms of their intensity and duration within the context of the entire catchment and its ability to absorb local impacts. Development projects are often planned for catchments which fall within more than one country, complicating conservation and management issues (see Case Study 5). Case studies below illustrate how stream conservation is tied to local versus regional water supplies (Case Study 1) and to both national and international politics (Case Study 2).

Case Study 1 – Local versus regional water needs in the Caribbean National Forest, Puerto Rico

In many areas of Latin America – particularly in the Caribbean – tourism and urbanization have both resulted in significant conflict between local and regional water needs. Water shortages can be particularly severe in 'ocean-locked' islands undergoing population explosions (Rice, 1986). Because most islands in the Caribbean have central governments, the water needs of a few centralized urban areas are typically considered before local needs. Water needs from government-sponsored tourist complexes are also often considered over local water demands or conservation needs.

The Caribbean National Forest (CNF) is an 11 269 ha US national forest in north-eastern Puerto Rico which has been protected since 1903 (Figure 2.4). However, at least 600 000 people are drawing water from the river systems that drain the forest. On an average day, more than 50% of river water draining the forest is abstracted for municipal water supplies (Naumann, 1994). Conditions are particularly severe during drought (Morris, 1994) and many streams have no water below their intakes for much of the year. Dams and associated water abstractions are threatening the biotic integrity of rivers and streams draining the CNF by causing significant mortality of migratory shrimps and fish which migrate between stream headwaters and coastal estuaries (Pringle, 1997a; March *et al.*, 1998; Benstead *et al.*, 1999).

The sheer magnitude of hydrological changes that are being proposed over a short period has kept resource managers busy counteracting poorly planned proposals. Conflicts between local water use and tourist complexes are aggravated by the fact that the dry season coincides with the tourist season. These conflicts can only be expected to increase as tourism expands: the number of tourists visiting the islands each year is now approaching the local population of more than 3.5 million (already a high population density of

396 inhabitants km^{-2}). Water conflicts in Puerto Rico are discussed in more detail in Pringle and Scatena (1999b) and Pringle (2000b).

Case Study 2 – National and international politics and the ecological integrity of the San Juan River of Central America

The San Juan River provides an excellent example of the conflict between preserving catchment integrity, international politics and national conflicts. The San Juan River is a transborder region situated between the two countries of Costa Rica and Nicaragua (Figure 2.5). It is also a corridor between the Caribbean and Pacific regions. The river's 41 600 km^2 catchment is the largest in Central America. Approximately 70% is located in Nicaragua and 30% in Costa Rica. The river system contains many interconnecting aquatic habitats which support a diversity of wildlife, from small lakes to freshwater and estuarine wetlands. The catchment lies in a 500 km long by 80 km wide rift valley and the 200 km long river connects Lake Nicaragua to the Caribbean. Since Lake Nicaragua is only separated from the Pacific Ocean by a 20 km wide divide (Figure 2.5), the lake and river complex have attracted many proposals to build canals. Since 1812, the plan to build a trans-isthmus canal across Nicaragua has been periodically revived (most recently in 1989 by the Japanese (Girot and Nietschmann, 1992)). Unfortunately, the perennial nature of these inter-oceanic canal proposals has acted to discourage other potential national projects and land-use schemes and the San Juan River area has remained isolated from institutions in Managua and San José (Girot and Nietschmann, 1992). The area has evolved as a transborder region with a bi-national identity, a distinct social network and economy, and large areas of tropical forest and wetland environments.

For the past 500 years, the San Juan River has been one of the most militarily and politically sensitive areas in Central America. Ironically, the history of military activity has acted indirectly to protect large areas of the San Juan River catchment. Since large areas of the remaining tropical forest are in transborder catchments, a bi-national programme called SIAPAZ (International System of Protected Areas for Peace) has been initiated by the governments of Costa Rica and Nicaragua to create a transboundary protected area. This project is attempting to integrate existing and new border area parks, reserves and wildlife refuges into a transboundary protected area (Figure 2.5 and see section later in text on 'Transboundary protected areas' for a more detailed description of the SIAPAZ Project).

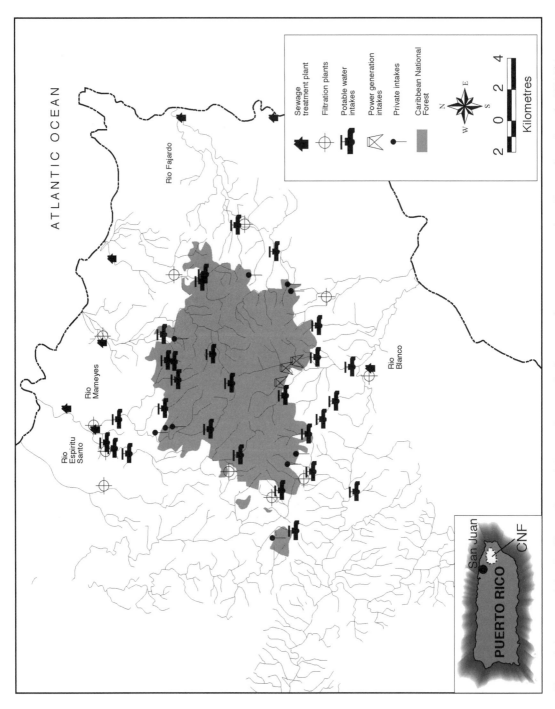

Figure 2.4   *Map of the Caribbean National Forest showing the location of major water intakes. Figure from Pringle (2000b)*

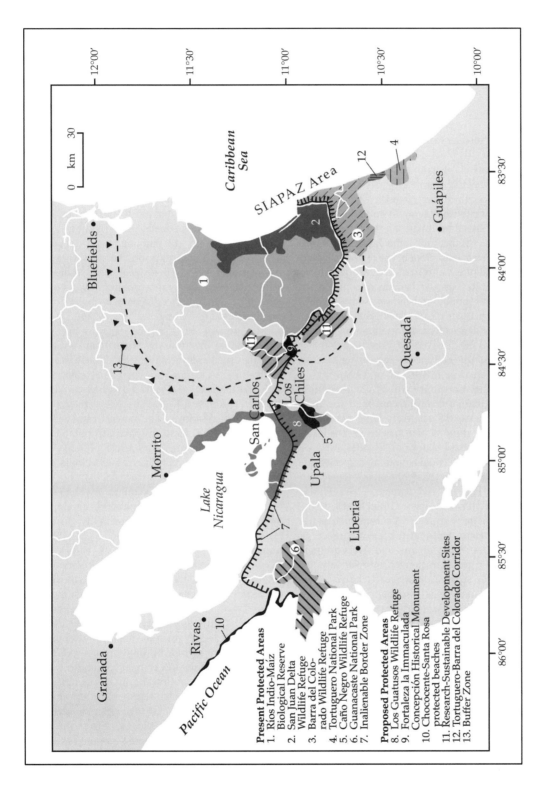

Present Protected Areas
1. Ríos Indio-Maíz Biological Reserve
2. San Juan Delta Wildlife Refuge
3. Barra del Colo- rado Wildlife Refuge
4. Tortuguero National Park
5. Caño Negro Wildlife Refuge
6. Guanacaste National Park
7. Inalienable Border Zone

Proposed Protected Areas
8. Los Guatusos Wildlife Refuge
9. Fortaleza la Immaculada Concepción Historical Monument
10. Chococente-Santa Rosa protected beaches
11. Research-Sustainable Development Sites
12. Tortuguero-Barra del Colorado Corridor
13. Buffer Zone

Figure 2.5 *Transboundary protected areas in the San Juan River catchment of Nicaragua and Costa Rica joined together under Project SIAPAZ (International System of Protected Areas for Peace). Figure from Girot and Nietschmann (1992)*

Nicaragua is promoting this project within the context of environmental security (i.e. the policy that protecting environments is a national security priority because environmental destruction is economically and politically destabilizing (Girot and Nietschmann, 1992)).

Information for conservation

There is a relatively small amount of published information in scientific journals that is available for river conservation in Latin America. The bulk of information is unpublished 'grey literature'. Much of the available published information is in journals from the USA, Europe (Netherlands) and Japan. Published information on freshwater biota of the region is largely of a taxonomic nature and is scattered throughout the systematic literature. However, in many regions (e.g. the Amazon), basic taxonomic information for most groups of organisms (from fish to algae) is lacking. A compelling illustration of the lack of basic information on tropical aquatic systems is the discovery of more than 240 new species of fish that were collected in deep-water habitats of the Brazilian Amazon in 1996 (Yoon, 1997).

In the Caribbean, more information for conservation is available for marine environments than for fresh water, in part due to the relatively large number of marine biological stations that occur in the area. Nonetheless, some basic information does exist on freshwater community composition and species life histories of freshwater Caribbean systems (for summaries see Harrison and Rankin, 1976; Hunte, 1980; Boon *et al.*, 1986; Covich and McDowell, 1996). The knowledge of large aquatic organisms such as fish and prawns is greater than that for aquatic insects or lower invertebrates. While streams and rivers draining Caribbean islands are characterized by relatively low diversity of larger aquatic organisms (relative to the mainland) which have wide distribution ranges, the diversity of smaller benthic invertebrates is often quite high (e.g. there are approximately 300 species of caddisflies (Trichoptera) described for the insular Caribbean).

There is an important need for inventories of aquatic biodiversity in freshwater habitats of the Caribbean (Alkins-Koo, 1998). Much of this may simply require collation of existing scattered data, although some field surveys of habitats will be necessary. Very few reliable data exist regarding pollution and contamination levels and their effects on aquatic biota and ecosystem functioning in Latin America and the Caribbean (van

Emden and Peakall, 1996). In contrast, there is a relative abundance of information on effects of water resource deterioration on human ecology (disease). While water quality standards are used voluntarily by many government agencies in the Caribbean, inadequate monitoring networks and laboratory facilities are common problems (UN ECLAC, 1990c).

In many cases, scientific data that are integral to management decisions have not been collected. For example, assessment of overall biodiversity effects of dams in Latin America has been hindered by the lack of pre-impoundment data. Construction programmes for the first five major dams constructed in the Amazon (coastal, central Amazon and Río Tocantins area) did not include broad-scale investigations of fish migrations before the impoundments were closed. The Coaracy–Nunes dam on the Río Araguari was completed in 1975 and no ecological studies of fish were conducted. General taxonomic surveys of species present before and after the dam was closed were not even made (Goulding *et al.*, 1996). With the construction of the last four dams in the Amazon, the National Institute of Amazonian Research in Manaus was contracted to do faunal surveys by Eletronorte. A comparison of gillnet catches before and after the Tucuruí (Figure 2.3) was closed revealed a 49% reduction in fish diversity below the dam and a 50% reduction above. The downstream decrease in diversity was at least in part attributable to the disruption of the migratory cycle of many species. Diversity in the reservoir fell by 55% (Goulding *et al.*, 1996).

Additional information on migratory biota is critical to ensuring that future hydroelectric development is pursued in a more 'environmentally sustainable' manner. As outlined by Barthem *et al.* (1991) with respect to long-distance migratory catfish (in the Tocantins Basin above the Tucuruí Reservoir in the Amazon), we do not know the answers to the following questions: (1) Can catfish eggs or fry reach the estuary when the water residence time in the reservoir is 51 days and as much as 130 days in the reservoir margins? (2) Can eggs and fry resist the impact of dropping from a barrage 80 m high or passing through a turbine? (3) Can they endure the low oxygen conditions within the reservoir (and just below it) in the summer? These questions need to be taken into account when considering proposals for hydroelectric dams. For example, how will hydroelectric effects on fisheries affect future benefits to regional economies?

For scientifically based catchment management, which considers aquatic biodiversity, Goulding *et al.*

(1996) recommend surveys from river headwaters to mouth in order to develop hypothetical models of fish migrations and movements within and out of the individual rivers involved. Such studies would provide important baseline data to evaluate development projects and would indicate the necessity of fish ladders or the installation of other devices in hydroelectric projects (see Pringle, this volume).

THE IMPORTANCE OF FIELD NOTES, MUSEUM COLLECTIONS AND LOCAL KNOWLEDGE

In many instances, the only information available for conservation may be old field notes and/or museum collections. This is an inexpensive way to gather data and utilize local knowledge. These sources of information can be extremely valuable as a frame of reference for evaluating how aquatic communities have changed between the past and present. Interviews with elders within a community can be an effective way to understand the general environmental history of an area, along with basic life-history patterns of fish species (e.g. Araujo-Lima and Goulding, 1997; Barthem and Goulding, 1997). Reznick *et al.* (1994) provide two examples where such sources can yield key information for aquatic conservation where: (1) visual censuses of fish communities by a scientist in freshwater streams in Trinidad over a 19 year period provided a qualitative index of change in fish communities accompanying anthropogenic changes in the habitat; and (2) fish collections, made in Costa Rican streams during the 1960s and 1970s and housed at the University of Costa Rica, were used to describe species abundance and diversity for entire watersheds, to describe the composition of the fish community at individual collecting sites, and to elucidate the life histories and ecology of resident species.

CONSERVATION, CLASSIFICATION, EVALUATION AND PRIORITIZATION SCHEMES

There are few conservation classification or evaluation schemes available for neotropical aquatic systems and this has hampered conservation activities in Latin America. Regional priorities for the conservation of freshwater biodiversity in Latin America and the Caribbean were assessed at a workshop in Bolivia in 1995 sponsored by the World Wildlife Fund and Wetlands International (Figure 2.6; Olson *et al.*, 1998). Prioritization was based on biological distinctiveness

(Table 2.2) and conservation status (i.e. high to medium levels of threat; Figure 2.7).

Areas of outstanding freshwater biodiversity that face significant and immediate threats were identified for the neotropics in a single document (Olson *et al.*, 1998) which includes tables and maps designed to inform conservation planners, investors, decision-makers and the public (e.g. Table 2.2, Figures 2.2, 2.6 and 2.7). Development of the freshwater prioritization scheme was based on: (1) representation of major freshwater habitat types; (2) the need to assess ecological and evolutionary phenomena as important elements of neotropical freshwater biodiversity; and (3) the importance of tailoring analytical criteria to particular patterns of biodiversity, ecological dynamics and responses to disturbances of different major habitat types.

Species richness for a wide range of taxa was estimated to be greatest for the ecoregions of the Amazon Complex and the Guiana Watershed (the Guayanan Highlands in particular). Ecoregions known for their large number of endemic fish species include the Guiana Watershed (Guayanan Highlands) and the Upper Amazon Piedmont, while regional centres of endemism (i.e. ecoregions with a high percentage of endemic species) include those ecoregions of the Río Grande/Río Bravo Complex, the Lerma/Santiago Complex, Papaloapam, Catemaco, Grijalvo-Usumacinta, Cuba, Hispaniola, Southern Orinoco and Río Negro ecoregions (Olson *et al.*, 1998).

In terms of overall conservation status, 82% of the freshwater ecoregions of Latin America and the Caribbean are considered to be either critically endangered, endangered òr vulnerable. Ten ecoregions were considered to be critically endangered, 44 endangered, 42 vulnerable, 19 relatively stable and two relatively intact. Critically endangered regions were found in the Caribbean lowlands and intermontane valleys of Colombia, the Maracaibo Basin, Lake Poopo, the delta of the Colorado River, coastal Sinaloa, Lerma and Patzcuaco of central Mexico, the Atacama/Sechura Deserts, the Panano–Platense Delta and the northern portions of the Mediterranean region of Chile. Endangered regions include much of the cerrado and Atlantic region of Brazil, northern and southern Mexico, higher elevation ecoregions of the northern Andes and the coastal deserts of Peru and Chile (Figure 2.7; Olson *et al.*, 1998).

Out of an original 117 ecoregions, a subset of 10 highest priority ecoregions and 36 high priority ecoregions were identified. Highest priority ecoregions

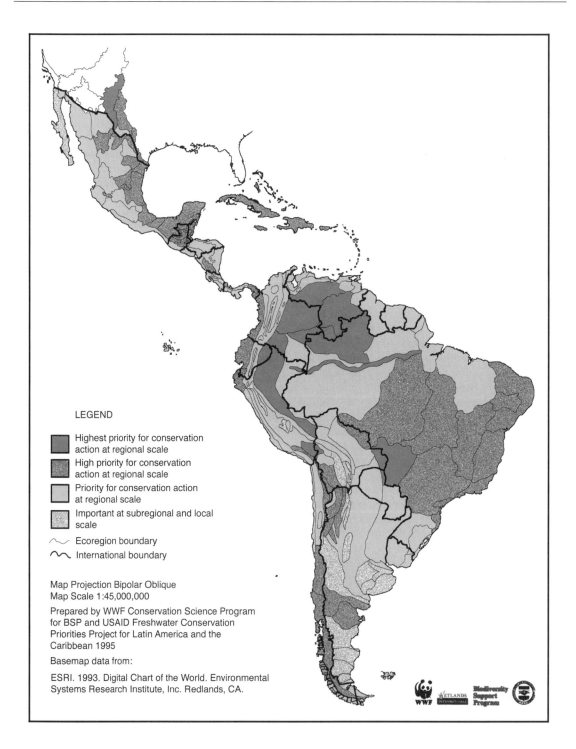

Figure 2.6   *Regional priorities for conservation action for freshwater ecoregions in Latin America and the Caribbean as determined at a special workshop on regional priorities for the conservation of freshwater biodiversity of Latin America and the Caribbean held in Santa Cruz, Bolivia, in 1995 (after Olson* et al.*, 1998)*

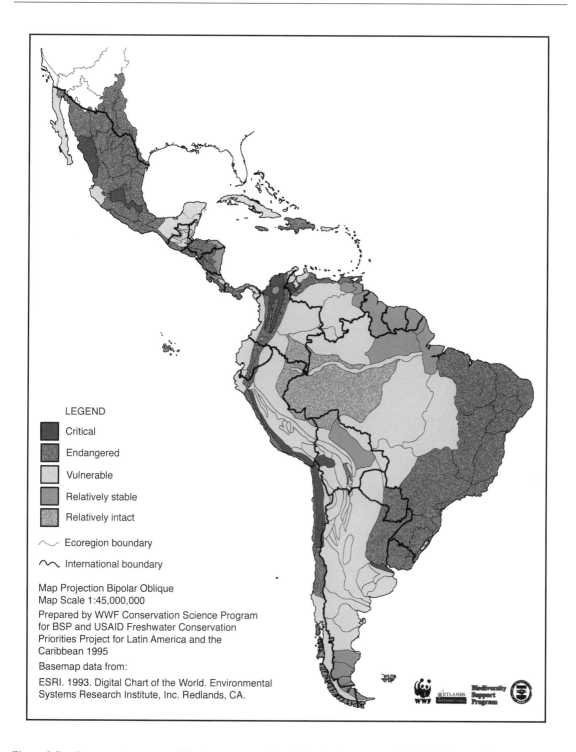

Figure 2.7 *Conservation status of freshwater ecoregions in Latin America and the Caribbean as determined at a special workshop on regional priorities for the conservation of freshwater biodiversity of Latin America and the Caribbean, held in Santa Cruz, Bolivia, in 1995 (after Olson* et al., *1998)*

Table 2.2 *Biological Distinctiveness Analysis – criteria and values*

Ecoregion	Biological Distinctiveness Indicators				Special Considerations		Biological distinctiveness
	Species richness (2–10)	Endemism (3–15)	Ecosystem diversity (1–5)	Total of indicator numbers (6–30)	Rarity of major habitat type	Unusual ecological/ evolutionary phenomena	
Baja California Complex							
1. Baja California	2	6	3	11			LI
Colorado River Complex							
2. Colorado Delta	2	6	2	10	RO[1]	RO[2]	RO
3. Sonoran	2	6	2	10			LI
Sinaloan Coastal Complex							
4. Sinaloan Coastal	4	9	3	16			RI
Río Bravo Complex							
5. Río Bravo	10	12	5	27			GO
6. Pecos	8	6	4	18			RO
7. Guzmán	6	6	4	16			RI
8. Mayoimi	4	9	3	16			RI
9. Cuatro Ciénagas	10	15	5	30		GO[3]	GO
10. Llanos El Salado	2	15	5	22			RO
11. Conchos	6	9	4	19			RO
12. Lower Río Bravo	8	12	4	24			RO
13. Río San Juan	6	9	4	19			RO
14. Río Salado	6	9	2	17			RI
Lerma/Santiago Complex							
15. Santiago	4	6	3	13			RI
16. Chapala	8	15	5	28	RO[4]	RO[5]	GO
17. Lerma	6	9	3	18			RO
18. Río Verde Headwaters	2	15	1	18			RO
19. Manantlan/Ameca	4	9	4	17			RI
Río Pánuco Complex							
20. Río Pánuco	6	12	4	22			RO
Balsas Complex							
21. Balsas	2	12	3	17			RI
Pacific Central Complex							
22. Tehuantepec	4	9	3	16			RI
Atlantic Central Complex							
23. Southern Veracruz	6	12	3	21			RO
24. Belizean Lowlands	6	9	3	18			RO
25. Central American Caribbean Lowlands	6	9	3	18			RO
26. Talamancan Highlands	2	3	3	8			LI
27. Catemaco	6	12	4	22			RO
28. Coatzacoalcos	6	9	3	18			RO
29. Grijalva-Usumacinta	8	12	3	23			RO
30. Yucatán	4	9	5	18			RO

Latin America and the Caribbean 61

Region						
31. Guatemalan Highlands	6	6	3	15		RO
32. Central American Karst Highlands	6	9	4	19		RO
33. Honduran/Nicaraguan Highlands	6	6	3	15		RI
34. Lake Nicaragua	8	9	4	21	RO[4]	RO
Isthmus Atlantic Complex						
35. Isthmus Atlantic	6	6	2	14		RI
Isthmus Pacific Complex						
36. Isthmus Pacific	6	9	2	17	RO[4]	RI
Bahama Archipelago Complex						
37. Bahamas	4	9	3	16		RI
Western Insular Caribbean Complex						
38. Cuba	8	12	3	23		RO
39. Hispaniola	6	12	3	21		RO
40. Jamaica	4	6	3	13		RI
41. Cayman Islands	2	6	3	11		LI
42. Florida Keys	2	3	3	8		LI
Eastern Insular Caribbean Complex						
43. Puerto Rico and Virgin Islands	8	12	3	23		RO
44. Windward & Leeward Islands	4	3	2	9		LI
Chocó Complex						
45. Chocó	10	12	2	24		RO
South American Caribbean Complex						
46. Magdalena	8	3	1	12		LI
47. Momposina Depression-Rio Cesar	8	3	2	13		RI
48. Ciénega Grande de Santa Marta	8	3	3	14		RI
49. Guajira Desert	8	3	2	13		RI
50. Maracaibo Basin	8	12	2	22		RO
High Andean Complex						
51. Páramos	8	12	3	23		RO
52. Peru High Andean Complex	8	6	2	16		RI
53. Bolivian High Andean Complex	4	3	2	9	RO[6]	LI
54. Arid Puna	6	9	5	20		RO
55. Subandean Pampas	6	9	3	18		RO
56. South Andean Yungas	6	9	3	18		RO
Inter-Andean Dry Valleys Complex						
57. Inter-Andean Dry Valleys	4	3	1	8		LI
North Andean Montane Complex						
58. North Andean Montane	6	9	1	16		RI
59. Humid Andean Yungas	4	9	2	15		RI
60. Chuquisaca and Tarija Yungas	4	6	1	11		LI
61. Salta and Tucumán Yungas	4	3	1	8		LI
62. Sierra de Córdoba	4	3	1	8		LI
Puyango-Tumbes Complex						
63. Puyango-Tumbes	10	9	2	21		RO
64. Atacama/Sechura Deserts	2	6	1	9		LI

continues overleaf

Table 2.2   *(Continued)*

Ecoregion	Biological Distinctiveness Indicators				Special Considerations		Biological distinctiveness
	Species richness (2–10)	Endemism (3–15)	Ecosystem diversity (1–5)	Total of indicator numbers (6–30)	Rarity of major habitat type	Unusual ecological/ evolutionary phenomena	
Pacific Coastal Desert Complex							
65. Pacific Coastal Deserts	6	6	3	15			RI
Lake Titicaca/Poopó Complex							
66. Lake Titicaca	8	15	5	28	GO[4]	GO[5]	GO
67. Lake Poopó	6	6	5	17			RI
Galápagos Complex							
68. Galápagos	2	9	1	12			LI
Mediterranean Chile Complex							
69. North Mediterranean Chile	6	9	4	19			RO
70. South Mediterranean Chile	4	12	4	20			RO
Juan Fernández Islands Complex							
71. Juan Fernández Islands	4	9	4	17			RI
Southern Chile Complex							
72. Valdivian	6	9	5	20			RO
73. Chiloé Island	6	9	4	19			RO
74. Chonos Archipelago	4	6	4	14			RI
75. Magallanes/Ultima Esperanza	4	6	4	14			RI
Subantarctic Complex							
76. Subantarctic	2	3	4	9			LI
Venezuelan Coast/Trinidad Complex							
77. Venezuelan Coast/Trinidad	4	3	1	8			LI
Llanos Complex							
78. Llanos	10	12	5	27	RO[7]	GO[3]	GO
Guiana/Orinoco Complex							
79. Eastern Morichal	6	9	5	20			RO
80. Orinoco Delta	10	6	4	20	RO[1]		RO
81. Southern Orinoco	10	15	5	30	GO[8]		GO
82. Guiana Watershed	8	15	4	27			GO
Amazon Complex							
83. Amazon Delta	8	6	5	19	RO[1]		RO
84. Amazon Main Channel	10	9	5	24	GO[8]	GO[2]	GO
85. Northern Amazon Shield Tributaries	6	12	4	22			RO
86. Rio Negro	8	15	5	28		GO[2]	GO
87. Upper Amazon Piedmont	10	12	4	26			GO
88. Western Amazon Lowlands	8	9	3	20			RO
89. Central Brazilian Shield Tributaries	8	9	3	20			RO
90. Tocantins-Araguaia	8	12	3	23			RO
Northeastern Atlantic Complex							
91. Maranhão	6	3	3	12			LI

Ecoregion				Total	Biological Distinctiveness
Mata-Atlantica Complex					
92. Northeast Mata-Atlantica	6	12	3	21	RO
93. East Mata-Atlantica	8	12	3	23	RO
94. Southeast Mata-Atlantica	8	12	3	23	RO
São Francisco Complex					
95. Caatinga	6	12	4	22	RO
96. Cerrado	6	12	4	22	RO
Upper Paraná Complex					
97. Upper Paraná	8	12	2	22	RO
Beni Complex					
98. Beni	8	6	3	17	RI
Paraguay-Paraná Complex					
99. Pantanal	8	6	4	GO[7] 18	GO
100. Lower Paraná	8	12	3	23	RI
Southern Atlantic Complex					
101. Jacuí Highlands	4	3	3	10	LI
102. Lagoa dos Patos Coastal Plain	4	3	3	10	LI
Chaco Complex					
103. Chaco	6	6	5	17	RI
Pampas Complex					
104. Parano-Platense Basin	7	3	3	13	RI
105. Rio Salado and Arroyo Vallimanca Basin	5	3	5	13	RI
106. Northwest Pampas Basins	2	3	5	10	LI
107. Pampas Coastal Plains	4	3	5	12	LI
108. Southwest Pampas Basins	4	3	3	10	LI
Patagonia Complex					
109. Rio Colorado	6	6	3	15	RI
110. Rio Lamay-Neuquen-Rio Negro	6	15	4	25	RO
111. Meseta Somuncura	4	15	3	22	RO
112. Rio Chubut-Rio Chico	6	12	3	21	RO
113. Rio Deseado	2	6	2	10	LI
114. Rio Santa Cruz-Rio Chico	2	6	3	11	LI
115. Rio Coyle	2	15	2	19	RO
116. Rio Gallegos	2	6	2	10	LI
117. Tierra del Fuego-Rio Grande	2	12	2	16	RI

Notes:

(a) Scoring for Biological Distinctiveness Indicators: higher numbers correspond to greater distinctiveness

(b) Biological Distinctiveness: Globally Outstanding (GO), Regionally Outstanding (RO), Regionally Important (RI), Locally Important (LI). GO = 26–30 points, RO = 18–25, RI = 13–17, LI = 6–12; total range 6–30 points

(c) Special Considerations: Special considerations automatically raised ecoregion distinctiveness to highest assigned category

Rarity of major habitat types and unusual ecological/evolutionary phenomena:

[1] Large river delta
[2] Large-scale fish migrations
[3] Very high beta diversity/unusual adaptations and radiations
[4] Large tropical lakes
[5] Pronounced radiations of fish in tropical lakes
[6] Tropical saline lakes
[7] Extensive flooded savanna
[8] Large river channels

for Latin America and the Caribbean include the western arc of the Amazon; the freshwater ecosystems of the Amazon main channel, particularly the várzea flooded forests; the wetland complexes of the Llanos and the Pantanal; the Southern Orinoco including the Guayanan Highlands; Cuatro Cienegas in the Chihuahuan Desert; Lake Titicaca; the upper Río Bravo; and Lake Chapala in Mexico (Figure 2.6; Olson *et al.*, 1998).

The framework for conservation and conservation in practice

Traditionally, countries within Latin America and the Caribbean have focused on water resources management from the perspective of irrigation, hydropower, drinking water and sanitation. Most countries have formulated plans for the management of water resources for these needs, some of which have completely dominated the process to the exclusion of other important issues (Griesinger and Gladwell, 1993). For example, irrigation and energy often dominate national and regional strategies for water planning. Thus, there is a critical need for comprehensive and systematic assessments of national and regional water resources and the formulation of integrated water management plans (UN ECLAC, 1985; Griesinger and Gladwell, 1993). There is also an important need for local catchment plans. In some cases, very broad and general plans are developed while plans at the intermediate and local scales are forgotten. The development of local catchment commissions (e.g. in the United States the Potomac River Basin Commission, the Tennessee Valley Authority and the Mississippi River Basin Commission) can be an effective method to deal with long-term planning issues.

Continuous monitoring protocols for both hydrological and water quality data are lacking in most areas throughout Latin America. In the Caribbean, the problem is compounded by the continuous decimation of limited facilities by hurricanes and flood damage (Alkins-Koo, 1998). While almost all islands have some long-term rainfall records, few have stream gauges or water quality stations. With the exception of Puerto Rico, Jamaica and Cuba, few islands have a systematic network of stream gauging stations. The Caribbean Environmental Health Institute (CEHI) was given the authority to collect data on pesticides in river waters, but no data have been collected due to lack of properly

functioning equipment. Protocols for a low-technology and low-cost biomonitoring programme are being developed at present by this agency (CEHI) for use within the Caribbean.

Environmental planning has rarely been a high priority in Latin American development projects. For example, two-thirds of Central America's rain forests have disappeared since 1945 as a result of the uncontrolled growth of cattle ranching, logging concessions and farming. International lending agencies (e.g. International Development Bank) have contributed to the problem by paying inadequate attention to watershed management in their rush to fund hydroelectric projects.

Redefining the development policies of 'foreign' governments outside of Latin America is an important step in improving water quality and conserving river systems in the region, since policies of outside countries influence many governments in the region and can shape development policy (see Pringle, this volume; Wishart *et al.*, this volume, Part II). Creation of a more open, peer-reviewed environmental review process for the development of loans is critical, especially for large agribusiness, dam construction and road-building projects. More international focus should be placed on the development of water basin commissions, national parks and reserves.

A major role is being played by agricultural colleges, universities and other local government institutions. For example, in Brazil, several state universities have been actively engaged in research on integrated pest management and implementation of lower-input technology (reduction of pesticide use, etc.; Gravena *et al.*, 1987).

PUBLIC AWARENESS AND ENVIRONMENTAL EDUCATION

Public awareness of environmental problems is strongly influenced by the socio-political/religious–cultural background of the many communities within the region. In Latin America, as in many developing and developed countries, there is a strong socio-cultural perspective of rivers as conduits to remove pollution. The United Nations sponsored Earth Summit in Rio de Janeiro had a big impact on the environmental awareness of peoples in Latin America. Local radio and television frequently present the need for environmental conservation to a large proportion of the population. However, very little of this information is transmitted to rural populations (van Emden and Peakall, 1996).

Community-based watershed projects are most effective when they interact with local government agencies. This helps build political support and gets policies into law. Community-based conservation can be most successful when it involves the participation of stakeholders in the research process and local resource management (White and Runge, 1995). It often involves the need to solve a resource or land-use problem, a local organization (e.g. NGO) that provides start-up funding and training, and the collaboration of scientists (Western and Wright, 1994). Project SIAPAZ, in the transboundary catchment of the Río San Juan between Costa Rica and Nicaragua, provides one such example of community-based catchment conservation (see Case Study 2).

The two case studies below provide examples of environmental outreach and community-based participation: (a) environmental outreach activities on water quality and implementation of a volunteer stream monitoring programme in lowland Costa Rica; and (b) grassroots efforts initiated by local communities themselves to manage fisheries of floodplain lakes in the Brazilian Amazon.

Case Study 3 – Implementation of 'Water-for-Life' Programme in Puerto Viejo de Sarapiquí, Costa Rica

The community of Puerto Viejo de Sarapiquí (population 10 000) in the lowlands on Costa Rica's Caribbean slope is facing severe water quality and quantity problems as a result of rapid population growth which has occurred largely as a result of the development of extensive banana plantations in the area. Local streams have become contaminated by faecal coliforms (introduced by livestock and domestic sewage), pesticides and herbicides from the banana plantations. Despite an average annual rainfall of near 4 m, the community must pipe in water from a distant location (i.e. a spring located 8 km away in a national park) to satisfy the needs of the rapidly growing population. The current success of banana culture in the province of Sarapiquí has had the effect of attracting workers and their families who caused the population to increase from 3000 to 30 000 in just three years. Such explosive growth has placed extreme demands on the municipality and water quality and supply issues promise to become even more significant in the coming years.

In response to these water quality problems, an environmental outreach programme on water quality was created for local high school students and adults in local communities of Sarapiquí (Pringle, 1997b, 1999).

The programme was initiated by aquatic scientists working at the Organization for Tropical Studies (OTS) La Selva Biological Station (located just 5 km from the town of Puerto Viejo) and is now administered by OTS staff who are involved in environmental education and community relations. This arrangement has provided an infrastructural support system to ensure the programme's longevity.

Environmental outreach materials on water quality include: (1) slide show presentations which describe the history of municipal water resources in Sarapiquí Province and threats to water resources (Vargas, 1995); (2) catchment protection posters (http://www.arches.uga.edu/~cpringle/wflproducts.html) which illustrate the importance of intact riparian zones, the dependence of local communities on forested catchments in the highlands for potable water supplies, and the importance of river corridors for migrating biota (Figure 2.8); and (3) the implementation of an 'Adopte Una Quebrada' (i.e. Adopt a Stream) programme at the local high school in Puerto Viejo, accompanied by the development of a manual in Spanish and English which provides details on how to initiate volunteer stream monitoring programmes, the sampling methodology to follow, and data interpretation (Laidlaw, 1996); and (4) an instructor's manual for local high school teachers on water quality and river conservation (Pohlman, 1998).

Case Study 4 – Fisheries management of Amazonian floodplain lakes by local communities

Local communities in floodplain areas of the Lower Amazon are becoming increasingly dependent on fishing for both cash income and family subsistence as a result of the collapse of floodplain agriculture and the expansion of ranching (Furtado, 1988). At the same time, commercial fishing has intensified, greatly reducing the catch of local fishermen (Goulding, 1983; Smith, 1985; Furtado, 1990). As competition for fish has increased, local communities have responded by attempting to control local floodplain lakes by excluding fishermen from outside of the community. These lake 'reserves' constitute a new strategy of resource management based on a traditional knowledge of the ecology of local floodplain fisheries (McGrath *et al.*, 1993).

Control of a given floodplain lake is usually based on a system of collective ownership of shorefront property by members of the community. The formation of a lake reserve typically involves not only the exclusion of outsiders but the regulation of fishing activity on the

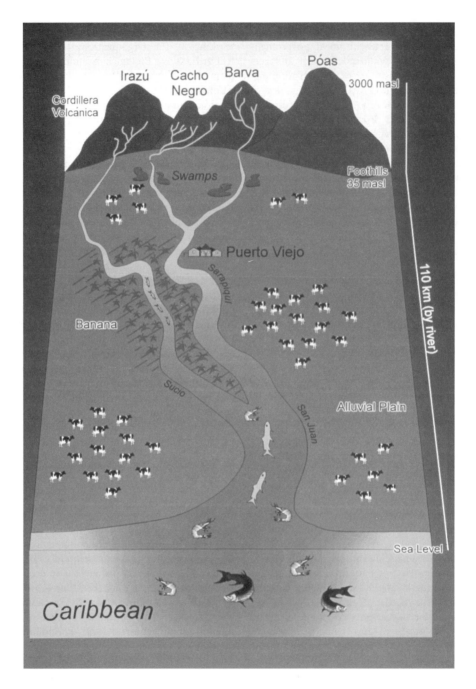

Figure 2.8 *Figure produced as part of an environmental outreach programme on water quality designed for the local community of Puerto Viejo de Sarapiquí, Costa Rica: illustration of land-use activities in Sarapiquí region (banana plantations, cattle grazing) with impacts on riverine ecosystems. This figure shows landscape linkages between stream headwaters in the mountains, lowland estuaries and the ocean. Migratory freshwater shrimp and fish create a critical linkage between highland streams in Costa Rica's Central Mountain Range and the Caribbean and must pass through a gauntlet of pesticides introduced into the stream system by the banana plantations. The 19 hydroelectric projects which exist or are planned for the Sarapiquí catchment will further fragment river systems*

lake. Regulatory measures are based on local understanding of the ecology of floodplain fisheries and widely held beliefs about the effects of different fishing practices on lake fish populations. One of the more common measures is to prohibit the use of gill nets during the low water season. As an example, in Ilha de São Miguel, gill nets are completely prohibited and commercial fishing is limited to the sale of pirarucu and catfish during a six-month season extending from June to December. The productivity of local fisheries in this region is high and most fishermen can meet family subsistence needs with relatively little effort, while seasonal commercial fishing provides an important cash supplement (McGrath *et al.*, 1993).

While these lake reserves are a hopeful strategy for managing lake fisheries on a sustainable basis, floodplain lakes are small elements of a complex riverine system where fisheries face many different threats – from overfishing by commercial fisheries to hydrological modifications which block the migrations of economically important fish. It thus remains to be seen whether individual lake fisheries will be sustainable in the long term. In the successful São Miguel example cited above, while managing individual lake systems appeared to improve the productivity of lake fisheries, lake reserves alone were not sufficient to revitalize the local economy (McGrath *et al.*, 1993).

STATUTORY FRAMEWORK, APPLICATION AND ENFORCEMENT OF LEGISLATION

Most countries in Latin America and the Caribbean have begun to develop a body of law on water pollution control that includes legislation which empowers public agencies to take steps to control water pollution (UN ECLAC, 1990c). (A general discussion of laws aimed at controlling water pollution in Latin America can be found in UN ECLAC, 1990c.) However, existing legislation is rarely applied, and immediate social needs often take precedence over long-term ecological and social needs until the extent of environmental deterioration becomes an immediate social concern. For example, El Salvador is the smallest and most densely populated country on the American continent. The relationship between environmental deterioration and poverty is evidenced by the relation between quality of life and public services. Only 41% of the population has access to potable drinking water and 61% to sanitary services. Ninety per cent of the surface water is contaminated with domestic, industrial and agricultural discharges. Also, the availability of water

(including groundwater) is unevenly distributed, presenting additional problems for distribution. El Salvador has a very low unitary discharge and availability of water per capita, with only 30 L s^{-1} km^{-2} and 3302 m^3 per capita, respectively. In comparison, the average Central American country has 501 L s^{-1} km^{-2} and 6800 m^3 per capita.

The river basin concept has not been widely applied in water management in Latin America and the Caribbean. While isolated application of the concept has occurred in various countries in the past, there are few contemporary examples (Lee, 1991). The application of the concept of integrated river basin management (e.g. management efforts which observe watershed boundaries and consider the entire river basin) is not even very common at the planning stage. However, Brazil and Chile have been recognized for their recent innovations in water management policy that may eventually lead to future creation of national water management systems that are based on the concept of integrated river basin management (e.g. see Lee, 1991).

Existing legislation related to water quality is often very dispersed and buried within many different laws, decrees, ordinances and regulations. For instance, the water sector in Honduras has 37 laws with 420 articles without any uniform criteria. There is a lack of discharge regulations for industry, agriculture, urban and rural housing and sewage treatment facilities in many countries throughout Latin America. Costa Rica established 'water discharge normatives' in 1995, while most Central American countries have none.

Environmental legislation is also urgently needed throughout Latin America to reform hydroelectric development projects. Legislation and/or legal requirements to clear vegetation from reservoir areas before they are submerged is particularly important in tropical areas since it can take many decades or even centuries for organic matter to decompose, causing severe deoxygenation of water and massive emission of hydrogen sulphide. For example, large-scale flooding in Surinam submerged 1500 km^2 of rain forest to create the Brokopondo Dam. Organic matter decomposition resulted in severe deoxygenation of the water and massive emissions of hydrogen sulphide. Workers at the dam wore masks for two years after the reservoir started to fill in 1964 and the cost of repairing damage to the dam's turbines by the acidic water totalled more than US $4 million (see Pringle, this volume). Despite a legal requirement to clear vegetation from all areas to be submerged, the Brazilian electricity utility Eletronorte cleared less than one-fifth of the 2250 km^2

of rain forest inundated by construction of the Tucuruí Dam (on the border between Argentina and Paraguay) and only 2% of the 3150 km^2 of forest inundated by the Balbina Dam on Brazil's Uatuma River, with similar disastrous consequences. The Uatuma River below the Balbina Dam is receiving almost totally deoxygenated water from the reservoir. Consumption of oxygen by decomposing vegetation in the Yacyreta Reservoir is believed to have killed more than 120 000 fish found downstream after the first test of the dam's turbines in 1994.

THE ROLE OF NON-GOVERNMENTAL ORGANIZATIONS

Non-governmental organizations (NGOs) are emerging as a major force in the conservation of aquatic resources in the Caribbean and Latin America. Efforts of NGOs to protect riverine ecosystems in Latin America and the Caribbean include establishing and nurturing protected areas; working with indigenous peoples and other traditional riverine groups; designing alternative development strategies for the region; lobbying to halt destructive projects; technical and scientific work to raise the level of knowledge about the ecological imperatives of the wetlands and river systems; environmental education to make the public aware of the importance of water resources and wetlands; valuing traditional economic activities versus industrialization and export models; and international campaigns to reform the environmental policies of multilateral development banks (e.g. IDB and World Bank).

Local, national and international NGOs have become increasingly active on a variety of different scales, as illustrated by the following case study.

Case Study 5 – Community-based initiatives expand into large-scale efforts coordinated by international NGOs – Ríos Vivos versus the Hidrovia Paraguay/Paraná Waterway Project

The Hidrovia is an ambitious project to convert 3400 km of the Paraguay and Paraná River systems into a shipping canal stretching from Caceres, Brazil, to the Atlantic Ocean (Figure 2.9). The canal would accommodate convoys of 16–20 barges which require an engineered river channel 90 m wide. The five governments of South America's La Plata Basin have approved the engineering design of the first phases of the waterway – for the purpose of improved navigation to spur trade. Funding for the project has been requested from the Inter-American Development Bank and the United Nations Development Programme. Additional funding would be provided by commercial investors of the five countries involved.

Hydrological changes associated with this scheme may increase downstream flooding which will adversely affect indigenous peoples who live along the river. The project is also predicted to have major environmental impacts on Brazil's Pantanal, the world's largest intact wetland (e.g. Bucher *et al.*, 1993). The Pantanal is an ecologically important highly diverse area as a result of its location at the confluence of the Amazons, Cerrado and Chaco ecoregions. It stretches from Corumba to Caceres and encompasses about 500 000 km^2 of wetlands in the Brazilian states of Mato Grosso and Mato Grosso do Sul. The Pantanal hosts 650 bird species, caiman, giant otters and more than 260 species of fish. The project involves dredging at 93 separate places on the river, including sites in the Pantanal. Millions of cubic metres of material (some of it toxic) must be disposed of. Maintenance dredging of the river, once the initial project is completed, is predicted to have negative effects on the habitat of many river-dwelling species. In addition, natural river meanders will be straightened at seven points. The plan involves extensive modifications of the river bed and bank and closing off of tributaries and lakes.

Recent studies indicate that: (1) the Hidrovia Project will have a substantial impact on the flood regime of the Upper Paraguay River; (2) the proposed removal of rock outcroppings in the river bed (via dynamite) 'will have an irreversible impact on the hydrology of the Upper Paraguay River' and the Pantanal region will becoming progressively drier; (3) channel straightening upstream of Corumba will intensify annual floods downstream and hasten the drying of the Pantanal; and (4) these changes will eventually alter the nature of the Pantanal, with a loss of soil nutrients leading to 'decreases in biotic productivity' (Ponce, 1995).

International efforts to protect the Pantanal were an outcome of a meeting of environmentalists, human rights workers, indigenous representatives and members of the women's and black people's movements held in the town of Chapado dos Guimares in Mato Grosso in 1994. Participants of the Chapado meeting summarized their resolutions in a 'Letter from Chapado', in which they detailed their concerns which included: questioning the validity of forecasts of increased agricultural production associated with the project; a marginal, if any, economic return; severe environmental impacts; and

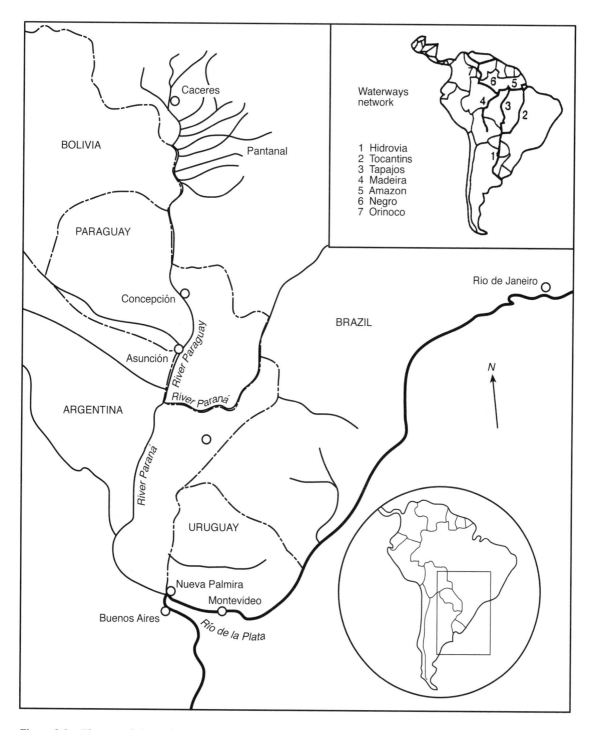

Figure 2.9 *The site of the Hidrovia Project in the Rio Paraguay–Paraná watershed. The plan is to turn the Paraguay–Paraná River system into a grand shipping lane; its implementation, say conservationists, could then spur the development of a waterways network stretching from the Orinoco in the north to the La Plata estuary in the south (modified from Heath, 1995)*

negative social impacts (Switkes, 1995). As a result of the Chapado meeting, more than 70 NGO representatives from Brazil, Argentina, Bolivia, Paraguay, Uruguay, Canada, Holland and the United States agreed to work together to ensure ongoing monitoring of project plans and participation in carrying out the Environmental Impact Study process. This group has now expanded to a 300 member coalition called 'Ríos Vivos'.

The International Rivers Network (IRN), based in Berkeley, California, USA, currently serves as the North American coordinator for the Ríos Vivos Coalition. IRN serves as a 'clearing house' for information; it monitors the project, disseminates information and works to improve public participation on the project (e.g. through videos and dossiers). It has also pushed international agencies to be proactive in the opening of Hidrovia studies to public scrutiny. A cause for optimism is that doubt about the project appears to be growing at official political levels. In 1998, the President of the Brazilian government's Environmental Institute announced that the country is abandoning its current plans to construct Hidrovia. The government of Paraguay is also wavering in its plans to dynamite rock passes along the Paraguay River owing to a study about the project's impact by the US Army Corps of Engineers (Switkes, 1998), Nonetheless, the other countries in the region (Argentina, Uruguay and Bolivia) still appear to be determined to carry out extensive engineering works along the Paraguay and Paraná and are pressing Brazil to reconsider its decision.

IN SITU CONSERVATION: PROTECTED AREAS

Over the last few decades many countries in Latin America and the Caribbean have taken major steps to create parks and conservation areas. While many of these newly established park systems rival those of developed countries in their relative size and ecological diversity, they are often very vulnerable – with no official budget allocations and inadequate protection. Most existing protected areas are becoming increasingly vulnerable to external factors such as expanding populations (Hunter, 1994), increased needs for water supply (Pringle, 1997b, 2000a; Pringle and Scatena, 1999b), poaching and fishing (Goulding et al., 1996), and pollution from adjacent areas (Burger et al., 1993, 1994). There is often a lack of buffer zones around conservation areas to absorb such development effects.

While protected reserves hold some promise as safe havens for river dolphins, they often face great threats themselves and are in need of stronger legislative protection (Carpino, 1994). For example, the Pacaya–Samiria National Reserve located in the Peruvian Amazon is a large protected wetland that lies between the Ucayali and Maranon Rivers (where they merge to form the Amazon River). The reserve provides habitat for the Amazon River dolphin (*Inia geoffrensis*) and grey dolphins (*Sotalia fluviatilis*) which are protected by law from poaching and habitat disruption in the region. It is operated by the Peruvian Foundation for the Conservation of Nature, with assistance from the Nature Conservancy and the US Agency for International Development. External threats to the reserve include oil companies which degrade forest areas adjacent to the reserve while exploring and drilling for oil; industrial activities and boat traffic which are radiating outward from the city of Iquitos and beginning to encroach on the reserve; and local poaching (Carpino, 1994).

Even when the conservation significance of certain regions is understood, a myriad socio-political factors may impede protection. This is well illustrated by Brazil's Pantanal, which was declared a priority area for research and conservation in 1985 (e.g. Alho et al., 1988). The Pantanal is currently threatened by hydroelectric development (see Case Study 5), deforestation, expanding agriculture, illegal hunting and fishing, and water pollution from herbicides and pesticides.

Protected areas can be created through a variety of innovative mechanisms including: (1) bi- or tri-national transboundary protection areas; (2) more broadly international cooperative agreements such as debt-for-nature swaps and 'joint implementation' programmes; and (3) by involving indigenous peoples in local management efforts. Below we discuss how some of these different mechanisms work to protect riverine ecosystems using examples from Latin America and the Caribbean.

Transboundary protected areas

Since river catchments do not conform to national boundaries, transboundary protected areas have great potential for riverine protection. Transboundary protected areas offer several advantages over conventional national projects including: (1) more international attention and funding is often focused on multinational environmental projects; (2) they provide mechanisms and policies for controlling transboundary destructive resource exploitation and contraband; (3) they can provide the means to improve relations between neighbour countries, and to resolve or reduce border disputes; and (4) they are an innovative means to secure demilitarization of war-torn border regions (Nietschmann, 1987; Thorsell, 1990; Girot and Nietschmann, 1992).

In Central America, since the late 1980s, several transborder protected-area projects have resulted in the protection of important riverine ecosystems. These include: La Amistad International Park between Panama and Costa Rica; the Trifinio international project between El Salvador, Guatemala and Honduras; and the SIAPAZ between Costa Rica and Nicaragua in the River San Juan region (see Case Study 2).

The SIAPAZ Project is attempting to join existing and new border area parks, reserves and wildlife refuges into an integrated transborder system of protected areas in the San Juan River catchment (Figure 2.5). The project was formally agreed to in 1991 by both Costa Rica and Nicaragua and it is being developed by a bi-national commission made up of representatives of each country's governments, rural organizations and NGOs. Protected areas total 11 500 km^2 and include: (1) the 5000 km^2 Río Indio–Río Maiz Biological Reserve, a vast stretch of intact tropical rain forest that extends southwards across the border and the San Juan River Delta to the Costa Rican 'protected' area of Barra del Colorado and Tortuguero; and (2) the Los Guatusos Wildlife Reserve, a large wetland area that extends from the southern shore of Lake Nicaragua to Caño Negro in Costa Rica, which is a key area for resident and migratory waterfowl. SIAPAZ is attracting ecotourism development projects, facilities for scientific research, a buffer zone for sustainable agriculture and re-training and employment opportunities for ex-military persons to work as guides and guards in protected areas (Girot and Nietschmann, 1992).

Challenges which must be addressed for the SIAPAZ Project to succeed include: the difficulty of administering such a huge geographic area; achievement of immediate economic benefits for local residents; continued political and economic support of the international community; overcoming historical differences between Costa Rican and Nicaraguan authorities; deforestation and development by impoverished people (Nicaragua) and lumber and banana companies (Costa Rica) in the northern and southern catchment which is causing riverine deterioration; and residual effects of military activity and drug trade (Thrupp, 1990; Girot and Nietschmann, 1992).

Debt-for-nature swaps

Debt-for-nature swaps have been important to overall conservation efforts in Latin America and the Caribbean, particularly for protected areas. This mechanism allows conversion of unpaid loans to indebted countries into funds for conservation activities in those countries: a certain amount of foreign debt is cancelled in exchange for local investments in conservation programmes. Brazil's first debt-for-nature swap helped in the acquisition of many private ranches in the Pantanal. Costa Rica is among the most aggressive 'debt-swappers'; the country has converted $80 million of debt to protect millions of acres of tropical habitat. For debt-for-nature swaps to be effective, a clear definition of land boundaries must be made along with effective mechanims to meet the needs of local residents.

The next step in debt-for-nature swaps, which is being promoted by the US Nature Conservancy, is the establishment of National Conservation Trust Funds. The purpose of such trust funds would be to generate a steady and reliable cash stream on a national scale that would allow for the conservation of protected areas in perpetuity (e.g. funding to support forest rangers, equipment, training and stewardship, educational programmes, community outreach and sustainable development).

Protection via a US Initiative on Joint Implementation

The US Initiative on 'Joint Implementation' is a provision of the Framework Convention on Climate Change that was signed by nearly 200 countries in 1992 at the Earth Summit Conference in Brazil. It was established to encourage private-sector incentives for reducing greenhouse gases. The Río Bravo Conservation and Management Area is a 92 000 ha site in north-western Belize which harbours more than 380 bird species and 12 endangered mammal species. In a recent private partnership, five US electric utility companies and organizations joined forces with the Nature Conservancy and the Programme for Belize to conserve and manage this reserve. The partnership is known as the 'Río Bravo Carbon Sequestration Pilot Project'. It is a voluntary effort which is aimed at illustrating that protection of tropical forests can help offset worldwide carbon release (Cutright, 1996). The Río Bravo Pilot Project is one of seven projects approved from around the world and it is the first private forest management and preservation project to be fully funded under 'Joint Implementation'. The partnership, formed between the Nature Conservancy, Programme for Belize and the five electric utility companies and organizations, is designed as a model for private international cooperation. Together they are contributing $2.6 million for the acquisition and management of a 500 ha parcel of endangered forested land that was slated to be converted into agricultural land. A goal of the project is to establish conservation and management procedures for a sustainable forestry programme that will make the Río Bravo Protection Area economically self-sustaining in 10 years.

Protection by local and indigenous people

Locally based management initiatives in parks and reserves have the potential to be an effective mechanism to protect river ecosystems. The Mamiraua Ecological Station in northern Brazil provides an example of a protected area whose success depends on the involvement of local peoples. This 11 000 km^2 reserve in northern Brazil (at the juncture of the Japura and Solimões rivers in the Upper Amazon) directly involves local people living in this várzea area by training them as park rangers and research assistants. The project will be used as the basis of a várzea management plan (Alexander, 1994). It is hoped that this effort will be successful in creating a model for saving other rain forest areas. The advantage of training local river communities as caretakers is clear given the scant resources that the Brazilian government allocates to enforcing national parks (e.g. there is an average of one ranger per 366 000 km^2 in Brazil; Alexander, 1994). Local communities are benefiting by receiving aid subsidies and almost all manual labour for the project is provided by employing local villagers. Villagers are also involved in fish censuses and growing native trees for commercial production.

BIOLOGICAL FIELD STATIONS AND INDIVIDUAL SCIENTIFIC RESEARCH INITIATIVES

There are a myriad biological stations throughout Latin America and the Caribbean which provide resources and infrastructural support for scientists who are studying aquatic systems. Research conducted at these biological stations can be critical to river conservation efforts (e.g. Alexander, 1994; Benstead *et al.*, 1999). In addition, biological stations can serve as excellent bases for environmental outreach activities on water quality (Pringle, 1997b, 1999; see Case Study

3). Many biological stations are associated with reserves and/or national parks; scientists often become active in conservation activities in these natural areas and adjacent developed areas. In some cases, scientific information collected at biological stations can help provide the impetus to establish adjacent protected catchments (e.g. Pringle, 1988; Terborgh, 1990).

REGIONAL THREATS AND THEIR MITIGATION

While the intensity and scale of threats to river systems vary among regions of Latin America and the Caribbean, several general regional patterns emerge:

- River systems draining coastal plains have traditionally suffered the greatest direct human impact, including channelization, water abstraction, sewage and industrial waste disposal, and elimination of riparian vegetation (UN ECLAC, 1990c; Pringle and Scatena, 1999a).
- A major factor threatening large river systems in Latin America is in-stream regulation, particularly for hydropower development (UN ECLAC, 1990c; Fearnside, 1995; McCully, 1996).
- Many montane streams have been damaged by intensive agricultural operations (e.g. coffee production) through sedimentation, eutrophication, pesticides, losses in riparian vegetation and water abstraction (Olson *et al.*, 1998).
- River systems in dry climates are highly threatened throughout Latin America and the Caribbean (Contreras-Balderas, 1978a,b; Contreras-Balderas and Escalante, 1984), largely because of use and pollution of water by humans, and human destruction of riparian vegetation (Olson *et al.*, 1998). Drier life zones are also very sensitive to climate change.
- Habitat types which are most endangered include: (a) large river floodplain habitats such as floating meadows and várzea forests, which are threatened by logging and conversion to pasture (e.g. Goulding *et al.*, 1996); (b) stream cataracts which have been lost over vast areas due to dams and water diversions (McCully, 1996; Olson *et al.*, 1998); and (c) coastal wetlands and estuaries which are typically modified by channelization and urban development, thus disrupting the life cycle of many freshwater organisms.

• Large freshwater species are particularly threatened throughout Latin America owing to intensive hunting and habitat fragmentation resulting from land-use activities and stream regulation. These taxa include: river otters (*Lutra longicaudis* and *Pteronura brasiliensis*), black caiman (*Melanosuchus niger*), migratory catfish (e.g. *Brachyplatystoma flavicans*, *B. filamentosum*), manatees (*Trichechus inunguis*), dolphins (*Inia geoffrensis*, *Sotalia fluviatilis* and *Pontoporia vlainvillei*), pirarucu (*Arapaima gigas*) and river turtles (e.g. *Podocnemis expansa*) (Perrin *et al.*, 1989; Vidal, 1993; Polisar and Horwich, 1994; Barthem and Goulding, 1997).

Environmental decline of water resources in Latin America and the Caribbean demands urgent attention and, in many cases, problems can be addressed through remedial and preventive measures. Development of catchment management plans and zoning regulations today will help reduce the need for expensive restoration and development of flood control structures later. To do this, however, will require addressing a spectrum of interests including the innovation and dissemination of less polluting technologies; concentrated efforts by local, national and international agents to clean up wastes and improve waste disposal (Bartone and Salas, 1984; IDB, 1984; PAHO, 1990a,b, 1991, 1992); conservation measures to protect marginal and common lands, to promote reforestation and to protect water quality and fisheries; clarification of land tenure and adoption of agrarian reform; measures to address poverty (Witt, 1984; Lee, 1988; United Nations Population Fund, 1991); environmental legislation and institutional strengthening; and education and civil responsibility.

Acknowledgements

CMP gratefully acknowledges National Science Foundation grant DEB-95-28434 which partially supported the writing of this chapter. We thank J. Benstead, J. March, A. Ramirez and B. Toth for their helpful comments on the manuscript.

References

Abramovitz, J.N. 1996. Imperiled waters, impoverished future: The decline of freshwater ecosystems. *Worldwatch Paper* 128: March.

Alexander, B. 1994. People of the Amazon fight to save the flooded forest. *Science* 265:606–607.

Alho, C.J.R., Lacher, T.E. and Goncalves, H.C. 1988. Environmental degradation in the Pantanal Ecosystem. *BioScience* 38:164–171.

Alkins-Koo, M. 1998. Status report on Caribbean riverine systems and coarse-scale assessment of conservation priorities. In: Olson, D.M., Dinerstein, E., Canevari, P., Davidson, I., Castro, G., Morriset, V., Abell, R. and Toledo, E. (eds) *Freshwater Biodiversity of Latin America and the Caribbean: A Conservation Assessment*. America Verde Publications, Biodiversity Support Program, Washington, DC, 43–44.

Alva, C.A., van Alphen, J.G., de la Torre A. and Manrique, L. 1976. Problemas de drenaje y salinidad en la costa Peruana. *International Institute for Land Reclamation and Improvement Bulletin* 16:28.

Andreoli, C. 1993. The influence of agriculture on water quality. In: *Prevention of Water Pollution by Agriculture and Related Activities*. Proceedings of the Food and Agriculture Organization (FAO) of the United Nations, Santiago, Chile. FAO, Rome, 53–65.

Araujo-Lima, C. and Goulding, M. 1997. *So Fruitful a Fish: Ecology, Conservation and Aquaculture of the Amazon's Tambaqui*. Columbia University Press, New York.

Barros, A.C. and Uhl, C. 1995. Logging along the Amazon River and estuary: patterns and problems and potential. *Forest Ecology and Management* 77:87–105.

Barthem, R.B. and Goulding, M. 1997. *The Catfish Connection: Ecology, Migration and Conservation of Amazon Predators*. Columbia University Press, New York.

Barthem, R.B., Lambert de Brito Ribeiro, M.C. and Petrere, M. 1991. Life strategies of some long-distance migratory catfish in relation to hydroelectric dams in the Amazon Basin. *Biological Conservation* 55:339–345.

Bartone, C. and Salas, H.J. 1984. Developing alternative approaches to urban wastewater disposal in Latin America and the Caribbean. *PAHO Bulletin* 18:323–336.

Bayley, P.B. and Petrere, M. 1989. Amazon fisheries: assessment methods, current status and management options. *Canadian Special Publications on Aquatic Sciences* 106:385–398.

Benstead, J.P., March, J.G., Pringle, C.M. and Scatena, F.N. 1999. Effects of a low-head dam and water abstraction on migratory tropical stream biota. *Ecological Applications* 9:656–668.

Best, R. and da Silva, V. 1989. Biology status and conservation of *Inia geoffrensis* in the Amazon and Orinoco river basins. In: *Biology and Conservation of the River Dolphins*. Occasional Papers of the International Union for the Conservation of Nature Species Survival Commission, No. 3, International Union for the Conservation of Nature, Gland, Switzerland.

Birdsey, R.A. and Weaver, P.L. 1987. *Forest Trends in Puerto Rico*. US Department of Agriculture Forest Service Southern Forest Experiment Station Research Note 30–331, New Orleans, Louisiana.

Boon, P.J., Jupp, B.P. and Lee, D.G. 1986. The benthic ecology of rivers in the Blue Mountains (Jamaica) prior to construction of a water regulation scheme. *Archiv für Hydrobiologie Supplement* 74:315–355.

Bucher, E.H., Bonetto, A., Boyle, T., Canevari, P., Castro, G., Huszar, P. and Stone, T. 1993. *Hidrovia: An Initial Environmental Examination of the Paraguay–Parana Waterway*. Special Publication of Wetlands for the Americas, Manomet, Massachusetts.

Bunkley-Williams, L., Williams, E.H., Lilystrom, G.G., Corvjo-Flores, I., Zerbi, A.J., Aliaume, C. and Churchill, T.N. 1994. The South American Sailfin Armored Catfish, *Liposarcus multiradiatus* (Hancock), a new exotic established in Puerto Rican freshwaters. *Caribbean Journal of Science* 30:90–94.

Burger, J., Rodgers, J.A. and Gochfeld, M. 1993. Heavy metals and selenium levels in endangered wood storks (*Mycteria americana*) from nesting colonies in Florida and Costa Rica. *Archives of Environmental Contamination and Toxicology* 24:417–420.

Burger, J., Marquez, M. and Gochfeld, M. 1994. Heavy metals in the hair of opossum from Palo Verde, Costa Rica. *Archives of Environmental Contamination and Toxicology* 27:472–476.

Bussing, W.A. 1994. Ecological aspects of the fish community. In: McDade, L.A., Bawa, K.S., Hespenheide, H.A. *et al.* (eds) *La Selva: Ecology and Natural History of a Neotropical Rainforest*. University of Chicago Press, Chicago, Illinois, 195–198.

Carpino, E.A. 1994. *River Dolphins: Can They Be Saved?* International Rivers Network, Working Paper 4, Berkeley, California.

Connell, D.W. and Hawker, D.W. 1992. *Pollution in Tropical Aquatic Systems*. CRC Press Inc, Ann Arbor, Michigan.

Contreras-Balderas, S. 1978a. Speciation aspects and man-made community composition changes in Chihuahuan Desert fishes. In: *Transactions of the First Symposium on Biological Research in the Chihuahuan Desert* (Alpine, Texas), 405–431.

Contreras-Balderas, S. 1978b. Environmental impacts in Cuatro Cienegas, Coahuila, Mexico: a commentary. *Journal of the Arizona–Nevada Academy of Science* 19:85–88.

Contreras-Balderas, S. and Escalante, M.A. 1984. Distribution and known impacts of exotic fishes in Mexico. In: Courtenay, W.R. and Stauffer, J.R. (eds) *Distribution Biology and Management of Exotic Fishes*. Johns Hopkins University Press, Baltimore, Maryland, 102–129.

Covich, A.P. and McDowell, W.D. 1996. Aquatic consumers. In: Reagan, D.P. and Waide, R.B. (eds) *The Food Web of a Tropical Rainforest*. University of Chicago Press, Chicago, Illinois, 433–459.

Cutright, N.J. 1996. Joint implementation: biodiversity and greenhouse gas offsets. *Environmental Management* 20:913–918.

Dinerstein, E., Olson, D.M., Graham, D.J., Webster, A.L.,

Primm, S.A., Bookbinder, M.P. and Ledec, G. 1995. *A Conservation Assessment of the Terrestrial Ecoregions of Latin America and the Caribbean*. The World Bank and World Wildlife Fund, Washington, DC.

Erdman, D.S. 1984. Exotic fishes in Puerto Rico. In: Courtenay, W.R. and Stauffer, J.R. (eds) *Distribution, Biology and Management of Exotic Fishes*. John Hopkins University Press, Baltimore, Maryland, 162–176.

Fearnside, P.M. 1995. Hydroelectric dams in the Brazilian Amazon as sources of 'greenhouse' gases. *Environmental Conservation* 22:7–19.

Food and Agriculture Organization of the United Nations (FAO) 1993. *Prevention of Water Pollution by Agriculture and Related Activities*. Proceedings of the FAO Expert Consultation, Santiago, Chile, 20–23 October 1992. FAO, Rome.

Fuentes, C.M. and Quiros, R. 1988. *Variacion de la composicion de la captura de peces en el rio Paraná, durante el periodo 1941–1984*. Instituto Nacional de Investigacion y Desarrollo Pesquero, Mar del Plata, Argentina. Serie Informes Tecnicos del Departamento de Aguas Continentales 6.

Furtado, L.G. 1988. Os caboclos pescadores do Baixo Rio Amazonas e o processo e mudanca social e economica. In: *Ciencias Socias e o Mar no Brasil*. II. Programa de Pesquisa e Conservação de Areas Unidas no Brasil. São Paulo.

Furtado, L.G. 1990. Caracteristicas gerais e problemas da pesca Amazonica no Para. *Boletim do Museu Paraense Emilio Goeldi. Antropologia* 6:41–93.

Gentry, A.H. and Lopez-Parodi, J. 1980. Deforestation and increased flooding of the upper Amazon. *Science* 210:1354–1356.

Girot, P.O. and Nietschmann, B.Q. 1992. The Rio San Juan. *National Geographic Research and Exploration* 8:52–63.

Gladwell, J.S. and Bonell, M. 1990. An international programme for environmentally sound hydrological and water management strategies in the humid tropics. In: *Proceedings of the International Symposium on Tropical Hydrology and Fourth Caribbean Islands Water Resources Congress*. American Water Works Association Technical Publication Series TPS-90-2, 1–10.

Glynn, P.W., Howard, L.S., Corcoran, E. and Freay, A.D. 1984. The occurrence and toxicity of herbicides in reef building corals. *Marine Pollution Bulletin* 15:370–374.

Goenaga, C. 1991. The state of coral reefs in the wider Caribbean. *Interciencia* 16:12–20.

Gomez-Sanchez, C.E. 1993. The influence of agriculture on water quality in Colombia. In: *Prevention of Water Pollution by Agriculture and Related Activities*. Proceedings of the Food and Agriculture Organization (FAO) of the United Nations, Santiago, Chile, 1992. FAO, Rome, 93–101.

Goulding, M. 1983. Amazonian fisheries. In: Moran, E. (ed.) *The Dilemma of Amazonian Development*. Westview Press, Boulder, Colorado, 189–210.

Goulding, M., Smith, N.J.H. and Mahar, D.J. 1996. *Floods of*

Fortune: Ecology and Economy Along the Amazon. Columbia University Press, New York.

Graham, N.A. and Edwards, K.L. 1984. *The Caribbean Basin to the Year 2000: Demographic, Economic and Resource Use Trends in Seventeen Countries*. Westview Press, Boulder, Colorado.

Grainger, A. 1990. *The Threatening Desert*. Earthscan, London.

Gravena, S., Pazini, W.C. and Fernandes, O.A. 1987. Centro de manejo integrado de pragas – DEMIP. *Laranja Cordeiropolis* 1:33–46.

Griesinger, B. and Gladwell, J.S. 1993. Hydrology and water resources of tropical Latin America and the Caribbean. In: Bonell, M., Hufschmidt, M.M. and Gladwell, J.S. (eds) *Hydrology and Water Management in the Humid Tropics: Hydrological Research Issues and Strategies for Water Management*. Cambridge University Press, New York, 84–98.

Guzman, H.M. and Jimenez, C.E. 1992. Contamination of coral reefs by heavy metals along the Caribbean coast of Central America (Costa Rica and Panama). *Marine Pollution Bulletin* 24:554–561.

Harrison, A.D. and Rankin, J.J. 1976. Hydrobiological studies of Eastern Lesser Antillean Islands. *Archiv für Hydrobiologie Supplement* 50:275–311.

Heath, R. 1995. Hell's highway. *New Scientist*, June 3:22–25.

Hunte, W. 1980. The laboratory rearing of larvae of the shrimp, *Macrobrachium faustinum*. *Caribbean Journal of Science* 16:1–4.

Hunter, J.R. 1994. Is Costa Rica truly conservation-minded? *Conservation Biology* 8:478–481.

Hunter, J.M. and Arbona, S.I. 1995. Paradise lost: an introduction to the geography of water pollution in Puerto Rico. *Social Science and Medicine* 40:1331–1355.

Hurtado, M.E. 1987. Agrotoxics: Blight on the next generation. *South* 77:97.

Inter-American Development Bank (IDB) 1984. Water resources of Latin America. *Water International* 9:26–36.

International Rivers Network (IRN) 1988a. IDB and Central America: Deforestation threatens big hydro. *World Rivers Review* 3:4–5.

International Rivers Network (IRN) 1988b. Rivers and rainforests. *World Rivers Review* 3:2.

Klinowska, M. 1991. *Dolphins, Porpoises and Whales of the World: The IUCN Red Data Book*. International Union for the Conservation of Nature, Gland.

Laidlaw, K. 1996. The implementation of a volunteer stream monitoring program in Costa Rica. Unpublished Masters Thesis, Institute of Ecology, University of Georgia, Athens, Georgia.

Lee, T.R. 1988. The evolution of water management in Latin America. *Water Resources Development* 4:160–168.

Lee, T.R. 1991. *Water Resources Management in Latin America and the Caribbean*. Studies in Water Policy and Management, No. 16. Westview Press, Boulder, Colorado.

Lowe-McConnell, R.H. 1987. *Ecological Studies in Tropical Fish Communities*. Cambridge University Press, New York.

March, J.G., Benstead, J.P., Pringle, C.M. and Scatena, F.N. 1998. Migratory drift of larval amphidromous shrimps in two tropical streams, Puerto Rico. *Freshwater Biology* 40:1–14.

Maser, C. and Sedell, J.R. 1994. *From the Forest to the Sea: The Ecology of Wood in Streams, Rivers, Estuaries and Oceans*. St Lucie Press, Delray Beach, Florida.

McCully, P. 1996. *Silenced Rivers: The Ecology and Politics of Large Dams*. Zed Books, Atlantic Highlands, New Jersey.

McGrath, D.G., de Castro, F., Futemma, C., de Amaral, B.D. and Calabria, J. 1993. Fisheries and the evolution of resource management on the lower Amazon floodplain. *Human Ecology* 21:167–195.

McNeeley, J.A., Miller, K.R., Reid, W.V., Mittermeier, R.A. and Werner, T.B. 1990. *Conserving the World's Biological Diversity*. International Union for Conservation of Nature and Natural Resources, Gland.

Mittermeier, R.A. 1989. Primate diversity and the tropical forest: case studies from Brazil and Madagascar and the importance of the megadiversity countries. In: Wilson, E.O., Francis, F. and Peter, M. (eds) *Biodiversity*. National Academy Press, Washington, DC, 145–154.

Morris, L.M. 1994. Ten concepts on water supply and drought in Puerto Rico. *Dimensions, Segundo Trimestre* 7–14.

Myers, N. 1988. Threatened biotas: 'hot-spots' in tropical forests. *The Environmentalist* 8:1–20.

Nash, L. 1993. Water quality and health. In: Gleick, P.H. (ed.) *Water in Crisis: A Guide to the World's Fresh Water Resources*. Oxford University Press, New York, 25–39.

Naumann, M. 1994. *A Water-use Budget for the Caribbean National Forest of Puerto Rico*. Special Report, USDA Forest Service, Rio Piedras, Puerto Rico.

Nietschmann, B. 1987. Conservation by conflict. *Natural History* November:42–49.

Olson, D.M., Dinerstein, E., Canevari, P., Davidson, I., Castro, G., Morisset, V., Abell, R. and Toledo, E. (eds) 1998. *Freshwater Biodiversity of Latin America and the Caribbean: A Conservation Assessment*. America Verde Publications, Biodiversity Support Program, Washington, DC.

Pan American Health Organization (PAHO) 1990a. *The Situation of Drinking Water Supply and Sanitation in the American Region at the End of the Decade 1981–1990 and Prospects for the Future*. Volume 1. Washington, DC.

Pan American Health Organization (PAHO) 1990b. Wastewater disposal in the Caribbean: status and strategies. *Bulletin of PAHO* 24:252–255.

Pan American Health Organization (PAHO) 1991. Drinking water supply and sanitation in the Americas: status and prospects. *Bulletin of PAHO* 25:87–96.

Pan American Health Organization (PAHO) 1992. Health and the environment. *Bulletin of PAHO* 26:370–378.

Pena-Torrealba, H. 1993. Natural water quality and agricultural pollution in Chile. In: *Prevention of Water Pollution by Agriculture and Related Activities*. Proceedings of the Food and Agriculture Organization (FAO) of the

United Nations, Santiago, Chile, 1992. FAO, Rome, 67–76.

Perrin, W.F., Brownell, R.L., Kaiya, Z. and Jiankang, L. (eds) 1989. *Biology and Conservation of River Dolphins*. Occasional Papers of the International Union for the Conservation of Nature Species Survival Commission, No. 3. International Union for Conservation of Nature and Natural Resources, Gland.

Pohlman, S. 1998. Towards implementation of community based conservation: 'The Water for Life Program' in Puerto Viejo de Sarapiqui, Costa Rica. Unpublished Masters Thesis, Institute of Ecology, University of Georgia, Athens, Georgia.

Polisar, J. and Horwich, R.H. 1994. Conservation of the large, economically important river turtle *Dermatemys mawii* in Belize. *Conservation Biology* 8:338–342.

Ponce, V. 1995. *Hydrologic and Environmental Impact of the Parana–Paraguay Waterway on the Pantanal of Mato Grosso, Brazil*. Special report available from Environmental Defense Fund, Berkeley, California.

Postel, S. 1984. *Water: Rethinking Management in an Age of Scarcity*. Worldwatch Paper, Washington, DC.

Postel, S. 1987. *Defusing the Toxics Threat: Controlling Pesticides and Industrial Waste*. Worldwatch Paper 79:11.

Pringle, C.M. 1988. History of conservation efforts and initial exploration of the lower extension of Parque Nacional Braulio Carrillo. In: Almeda, F. and Pringle, C.M. (eds) *Tropical Rainforests: Diversity and Conservation*. Allen Press, Lawrence, Kansas, 225–241.

Pringle, C.M. 1997a. Exploring how disturbance is transmitted upstream: going against the flow. *Journal of the North American Benthological Society* 16:425–438.

Pringle, C.M. 1997b. Expanding scientific research programs to address conservation challenges in freshwater ecosystems. In: Pickett, S.T.A., Ostfeld, R.S., Shachak, M. and Likens, G.E. (eds) *Enhancing the Ecological Basis of Conservation: Heterogeneity, Ecosystem Function and Biodiversity*. Proceedings of the Sixth Cary Conference, Institute of Ecosystem Studies. Chapman & Hall, New York, 305–319.

Pringle, C.M. 1999. Changing academic culture: interdisciplinary, science-based graduate programmes to meet environmental challenges in freshwater ecosystems. *Aquatic Conservation: Marine and Freshwater Ecosystems* 9:615–620

Pringle, C.M. 2000a. Riverine connectivity: conservation and management implications for remnant natural areas in complex landscapes. *Verhandlungen der Internationalen Vereinigung für theoretische und angewandte Limnologie* (in press).

Pringle, C.M. 2000b. Threats to U.S. public lands from cumulative hydrological alterations outside of their boundaries. *Ecological Applications* (in press).

Pringle, C.M. and Benstead, J.P. 2000. Effects of logging on tropical riverine ecosystems. In: Fimbel, R., Grajal, A. and Robinson, J. (eds) *Conserving Wildlife in Managed Tropical Forests*. Columbia University Press, New York (in press).

Pringle, C.M. and Scatena, F.N. 1999a. Aquatic ecosystem deterioration in Latin America and the Caribbean. In: Hatch, L.U. and Swisher, M.E. (eds) *Managed Ecosystems: The Mesoamerican Experience*. Oxford University Press, New York, 104–113.

Pringle, C.M. and Scatena, F.N. 1999b. Freshwater resource development: case studies from Puerto Rico and Costa Rica. In: Hatch, L.U. and Swisher, M.E. (eds) *Managed Ecosystems: The Mesoamerican Experience*. Oxford University Press, New York, 114–121.

Pringle, C.M., Rowe, G.L., Triska, F.J., Fernandez, J.F. and West, J. 1993. Landscape linkages between geothermal activity, solute composition and ecological response in streams draining Costa Rica's Atlantic Slope. *Limnology and Oceanography* 38:753–774.

Quiros, R. 1993. Inland fisheries under constraint by other uses of land and water resources in Argentina. In: *Prevention of Water Pollution by Agriculture and Related Activities*. Proceedings of the Food and Agriculture Organization (FAO) of the United Nations, Santiago, Chile, 1992. FAO, Rome, 29–94.

Reeves, R. and Leatherwood, S. 1994. *Dolphins, Porpoises and Whales: 1994 Action Plan for the Conservation of Cetaceans*. International Union for Conservation of Nature and Natural Resources/Species Survival Commission Specialist Group, Gland.

Reid, R. 1978. The Caribbean Region water resources management problems. *Water Quality Bulletin* 3:4.

Reiff, F. 1993. Health impacts related to irrigated agriculture in Latin America. In: *Prevention of Water Pollution by Agriculture and Related Activities*. Proceedings of the Food and Agriculture Organization (FAO) of the United Nations, Santiago, Chile, 1992. FAO, Rome, 327–340.

Reznick, D., Baxter, R.J. and Endler, J. 1994. Longterm studies of tropical fish communities. The use of field notes and museum collections to reconstruct communities of the past. *American Zoologist* 34:452–462.

Rice, B. 1986. The fresh connection. *Audubon* 88:104–107.

Richey, J.E., Nobre, C. and Deser, C. 1989. Amazon River discharge and climate variability: 1903–1985. *Science* 246:101–103.

Rose, M. 1987. The cost of Mexico's filthy riches: catalog of devastation. *South* 80:106–107.

Schemo, D.J. 1997. More fires by farmers raise threat to Amazon. *New York Times International* 2 November.

Schmidt-Lynch, C. 1994. Myth or mammal: in search of the Amazon god of the water. *Nature Conservancy* pp. 18–22.

Seeliger, U.L., de Lacerda, D. and Patchineelam, S.R. (eds) 1988. *Metals in Coastal Environments of Latin America*. Springer-Verlag, New York.

Seva, O. 1990. Works on the Great Bend of the Xingu – a historic trauma? In: Stanos, L.A. de O., Andrade. L.M.M. de (eds) *Hydroelectric Dams on Brazil's Xingu River and Indigenous Peoples*. Cultural Survival Report 30. Cultural Survival, Cambridge, Massachusetts, 19–35.

Shukla, J.C.N. and Sellers, P. 1990. Amazon deforestation and climate change. *Science* 247:1322–1325.

Smith, N. 1985. The impact of cultural and ecological change on Amazonian fisheries. *Biological Conservation* 32:355–373.

Switkes, G. 1995. Tiete River cleanup: IDB flushes hundreds of millions down the drain in 'Environmental Boondoggle'. *World Rivers Review. Newsletter of the International Rivers Network*, May 10: 3.

Switkes, G. 1998. Brazilian rejection of Hidrovia elicits cautious optimism. *World Rivers Review. Newsletter of the International Rivers Network*, April, 13.

Terborgh, J. 1990. An overview of research at Cocha Cashu Biological Station. In: Gentry, A.H. (ed.) *Four Neotropical Rainforests*. Yale University Press, New Haven, Connecticut, 48–59.

Thorsell, J. 1990. *Parks on the Borderline: Experience in Transfrontier Conservation*. International Union for Conservation of Nature and Natural Resources, Gland.

Thrupp, L.A. 1990. Environmental initiatives in Costa Rica: a political ecology perspective. *Society and Natural Resources* 3:243–256.

United Nations Economic Commission for Latin America and the Caribbean (ECLAC) 1985. *Water Resources of Latin America and their Utilization*. Estudios e informes No. 53, Santiago, Chile.

United Nations Economic Commission for Latin America and the Caribbean (ECLAC) 1990a. *Latin America and the Caribbean: Inventory of Water Resources and their Use*, Volume I: *Mexico, Central America and the Caribbean*. Santiago, Chile.

United Nations Economic Commission for Latin America and the Caribbean (ECLAC) 1990b. *Latin America and the Caribbean: Inventory of Water Resources and their Use*, Volume II: *South America*. Santiago, Chile.

United Nations Economic Commission for Latin America and the Caribbean (ECLAC) 1990c. *The Water Resources of Latin America and the Caribbean: Planning, Hazards and Pollution*. Santiago, Chile.

United Nations Population Fund (UNPF) 1991. *Population, Resources and the Environment: The Critical Challenges*. London.

van Emden, H. and Peakall, D.B. 1996. *Beyond Silent Spring: Integrated Pest Management and Chemical Safety*. Chapman & Hall, New York.

Vargas, R.J. 1995. History of municipal water resources in Puerto Viejo, Sarapiqui, Costa Rica: A socio-political perspective. Unpublished Masters Thesis, Institute of Ecology, University of Georgia, Athens, Georgia.

Vidal, O. 1993. Aquatic mammal conservation in Latin America: Problems and perspectives. *Conservation Biology* 7:788–795.

Western, D. and Wright, M. (eds) 1994. *Natural Connections: Perspectives in Community-based Conservation*. Island Press, Washington, DC.

White, T.A. and Runge, C.F. 1995. The emergence and evolution of collective action: lessons from watershed management in Haiti. *World Development* 23:1683–1698.

Witt, V.M. 1984. The water and sanitation decade in Latin America and the Caribbean: strategies for its success. *Water Quality Bulletin* 9:188–194.

Witt, V.M. and Reiff, F.M. 1991. Environmental health conditions and cholera vulnerability in Latin America and the Caribbean. *Journal of Public Health Policy* 12:450–463.

World Conservation Monitoring Centre (WCMC) 1992. *Global Biodiversity: Status of the Earth's Living Resources*. A report compiled by the World Conservation Monitoring Centre. Chapman & Hall, New York.

World Meteorological Organization (WMO) 1993. *Report of WMO/WHO/UNEP Workshop on Regional Needs for Water Quality Monitoring in the Caribbean*. Switzerland.

World Resources Institute (WRI) 1990. *World Resources 1990–91*. Oxford University Press, New York.

Yoon, C.K. 1997. Amazon's depths yield strange new world of unknown fish. *New York Times* 18 February.

3

River conservation in the European Community, including Scandinavia

T.M. Iversen, B.L. Madsen and J. Bøgestrand

Introduction

Rivers are an important part of the European landscape and of great significance for European biodiversity. The intensive land use in most of Europe and the intensive use of rivers for a variety of purposes has led to a significant degradation of the rivers all over the region during the last century. The need for river conservation has therefore become more and more evident. It has long since been recognized that the interactions between a river and its valley are of significant importance for the ecological functioning of a river (Hynes, 1975). This chapter will therefore deal with river conservation as the conservation of rivers and their riparian areas.

Conservation can be defined as 'the prevention of waste, loss or damage, especially the official protection or care of forests, rivers, waterpower, etc.' (Hornby *et al.*, 1960). Following this definition the chapter will cover the protection of pristine rivers as well as management of impacted rivers in order to improve their ecological quality. As the region contains a wealth of examples of damaged rivers (in terms of water quality and quantity, physical quality), river restoration or rehabilitation will also be included as a necessary measure for significant improvements.

The regional context

The region includes 17 main countries (Figure 3.1) and a total land area of 3.7×10^6 km². The countries vary widely in size from Luxembourg (2568 km²) to France (543 965 km²) (Table 3.1). In addition, the region contains eight small countries: Andorra, Faroe Islands, Gibraltar, Liechtenstein, Malta, Monaco, San Marino and Vatican City. They have a total area of about 2400 km², but will not be dealt with further due to lack of information. The region covers about 37 degrees of latitude from Norway in the north to Greece in the south and about 52 degrees of longitude from Iceland in the west to Finland and Greece in the east. In this chapter, the countries of the region will be referred to as Nordic, Western and Southern European countries (Table 3.2) following Kristensen and Hansen (1994).

In the region there are 13 rivers with watersheds >50 000 km², of which all or most of the watershed is within the region, and another three with the major part of the watershed outside the region (Table 3.3). Of these, five rivers are shared by two to eight countries within the region and eight of these large rivers are transnational. The catchment areas of the main river in each country naturally vary widely (Table 3.4). In four countries the watershed of the main river covers 44–96% of the country. From these data it is obvious that, especially in Western countries, river conservation also demands international cooperation.

Morris and Kronvang (1994) have compiled available estimated total river lengths based on 1:50 000 maps (Table 3.1). The region has about 2.4×10^6 km of rivers and about 0.6 km river km⁻². Specific river lengths vary from about 0.3 km river km⁻² in Spain to about 1 km river km⁻² in France, partly reflecting variations in water flow (see below).

Global Perspectives on River Conservation: Science, Policy and Practice.
Edited by P.J. Boon, B.R. Davies and G.E. Petts. © 2000 John Wiley & Sons Ltd.

Figure 3.1   *Map of the region. Country codes as on Table 3.4*

The seasonal pattern of natural water flow varies significantly from north to south (Figure 3.2) due to variations in temperature and precipitation. Similarly there are great variations in specific water flow. For the 16 large rivers of Table 3.3 the mean specific water flow varies from about 1 L s^{-1} km^{-2} in the Spanish Guadalquivir to about 20 L s^{-1} km^{-2} in the French Rhone. The low specific water flow in Guadalquivir also reflects human impacts such as irrigation.

River gradients have an overall impact on the ecology of rivers. The majority of rivers in the region have steep gradients – for example, in the mountainous regions of Norway, Sweden and Central Europe. Low-gradient rivers are especially common in the Baltic region.

DEVELOPMENT PRESSURES AND HUMAN IMPACT

The main pressures affecting rivers can be related to the density of population and households, industrialization and land use (agriculture and forestry).

The total population of the region in 1995 was about 376 × 10^6 (Table 3.1). Over the region the population density varies widely from <20 km^{-2} in Iceland, Norway, Finland and Sweden to >200 km^{-2} in Germany, United Kingdom and Belgium, with a maximum of 373 km^{-2} in The Netherlands. In 1950 the total population was about 300 × 10^6. Thus, between 1950 and 1995 the population has increased by 25%. At

Table 3.1　*Country area, population size and river length as estimated on a 1:50 000 map.*
Data from European Commission (1998) and Morris and Kronvang (1994). nd – no data

	Country area km^2	Population nos $\times 10^6$ (1995)	Population density (nos km^{-2})	River length (km $\times 10^3$)
Austria	83 858	8.045	96	47
Belgium	30 518	10.127	332	23
Denmark	43 094	5.223	121	28
Finland	338 145	5.107	15	159
France	543 965	58.104	107	563
Germany	356 970	81.594	229	179
Greece	131 957	10.454	79	nd
Iceland	103 000	0.269	3	nd
Ireland	70 285	3.546	50	34
Italy	301 323	57.204	190	136
Luxembourg	2 568	0.407	157	1
The Netherlands	41 526	15.482	373	20
Norway	323 880	4.332	13	201
Portugal	91 905	9.815	107	172
Spain	505 990	39.627	78	172
Sweden	449 964	8.788	20	315
United Kingdom	244 101	58.079	238	171
Total	3 666 049	376.203	103	2300–2400

Table 3.2　*Countries comprising the three groups widely dealt with in this chapter. Modified from Kristensen and Hansen (1994)*

Nordic countries (area: 1.2 $\times 10^6$ km^2)	Western countries (area: 1.4 $\times 10^6$ km^2)	Southern countries (area: 1.0 $\times 10^6$ km^2)
Finland	Austria	Greece
Iceland	Belgium	Italy
Norway	Denmark	Portugal
Sweden	France	Spain
	Germany	
	Ireland	
	Luxembourg	
	The Netherlands	
	United Kingdom	

the same time urban development has increased, and a proportionately larger part of the population now lives in urban areas.

The effects of population pressures on rivers can be categorized in terms of river water quality, river water quantity, the physical river environment and the riparian areas. Wastewater loading reduces water quality, changes the ecological structure and functioning of rivers, and is a major problem in many parts of the region. Wastewater production from households is mainly caused by population density, household numbers and size, disposable income and consumer spending (Stanners and Bourdeau, 1995). Waste water is also produced in industry, and industrial production has almost doubled in the period 1970–1990. Food and agro-product processing industries especially are large producers of waste organic matter. In contrast, industry is the main source of synthetic organic chemical substances, heavy metals and several micropollutants, as well as being a main user of water. Although exact data are scarce there is no doubt that the amount of waste water produced in the region increased tremendously during the 20th century.

Wastewater treatment varies widely from country to country within the region, although there are some gaps in data (Table 3.5). Generally, less than 50% of the population is served by wastewater treatment plants in Belgium, Greece, Iceland, Ireland and Portugal, whereas 80% or more of the population is served in Denmark, Germany, Luxembourg, the Netherlands, Sweden and the United Kingdom. The available data

Table 3.3 *Nordic, Western and Southern European rivers with catchment areas >50 000 km².*
For country code see Table 3.4. Modified from Kristensen and Hansen (1994)

River	Country	Catchment area (km² × 10³)	Mean discharge (km³ yr⁻¹)	Length (km)
Danube[b]	DE, AU	817	205	2850
Neva[b]	FI	281	79	75
Rhine[a]	AU, DE, FR, NL, IT, LU, BE	185	69	2200
Elbe[a]	DE, AU	148	24	1140
Oder[b]	DE	119	16	850
Loire	FR	118	32	1010
Douro	ES, PO	98	20	790
Rhone[a]	FR	96	54	810
Garonne	FR	85	21	575
Ebro	ES	84	17	910
Tajo	ES, PO	82	6	1010
Seine	FR	79	16	780
Guadiana	ES, PO	72	6	800
Po	IT	69	46	670
Guadalquivir	ES	57	2	675
Kemijoki	FI	51	17	510

[a] minor and [b] major part of the catchment area outside the region

Table 3.4 *The largest river in each country of the region and its*
catchment area. Modified from Kristensen and Hansen (1994)

Country (code)	River	Catchment area km² × 10³	% of country
Austria (AU)	Danube	80.6	96
Belgium (BE)	Meuse	13.5	44
Denmark (DK)	Gudenå	2.6	6
Finland (FI)	Vuoksa[a]	61.1	18
France (FR)	Loire	117.5	21
Germany (DE)	Rhine	102.1	29
Greece (GR)	Aliákmon	9.5	7
Iceland (IS)	Jökulsá-á-Fjöllum	7.8	8
Ireland (IR)	Shannon	~14.0	~20
Italy (IT)	Po	69.0	23
Luxembourg (LU)	Sûre[b]	~2.0	~77
Netherlands (NL)	Rhine	~25.0	~60
Norway (NO)	Glomma	41.4	13
Portugal (PO)	Tajo	24.9	27
Spain (ES)	Ebro	84.2	17
Sweden (SE)	Göta älv	42.8	10
United Kingdom (UK)	Thames	15.0	6

[a] tributary to Neva
[b] tributary to the Rhine

indicate a significant improvement in wastewater treatment all over the region, highlighted by the proportion of different treatment processes with different efficiencies (Table 3.6). By 1990, tertiary treatment was most commonly used in Sweden (serving 84% of the population), Finland (76%) and Norway (43%), but in many of the other countries only 10–20% of the population is served by tertiary wastewater treatment (European Commission, 1995a). By 1970,

tertiary treatment of waste water had only been reported from Norway (20%), Sweden (3%) and Finland (1%).

The implication of the wastewater data in Table 3.5 for rivers, however, must be considered with care. For countries such as Austria and Luxembourg all waste water is emitted to inland waters and especially to rivers. For all other countries much waste water is emitted directly to the marine environment. Data for

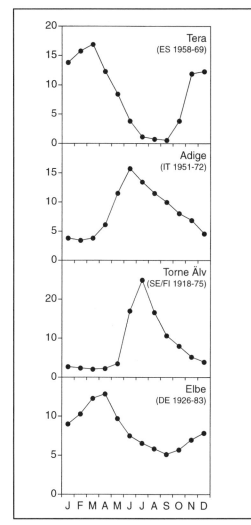

Figure 3.2 *Long-term annual flow regimes (percentage of annual flow) in the Tera[a], Adige[b], Torne Älv[c], and Elbe[d] rivers (revised from [a]Cassado et al. (1989); [b]Duzzin et al. (1988); [c]Petersen et al. (1992); [d]Wassergütestelle Elbe (1990)). Country codes as on Table 3.4*

Table 3.5 *The population in the region served by wastewater treatment plants (1970–1990). Data from European Commission (1998). nd – no data*

	1970	1980	1990
Austria	17	38	76
Belgium	4	23	nd
Denmark	54	nd	99
Finland	16	65	77
France	19	62	77
Germany	nd	80	89
Greece	nd	1	34
Iceland	nd	nd	4
Ireland	nd	11	45
Italy	14	30	61[a]
Luxembourg	28	81	88
The Netherlands	44	73	96
Norway	21	34	67
Portugal	nd	2	21[a]
Spain	nd	18	48
Sweden	63	82	95
United Kingdom	nd	82	86

[a] data from 1995

Table 3.6 *The reduction efficiency of different commonly used treatment processes. (Modified from Miljøstyrelsen 1990)*

Treatment process	BOD (%)	P (%)	N (%)
Primary	20–40	10–30	10–20
Secondary	70–90	20–40	20–40
Tertiary	>95	85–97	20–40[a]

[a] N-reductions increase to 85–95% with nitrification and denitrification processes included

instance from Denmark clearly indicate that waste water to inland waters is treated more efficiently than waste water to the marine environment.

The main wastewater pollution indicators in rivers are organic matter (Biochemical Oxygen Demand – BOD, Chemical Oxygen Demand – COD), phosphorus and ammonium. Excess organic matter is decomposed in the river, causing oxygen deficiency and significant impacts on biological structure and processes. Excess phosphorus may increase benthic algal biomass and production in low-order rivers, whereas excess ammonium increases the probability of high concentrations of ammonia, which is toxic to fish.

An overview of these indicators is given in Kristensen and Hansen (1994). The main conclusions are:

- concentrations of organic matter and phosphorus in river water increase with increasing population density;
- concentrations of organic matter and phosphorus in river water are lowest in the Nordic countries, higher in Southern countries, and highest in the Western countries.

Emissions to the atmosphere may cause deterioration in river water quality through acid precipitation. Emissions of sulphur oxides (SO_x) from the combustion of sulphur-containing fossil fuels and from certain industries, and of nitrogen oxides (NO_x)

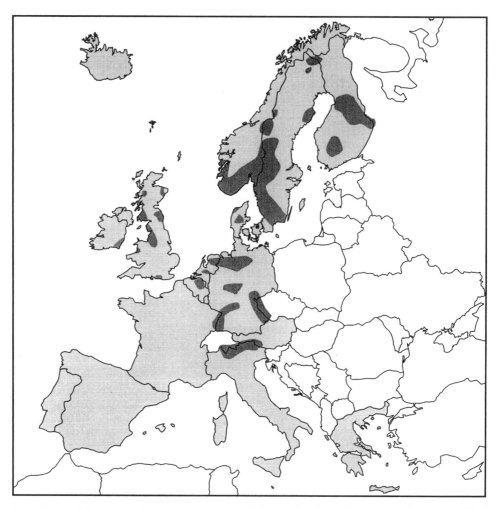

Figure 3.3   *Areas of the region where surface water acidification has been observed. Redrawn from Skjelvåle and Wright (1990) and Merilehto* et al. *(1988). Light shading, countries of the regions; dark shading, areas of acidification*

from electricity production and from road transport, are deposited due to transboundary air pollution causing acidification. Generally emissions have increased since the last World War until 1985, whereafter emissions, especially of sulphur dioxide, have been reduced (European Environment Agency – EEA, 1998).

Catchments with soils or bedrocks poor in lime and in other easily weatherable minerals that buffer against deposition of acidifying substances are sensitive to acidification. Such catchments are found throughout the region apart from Iceland and Greece (Merilehto *et al.*, 1988; Skjelkvåle and Wright, 1990). River acidification is a main problem in the Nordic

countries, where it has increased slowly between 1915 and 1950 and rapidly since 1950 (Brodin and Kuylenstierna, 1992). Surface water acidification in the region has also been documented from Western countries (Figure 3.3), but not generally from Southern countries.

The development of agriculture has significantly affected the rivers of the region, not least their riparian areas. Since 1960 there has been an overall increase in agricultural production levelling out in the 1980s, even though the area of agricultural land has decreased (Stanners and Bourdeau, 1995). This means that there has been an even higher increase in agricultural production in terms of yields in t ha^{-1}. This increase has been achieved through a number of changes in agricultural practice which may be summed up in the term 'intensification'. Intensification has been particularly developed in Western countries such as the Netherlands, southern UK, northern France, Belgium, Denmark and parts of northern Germany, and includes issues such as specialization and concentration of crops and livestock, the application of fertilizer and pesticides, water abstraction and irrigation, and land drainage.

One aspect of agricultural intensification is the application of fertilizers and pesticides, which affects river water quality. Nitrate concentrations in rivers throughout the region have increased, but the ecological effects of nitrate in rivers are generally small although some evidence indicates that *Ranunculus fluitans* is favoured by high nitrate concentrations (Grasmück *et al.*, 1995). Pesticides are recognized as a major environmental problem in groundwater in areas with intensive agriculture. Although there is a shortage of data on pesticide concentrations in rivers, there is substantial evidence that pesticides also affect the biological structure and processes especially of small rivers. As an example it is believed that about 10% of the rivers in the Danish county Funen do not meet quality objectives because of pesticides (Wiberg-Larsen *et al.*, 1991).

Another aspect of agricultural intensification is water abstraction for irrigation, which increases evapotranspiration and reduces water flow in rivers. In Portugal and Spain about 70% of total water supply is used for agriculture (European Commission, 1998) and similar figures are likely in other Southern countries. In Denmark and the Netherlands, too, a substantial part of the water abstracted is used for agriculture. A reduced water flow implies reduced water velocity, which may significantly affect benthic communities and fish. In severe cases rivers dry up with profound ecological effects (Iversen *et al.*, 1978). In Southern countries reservoirs are often built to store water for irrigation, and this also affects rivers (see below).

The third major aspect of agricultural intensification is drainage of wetlands and flooded riparian areas. The main purpose is to lower the groundwater level to facilitate crop production. In lowland areas of the Western countries the agricultural incentive for removing excess water from the fields has led to large-scale channelization of previously meandering rivers, which increases river slope, water velocity and thus removal of water from upstream parts of the watershed. This has furthermore been enhanced by

widening and deepening of rivers, and by river maintenance schemes including weed cutting and sediment removal. It is estimated that 80–98% of all natural rivers in Denmark have lost their natural physical properties due to channelization (Iversen *et al.*, 1993), 96% in the Netherlands (L.W.G. Higler, Institute for Forestry and Nature Research, the Netherlands, personal communication) and a similar picture is seen in other lowland Western countries (Petts, 1988) and Southern countries (Garcia de Jalon, 1987). Together with draining, these measures have turned significant areas of riparian wetland into agricultural land and changed the landscape.

Flood control and navigation have long since been the reason for physical modifications of lowland rivers and their riparian areas. In the River Rhine valley in the Netherlands the development has been described by Postma and van de Kamer (1995). Figure 3.4 shows how the natural lowland river valley was settled and the settlements protected through dikes in the Middle Ages. Development in the 20th century included extensive agricultural use of the riparian areas between minor summer dikes and the winter dikes, and further draining, agricultural use and settlements in the previous marshland. Furthermore, the river was deepened for navigation and groynes were constructed to prevent ice dams. The net result of this process is a cultural landscape with a deep river-bed, steep banks, and a regularly flooded zone which has been reduced from a width of several kilometres to one of some hundred metres. Very little is left of the ecological functioning of the River Rhine.

European rivers are important arteries for transport, and in several Western countries they have for centuries formed a dense network, connecting the great North Sea harbours (e.g. Rotterdam) and inland Europe. The Main–Danube Canal now provides passage between the Black Sea and the North Sea. The EU authorities are concerned about the environmental impact of the increasing truck traffic, so it is likely that in future huge Eurobarges carrying the equivalent of 80 truck trailers will be important carriers of bulk cargo such as coal and grain between the emerging Eastern market and the rest of the world (Bryson, 1992).

In forested areas of the Nordic countries many rivers (up to the 1960s) were modified to ease the floating of timber. In northern and eastern Finland almost all rivers have been dredged by removing the stones and boulders from riffles (Jutila, 1985). In the River Simijoki the density of salmonid parr in the dredged

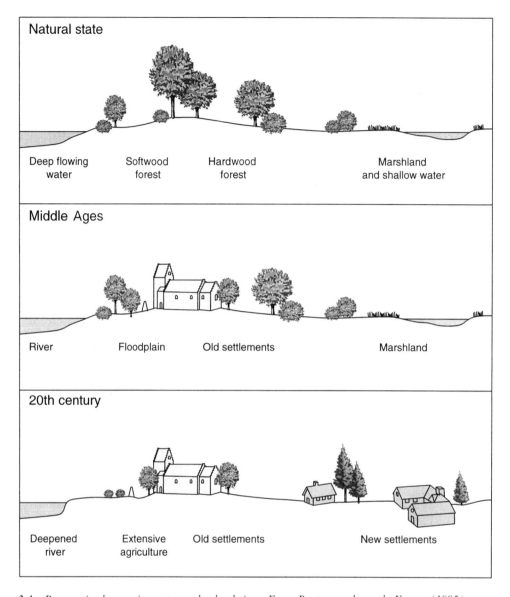

Natural state

Deep flowing water | Softwood forest | Hardwood forest | Marshland and shallow water

Middle Ages

River | Floodplain | Old settlements | Marshland

20th century

Deepened river | Extensive agriculture | Old settlements | New settlements

Figure 3.4 *Progressive human impact on a lowland river. From Postma and van de Kamer (1995)*

areas decreased to 1–4 parr per 100 m^2 compared with 6–9 in undisturbed areas. This caused an annual loss of 10–20 t in the annual salmon catches in the area. In some dredged rivers the local stock of salmonids has become extinct because the spawning grounds and fish cover have been destroyed. After the 1960s the need for navigable rivers ceased when truck transportation took over from floating.

Establishing reservoirs for power production is a long tradition in Europe. Since the invention of the water wheel, innumerable watercourses have been dammed to harness their energy, but with potential damage to salmonid fisheries, through loss of spawning and feeding grounds. The statistics in the Domesday Book (1086) gives an impression of the potential impact: in southern England there were 5624 water-mills, about one mill for each 50–60 inhabitants. Soon the larger rivers fell victim to this fate too. Around the year 1200, the River Arbonne in southern France was dammed at three sites in order to control the water flow

(Reynolds, 1983). Most water-mills have long since gone, but even if the water wheel has perished, the dam and mill pond still remain, preventing the passage of migrating fish.

The number of reservoirs in the region increased until the 1980s (Boon, 1992), but in recent years there has been a stagnation. In Norway the potential for hydroelectric power is 177.7 TWh of which 112.3 TWh is utilized (1995 data: Berntsen, 1996). Rivers with a potential of 35.3 TWh have been protected leaving very little room for further development. Out of 42 Norwegian rivers with flow rates >40 m^3 s^{-1}, 38 have been regulated. The decline in reservoir building seen all over the region is believed to be caused by a lack of suitable sites (Williams and Musco, 1992), but evidence from Norway suggests that the growing concern for environmental issues has also been important.

Today reservoirs usually have multiple purposes including hydroelectric power generation, irrigation, domestic and industrial power supply and flood control. However, in all cases building reservoirs interferes with the hydrological cycle and significantly disrupts the river continuum (*sensu* Vannote *et al.*, 1980) affecting the biological structure and processes of natural rivers.

Species introductions impose a very different pressure on rivers. The nematode *Anguillacola* spp infects the swim bladder of eel and occurs naturally in Asian and Pacific waters, where the eel populations apparently have evolved a certain degree of resistance. The parasite has been introduced to European waters with live Asian eels. It was first observed in an Italian fish farm in 1982 and in 1987 in Sweden. In Denmark it has been spread through stocking schemes before strict control was introduced and it has infected the natural eel population, spreading with small eels migrating into freshwater systems. It is unclear whether or not strict control of eel stocking can solve the problem in the foreseeable future.

In Norway, the cool and clean waters around its coastline are very attractive for salmon farming. An unwanted side-effect is the disease that has spread catastrophically among Norwegian wild salmon. The salmon parasite *Gyrodactylus salaris* was first observed mass-infecting Atlantic salmon parr in the River Lakseelv in 1975. It caused heavy mortality, and within two years salmon parr had disappeared from the river. By 1985 the parasite had spread to 26 Norwegian rivers, infecting the natural populations too. The origin of the disease is most probably imported Baltic salmon, which are resistant to the parasite. The local spread came from stocking rivers with infected salmon from the Norwegian hatcheries (Johnson and Jensen, 1986). From the stocked rivers the parasite has been spread to nearby rivers by migrating salmon and to more remote rivers, probably by illegal stocking. Transport by birds cannot be excluded and additional vectors may be the numerous tourists holidaying in mobile homes. Despite warning signs they empty drinking water containers, originating from yesterday's camping site, into the next river, from which they make a refill. The spread of the parasite to a river system often results in sudden outbreaks of the disease, leading to drastic reduction or near extermination of the salmon population. It has led to catastrophic declines in the catches of salmon, and drastic measures have been applied such as using rotenone, eradicating all fish in infected river systems. When the size of the rivers and their discharge is taken into consideration this is a serious human impact in addition to the salmon decline in pristine rivers in the Northern regions.

CATCHMENT VERSUS LOCAL ISSUES

There is a growing recognition that water management should be integrated and implemented at the catchment level, and that river conservation should be one component of integrated water management (e.g. see Hooper and Margerum, this volume). This reflects the fact that river flow is one component of the hydrological cycle, which may be affected by human impacts on other components, such as increased evapotranspiration due to irrigation, which may reduce river flow and cause drought. Similarly, human impacts on upstream water quality may also affect downstream water quality.

The necessity for catchment management is illustrated by the international cooperation on the River Rhine. The River Rhine in the Swiss Alps and the catchment area of 185 000 km^2 is spread over eight countries. The Rhine catchment has become the most important industrial agglomeration in Europe, with a number of large cities and a mean population density of 294 persons km^{-2}. It accounts for 20% of the world's chemical production and boasts several steel industries, coal mines and power stations (Lelek, 1989).

The pollution of the River Rhine increased after World War II and in the 1960s and 1970s the river was severely polluted (Figures 3.5 and 3.6). In 1963 an agreement was signed in Berne regarding the International Commission for the Protection of the Rhine, and in 1977 an EC Council Decision was made regarding chemical pollution. The coordinated

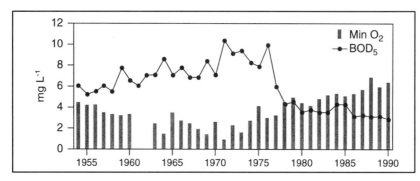

Figure 3.5   *Annual mean BOD₅ and minimum oxygen concentrations in the River Rhine at Lobith near the border with the Netherlands. Redrawn from Umweltbundesamt (1994)*

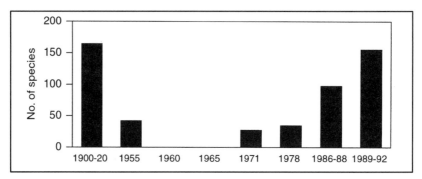

Figure 3.6   *Number of selected invertebrate species in the main course of the River Rhine. Redrawn from Umweltbundesamt (1994)*

measures in the Rhine watershed have been a major success, lowering the human impact on the river and re-establishing major ecological functions (Figures 3.5 and 3.6). The severe floods in the Rhine valley in early 1995 clearly stressed the need for management at catchment level. Loss of floodplains through draining increases peak flows and such human interventions in the hydrological cycle in the upper parts of the Rhine Basin have reduced its water retention capacity, thus magnifying the floods in Germany and the Netherlands and increasing the need for flood defence.

Other international agreements involving binational or multinational watershed cooperation within the region include the River Moselle (1961), the River Meuse (1994) and the River Schelde (1994) (Nixon *et al.*, 1996b). Similarly several agreements between countries in the region and Eastern countries have been established. In 1995 the Council of Ministers and the Environment Committee of the European Parliament called for a fundamental review of Community Water Policy. At present a Water Framework Directive is being discussed, which among other things will require integrated water management planning on a catchment basis.

In most of Europe the pressures on rivers are still high. Although a general improvement in water quality and ecological quality can be documented for the region, there are many examples of deterioration still taking place, not least in the physical properties of rivers (Holmes, 1997). Three main issues can be identified:

- Areas subject to acidification, which can only be improved through efficient international reductions in emissions of pollutants to the atmosphere;
- Areas in Southern countries, where human and industrial water consumption, together with irrigation agriculture, are given priority over environmental conservation;
- Intensive agriculture in Western countries where drainage and physical disruptions of rivers and their riparian areas are still dominant features.

Table 3.7   *Biological macroinvertebrate assessment methods for rivers currently used in Europe (Nixon* et al., *1996a)*

	Index method	B/S	Standard
Austria	Saprobic	B/S	National
Belgium	Belgium Biotic Index	B	National
Denmark	Danish Fauna Index	B	National
Finland	Various (e.g. BMWP)	B	Regional
France	Global Biotic Index	B	Regional
Germany	BEOL/Saprobic	S	Regional
Greece	None		
Iceland	None		
Ireland	Quality Rating System	B	National
Italy	Extended Biotic System	B	Regional
Luxembourg	Biotic Index	B	National
The Netherlands	Quality Index K 135	B	Regional
Norway	No index		Regional
Portugal	Belgium Biotic Index	B	Regional
Spain	BMWP Spanish Modf.	B	Regional
Sweden	No index used	–	Regional/National
United Kingdom	BMWP-Score	B	National

B biotic
S saprobic

Information for conservation

CLASSIFICATION AND EVALUATION SCHEMES

Classification and evaluation schemes are a necessity for any river conservation strategy. For a European strategy, consensus about the framework (but not necessarily the details) is important and a discussion on river classification in Europe has continued since the early 20th century (Steinmann, 1907). This partly reflects real disagreements and partly the fact that classifications have been established for different purposes.

The classification of rivers based on stream order (Horton, 1945; Strahler, 1957) is recognized as useful for a number of purposes. However, stream order has limited value for classifying running-water habitats over a wider region and, therefore, limited value for conservation. Hawkes (1975) and Naiman *et al.* (1992) reviewed a wide range of river zonation schemes. In reality, rivers display a continuum of ecological changes from source to mouth (Vannote *et al.*, 1980), and there is no reason to believe that discrete zones exist. Illies (1961) proposed a river classification for worldwide application with two primary zones – the rhithron and the potamon – which he characterized faunistically. In the Netherlands Peeters *et al.* (1994) have proposed six river types: the upper, middle and lower reaches of upland and lowland rivers. A similar

classification has been proposed in Germany by Braukman and Pinter (1995).

The CORINE Biotope classification represents a standardized classification of both terrestrial and aquatic ecosystems. The main categories of rivers are rivulets, trout zone, grayling zone, barbel zone, bream zone and intermittent rivers. The fish zones are based on the work of Huet (1959). There are further subdivisions based on plant communities. Whether or not this classification is useful in river conservation remains to be seen.

Classification and evaluation of river 'quality' raises the question of definition. This question has been thoroughly discussed in Boon and Howell (1997). In the following paragraphs 'river quality' is defined as 'the totality of the features and characteristics of the river and its riparian areas that bear upon its ability to support an appropriate natural flora and fauna and to sustain legitimate uses' (modified from Pugh, 1997).

Evaluation methods for assessing river quality have been reviewed by Nixon *et al.* (1996a). Physical and chemical parameters have been used all over the region as well as methods based on a wide variety of organisms including benthic algae, phytoplankton, mosses, angiosperm macrophytes, and fish. Throughout the region, however, methods based on benthic macroinvertebrates are clearly the most common (Table 3.7). Early schemes were based on the Saprobic System (Kolkwitz and Marsson, 1902;

Sladecek, 1973), but today most biotic indices are based on the approach of Woodiwiss (1964). A crucial element in evaluation methods is the reference situation with which the actual situation is compared as, at least in some countries, true reference situations cannot be found any more. In most of the schemes the highest quality class is the reference situation, although it is recognized that the reference situation in higher order streams may not reach the highest class.

Predictive systems such as RIVPACS (River InVertebrate Prediction And Classification System) used in the UK (Wright *et al.*, 1989) represent a different and more stringent approach. Using key physical, chemical and geographical variables the macroinvertebrate community of a given site in its natural state can be predicted and compared with the community recorded. A similar rationale underpins the HABSCORE system (Milner *et al.*, 1998) for assessing the suitability of salmonid habitat, and the River Habitat Survey (RHS), developed in the UK for assessing and predicting a wider range of habitat features, also uses a modelling aproach (Raven *et al.*, 1997).

In contrast to the schemes and methods described above, SERCON (System for Evaluating Rivers for Conservation) developed in the UK (Boon *et al.*, 1997) is the only approach which more holistically also includes banks, riparian zones and associated floodplains. It is specifically designed for conservation assessment using six generally accepted conservation criteria: physical diversity, naturalness, representativeness, rarity, species richness and a category known as 'special features'.

The proposed EC Water Framework Directive will no doubt increase the needs for R&D to develop both comparable classification and evaluation schemes for a number of taxonomic groups, as well as more broadly based holistic schemes.

The concept of a Red List of Threatened Species (International Union for Conservation of Nature and Natural Resources: IUCN, 1990) is quite another approach with particular relevance for biodiversity conservation. Species are assigned to categories using information on previous and present distribution. The categories and criteria have recently been revised (IUCN, 1996) and are now described as:

Threatened Species
Lower Risk: conservation dependent
Lower Risk: near threatened
Extinct, and Extinct in the Wild

Red lists have been prepared globally (United Nations, 1991) as well as for specific countries or smaller regions, but mainly for larger species. This type of

approach suffers from substantial subjectivity by those making the designation, but the concept is easy to understand and puts public focus on the need to protect species and their habitats.

AVAILABILITY AND DISSEMINATION

Political decisions on conservation must be based on comparable and reliable data, whose scientific quality cannot be questioned. In scientific journals river conservation has never been adequately covered (Boon, 1992), although the appearance of journals such as *Regulated Rivers: Research & Management* and *Aquatic Conservation: Marine and Freshwater Ecosystems* has significantly improved the situation. Nevertheless, compared with terrestrial habitats and ecosystems, scientific data on European rivers are rather well established. Scientific publications on the region-wide distribution of different taxonomic groups include those on freshwater fish (Maitland, 1991), aquatic macrophytes (Haslam and Wolsley, 1981) and freshwater invertebrates (Illies, 1967, 1978). In addition, several publications describe the distribution of specific taxonomic groups within a country or a sub-set of countries within the region. However, the availability of data clearly decreases with the size of the organisms.

In the 'grey literature' huge amounts of data of relevance for river conservation are available within the region although there are large geographical variations. River monitoring is the most important topic for which such unpublished data are available, and from a recent review of surface water monitoring in the region (Kristensen and Bøgestrand, 1996) the following conclusions can be drawn:

- national monitoring programmes for physical and chemical assessment of water quality exist in nearly all countries;
- it is mainly larger rivers that are monitored; streams draining small catchments are monitored only in Denmark, Finland and Sweden;
- most monitoring focuses on organic pollution and nutrients, with metals included in about 40% of the programmes;
- national biological assessment programmes exist in Austria, Belgium, Denmark, Germany, Luxembourg, Ireland, The Netherlands, Spain and the UK;
- biological assessment is mainly based on macroinvertebrates, but in a few countries algae, macrophytes and fish are included;

- most national monitoring programmes are based on information collected regionally;
- there are large variations between countries in the region; e.g. national monitoring activities in Greece and Italy are scarce or absent.

In contrast with the information on rivers, the state of riparian areas and floodplains is generally not monitored in the region.

Most of the information from monitoring is published in the native language, which is ideal for the public of that particular country, but this restricts its wider use on a European scale. In recognition of this, the EU established a European Environment Agency (EEA) in 1993 located in Copenhagen. A main purpose of the EEA is to put the huge amount of existing data into work to support European policy. A major product has been the publication of *Europe's Environment: The Dobříš Assessment* (Stanners and Bourdeau, 1995), the preparation of which started in 1991 by the Task Force preparing the EEA, and later finished by the EEA. The Dobříš Assessment was the first comprehensive report on the environment of the whole of Europe, and this has recently been followed by a second assessment (EEA, 1998). The EEA has also started a series of monographs, the first of which was *European Rivers and Lakes* (Kristensen and Hansen, 1994), which was a background report for the Dobříš Assessment.

PUBLIC AWARENESS AND EDUCATION

The value of involving the public in environmental issues is increasingly being recognized. Public opinion influences the political will to initiate and support conservation programmes at all levels, from municipal authorities to the European Commission and the European Parliament.

An example of the importance of public awareness was discussed by Boon (1992) in the case of the willingness of Norwegian households to pay an increase of 500–1000 NOK in yearly electricity rates as a consequence of protecting 50 watercourses from hydroelectric power development. Of households 'concerned about nature conservation' 53.5% were willing to pay and 12.6% were not. In contrast, only 11.5% of households 'little concerned about nature conservation' were willing to pay and 37.0% were not.

A good model for increasing public awareness on lake conservation is described by Shapiro *et al.* (1975) reporting on a programme involving volunteers in monitoring lake transparency. In Denmark, the Biology Teachers Association has organized the 'Blue Stream' programme involving school classes all over the country in studying stream quality and reporting to a database. Recently, Tent (1998) described schoolchildren actively taking part in the practical work of a stream restoration project in Germany by planting trees. Similarly Nielsen (1997) has emphasized the significance of awareness and involvement of the local population, and not least, the landowners, in the success of river restoration projects. In fact, some of the Danish river restoration schemes have been initiated by the public.

At a European level the EEA is expected to raise public awareness. The main language is English but 'public awareness of the region' requires the use of a large number of languages. The cost of translation is a major hindrance for disseminating information and creating public awareness on European environmental issues, including river conservation. Whilst the EEA clearly has an important role in publishing environmental data, national authorities have a prime responsibility for raising public awareness.

Education is the foundation for public awareness, and environmental issues should be integrated within the educational systems at all levels. There is no overview as to the extent to which this has happened. In some countries there has been significant progress, but throughout the region there are undoubtedly large differences. There has been a shift from natural history to an ecological approach and in some countries conservation issues are dealt with both locally (e.g. rivers) and globally (e.g. rain forests). The education of practitioners has a major influence on river quality. Operators of sewage treatment plants influence discharges to the rivers and thereby river quality. Today a formal technical training is now compulsory for Danish operators of sewage treatment plants, ensuring optimum operation of the system.

River maintenance schemes in Denmark are run by river keepers, who play a crucial role in the conservation of Danish streams. By tradition they must ensure that stream discharge capacity is maintained, for draining arable fields. However, new legislation states that physical characteristics are as important as chemical constituents in determining water quality. This gives river keepers a dual role: when cutting weed or dredging they must ensure that the ecological requirements of the river are also fulfilled. This demands a special skill and an understanding of basic river ecology as well as an intimate knowledge of the legislative and administrative framework of river management. This is important since the river keepers meet the landowners, who may not be sympathetic to ecological river management. If a river keeper can give a sufficient explanation to the landowner, a complaint to the next administrative level may be avoided.

Danish river keepers are offered education and courses, often run by the local county river authorities. This ensures that an emphasis is put on the regional characteristics of rivers such as precipitation, geology and land use. During the courses the river keepers hone their skills in weedcutting, in understanding the properties of meandering currents as they pass through weed, and the relationships between discharge and water level, etc. Weedcutting and dredging are their main task, but often the river keepers are taught about restoration methods, such as how to construct a bypass at a weir. Their basic manual is the *River Keepers Field Book* (Madsen, 1995a), but a wide selection of printed material and videos is available.

The framework for conservation

THE SOCIO-POLITICAL/RELIGIOUS–CULTURAL BACKGROUND

After World War II the policy throughout the region was dominated by efforts to become self-sufficient in agricultural products. The experiences during the war of rationing and hunger were fresh in the minds of people. Intensification of agriculture accelerated, and with it, draining, river channelization and rigorous river maintenance to remove excess water from the fields to increase production.

Another important historical event was the establishment of the European Community in 1957 with the Treaty of Rome, starting with six countries and now including 15 countries. Today in the region only Norway and Iceland are not within the EC. The policies of the EC have had a significant impact on the rivers of the region and much EC legislation is also implemented in Norway. The agricultural policy of the EC continued the post-war trend for Europe described above until the 1980s. Recently the over-production of agricultural products and the consequent financial drain on EC budgets have forced a change leading to set-aside schemes and other measures. Although unintended at the beginning, these measures may have significant environmental benefits also for rivers. Environmental

issues including conservation have been on the EC political agenda, not least in the Environmental Action Programme Towards Sustainability (European Commission, 1992) and the EC is a co-signatory of several international agreements, conventions, etc.

A main characteristic is that, despite EC efforts, there are still very significant differences in the standard of living throughout the region, with much lower levels in Southern countries. Consequently priorities are different and in these areas conservation is given lower priority than in the rest of the region. To some extent EC policy has supported this and in several cases the EC structural funds have been used in a way which clearly conflicts with conservation interests. However, although conservation awareness generally has increased in recent years, there is certainly no guarantee that conservation will remain high on the agenda in the region as a whole, or in specific countries.

STATUTORY FRAMEWORK

The international statutory framework in the region consists of international conventions and EC legislation, mainly Directives. Whereas participation in conventions is voluntary, EC Directives are binding and must be implemented in national legislation by Member States.

Three conventions are broadly aimed at nature conservation, and are relevant to rivers. The Ramsar Convention (1971) is a global agreement to protect 'wetlands' (defined to include running waters) of international importance. The participating countries are obliged *inter alia* to designate at least one wetland and to promote the sustainable use of wetlands. At present all 17 countries of the region participate in the convention, which also includes marine wetlands. The Berne Convention (1982) is a European initiative developed by the Council of Europe, but also includes African countries. The aim of the convention is to protect flora and fauna, especially endangered species and their habitats, not least migratory species. In total, 122 species of freshwater fish are included, several of them important components of river ecosystems. The Convention on Biological Diversity (1992) aims at conserving biodiversity through establishing a system of protected areas or areas where specific measures have to be taken. It has been signed by all 17 countries in the region and follow-up initiatives are under way.

The EC Council Directive 92/43/EEC of 21 May 1992 on the conservation of natural habitats and wild fauna and flora (generally known as the 'Habitats

Directive') is the main Directive dealing broadly with nature conservation. Annex I lists a wide variety of habitats identified as being important at the EC level. Eight of the 175 habitats are river types:

- Alpine rivers and the herbaceous vegetation along their banks
- Alpine rivers and their ligneous vegetation with *Myricaria germanica*
- Alpine rivers and their ligneous vegetation with *Salix elaegnos*
- Constantly flowing Mediterranean rivers with *Glaucium flavum*
- Floating vegetation of *Ranunculus* of plain, submountainous rivers
- *Chenopodietum rubri* of submountainous rivers
- Constantly flowing Mediterranean rivers: *Paspalo-Agrostidion* and hanging curtains of *Salix* and *Populus alba*
- Intermittently flowing Mediterranean rivers

Annex II lists a number of species of animals and plants whose protection may require the designation of Special Areas of Conservation (SACs). There are relatively few riverine species in Annex II; these include mammals such as the European beaver (*Castor fiber*), the Eurasian otter (*Lutra lutra*) and the European mink (*Mustela lutreola*) as well as several species of fish (e.g. Atlantic salmon *Salmo salar*) and invertebrates (e.g. freshwater pearl mussel *Margaritifera margaritifera*). The only true riverine macrophyte is *Oenanthe conoides*. By the end of 1996 each country was supposed to have identified a number of protected habitats according to the directive but this deadline has not been met by most countries. The key aim of the Habitats Directive is to develop a coherent ecological network of SACs, which are to be maintained at 'favourable conservation status'.

The 5th Environmental Action Programme 'Towards Sustainability' (European Commission, 1992) has identified five major environmental issues of special concern, three of which are highly relevant to rivers: loss of biodiversity, integrated water management, and acidification. In recognition of the existing damage to European ecosystems, one of the measures identified to enhance biodiversity is restoration. To facilitate this process the EU LIFE-programme has financially supported river restoration, for instance in Denmark and the UK (Holmes and Nielsen, 1998).

Two Directives deal more specifically with issues related to river (and lake) conservation. The Surface Water Directive (75/440/EEC) classifies sources of surface water for abstraction of drinking water. One purpose of the Directive is to improve rivers used as sources for drinking water but focuses only on water quality. Of greater importance has been the Freshwater Fish Directive (78/659/EEC). This sets quality objectives for designated fresh waters in order to support fish life, and contains two categories: salmonid waters, and cyprinid waters. The implementation of the Directive has varied widely between countries. In Denmark 21 000 km out of 35 000 km of rivers with quality objectives have been identified as fish waters, and in the United Kingdom 54 771 km (mainly in Scotland) have been identified, 88% of them being salmonid waters (European Commission, 1995b). By contrast, in Italy only 403 km in six rivers in one province have been identified as salmonid waters and in 1998 the European Court decided that Italy had not implemented the Directive correctly (European Commission, 1995b). In Portugal the Directive had by 1995 not been implemented at all. A major step forward in the Freshwater Fish Directive was the introduction of water quality objectives and formal requirements for monitoring adherence to these objectives.

In the proposed Water Framework Directive these ideas are further developed. The overriding objective is to ensure good ecological quality of all surface waters in the EC. Those aspects of quality under discussion include physical, chemical and biological elements:

- dissolved oxygen;
- concentrations of toxic and other harmful substances in water, sediment and biota;
- levels of disease in animal life, including fish and in plant populations due to human influences;
- diversity of invertebrate communities (planktonic and bottom-dwelling) and key species/taxa normally associated with the undisturbed condition of the ecosystem;
- diversity of aquatic plant communities, including key species/taxa normally associated with the undisturbed condition of the ecosystem, and the extent of macrophyte or algal growth due to elevated nutrient levels of anthropogenic origin;
- diversity of the fish population and key species/taxa normally associated with the undisturbed condition of the ecosystem. Passage of migratory fish, in so far as it is influenced by human activity;
- diversity of vertebrate communities (amphibians, birds and mammals);

Table 3.8 *Quality objectives for Danish rivers (Miljøstyrelsen, 1983). The Roman numerals refer to a system for describing the effect of organic pollution with I the least and IV the most polluted water*

		Quality objective	Maximum stream pollution index
Rigorous objectives	A	Area of special scientific interest	II
General objectives	B₁	Salmonid spawning and fry production area	II
	B₂	Salmonid water	II
	B₃	Cyprinid water	II
Eased objectives	C	Streams with only run-off interests	II–III
		Streams affected by:	
	D	waste water	II–III
	E	groundwater abstraction	II–III
	F	ochre	II–III

- the structure and quality of the sediment and its ability to sustain the biological community of the ecosystem;
- the riparian and coastal zones, including the biological community and the aesthetics of the ecosystem;
- water quantity.

No doubt all these elements cannot be made operational within a few years, but the holistic ecosystem approach is promising.

At the national level there are not only large differences in statutory frameworks, but also similarities. Early in the 20th century nature preservation appeared on the political agenda, focusing on undisturbed habitats and landscapes or on threatened species. A good example is the establishment of the first national parks in Sweden in 1909 and in Spain in 1918. One of the main aims of river legislation before about 1980 was to remove excess water from lowland areas, including fields. The Land Drainage Act 1930 in England (Newson, 1992) and the Danish Watercourse Act (1949, 1962) are illustrative, ecological river quality not being mentioned at all.

Another general picture is the lack of coordination in river management, many Acts and regulations dealing with different aspects of river functioning. For instance in Denmark seven pieces of legislation are involved (Iversen *et al.*, 1993). The use of quality objectives as part of Danish environmental regulation was established in the 1970s with eight different quality objectives (Table 3.8). The counties are responsible for setting the politically determined quality objectives and for taking the necessary measures to fulfil them. An index based on macroinvertebrates is used as an operational indicator and the counties are responsible for monitoring whether or not quality objectives are met. In the region quality objectives are used at least in

connection with the Freshwater Fish Directive, but with significant differences between countries, and in Norway and Sweden quality objectives are under consideration (Wiederholm, 1997). No doubt the future Water Framework Directive will promote the use of quality objectives all over the region.

It is the stated intention of the EC 5th Environmental Action Programme to integrate environmental issues into sectoral policies, not least agriculture and forestry. Until now the results have been rather discouraging.

ROLE OF VOLUNTARY ORGANIZATIONS

There is a long tradition in several countries in the region for Non-Governmental Organizations (NGOs) to play a significant role. In the UK river management with the clear purpose of increasing the catch of trout and salmon has been practised for more than a century. Although some of the remedies (e.g. removal of pike and burbot) oppose the general view that conservation should benefit a diverse animal and plant life, these activities probably saved many salmon rivers from being ruined. Some of the old handbooks for the prudent management of such streams contain information that is still used in today's stream management (e.g. Bund, 1899) with detailed advice on the proper conservation of spawning beds. Several books from continental Europe, written for commercial river fishermen, have the same message (e.g. Walter, 1912). These people were the fathers of river conservation and their wisdom is still viable.

Anglers' associations in Denmark have had influence in lobbying for Danish environmental legislation, for example the revision of the Water Course Act where they helped initiate non-aggressive weedcutting and restoration. Recently they have influenced measures for ensuring a minimum water flow bypassing Danish trout farms, where for years stream continuity has been disrupted during times of low flow by diverting all the water through the trout farm, leaving a 'dead' river stretch (Iversen, 1995). On local issues, anglers have actively been involved in practical conservation works. A good example is the restoration of the breeding streams for lake trout living in the Danish Lake Hald Sø (Madsen, 1995b).

Voluntary organizations have also played a role in enhancing the distribution of native species such as the otter. This species was once regarded as a nuisance, but is now endangered and is protected in the region (Foster-Turley *et al.*, 1990). In the UK the Otter Trust was founded in 1971 with the goal of improving the chances of otter survival. The Trust has demonstration sites (Wayre, 1979), initiates research and develops methods for protecting and enhancing otter populations. One of the most important activities is a breeding programme. Up to 1994 the Trust had bred 178 otters in 90 litters, and by April 1995 they had released 72 otters in East Anglia and other regions in lowland England. Their influence on public attitudes to otters has been considerable, and a very important activity has been to involve the public in surveying signs of otter. Similar activities have spread to other countries in the region, such as 'Action Fischotterschutz' established in 1977 in Germany.

Conservation in practice

IN SITU CONSERVATION

Over the last century a wide range of areas have been protected according to different national initiatives and international conventions. IUCN (1993) has categorized different types of protected areas according to their purpose and management measures used (Table 3.9). About 200 000 km^2 in the region have been protected in some way including marine areas (Table 3.10) with large differences between countries in total area protected and the degree of protection. However, these areas may still experience human pressure. For instance, dams for hydroelectric power were established in a Swedish national park, and illegal builders are a major problem in a Portuguese national park (Figuerido, 1995). Seen in a river conservation perspective some of these areas are valuable, but their overall impact on European river conservation is limited.

River conservation in its widest sense includes water quality, water quantity, physical features and riparian

Table 3.9 *Management categories of protected areas established by the Commission on National Parks and Protected Areas (IUCN, 1993; Stanners and Bourdeau, 1995). Categories I–V have been in use since 1978, category VI since 1992*

I	Strict Nature Reserve/Wilderness Area: protected area managed mainly for science or wilderness protection
II	National Park: protected area managed mainly for ecosystem protection and recreation
III	Natural Monument: protected area managed mainly for conservation of specific features
IV	Habitat/Species Management Area: protected area managed mainly for conservation through management intervention
V	Protected Landscape/seascape: protected area managed mainly for landscape/seascape conservation and recreation
VI	Managed Resource Protected Area: protected area managed mainly for the sustainable use of natural resources

Table 3.10 *Major protected areas (km^2) in the region according to the IUCN categories (Table 3.9). Data from European Commission (1995a)*

	I	II	III	IV	V
Austria				948	14 991
Belgium				40	679
Denmark	89		63	787	3 289
Finland	1 512	3 541			3 020
France	171	2 613		1 294	43 709
Germany		131		1 695	47 715
Greece		526	180	106	224
Iceland	3	1 801	386	520	6 450
Ireland		225		43	
Italy		1 259		2 775	8 972
Luxembourg		4		4	
The Netherlands	42	54	2 206	1 248	
Norway	26 373	19 102		187	1 962
Portugal		994	39	732	2 771
Spain		1 128		15 710	18 173
Sweden		5 892		10 618	1 074
United Kingdom		41		1 332	45 018
Total	28 190	37 411	2 874	38 039	198 044

areas. Since the 1960s the main focus has been on measures to improve water quality. Due to national or local legislation and regulation vast investments have been made in sewage treatment technology, especially in densely populated areas. During the last two decades these measures have in general resulted in a significant decrease in concentrations of organic matter, phosphorus and ammonium in rivers (Kristensen and Hansen, 1994). The decrease has been greatest in Western countries, where the River Rhine is a good

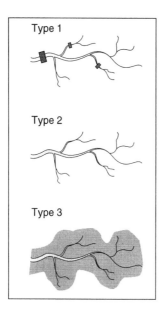

Figure 3.7 *Schematic definition of three types of river rehabilitation project. Type 1: Rehabilitation of watercourse reaches. Type 2: Restoration of continuity between watercourse reaches. Type 3: Rehabilitation of river valleys*

example (Figure 3.5), but the pattern is the same all over the region. However, despite the improvements, poor river quality is still a major problem, not least in small rivers in densely populated areas.

Reduced river water flow due to irrigation and consumption in households and industry is also a serious problem, not least in Southern countries. The pressure on water resources in these regions is certainly not declining, and this is of major concern for river conservation.

The loss of the natural physical properties of rivers and of their riparian areas can only be improved through rehabilitation measures, often called 'river restoration' (see Petts *et al.*, this volume). The recently established European Centre for River Restoration, with a secretariat at the National Environmental Research Institute in Denmark, collects and disseminates data and information on practical river restoration in the region. The various measures for restoration may be classified as follows (Hansen *et al.*, 1996, Figure 3.7):

- rehabilitation of river reaches
- re-establishing continuity between river reaches
- rehabilitation of river valleys

During the last 10–15 years physical rehabilitation of rivers has taken place especially in the Nordic and Western countries (Jutila, 1985; Brookes, 1992; Petersen *et al.*, 1995; Iversen *et al.*, 1993; Kern, 1995). A review of practical examples is given by Brookes and Shields (1996), Hansen (1996) and Holmes (1997).

In Austria a 1.5 km section of the channelized River Romaubach was given new meanders in 1988 (Schlott, 1995). The project originated from farmers wanting to dredge the silted channel, but the state authorities were only prepared to fund a more environmentally sensitive option, mirroring the attitudes slowly emerging in parts of the region. The limited width of the available river corridor constrained the dimensions of the bends, so that the new meandering stretch had increased in length by only 10%. Within three years, the river and banks were revegetated with a rich variety of plants. No significant change in fish populations occurred but the overall landscape diversity and aesthetics had improved, which is important for convincing the local communities of the value of such projects. In the fifth-order Melk River in Austria, Jungwirth *et al.* (1995) demonstrated that re-creation of in-stream river-bed structures significantly increased fish diversity, density and biomass. Similarly newly created riparian zones provided important refuge areas during flooding, and nursery grounds for fish fry.

The German Landesamt Niedersachsen with its 180 000 km of streams and rivers (the Weser River system) has initiated a regional restoration programme (Kairies and Dahlman, 1995). The goal is to re-establish their inherent, natural qualities. Measures are:

- removing obstacles in streams
- establishing public ownership of stream margins
- re-establishing the natural vegetation along the streams
- changing steep and deep stream profiles to a more natural shape
- re-establishing flooding processes in the valleys

The progress of the programme is shown in Figure 3.8.

River valley restoration aims at re-establishing natural hydrology and hydrological processes. Re-establishing the water retention capacity through raising the water table, re-meandering and re-establishing pools and lakes reduces the risk of downstream floods. The loss of natural hydrological processes, however, has also greatly reduced wetland nutrient retention capacity. The increased nitrogen leaching due to intensification of agriculture causes

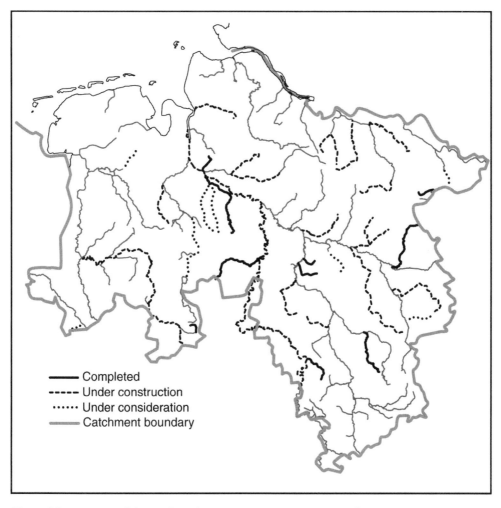

Figure 3.8 *Progress of the Niedersachsen restoration programme as of January 1995. From Kairies and Dahlman (1995)*

marine eutrophication with oxygen deficiency, etc. (Kronvang *et al.*, 1993). Undisturbed wetlands have been reported to have a nitrogen removal capacity of 42–875 kg N ha^{-1} yr^{-1} with an efficiency of 56–99%, and re-established previously drained areas have the same capacity (Iversen *et al.*, 1995). The main measure needed to reduce marine eutrophication is reduced nitrogen leaching from agriculture through a more appropriate use and handling of manure and fertilizer. However, river valley restoration may contribute significantly and will have an immediate effect (Iversen *et al.*, 1995).

In Denmark river quality in many rivers in Jutland is significantly reduced by ochre pollution (Jensen, 1996). Draining wetlands with pyrite causes oxygenation and

leaching of acid water with ferrous ions. River valley restoration to raise the water table will stop this process and, furthermore, flooded meadows retain a significant amount of particulate and soluble iron (Jensen, 1996).

EX SITU CONSERVATION

Ex situ conservation measures such as the use of gene banks for forest trees or for husbandry does not exist for riverine species. However, the management of fish populations either for conservation purposes (threatened populations) or for recreational use (angling) contains some *ex situ* elements. In several countries fish stocking programmes are actively pursued in the belief that the natural fish populations are far below carrying capacity in many rivers. The fish being stocked are reared *ex situ* in hatcheries or fish farms. In the past, it was common for the eggs to come from stocks reared and maintained in fish farms. In recognition that this may produce fish that are genetically very different from natural populations it is now common to rear eggs from female fish caught in the river to be stocked, and fertilized with sperm from males from the same river.

The houting *Coregonus oxyrinchus* is a globally endangered species. It lives in shallow coastal waters and migrates into rivers in the spawning season in winter. Its area of distribution has been significantly reduced (Holčik, 1990). It was common in the Waddensee along the Dutch, German and Danish North Sea coast, but from early in the 20th century the population began plummeting, the cause assumed to be the destruction of its breeding grounds, and dams and weirs preventing upstream migration. In the 1970s only a very small population, facing extinction, persisted in the Danish Waddensee, breeding in the River Vidå at the Danish–German border (Ejbye-Ernst *et al.*, 1988).

From the River Vidå population a breeding scheme was initiated in 1981 and from 1982 onwards fry were released in the Vidå and the other rivers running into the Danish Waddensee. In addition, the rivers were restored to make them suitable breeding habitats, i.e. riffles replaced impassable weirs, new spawning gravel was introduced, and programmes for reducing ochre, which otherwise clogs the spawning banks, were initiated. The stocking programme was terminated in 1992. Monitoring the houting populations in the rivers has shown that the project has so far been a success, probably averting the extinction of the population. In 1994–1995 mature houting were found at an age that indicates they originated from eggs laid in the rivers, especially in the River Vidå (Ejbye-Ernst, the County of Ribe, Denmark, personal communication). One of the lessons learned from the project is that a detailed knowledge of the biology of the species is essential, so that the subtle interaction between the river and the species can be taken into consideration in matching the needs of a species to a particular type of restoration.

In other cases fish populations have totally disappeared from a watershed. This has been reported, for example, for the Atlantic salmon *Salmo salar*, native to the Atlantic region from Iceland to Norway, owing to the prevention of upstream migration by dams, poor water quality, acidification and overfishing. The River Rhine was probably the most productive European salmon river in the 19th century, but due to human impact the salmon became extinct in 1957. The salmon has now returned to a number of rivers including the Rhine. An important measure has been large-scale stocking: Kennedy (1988) reported a release of more than 38 million juvenile salmon in the North Atlantic region. However, the new populations are genetically different from those that became extinct. For instance, the re-established population in the Danish River Gudenå is based on salmon from Ireland and Scotland. Thus, whilst the species has been re-established, there has been a loss in genetic biodiversity.

Ex situ conservation may be problematic. The Baltic salmon (*Salmo salar*) has lost its most important spawning grounds in the large rivers of Finland and Sweden because of the construction of reservoirs. A large stocking programme was established because of the profitable commercial salmon fisheries in the Baltic Sea. Millions of fry were reared from upstream migrating mature males and females, and released as smolt in the Baltic. From Sweden alone nearly 2 million smolt were released. The catches consisted of 90% reared fish, while the remaining 10% was wild salmon. From 1974 an undiagnosed disease appeared in the rearing plants killing the alevins (Karlsson and Karlström, 1994). In 1993 the disease killed 90% of the alevins in the Finnish hatcheries at the River Torne Älv, and in 1994 70% of the alevins in the Swedish hatcheries. This not only threatens the stocking programme and the commercial fisheries, but the disease has spread to small, wild populations too and ultimately the stocks may be decimated to near extinction.

The European beaver *Castor fiber* has successfully been reintroduced in a number of countries in the region. From being widespread in Europe, only a few hundred animals remained around 1900 in Norway, Germany, France and the former Soviet Union (Valeur, 1990). In 1922 the European beaver was re-introduced from Norway to Sweden and the Swedish population has expanded from 80 animals in 1922–1939 to about 100 000 animals in 1992 mainly due to restricted hunting. Similar re-introductions have taken place in Finland, Germany, France and Austria with varying success, and a pilot re-introduction is being considered in Scotland (Scott Porter Research & Marketing Ltd, 1998). In 1937 not only the European beaver but also the Canadian beaver (*Castor canadensis*) was introduced to Finland. Where both species were introduced the Canadian beaver has had a higher rate of increase (Lathi and Helminen, 1974). Introducing alien species (as also seen for many fish species) reduces the naturalness of ecosystems, thus conflicting with an important principle of river conservation.

FUTURE OUTLOOK

The main pressures described in this chapter were domestic, industrial and agricultural. It has been shown that efficient measures (e.g. for wastewater treatment and physical rehabilitation) can significantly reduce human impact. Similarly acidification, the result of transboundary atmospheric pollution, can be reduced by limiting emissions. If there is the necessary political will, rivers can be restored – not back to a pristine state, but certainly to a state of good ecological quality.

Recently there have been examples of environmental deterioration where the causes are more elusive. For example, sea trout (*Salmo trutta*) catches on the west coast of Ireland have collapsed since 1989 (Whelan and Poole, 1996) and there is no obvious reason. The dramatic decline in the distribution of freshwater pearl mussel (*Margaritifera margaritifera*) in inland waters of Ireland (Chesney, 1996) cannot be explained by sewage pollution or by acid rain, since it has occurred in a region not suffering from such impacts. One may ponder why such a creature, with a complex life cycle involving another species in taking its progeny upstream, has been able to sustain very dense populations during centuries, but whch is now clearly facing extinction not only in pollution-prone continental waters but also in remote Irish waters. These examples clearly indicate that there is still much to learn for river conservation to be effective.

Global climate change may significantly affect ecosystem structure and processes. Global mean temperature has increased since the first reliable measurements around 1860, especially between 1910 and 1940 and since the mid 1970s. The Intergovernmental Panel on Climate Change (IPCC, 1996) concludes that 'the balance of evidence suggests a discernible human influence on global climate'. Estimates for the increase in global mean temperature by the year 2100 vary between 1.0 and 3.5°C. For the EC countries including Scandinavia the evidence suggests an overall increase in temperature and precipitation with significant differences between countries and between seasons (Colman *et al.*, 1995; Mitchell *et al.*, 1995). Similarly an increase is expected in the frequency of extreme events such as floods and droughts. Climate change will affect biogeography as well as species distribution and biodiversity in rivers,

and hydrological changes will increase the complexity of future river management.

The challenges for river conservation as for nature conservation in general will certainly not decrease. The known impacts will increase and unknown or only vaguely known impacts will appear. Solutions can only be found through implementing political decisions that may have considerable economic impact. The scientific community has the obligation to provide politicians with scientifically based concepts and arguments. Throughout the region there are positive signs of improvement but also of deterioration and increase in pressures. It is to be hoped that joint effort will bring the case forward. Rivers are important by themselves but they are even more important for the overall diversity of landscapes and biota.

Acknowledgements

We wish to thank Dr Phil Boon, Dr Peter Kristensen and Dr Steve Nixon for valuable help and comments on the manuscript.

References

Berntsen, T. 1996. *Miljøvernpolitisk redegjørelse: Miljøstatus* (in Norwegian) [Review of the policy for environmental protection: environment status]. Miljøverndepartetmentet, Oslo.
Boon, P.J. 1992. Essential elements in the case for river conservation. In: Boon, P.J., Calow, P. and Petts, G.E. (eds) *River Conservation and Management*. John Wiley, Chichester, 11–33.
Boon, P.J. and Howell, D.L. (eds) 1997. *Freshwater Quality: Defining the Indefinable?* The Stationery Office, Edinburgh.
Boon, P.J., Holmes, N.T.H., Maitland, P.S., Rowell, T.A. and Davis, J. 1997. A system for evaluating rivers for conservation ('SERCON'): development, structure and function. In: Boon, P.J. and Howell, D.L. (eds) *Freshwater Quality: Defining the Indefinable?* The Stationery Office, Edinburgh, 299–326.
Braukmann, U. and Pinter, I. 1995. *Concept for an Integrated Ecological Evaluation of Running Waters*. Strategy paper. Landesanstalt für Umweltschutz, Baden-Württemberg.
Brodin, Y.W. and Kuylenstierna, J.C.J. 1992. Acidification and critical load in the Nordic countries, a background. *Ambio* 21:332–328.
Brookes, A. 1992. Recovery and restoration of some engineered British river channels. In: Boon, P.J., Calow, P. and Petts, G.E. (eds) *River Conservation and Management*. John Wiley, Chichester, 337–352.
Brookes, A. and Shields, D. 1996. *River Channel Restoration.*

Guiding Principles for Sustainable Projects. John Wiley, Chichester.

Bryson, B. 1992. Main–Danube canal: Linking Europe's waterways. *National Geographic* 182:3–31.

Bund, J.W.W. 1899. *A Handy Book of Fisheries Management.* Lawrence and Bullen, London.

Casado, C., de Jalon, D.G., del Olmo, C.M., Barcelo, E. and Menes, F. 1989. The effect of an irrigation and hydroelectric reservoir on the downstream communities. *Regulated Rivers: Research & Management* 4:275–284.

Chesney, H.C.G. 1996. Irish pearl mussels: going, going, gone. In: Reynolds, J.D. (ed.) *The Conservation of Aquatic Systems.* Royal Irish Academy, Dublin, 142–150.

Colman, R.S., Power, B., MacAvaney, B. and Dahlni, R. 1995. A non-flux corrected transient CO_2 experiment using the BMRC coupled A/OGCM. *Geophysics Research Letters* 22:3047–3050.

Duzzin, B., Pavoni, B. and Donazzolo, R. 1988. Macroinvertebrate communities and sediments as pollution indicators for heavy metals in the river Adige (Italy). *Water Research* 22:1353–1363.

Ejbye-Ernst, M., Grøn, P.N., Larsen, L.K., Møller, B. and Nielsen, M.B. 1988. *Snæbelen – en truet fiskeart* (in Danish) [The houting – a threatened fish]. Ribe and Sønderjyllands Amtsråd.

European Commission 1992. *Towards Sustainability. A European Community Programme of Policy and Action in Relation to the Environment and Sustainable Development.* COM(92)23 – vol II. Commission of the European Communities, Brussels.

European Commission 1995a. *Europe's Environment: Statistical Compendium for the Dobříš Assessment.* Office for Official Publications of the European Communities, Luxembourg.

European Commission 1995b. *Kvaliteten af ferske vande og skaldyrsvande* (in Danish) [The Quality of Fresh Waters and Shellfish Waters]. Office for Official Publications of the European Communities, Luxembourg.

European Commission 1998. *Europe's Environment: Statistical Compendium for the Second Assessment.* Office for Official Publications of the European Communities, Luxembourg.

European Environment Agency (EEA) 1998. *Europe's Environment: The Second Assessment.* Office for Official Publications of the European Communities, Luxembourg. Elsevier Science Ltd, Oxford.

Figuerido, J. 1995. Nature restoration in Portugal. *Proceedings for Nature Restoration in the European Union.* Ministry of Environment and Energy, the National Forest and Nature Agency, Copenhagen, 126–128.

Foster-Turley, P.S., Macdonald, P.S. and Mason, C. (IUCN/SSC Otter Specialist Group) 1990. *Otters, An Action Plan for their Conservation.* International Union for Conservation of Nature and Natural Resources, Gland.

Garcia de Jalon, D. 1987. River regulation in Spain. *Regulated Rivers: Research & Management* 1:343–348.

Grasmück, N., Haury, J., Léglize, L. and Muller, S. 1995.

Assessment of the bio-indicator capacity of aquatic macrophytes using multivariate analysis. In: Balway, G. (ed.) *Space Partition within Aquatic Ecosystems. Hydrobiologia* 300/301:115–122.

Hansen, H.O. (ed.) 1996. *River Restoration – Danish Experience and Examples.* Ministry of Environment and Energy, National Environmental Research Institute. Silkeborg, Denmark.

Hansen, H.O., Kronvang, B. and Madsen, B.L. 1996. Classification system for water course rehabilitation. In: Hansen, H.O. (ed.) *River Restoration – Danish Experience and Examples.* Ministry of Environment and Energy, National Environmental Research Institute, Silkeborg, Denmark, 73–79.

Haslam, S.M. and Wolsley, P.A. 1981. *River Vegetation: Its Identification, Assessment and Management.* Cambridge University Press, Cambridge.

Hawkes, H.A. 1975. River zonation and classification. In: Whitton, B.A. (ed.) *River Ecology.* Blackwell, Oxford, 312–374.

Holčik, J. (ed.) 1990. *The Freshwater Fishes of Europe.* AULA Verlag, Wiesbaden.

Holmes, N.T.H. 1997. The UK experiences on river restoration. In: Hansen, H.O. and Madsen, B.L. (eds) *River Restoration 96 – Plenary Lectures.* International conference arranged by the European Centre for River Restoration. Ministry of Environment and Energy, National Environmental Research Institute, Silkeborg, Denmark, 57–84.

Holmes, N.T.H. and Nielsen, M.B. 1998. Restoration of the Rivers Brede, Cole and Skerne: a joint Danish and British EU-LIFE demonstration project: I – Setting up and delivery of the project. *Aquatic Conservation: Marine and Freshwater Ecosystems* 8:185–196.

Hornby, A.S., Gatenby, E.V. and Wakefield, H. 1960. *The Advanced Learners Dictionary of Current English*, 12th impression. Oxford University Press, London.

Horton, R.E. 1945. Erosional development of streams and their drainage basins: hydrophysical approach to quantitative morphology. *Geological Society of America Bulletin* 56:275–370.

Huet, M. 1959. Profiles and biology of Western European streams related to fish management. *Transactions of the American Fish Society* 88:155–163.

Hynes, H.B.N. 1975. The stream and its valley. *Verhandlungen der Internationalen Vereinigung für theoretische und angewandte Limnologie* 19:1–15.

Illies, J. 1961. Versuch einer algemein biozönotischen Gliederung der Fliessgewässer [An attempt at a general biocentric classification of rivers]. *Internationale Revue der gesamten Hydrobiologie* 46:205–213.

Illies, J. 1967. *Limnofauna Europaea.* Gustav Fischer, Stuttgart.

Illies, J. 1978. *Limnofauna Europaea.* Gustav Fischer, Stuttgart.

Intergovernmental Panel on Climate Change (IPCC) 1996. *Climate Change 1995. Impacts, Adaptations and Mitigations of Climate Change: Scientific and Technical Analyses* (eds Watson, R.T., Zinyowera, M.C. and Moss, R.H.). Cambridge University Press, Cambridge.

International Union for Conservation of Nature and Natural Resources (IUCN) 1990. *1990 IUCN Red List of Threatened Animals*. Gland.

International Union for Conservation of Nature and Natural Resources (IUCN) 1993. *Action Plan for Protected Areas in Europe. Second Draft. November 1993*, Gland.

International Union for Conservation of Nature and Natural Resources (IUCN) 1996. *Red List of Threatened Animals*. Gland.

Iversen, T.M. 1995. Fish farming in Denmark. Environmental impact of regulative legislation. *Water Science and Technology* 31:73–84.

Iversen, T.M., Wiberg-Larsen, P., Hansen, S.B. and Hansen, F.S. 1978. The effect of partial and total drought on the macroinvertebrate communities of three small Danish streams. *Hydrobiologia* 60:235–242.

Iversen, T.M., Kronvang, B., Madsen, B.L., Markmann, P. and Nielsen, M.B. 1993. Re-establishment of Danish streams: restoration and maintenance measures. *Aquatic Conservation: Marine and Freshwater Ecosystems* 3:73–92.

Iversen, T.M., Kronvang, B., Hoffmann, C.C., Søndergaard, M. and Hansen, H.O. 1995. Restoration of aquatic ecosystems and water quality. *Proceedings from Nature Restoration in European Union*. Ministry of Environment and Energy, The National Forest and Nature Agency, Copenhagen, 63–69.

Jensen, P.S. 1996. Rind stream at Herning. In: Hansen, H.O. (ed.) *River Restoration – Danish Experience and Examples*. Ministry of Environment and Energy, National Environmental Research Institute, Silkeborg, Denmark, 32–34.

Johnson, B.O. and Jensen, A.J. 1986. Infestations of Atlantic salmon, *Salmo salar*, by *Gyrodactylus salaris* in a Norwegian river. *Journal of Fish Biology* 29:233–241.

Jungwirth, M., Muhar, S. and Schmutz, S. 1995. The effects of recreated instream and ecotone structures on the fish fauna of an epipotamal river. In: Schiemer, F., Zalewski, M. and Thorpe, J.E. (eds) *The Importance of Aquatic–Terrestrial Ecotones for Freshwater Fish. Hydrobiologia* 303:195–206.

Jutila, E. 1985. Dredging of rapids for timber-floating in Finland and its effect on river-spawning fish stocks. In: Alabaster, J.A. (ed.) *Habitat Modification and Freshwater Fisheries*. Butterworths, London, 104–108.

Kairies, E. and Dahlman, I. 1995. Fliessgewässerrenaturierung in Niedersachsen – Grundlagen und Erfahrungen [River restoration in Niedersachsen – foundation and experiences]. In: *Fliessgewässerrenaturierung in Praxis*. Niedersächsisches Landesamt für Ökologie, Hildesheim, 119–122.

Karlsson, L. and Karlström, Ö. 1994. The Baltic Salmon (*Salmo salar*): its history, present situation and future. *Dana* 10:61–85.

Kennedy, G.J.A. 1988. Stock enhancement of Atlantic salmon (*Salmo salar*). In: Mills, D. and Piggins, D. (eds) *Atlantic*

Salmon: Planning for the Future. Croom Helm, London, 345–372.

Kern, K. 1995. Bettbildung und Morphodynamik von Fliessgewässern [River bed formation and morphodynamics of rivers]. In: *Fliesswasserrenaturierung in der Praxis*. Niedersächsisches Landesamt für Ökologie, Hildesheim, 129–138.

Kolkwitz, R. and Marsson, M. 1902. Grundsätze für die biologische Beurteilung des Wasser nach seine Flora und Fauna [Principles of biological assessment of water using flora and fauna]. *Mitteilungen aus der Königlichen Prüfungsanstalt Wasserversorgung Abwasserbeseitigung zu Berlin* 1:33–77.

Kristensen, P. and Bøgestrand, J. 1996. *Surface Water Quality Monitoring*. Topic report 2/96 (Inland Water). European Environment Agency, Copenhagen.

Kristensen, P. and Hansen, H.O. (eds) 1994. *European Rivers and Lakes*. EEA Environmental Monographs 1. European Environment Agency, Copenhagen.

Kronvang, B., Ærtebjerg, G., Grant, R., Kristensen, P., Hovmand, M. and Kirkegaard, J. 1993. Nationwide monitoring of nutrients and their ecological effects: state of the Danish aquatic environment. *Ambio* 22:176–187.

Lathi, S. and Helminen, M. 1974. The beaver *Castor fiber* L. and *Castor canadensis* (Kuhl) in Finland. *Acta Theriologica* 19:177–189.

Lelek, A. 1989. The Rhine river and some of its tributaries under human impact in the last two centuries. In: Dodge, D.P. (ed.) *Proceedings of the International Large River Symposium. Canadian Special Publication of Fisheries and Aquatic Sciences* 106:469–487.

Madsen, B.L. 1995a. *A River Keepers Field Book*. Ministry of Environment and Energy, Danish Environmental Protection Agency, Copenhagen.

Madsen, B.L. 1995b. *Danish Watercourses. Ten Years with the New Water Course Act*. Miljønyt No 11. Danish Environmental Protection Agency, Copenhagen.

Maitland, P.S. 1991. *Conservation of Threatened Freshwater Fish in Europe*. Nature and Environment Series, No. 46. Council of Europe, Strasbourg.

Merilehto, K., Kenttämies, K. and Kämäri, J. 1988. *Surface Water Acidification in the ECE Region*. Miljörapport 1988: 14. Nordic Council of Ministers, Copenhagen.

Miljøstyrelsen 1983. *Vejledning fra Miljøstyrelsen, Vejledning i recipientkvalitetsplanlægning, Del I vandløb og søer* (in Danish) [Guidelines from the Environmental Protection Agency. Guidelines for planning the quality of recipients. Part 1 – rivers and lakes]. Miljøstyrelsen Vejledning nr. 1/1983, Copenhagen.

Miljøstyrelsen 1990. *Lavteknologisk spildevandsrensning i danske landsbyer* (in Danish) [Low technological purification of waste water in Danish villages. Sewage water research from the Environmental Protection Agency]. Spildevandsforskning fra Miljøstyrelsen 5, Copenhagen.

Milner, N.J., Broad, K. and Wyatt, R.J. 1998. HABSCORE – applications and future development of related habitat models. *Aquatic Conservation: Marine and Freshwater Ecosystems* 8:633–644.

Mitchell, J.F.B., Johns, T.C., Gregory, J.M. and Tett, S.F.B. 1995. Climate response to increasing levels of greenhouse gases and sulphate aerosols. *Nature* 376:501–504.

Morris, D.G. and Kronvang, B. 1994. *Report on a Study into the State of River and Catchment Boundary Mapping in the EC and the Feasibility of Producing an EC-wide River and Catchment Boundary Database.* Report to the EEA-TF. Institute of Hydrology, Wallingford.

Naiman, R.J., Lonzarich, D.G., Beechie, T.J. and Ralph, S.C. 1992. General principles of classification and the assessment of conservation potential in rivers. In: Boon, P.J., Calow, P. and Petts, G.E. (eds) *River Conservation and Management.* John Wiley, Chichester, 93–124.

Newson, M.D. 1992. River conservation and catchment management: a UK perspective. In: Boon, P.J., Calow, P. and Petts, G.E. (eds) *River Conservation and Management.* John Wiley, Chichester, 385–396.

Nielsen, M.B. 1997. Danish experience on river restoration III: From idea to completion. In: Hansen, H.O. and Madsen, B.L. (eds) *River Restoration 96 – Plenary Lectures.* International conference arranged by the European Centre for River Restoration. Ministry of Environment and Energy, National Environmental Research Institute, Silkeborg, Denmark, 47–56.

Nixon, S.C., Mainstone, C.P., Iversen, T.M., Kristensen, P., Jeppesen, E., Friberg, N., Papathanassion, E., Jensen, A. and Pedersen, F. 1996a. *The Harmonized Monitoring and Classification of Ecological Quality of Surface Waters in the European Union.* Report to the European Commission, Medmenham, UK.

Nixon, S.C., Rees, Y.J., Gendebien, A. and Ashley, S.J. 1996b. *Requirements for Water Monitoring.* Topic Report 1/1996 (Inland Water). European Environment Agency, Copenhagen.

Peeters, E.T.H.M., Gardeniers, J.J.P. and Tolkamp, H.H. 1994. New methods to assess the ecological status of surface waters in the Netherlands. Part 1: running waters. *Verhandlungen der Internationalen Vereinigung für theoretische und angewandte Limnologie* 25:1914–1916.

Petersen, R.C., Gislason, G.M. and Petersen, L.B.-M. 1995. Rivers of the Nordic countries. In: Cushing, C.E., Cummins, K.W. and Minshall, G.W. (eds) *Ecosystems of the World 22. River and Stream Ecosystems.* Elsevier. Amsterdam, 295–342.

Petts, G.E. 1988. Regulated rivers in the United Kingdom. *Regulated Rivers: Research & Management* 2:201–220.

Postma, R. and van de Kamer, S. 1995. Ecological river restoration and floodings. *Proceedings from Nature Restoration in the European Union.* The National Forest and Nature Agency, Copenhagen, 56–62.

Pugh, K.B. 1997. Organizational use of the term 'freshwater quality' in Britain. In: Boon, P.J. and Howell, D.L. (eds) *Freshwater Quality: Defining the Indefinable?* The Stationery Office, Edinburgh, 9–23.

Raven, P.J., Fox, P., Everard, M., Holmes, N.T.H. and

Dawson, F.H. 1997. River Habitat Survey: a new system for classifying rivers according to their habitat quality. In: Boon, P.J. and Howell, D.L. (eds) *Freshwater Quality: Defining the Indefinable?* The Stationery Office, Edinburgh, 215–234.

Reynolds, T.S. 1983. *Stronger Than a Hundred Men. A History of the Vertical Water Wheel.* The Johns Hopkins University Press, Baltimore, Maryland.

Schlott, G. 1995. Restoration of a channelised stream: the Romaubach restoration project, Austria. In: Eiseltová, M. and Biggs, J. (eds) *Restoration of Stream Ecosystems: An Integrated Catchment Approach: A Training Handbook.* International Waterfowl and Wetlands Research Bureau, Slimbridge, 112–118.

Scott Porter Research & Marketing Ltd. 1998. *Re-introduction of the European Beaver to Scotland: Results of a Public Consultation.* Research, Survey and Monitoring Report 121, Scottish Natural Heritage, Edinburgh.

Shapiro, J., Lundquist, B. and Carlson, R.E. 1975. Involving the public in limnology – an approach to communication. *Verhandlungen der Internationalen Vereinigung für theoretische und angewandte Limnologie* 19:866–874.

Skjelkvåle, B.L. and Wright, R.F. 1990. *Overview of Areas Sensitive to Acidification: Europe.* Acid Rain Research, Report 20/1990. Norwegian Institute for Water Research, Oslo.

Sladecek, V. 1973. System of Water Quality from the Biological Point of View. *Ergebnisse der Limnologie, Archiv für Hydrobiologie,* Beiheft 7.

Stanners, D. and Bourdeau, P. (eds) 1995. *Europe's Environment: The Dobříš Assessment.* European Environment Agency. Office for Official Publications of the European Communities, Luxembourg.

Steinmann, P. 1907. Die Tierwelt der Gebirgsbäcke. Eine faunastisch–biologische Studie [The fauna of mountain brooks. A faunistic–biological study]. *Annales de biologie lacustre* 2:30–150.

Strahler, A.N. 1957. Quantitative analysis of watershed geomorphology. *American Geophysical Union Transactions* 38:913–920.

Tent, L. 1998. Reconstruction versus ecological maintenance – improving lowland rivers in Hamburg and Lower Saxony. In: Hansen, H.O. and Madsen, B.L. (eds) *River Restoration 96 – Session Lectures Proceedings.* International conference arranged by the European Centre for River Restoration. Ministry of Environment and Energy, National Environmental Research Institute, Silkeborg, Denmark, 170–174.

Umweltbundesamt 1994. *Daten zur Umwelt 1992/93.* Erich Schmidt Verlag GmbH & Co, Berlin.

United Nations 1991. *European Red List of Globally Threatened Animals and Plants and Recommendations on its Application as Adopted by the Economic Commission for Europe at its Forty-sixth Session (1991) by Decision D (46).* United Nations, New York.

Valeur, P. 1990. *Beverfamilien* (in Norwegian) [The beaver

family]. In: *Norges dyr, pattedyrene.* J.W. Cappelen, Oslo, 97–120.

Vannote, R.L., Minshall, G.W., Cummins, K.W., Sedell, J.R. and Cushing, C.E. 1980. The river continuum concept. *Canadian Journal of Fisheries and Aquatic Sciences* 37:130–137.

Walter, E. 1912. *Die Bewirtschaftung des Forellenbaches* [The biological communities of trout rivers]. Verlag J. Neumann, Neudamm.

Wassergütestelle Elbe 1990. *Nährstoffstudie der Elbe* [Nutrient studies of the Elbe]. Wassergütestelle Elbe, Hamburg.

Wayre, P. 1979. *The Private Life of the Otter.* Batsford, London.

Whelan, K.F. and Poole, W.R. 1996. The sea trout collapse 1899–92. In: Reynolds, J.D. (ed.) *The Conservation of Aquatic Systems.* Royal Irish Academy, Dublin, 101–109.

Wiberg-Larsen, P., Adamsen, N.B., Knudsen, J. and Larsen, F.G. 1991. Pesticides threaten Funen water courses (in Danish). *Vand & Miljø* 8:371–374.

Wiederholm, T. 1997. Assessing the nature conservation value of fresh waters: a Scandinavian view. In: Boon, P.J. and Howell, D.L. (eds) *Freshwater Quality: Defining the Indefinable?* The Stationery Office, Edinburgh, 353–368.

Williams, D. and Musco, H. 1992. *Research and Technological Development for the Supply and Use of Freshwater Resources.* SAST Project No. 6. Strategic Dossier. Commission of the European Communities, DGXII, Brussels.

Woodiwiss, F.S. 1964. The biological system of stream classification used by the Trent Water Board. *Chemistry and Industry*, March 1964, 444–447.

Wright, J.F., Armitage, P.D., Furse, M.T. and Moss, D. 1989. Prediction of the invertebrate communities using stream measurements. *Regulated Rivers: Research & Management* 4:147–155.

4

River conservation in central and eastern Europe (incorporating the European parts of the Russian Federation)

P.A. Khaiter, A.M. Nikanorov, M.G. Yereschukova, K. Prach, A. Vadineanu, J. Oldfield and G.E. Petts

Introduction

The area covered in this chapter includes the countries of central and eastern Europe, the post-Soviet states of Ukraine, the Baltic States, Belarus and Moldova, and the European part of the Russian Federation west of the Ural Mountain chain (Figure 4.1). Until recently these countries were united under communism and many of the attitudes and approaches towards river management and nature conservation were mirrored throughout the region. However, the societal changes that have taken place since the late 1980s have increased the internal heterogeneity. The pace of economic, social and political change has varied substantially from country to country. Furthermore, while many of the countries situated in central and eastern Europe (and the Baltic States) are now lining up to join the European Union, the post-Soviet countries of Russia, Belarus, Moldova and Ukraine have formed the Commonwealth of Independent States together with the majority of other former Soviet Republics.

The region is characterized by a number of large rivers, many of which flow through more than one country (Table 4.1). The Volga and Danube river basins are the dominant river systems possessing substantial catchment areas and flowing for thousands of kilometres. The Danube catchment area encompasses 13 European countries which presents a serious obstacle

to any effective management activity (Margesson, 1997). In contrast, the Volga catchment area lies almost entirely within the borders of the Russian Federation. Approximately 1300 km^2 (0.1%) of the Volga Basin is situated within Kazakhstan. The flow regime of these rivers is dominated by the spring–summer snowmelt flood.

The regional context

THE HISTORICAL CONTEXT

Many of the countries of central and eastern Europe and the former Soviet Union have a long history of environmental change, although the pace of change was generally more rapid during the 20th century. Ancient human populations prior to the Neolithic Age (*ca* 6500 BC) tended to use wooded river corridors only as migration route-ways and for fishing. Over time, permanent settlements were established and forests were cleared for agricultural land, but most river floodplains remained forested until the 9th century. During the Middle Ages, large portions of floodplains were converted from woodland to secondary grasslands and used for hay-making and occasional grazing. Large-scale deforestation of the surrounding uplands and mountains led to sedimentation dominated by silt and organic sediments, and to

Global Perspectives on River Conservation: Science, Policy and Practice.
Edited by P.J. Boon, B.R. Davies and G.E. Petts. © 2000 John Wiley & Sons Ltd.

Figure 4.1 *Major rivers of central and eastern Europe and the European parts of the Russian Federation (Yugoslavia now comprises Montenegro and Serbia). See text for case studies on the Ludnice River (Czech Republic) and the Lower Danube*

Table 4.1   *Major rivers within central and eastern Europe*

	Catchment area (km² × 10³)	Average annual discharge (m³ s⁻¹)	Length (km)	Country
Volga	1360	8000	3530	Russian Federation
Danube	817	6450	2850	Germany, Austria, Slovakia, Hungary, Croatia, Yugoslavia (Serbia-Montenegro), Romania, Bulgaria, Ukraine[a]
Dniepr	558	1660	2270	Russian Federation, Belarus, Ukraine
Don	422	930	1870	Russian Federation
Northern Dvina	358	3560	740	Russian Federation
Pechora	322	4060	1810	Russian Federation
Neva	281	2530	75	Russian Federation
Vistula	199	1100	1095	Poland
Oder	119	580	850	Czech Republic, Poland, Germany
Dniestr	72	310	1350	Ukraine, Moldova

Source: Stanners and Bourdeau (1995, pp. 76, 78), Margesson (1997, p. 146)
[a] riparian states

changes in river floodplain geomorphology and flood regime (Opravil, 1983). Nearly all rivers in central Europe have been regulated since the end of the 18th century. At first, large-scale regulation schemes were carried out – for example, on the Elbe, Danube and Morava Rivers. Recently, there were two periods of intensive river regulation, in the 1950s and the 1970s, when substantial areas of existing floodplains were converted to arable land. At the same time, eutrophication of the rivers and floodplain backwaters was accelerated by industrial and municipal waste water and run-off from over-fertilized fields (Drbal and Rauch, 1996). There were some early attempts (between the World Wars) to conserve both remnants of primary alluvial forests and the species-rich mosaic of secondary grasslands associated with oxbows and permanent or periodic pools (Marsakova *et al.*, 1977).

Eastern European rivers drain relatively arid basins dominated by continental climates, and in such environments river wetlands and deltas are especially important for birds and wildlife. For example, the Danube supports important breeding, migrating and wintering bird populations (about 275 species), including several globally threatened species, such as red-breasted goose (*Branta ruficollis*) and pygmy cormorant (*Phalacrocorax pygmens*). The Danube Delta is an important refuge for the European mink (*Mustela lutreola*), wild cat (*Felis sylvestris*) and otter (*Lutra lutra*) (Wilson and Moser, 1994). Similarly, the Don River ecosystem includes 130 species of waterfowl as well as two mammals (Russian desmon, *Desmana moschata* and European beaver, *Castor fiber*), three species of reptile, six species of amphibian and one

lamprey (*Lampetra mariae*). The delta regions of the Danube, Don and Volga provide nesting sites for 90% of the world's population of two pelican species (*Pelecanus crispus* and *P. onocrotalus*). Both birds are listed in the Red Data Books of the former USSR and Newly Independent Republics, and their populations in the Volga Delta have declined since the 1930s (Krivonosov *et al.*, 1994). Until the 1940s, pelicans also nested in the Dniestr Delta, but intensive fisheries, pollution, habitat destruction and hunting have caused their disappearance. In Poland, approximately two-thirds of the bird fauna (195 species) is associated with the Vistula (Kajak, 1992).

SOCIO-POLITICAL AND CULTURAL BACKGROUND

River management and conservation issues within the region must be placed within the context of societal change and transformation. The fall of communism initiated a drive towards liberal democracy in varying degrees throughout the region. The emergence of free market mechanisms and the increased openness towards foreign ideas, combined with the changing political and social situation, will all have a substantial impact upon future environmental policies. Nevertheless, the present period of transition is imbued with the legacies of the communist era. Approaches towards resource management cannot be changed overnight, especially considering the apparent ability of many former communist leaders and managers to maintain themselves in positions of power and influence. The environmental legacy of the region testifies to the often antagonistic relationship between communism and the wider environment (Weiner, 1990; McCannon, 1995). Water resources were not exempt from the aggressive approach of Soviet society. Within the Soviet Union many rivers were subject to extensive management strategies involving large reservoirs and hydroelectric production facilities (see 'Engineering Projects'). In addition, several immense water development schemes were formulated of which the Sibaral project is perhaps the best known (Pryde, 1991; Micklin, 1992; Peterson, 1993; Petts, 1994). The scheme proposed to reverse the flow of two large Siberian rivers in order to supply the southern dryland regions of Central Asia. This was to compensate for massive abstractions from the Amu Darya and Syr Darya to support the extensive cotton irrigation infrastructure within Turkmenistan and Uzbekistan. The political upheavals of recent years have resulted in the abandonment of the project.

The ongoing societal changes are helping to modify the prevailing attitudes within these countries towards environmental issues such as river conservation. The greater availability of data regarding river quality, and the growth of Non-Governmental Organizations (NGOs), has been instrumental in highlighting environmental issues for the public. Nevertheless, financial problems are proving a major obstacle to effective river management. The acute shortage of available finance for conservation projects in general is encouraging academics, scientists and government officials to be more receptive towards Western collaborative initiatives. As a consequence, foreign governments, lending institutions and NGOs are becoming increasingly involved in river conservation and management issues. This activity provides a conduit for Western technical know-how and can therefore play an influential role in the resultant conservation practices adopted within the region. For example, the World Wide Fund for Nature (WWF) has established a Danube programme (Green Danube) with the aim of protecting ecologically important sections of the river. A considerable amount of attention has been focused on the delta region of the river and WWF has joined forces with both Romanian and Ukrainian NGOs in order to protect the area's future. In addition, WWF has set up conservation projects in countries such as Bulgaria and Slovakia. The European Bank for Reconstruction and Development (EBRD) has provided financial assistance in a number of cases in order for improvements to be made to water treatment infrastructure. A great deal of attention has been directed at the level of pollution within the Baltic Sea basin. One of the EBRD's current environmental

infrastructure projects aims to improve the wastewater treatment facilities of Kaunas in Lithuania (EBRD, 1995a). This project will help significantly to reduce pollution discharges into the Neumunas River on which the city is situated. Similar investment projects have also been initiated in Estonia (EBRD, 1995b) and the Latvian capital Riga (EBRD, 1997). Riga is responsible for approximately 60% of the country's overall municipal water pollution and a high percentage of this is discharged directly into the adjacent Daugava River (EBRD, 1997).

DEVELOPMENT PRESSURES AND HUMAN IMPACTS

Many of the contemporary environmental problems associated with the rivers of the region can be traced back to the communist era. The precise dates for this period vary throughout the region. The communist regime within the former Soviet Union was in power from 1917 to 1991; in contrast, the communist regimes within central and eastern Europe were not established until after World War II.

Water quality

The extensive industrial development policy characteristic of communism encouraged a range of pollution problems and a significant literature has built up dealing with the resulting environmental legacy (Pryde, 1991; Manser, 1993; Peterson, 1993; Jancar-Webster, 1995; Carter and Turnock, 1996). Industry and population centres proved substantial polluters. The available water treatment infrastructure tended to be of poor quality and often performed inefficiently. As a consequence, the transition countries have inherited a wastewater treatment infrastructure in serious need of repair and replacement. Unfortunately, these countries are in no position to invest large amounts of capital into wastewater projects. Water and sewage works continue to deteriorate and this results in greater pressures being placed on local river systems. For example, in the Russian Federation it is estimated that 60% of the existing sewage treatment system is overloaded and approximately 40% of installations are in need of repair (*Zelenyi Mir*, 1997b).

The legacy of contaminated land and polluted groundwater is now a major problem for rehabilitation schemes (Somlyody, 1994). At the present time more than 50% of the length of many rivers belongs to the poorest water quality class. High levels of Biochemical Oxygen Demand (BOD), bacterial contamination,

Table 4.2 *Status of the Volga reservoirs based on routine observations over the period 1979–1991 obtained by the State Monitoring System of Russia*

Reservoir	Phytoplankton change[a]	Zooplankton change[b]	Oligochaeta change[a]	Concentration of total Phosphorus[c], Nitrate-N[d], Ammonium-N[e]		
				P[c]	Nitrate-N[d]	Ammonium-N[e]
Ivankovo	+					
Uglich	+					
Rybinsk	–		+			High
Gor'kiy	–	A	+	High		
Cheboksarsk	+					High
Kuybyshev	–	C	+	High	High	
Saratov	–	B,C	+	High	High	High
Volgograd	–		+	High		High

Sources: Bryzghalo and Khaiter (1993), Khaiter (1995)
[a] Change of numbers
[b] A = decrease of species diversity; B = intensive development of Rotatoria; C = intensive development of Cladocera
[c] Total phosphorus. High = >0.60 mg L^{-1}
[d] Nitrate – N. High = >2.00 mg L^{-1}
[e] Ammonia and nitrogen. High = >2.00 mg L^{-1}

ammonia and nutrients, and serious heavy metal contamination are common problems (Somlyody, 1994). For example, the Volga is classified as 'polluted' from its source to the delta with high levels of BOD, copper and phenol. Ecological degradation of the Volga's reservoirs is illustrated in Table 4.2. The changes include an overall reduction in species diversity and the development of an exceptionally high abundance of a few common species. For example, the phytoplankton became dominated by blue-greens and the zoobenthic community by oligochaetes. Similar studies of the 22 largest water bodies of Russia, Ukraine, Belarus and the Baltic States provided further evidence of widespread ecological degradation related to pollution (Bryzghalo and Khaiter, 1993).

Engineering projects

Large-scale engineering projects were very much a feature of the communist period. Within the Soviet Union large sections of the Volga and other rivers were transformed into reservoirs resulting in the loss of large areas of land (Pryde, 1972). For example, dam construction on the River Volga between 1937 and 1981 turned it into a series of eight shallow lakes between Rybinsk and Volgograd, facilitating navigation and hydroelectric power production. The lakes have a combined volume of more than 150 km^3 and a surface area of nearly 20 000 km^2. This type of activity can have an immense impact on the functioning of a river system (Petts, 1984). The most valuable fish species, such as sturgeon (*Acipenser sturio*), salmon (*Salmo salar*), sea trout (*Salmo trutta* m. *trutta*) and vimba (*Vimba vimba*), have disappeared or at least decreased significantly from major rivers in the region as a result of pollution and dam construction. The Danube still supports more than 75 species of fish belonging to 22 families. However, sturgeon has virtually disappeared and the annual catch of *Acipenser sturio* declined from 1000 t in the early parts of the 20th century to 20 t in 1989. On the Don, the Tsimlyansk Dam eliminated 80% of the traditional spawning habitats of the Azov Russian sturgeon (*Acipenser guldenstadti tanaicus*) and vimba (*Vimba vimba natio carinata*) and 100% of the spawning grounds of the Azov great sturgeon (*Huso huso maeoticus*), the population of which has crashed to 5–15% of the levels in the early 1950s. Many fish are contaminated with pollutants such as heavy metals and DDT. The economic losses from pollution, fisheries decline and a reduction in tourism revenue is estimated at over US$500 million (Wilson and Moser, 1994).

Although it is widely recognized that dam construction causes problems, it continues to be an issue in the region. A case in point is the Gabcikovo–Nagymaros Dam project situated on the Danube (Boucher, 1990). This was to be a joint project between Czechoslovakia and Hungary until Hungary pulled out towards the end of the 1980s, largely in response to public protest at the ecological repercussions of the scheme (Jancar-Webster, 1995; WWF Press Release, 18

February 1998). However, Slovakia continued to work on the Gabcikovo Dam part of the project and its subsequent opening in 1992 has encouraged Hungary to consider restarting construction work on the Nagymaros Dam. In addition to the attendant environmental ramifications of such a project, the dam would also be situated within the newly created Danube–Ipoly National Park (WWF Press Release, 18 February 1998). There is a great deal of opposition to the proposed dam construction at both the local and international level; indeed, the implementation of the project may jeopardize Hungary's attempt to join the European Union (WWF Press Release, 18 February 1998).

MAIN AREAS AT RISK

It is a commonly held perception that the current period of societal transformation should help to reduce the pollution output of the countries within the region (World Bank, 1996). Societal restructuring has resulted in substantial falls in the level of both industrial and agricultural activity throughout the former communist countries. This has led to a decline in levels of drainage discharge and therefore reduced the pollution load in many of the region's rivers. The decline of industrial pollution and related wastewater discharges has improved the environmental state of some heavily polluted rivers. Large increases of both fertilizer and pesticide prices have also resulted in reduced levels of application. This has led to a concomitant reduction in non-point-source pollution (Kindler, 1994). As a consequence, a number of rivers have registered an improvement in quality. Similarly, falling levels of economic activity within the former Yugoslavia, caused by a combination of war and UN sanctions, have encouraged a concomitant improvement in the water quality of rivers within the country (Krizan and Vojinovic-Miloradov, 1997).

It would be wrong to suggest that the overall pressure upon water resources has declined drastically in response to the fall in levels of economic activity. This is due to a number of factors. The relative scarcity of available investment is leading to a deterioration in water treatment infrastructure. Furthermore, there is evidence of a relaxation in the application of environmental procedures and regulations within some countries in response to the economic and legal confusion generated by the transition period. Within the Russian Federation the majority of large rivers remain heavily polluted (*Zelenyi Mir*, 1997a). Rivers running through agricultural regions are subjected to

Table 4.3   *Changes in the volume of drainage discharge emissions, Moscow city (m³ × 10⁶)*

	1992	1993	1994	1995	% change 1992–95
Drainage discharge	2991	3032	2889	2809	–6.1
Including:					
Industrial enterprises and the energy sector	704	652	545	432	–38.6
Municipal economy	2287	2380	2344	2377	+3.9

Source: Moskompriroda (1996, p. 112)

discharges of polluted run-off, sediment deposition resulting from soil erosion, and infringement of the water protection zones surrounding water bodies. Unfortunately, during the Soviet period, many farms were located within these zones and are therefore able to exercise a considerable influence on the adjacent water source. In addition, small rivers situated near industrial centres continue to be placed under substantial pressure from production and municipal waste. Effluent from municipal sewage systems changed little during the 1990s despite the socio-economic upheavals. Indeed, in Moscow city available data suggest that the volume of municipal drainage water discharged to surface waters has actually increased slightly in recent years. This rise has partially compensated for the noted fall in discharge levels attributable to industrial enterprises and the energy sector (Table 4.3).

CATCHMENT MANAGEMENT VERSUS LOCAL ISSUES

The effective management of a river system requires the whole catchment area to be taken into consideration. The complexity of the region under study ensures that management at the catchment level is difficult to administer because many river systems cross several international borders. The Russian Federation provides the one real exception since a number of large European rivers are contained entirely, or almost entirely, within its borders. Nevertheless, Russia is struggling to come to terms with its new Federal structure and the automatic compliance of individual Federal regions is not guaranteed. The Volga Basin unites 38 of the country's 89 Federal units, eight of which have Republic status (*Ekos-Inform*, 1996, p. 19).

Considering the poor environmental record of the former communist countries it is often assumed that effective environmental management approaches were rarely practised. It is certainly true that conservation aims were often secondary to considerations of productive potential. However, it would be a gross oversimplification to argue that more holistic management approaches were not practised. During the Soviet period, the constituent Republics of Ukraine and Moldova cooperated over the regulation of the Dniester River Basin (Jancar-Webster, 1995). In addition, the Danube Commission provided a point of focus for the activities of the riparian states over a number of decades (Margesson, 1997). Cooperation between states has, if anything, increased in recent years. For example, the Elbe Commission was established in 1990 to address the marked environmental problems of the Elbe River (Jancar-Webster, 1995). A number of initiatives have been directed towards the Danube in recognition of its substantial importance within the region (Margesson, 1997). Furthermore, river management issues are increasingly being considered in the context of the wider environment. For example, the Danube is recognized as an integral part of the Black Sea water system. In a similar vein, the pollution levels of the Vistula and Oder Rivers are considered in relation to the Baltic Sea environment (Kajak, 1992; Jancar-Webster, 1995).

Unfortunately, the existing socio-economic and political situation is hardly conducive to effective cooperation in catchment management. Other issues, such as the unrest in the former Yugoslavia, present further barriers to effective action.

Information for conservation

SCIENTIFIC DATA AND EVALUATION SCHEMES

During the communist period, many of the countries of central and eastern Europe and the former Soviet Union operated a reasonably comprehensive environmental monitoring system (Carter and Turnock, 1996). This is reflected in the extent of the post-Soviet monitoring networks. For example, in 1994 the State Monitoring System of Russia had a network of 1172 hydrochemical monitoring stations located on rivers throughout the country. Hydrological and hydrophysical characteristics such as flow rate and temperature were assessed in addition to an extensive range of hydrochemical and hydrobiological traits. These observations were carried out along more than 1100 rivers (including both European and non-

European parts of the Russian Federation). Rivers were then graded according to the level of pollution. For example, in 1996 the River Neva, located in the north-west region of European Russia, was classed as moderately polluted according to the national index of pollution, whereas the basin of the River Dniepr ranged from weakly polluted to polluted (*Zelenyi Mir*, 1997a). The monitoring network includes records at some stations dating back to 1936–37. Since 1974, these data have been complemented by the National Hydrobiological Network of Observations which includes about 25% of the sites monitored by the Hydrochemical Network. The list of criteria recorded by these networks is given in Table 4.4. Table 4.2 provides an illustration of how the data are applied in order to classify levels of ecological degradation. Other data include regular bird counts (e.g. Krivonosov *et al.*, 1994), especially of Red Data Book species. However, during the communist periods, environmental data were rarely released for general viewing and much of the information that was generated remained 'in-house'.

Environmental data have become more openly available since the collapse of the Soviet Union and the fall of communist regimes throughout central and eastern Europe. State statistical bodies of former Soviet Republics such as Russia, Ukraine, Belarus and Moldova publish details of water use and drainage discharge together with a breakdown of major pollutants. Furthermore, the previously extensive Soviet monitoring system has been largely maintained and this enables a wide range of data to be collected.

Nevertheless, channels for the dissemination of data remain poor and a substantial volume of data is therefore under-utilized. Data quality continues to be a major problem within the region. The equipment and standards used for monitoring can vary, not only from one country to another, but also within countries. In many cases the existing environmental monitoring systems are in need of modernization and restructuring work, a situation which is aggravated by financial difficulties. Russian water quality data are often presented in relation to the accepted norms rather than absolute values (International Monetary Fund – IMF *et al.*, 1991). In addition, accepted quality norms do not necessarily coincide with those used in the West, and this can make comparative work extremely difficult. Those countries wishing to join the European Union obviously have a greater incentive to bring their assessment procedures into line with those of western Europe.

Table 4.4   *Characteristics recorded by (a) hydrochemical and (b) hydrobiological networks established by the State Monitoring System of Russia (a) since 1936–37 and (b) since 1974*

(a)   Hydrochemical

Hydrological
- Water discharge ($m^3 s^{-1}$)
- Flow ($m s^{-1}$)

Hydrophysical
- Temperature (°C)
- Transparency (cm)
- pH
- Redox potential (Eh)

Hydrochemical
- Concentration of dissolved gases: O_2, CO_2 ($mg L^{-1}$)
- Concentration of suspended matter ($mg L^{-1}$)
- Concentration of main ions (Cl^-, SO_4^{2-}, HCO_3^-, Ca^{2+}, Mg^{2+}, Na^+, K^+) and a sum of ions ($mg L^{-1}$)
- COD ($mg L^{-1}$)
- BOD ($mg L^{-1}$)
- Concentration of nutrients: NH_4^+, NO_2^-, NO_3^-, phosphates, Fe_{tot}, silicon ($mg L^{-1}$)
- Concentration of pollutants: oil ingredients, phenols and heavy metal compounds ($mg L^{-1}$)

(b)   Hydrobiological

Phytoplankton
- Total number of cells (10^3 cells cm^{-3} or cells ml^{-1})
- Total number of species
- Total biomass ($mg L^{-1}$)
- Number of the primary groups of biological species (10^3 cells cm^{-3} or cells ml^{-1})
- Biomass of the primary groups ($mg L^{-1}$)
- Abundant species and 'key' or indicator species (%)

Zooplankton
- Total number of organisms (specimens m^{-3})
- Total number of species
- Total biomass ($mg m^{-3}$)
- Number of the primary groups (specimens m^{-3})
- Biomass of the primary groups (specimens m^{-3})
- Number of species in a group
- Abundant species and 'key' indicator species (%)

Zoobenthos
- Total number (specimen m^{-2})
- Total biomass ($g m^{-2}$)
- Total number of species
- Number of species in a group
- Number of the primary groups (specimen m^{-2})
- Biomass of the primary groups ($g m^{-2}$)
- Abundant species and 'key' or indicator species (%)

Periphyton
- Total number of bacteria (10^6 cells cm^{-3} or cells ml^{-1})
- Number of saprophyte bacteria (10^3 cells cm^{-3} or cells ml^{-1})

Intensity of phytoplankton photosynthesis and destruction of organic matter
- Photosynthesic intensity according to oxygen and carbon units ($mg L^{-1} d^{-1}$)
- Destruction of organic matter according to oxygen and carbon units ($mg L^{-1} d^{-1}$)
- Photosynthetic intensity – organic matter destruction ratio
- Chlorophyll content ($mg L^{-1}$)

Macrophytes
- Projective covering ($100 m^2$)
- Total number of species
- Dominant species

Toxicological indicators
- Toxic biotesting using *Daphnia*
- 50% mortality of *Daphnia*
- Biotesting using phytoplankton (50% reduction of cells)

PUBLIC AWARENESS AND EDUCATION

During the communist period public awareness of environmental issues was limited owing to the lack of available information. Nevertheless, environmental problems played a not inconsiderable part in the fall of the political regimes within the region. Public attention focused primarily on major environmental issues such as the Chernobyl explosion, the fouling of Lake Baikal (East Siberia) and high pollution levels along the Volga and Danube Rivers. The greater freedom afforded to the media and local NGOs in recent years has certainly helped to increase public awareness of environmental issues in general. For example, NGOs such as the International Association for Danube Research, the Association of Danube Cities and the World Wide Fund for Nature are all involved in conservation work within the Danube Basin.

The communist regimes of central and eastern Europe and the former Soviet Union paid little attention to environmental education. As a result, although the countries possess a wealth of technical expertise in issues of environmental management, they tend to lack personnel with strengths in resource management and conservation issues. The transition period has seen an improvement in the status of environmental education. The Russian government, for example, has made a concerted effort to integrate environmental issues into the existing school curriculum (Ministry for Environment Protection and Natural Resources of the Russian Federation – MEPNR, 1994). Nevertheless, the prevailing socio-economic situation within the region ensures that environmental issues are often given a low priority both by governments and the general public.

The statutory framework for conservation

The framework for environmental law is in a state of flux throughout the region (Jancar-Webster, 1995). Environmental laws have been rewritten and modified in order to take account of the new social, political and economic situation. A major environmental law was formulated towards the end of the Soviet period in the Russian Federation and this has encouraged the emergence of a host of related laws. Many of the countries within the region have concentrated on bringing their environmental regulations and standards into line with those of the European Union in preparation for them becoming member states. Nevertheless, the difficulties of transition ensure that

economic, social and political concerns often take precedence (Jancar-Webster, 1995).

International environmental management programmes play an important role in stimulating, directly and indirectly, improvements in river quality from both chemical and biological perspectives. For example, the Helsinki Commission has organized a programme for Baltic Sea protection (Tonderski *et al.*, 1994) which is seeking to reduce nutrient loads from major rivers such as the Vistula by controlling point-source wastewater discharges. The United Nations Development Programme/United Nations Environment Programme (UNDP/UNEP), World Bank, European Community and national governments of central and eastern Europe are cooperating in an Environmental Programme for the Danube River Basin (Rodda, 1994; Margesson, 1997). This also includes a 300 000 ha Biosphere Reserve within the Romanian part of the Danube Delta which has been registered as being of international importance and included on the World Heritage List. The Ukrainian part of the Danube Delta (the Dounaiski plavni) is currently subject to a US$1.5 million grant to establish a Biosphere Reserve (Wilson and Moser, 1994). These initiatives obviously encourage a more holistic approach towards the Danube's water management issues.

Conservation in practice

In western Europe, the restoration of river systems generally followed the sequence: (i) pollution control for public health; (ii) fish stocking; (iii) water quality improvement; (iv) physical habitat management; and (v) water resource allocation to meet ecological needs. In common with developing nations, the countries of central and eastern Europe and the former Soviet Union have only recently come to address issues of environmental degradation, as public health concerns have generally been the top priority. In several of these countries the *per capita* cost of controls to meet surface water quality standards would exceed the *per capita* gross domestic product! One example of a positive development is the least-cost water quality control programme for the Nitra River, Slovakia (Somlyody *et al.*, 1994) which receives numerous municipal and industrial discharges and has a low level of wastewater treatment.

A first step in the process of river restoration is to classify river reaches in relation to impacts and realistic targets. Within the region under study, the water quality criteria are often based on hydrochemical characteristics (L'vov and Kuzin, 1987) as considerable effort has been put into developing water quality models (Strashkraba and Gnauk, 1985). However, countries also utilize hydrobiological approaches (Schiemer and Waidbacher, 1992; Welcomme, 1992). Thus, Abakurnov (1979) and Abakurnov and Sushenya (1991) applied the concept of ecological modification which uses relationships between biodiversity, structural and functional organization of aquatic communities and the stability of hydro-biocoenoses (Alirnov, 1982; Magurran, 1988) to classify rivers and to define their ecologically permissible state (Izrael *et al.*, 1981). Such an approach has been used, for example, on the Volga (Table 4.2).

The general confusion and dislocation which has accompanied the structural transformation of countries in central and eastern Europe has had a significant impact on nature conservation work within the region. Rivers such as the Lower Dniepr, Northern Dvina, Pechora and Neva are at present unprotected. In contrast, the Volga Delta forms part of the 63 400 ha Astrakhan State Reserve, one of the oldest protected areas in the former USSR, established in 1919. A large area of the River Don Basin is incorporated into four wildlife conservation areas: Voronezh State Reserve (31 053 ha), Hoper State Reserve (16 178 ha), Rostov State Steppe Reserve and the Don Fish Sanctuary (68 000 ha, which includes 28 000 ha of the Don Delta and 40 000 ha of the eastern part of the Taganrog Bay of the Sea of Azov). Other rivers, such as the Dniestr Delta, have been the subject of conservation proposals. The recreational and wildlife characteristics of the Middle and Lower Vistula have prompted the concept of a Vistula Landscape Park. This would help to balance the needs of wildlife conservation with the pressures of recreation activities and tourism. The Landscape Park would incorporate a range of other nature protection areas, such as national parks, reserves, areas of protected landscape and interconnecting corridors.

CASE STUDY 1: THE LUDNICE RIVER, CZECH REPUBLIC

The Ludnice River on the border between Austria and the Czech Republic (Figure 4.1) is one of the last rivers in central Europe to remain in a semi-natural state. The wet, clayey, nutrient-poor soil coupled with the marginal geographical position of the river ensured that the whole countryside was colonized later (in the

13th century) and less intensively than surrounding regions. Thus the human impact on the river and its floodplain started later and was not so intensive as elsewhere. Declaration of the surrounding area of the Trebon Basin (*ca* 700 km^2) as a Biosphere Reserve under the UNESCO scheme in 1977 and as a Protected Landscape Area by a local law in 1979, helped to stop many attempts in the late communist era to regulate the remaining untouched parts of the river. At present, there are about 30 km of unregulated river. The total length of the river is 200 km and its altitude ranges between 347 and 990 m above sea level. The average discharge at the mouth of the river is 24.3 m^3 s^{-1}. The river forms part of the Elbe River system and has a drainage area of 4225 km^2. Thus the river is small in comparison with others in central Europe. The best-preserved part of the Ludnice River floodplain was subjected to an intensive ecological study between 1986 and 1993, when geomorphology, hydrology, soils, nutrient balance, biotic communities and human impact were investigated (Prach *et al.*, 1996). Results of the project helped to determine which parts of the river to conserve or restore. Importantly, the study defined the scientific requirements needed to underpin a conservation/restoration strategy specifically highlighting the need to understand rates of successional transitions.

Management of the Ludnice floodplain

A management scheme was constructed on the basis of: (i) comparisons of floodplain segments having been abandoned for different periods ranging from 5 to 50 years; (ii) personal observations by Karel Prach over a period of 20 years; (iii) investigations using permanent plots and transects between 1986 and 1996; and (iv) experimental re-establishment of the regular cutting regime in part of the floodplain, previously neglected for more than 20 years. The dynamics of the grassland vegetation in the Ludnice floodplain are summarized in Figure 4.2. This compares vegetation units under regular management with other units totally neglected for a period of nearly 50 years in relation to water-table levels and nitrogen concentrations. There is an obvious difference in species diversity between managed and neglected meadows; diversity is also seen to decrease towards wet and nutrient-rich sites. The opposite trends are seen in the case of biomass production. However, regularly mown *Alopecurus* meadows exhibit relatively high production especially if cut two or three times a year. Such management techniques can have economic benefits too (Kvet *et al.*, 1996).

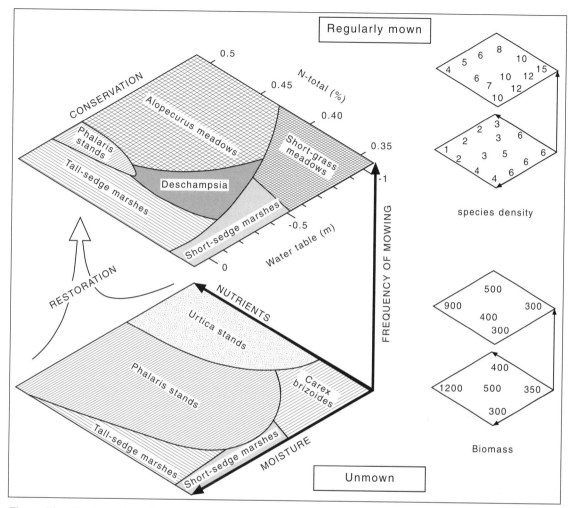

Figure 4.2 *The dynamics of grassland vegetation in the Ludnice floodplain comparing neglected vegetation units (unmown) with those under regular management (mown), and including the number of species and biomass (g m^{-2})*

Results of the management scheme demonstrated the rapid restoration of the degraded alluvial meadow. In only four years, the previous dominants *Phalaris arundinacea* (nearly) and *Urtica dioica* (completely) disappeared, while typical meadow species expanded. The total number of species increased from 28 to 79, and average species density increased from 4.0 to 8.2 m^{-2}. There are apparently two main causes of the rapid restoration: (i) species typical of regularly cut meadows survived, at least in some cases, during the degraded stages; and (ii) transport of diaspores was helped by floods from nearby sources upstream, where regularly cut meadows still exist. If these two conditions are ensured in any given floodplain, the restoration of degraded meadows may be comparatively easy. This issue was discussed in relation to other rivers within the country in Straskrabova *et al.* (1996).

As a result of the major research project, the Ludnice River floodplain is probably the best known floodplain in central Europe with regard to its ecological functioning. It is useful, therefore, to try to apply the results of the research to other river corridors. The main conclusions (Table 4.5) seem to be generally valid (see also Straskrabova *et al.*, 1996).

Conservation of the Ludnice River

The first nature reserve protecting a part of the Ludnice River was declared in 1956. The stretch of river

Table 4.5 *Scientific benefits of the Ludnice River floodplain studies*

Effective conservation and restoration schemes must:

i. preserve/restore natural water-table fluctuations;
ii. preserve/restore physical habitat diversity (e.g. meanders, oxbow lakes, permanent and ephemeral floodplain pools);
iii. conserve the connectivity between river and floodplain which supports the self-purification of the river, as well as high biotic diversity;
iv. ensure that three main environmental gradients (moisture, nutrients and disturbance intensity, see also Day *et al.*, 1988) interact to determine biotic communities;
v. recognize that nutrient enriched or contaminated floodplains require special consideration e.g. as in the Ludnice floodplain.

Selected findings relevant to (v) are:

● floodplains are nutrient sinks; high levels of nutrients transported by the river will not decline rapidly;
● natural succession to alluvial forest will be restricted because establishment of woody species is limited by competition from the herb layer under high nutrient levels;
● artificial maintenance is necessary – meadows must be regularly cut to avoid rapid degradation into monotonous swards of competitively strong herb species; regular cutting is also the best way to prevent invasions of alien organisms (Pysek and Prach, 1993; de Waal *et al.*, 1994);
● rates of successional transition between vegetation types must be quantified to define the most appropriate management practice;
● large-scale restoration of the natural alluvial forest by succession is impossible.

situated within the Trebon Biosphere Reserve received some protection from the late 1970s onwards. However, during the communist period, this protected status did little to prevent the channelization of the river, the extraction of sand and gravel from the river channel, or the conversion of meadows to arable land. In addition, other meadows were poorly managed or abandoned. During the early 1990s two sections of the floodplain were declared as nature reserves and this included the area where the research project was carried out. Furthermore, many valuable parts of the floodplain located within the Czech Republic were included in a system of Important Landscape Elements and highlighted in Territorial Systems of Ecological Stability (TSES). The TSES included land-use analysis, evaluation of individual segments and the identification, protection and suitable management of the most valuable ecosystems within the various regions of the country. As a consequence of

the improved management of the floodplain region, the threat of channelization has been removed and the volume of nutrients has tended to fall in recent years (Drbal and Rauch, 1996). At present, a range of activities is prohibited within the floodplain. These include conversion to arable land; the use of pesticides, artificial fertilizers and slurry wastes; new building activities; and the introduction of non-native plants and animals. However, the present nature conservancy faces other problems, the most serious of which concerns the decline of hay-making activity within the region and degradation of grasslands. Unfortunately, the floodplain has little economic value at present and therefore government money is required in order to stimulate proper management within the floodplain.

CASE STUDY 2: THE LOWER DANUBE RIVER SYSTEM

The Danube catchment of 817 000 km^2 is an international basin, draining all or part of 13 countries in central and eastern Europe (Figure 4.1, Figure 4.3). Annual run-off of 200 km^3 yr^{-1} (a mean flow of about 6000 m^3 s^{-1}) represents 65% of the freshwater input to the Black Sea. The lower floodplain of the Danube is one of the largest and most complex freshwater systems in Europe (Figure 4.3, Table 4.6). This reach is very sensitive to catchment processes and provides an important control on the Danube Delta into which it feeds (Banu, 1967; Ianovici *et al.*, 1969; Cousteau, 1993; Gastescu, 1993).

Conservation areas

The Lower Danube River System (LDRS) is one of the most important wetland areas of Europe because of its ecological value, considerable surface area (82% in Romania and 18% in the Ukraine) and geographical location. The delta is a heterogeneous and dynamic complex of natural and semi-natural ecosystems of various types and in different successional stages. For example, it includes lakes, swamps, channels, river branches, reed wetlands, grasslands, dunes and forests (oak and older willow), as well as human-influenced and controlled ecosystems (30%) based upon activities such as agriculture, forest plantations (mainly poplar) and fish farming. Upstream of the delta system, along a stretch of more than 700 km, the Danube River has developed an extensive wetland system of approximately 500 000 ha. More than 83% of this system is located within Romania and is characterized by a diverse array of natural formations such as river islands, lakes,

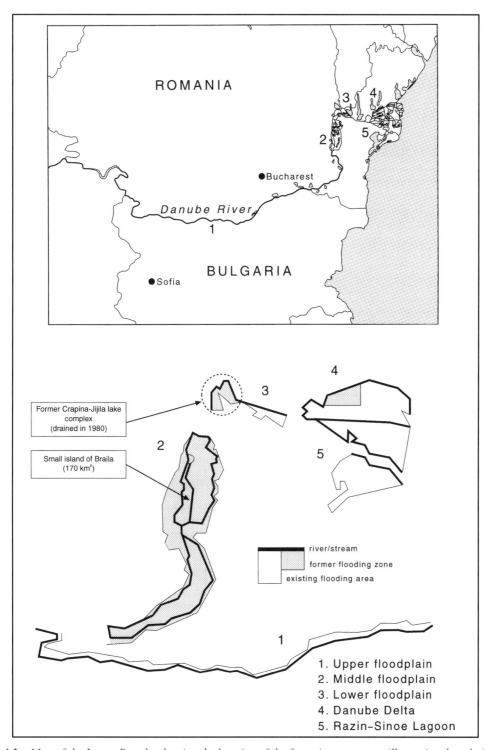

Figure 4.3 *Map of the Lower Danube showing the location of the five primary sectors, illustrating the relative areas of former and existing floodplains*

Table 4.6   *The primary hydrogeomorphological units of the Lower Danube River System (see Figure 4.3)*

Characteristic	1. Upper Floodplain	2. Middle Floodplain	3. Lower Floodplain	4. Danube Delta	5. Razim–Sinoe Lagoon
Position (river km)	840–370	370–170	170–70	70–0	
Surface area (km²)	2220	3250	525	2544	1010
Altitude (m)	49–13	13–7	7–5	5–0	15–3
Water area (%)	8.4	7.6	39.3	68.3	75
Forestry (%)	10.4	8.7	11	4.7	0
Pasture (%)	3.2		10.4	6.2	7.6
Agriculture (%)	68.7	76.8	30.4	17.1	9
Other uses (%)	9.3	6.9	8.9	3.7	8.4

marshes, ponds, channels, oxbows, grasslands and alluvial forest (both natural and semi-natural). A high proportion of this area (80%) is dominated by human activities: crop management (64%) and wood production (8%) as well as fish farming (3%) and pastures (12%).

The ecological structure of the LDRS has experienced three main waves of human-induced structural changes in the last century:

1. Large river engineering works to improve the navigation potential and to allow maritime shipping to gain access to large river harbours such as Reni, Galati and Braila. These changes were extended during the early decades of the 20th century by additional waterways for improving freshwater circulation mainly within the delta and the Razim–Sinoe Lagoon.

2. The second phase of structural changes was implemented during the 1960s and 1970s. River engineering works were carried out in order to establish large enclosures within the delta where reed growing and harvesting or fish breeding could be managed in order to increase the level of productivity. In addition, land reclamation activity was carried out higher up the river's course.

3. The third phase of structural changes was partially accomplished during the 1980s with an increase in the area of polders in the delta from 3 to 25% and the decline in natural or semi-natural areas from 88 to 62%. Fish farming and forestry also showed slight increases.

The technical and financial efforts were mostly focused on land reclamation for intensive agricultural use, fish farming, cattle-breeding and poplar plantations within the Danube Delta. The scale and the rate of land reclamation for crop production and forest plantations have been almost identical on the Bulgarian side of the LDRS. However, there was a very

clear imbalance between the target set and the scale of the actual changes. Fortunately, the established parameters were not achieved before the very significant political and strategic changes that occurred at the end of 1989.

Even after the long history of human-induced structural changes in the LDRS, the delta system still possesses many of the structural and functional characteristics of the natural system. In addition, a series of relatively small and disconnected natural and semi-natural wetlands survived, mainly along the upstream stretch of the river. Among the remnant parts of the upstream Danube Delta wetlands, the Small Island of Braila (Figure 4.3) plays an important role in the actual and potential rehabilitation of the LDRS. An analysis of historical data and of information provided by the ongoing process of updating and improving the description of biological diversity, shows clearly that the delta system contains very rich plant and animal communities (Table 4.7). Despite severe fragmentation, the species richness and habitat diversity proved to be well preserved in the remaining upstream wetlands (Baboianu, 1998; Cristofor *et al.*, 1998; Vadineanu *et al.*, 1998).

The Small Island of Braila, a relatively small wetland of 23 000 ha, contains 14 types of habitat listed in the EU Habitats Directive, 34 bird species listed in the EU Birds Directive and 16 other wildlife species listed in both the Habitats Directive (Annex II) and the Berne Convention. However, it is important to realise that since 1938 the process of anthropogenic change has been accompanied by a series of specific measures to conserve some rare or endangered species (e.g. spoonbill – *Platalea leucorodia*; shelduck – *Tadorna tadorna*; great white egret – *Egretta alba*; white pelican – *Pelecanus onocrotalus*; Dalmatian pelican – *Pelecanus crispus*; black-winged stilt – *Himantopus himantopus*; avocet – *Recurvirostra avosetta*; mute swan – *Cygnus olor*). Some characteristic terrestrial and aquatic ecosystems within the Danube Delta,

Table 4.7 *Species richness in the Danube Delta (from Baboianu, 1998)*

Category	Total	New for DDBR[a]	New for Romania	New for science
Flora				
Algae	562	2	–	–
Fungi	47	5	–	–
Marine fungi	14	–	14	–
Cormophyta	945	409	14	–
Total flora	1668	416	28	–
Fauna				
Invertebrates				
Worms	446	101	37	4
Molluscs	106	–	–	–
Spiders	240	66	17	1
Crustaceans	146	10	–	–
Myriapods	34	11	1	–
Insects	2419	312	48	14
Total invertebrates	3391	500	102	19
Vertebrates				
Fish	82	–	2	1
Amphibians	9	1	–	–
Reptiles	12	–	–	–
Birds	200	–	–	–
Mammals	41	–	–	–
Total vertebrates	344	1	2	1
Total fauna	3735	501	104	20

[a] DDBR – Danube Delta Biosphere Reserve

Table 4.8 *Environmental change along the Lower Danube (data from Gastescu, 1993; Bondar, 1994; Vadineanu et al., 1998)*

Channel works

Flood attenuation capacity reduced by 4.5 km^3 as a result of regulation (embanking)

Two large reservoirs built in the lower floodplain sector

Danube–Black Sea Canal cut connecting the Middle floodplain sector to the sea, diverting a daily flow of about 320 000 m^3

Land-use change

80% (3500 km^2) of the former floodplain drained

15% of the delta changed to polders (36 000 ha) and fish ponds (32 000 ha)

60% of the former wetlands along main tributaries reclaimed for agriculture

Water quality

Transport of suspended solids decreased from 67.5×10^6 t yr^{-1} at about 1900 to 53×10^6 t yr^{-1} in the 1950s and to 30×10^6 t yr^{-1} in the late 1980s

Rapid increase of nutrient loads from 5–30 μg P L^{-1} and 1200 μg N L^{-1} in 1980 to 70–280 μg P L^{-1} and > 2200 μg N L^{-1} in 1990

Slow increase of other contaminants including heavy metals and pesticides

which in 1962 covered almost 41 500 ha, were also protected. It was in this context that the first Romanian Biosphere Reserve was established in 1979 in the north-eastern part of the Danube Delta – the 'Rosca–Letea'.

The profound changes in the spatio-temporal organization of the LDRS (and, in particular, of the upstream delta sector) which occurred before 1990, have led to a series of dramatic environmental and ecological changes. The buffering capacity of the whole system has been reduced, spawning and feeding areas for many migratory fish species have been lost and the annual fish yield has declined from $15–10 \times 10^3$ t yr^{-1} to less than 5×10^3 t yr^{-1}. Furthermore, the vulnerability of ecological units and species within the delta and lagoon system has increased and there has been a negative impact on biological diversity and renewable resources within both north and western parts of the Black Sea.

The protection of these areas has occurred in the face of rapid ecological change caused by regulation works and eutrophication (Table 4.8). Ecological changes are described in Figure 4.4 using between 25 and 35 variables selected to represent water quality and ecological compartments (i.e. pelagic and benthic plant and animal populations, birds, target species and renewable resources). Despite the impact on many fish and bird populations, the main river and the Small Island of Braila lakes have an ecological structure that is similar to the reference system, which was established using data covering the period 1952–1965. Biodiversity has declined in both the Danube Delta and the Razim–Sinoe Lagoon system with the loss of some community components in most of the lake ecosystems (i.e. submerged macrophyte–epiphyte complex, phytophylous fauna, and a rich benthic community dominated by filter feeders). A decline of species richness is also associated with reorganization within community components (e.g. cyanobacterial dominance of phytoplankton, invasive species in aquatic weed structure, etc.).

Protection has also been achieved in the face of intensive human pressure. About 2.2 million people are directly dependent upon the LDRS as a water source for domestic, industrial and agricultural uses. More than 1.5 million people within Romania live adjacent to the LDRS and depend upon the river for renewable natural capital. Southern Romania encompasses a large area of drought steppe (1.9×10^6 ha) and the region is dependent upon the Danube for irrigation. Most of the drained floodplain area is devoted to agriculture; there is an important chemical industry and a wide range of metal, building

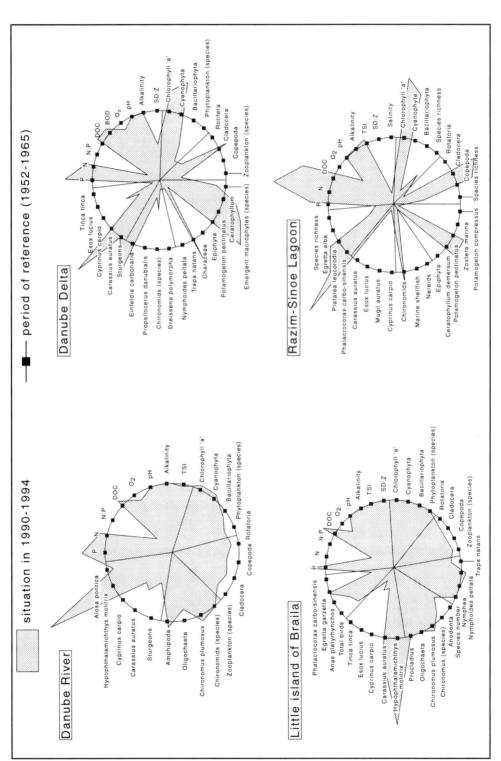

Figure 4.4 *Ecological changes within four representative aquatic systems of the Lower Danube River System. Data based on the AMOEBA method (Ten Brink et al., 1991) were derived from three to eight stations in each of the following areas: Razim–Sinoe lagoon (two lakes), Danube Delta (eight lakes), Small Island of Braila (two lakes) and LDRS (four reaches). During the period 1990–1994 data relate to monthly surveys of water chemistry, phytoplankton and zooplankton; from annual surveys of submerged macrophytes, birds and fish TSI = trophic state index; SD: Z = relative water transparency; Z = mean depth; SD = Secchi depth*

materials and processing industries. Marine navigation is practised on the lower 200 km river stretch and river navigation connects the Black Sea to central and western Europe.

By the beginning of the 1990s, it was becoming increasingly evident to the academic community, the newly established political institutions and the general public that the previous management plans had been inadequate. The emerging understanding tended to correspond with the newly established western European and UN strategies for dealing with the conflict between the environment and development and, in particular, with those focused on pollution abatement and biodiversity conservation. Consequently, the commitment of the newly established governmental authorities to change the trend of human impacts on the LDRS, and to declare its main components as protected areas, was very well received by both national and international bodies. The government decision to manage the most important components of the LDRS (the delta and Small Island of Braila) as special areas and pilot regions for sustainable development had two main aims. First, it would enable the pace of environmental degradation to be slowed down. Second, it prepared the ground for establishing a framework for the long-term and large-scale rehabilitation of the LDRS and north-western areas of the Black Sea.

The Romanian authorities submitted a proposal in September 1990 to integrate the whole Romanian part of the Danube Delta and Razim–Sinoe Lagoon system, together with the Black Sea buffer zone consisting of the littoral waters up to the isoline of 20 m depth (580 000 ha), into the UNESCO network of Biosphere Reserves. In addition, they sought to obtain World Heritage Site and Ramsar Site status for the region. This received full international support and the process was completed in less than a year.

The newly established unit, known since September 1991 as the Danube Delta Biosphere Reserve (DDBR), has played a major role in the development and implementation of two international programmes within the Danube River Basin and the Black Sea. These focus on building the scientific and managerial background for the sustainable management and socio-economic transition of the region's environment.

Management

Management of the LDRS involves five national administrations: Romania, Bulgaria, Ukraine, Yugoslavia (see Table 4.1) and Moldavia. The

Romanian administration is responsible for the largest part, incorporating an area of 10 400 km^2. This includes more than 30 islands, a large floodplain, the delta (with the exception of 732 km^2 under Ukrainian administration) and the lagoon complex. Within Romania the administration of the LDRS involves 12 counties, 7 cities, 11 towns and 55 village councils as well as a special administration for the protected areas: the Danube Delta Biosphere Reserve Authority and the Administrative Council of the Small Island of Braila. Romania is also party to a series of international conventions regarding the management of the Danube River: (i) Convention for the Protection and Rational Utilization of the Danube River (Sofia, 1995); (ii) International Agreement for Navigation on the Danube River; and (iii) Convention for Fishing on the Danube River.

Within this framework, three categories of management programmme are in different stages of development: (i) management projects specific to the protected areas; (ii) local and/or sectoral programmes for different zones and resources (forestry, fishery, fish and wildlife, agriculture, tourism, etc.); and (iii) integrated management of the LDRS requiring development on the basis of the Danube River Ecological Network's recommendations. Implementation of international conventions and national regulations, laws, government decisions, decrees and orders in Romania is the responsibility of specific national institutions (e.g. Ministry of Water, Forests and Environmental Protection; Ministry of Agriculture and Food Industry; Department of Fluvial Navigation of the Ministry of Transport).

The Management Plan for the Danube Delta Biosphere Reserve was elaborated in 1994 following its new legal status as a Biosphere Reserve, a World Heritage Site and a Ramsar Site. Four zones are delineated: (i) strictly protected areas; (ii) marine and coastal buffer zones; (iii) terrestrial and freshwater buffer zones; and (iv) areas for sustainable socio-economic development (DDBRA, 1995). The Management Plan comprises 35 Management Objectives and 87 Management Projects (Table 4.9).

An integrated management plan for the Nature Reserve 'Small Island of Braila' was proposed under the European Union's (DG XI) Life – Nature 98 Programme. The development of this plan is one objective of Romania's National Strategic Action Plan for Biodiversity Conservation and will be supported by the Ministry of Water, Forestry and Environmental Protection. Six key objectives and activities are proposed (Table 4.10).

Table 4.9   *Key projects in the Management Plan for the Danube Delta Biosphere Reserve*

- Identify, maintain and protect feeding, spawning, resting and breeding areas
- Restore sturgeon populations, including fingerling production
- Assess the environmental impact and cost/benefit of creating a new canal to discharge Danube sediment
- Restore and/or create habitat in abandoned polders
- Improve the monitoring network for hydrological, chemical and meteorological conditions
- Investigate alternative options for disposal of dredged materials
- Investigate the potential for extending reserve boundaries
- Develop a plan for tourism
- Explore opportunities for fisheries and aquaculture
- Evaluate options for a water distribution system
- Examine ways to sustain traditional practices in fishing and agriculture

Table 4.10   *Primary objectives of the Small Island of Braila Management Plan*

- Develop and implement action plans for birds (Birds Directive 79/409/EEC)
- Rehabilitate the former alluvial forests (Habitats Directive 92/43/EEC)
- Maintain local genetic resources
- Develop and support traditional economic activities
- Advance a Public Awareness Campaign aimed at hunters, fishermen, local communities, NGOs and decision-makers
- Advance economic and tourism activities by individuals and legal entities in accordance with the reserve's designation as a Ramsar Site

Public support, scientific assessment and recommendations, as well as the actions of NGOs, have been important in the development of the Protection Areas and Management Plans. The sustainable development of the LDRS requires the integration of several elements: (i) science (especially the restoration of floodplains for nutrient and diffuse-source pollution control); (ii) natural capital in terms of 'goods' (e.g. industrial, manufacture and medicinal plants, forestry, fish and wildlife) and 'services' (e.g. water quality buffer, tourism, water transport); (iii) socio-economic capital adapted to the carrying capacity of the natural system (e.g. recognizing the value of traditional land-use practices); and (iv) the geopolitical location of the LDRS connecting the Ponto–Caspian region to central and western Europe and the Baltic region (e.g. the important challenge to manage the potentially significant oil deposits to be transported by the Black Sea–Danube River).

Many local, national and international NGOs have been involved in the protection of the LDRS. This has involved specific action at many levels: public awareness and training; waste recycling; monitoring of pollution; lobbying local and national authorities, and informal education. A significant role is played by the international professional NGOs focused especially on the coordination of sectoral efforts of GOs and NGOs towards real solutions, including knowledge transfer towards users. Specific organizations such as the International Association for Danube Research (IAD), the Association of the Danube Cities, Earth Voice Romania and Earth Voice International, in addition to large organizations like IUCN and WWF, have been active partners with the academic community, the general public and decision-makers. One recent outcome has been the development of a memorandum of cooperation with HSUS[1]/Wildlife Trust for developing the institutional infrastructure and management plan of the Small Island of Braila Nature Reserve and its implementation.

RESTORATION IN PRACTICE

A range of restoration projects is being implemented to reverse the ecological degradation of the past 20 years. These include research on fish movement at the Iron Gate dams, and coastal engineering works to prevent erosion of the south sea side of the Danube Delta. However, the most important component of the restoration strategy is the wetland restoration programme. This involves returning polders to wetlands within the delta, returning parts of the agricultural diked areas – currently including large areas of salinized and degraded soils – to floodplain wetlands and the rehabilitation of open waters affected by eutrophication.

Two agricultural polders in the Danube Delta with a surface area of 3680 ha were successfully restored to wetland in 1993. Two other polders with a combined surface area of more than 10 000 ha, as well as 12 other fish polders with a combined area of over 27 000 ha, have also been identified for wetland restoration. Upstream in the floodplain zone of the lower river, about 150 000 ha of reclaimed floodplain has been abandoned because of salinization. Restoration of these floodplain areas is considered to offer multiple benefits including the improvement of spawning,

---

[1]   HSUS is the Human Society of the United States, a non-governmental organization involved in animal protection and biodiversity conservation.

nesting and feeding areas as well as increasing the buffering capacity for protecting downstream ecosystems – the delta and the Black Sea – from poor quality catchment run-off.

Given the intensity of human use within the LDRS, a range of technologically advanced solutions are required to tackle specific problems, in addition to the restoration works outlined above. Some of these are large-scale, such as the operational management of the main navigation channels, others are small-scale including, for example, the development of 'screens' to prevent fish-fry mortalities at pumping-station intakes. However, advanced engineering solutions are very expensive and considered to be difficult to appraise using success criteria on an acceptable scale. The stimulation of re-naturalization processes in order to determine a new level of ecological equilibrium is considered to be the best strategy for future schemes.

The realization of the public and political will for Romania to join the EU depends largely on the capacity of the country to implement strategies that fit into the new objectives of the EU. Achieving rehabilitation and sustainable management of the LDRS requires significant financial resources and stronger policies, and an awareness by decision-makers and the public of the long-term benefits. The successful design and implementation of such a management plan requires a holistic approach with appropriate time-scales.

A future for river conservation?

The history of environmental change in the region is dominated by the legacy of communism. For the countries of the former Soviet Union, this legacy extends back to the early part of the 20th century. However, for the remaining countries, the era of communism was ushered in during the aftermath of World War II. In recent years, this latter group of countries has tended to strengthen links with the EU and, as a consequence, compliance with environmental objectives is an increasingly important driver of political change. In contrast, the former Soviet countries (with the exception of the Baltic States) are more removed, both physically and culturally, from the influence of the EU.

Bearing in mind the dangers of generalization, it would appear that the countries of central and eastern Europe and the former Soviet Union are characterized by severe economic constraints, weak institutional arrangements and fragmented political structures. Economic recovery and development, not

environmental improvement, are considered as main objectives. Much of the euphoria and expectation in the environmental sphere, which existed throughout the region at the fall of communism, has dissipated under the reality of societal transformation. Public opinion polls tend to reinforce this conclusion. Similarly, governments within the region are preoccupied with the problems of structural transformation, ensuring a low priority for environmental issues. Even when efforts are made to address key issues, little progress is made. For example, a recent programme proposed by the Russian State Committee for Water laid the foundations for an integrated regional approach to the management of the entire Volga Basin (*Ekos-Inform*, 1996). However, financial support for the programme is unlikely to be found, not least because the West appears to have little impetus to become involved in river management issues in Russia where most of the river systems are within the country's borders.

Those countries adjacent to the EU and which share 'ownership' of international river basins are better placed to obtain help and assistance with river management and conservation issues and there is evidence of growing involvement of international lending agencies and NGOs. As suggested above, most of these countries appear eager to provide evidence of compliance with the methods and modes of operation now functioning within the EU in order to strengthen their membership prospects. Certainly, for those who aspire to join the EU investment in conservation issues is seen as an important step in the process of integration. The case study of the Danube Delta in Romania provides a good example of what can be achieved.

Because of the extensive nature of environmental degradation and in many cases its severity, long-term, catchment-scale environmental management plans are required to restore river ecosystems. However, both case studies presented here illustrate the value of the 'nature reserve concept' – that is, the protection of key sites which retain high ecological quality. These sites have great short-term benefit in providing refuges for flora and fauna. The protection and restoration of these sites can also have a high political profile and can attract the interest of international organizations, often with benefits for the development process. Nevertheless, it is unlikely that major, sustainable improvements will be made within the region of study unless socio-economic growth and improvements in environmental education are realized, and this will require substantial international investment.

References

Abakurnov, V.A. 1979. General directions of aquatic biogeocenosis reconstruction under pollution of the environment. *The Problem of Ecological Monitoring and Modelling of Ecosystems* 2:37–47 (in Russian).

Abakurnov, V.A. and Sushenya, L.M. 1991. Hydrobiological monitoring of the state of freshwater ecosystems and means for its improvement. In: *Ecological Modification and Criteria for Ecological Standardisation*. Gidrometeoizdat, St Petersburg, 33–34.

Alirnov, A.F. 1982. Structural and functional approaches to studying the communities of water animals. *Ecology* 3:45–51 (in Russian).

Baboianu, G. 1998. The Danube Delta Biosphere Reserve: nature protection and sustainable development. *Proceedings of the 2nd International Seminar for Managers of Biosphere Reserves of the Euro MaB Network*. Stara Lesna, Slovakia, 23–27 September 1996, 49–54.

Banu, A.C. (ed.) 1967. *Limnology of the Danube River Romanian Stretch: Monographic Study*. Bucharest, Academy RSR.

Bondar, C. 1994. The hydrobiology of the Danube Delta – sectoral study. *The Management Program of the Danube Delta Biosphere Reserve*. EBRD (European Bank for Reconstruction and Development) Report, 17–26.

Boucher, K. 1990. Landscape and technology: the Gabcikovo–Nagymaros scheme. In: Cosgrove, D. and Petts, G. (eds) *Water Engineering and Landscape: Water Control and Landscape Transformation in the Modern Period*. Belhaven Press, London.

Bryzghalo, V.A. and Khaiter, P.A. 1993. A computer application for investigating the structural transformation of anthropogenically impacted aquatic ecosystems. *Hydrological, Chemical and Biological Processes of Transformation and Transport of Contaminants in Aquatic Environments*. IAHS Publication 219, Wallingford, 175–183.

Carter, F.W. and Turnock, D. (eds) 1996. *Environmental Problems in Eastern Europe*, Updated edition. Routledge, London.

Cousteau, J.I. (ed.) 1993. *The Danube Delta: For Whom and for What?* Equip Cousteau and EBRD (European Bank for Reconstruction and Development), Paris/London.

Cristofor, S., Vadineanu, A., Sarbu, A., Postolache, C., Ignat, Gh. and Ciubuc, C. 1998. Aquatic macrophyte to the floodplain functions in the Lower Danube River System. In: Ferreira, T. and Wade, M. (eds) *Proceedings of the 10th EWRS Symposium on Aquatic Weeds*, Lisbon, 175–178.

Danube Delta Biosphere Administration (DDBRA) 1995. *Management Objectives for Biodiversity Conservation and Sustainable Development in the Danube Delta Biosphere Reserve, Romania*. Information Press, Oxford.

Day, R.T., Keddy, P.A. and McNeill, J. 1988. Fertility and disturbance gradients: a summary model for riverine marsh vegetation. *Ecology* 69:1044–1054.

de Waal, L.C., Child, L.E., Wade, P.M. and Brock, J.H. (eds) 1994. *Ecology and Management of Invasive Riverside Plants*. John Wiley, Chichester.

Drbal, K. and Rauch, O. 1996. Water chemistry. In: Prach, K., Jenik, J. and Large, A.R.G. (eds) *Floodplain Ecology and Management. The Ludnice River in the Trebon Biosphere Reserve, Central Europe*. SPB Academic Publishing, Amsterdam, 47–51.

European Bank for Reconstruction and Development (EBRD) 1995a. Kaunas Water and Environment Project. *Environments in Transition: The Environmental Bulletin of the EBRD*, Autumn edition, 8–9.

European Bank for Reconstruction and Development (EBRD) 1995b. Estonia Small Municipalities Environment Project. *Environments in Transition: The Environmental Bulletin of the EBRD*, Autumn edition, 13–14.

European Bank for Reconstruction and Development (EBRD) 1997. Developments in environmental infrastructure. *Environments in Transition: The Environmental Bulletin of the EBRD*, Spring edition, 2–4.

Ekos-Inform 1996. Concept and Introduction to the restoration of the Volga programme. *Ekos-Inform* 8–9, 5–150 (in Russian).

Gastescu, P. 1993. The Danube Delta: geographical characteristics and ecological recovery. *GeoJournal* 29:57–67.

Ianovici, V., Mihailescu, V., Badea, L., Morariu, T., Tufescu, V., Iancu, M., Herbst, C. and Gramazescu, H. (eds) 1969. *Geography of the Romanian Danube Valley*. Bucharest, Ed. Acad. RSR (in Romanian).

International Monetary Fund (IMF), World Bank, Organization for Economic Co-operation and Development (OECD), European Bank for Reconstruction and Development (EBRD) 1991. *A Study of the Soviet Economy* (Volume 3). OECD, Paris.

Izrael, Yu.A., Gasilina, N.K. and Abakurnov, V.A. 1981. Hydrobiological service of surface water observation in the USSR. In: *Scientific Backgrounds of Surface Water Quality Control by Hydrobiological Data*. Gidrometeoizdat, Leningrad, 7–16 (in Russian).

Jancar-Webster, B. 1995. Environmental degradation and regional instability in Central Europe. In: De Bardeleben, J. and Hannigan, J. (eds) *Environmental Security and Quality after Communism: Eastern Europe and the Soviet Successor States*. Westview Press, Boulder, Colorado, 43–68.

Kajak, Z. 1992. The River Vistula and its floodplain valley (Poland): its ecology and importance for conservation. In: Boon, P.J., Calow, P. and Petts, G.E. (eds) *River Conservation and Management*. John Wiley, Chichester, 35–49.

Khaiter, P.A. 1995. Hydrobiological responses to eutrophication: Volga River reservoirs, Russia. In: *Man's Influence on Freshwater Ecosystems and Water Use*. IAHS Publication 230, Wallingford, 185–190.

Kindler, J. 1994. Some thoughts on the implementation of water quality management strategies for Central and Eastern Europe. *Water Science & Technology* 3:15–24.

Krivonosov, G.A., Rusanov, G.M. and Gavrilov, N.N. 1994. Pelicans on the northern Caspian Sea. In: Crivelli, A.J., Krivenko, V.G. and Vinogradov, V.G. (eds) *Pelicans in the Former USSR*. IWRB Publication 27, 25–33.

Krizan, J. and Vojinovic-Miloradov, M. 1997. Water quality of Yugoslav rivers (1991–1995). *Water Research* 31:2914–2917.

Kvet, J., Tetter, M., Klimes, F. and Suchy, K. 1996. Grassland productivity as a basis for agricultural use of the Ludnice floodplain. In: Prach, K., Jenik, J. and Large, A.R.G. (eds) *Floodplain Ecology and Management. The Ludnice River in the Trebon Biosphere Reserve, Central Europe*. SPB Academic Publishing, Amsterdam, 247–249.

L'vov, A.V. and Kuzin, A.K. 1987. Strategy of water quality planning and management. In: Ryans, R.C. (ed.) *Protection of River, Basin, Lakes and Estuaries*. American Fishing Society Press, Bethesda, Maryland, 5–19.

Magurran, A.E. 1988. *Ecological Diversity and its Measurement*. Croom Helm, London.

Manser, R. 1993. *The Squandered Dividend: The Free Market and the Environment in Eastern Europe*. Earthscan, London.

Margesson, R. 1997. Environment and international water management: dealing with the problems of the Danube Delta. *Environmental Impact Assessment Review* 17:145–162.

Marsakova, M., Mihalik, S. *et al.* 1977. *National Parks, Nature Reserves and Other Protected Natural Sites in Czechoslovakia*. Academia, Prague (in Czech).

McCannon, J. 1995. To storm the Arctic: Soviet polar exploration and public visions of nature in the USSR, 1932–1939. *Ecumene* 2:15–31.

Micklin, P.P. 1992. Water management in Soviet Central Asia: problems and prospects. In: Massey Stewart, J. (ed.) *The Soviet Environment: Problems, Policies and Politics*. Cambridge University Press, Cambridge, 88–114.

Ministry for Environment Protection and Natural Resources of the Russian Federation (MEPNR) 1994. *State of the Environment of the Russian Federation 1993: National Report*. Moscow.

Moskompriroda 1996. *State Report Concerning the State of the Environment of Moscow City in 1995*. REFIA, Moscow (in Russian).

Organization for Economic Co-operation and Development (OECD) 1996. *Environmental Information Systems in the Russian Federation: An OECD Assessment*. Paris.

Opravil, E. 1983. Ancient river floodplains. *Studia Archeologica Instituto, Brno* 11:1–77.

Peterson, D.J. 1993. *Troubled Lands: The Legacy of Soviet Environmental Destruction*. Westview Press, Boulder, Colorado.

Petts, G.E. 1984. *Impounded Rivers: Perspectives for Ecological Management*. John Wiley, Chichester.

Petts, G.E. 1994. Large-scale river regulation. In: Roberts, N. (ed.) *The Changing Global Environment*. Blackwell, Oxford, 262–284.

Prach, K., Jenik, J. and Large, A.R.G. (eds) 1996. *Floodplain Ecology and Management. The Ludnice River in the Trebon Biosphere Reserve, Central Europe*. SPB Academic Publishing, Amsterdam.

Pryde, P. 1972. *Conservation in the Soviet Union*. Cambridge University Press, Cambridge.

Pryde, P. 1991. *Environmental Management in the Soviet Union*. Cambridge University Press, Cambridge.

Pysek, P. and Prach, K. 1993. Plant invasions and the role of riparian habitats – a comparison of four species alien to central Europe. *Journal of Biogeography* 20:413–420.

Rodda, D.W. 1994. The environmental programme for the Danube river basin. *Water Science & Technology* 30:135–145.

Schiemer, F. and Waidbacher, H. 1992. Strategies for conservation of a Danubian fish fauna. In: Boon, P.J., Calow, P. and Petts, G.E. (eds) *River Conservation and Management*. John Wiley, Chichester, 363–382.

Somlyody, L. 1994. Quo vadis water quality management in Central and Eastern Europe? *Water Science & Technology* 30:1–14.

Somlyody, L., Kularathna, M. and Masliev, I. 1994. Development of least-cost water quality control policies for the Nitra river basin in Slovakia. *Water Science & Technology* 30:69–78.

Stanners, D. and Bourdeau, P. (eds) 1995. *Europe's Environment: The Dobříš Assessment*. European Environment Agency, Copenhagen.

Strashkraba, M. and Gnauk, A.H. 1985. *Freshwater Ecosystems. Modelling and Simulation*. Elsevier, Amsterdam.

Straskrabova, J., Prach, K., Joyce, Ch. and Wade, M. 1996. Floodplain meadows – ecological functions, contemporary state and possibilities for restoration. *Priroda* 4:1–176.

Ten Brink, E.J.B., Hosper, H.S. and Colyn, F. 1991. A quantitative method for description and assessment of ecosystems. The AMOEBA approach. *Marine Pollution Bulletin* 23:265–270.

Tonderski, A., Grimvall, A., Sundblad, K. and Stalnacke, P. 1994. An East–West perspective on riverain loads of nutrients in the Vistula and Rhine Basins. *Water Science & Technology* 30:121–130.

Vadineanu, A., Cristofor, S., Sarbu, A., Romanca, G., Ignat, Gh., Botnariuc, N. and Ciubuc, C. 1998. Changes of the biodiversity along the Lower Danube River System. *International Journal of Ecology and Environmental Sciences* 24:315–332.

Weiner, D.R. 1990. Prometheus rechained: ecology and conservation. In: Graham, L.R. (ed.) *Science and the Soviet Order*. Harvard University Press, Cambridge, Massachusetts, 71–93.

Welcomme, R.L. 1992. River conservation – future prospects. In: Boon, P.J., Calow, P. and Petts, G.E. (eds) *River Conservation and Management*. John Wiley, Chichester, 453–462.

Wilson, A.M. and Moser, M.E. 1994. *Conservation of Black Sea Wetlands: A Review and Preliminary Action Plan*. IWRB Publication 33.

World Bank 1996. *World Development Report: From Plan to Market*. Oxford University Press, Oxford.

World Wide Fund for Nature (WWF) 1998. Press release, 18 February. New Danube dam could threaten Hungary's accession to EU. http://www.panda.org/news/press/news_184.htm

Zelenyi Mir 1997a. State report concerning the state of the environment of the Russian Federation in 1996. *Zelenyi Mir* 24:4 (in Russian).

Zelenyi Mir 1997b. State report concerning the state of the environment of the Russian Federation in 1996. *Zelenyi Mir* 28:8 (in Russian).

5

River Conservation in North Africa and the Middle East

M.J. Wishart, J. Gagneur and H.T. El-Zanfaly

Introduction

Commonly referred to as the 'Arab Region', the countries of North Africa and the Middle East cover roughly 14×10^6 km^2 and include the Middle Eastern countries of the Arabian Peninsula (Bahrain, Kuwait, Oman, Qatar, Saudi Arabia, United Arab Emirates (UAE) and the Republic of Yemen (comprising the former Arab Republic of Yemen and the People's Democratic Republic of Yemen which joined in 1990)), Iraq, Israel, Jordan, Lebanon, Palestine and Syria, along with those countries of continental Africa north of the Sahara, Algeria, Egypt, Libya, Morocco, Sudan and Tunisia. Characterized by various topographic and geomorphological features, the region is generally distinguished by its long coastal boundaries, totalling *ca* 22 870 km along the Mediterranean and Red Seas, the Arabian Gulf and the Indian Ocean. Although encompassing a wide range of different climatic conditions, the greater part of the region is arid to semi-arid, with much of the area below the 200 mm isohyet, making it one of the most arid regions in the world.

The recorded history of the Middle East and North African region dates back more than three millennia, with the basin of the Tigris and Euphrates Rivers considered to be the cradle of modern Western civilization. For centuries communities in these regions have practised water-harvesting techniques and the conjunctive use of surface water and groundwater for dryland irrigation and rain-fed agriculture. Throughout this period water has been a key resource issue. With a rapidly increasing population and concomitant pressures on the region's resources, traditional technologies related to resource utilization have given way to more modern approaches. The result is that most of the states of the Arabian Peninsula, along with Jordan, Israel, Libya, Oman and Palestine, currently consume more water than their annual renewable supply. Similar scenarios face Egypt, Syria, Morocco, Tunisia and the Sudan. Today, water continues to be a key resource within a region plagued by complex political uncertainty and tensions. The situation is such that in 1996 the Vice President of the World Bank, Ismail Serageldin, observed that while 'many of the wars of this century were about oil . . . wars of the next century will be over water'. This view echoed that of the deputy prime minister of Egypt (then Boutros Boutros-Ghali), who four years earlier, stated that 'the next war in the Middle East will be over water, not politics'.

The regional context

THE MAIN RIVERS

The Middle East and North Africa has seen a complex tectonic history (e.g. Por, 1986). This complexity is reflected in the spatial arrangement, biogeography and characteristics of the region's river systems. Generally characterized by long coastal boundaries, most of the region is of low topography and covered with vast deserts that extend between the Atlantic Ocean and the Arabian Gulf. The climate of the Arab Region has, however, undergone several important shifts and

Global Perspectives on River Conservation: Science, Policy and Practice.
Edited by P.J. Boon, B.R. Davies and G.E. Petts. © 2000 John Wiley & Sons Ltd.

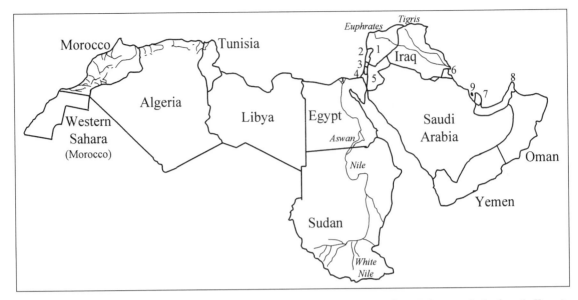

Figure 5.1   *North Africa and the Middle East. 1, Syria; 2, Lebanon; 3, Israel; 4, Palestine; 5, Jordan; 6, Kuwait; 7, Qatar; 8, United Arab Emirates; 9, Bahrain*

transitions. For example, 20 000 to 15 000 years BP, conditions around the Nile Basin were very arid as the current rainfall belt moved 450 km southwards. This extended the desert sand regions to around latitude 10°N. This was followed about 12 000 to 7000 years BP by a wetter period during which the rainfall belt moved 400 km northward such that the levels of precipitation recorded in the Sudan were two to three times greater than at present. The more recent history, around 7000 to 6000 years BP, was characterized by a period of climatic fluctuations which 3000 years BP became increasingly drier (Rzsóka, 1976b). Analysis of more recent historical flood-level time series of the River Nile (AD 622–1470) highlights a number of periods of abrupt climate change. Global estimates reveal three distinct epochs, AD 622–1078, 1079–1325 and 1326–1470, which coincided with larger-scale climate changes. These epochs are a relatively cool age, the Little Climatic Optimum of the Middle Ages and an interim period before the Little Ice Age (Fraedrich *et al.*, 1997).

Today, with the exception of the Mediterranean coastal margins and those along the Atlantic Ocean, temperatures and the degree of aridity typically increase inland, with the region now experiencing a general climatic state of hyper-aridity. The average annual surface run-off also decreases sharply from the extreme northern and southern boundaries of the region to the central desert regions. Located within this

arid and semi-arid climate, the Arab Region includes some of the most water-stressed countries in the world, characterized by high temperatures and low rates of precipitation. In the Hashemite Kingdom of Jordan, for example, only 2% of the land area receives more than 350 mm of rainfall per year, with more than 95% of the land areas receiving <200 mm. In Libya, rainfall ranges from 150 to 400 mm yr^{-1}, with most of it falling in the coastal region from September through to April. Due to strong winds and high temperatures, most of the available water in Libya is lost through evaporation, while only 5–10% recharges groundwater, and 30–40% reaches the sea as surface run-off (El Asswad, 1995). As a result of this aridity, the region includes only a few principal streams along with a number of smaller tributaries, with roughly 80 perennial rivers across the entire region. Thus, one of the most striking and definitive characteristics of the entire Arab Region is the paucity of perennial sources of water (Figure 5.1). Almost 25% of the population of the Arab Region live in countries with no perennial surface supplies, while countries such as the oil-rich state of Kuwait have no permanent rivers or streams, with surface run-off taking place only after exceptional rainfall.

With aridity typically increasing toward inland areas, perennial streams are largely confined to the coastal margins that experience more Mediterranean climes. Those rivers of significant size are limited largely to the basins of the Jordan, Nile and Tigris–Euphrates systems (Figure 5.1). The Mediterranean climate of Morocco results in a large number of smaller perennial streams, while more than 90% of Algeria's annual mean stream discharge comes from springs in the north-eastern mountains of the country (Zouini, 1997). As is the case with most of the region's principal streams, the major perennial basins of the Jordan, Nile and Tigris–Euphrates all cross international boundaries or share riparian rights between more than one country. Two-thirds of the population in the Arab countries depend on rivers flowing from 'non-Arab' countries. Furthermore, the Jordan River is the only one of these large systems for which recharge and discharge is confined to countries within the region. Both the Nile and Euphrates–Tigris receive most of their recharge from outside the borders of the region. Table 5.1 lists the most important perennial rivers (principal streams and tributaries) along with their basin areas, length of the principal streams and the average flow.

Local estimates derived from historical flood-level time-series of the River Nile (AD 622–1470) show that the reference time of abrupt changes can be clearly identified and that the associated time-scale coincides with the persistent anomaly period. Fraedrich *et al.* (1997) identified approximately eight almost synchronous abrupt changes in the minimum and maximum River Nile flood levels; many of them are associated with 35–45 year persistence time-scales. Although there is no information available on phase coherence, Fraedrich *et al.* (1997) suggested an association of these short time-scales with those of interdecadal variability reported for the mid- and high-latitude sea-surface temperature of the North Atlantic. Results from the analysis of two extensive data sets describing sea-surface temperature of the Pacific Ocean and the flow of water in the Nile River (Egypt) suggest that 25% of the natural variability in the annual flow of the Nile is associated with El Niño oscillations (Eltahir, 1996).

Assuming increasing importance across the Arab Region is the development of artificial rivers, such as Libya's Great 'Man-made' River (MMRs) project (Alghariani, 1997). Depending on the definition of what exactly constitutes a river, there are increasing moves toward developing a number of MMRs, in an attempt to reconcile the disparities between the distribution of water across the region and the distribution of demand. Alghariani (1997) points to a number of project developments aimed at achieving 'a more desirable distribution of available water

supplies', considered both 'technically feasible and economically rational' in light of the uneven distribution of water and its overall scarcity, and proposes that similar future projects may be constructed to serve other water-scarce regions such as the other countries of the Middle East, North Africa and the south-western USA. Similarly, many of the rivers that flow today are little more than conduits for sewage and other effluents. For example, the upstream abstractions in the River Yarqon have resulted in base flows of less than 10% of the average annual discharge (Gasith, 1992) and were it not for the input of sewage and effluents the River Yarqon would be largely intermittent. Even the artificial canals and drainage systems within the Nile Delta support a large array of different species (i.e. Shaltout *et al.*, 1994).

DEVELOPMENT PRESSURES AND HUMAN IMPACTS

Population growth rates for North Africa and the Middle East once ranked among the highest in the world. Given the arid climatic conditions, the scarcity of available water resources for human consumption and this burgeoning population, the greatest pressures upon the rivers of the region have historically come from the need to secure water supplies suitable for human utilization. The region has the highest concentration of water-scarce countries in the world. While recent records show that the amount of water per person in the region is about 1000 m^3 yr^{-1}, it is anticipated that this figure will fall to about 665 m^3 yr^{-1} by the year 2025. With only 432 m^3 of water available per person per year, Tunisia already rates as one of the most water-scarce countries in the world, yet this amount is expected to drop to 285 m^3 by 2020 (Matoussi, 1996). In order to respond to large and expanding population demands, traditional practices that have developed over centuries have been, and continue to be, abandoned in favour of major irrigation schemes in association with large-scale, pharaonic water-supply systems.

While the implementation of modern technologies to overcome these problems, such as the development of large dams and water-supply and storage schemes, is obviously confounded and contradictory to the arid climatic conditions of the region, this has not impeded their construction and development. For example, evaporation rates in the Nasser Reservoir behind the Aswan High Dam are on average *ca* 11.2 km^3 yr^{-1}. This is equivalent to roughly 10% of the reservoir's

Table 5.1 List of permanent rivers in the North African and Middle Eastern Region (from Shahin, 1985; United Nations Educational, Scientific and Cultural Organization/Arab Centre for the Study of Arid Zones and Dry Lands – UNESCO/ACSAD, 1988)

River or tributary	Basin area (km² × 10³)	Principal stream length (km)	Average flow (m³ × 10⁹ yr⁻¹)
Nile	2800	4800	85.8 (At Aswan)
White Nile		2380	21.4 (At Manjala)
Bahr El-Jabal (Upper White Nile)			
Bahr El-Ghazal	528	1460	
Bahr El-Arab	209	820	
River Loul	82		4.0
River Bonjo	70.2		0.5 Total 14.1
River Jor	64		5.0
River Toubekh	27		1.5 (Upper dams reigon)
River Maridi	22		
River El-Naam	16		0.52
River Yay	25		2.0
River Jel (Tabari)	12.8		0.55
El-Soubat	225.0	600	12.99 (Hallet Doulib)
El-Barou	41.4	400	11.6 (Upper Jambila)
El-Bibour	10.9	400	3.1 (Bibour Port)
Blue Nile	324.5	1650	49.2 (El Rous-aires)
El-Dandar[a]	35.6	750	3.0 (Hallet Idris)
El-Rahed	34.7	800	1.1 (At outlet)
El-Atbare	112.4	880	11.9 (At outlet)
Satit	68.8	330	
Tigris	258.0	1718	48.70
The Large Zab	26.0	260	12.18
The Small Zab	21.5	380	7.17
El-Azim	13.0	210	0.79
Davali	32.0	440	5.74
El-Karkha	46.0	780	6.30
El-Tayeb	5.0	80	1.00
Douyesrej	5.0	110	1.00
Euphrates	444.0	2330	29.00
El-Khabour	36.9	430	1.50
El-Balikh	14.4	2.2	0.15
El-Sajour	2.35	1.8	0.125
Shatt al-Arab	702.0	190	32.5
Kazoun	58.0	400	24.7
Khour El-Karkhah	46.0	780	
Barada	1.45	79	0.35
El-A'waj	1.21	91	0.10
Kowaik	4.21	126	0.095

River			
El-Sin		6	0.315
El-Kabir	1.06	80	0.325
El-Shamali			
El-Assi	16.90	571	2.00
Afrin	2.68	149	0.28
El-Kabir	0.98	50	0.320
El-Janoubi			
Ostwan	0.22	40	
Arka	0.13		
El-Bared	0.19	35	
Abu Ali	0.29	40	
El-Joz	0.15	33	
Ibrahim	0.27	42	
El-Kalb	0.25	25	Total flow of rivers = 3.00
Beirut	0.19	38	
Damour	0.39	35	
Awwali	0.25	50	
Zahrani	0.09	28	
Abu Aswad	0.22	23	
Litani	1.94	170	
Abu Zibel		20	
Jowaya		30	
Jordan		225	
Yarmouk	9.30	120	0.80
Banias			
Hasbani			
Zarqua		120	1.00
Mujaraada	24.00	380	0.05
Malian	2.28	110	1.30
Um El-Rabi		600	1.20
Sabou		500	
El-Lakous		100	
Abou Raqraq		250	
El-Malwiya		450	
Dea'a		1200	
Transifet		270	
Ziz		270	
Sous			
El-Shalif		490	
Fienna		170	

[a] It may turn into a seasonal river during the year when rainfall falls much below average.

stored water and, interestingly, is equal to the total withdrawal of water for residential and commercial use throughout Africa (Afifi and Osman, 1993; Ezzat, 1993; Shiklomanov, 1993; McCully, 1996). Between 1970 and 1990 the construction of large-scale water-supply and irrigation systems resulted in a dramatic increase in irrigated areas, helped to stimulate economic growth and alleviated some of the stresses caused by drought. There are, however, few opportunities left for further development of the region's rivers. Most of the available sites 'suitable' for dam construction have been exploited and more recent threats to these systems now come from upstream basin states, such as Ethiopia and Turkey (Allen, 1994; Hillel, 1994) which, in the absence of adequate water agreements between the riparian countries, will pose a crucial challenge to future stability of the region.

Within the broader region, some rivers remain relatively unexploited by some basin states. For example, while the Blue Nile Basin contributes 86% of the total annual discharge of the Nile, Ethiopia has yet to build any large hydroelectric power installations or modern dams in the basin (Woube, 1997) and a considerable quantity of water flows into the El-Kebir Wadi Basin. Water supply and management projects are being developed in this basin, including dams for irrigation and supplying water for urban and industrial uses (Zouini, 1997). New dam proposals are now also being increasingly focused toward temporary rivers. For example, there are proposals for the development of a dam on the highly seasonal intermittent Al Authaim River, near Jabal Hemrin, a tributary in the Tigris River Basin. Although these systems are temporary and experience annual or greater periods of no flow, periods of surface flow are crucial to the maintenance of the region's aquatic environments. The implications of such developments are highlighted by the construction of a dam on the Wadi Rajil in 1991, which has resulted in reduced flood peaks and which has affected one of Jordan's most important inland wetlands, the Azraq Oasis. This oasis comprises a complex system of spring-fed marshes adjacent to a seasonally flooded mudflat, the Qa Al Azraq. The system was already suffering as a result of reduced stream flow from the springs that feed the marsh, but the dam has further exacerbated the situation by cutting off the single most important source of water supplying the system. Several of the springs ceased to flow during August 1992, primarily as a result of over-exploitation of groundwater supplies for irrigation and domestic supply, highlighting the inextricable linkages and multi-dimensional nature of the hydrological regime.

One of the continuing effects of the intense level of river regulation throughout the region has been the reduction, or elimination, of the floods that have for centuries provided the fertile floodplains with nutrient-rich and fertile sediments upon which much of the agricultural sector has relied. Dam building on the Euphrates in Turkey and Syria, and the increasing utilization of the Euphrates and the Tigris for irrigation in upper and middle Iraq have, for example, reduced seasonal flood peaks. It has been estimated that by 1993, dams and barrages on the Tigris and Euphrates had resulted in a flow reduction of 44% in the Tigris at Kut, and a 21% reduction in the Euphrates at Hindiya. While there a few suitable sites for new developments, by the time all of the proposed constructions and developments for the system are completed and fully operational, flow will be reduced by 70% in the Tigris, and by 60% in the Euphrates (Maltby, 1994). These developments are primarily situated in Iraq, in the case of the Tigris, while for the Euphrates, it is mostly work in Turkey.

Even some of the region's larger perennial rivers are under threat from over-abstraction and heavy utilization such that they face the threat of the cessation of surface flow. Earlier this century Winston Churchill commented that the Nile would one day cease to flow into the Mediterranean. While the Nile still maintains perennial flows, heavy utilization and over-abstraction of other rivers across the region has resulted not only in the reduction of surface flows but has disrupted the perennial nature of these systems. With an annual average discharge of 50–60×10^6 m^3 the Yarqon River in Israel was considered second only to the Jordan (Avitsur, 1957; see above). With most of the management protocols derived from northern temperate systems, there is a need for the development of dryland protocols that address the inherent and driving variability that is characteristic of these systems. This feature is usually in contrast, and contradictory, to the needs and objectives of management strategies concerned with the supply of water during natural low-flow periods.

With a lack of suitably available sites for development the issue is increasingly becoming one not confined to water quantity, but also one of suitable quality (e.g. Basha and Touma, 1984; Al-Layla and Al-Rawi, 1988; DouAbul et al., 1988; Khorasani and Pooryadegar, 1988; Gabbay, 1989). While reduction of water consumption is one way to reduce the demand on the region's rivers, one of the problems with processes such as recycling is the quality of the return flows (Khouzam, 1996). With dwindling resources, water quality has been identified as being of primary importance by a number of nations (Bahrain, Jordan, Lebanon, Syria and the UAE). These problems were identified as originating from the over-abstraction of aquifers, agricultural run-off, human and industrial discharge of waste water and loss of habitat (Brooks, 1994; Lonergan and Brooks, 1994). In view of rapid industrial development in urban areas, mostly situated along perennial streams, the causes of water quality deterioration are becoming uncontrolled within acceptable economic standards (Khouzam, 1996). The declining water quality caused by pollution from the various industrial, agricultural and domestic activities is affecting public health, the productivity of resources and the quality of life, besides inducing steep increases in the cost of pollution control.

Other development pressures arise from the historical reliance on agriculture. For thousands of years agriculture has provided a staple means of existence for the region's people. With increasing demands and population growth, the advent of more modern technologies has resulted in large-scale hydrological engineering works typically aimed at enhancing agricultural production. An average of 60 to 90% of the water consumption in the region is diverted to agriculture. In Libya agriculture consumes nearly 99% of available water, with demand for water having doubled over the last 17 years (El Asswad, 1995). The natural subterranean springs on the islands of Bahrain that have supported plantations and agriculture for centuries have been decreasing, or becoming increasingly more saline, since the early 1970s as a result of lowering of the water table through the depletion and over-utilization of aquifers (Louri, 1990).

While irrigation is the largest consumer of water, proportionally it contributes little to most national economies. There are important exceptions though, for in Egypt the agricultural sector accounts for just under 50% of employment and about 80% of total export earnings. Irrigation and agricultural development in Egypt invariably means that the Nile and all but a small fraction of that country's agriculture is heavily reliant upon water from the River Nile. Traditionally irrigation and agriculture have been dependent upon the development and construction of canals from the river (Abu-Zeid, 1983). Technological developments since the start of the 20th century, however, have seen an increase in the construction of large-scale impoundments and water storage in dams such as Aswan, the Gabal El Awlia Reservoir and the Aswan High Dam (Samaha, 1979; Abu-Zeid, 1983). The Aswan

Dam was the first of these projects, constructed on the Nile in Egypt in 1902 with an initial storage capacity of 1×10^9 m^3. In 1913 this capacity was subsequently increased to 2.5×10^9 m^3. The Gabal El Awlia Reservoir was built in the 1930s but, while it is situated on the White Nile in Sudan, it provides water to Egypt. More recent developments south of the old Aswan Dam site have seen the construction of the Aswan High Dam, which has a storage capacity of 162×10^9 m^3, making it the fifth largest storage reservoir in the world (McCully, 1996).

The Aswan High Dam is often considered to be an environmental catastrophe, trapping the annual floods and resulting in the retention of almost the entire sediment load of the Nile (Entz, 1976; McCully, 1996). Prior to construction of the Aswan High Dam the river carried an average of *ca* 124×10^6 t of sediment onto the floodplain and into the delta. Today, more than 98% of this sediment falls to the bottom of the Nasser Reservoir (Lavergne, 1991). While government sources claim that the construction of the new dam has resulted in the 'reclamation' of 690 000 ha of land from the desert (Shalaby, 1993) information from the Food and Agricultural Organization of the United Nations (FAOUN) suggests that the actual area of irrigated land in Egypt remained largely unchanged between 1969, when construction of the dam commenced, and 1989. Similarly, the Assad Reservoir behind Syria's Tabqua Dam on the Euphrates River drowned 31 000 ha of irrigated land. Almost two decades later, in 1990, less than 83 000 ha had been recovered and received water from the Assad Reservoir. As much as 60% of this had already been irrigated by smaller-scale private pump schemes prior to construction of the dam. Initially the dam was proposed to have provided 640 000 ha of irrigated land (Scheumann, 1993).

Others argue, however, that the Aswan High Dam has in fact been overwhelmingly positive and that the negative image is based on predictions made some two decades ago (Abu-Zeid, 1990; Biswas, 1992: although see McCully, 1996). They argue that the huge costs of building the dam were recovered within only two years through increases in agricultural production and hydropower generation. They argue further that the dam has also proved its value to the country's economy through years of sustained drought and potentially catastrophic floods. Further, although there are environmental problems resulting from the dam's construction many – including river-bed erosion and the effects on fishing – have proved much less severe than first feared. Increases in waterborne diseases and other problems often ascribed to the dam, it is argued, are due mainly to lack of sanitation and the Aswan High Dam deserves great credit for its remarkable contribution to Egypt's socio-economic development (Abu-Zeid, 1990; Biswas, 1992).

In an attempt to provide a solution for the disparities between the distribution of water and demand there is an increasing propensity towards the development of interbasin water transfers (IBTs). Such developments have a long history throughout the region. For example, in 1862 the Nile was connected to the Suez Canal by the construction of the Ismailia Canal in Egypt (Abu-Zeid, 1983). This 128 km long canal created a navigable route between the Nile and the Suez Canal, providing irrigation and potable water along its route. The Ibrahimia Canal was built a few years later, in 1873, along the left bank of the Nile and carries water into, among others, a shallow basin, called the Fayoum Depression, where water can be stored. More recent developments have seen construction of the Jonglei Canal in southern Sudan (see also Pacini and Harper, this volume). The Jonglei diverts 20% of the flow in the White Nile around the Sudd Swamps to irrigate cultivated areas in Egypt and Sudan (Critchfield, 1978; Charnock, 1983). Completed in the early 1980s, the canal is 360 km long and can deliver 4×10^9 m^3 yr^{-1}, some of which is returned to the White Nile further downstream (Bailey and Cobb, 1984). The rationale for this development was the concern expressed in Sudan and Egypt that much of the water in the White Nile was being lost through evaporation from the Sudd swamps; hence, by bypassing these swamps, this water could be put to greater use elsewhere (Bailey and Cobb, 1984).

In order to alleviate water deficits in the northern, coastal regions of Libya, where most of the irrigated agriculture occurs, the government has undertaken an ambitious IBT scheme (El Asswad, 1995). It comprises three stages and will eventually transfer 2.25×10^9 m^3 d^{-1} through 4500 km of concrete piping from the regions of Alkufrah, Assarir and Fezzan, in the south of Libya, northwards to the coast. Costing a total of US\$900 000, the first stage is complete, transferring 2×10^6 m^3 d^{-1} for irrigation. When complete the scheme is expected to meet demands for the next 50 years (El Asswad, 1995). Despite the fact that concerns have been raised around the environmental, social and political implications of such transfers, there is no available information pertaining to the effects of this project (see Snaddon *et al.*, 2000).

The Jordan River was once the only perennial source of water in Jordan. The National Water Carrier, constructed in the mid 1960s, is the transfer route of the Jordan River Project in Israel, which transfers water from the Sea of Galilee (Lake Kinneret) to the central and southern coastal plain (Overman, 1976; Ambroggi, 1977; Gavaghan, 1986). Developed to supply water to Tel Aviv and the Negev region, where it is needed for agricultural purposes, the National Water Carrier has diverted most of the Jordan River's 1200×10^6 m^3 annual flow for irrigation in Israel (Pearce, 1995). Further proposed IBT developments within the basin involve the diversion of water from the Litani River in southern Lebanon to the headwaters of the Jordan (Ambroggi, 1977; Pearce, 1995). This proposal suggests that transferred water should be kept in a 'water bank', to be managed jointly by the governments of the Middle East region. Water would then be allocated from the 'bank' to countries in need. Aside from the obvious political and ecological implications of such transfers there are a myriad other effects. Diversion of the Jordan River water away from the country of Jordan and the Dead Sea through the National Water Carrier has resulted in a reduction of approximately 300 km^2 in surface area, representing 30% of the original area of the Dead Sea. This dramatic reduction has left factories, that extracted potash and other salts from the sea water, stranded several kilometres from the edge of the sea (Pearce, 1995).

Concerns have also been raised about the spread of various aquatic taxa and their parasites by IBTs. Other problems have arisen through the regulation of surface flows and the elimination of the intensity and duration of flood peaks. For example, in Khartoum during the 1950s and 1960s, mass emergence of adult chironomids resulted in huge swarms that entered the suburbs of Khartoum causing allergic problems (Rzsóka, 1976a). The huge densities of *Cladotanytarsus lewisi* between November and March disappeared in April, just before the June floods that resulted in the removal of most of the remaining biomass.

The Tigris and Euphrates rivers support a vast network of marshes, the Mesopotamian Marshes, which cover about 15 000 km^2 of Iraq. With elevation changing only 4 cm km^{-1} over the last 300 km of the Euphrates and 8 cm km^{-1} along the Tigris, these floodplain areas are subject to flooding and inundation, to the order of 1.5 to 3 m, and comprise about 25% of Iraq's surface area. These regions are now threatened by the almost total diversion of the flow of the Euphrates and many of its tributaries. Intensive regulation by dams upstream on both rivers, particularly on the Euphrates in Turkey and Syria, has resulted in a

Table 5.2 Inter-basin water transfer schemes in North Africa and the Middle East (taken from Snaddon et al., 2000)

Country	Scheme	Donor	Recipient	Annual volume transferred ($m^3 \times 10^6 \ yr^{-1}$)	Average transfer rate ($m^3 \times 10^6 \ d^{-1}$)
Egypt	Jonglei Project	Atem (Bahr El Jebel)	White Nile	–	20
Libya	Great Man-Made River Project	Alkufrah/Assarir/ Fezzan regions	Coastal region	–	2
Israel	Jordan River Project	Jordan River (Sea of Galilee)	Central and southern coastal plain	1200	–
Jordan	Red–Dead Canal	Red Sea	Dead Sea	–	–
Iraq	Tharthar Development Project	Tigris	Tharthar Depression	–	–
		Tigris	Euphrates	500	–
		Tharthar Depression	Tigris	600	–

decrease in the magnitude of the seasonal floods. In addition, abstractions from the Euphrates by Turkey, in order to fill the relatively recently completed Ataturk Dam, have demonstrably reduced the flow of the river to the marshes by 10% between 1985 and 1990 (Pearce, 1993a). Eventual implications will be a loss in productivity as these floods fail to deliver sediments and nutrients and provide the appropriate stimulus for species to breed and to migrate.

Designed by British engineers in 1951, the Tharthar Development Project (Table 5.2) has led to the partial diversion of the waters of the Euphrates and Tigris rivers in Iraq (Pearce, 1993a). Before diversion, snowmelt from the Turkish headwaters annually flooded the low-lying marshes near the coast, bringing nutrient-laden silt to major wetlands, known as the Howeiza, Amara and Hammar marshes and the Shatt-al Arab, which is home to the Ma'dan or 'Marsh Arabs'. The marshes are also regarded as the most important wetland bird habitat in Eurasia, with millions of waterfowl feeding on the fish and invertebrates every year. The area is not declared under the Ramsar Convention and its loss will be measured on a global scale. The scheme was originally designed to 'reclaim' land from these vast marshes that naturally link the two rivers. Construction of the scheme began in 1953, with a 20 km canal being built which drained the saline and waterlogged farmlands in the marshes. In December 1992, the canal was extended to 560 km, with a width of up to 1.2 to 2 km in places. This canal, known as the 'Third River', cuts off more than 40 tributaries from the marshes and has allowed the 'reclamation' of some 1.5×10^6 ha of land for farming. The canal now cuts to the west of the Hammar Marsh, crosses the Euphrates in three pipes and eventually drains through a 90 m wide channel into the Persian Gulf along the western edge of the Shatt al-Arab. Furthermore, a series of lock

gates on the easterly Tigris can halt or lower the flow of this river and, together with the construction of levees and an additional 90 km-long canal, the water can be diverted away from the eastern Howeiza Marsh (Pearce, 1993b).

CATCHMENT VERSUS LOCAL ISSUES

The international nature of many of the region's perennial watersheds means that there is often conflict over development rights, stimulated by scarcity of resources and rapidly increasing population pressures. Resulting from its scarcity and important strategic position, water is a crucial resource and key factor in the economic development of the region. The Tigris–Euphrates, Nile and Jordan systems, constituting the region's only large perennial sources of water, all cross international boundaries. The Nile, with its tributaries, has a large basin that covers most of the eastern part of Africa and includes nine different nation states. Obviously with water being such an important resource there are a number of issues that need to be considered. For example, there are local and national state considerations (such as water supply, drainage and waste disposal, transport, fisheries, ecology, leisure) which contrast with international issues related to the upstream and downstream needs of neighbouring states. The absence of rational cooperation and multilateral policy agreements governing the development issues and utilization of these systems will not facilitate maximization of the national welfare (Durth, 1996). Durth (1996) points out that this is complex enough given a homogeneous region in which there already exists a large degree of cooperation and exchange. In heterogeneous social, economic and political regions in which concerns are often conflicting, there is the potential, in the absence of any peace agreements, for this to become largely confounded and confrontational.

There have been a number of research projects, analyses, workshops, discussions and treaty agreements over the shared basins in the region (e.g. Wolf, 1993; Durth, 1996; Kibaroglu, 1996). Many of these still require additional detail and consolidation before agreement can be reached. For example, Kibaroglu (1996) proposed a basis for cooperation between Turkey and Syria over the management and allocation of the waters in the Euphrates–Tigris Basin. Identifying a number of rules and procedures for the management and allocation of transboundary watercourses, Kibaroglu (1996) pointed out the need for further elaboration, but suggested that Turkey could act as a leader for such river basin management (see below: socio-political environment and discussion on the headwaters of the Euphrates–Tigris Basin). Although an apparently capital-intensive technique requiring a high number of skilled individuals, it is proposed that each water dispute could be dealt with in an extensive case-by-case manner. Given the value of the region's rivers and the reported threat to stability in the region, Kibaroglu (1996) argued that such a price may be considered by many to be easily affordable and, indeed, essential. The International Law Commission (ILC) rules have also attempted to deal with the issue and to accommodate incorporation of equitable utilization and the rule of 'no significant harm'. In so doing though, the ILC has run the risk of losing the concepts of reasonableness and the equitable consideration of all factors (Utton, 1996). Utton (1996) points to the Nile and the Tigris and Euphrates (Iraq) river systems as two of four systems in the world where agreements on the cooperative use of international water systems have not been reached, the other two being the Ganges (Ganga; see Gopal *et al.*, this volume) and the Paraná River in South America (e.g. Pringle *et al.*, this volume).

As the region's largest perennial source of water set within the region's arid constraints, the pressures on the Nile are arguably greater than those experienced anywhere else in the world. With its origins deep in the southern parts of the Sudan, the headwaters of the Nile rise along the border with the Central African Republic (CAR), the Democratic Republic of Congo (DRC), Uganda and Ethiopia (see also Pacini and Harper, this volume). These tributaries join at Malakal to form the White Nile. Just upstream of the Sudanese capital of Khartoum, the White Nile is joined by the Blue Nile flowing from Ethiopia. For the remainder of its length, the Nile then flows through the Sudan and into Egypt where it enters the Mediterranean Sea. Along with the history of regional development, the fact that it covers such huge distances and crosses so many countries creates a number of inherent problems. For example, Egypt is the most downstream basin state of the Nile and relies heavily on the river which provides 97% of Egypt's water. As resources dwindle, upstream development pressures are likely to reduce downstream discharge and deprive Egypt of at least some of that which it already receives. In the absence of any basin-wide agreements, reconciliation of the problems surrounding the sustainable and equitable utilization of the Nile seems unlikely. However, this fact has been recognized and acknowledged by the basin states and there are initiatives toward seeking agreement on the shared and integrated development of the Nile (see 'Statutory Framework' below).

While the Nile Basin encompasses and captures the international arena, local issues are focused more on dealing with the day-to-day struggle of securing sufficient resources to sustain agricultural production and to provide potable water supplies. The Nile Basin at present includes five of the world's 10 poorest, or least developed, nations. So, while strengthened cooperation and collaboration are needed to remove the current impediments and to provide mechanisms for equitable allocation and use of the Nile waters, these are often difficult to reconcile with more immediate local concerns. This is true of nearly all of the North African and Middle Eastern basins where local point-source consumption is forced to operate within the constraints of the entire basin. The provision and security of local supplies are inevitably and inextricably linked to the equitable distribution of finite volumes within each of the basin states. Without these international regional agreements in place, the local catchment issues cannot suitably be addressed. While invaluable and essential, local initiatives at reducing consumption through better training and management practices can only serve to postpone the inevitable arrival of constraints that will impinge upon the sustainability and success of local communities. Issues of distribution and access at the local level may also have implications for internal stability, particularly as IBTs assume greater importance. The perceived disparities in the distribution of water and the resulting redistribution from one catchment to another has the potential to increase conflict within and between catchments, as the constraints imposed by diminishing water supplies intensify (see Snaddon *et al.*, 1998).

MAIN AREAS AT RISK

Given the socio-economic realities of the North African and Middle Eastern region, preservation-centred arguments towards conservation are difficult to develop. Faced with increasing and necessary demands, all as yet uncommitted surface waters are under threat of development. In-stream and environmental allocations are difficult to justify given the extant problems of sufficient quantity. Most at risk from future developments are existing dependants in downstream basin states. For example, as pointed out earlier, while Egypt has extensive developments inextricably linked to the Nile River, Ethiopia has no large hydroelectric power installations or modern dams in the Blue Nile Basin. With its origin in the Ethiopian highlands, the Blue Nile passes through several geographic zones and provides 86% of the annual discharge of the Nile (Woube, 1997). Efforts in Ethiopia to divert its waters have brought confrontations with the lower Nile riparian states and there is an increasing perception that 'the river carries the country's valuable natural resources to the lower riparian countries while the Ethiopian people suffer from recurrent drought and famine' (Woube, 1997).

While Egypt encourages countries upstream of its borders to harness the power of the Nile River (which reduces the siltation of Nile water management projects), it uses a vast quantity of the water for agricultural, industrial and domestic use such that only 2% of the flow reaches the sea (Swain, 1997). The future availability of water in Egypt is affected at three different scales (Conway *et al.*, 1996): global (climate change), regional (land-use change) and river basin (water-resources management) scales. All have implications for Egypt's future development. The combined effects of these driving forces on future water availability in Egypt range from large water surplus to large water deficit by the year 2050. The future status is dependent more upon Egypt's approach to population growth, agricultural policy and human aspirations for greater water use in the future (Conway *et al.*, 1996).

Given the interlinked and interconnected nature of large river–floodplain systems, the marshes and wetlands reliant upon annual or asecular flooding, and the wildlife dependent upon them, represent some of the region's most vulnerable environments (see Scott, 1995). Seasonal vegetation providing valuable fodder and grazing for domestic and wild animals has given way to more permanent established vegetation comprising mainly mono-specific agriculture. The cycle

of flooding and processes of sediment transport have provided these areas with fertile and highly productive soils, which has made them attractive options for reclamation (see also Pacini and Harper, this volume). These floodplain environments are heavily reliant upon the input of flood-derived sediments for maintenance of productivity and, while traditional societies operated with migrating floodplain agriculture, more technical solutions are required in the absence of flooding.

Information for conservation

SCIENTIFIC DATA

Although there is a long history of learned societies associated with the North African and Middle Eastern region, formalized scientific research into the aquatic ecosystems of parts of North Africa and the Middle East only began in the later part of the 19th and in the early 20th centuries (e.g. Gervais and Jasienski, 1879; De Chaignon, 1904). These early works typically were taxonomic accounts focused on species distribution and occurrence, and limited primarily to larger vertebrate species such as fish (Gervais and Jasienski, 1879; De Chaignon, 1904; Pellegrin, 1921; Seurat, 1940). There were, however, a few early investigations into some of the larger invertebrate species (i.e. the gastropod fauna – Bourguignat, 1868; Bourguignat and Letourneux, 1887; Pallary, 1923) and flora (Gauthier-Lièvre, 1931). Subsequently there was little research across the region. This is highlighted by the fact that between 1880 and 1940 there were roughly 100 referenced taxonomic works for Tunisia, while for the period between 1940 and 1970 there are a few more than 10 referenced works (Zaouali, 1995). Little work was carried out in Tunisia until the 1970s, at which time earlier works were expanded upon by Khallel (1974), Boumaiza (1984) and Kraiem (1993) (see Zaouali, 1995).

As a result of this early work and continued research interests in the region, comprehensive species lists and taxonomic information exists for some areas and for certain taxa (i.e. Wagner, 1992; Fadl *et al.*, 1993; Gussev *et al.*, 1993; Salman, 1993; Gheit, 1994; Smit, 1995). The emphasis of aquatic research, however, has been directed historically toward lentic systems; despite the long history of river use, rivers have received attention only relatively recently. For example, Moubayed and Laville (1983) produced the first faunistic inventory for Chironomidae of the Lebanon. A total of 142 taxa have been identified from the Oronte and Litani rivers in the

Bekaa Plain and from a river near Beyrouth. Of the recorded species, 81 are new records for the eastern Mediterranean while 131 are new records for the Middle East. The only previous records were for lake species from Israel. Similarly, a recent analysis of data from the literature between 1955 and 1986 indicates that 65 species of chironomids have been recorded and described from Morocco (Azzouzi and Laville, 1987). More recent collections of chironomid species from various rivers flowing through the towns of Fes, Meknes, Khenifra and Marrakech have now increased the number to 134. These included 16 Tanypodinae, eight Diamesinae, 42 Orthocladiinae, 45 Chironomini and 23 Tanytarsini. Of these, 11 species are of Ethiopian origin and six are considered indigenous to North Africa: *Rheopelopia* n.sp., *Telopelopia maroccana*, *Rheomus alatus*, *R. yahiae*, *Stempellina almi* and *Virgatanytarsus ansatus* (Azzouzi and Laville, 1987).

Studies on other groups such as the Ephydridae (Vitte, 1988) and the Simuliidae (Clergue-Gazeau *et al.*, 1991) have similarly revealed a number of new species and distribution records. From these it is apparent that there remains a large amount that is unknown about the invertebrate fauna of the North African and Middle Eastern region (e.g. Gheit, 1994; Smit, 1995). There are numerous other examples from the recent literature; for example, Lohmann (1990) described what would appear to be the first record of *Enallagma cyathigerum* (Charp.) (Zygoptera: Coenagrionidae) from North Africa. There have also been other recent major publications, for example the *Freshwater Ichthyogeography of the Levant* (Krupp, 1985).

The biogeography of the North African and Middle Eastern fauna seems to reflect the complex geological history of the region (e.g. Rzóska, 1976b; Dumont, 1979, 1986a,b), resulting in greater affinities with the Palaearctic fauna of the European continent than the sub-Saharan African fauna (e.g. Stauder and Meisch, 1991; Vitte, 1991). The North African fauna, for example, is predominantly of Palaearctic origins, moving east into Egypt. However, the Nile provides a channel for the spread of African species (e.g. Rzóska, 1952; Dumont, 1986b) which is reflected in the greater proportion of African species in Egypt. Aside from the Nile, the Sahara has provided an effective barrier to the northward spread of the African fauna (Rzóska, 1976b). The geographical distribution of chironomids of the Lebanon region, for example, indicates that this region is placed at the cross-roads of Palaearctic, Ethiopian and Oriental influences. The Simuliidae of the Maghreb also show what is essentially a Palaearctic fauna of a west or circum-Mediterranean distribution,

reaching its southern geographical limit. In the Simuliidae of the Maghreb region, however, the Ethiopian and Oriental elements are poorly represented while some North African species are endemic or penetrate weakly in to Europe through the south-west (Clergue-Gazeau *et al.*, 1991). Within the crustacean fauna currently inhabiting the region there are essentially six historico-biogeographical 'strata' that can be recognized (Por, 1986). These are: (1) subterranean and interstitial forms (see also Dumont, 1986b) which appear to extend back to the Permo-Triassic when Pangaea was still intact; (2) subterranean and interstitial species with an origin in the Mesozoic when the Tethys Sea was present; (3) euryhaline species related to the Miocene Messinian salinity crisis; (4) marine forms which originated with the Pliocene recolonization of the Mediterranean and Red Seas; (5) wet-tropical and wet-temperate species which migrated through adjacent river systems during the Pleistocene climatic fluctuations; and (6) Lessepsian migrants between the Mediterranean and Red Seas after the opening of the Suez Canal.

Increasingly, molecular techniques are being used to elucidate and identify patterns in biogeography and dispersal as well as to resolve some taxonomic affinities (Machordom and Doadrio, 1993; Urbanelli *et al.*, 1996). Molecular investigations of Iberian barbel species suggest that isolation of the Iberian Peninsula from Europe since the Oligocene–Miocene may offer an explanation of the genetic affinities of the Iberian barbels with those of North Africa rather than with the European group (Machordom *et al.*, 1995). Biosystematic and evolutionary studies of the sub-genera *Barbus* and *Labeobarbus* from Morocco show that *Barbus* is probably of Asian origin, resulting from allopatric radiation, while the sub-genus *Labeobarbus* has strictly African distribution. This is supported by biogeographical studies of the parasitic fauna of the genus *Barbus* which indicates exchanges between Africa and the Iberian Peninsula. The existence, in the latter region, of parasitic forms from Asia, that are absent from North Africa, underlines the enigma of the origin of Iberian *Barbus* species (El Gharbi *et al.*, 1993).

Given the increasing propensity towards the redistribution of water between river catchments, these recent investigations give reason for concern. Biometrical investigations on populations of *Barbus callensis* collected in different streams, rivers and reservoirs from across Tunisia showed that the north-western population, belonging to an Algerian catchment, was ecologically and genetically isolated from the other Tunisian populations, which appeared more uniform (Kraiem, 1993). Similarly, enzyme electrophoresis on horizontal starch gel carried out on 278 individuals, again of the species *Barbus callensis*, from 10 rivers in Tunisia, combined with an ecological investigation, showed clear differentiation of the two samples from north-western Tunisia (Berrebi *et al.*, 1995). This differentiation was only partly correlated with the ecological characteristics of the rivers they inhabit. Such results do not indicate any genetic cline, but rather a discontinuity between populations in the north-westernmost watershed and the other Tunisian populations. This differentiation probably has a palaeohistoric origin not only related to adaptation to ecological conditions, but also to difficulties in colonizing the watersheds. It would appear from similar molecular investigations that there is some sort of historical impediment to the migration of other fish species between Morocco and the western Algerian highlands and those river basins to the east, in eastern Algeria and Tunisia (Machordom and Doadrio, 1993).

According to the definition of the International Union for the Conservation of Nature and Natural Resources/United Nations Environment Programme/World Wide Fund for Nature (IUCN/UNEP/WWF, 1991) – there are three specific objectives of conservation: (1) to maintain essential ecological processes and life support systems; (2) to ensure that the utilization of species and ecosystems is sustainable; and (3) to preserve genetic diversity. The very nature of river basins provides natural isolating mechanisms that safeguard genetic diversity. The nature of the information presented above would suggest that to transport or transfer water between these river catchments would risk the breakdown of the natural biogeographic barriers and compromise the genetic integrity of the natural populations. Such precautions have not previously been considered in the development and construction of IBTs (e.g. see Davies *et al.*, this volume).

While information relating to taxonomy and species distribution is relatively well documented, information relating to the ecological functioning of the rivers of the region is lacking. Harrison (1995) points out that much of the work carried out thus far in the north-eastern parts of the African continent has been exploratory in nature, although information for a few systems is often very good and very detailed. For example, the data for the Nile system on its discovery and exploration, history, hydrology, ecology and management have been the subject of numerous publications already (e.g. Rzóska, 1976a,b). Much of the information pertaining to other systems is rarely

quantitative, however, typically relying on qualitative or descriptive speculation (although see Gagneur, 1994; Gagneur and Thomas, 2000). Given the nature of the region's river systems, and the dominant derivation of theoretical paradigms from the northern temperate regions, such speculations may be misplaced.

Maamri *et al.* (1994) examined leaf litter input under the warm Mediterranean climate of northern Morocco and found that peak inputs were delayed by about one month, when compared with the earlier (summer) litter fall generally mentioned in the Southern Hemisphere under similar climates, and included a greater proportion of leaf material. A comparison of inputs on the banks and in the water under the closed canopy showed that much litter presumably drifted downstream and that downstream reaches received partly processed matter from upstream. This pattern of organic processing was reflected by the aquatic invertebrates which upstream were predominantly shredders (mainly gastropods), with peak numbers in October–November and June, and collectors (mainly chironomids) downstream. Evidence from other rivers within the region, however, shows a lack of conformation with the predictions of the River Continuum Concept (RCC) of Vannote *et al.* (1980). The River Dan, for example, is the largest of the headwater rivers of the Jordan. The karstic exsurgence of the Dan has a seasonally stable output in which Por *et al.* (1986) found a total of 156 taxa. Most of these were almost exclusively of Palaearctic origin and about half were limited to the northernmost part of Israel. The faunal complex described by Por *et al.* (1986) does not present any longitudinal zonation, nor does it present any seasonal changes in species composition. Furthermore, the majority of rivers throughout the Arab Region experience at least annual, if not more prolonged, cessations in surface stream flows. The implications of this and of the discontinuity in the river continuum have not yet been properly addressed (although see Stanley *et al.*, 1994; Boulton *et al.*, this volume). These streams typically lack the well-canopied headwaters and longitudinal linkages required to fulfil the predictions of concepts such as the RCC. Small-scale secular and aseasonal transitions in flow, between the presence and absence of surface waters, create transitions between the terrestrial and aquatic condition similar to those outlined for the Flood-Pulse Concept of large tropical and sub-tropical rivers (Junk *et al.*, 1989; see Puckridge *et al.*, 1998). Despite some exceptions and excluding many of the larger rivers, these systems would not appear to adhere to the predictions of longitudinal paradigms of river ecosystem functioning and are particularly susceptible to water diversions and flow regulation. Such activities reduce the intensity and frequency of floods and base flows, alter the normal stream–floodplain interaction (Boulton and Lloyd, 1992; Boulton *et al.*, this volume) and change water quality conditions (e.g. Tuch and Gasith, 1989; Resh and Gasith, 2000).

While the effects of river regulation are fairly well documented, most of the empirical data and information have been derived from perennial systems in temperate and tropical regions. The effects of regulation and the elimination of the seasonal flow of water in intermittent systems has yet to be examined. Many of the region's rivers are highly variable and unpredictable, inhabited by an opportunistic fauna that may be able to tolerate such impositions. However, organisms adapted to, and having evolved to, the seasonal cues provided by more predictable patterns in discharge, such as a seasonal hydrograph, will suffer reductions in growth, productivity and survivorship. The effects of impoundment and river regulation on temporary river systems have not really been addressed. With the development of dams now directed toward intermittent systems, such as the highly seasonal Al Authaim River, there is an urgent need for the development of dryland protocols that recognize and address the functioning and maintenance of dryland rivers. The extrapolation of northern, temperate limnological paradigms needs to be critically examined within the climatic constraints of the Arab Region and protocols for the conservation and management of river systems need development within the region's own constraints. One of the problems now, though, is that given the history of the region and the intense level of utilization, there are few, if any, systems that exist in a pristine condition and much of the damage may already be irreversible.

Initial perceptions of the situation of research in North Africa and the Middle East can be misleading. There are a number of institutions throughout the region that have at some stage or another published scientific articles and research papers pertaining to lotic ecosystems. A survey of the National Information Services Corporation (NISC) Aquatic Biology, Aquaculture and Fisheries Resources database between 1970 and June 1998 recovered only 142 titles relating to river ecosystems in North Africa and the Middle East. Of these 30% were by authors from institutions in North African countries, 28% from Middle Eastern authors and 42% from institutions outside the Arab Region. The dominance of

publications from external authors is a trend common to many countries of the 'Third World' (see Wishart and Davies, 1998; Wishart *et al.*, this volume, Part II). These institutions include various departments at universities such as (among others) the Hebrew University in Jerusalem, Rabat, Hassan II and Tetouan universities in Morocco, the Lebanese University, Tanta University and the University of Cairo in Egypt, Saudi Arabia's King Fasal University, Alfateh University in Libya and the Université de Annaba in Algeria. The list of publications also included those from a number of government departments and institutes such as Algeria's Institut National Agronomique El-Harrach, the Lebanese National Council for Scientific Research and the Biological Research Centre in Iraq. In a survey of 22 institutions from across the Arab Region, 46% of respondents replied that they received their funding via government sources, with a similar percentage originating from international funding agencies, while 31% reported support of some sort from universities (Ghezawi, 1997).

The problem of research throughout the region remains one that is common to most countries in the developing world (Wishart and Davies, 1998; Wishart *et al.*, this volume, Part II; Denny, 2000). In the absence of a substantial critical mass it is difficult to motivate and establish long-term coordinated research programmes. Individual initiatives go some of the way towards providing the scientific foundations but are often restricted in their geographical and disciplinary coverage. While the ecological information relating to the region's river systems may be limited in scope, the dominant feature of the research is now focused on using and maximizing the scarce water resources they provide. While more recent developments in research have seen the shift away from biotic processes, the region is a leader in the development of Water Demand Management (WDM). Over 77% of reported activities in water research institutions involve some form of WDM work. While this has been due to necessity, the reduction of demand is perhaps the most direct method towards ensuring at least partial conservation of the region's rivers.

CLASSIFICATION AND EVALUATION SCHEMES

Although there have been several attempts to provide qualitative assessments of some of the region's aquatic environments (e.g. Scott, 1995), there are few examples of conservation-oriented classification and evaluation systems. Various procedures have been developed for using correlations between the flow regime and El Niño oscillations to improve the predictability associated with the Nile flood (Eltahir, 1996). Other projects, such as the *Directory of Wetlands in the Middle East*, sponsored by IUCN, WWF, the International Waterfowl and Wetlands Research Bureau, Birdlife International and the Ramsar Convention Bureau, provide valuable assessments and syntheses of available information. The emphasis of the directory, however, was primarily on bird species, but provides a country by country account and a detailed network of individuals and organizations.

Despite the advent of in-stream flow methodologies elsewhere in the world, there appears to have been no development or implementation of such technologies in North Africa or the Middle East (Dunbar *et al.*, 1998). Many of the rivers around the region are temporary, experiencing the cessation of surface stream flow. While in-stream methodologies would appear to be appropriate and applicable to perennial river systems, their relevance to temporary systems remains unknown, as the cessation of surface flow typically coincides with peak demand. Water augmentation management practices often invoke perennial flows in historically temporary systems, the implications of which remain unassessed (although see the Fish River in South Africa: O'Keeffe and De Moor, 1988). The development of dryland-based protocols for the management of these highly variable and often unpredictable and temporary systems is an important priority. Such management protocols would have to assess the implications and importance of the dry phase for the maintenance of ecological processes and the sustainability of the biota.

AVAILABILITY AND DISSEMINATION

While there has been a fair amount of research carried out in some parts of the region, much of it is confined to the local literature, published in local journals belonging to individual research stations or institutions, or remaining as unpublished theses (e.g. see references in Scott, 1995). These studies have covered a wide range of topics, from the general ecology of river systems to the specific functioning of gastropod populations (Al-Dabbagh and Daod, 1985; Al-Dabbagh and Luka, 1986a,b) and those of other invertebrates (Ali, 1976, 1978a,b; Al-Saboonchi *et al.*, 1982, 1986; Salman *et al.*, 1990), as well as studies on aquatic vegetation (Antoine, 1984; Al-Saadi and Al-

Mousawi, 1988) and parasites (Mhaisen *et al.*, 1990). The other problem in disseminating information from the region is in the use of language. From 142 titles relating to river ecosystems recovered in a database search three were published in German, eight in Persian, nine in Arabic, 21 in French and 99 in English. While many papers are abstracted in English (e.g. Vitte, 1988; Wagner, 1992; El Gharbi *et al.*, 1993), a large amount of information is contained within publications written in languages other than English (e.g. see Azzouzi and Laville, 1987; Clergue-Gazeau *et al.*, 1991; Kraiem, 1993; Alouf *et al.*, 1996). A general challenge facing science and conservation initiatives is bridging the disparity between English and other language literature to develop a more cohesive and inclusive forum for the presentation and discussion of information. While local journals provide a valuable forum for discussion and documentation of local issues and research that may not be otherwise accepted into the international literature, there is obviously a problem with accessibility to the broader community (e.g. Tawfik *et al.*, 1990; Alouf, 1991; Slim, 1991).

Access to and dissemination of information is a problem common to many countries of the developing world (e.g. Wishart and Davies, 1998; Wishart *et al.*, this volume, Part II; Denny, 2000) and remains one of the biggest problems facing the region. The development of the Internet has provided a mechanism to overcome some of those inherent problems and to facilitate not only access, but also dialogue between those both within and outside the region. For example, initiatives within the Arab world, such as the Arab Water Information Network (AWIN), based in Lebanon, aims to provide a forum for discussion and interaction through the Internet and the World Wide Web (WWW) on issues pertaining to water. There is also the Egyptian based Centre for Environment and Development for the Arab Region and Europe (CEDARE) Water Information Network which also utilizes the Internet via the WWW and deals with water issues in relation to environmental development. The Middle East and North Africa (MENA) Economic Summit Waternet from Jordan, provides a forum for discussion of water in relation to Arab countries' economic summit processes. There is also the Inter-Islamic Network on Water Resources Development and Management, based in Jordan. This is an information exchange, data bank, research umbrella organization and training institute, which focuses on water management research and implementation. It involves 12 member states, and has been active since 1987.

PUBLIC AWARENESS AND EDUCATION

A number of conferences and workshops have addressed the myriad problems associated with river basins in the Arab Region (International Irrigation Management Institute – IIMI, 1995; Schiffler and Libiszewski, 1995; Green Cross International – GCI, 1998). However, little information is available relating to community-based, public-education initiatives.

The framework for conservation

THE SOCIO-POLITICAL/RELIGIOUS– CULTURAL BACKGROUND

The history of the region is ancient. For thousands of centuries people have lived within the dryland constraints of the area in a delicate balance reliant upon cycles of prosperity interspersed with drought. Many of the world's earliest societies developed around the rivers of the Middle East and North African region. Civilization was well established in Mesopotamia by the 4th millennium BC, and was largely reliant upon sophisticated irrigation systems. As with many early civilizations, the people of this region operated around a series of myths and legends. The flood myth, surviving today in three different versions, but all based on a common prototype, reflects the unpredictable nature of the Tigris–Euphrates River systems. It also highlights the fact that the stresses and problems being experienced today have been common to civilizations throughout the centuries. As the story goes, humanity was created to serve the gods and to relieve them of the necessity of labour. The rapid growth of the human population results in noise which disturbs the gods and the deity Enlil then seeks to reduce their numbers, first through plague and then through successive droughts. When this fails, due to the intervention of another deity, Enki, Enlil then enlists the help of the other deities and they send a great flood. In order to make a fresh start for humankind, Enki spares one family. Today the Sumerian flood myth survives in only a fragmentary state, but it is clearly the origin of all other Mesopotamian versions (Porter, 1993). In it, it offers insights into the early understandings not only of the institutions of Sumerian civilizations and kingship, but also into the understandings of the natural order and irrigation practices.

For three millennia water has been of central importance to life in the Arab world. Remains of the earliest known dam are to be found in modern-day Jordan. Built around 3000 BC the dam, which included a 200 m-wide weir and diverted water via a canal into 10 small rock and earth reservoirs, formed part of a water supply system for the town of Jawa. The largest dam was more than 4 m high and 80 m long. Around the time of the first pyramids, some 400 years later, the Egyptians constructed Sadd el-Kafra, the 'Dam of the Pagans', across a seasonal stream near what is now the city of Cairo. The 14 m-high and 113 m-long mass of sand, gravel and rock, which was retained by 17 000 cut stone blocks, was washed away a decade after commencing construction and before it could be completed. It is thought that because of the annual flooding of the Nile, the dam was probably not built for irrigation or agricultural purposes, but probably rather to supply the quarries. Although there is no evidence of any dams, 8000-year-old irrigation canals have been found in the Tigris and Euphrates basins. It is also known that by 6500 the Sumerians of the region had established a network of irrigation canals that criss-crossed the plains of the lower reaches and were no doubt controlled by small earth-wall dams.

Traditionally issues of water in North Africa and the Middle East have focused on scarcity, but increasingly water quality and, more so, equity, are assuming the mantle as one of the key issues of the region. Most countries within the region are experiencing rapid population growth (Table 5.3). With a mean value of about 2.59%, population growth rates are higher than the mean value of 1.7% for the world's less developed regions (Marshall, 1998). While actual population densities are not considered high by world standards, densities ha^{-1} of arable land are some of the highest in the world. In Bahrain, for example, the density is *ca* 7000 people arable ha^{-1}. Egypt and Kuwait have densities of about 2000, while Israel, Jordan and Lebanon are at about 500 (Rogers, 1994). For comparison, the United States of America has a ratio of fewer than two people arable ha^{-1}. The scarcity and high densities have necessitated the development of good infrastructure and Jordan, Lebanon and Tunisia all have very high levels of access to safe potable water (Table 5.3: Marshall, 1998). These statistics indicate the percentage of the population with access to an adequate amount of safe drinking water located within a convenient distance from the user's dwelling and are derived from data obtained between 1990 and 1994 by the World Health Organization (WHO) and the United Nations Children's Fund (UNICEF). Other countries though, such as the Sudan and Iraq, still have more than half of their population without access to safe potable

Table 5.3 *Demographic statistics for the North Africa and Middle Eastern Region (from Marshall, 1998)*

Country	Population (millions: 1998)	Av. pop. growth (%: 1995–2000)	Access to safe water (% pop.)
Algeria	30.2	2.3	?
Bahrain	0.594	2.7	?
Egypt	65.7	1.9	64
Iraq	21.8	2.8	44
Israel	5.9	1.9	?
Jordan	6.0	3.3	89
Kuwait	1.8	3.0	?
Lebanon	3.2	1.8	100
Libya	6.0	3.3	?
Morocco	28.0	1.8	52
Oman	2.5	4.2	63
Qatar	0.579	2.1	?
Saudi Arabia	20.2	3.4	?
Sudan	28.5	2.2	50
Syria	15.3	2.5	85
Tunisia	9.5	1.8	99
United Arab Emirates	2.4	2.0	?
Yemen	16.9	3.7	?

water. Furthermore, given the scarcity of water across the region, it is somewhat surprising to find that some of the Gulf States rate among the highest per capita users of water in the world (Brooks, 1996).

Today, the issues surrounding the politics of water as a resource and key to the peace process of the Middle East has been the focus of much research and discussion (Cooley, 1984; Wishart, 1990; Starr, 1991; Hillel, 1994; Isaac and Shuval, 1994; Elmusa, 1995; Farinelli, 1997). It would be impossible to try to summarize or to propose solutions to the situation in the Arab Region. The aim of the following discussion is simply to introduce some of the complexities and to show how water is often proposed as being central and intertwined with the political situation. Many centres in the region have been damaged by civil and other conflicts. For example, the outbreak of civil war in Lebanon in 1975, between rival political and religious factions seeking to gain control, effectively destroyed what was then one of the most important financial and commercial centres within the region. The Gulf War, the recent American-led attack on the Sudanese capital of Khartoum, and the Israel–Palestinian conflict are yet other examples, highlighting many of the seemingly intractable problems besetting the region. Although there are continued signs of hope, they seem far from resolute.

Given the apparent hostile political climate and the importance of securing scarce resources, it is often postulated that war over water in the Middle Eastern region is more or less inevitable (e.g. Cooley, 1984;

Bulloch and Darwish, 1993; Homer-Dixon, 1994; International Committee of the Red Cross – ICRC, 1994). It is true that Iraq destroyed much of the water-desalinization capacity in Kuwait during the Gulf War and that Israel bombed the partially completed Yarmouk Dam late during the war that began in 1967. The Yarmouk River constitutes part of the tri-nation border between Syria, Palestine and Jordan and for most of its length (80%) flows through Syria. Prior to the 1960s water from the river had been amicably shared between the basin states following an agreement reached in 1948. After a summit held in 1964 in response to Israeli diversion of waters a decision was taken that allowed Syria, Jordan and Lebanon to channel water from the Jordan River, through Syria and into the Yarmouk tributary. This would have effectively granted control of these waters to Jordan and deprived Israel of valuable resources. It is argued that this dispute was a key factor in the war that broke out in 1967 between Israel and Jordan (Hof, 1995). This region is still contested and key to political peace processes within the region, with some believing that 'the conflict over these rivers will determine the security in the region especially after the Israeli occupation of the Syrian Golan and the 12 kilometres of the Yarmouk' (Isaac and Shuval, 1994). Dam constructions on tributaries of the Jordan River, such as the Yarmouk River, continue to create tension between member basin states. Jordan claims, for example, that Syria is removing more than its share of water from the tributary.

The arguments, the issues and their interpretation, are complex and extremely intricate, and are disguised by a long history of political tension. To suggest that tensions leading from the allocation and utilization of water could initiate such a scenario is to risk ignoring the suite of other factors at play. In simple economic terms researchers from Harvard University have shown that in the Jordan Valley the Middle East Water Project, perhaps arguably the most volatile and disputed point of debate, is worth no more than \$600 million CAD yr^{-1} – an annual cost through water loss well under the daily cost of modern warfare. Such speculations also fail to recognize available alternatives, such as drip irrigation and shifts towards more water-conservative crops. Such technologies would all provide simpler, more feasible alternatives to war. 'Water shortages will aggravate tensions and unrest within societies' but, as opposed to outright warfare, 'internal civil disorder, changes in regimes, political radicalisation and instability' are the more likely consequences (Homer-Dixon *et al.*, 1993).

Invariably though, many water development projects across the region serve dual purposes. Turkey's Southeast Anatolia Project (GAP, from its Turkish acronym), for example, is a series of 22 dams in the headwaters of the Tigris and the Euphrates Rivers. At a cost of \$32 billion the project will provide internal benefits through irrigation and hydropower generation. However, there are also political overtones, both internationally and internally. The project will help to secure Turkey's political position in a poverty-stricken and remote area that is currently the stronghold of the militant separatist Kurdistan Workers Party (PKK). Upon completion the project will also give Turkey's political structures control over large proportions of Syria and Iraq's water supply. Perhaps the political objectives, although not stated at government levels, are best summed up with this quote from the site manager of the Ataturk Dam to an American reporter in 1993. 'We can stop the flow of water into Syria and Iraq for up to eight months without the same water overflowing our dams, in order to regulate their political behaviour' (Barham, 1994; Kaplan, 1994; both in McCully, 1996).

STATUTORY FRAMEWORK (NATIONAL AND INTERNATIONAL LEGISLATION)

Despite the political tensions that still dominate much of the region, many countries of the region are now entering into bi-lateral or multi-lateral agreements in order to secure water supply and development rights to the region's many shared basins. While such agreements are set within a fairly tense political environment, today's negotiations and agreements reflect the fruition of years of negotiation. It should also be noted though that despite peace agreements reached in 1994 between Israel and neighbouring states, recent joint proposals by Jordan and Syria, addressing cooperation relating to dam construction on the Yarmouk, have met with opposition from Israel. Although military intervention can no longer realistically be considered an appropriate action, Israel continues to lobby the World Bank and the USA to stop consideration of the project (Cooley, 1984; Lowi, 1993; Hof, 1995). Such disputes continue to highlight not only the value of water as a resource, but also the regional issues that are brought about through scarcity and differences in political ideologies.

The Convention on the Law of the Non-Navigational Uses of International Watercourses adopted by the General Assembly of the United Nations on 21 May 1997 has 11 Signatories and four Parties. Of the North African and Middle Eastern countries Jordan still has to ratify and accept the signatory agreement that it made on 17 April 1998. The Syrian Arab Republic was the first country in the world to sign, on 11 August 1997, with the agreement being ratified on 2 April 1998. The agreement remains open to all States and regional economic-integration organizations for signature until 21 May 2000. Meanwhile the Syrian Arab Republic lodged reservations, stating that its acceptance of the Convention and its ratification by the Government shall not under any circumstances be taken to imply recognition of Israel and shall not lead to its entering into relations therewith that are governed by its provisions. In regard to the reservation made by the Syrian Arab Republic upon ratification, Israel also lodged objections on 15 July 1998.

'In the view of the Government of the State of Israel such reservation, which is explicitly of a political nature, is incompatible with the purposes and objectives of this Convention and cannot in any way affect whatever obligations are binding upon the Syrian Arab Republic under general international treaty law or under particular conventions. The Government of the State of Israel will, in so far as concerns the substance of the matter, adopt towards the Syrian Arab Republic an attitude of complete reciprocity.'

There are other more positive developments within the region though, such as recent developments around one of the region's largest and most contested basins. The following is taken from a report on the history and progress of the Nile Basin Initiative. Launched in 1992 by the Council of Ministers (COM) of Water Affairs of the Nile Basin States, the Nile Basin Initiative was to promote cooperation and development within the Basin. The COM provides, among others, a forum in which all the Nile River Basin states can deliberate issues of mutual interest pertaining to the development and management of the Nile Water resources, on an equitable basis. Six of the riparian countries (the DRC, Egypt, Rwanda, Sudan, Tanzania and Uganda) formed the Technical Cooperation Committee for the Promotion of the Development and Environmental Protection of the Nile Basin (TECCONILE). The other four riparian states participated as observers. Within this framework, the Nile River Basin Action Plan (NRBAP) was prepared with support from the Canadian International Development Agency (CIDA).

At the Sixth meeting of the COM responsible for water affairs in the Nile Basin countries, held on 2 March 1998, in Tanzania, one of the main points on the agenda was to deliberate and to reach agreement on the outcome of the report of the revised NRBAP for basin-wide development. This two-day meeting was attended by eight of the riparian states, with seven Ministers, from Egypt, Ethiopia, Kenya, Rwanda, Sudan, Uganda and the host country Tanzania, attending. Burundi was represented. The DRC and Eritrea were the only countries not to attend. This work was being prepared jointly with technical assistance from the World Bank, CIDA and the United Nations Development Programme (UNDP) in consultation with experts from all Nile Basin countries. The report, among others, points out that:

- 'The Nile River constitutes key natural resources in our respective countries and whose potential largely remains underdeveloped in most of the countries. There are emerging conflicts over access and use of the resource among the countries.'
- 'Five out of the 10 poorest or least developed countries in the world are found in the Nile Basin. Strengthened co-operation and collaboration is needed to remove the current impediments including inequity in the access and use of the resource and set a framework and mechanism to promote equitable allocation and use of Nile Waters

for socio-economic development of the riparian countries.'

The meeting overcame a number of impediments related largely to representation, with record participation, and introduced discussion on issues that had previously barred progress. The most important product of the meeting was the agreement reached on transitional institutional arrangements in which all riparian states were keen to be full members. Other areas where consensus was reached included:

- recognition that a solution must be found to managing and sharing the resources of the basin;
- willingness to accommodate the various views and concerns that were aired;
- a belief that a shared vision can only be legitimized by action on the ground; actions that benefit the poor and the disadvantaged in particular.

As an outcome of this meeting the COM agreed on a transitional mechanism comprising a Council of Ministers responsible for Water Affairs, a Technical Advisory Committee (TAC) and a Secretariat.

Currently, the UNDP is supporting a project aimed at developing a framework acceptable to all riparian states for basin-wide cooperation, while the World Bank, together with CIDA and UNDP, is assisting in reviewing the NRBAP. Other External Support Agencies' representatives present at the meeting included those from the FAOUN, UNEP, Swedish Agency for Development Cooperation (SIDA) and United States Agency for International Development (USAID).

In July 1998 senior officials from eight Nile Basin countries (Burundi, DRC, Egypt, Kenya, Sudan, Uganda, Tanzania and Ethiopia) met in Dar es Salaam for the first meeting of the Nile Technical Advisory Committee (Nile-TAC) which was created by the Nile COM for water affairs of the Nile Basin states. Following agreement on the terms of reference and *modus operandi* of the Nile-TAC there was agreement on the establishment of a secretariat followed by preparation of a set of policy guidelines for approval by the Nile-COM. The consensus achieved and the agreement reached on a shared vision for the development of the Nile Basin to the benefit of all its peoples was felt to be a reflection of the high degree of cooperation between the countries of the Nile Basin.

So far one of the projects, the objective of which is to develop a cooperative framework for management of the Nile, is under implementation with UNDP funding

Table 5.4 *North African and Middle Eastern signatories to some of the aquatic-related international conventions. The numbers indicate the year of signing (or acceptance (Ac)) with those in parentheses indicating the number of listed World Heritage sites*

	Ramsar	World Heritage	Biodiversity	Bonn CMS[a]
Algeria	84 (2)	74 (7)	95	–
Bahrain	98 (2)	91	96	–
Egypt	88 (2)	74 (5)	94	83[b]
Ethiopia	–	77 (7)	94	–
Iraq	–	74 (Ac) (1)	–	–
Israel	97 (2)	–	95	83
Jordan	77 (1)	75 (2)	93	–
Kuwait	–	–	92[b]	–
Lebanon	–	83 (4)	94	–
Libya	–	78 (5)	92[b]	–
Morocco	80 (4)	75 (6)	95	93[b]
Oman	–	85 (Ac) (3)	95	–
Qatar	to be ratified	84 (Ac)	96	–
Saudi Arabia	to be ratified	78 (Ac)	–	91
Somalia	–	–	–	86
Sudan	–	74	95	–
Syria	98 (1)	75 (Ac) (4)	96	–
Tunisia	–	75 (8)	93	87
United Arab Emirates	–	–	92[b]	–
Yemen	–	80 (3)	96	–

[a] Bonn CMS = Bonn Convention on Migratory Species
[b] signed but not ratified

(Project D3, the Cooperative Framework). The World Bank has also been asked by the COM to play a lead role coordinating the inputs of external agencies to finance and implement the NRBAP.

Many North African and Middle Eastern countries are signatories to a number of international agreements (Table 5.4). There are also other regional agreements, such as the Convention for the Conservation of the Red Sea and Gulf of Aden Environment, to which several countries have agreed. Iraq, for example, is signatory to a number of treaties and international agreements, although notable perhaps, it is not party to the Bonn Convention on Migratory Species or the Ramsar and Biodiversity conventions. This is significant in light of the importance of Tigris–Euphrates wetland systems. Many of these agreements are focused on the migratory bird species, with the region's wetlands and estuarine environments providing an important staging post for migratory waders.

At national level, most countries have centralized water management sectors. Given the importance of the river systems to irrigation and agricultural practices, these sectors typically share very close ties to the agricultural sector. This is particularly true for

the Middle East (Brooks, 1996). Most countries also have policies developed to cover the fishing and hunting of wildlife, but little enforcement- or conservation-directed policy. The Iraqi Government, which has no national conservation strategy, has taken no special measures nor provided legal protection for any wetland system. Again, legislation has been introduced to protect fish during spawning through a ban on fishing in the spawning season, but no steps have been taken to put in place structures to ensure implementation and it is, therefore, largely disregarded. Similarly, there is no legislation in Jordan that specifically protects wetland areas. The development of the National Environmental Strategy Plan (1988–91) in Jordan was coordinated by the Department of Environment and included a review of the national legislation. The National Environmental Strategy was then ratified by the government in 1992 and the Law of Environment presented to government. The current status of this amendment is not known.

ROLE OF VOLUNTARY ORGANIZATIONS

A range of international aid organizations operate throughout the region, such as the International

Development Research Centre in Canada which has been helping to identify and develop capacity throughout Africa. An example of their work is the Dryland Water Management Program Initiative (DWMPI) which is aimed at encouraging greater efficiency in the use of water resources through the management of water. Other organizations, such as Birdlife International, with a regional office situated at the Royal Society for the Conservation of Nature (RSCN) in Amman, Jordan, are involved in more environmental pursuits. Established in 1944, Birdlife International monitors the status of important bird areas throughout the Middle East and is involved in the management of protected areas.

EcoPeace is a Middle Eastern Environmental Non-governmental Organization (NGO) forum that has formed a consortium of Egyptian, Israeli, Jordanian and Palestinian environmental NGOs. The consortium aims to work together to promote environmentally sustainable development in the region and plans to make climate change a major issue within the Middle Eastern NGO community. There are a number of other initiatives aimed at further consolidating regional networks, such as the Middle East Water Information Network (MEWIN), based at the University of Pennsylvania. MEWIN is an international, non-profit, professional association founded with the assistance of the Ford Foundation in 1994 by a group of leading water specialists from nations of the Middle East, Europe and North America. Its purpose is to improve the management and conservation of water resources in the Middle East, to promote their peaceful, cooperative use, motivate sound environmental planning in the region, and to encourage the sharing and exchange of information and data which is deemed essential to the achievement of all of MEWIN's goals. The network has produced the 'MEWIN Directory of Individuals and Organizations Specializing in Middle East Water Resources' which provides contact and specialized information on approximately 400 individuals and 150 organizations.

Conservation in practice

APPLICATION AND POLICING OF LEGISLATION

Although many countries are signatory to international agreements they face the problems common to many of the world's developing regions. Governments are

hampered by lack of finances and technical staff, administrative backlogs, inappropriate or expensive institutional structures and poor legislative staff (see Scott, 1995; Wishart *et al.*, this volume, Part II). Such problems are common to countries throughout the developing or Third World and need to be contextualized within a broader global framework. Interestingly, the United Nations has estimated that in 1988 there were 70 000 Africans working abroad and 80 000 expatriates working in Africa. Due to innate shortcomings, policies in a number of countries to enhance water quality have not been fully successful in achieving their goals (Khouzam, 1996). Khouzam (1996) presents an alternative policy that aims to avoid pollution generation at its source and comprises an array of options to replace polluting inputs or techniques.

During 1995, the World Bank undertook a number of shifts in aquatic resource management investments as part of its long-term strategy for managing water resources that underlies its Water Resources Management Policy (Lintner, 1996). The first shift emphasized the need to be more comprehensive in its approach. As a result, the decision was taken to continue providing funding for Environment Programmes, not only for marine areas but also for regional and basin-level programmes in freshwater management. The focus of this funding was to promote integrated management in, among others, Africa and the Middle East. The second shift was towards funding programmes that lead to the adoption of preventative measures for pollution control in marine environments. Strategic water-management investment, with the coupling of complementary actions of regional environment programmes, national water resource strategies and management programmes for river basins, lakes and coastal areas provided the third shift. The fourth initiative was to take pilot activities and to apply them more broadly (i.e. bringing them into the mainstream). This is reflected in some of the more recent activities in the North African region that has seen the World Bank involved in reviewing the NRBAP. The World Bank has also been asked by the COM of Water Affairs of the Nile Basin States to play a leading role in coordinating the inputs of external agencies to finance and implement the NRBAP.

RESTORATION AND REHABILITATION PROGRAMMES

Given the existing scientific framework there is very little information that could feed into any restoration or rehabilitation programmes. There is, however, an increasing recognition of the need to conserve water resources across the region, and all three countries of the Maghreb Region – Morocco, Tunisia and Algeria – have launched activities to promote soil and water conservation (Matoussi, 1996). These measures and their objectives include the reforestation of watersheds, construction of new, smaller dikes and dams to retain water higher in the hills and mountains, evaluation of additional resources and optimization of conjunctive management of surface water and groundwater. There is also a growing awareness that if the water resources of the Nile Basin are to be utilized on a sustainable basis, by the lower and upper Nile riparian countries alike, then there is an urgent need for political stability, understanding, and environmental rehabilitation (Woube, 1997). Deterioration of the Blue Nile Basin has resulted in losses in the amount of water and soil nutrients available to the Sudan and Egypt. The proposed implementation of a number of new water-resource development projects on the Blue Nile in Ethiopia has recently emphasized development through conservation-based sustainability measures (Woube, 1997).

With high levels of utilization and a heavy reliance upon all of the water resources of the region there is a need to assess the role and importance of perennial systems in maintaining the aquatic biota. In order to manage temporary river systems effectively, both for conservation and human utilization, there is also an urgent need to assess the faunal similarity of perennial and temporary systems. In regions like North America, where there is little similarity (Williams and Hynes, 1976, 1977; Wright *et al.*, 1984; although see Delucchi, 1988), temporary systems require management directed at the maintenance and preservation of the fauna which is unique and characteristic. Relatively greater disparity in the amplitude of the physical and chemical constraints imposed upon the fauna in the temporary and perennial streams of more temperate climates suggests independent adaptation to the conditions in these different systems. It would also suggest that perennial systems do not provide a seeding source of new individuals. In contrast, in regions such as Australia, where there is a large degree of faunal similarity between perennial and temporary streams, the implications of inducing or increasing the period of flow cessation may not be as adverse as in northern, temperate systems.

The levels of faunal similarity throughout would suggest that many species can tolerate the absence of surface stream flow, either through recolonization from perennial sources or through the adaptation of resistant stages. That said, many species in fact require nearby permanent sources of water from which to recolonize. For example, work on the Sabie River in Mpumalanga, South Africa, has shown that permanent tributaries act as refuges from which fish can recolonize areas of temporary streams following the onset of flow (O'Keeffe *et al.*, 1996). If these tributaries dry up the refuge is effectively removed, with important implications for the aquatic biota. Obviously the need arises to ensure that a balance is maintained or, at least, that some core areas are preserved in which perennial flows are guaranteed, providing refugia and serving as sources for recolonization following the onset of flow. The River Dan, for example, is considered to be a post-Pleistocene river which has provided an important refugium for a wide area of aquatic water bodies, including the presently drained Lake Hula (Por *et al.*, 1986).

FUTURE THREATS AND THEIR MITIGATION

Water resources in North Africa and the Middle East are central to the success of the region's future prosperity. Given the scarcity and the disparities in the distribution of water resources, the challenges facing conservation and sustainable use seem daunting, if not insurmountable. Without doubt the greatest threat, both now and in the future, is the framework of human consumption set against a finite resource. Under immediate threat are those sites suitable for dam construction, but which as yet are undeveloped. With many of the larger potential dam sites already exploited, or earmarked for development, water-resource developers are now directing their attention to more numerous small-scale projects. For example, Tunisia has embarked on a 10-year programme for the construction of 1000 small storages in the hills and 200 small mountain dams, with the aim of 'recovering' 160×10^6 m^3 of water yr^{-1} (Matoussi, 1996). Unless investigations and analyses of the viability, suitability and implementation of demand-orientated management solutions accompany such developments they will be of little long-term sustainable benefit.

For future conservation efforts to be successful they will need to be directed through programmes aimed at minimizing the impacts of human development. The implementation of water demand-management strategies, such as realistic water pricing and step tariffs, introduction and establishment of sustainable irrigation technologies and strategies, increased public awareness and research, and the integration and decentralization of management, have all been undertaken in some form or other within the region (Matoussi, 1996). Other alternatives, such as water recycling, provide viable options to meet shortages and to provide return flows, but are all dependent on the quality of the returning water (Khouzam, 1996). In response to declining water quality and in an effort to improve the quality of return waters, there is increasing utilization of constructed wetlands for the treatment of secondary effluents to ensure safe river discharge (Green *et al.*, 1996). Results from experiments carried out in constructed wetlands on the Alexander River in Israel have shown very efficient reductions in BOD and TSS. Removal efficiency of nitrogen and phosphorus compounds, however, displayed highly variable responses within a very wide range from 95% to zero. Currently, the re-use of water in the region is most advanced in Israel, where 70% of sewage is treated and used to irrigate 19 000 ha of cropland. It is predicted that by the year 2010, re-used and recycled water will provide an estimated one-fifth of Israel's total water supply and one-third of its irrigation (Watzman, 1995).

Many of the features of the North African and Middle Eastern region are shared with the countries of southern Africa. Although there is as yet little formalized communication between the member states, many of the scientific principles and policy options, including demand management, decentralization of water management to communities and privately owned local structures, water harvesting and small-scale agriculture, could benefit from an increased dialogue and exchange given the similarities (Rached, 1996). The establishment and collaboration, for example, of training programmes such as those carried out in Egypt and Morocco has improved on-farm management skills. Introduced through farmers' organizations they have resulted in improved irrigation efficiencies of up to 10–15% overall and as much as 30% in productivity (Xie *et al.*, 1993). Both North Africa and the Middle East, and southern Africa, represent dryland regions faced with burgeoning population growth and incumbent pressures, and both could benefit greatly from mutual cooperation in developing scientific principles upon which to base management and conservation policies. Southern Africa is moving towards greater formalized regional structures, such as the Environment and Land Management Sector of the Southern African Development Community, in recognition of the challenges that lie ahead (Davies and Wishart, this volume). The political problems and tensions facing the North African and Middle Eastern region have long

since provided an impediment to such dialogue and regionalization. However, with continued negotiations and an increasing realization of the necessity for cooperative management of the region's rivers it is hoped that similar initiatives will soon reach fruition.

Acknowledgements

We would like to thank Phil Boon and Bryan Davies for the invitation and opportunity to present this work and for their continued support throughout. Special thanks also go to all those individuals and agencies who provided information, critical discussion and thought-provoking comment.

References

Abu-Zeid, M. 1983. The River Nile: main water transfer projects in Egypt and impacts on Egyptian agriculture. In: Biswas, A.K., Dakang, Z., Nickum, J.E. and Changming, L. (eds) *Long-Distance Water Transfer. A Chinese Case Study and International Experiences*. Tycooly International and United Nations University, Shannon, 15–34.

Abu-Zeid, M. 1990. Environmental impacts of the High Aswan Dam: a case study. In: Thanh, N.C. and Biswas, A.K. (eds) *Environmentally-Sound Water Management*. Oxford University Press, Delhi.

Afifi, A.K. and Osman, H. 1993. *Water Losses from Aswan High Dam*. In: Egyptian National Committee on Large Dams (ed.). *High Aswan Dam Vital Achievement Fully Controlled*, ENCOLD, Cairo, 153.

Al-Dabbagh, K.Y. and Daod, Y.T. 1985. The ecology of three gastropod molluscs from Shatt-Al Arab. *Journal of Biological Sciences Research* 16:155–167.

Al-Dabbagh, K.Y. and Luka, J.K. 1986a. Respiration studies and population metabolism of the gastropod *Theodoxus jordani*. *Freshwater Biology* 16:449–453.

Al-Dabbagh, K.Y. and Luka, J.K. 1986b. Population dynamics of the gastropod *Theodoxus jordani* in Shatt-Al Arab River. *Freshwater Biology* 16:443–449.

Alghariani, S.A. 1997. Man-made rivers: a new approach to water resources development in dry areas. *Water International* 22:113–117.

Ali, H.A. 1976. Preliminary study on the aquatic beetles of Iraq (Haliplidae, Coleoptera). *Bulletin of the Basrah Natural History Research Centre* 3:89–94.

Ali, H.A. 1978a. A list of some aquatic beetles of Iraq (Coleoptera: Dytiscidae). *University of Baghdad: Bulletin of the Natural History Research Centre* 7:11–13.

Ali, H.A. 1978b. Some taxonomic studies on the aquatic beetles of Iraq (Coleoptera: Gyrinidae). *University of Baghdad: Bulletin of the Natural History Research Centre* 7:15–20.

Al-Layla, M.A. and Al-Rawi, S.M. 1988. Impact of Mosul Textile Factory effluents on Tigris River water quality. *Journal of Environmental Science and Health, Part A. Environmental Science and Engineering* 23:559–568.

Allen, T. 1994. Water: a substitutable resource? Department of Geography, School of Oriental and African Studies, University of London, London. Unpublished paper.

Alouf, N. 1991. *Invertebrate Study on Some Lebanese Rivers. Ten Years Report (1975–1984).* Beirut.

Alouf, N., Dia, A., El Zein, Gh., Hamze, M. and Slim, K. 1996. *Study of Biological Diversity of Lebanon. Aquatic Fauna and Flora [Étude de la Diversité Biologique du Liban. Faune et Flore Aquatiques].* Beyrouth (Liban) 6:1–65.

Al-Saadi, H.A. and Al-Mousawi, A.H. 1988. Some notes on the ecology of aquatic plants in the Al-Hammar Marsh, Iraq. *Vegetatio* 75:131–133.

Al-Saboonchi, A.A., Mohamed, A.R.M. and Barak, N.A. 1982. A study of phytoplankton in the Garma Marshes, Iraq. *Iraqi Journal of Marine Science* 1:67–78.

Al-Saboonchi, A.A., Barak, N.A. and Mohamed, A.R.M. 1986. Zooplankton of Garma Marshes, Iraq. *Journal of Biological Sciences Research* 17:33–40.

Ambroggi, R.P. 1977. Freshwater resources of the Mediterranean Basin. *Ambio* 6:371–373.

Antoine, S.E. 1984. Studies on the bottom sediments of Al-Hammar Marsh area in southern Iraq. *Limnologica* 16:25–28.

Avitsur, S. 1957. *The Yarkon, the River and its Environment.* Hakibbutz Hameuchad Publishing House Ltd, Israel.

Azzouzi, A. and Laville, H. 1987. First faunistic inventory of chironomids (Diptera, Chironomidae) of Morocco. *Annales de Limnologie* 23:217–224.

Bailey, R.G. and Cobb, S.M. 1984. A note on some investigations carried out in the area of the Sudan Plain to be affected by the Jonglei Canal. *Hydrobiologia* 110:45–46.

Barham, J. 1994. Demirel raises stakes in tense regional game. *Financial Times,* 10 November 1994.

Basha, R. and Touma, A. 1984. *Study of Pollutant Factors on Litani River. Ten Years Report (1975–1984).* Beirut.

Berrebi, P., Kraiem, M.M., Doadrio, I., El-Gharbi, S. and Cattaneo-Berrebi, G. 1995. Ecological and genetic differentiation of *Barbus callensis* populations in Tunisia. *Journal of Fish Biology,* 47:850–864.

Biswas, A.K. 1992. The Aswan High Dam revisited. *Ecodecision* 6:67–69.

Boulton, A.J.F. and Lloyd, L.N. 1992. Flooding frequency and invertebrate emergence from a dry floodplain. *Regulated Rivers: Research & Management* 7:137–151.

Boumaiza, M. 1984. Contribution à la limnologie de la Tunisie: étude physico-chimique. *Archives de l'Institut Pasteur de Tunis* 61:205–246.

Bourguignat, J.R. 1868. *Histoire malacologique de la Régence de Tunis.* Imprime Nationale, Paris.

Bourguignat, J.R. and Letourneux, 1887. *Prodrome de la malacologie terrestre et fluviatile de la Tunisie.* Imprime Nationale, Paris.

Brooks, D.B. 1994. Economics, ecology and equity: lessons from the energy crisis in managing water shared by Israelis and Palestinians. In: Isaac, J. and Shuval, H. (eds) *Water and Peace in the Middle East.* Elsevier Scientific, Amsterdam, 441–450.

Brooks, D.B. 1996. Between the great rivers: water in the heart of the Middle East. In: Rached, E., Rathgeber, E. and Brooks, D.B. (eds) *Water Management in Africa and the Middle East: Challenges and Opportunities.* International Research Development Centre, Ottawa.

Bulloch, J. and Darwish, A. 1993. *Water Wars: Coming Conflicts in the Middle East.* Victor Gollancz, London.

Charnock, A. 1983. A new course for the Nile. *New Scientist,* 27 October, 281–288.

Clergue-Gazeau, M., Lek, S. and Lek, S. 1991. The blackflies of North Africa. New data on the distribution of the fauna of Morocco and biogeography of the Maghrebian species (Diptera, Simuliidae) [Les simulies d'Afrique du Nord. Nouvelles données sur la repartition de la faune du Maroc et biogéographie des espèces maghrebines (Diptera, Simuliidae)]. *Revue d'hydrobiologie tropicale* 24:47–59.

Conway, D., Krol, M., Alcamo, J. and Hulme, M. 1996. Future availability of water in Egypt: the interaction of global, regional, and basin scale driving forces in the Nile Basin. *Ambio* 25:336–342.

Cooley, J.K. 1984. The war over water. *Foreign Policy* 54:3–26.

Critchfield, R. 1978. Crocodiles, cattle and the Jonglei Canal. *International Wildlife* 8:20–25.

De Chaignon, H. 1904. Contribution à l'histoire naturelle de la Tunisie. *Bulletin de la Société d'Histoire Naturelle de Autun* 17:1–162.

Delucchi, C.M. 1988. Comparison of community structure among streams with different temporal flow regimes. *Canadian Journal of Zoology* 66:579–586.

Denny, P. 2000. Limnological research and capacity building in tropical developing countries. *Verhandlungen der Internationalen Vereinigung für theoretische und angewandte Limnologie* (in press).

DouAbul, A.A., Al-Saad, H.T., Al-Timari, A.A. and Al-Rekabi, H.N. 1988. Tigris–Euphrates Delta: a major source of pesticides to the Shatt al-Arab River (Iraq). *Archives of Environmental Contamination and Toxicology* 17:405–418.

Dumont, H.J. 1979. Limnologie von Sahara en Sahel: bÿdrage tot een beter begrip ran de klimaatsveranderingen van het laat-Pleistocen en Holoceen. Unpublished PhD Thesis, Rijksuniversiteit, Gent.

Dumont, H.J. 1986a. The Nile River system. In: Davies, B.R. and Walker, K.F. (eds) *The Ecology of River Systems, Monographiae Biologicae* 60, Dr W. Junk, Dordrecht, 61–74.

Dumont, H.J. 1986b. Zooplankton of the Nile system. In: Davies, B.R. and Walker, K.F. (eds) *The Ecology of River Systems, Monographiae Biologicae,* 60, Dr W. Junk, Dordrecht, 75–88.

Dunbar, M.J., Gustard, A., Acreman, M. and Elliot, C.R.N.

1998. *Overseas Approaches to Setting River Flow Objectives.* Institute of Hydrology R&D Technical Report, W6B(96)4. Environment Agency, London.

Durth, R. 1996. Supra-border environmental problems and regional integration. The political economy of upstream–downstream problems in international rivers. *Grenzuberschreitende Umweltprobleme und regionale Integration. Zur Politischen Okonomie von Oberlauf–Unterlauf-Problemen an internationalen Flussen,* Germany. Nomos Verlagsgesellschaft. Schriftenreihe des Europa-Kollegs Hamburg zur Integrationsforschung No. 10.

El Asswad, R.M. 1995. Agricultural prospects and water resources in Libya. *Ambio* 24:324–327.

El Gharbi, S., Lambert, A. and Berrebi, P. 1993. The genus *Barbus* (subgenera *Barbus* and *Labeobarbus*) in Morocco. Genetics and parasitology. In: Poncin, P., Berrebi, P., Philippart, J.C. and Ruwet, J.C. (eds) *Biology of the European, African and Asiatic Barbus,* 13, 223–226.

Elmusa, S.S. 1995. Dividing common water resources according to international water law: the case of the Palestinian–Israeli waters. *Natural Resources Journal* 35:223–242.

Eltahir, E.A.B. 1996. El Niño and the natural variability in the flow of the Nile River. *Water Resources Research* 32:131–137.

Entz, B. 1976. Lake Nasser and Lake Nubia. In: Rzóska, J. (ed.) *The Nile, Biology of an Ancient River.* Dr W. Junk, The Hague, 271–298.

Ezzat, M.N. 1993. *Nile Water Flow, Demand and Water Development.* In: Egyptian National Committee on Large Dams (ed.) *High Aswan Dam Vital Achievement Fully Controlled.* ENCOLD, Cairo.

Fadl, H.H., El-Sherif, L.S. and Shaarawi, F.A. 1993. Hydrophilidae of Egypt (subfamilies: Helophorinae, Hydrochinae and Spercheinae) Coleoptera. *Bulletin of the Entomological Society of Egypt* 71:101–108.

Farinelli, X.H. 1997. Freshwater conflicts in the Jordan River Basin. Green Cross International, June 1997. http://www4.gve.ch/gci/

Fraedrich, K., Jiang Jian Min, Gerstengarbe, F.W. and Werner, P.C. 1997. Multiscale detection of abrupt climate changes: application to River Nile flood levels. *International Journal of Climatology* 17:1301–1315.

Gabbay, S. 1989. Annual Environment Report. *Israel Environmental Bulletin, Ministry of the Environment* 12:7–9.

Gagneur, J. 1994. Flash floods and drying up as a major disturbance upon benthic communities in North-African wadis. *Verhandlungen der Internationalen Vereinigung für theoretische und angewandte Limnologie* 25:1807–1811.

Gagneur, J. and Thomas, A.G.B. 2000. Factors influencing the ecological repartition of aquatic Diptera families in a catchment of north-western Algeria. *Verhandlungen der Internationalen Vereinigung für theoretische und angewandte Limnologie* (in press).

Gasith, A. 1992. Conservation and management of the coastal streams of Israel: an assessment of stream status and prospects for rehabilitation. In: Boon, P.J., Calow, P. and

Petts, G.E. (eds) *River Conservation and Management.* John Wiley, Chichester, 51–64.

Gauthier-Lièvre, L. 1931. Recherches sur la flore des eaux continentales de l'Afrique du Nord. *Bulletin de la Société Histoire Naturelle d'Afrique Noire, Mémoire H.S.*

Gavaghan, H. 1986. A saline solution to Israel's drought. *New Scientist,* 10 July, 26–27.

Gervais, H. and Jasienski, H. 1879. *Les poissons de Tunisie: énumération et description valeurs alimentaire et commerciale.* Saint-Denis Imprime, C. Lampert, Paris.

Gheit, A. 1994. New water-bugs recorded for the Moroccan fauna (Heteroptera, Hydrocorisae). *Bulletin de la Société Entomologique de France* 99:515–516.

Ghezawi, A. 1997. Water demand management networking in the Middle East and North Africa. In: Brooks, D.B., Rached, E. and Saade, M. (eds) *Management of Water Demand in Africa and the Middle East: Current Practices and Future Needs.* International Research Development Centre, Ottawa.

Green, M., Safray, I. and Agami, M. 1996. Constructed wetlands for river reclamation: experimental design, start-up and preliminary results. *Bioresource Technology* 55:157–162.

Green Cross International (GCI) 1998. *Averting a Water Crisis in the Middle East: Make Water a Medium of Cooperation Rather Than Conflict.* Report of a Workshop Held in Paris on 18 March 1998. GCI, Switzerland.

Gussev, A.V., Ali, N.M., Abdul-Ameer, K.N., Amin, S.M. and Molnar, K. 1993. New and known species of *Dactylogyrus* Diesing, 1850 (Monogenea, Dactylogyridae) from cyprinid fishes of the River Tigris, Iraq. *Systematic Parasitology* 25:229–237.

Harrison, A.D. 1995. Northeastern Africa rivers and streams. In: Cushing, C.E., Cummins, K.W. and Minshall, G.W. (eds) *Ecosystems of the World 22. River and Stream Ecosystems.* Elsevier, Amsterdam, 507–517.

Hillel, D. 1994. *Rivers of Eden: The Struggle for Water and the Quest for Peace in the Middle East.* Oxford University Press, Oxford.

Hof, F. 1995. The Yarmouk and Jordan Rivers in the Israel–Jordan Peace Treaty. *Middle East Policy,* April, 47.

Homer-Dixon, T.F. 1994. Environmental scarcities and violent conflict: evidence from cases. *International Security* 19:5–40.

Homer-Dixon, T.F., Boutwell, J. and Rathjens, G. 1993. Environmental changes and violent conflict. *Scientific American* 268:38–45.

International Committee of the Red Cross (ICRC) 1994. *Water in Armed Conflicts.* ICRC, Geneva.

International Irrigation Management Institute (IIMI) 1995. *Nile Water Resources Management in Egypt, in 2010: Achieving a Common Vision.* Alexandria, IIMI, March. Colombo, Sri Lanka.

International Union for Conservation of Nature and Natural Resources, United Nations Environment Programme, and World Wildlife Fund (IUCN/UNEP/WWF) 1991. *Caring for the Earth: A Strategy for Sustainable Living.* Gland.

Isaac, J. and Shuval, H. (eds) 1994. *Water and Peace in the Middle East.* Elsevier, New York.

Junk, W.J., Bayley, P.B. and Sparks, R.E. 1989. The flood pulse concept in river–floodplain systems. In: Dodge, D.P. (ed.) *Proceedings of the International Large River Symposium. Canadian Special Publication of Fisheries and Aquatic Sciences* 106: 110–127.

Kaplan, R. 1994. The coming of anarchy. *Atlantic Monthly*, February, 67.

Khallel, M.R. 1974. *Monographie de la Mejerda.* Publication of the Ministry of Agriculture, Tunis.

Khorasani, N. and Pooryadegar, V. 1988. Study of Djadjroud River pollution and its role on the water quality. *Iranian Journal of Natural Resources Karaj* 42:45–61.

Khouzam, R. 1996. A water quality management strategy study case: the Nile River. Water policy: allocation and management in practice. In: Howsam, P. (ed.) *Water Policy: Allocation and Management in Practice.* Proceedings of International Conference, Cranfield University, 23–24 September. E. & F.N. Spon Ltd, London.

Kibaroglu, A. 1996. Prospects for cooperation in the Euphrates–Tigris basin. In: Howsam, P. (ed.) *Water Policy: Allocation and Management in Practice.* Proceedings of International Conference, Cranfield University, 23–24 September 1996. E & F N Spon Ltd, London, 31–38.

Kraiem, M.M. 1993. Variability analysis in barbel populations (*Barbus callensis*) of Tunisia. In: Poncin, P., Berrebi, P., Philippart, J.C. and Ruwet, J.C. (eds) *Biology of the European, African and Asiatic Barbus. Proceedings of the International Round Table Barbus II.* 6–8 July 1993, 13:159–162.

Krupp, F. 1985. *Freshwater Ichthyogeography of the Levant. Proceedings of the Symposium on the Fauna and Zoogeography of the Middle East*, Mainz, 1985, 229–237.

Lavergne, M. 1991. *The Seven Deadly Sins of Egypt's High Aswan Dam. The Social and Environmental Impacts of Large Dams*, Vol. 2: *Case Studies.* Wadebridge Ecological Centre, Cornwall.

Lintner, S. 1996. Strategically managing the world's water. *World Bank and the Environment* 1996:48–49.

Lohmann, H. 1990. *Enallagma cyathigerum* (Charp.) in Morocco: the first record from North Africa (Zygoptera: Coenagrionidae) [*Enallagma cyathigerum* (Charp.) in Marokko: Erstnachweis fuer Nordafrika (Zygoptera: Coenagrionidae)]. *Notulae Odonatologicae Utrecht* 3:76–77.

Lonergan, S.C. and Brooks, D.B. 1994. *Watershed: The Role of Fresh Water in the Israeli–Palestinian Conflict.* International Research Development Centre, Ottawa.

Louri, E.J. 1990. The water resources of Bahrain. *Wildlife in Bahrain Biennial Report*, Bahrain Natural History Society, 67–76.

Lowi, M.R. 1993. *Water and Power: The Politics of a Scarce Resource in the Jordan River Basin.* Cambridge University Press, London.

Maamri, A., Chergui, H. and Pattee, E. 1994. Allochthonous input of coarse particulate organic matter to a Moroccan mountain stream. *Acta Oecologica* 15:495–508.

Machordom, A. and Doadrio, I. 1993. Phylogeny and taxonomy of North African barbels. In: Poncin, P., Berrebi, P., Philippart, J.C. and Ruwet, J.C. (eds) *Biology of the European, African and Asiatic Barbus. Proceedings of the International Round Table Barbus II.* 6–8 July 1993, 13:218.

Machordom, A., Doadrio, I. and Berrebi, P. 1995. Phylogeny and evolution of the genus *Barbus* in the Iberian Peninsula as revealed by allozyme electrophoresis. *Journal of Fish Biology* 47:211–236.

Maltby, E. (ed.) 1994. *An Environmental and Ecological Study of the Marshlands of Mesopotamia.* Draft Consultative Bulletin. Wetland Ecosystems Research Group, University of Exeter. Published by The AMAR Appeal Trust, London.

Marshall, A. 1998. *The State of the World Population.* United Nations Population Fund, New York.

Matoussi, M.S. 1996. Sources of strain and alternatives for relief in the most stressed water systems of North Africa. In: Rached, E., Rathgeber, E. and Brooks, D.B. (eds) *Water Management in Africa and the Middle East: Challenges and Opportunities.* International Research Development Centre, Ottawa.

McCully, P. 1996. *Silenced Rivers: The Ecology and Politics of Large Dams.* Zed Books, London.

Mhaisen, F.T., Khamees, N.R. and Al-Sayab, A.A. 1990. Flat worms (Platyhelminthes) of two species of gull (*Larus ichthyaetus* and *L. canus*) from Basrah, Iraq. *Zoology in the Middle East*, 4:113–116.

Moubayed, Z. and Laville, H. 1983. The chironomids (Diptera) of Lebanon 1. First faunistic inventory. *Annales de Lumnologie Toulouse* 19:113–114.

O'Keeffe, J.H. and De Moor, F.C. 1988. Changes in the physico-chemistry and benthic invertebrates of the Great Fish River, South Africa, following an interbasin transfer of water. *Regulated Rivers: Research & Management* 2:39–55.

O'Keeffe, J.H., Weeks, D.C., Fourie, A. and Davies, B.R. 1996. *A Pre-impoundment Study of the Sabie–Sand River System, Mpumalanga with Special Reference to Predicted Impacts on the Kruger National Park.* Volume Three: *The Effects of Proposed Impoundments and Management Recommendations.* Report No. 294/3/96, Water Research Commission, Pretoria.

Overman, M. 1976. *Water: Solutions to a Problem of Supply and Demand*, revised edition. Oxford University Press, Oxford.

Pallary, P. 1923. Faune malacologique des eaux douces de la Tunisie. *Archives de l'Institut Pasteur de l'Afrique du Nord* 3:22–47.

Pearce, F. 1993a. Draining life from Iraq's marshes. *New Scientist*, 17 April:11–12.

Pearce, F. 1993b. Greek dam dooms Byron's wetland. *New Scientist*, 10 July:4.

Pearce, F. 1995. Raising the Dead Sea. *New Scientist*, 22 July:32–37.

Pellegrin, J. 1921. Les poissons des eaux douces l'Afrique du Nord française. *Mémoire de la Société des Sciences Naturelles et Physique du Maroc*, 1.

Por, F.D. 1986. Crustacean biogeography of the Late Middle Miocene Middle Eastern landbridge. *Crustacean Issues* 4:69–84.

Por, F.D., Bromley, H.J., Dimentman, C., Herbst, G.N. and Ortal, R. 1986. River Dan, headwater of the Jordan, an aquatic oasis of the Middle East. *Hydrobiologia* 134:121–140.

Porter, J.R. 1993. The Middle East. In: Willis, R. (ed.) *World Mythology: The Illustrated Guide*. Duncan Baird Publishers, London, 56–67.

Puckridge, J.T., Sheldon, F., Walker, K.F. and Boulton, A.J. 1998. Flow variability and the ecology of large rivers. *Marine and Freshwater Research* 49:55–72.

Rached, E. 1996. Introduction. In: Rached, E., Rathgeber, E. and Brooks, D.B. (eds) *Water Management in Africa and the Middle East: Challenges and Opportunities*. International Research Development Centre, Ottawa, 3–6.

Resh, V. and Gasith, A. 2000. Streams in Mediterranean climates: abiotic influences and biotic responses to predictable seasonal events. *Annual Review of Ecology and Systematics* (in press).

Rogers, P. 1994. The agenda for the next thirty years. In: Rogers, P. and Lydon, P. (eds) *Water in the Arab World: Perspectives and Prognoses*. Harvard University Press, Cambridge, Massachusetts.

Rzóska, J. 1976a. The Joint Nile in the Sudan. In: Rzóska, J. (ed.) *The Nile, Biology of an Ancient River*. Dr. W. Junk, The Hague, 257–309.

Rzóska, J. (ed.) 1976b. *The Nile, Biology of an Ancient River*. Dr. W. Junk, The Hague.

Salman, A.L.Y. 1993. The Sphaeridiinae of Egypt (Coleoptera: Hydrophilidae). *Bulletin of the Entomological Society of Egypt* 71:91–100.

Salman, S.D., Ali, M.H. and Al-Adhub, A.H.Y. 1990. Abundance and seasonal migrations of the penaeid shrimp *Metapenaeus affinis* (H. Milne-Edwards) within Iraqi waters. *Hydrobiologia* 196:79–90.

Samaha, M.A.H. 1979. The Egyptian Master Water Plan. *Water Supply & Management* 3:251–266.

Scheumann, W. 1993. New irrigation schemes in southeast Anatolia and Northern Syria: more competition and conflict over the Euphrates? *Quarterly Journal of International Agriculture*, July–September.

Schiffler, M. and Libiszewski, S. 1995. *Water in the Middle East. Interdisciplinary Academic Conference*. German Development Institute, 17–18 June 1995, Berlin, Germany.

Scott, D.A. (ed.) 1995. *A Directory of Wetlands in the Middle East*. IUCN, Gland and IWRB, Slimbridge.

Seurat, L.G. 1940. Peuplement des eaux continentales de Cap Bon et de Zembra (Tunisie septentrionale). *Compte Rendu des Séances de la Société de Biogéographie* 27–29.

Shahin, M. 1985. *Hydrology of the Nile Basin*. Developments in Water Science No. 21, Elsevier Scientific Publishers, New York.

Shalaby, A.M. 1993. The role of High Aswan Dam in horizontal and vertical land expansion and yield promotion. In: Egyptian National Committee on Large Dams (ed.) *High Aswan Dam Vital Achievement Fully Controlled*. ENCOLD, Cairo, 153.

Shaltout, K.H., El-Din, A.S. and El-Sheikh, M.A. 1994. Species richness and phenology of vegetation along irrigation canals and drains in the Nile Delta, Egypt. *Vegetatio* 112:35–43.

Shiklomanov, I.A. 1993. World freshwater resources. In: Gleick, P. (ed.) *Water in Crisis: A Guide to the World's Freshwater Resources*. Oxford University Press, Oxford, 19–20.

Slim, K. 1991. *Study of Algae in Lebanese Rivers. Ten Years Report (1975–1984)*. Beirut.

Smit, H. 1995. New records of water mites from Morocco, with the description of one new subspecies (Acari, Hydrachnellae). *Aquatic Insects* 17:17–24.

Snaddon, C.D., Wishart, M.J. and Davies, B.R. 1998. Some implications of inter-basin water transfers for river ecosystem functioning and water resources management in southern Africa. *Aquatic Ecosystem Health and Management* 1:159–182.

Snaddon, C.D., Davies, B.R. and Wishart, M.J. 2000. *The Ecological Effects of Inter Basin Transfer Schemes, with a Brief Appraisal of their Socio-economic and Socio-political Implications, with an Outline of Guidelines for their Management*. Contract Report, Water Research Commission, Pretoria.

Stanley, E.H., Buschman, D.L., Boulton, A.J., Grimm, N.B. and Fisher, S.G. 1994. Invertebrate resistance and resilience to intermittency in a desert stream. *American Midland Naturalist* 131:288–300.

Stanley, E.H., Fisher, S.G. and Grimm, N.B. 1997. Ecosystems expansion and contradiction in streams. *Bioscience* 47:427–435.

Starr, J.R. 1991. Water wars. *Foreign Policy* 82:17–36.

Stauder, A. and Meisch, C. 1991. Freshwater Ostracoda (Crustacea) collected in a stream of the island of Madeira. *Bocagiana Funchal* 147.

Swain, A. 1997. Ethiopia, the Sudan, and Egypt: the Nile River dispute. *Journal of Modern African Studies* 35:675–694.

Tawfik, M.F.S., Bakr, H.A., Hemeida, I.A. and Agamy, E. 1990. Corixids and their natural habitats in Egypt (Hemiptera, Corixidae). *Bulletin de la Société Entomologique d'Egypte* 69:209–215.

Tuch, A. and Gasith, A. 1989. Effects of an upland impoundment on structural and functional properties of a small stream in a basaltic plateau (Golan Height, Israel). *Regulated Rivers: Research & Management* 3:153–167.

United Nations, Educational, Scientific and Cultural Organization/Arab Centre for the Study of Arid Zones and Dry Lands (UNESCO/ACSAD) 1988. *Water*

Resources Assessment in the Arab Region. Paris, Damascus, Delft.

Urbanelli, S., Sallicandro, P., Vito, E., De Colonnelli, E. and Bullini, L. 1996. Molecular re-examination of the taxonomy of *Ochthebius* (*Calobius*) (Coleoptera: Hydraenidae) from the Mediterranean and Macaronesian regions. *Annals of the Entomological Society of America* 89:623–631.

Utton, A.E. 1996. Regional cooperation: the example of international waters systems in the twentieth century. *Natural Resources Journal* 36:151–154.

Vannote, R.L., Minshall, G.W., Cummins, K.W., Sedell, J.R. and Cushing, C.E. 1980. The river continuum concept. *Canadian Journal of Fisheries and Aquatic Sciences* 37:130–137.

Vitte, B. 1988. Studies on Diptera Ephydridae from Morocco. The Ephydridae from lakes and streams from the northern part of Moroccan Middle Atlas (Diptera, Brachycera) [Etude des diptères Ephydridae du Maroc. Les ephydrides des lacs et des ruisseaux du nord du Moyen-Atlas marocain (Diptera, Brachycera)]. *Nouvelle Revue d'Entomologie* 5:389–395.

Vitte, B. 1991. *Rhithrogena mariae* n. sp., a new Ephemeroptera species from Moroccan Rif Mountains (Ephemeroptera, Heptageniidae) [*Rhithrogena mariae* n. sp., Ephemeroptère nouveau du Rif marocain (Ephemeroptera, Heptageniidae)]. *Nouvelle Revue d'Entomologie* 8:89–96.

Wagner, R. 1992. On some species of Psychodidae and aquatic Empididae from the Middle East. *Spixiana* 5:15–17.

Watzman, H. 1995. Sewage slakes Israel's thirst for water. *New Scientist* 23/30 December.

Williams, D.D. and Hynes, H.B.N. 1976. The ecology of temporary streams. I. The faunas of two Canadian streams.

Internationale Revue der gesamten Hydrobiologie 61:761–787.

Williams, D.D. and Hynes, H.B.N. 1977. The ecology of temporary streams. II. General remarks on temporary streams. *Internationale Revue der gesamten Hydrobiologie* 62:53–61.

Wishart, D.M. 1990. The breakdown of the Johnston negotiations over Jordan Waters. *Middle Eastern Studies* 26:536–546.

Wishart, M.J. and Davies, B.R. 1998. The increasing divide between First and Third Worlds: science, collaboration and conservation of Third World aquatic ecosystems. *Freshwater Biology* 39:557–567.

Wolf, A. 1993. Water for peace in the Jordan River Watershed. *Natural Resources Journal* 33:797–839.

Woube, M. 1997. The Blue Nile river basin: the need for new conservation-based sustainability measures. *Sinet: Ethiopian Journal of Science* 20:115–131.

Wright, J.F., Hiley, P.D., Cooling, D.A., Cameron, A.C., Wigham, M.E. and Berrie, A.D. 1984. The invertebrate fauna of a small chalk stream in Berkshire, England, and the effect of intermittent flow. *Archiv für Hydrobiologie* 99:179–199.

Xie, M., Kuffner, U. and Le Moigne, G. 1993. *Using Water Efficiently.* World Bank Technical Paper 205, World Bank, Washington, DC.

Zaouali, J. 1995. Limnology in Tunisia. In: Gopal, B. and Wetzel, R.G. (eds) *Limnology in Developing Countries.* International Association for Theoretical and Applied Limnology, New United Press, India, 41–61.

Zouini, D. 1997. Surface water resources for water management projects in the El-Kebir Wadi Basin (northeastern Algeria). *Sécheresse* 8:13–19.

6

River conservation in central and tropical Africa

N. Pacini and D.M. Harper

Introduction

Despite a century and a half of European colonization of the continent, the ecology of the rivers of tropical Africa is still relatively unknown. The major rivers, and the lure of their sources, were the focal point for early exploration and, as a result, the knowledge that exists is unevenly distributed. A large amount of information and understanding exists for the Nile, for example (Rzóska, 1976) when compared with the Congo. Human uses of rivers though, are as old as our species and will continue to affect rivers as long as humankind inhabits the earth. Early Egyptian civilizations, 5000 years BC, first regulated the Nile, creating basins for flood waters and irrigation canals to build their cities. In the middle of the 20th century, the building of the High Dam at Aswan definitively changed the ecology of the entire region. The future, however, may produce even greater surprises following the launching in 1997 of the Tuzhka Project which, within some 25 years, could build an artificial channel across western Egypt to replicate the ancient river (Tadesse, 1998).

The colonization of Africa brought with it the expertise of European-style civil service, and the application of new scientific ideas (Worthington, 1958), but the development of ecological knowledge was still uneven. It was focused upon the development of fisheries (and more oriented towards lakes than rivers), upon the ecology and control of disease vectors, the control of exotic species, and the supply of drinking water. The development of concepts of conservation on land, with the growth of National Parks and other protected areas, was not mirrored by any form of aquatic conservation except where floodplains contained large mammal herds. Since the middle of the 20th century, much money and effort has been spent understanding the ecological impact of large reservoirs on major river systems such as Kariba on the Zambezi, Volta in Ghana, and Aswan on the Nile, but almost all were conducted after the event, and had little or no influence on management decisions. Even up to the present day, the occurrence of serious environmental evaluation of water resource development schemes is rare and conservation is very much a 'catching-up' activity, usually after the expatriate engineers, technologists and economists have gone home (Pacini *et al.*, 1999). Capital aid programmes are rarely linked to assistance for subsequent running costs and, as a result, many large projects – from unserviced University equipment to unfunded resettlement schemes – have failed to deliver environmental improvements (see also Wishart *et al.*, this volume, Part II).

The conservation of tropical aquatic habitats has been overshadowed by the mobilization of public interest and support generated by the urgent need to conserve tropical rain forests. Running waters in particular still have a low profile within the new nations of tropical Africa. National parks and protected areas are often delimited by rivers instead of being set around them, thus disregarding the link between rivers and their catchments. This link is particularly crucial in Africa where water is more precious than in many other parts of the world. Indeed, current estimates forecast significant decreases in water

Global Perspectives on River Conservation: Science, Policy and Practice.
Edited by P.J. Boon, B.R. Davies and G.E. Petts. © 2000 John Wiley & Sons Ltd.

availability in much of Africa. Populous countries undergoing rapid industrialization, such as Kenya and Nigeria, are subject to the highest water stress and may lose half of their water supply during the next decade (Rached, 1996). Water stress should not, however, push conservation to a concern of secondary importance but, on the contrary, should underline the urgency of preserving Africa's rivers as the most valuable resources of the continent. River conservation efforts can only succeed if they are integrated as sustainable water-development programmes.

The regional context

THE RIVER SYSTEMS, THEIR CATCHMENTS AND THEIR RECENT HISTORIES

The major rivers are the Nile, flowing from the East African highlands northwards to the Mediterranean; the Congo, rising in the central East African highlands and flowing east and south to the Atlantic; and the Niger, rising in the West African highlands of Guinea and flowing in a large loop northwards and then south to the coast. Smaller rivers such as the Senegal, the Tana and the Ruaha flow west and east to the oceans while the Chari system flows centrally to Lake Chad (Figure 6.1).

Tropical Africa is characterized by rivers with seasonal floodplains and wetlands, often so extensive that they are referred to as 'inland deltas' in their widest locations. The Niger in the west, Lake Chad in the centre and the Nile 'Sudd' in the east are examples: each ecologically different but dependent upon the same hydrological phenomenon – a highly seasonal rainfall pattern. They are also characterized by considerable inter-annual variation, which has been variously interpreted in terms of short periodic cycles, global climatic events, and severe human intervention. For instance, there has been considerable variation in river discharge in almost every decade this century, and attempts have been made to interpret records in terms of periodic oscillations, such as a 5–6 year cycle in rainfall (Nicholson, 1996), an 11 year cycle in Lake Victoria levels, or 7 year cycles in Sahel rivers (Grove, 1996). Most 'cycles' however, have not been strongly evident over the last 30 years, a period characterized by relatively large fluctuations. In East Africa high rainfall and river discharges were typical of the early 1960s, leaving Lake Victoria, for example, at a higher level than at any time since the 1870s, and the discharge of the Tana River, in November 1961, five to 10 times its

normal values for the month (Grove, 1996). A second rainfall peak occurred at the end of the 1970s, but the general trend from *ca* 1965 to the present has been a decline, although in the first half of 1998 river discharges in eastern Africa again showed a large pulse as they responded to the 'El Niño' oscillation (see below). Analyses of river discharges in the west have shown that, since the mid 1970s, the Sahel has undergone a period of serious rainfall shortage, which has affected water availability in the floodplains of the Senegal and Niger, and extended its influence to the northern tributaries of the Congo (Olivry, 1987; Briquet *et al.*, 1996; Laraque and Olivry, 1996). Lake Chad declined in area by about 90% during this time and fish populations decreased dramatically in the lake as in all Sahelian floodplains (Lévêque, 1994, 1995).

The widespread similarity of the main climatic events across tropical Africa are a consequence of global-scale controls which are superimposed upon regional variability caused by mountain masses and large lakes. The regional pattern is driven by the mixing of three continental-scale air streams at convergence zones, the most important of which is the Inter-Tropical Convergence Zone, ITCZ. This is driven by global-scale events, linked to the temperature fluctuations in adjacent oceans (higher than average Indian Ocean temperatures in 1961 for example), themselves linked to the periodicity of the 'El Niño' warm-water upwellings off the South American west coast (Nicholson, 1996).

The extent to which these events will change with global warming is difficult to forecast. Models based upon a predicted 2–4°C rise over the African Great Lakes region produce scenarios which generally show declining water levels (Cohen *et al.*, 1996), and large-scale habitat loss due to flooding and seawater intrusions along the West African coastline as a consequence of sea-level rise (Nicholls *et al.*, 1993; French *et al.*, 1994). Widespread human impact through deforestation has not been shown to have a regional effect similar to that demonstrated for the Amazon, but undoubtedly it has catchment effects. The largest catchment effects, however, are produced by the construction of dams. Every major river in tropical Africa, with the exception of the Congo, is impounded by structures whose impact on downstream discharges is dramatic. Additional floodplain developments such as irrigation schemes have further modified the hydrographs. These impacts have had greater ecological effects than past climatic changes, but an understanding of conservation and mitigation

Figure 6.1   *The river systems of tropical Africa*

possibilities has to be developed against a background of the region's natural variability.

CLIMATE AND HYDROLOGY

The huge seasonal variations in flow are related to a relatively regular annual alternation of rainy and dry seasons, defined by the movements of the ITCZ. This atmospheric region separates the two monsoons and marks the boundary between moist air masses of Atlantic origin and hot dry continental air. The migration of the ITCZ about the Equator occurs in conjunction with monsoon winds and is the most significant climatic feature of the continent. In West Africa rains coincide with the monsoon which drives

humid air masses from the ocean towards the land. Adiabatic cooling then causes precipitation. The dry season, called 'harmattan', lasts from November to April, when hot and dry south-westerly winds blow from the Sahara across the west coast.

North of the Equator the annual rainfall peaks in April–July, while to the south maxima occur in October–January. In the northern and southern extremes, rainfall seasonality is unimodal with maxima occurring during 'high-sun' in the respective hemisphere (Nicholson, 1996). Hydraulic regimes follow, after a short lag. Rivers which receive most of their waters from the northern or southern tropics have single annual floods. For example, the Senegal River originating at about 10°N exhibits an annual flood

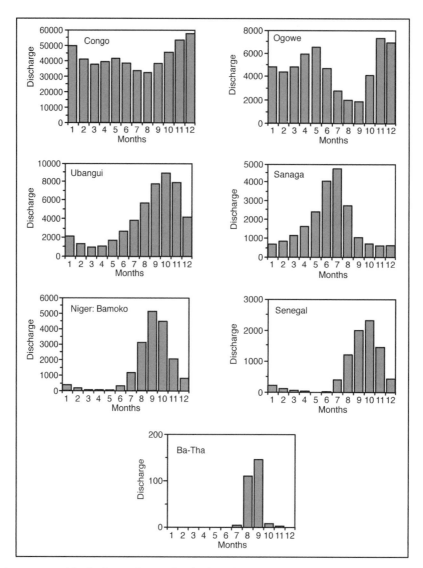

Figure 6.2   *Average monthly discharge data (m³ s⁻¹) for selected rivers of central and tropical Africa. From Lévêque (1997)*

peaking in September–October, about a month after the July rains at its sources in the Fouta D'jallon (Guinea) (Figure 6.2). In East Africa two rainy seasons of unequal duration are experienced during inter-monsoon periods. During the monsoons dry winds blow towards the Indian Ocean and associated high evaporation rates are typical of the dry-season savannah climate which characterizes the East African lowland plains.

Equatorial regions are characterized by two rainy and two dry seasons and are moister due to their proximity to the Atlantic Ocean and to the fact that the ITCZ annually passes twice over the Equator (Figure 6.3). The Tana in Kenya, for example, has a dual flood due to the two distinct rainy seasons in its equatorial catchment (Figure 6.4).

The hydrographs of larger rivers may show modified patterns; the Nile and the Niger, for example, have two floods along parts of their course due to tributaries originating in different geographical areas (Thompson, 1996). However, as the White Nile enters the Sudd, it divides into shallow lagoons and side channels, and its

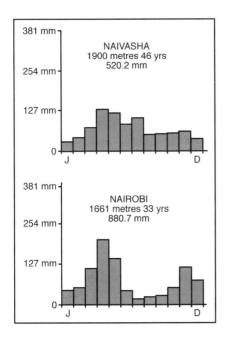

Figure 6.3 *Mean monthly rainfall patterns recorded for two meteorological stations in the Kenya Rift Valley; Naivasha and Nairobi MOW*

flood is attenuated. Below the Atbara one annual flood (July to November) is dominated by the Blue Nile and Atbara, both of which rise in the Ethiopian Highlands.

Floods are the most distinctive characteristic of the large African rivers. Due to the distance between mountain tributaries and the plains, flood passage may be delayed, reaching the plains often in the middle of the dry period. Mean velocities in the Senegal and the Niger are typically 17 km d^{-1}, 18 km d^{-1} in the Chari, but only 13 km d^{-1} in the broad, well-vegetated central delta of the Niger (Welcomme, 1989). The effects of floods are therefore expected to occur with a significant lag-time, but 'pre-flooding' – filling of depressions by local rains before the river flood surge – has been reported (Lévêque, 1995), extending moisture availability in floodplains and maximizing the effects of the incoming flood.

The hydrology of large floodplain rivers is further 'regulated' by numerous tributaries, by floodplain swamps and large aquifers, and for the last half-century, by dams. These rivers are often referred to as 'reservoir rivers' as opposed to their tributaries and other smaller rivers that exhibit irregular discharges and which are influenced by heterogeneous local climates. In East Africa, sharp gradients are

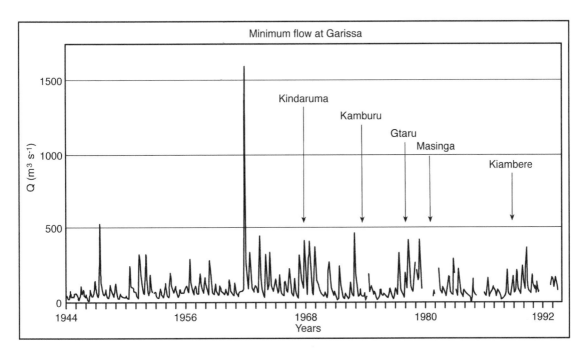

Figure 6.4 *Hydrograph of the Tana River at Garissa, northern Kenya. Names of upper Tana dams are inserted at dates of dam closure*

associated with steep changes in altitude from the dry lowlands to the perennial snows of Kilimanjaro, Mt Kenya and the Ruwenzoris. In the Sahelian region, from northern Mali to Somalia and the northern Kenyan coast, the timing and quantity of the precipitation is less predictable at lower altitudes: a direct relationship exists between degree of aridity and the spatial and temporal variability of rainfall (Thompson, 1996).

BIOGEOCHEMICAL SETTING

The alternation of dry and wet periods creates dynamic hydrological patterns which, together with high weathering rates and high primary productivity, results in a high efficiency of exchange between land and water and close coupling of land–water interactions with important consequences for nutrient cycling. Locally, the extent of land–water interactions is controlled by the integrity of ecotones. These zones of gradual passage between the riverine and the terrestrial environment are preserved in tropical regions of low human impact. The degree of hydraulic connectivity between the river channel, the floodplain and groundwater provides a qualitative measure of ecotone functioning. Geochemical and microbial processes within riparian zones and at the groundwater boundary comprise self-sustaining entrapment mechanisms which bind materials carried in solution and in suspension down the channel. The ecotonal surface available for exchange between the water column and these reactive sites is proportional to the extent and the geomorphological complexity of the floodplain. The chemical composition of tropical rivers reflects this efficient solute retention which has produced deeply weathered soils and dilute waters with little influence of the parent material (Viner, 1975).

Through its action on the physical disintegration of rocks, the chemical weathering of soil constituents, and its selective pressure on biological processes, temperature accounts for large-scale variation in soils and the chemical composition of run-off. Tropical soil and water temperatures are responsible for accelerated chemical denudation rates. The ionizing power of water increases steeply with temperature (at 24–25°C it is four times higher than at 10°C) causing chemically aggressive interstitial solutions able to penetrate deep into the bedrock (Faniran and Areola, 1978). The solubility of silica, the main constituent of primary rock structure, also increases with temperature (Viner, 1975; Meybeck, 1986). In the tropics the selective leaching of silica tends to degrade clay structure, contributing large amounts of silica to run-off and leading to the formation of silica-poor clays (Faniran and Areola, 1978; Meybeck, 1981). The baking and successive wetting and cooling of such clays causes physical stress and a high rate of swelling and cracking in tropical soils. Under these conditions, nutrients are easily mobilized (Twinch, 1987), while organic matter may be chemically oxidized. These processes should be conducive to high nutrient concentrations in solution. However, efficient nutrient retention mechanisms in African floodplains and high rates of gross primary production (both unicellular and higher plants) leads to a high demand for nutrients and contribute to the maintenance of oligotrophic waters.

A direct relationship on a global scale between water chemistry and temperature was suggested by Walling and Webb (1986) who showed that temperature has a determining impact on the ionic balance of streamwater in the rivers of the arid zone compared with those of other morpho-climatic regions. High temperatures coupled with high irradiation are also responsible for the direct volatilization of several aromatic toxic compounds such as DDT and its derivatives.

DEVELOPMENT PRESSURES AND HUMAN IMPACTS

With the possible exception of the equatorial forests of the Congo Basin, the greatest ecological impact has come from river regulation by dams, chiefly for hydro-electricity generation. The most widespread pressure on rivers is from human subsistence requirements for drinking water, washing and cooking, and agriculture and livestock watering. Pollution by industrial- or sewage-derived wastes associated with urban settlements, can be intense but remains localized and is of lesser ecological significance at a regional scale, but can significantly affect rapidly increasing urban populations. Due to technological inefficiencies the intensity of pollution in tropical countries is often higher than in the temperate regions, as documented by reports of pollution associated with mining (e.g. Chukwuma, 1997) and oil exploitation (Moffat and Lindén, 1995).

Urban stream pollution is increasing across equatorial Africa as settlements expand rapidly, and comprises mainly untreated sewage. In some industrial centres such as Nairobi, industrial pollution comprises heavy metals (Kinyua and Pacini, 1991), machine oils discharged by open-air garages, factory pollutants and acidic wastes from breweries and tanneries. As a consequence the Nairobi River crossing the city is heavily polluted. Known cases of industrial pollution are related to mining in Ghana (Amonoo-Neizer *et al.*, 1996), Tanzania (Bowell *et al.*, 1995; Ikingura and Akagi, 1996), Mali and in the Democratic Republic of Congo (DRC), and to mineral oil exploitation plants along the West African coast, in Angola and in Gabon. Recent analyses highlight the impact of secondary effects of oil exploitation that do not derive directly from oil spills but which have important long-term impacts (Moffat and Lindén, 1995). The significance of the environmental impact of industrial plants in the tropics is still poorly documented in the international scientific literature and too little is known in terms of the ecotoxicological effects on single species.

Long-term environmental impacts have been caused by the extensive use of pesticides targeting water-borne parasites. Schistosomiasis, onchocerciasis, malaria and dracunculiasis are some of the most widespread water-borne human diseases which have been tackled with synthetic pesticides (Dejoux, 1988). Organochlorines such as DDT were the first products to be introduced on the African market and remain in wide use. Various cocktails of chemicals have been experimented with following the spread of resistance to pesticides among target organisms. The degradation of many synthetic chemical compounds increases steeply at higher temperatures (Koeman *et al.*, 1972). However, so does their toxicity (Cairns *et al.*, 1975). This observation predicts major differences in the impact of pollutants in temperate and tropical regions (Ekweozor, 1989). Today regional programmes of pesticide application may choose from hundreds of commercial products for a variety of purposes. Ecotoxicological tests have shown that means of dispersal and physical phases (dry powders, hydrophilic or hydrophobic solutions) are decisive in determining the impact on aquatic systems (Dejoux, 1988), while dangers linked to the activity of specific active components are less relevant than the actual use made of products (dose applied, time of year, etc.). In Kenya, fish kills and cattle deaths due to improper pesticide use have been reported at Lake Naivasha and in the lower Tana, each in proximity to intensive agricultural plots. In West Africa lindane (a widespread organochlorine phytotoxic compound), as well as other toxins, is often used in fishing by poisoning entire water courses (Lévêque, 1995).

Human subsistence impacts on rivers (extraction of water for drinking and cooking) are usually severe, localized but widespread, yet almost nothing quantitative is known about them. Bretschko (1995) described the intensive use of water made in a typical small river in the Rift Valley in Kenya. Each family abstracts water from the stream for daily cooking and washing, mainly by children and women. Cattle and donkeys are also watered in large numbers. Disturbance and water removal are the main impacts, although pollution from livestock and nutrient enrichment from clothes washing are also unquantified potential threats. Along the whole river, the impacts of fuelwood collection and livestock grazing and watering were apparent, although the riparian woodland retains some important elements of natural vegetation. Substantial impoverishment of benthic fauna occurred compared with less frequently visited control sites (Bretschko, 1996). At this small scale, but with high frequency, subsistence farmers affect the hydrology of headwater wetlands (*dambos*) that are common features of African streams. The creation of footpaths, cattle tracks and roads is mainly associated with agriculture and, in particular, with the transport of produce; detailed erosion monitoring showed that they significantly increase erosion (Dunne, 1979). Ogbeibu and Victor (1989) analysed the consequences of track construction on the benthic community of a suburban stream in the Bendel State of Nigeria. High silt concentrations decreased the number of taxa and biomass, and a general worsening of the ecological state of the river was illustrated by several biotic indices used by the authors. Silt had an abrasive effect on the benthos and interfered with feeding and respiration.

Seasonal changes in water quality in tropical rivers are mainly related to high suspended solids during the rainy seasons. The properties of tropical soils, erosivity of rainfall and high population growth rates, have all been identified as the recurring causes of soil erosion. High sediment loads and very low transparencies are now a common feature of tropical streams. Deforestation, ranching and erosion by agriculture in riparian areas have been indicated as specific mismanagement issues leading to elevated suspended solids loads (Timberlake, 1985).

Fisheries exploitation is becoming a major concern in terms of the survival of endemic species and the maintenance of sustainable stocks. Case studies of lake fisheries are better known as they have more reliable records. In Lake Victoria, for example, changes to the fish community induced by the introduction of the Nile perch, *Lates niloticus*, led to the loss of more than 200 species of haplochromine cichlids. A reduction in fish yields induced fishermen fishing for haplochromines and for 'dagaa' (*Rastrineobola argentea*, a small-bodied but appreciated commercial species) to reduce net mesh sizes to 3 mm with a dramatic increase in fishing pressure. Information on the impacts of overfishing on the Niger River is provided by Malvestuto and Meredith (1989).

HEALTH AND DISEASE

Three major, debilitating and mortality-causing, waterborne parasitic diseases are rife throughout much of the region. The most widespread, malaria, is common throughout the tropics on all continents, as well in parts of the sub-tropics. Caused by one of several species of the protozoan genus *Plasmodium*, its vector is the female mosquito of the genus *Anopheles*, which requires still or slowly moving water for egg-laying and larval development. The second widespread disease of all tropical continents (500 million cases worldwide) is schistosomiasis (bilharzia). Larvae of the trematode worm of the genus *Schistosoma* infect humans in shallow waters directly through the skin. The intermediate host is a species of two genera of gastropod snail: *Bulinus* and *Biomphalaria*. Both species of human *Schistosoma* (*S. mansoni* and *S. haematobium*) live in the hepatic portal system. While the eggs of *S. mansoni* invade mesenteric and colonic veins causing rectal bleeding (intestinal bilharzia), those of *S. haematobium* invade the capillary vessels of the urinary bladder (haemoglobinurea). The cycle is maintained by defecation or urination in or near water.

The third disease in order of importance in tropical Africa is river blindness, which affects around 18 million people in Central and Western Africa, as well as having foci in South America. It is caused by the nematode worm, *Onchocerca volvulus*, which lives under the human skin, causing thickening and the formation of cysts. During their life within the human body (up to 15 years), females produce immature, microfilaria larvae which migrate below the skin causing skin problems and ocular lesions that may lead to blindness. Transmission is effected by the bite of an adult female *Simulium* blackfly (usually *S. damnosum* in Africa) whose larvae thrive on solid surfaces in turbulent river reaches. Onchocerciasis is endemic in the West Africa interior, with hyperendemicity in northern Ghana, where infection exceeds 90% of the population over 30 years of age (Dejoux, 1988). Further important foci occur in Central and Southern Africa; in East Africa it occurs in Tanzania and in Uganda.

From the foregoing, it is clear that for river conservation in Africa to succeed it cannot be separated from human health and economic development; if it is, then ordinary people will have no stake in it. Second, economic development of the tropics cannot rely on capital-intensive, large-scale engineering projects which have only negative impacts upon the rural and urban poor; they are irreversibly destructive. Third, education is essential to break the cycle of poverty, ill-health and disease (see also Wishart *et al.*, this volume, Part II).

Information for conservation

RIVERINE BIODIVERSITY

Threats to the biodiversity of tropical inland ecosystems are evident from the documented disappearance of an increasing number of animals and plants. While it is not possible to consider all tropical flora and fauna, it has been suggested that a focus for the efforts of freshwater conservation should be tropical fish (Lévêque, 1994). The advantage is that fish represent an important economic resource and their taxonomy is well documented in relation to other tropical animal groups. Fish are also profoundly tied to water quality, ecosystem functioning and to human activities, and their protection may result in the preservation of other groups of animals and plants as well. At present African inland fish number 75 families and more than 3000 species (Lévêque, 1994). Direct proportionality between numbers of species and basin area, and discharge, have been reported respectively by Welcomme (1989) and by Hugueny (1989). Some authors, however, still maintain that less than 50% of African freshwater fish have been described to date (Ribbink, 1994). This figure is partly due to the explosive speciation of stenotope (*K*-selected, sedentary species of small body size; habitat specialists residing in highly specialized habitats) cichlids in the large East African lakes (Lowe-McConnell *et al.*, 1994). In comparison with lakes, tropical rivers are believed to shelter far fewer fish species. Although rivers include heterogeneous habitats, cover large basins and (some) may be as ancient as the great lakes (Tanganyika and Malawi), due to seasonal hydrology, they are considered to be more suited to opportunistic and slow-speciating eurytope (*r*-selected, mobile, widely distributed, large body size) fish and have, as a result, a moderate species richness (Ribbink, 1994) (Figure 6.5). A comparison between rivers substantiates this

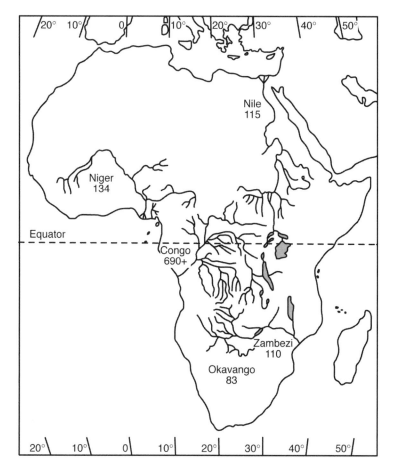

Figure 6.5   *The major rivers of Africa showing the approximate number of fish species in each and the geographical area each covers relative to the lakes; from Ribbink (1994)*

hypothesis. The Congo, an equatorial river fed by both northern and southern tributaries, is characterized by higher hydrological stability than 'Sahelian-type' tropical rivers, which undergo wide variations in discharge. The Congo contains a far larger number of species than other basins and its endemics stand at >80% (Beadle, 1981). The ecological stability of this system is related to its 2000 km-long, slow-flowing, potamon section between Kisangani and Kinshasa (mean slope 0.05 m km^{-1}) which has little seasonal or interannual change in water level, rendering the river more like an elongated lake (Beadle, 1981; Laraque and Olivry, 1996). Its forested catchment has very low erosion rates, producing nearly constant suspended sediment concentrations. These features derive from the large number of tributaries originating in different climatic regions (Laraque and Olivry, 1996).

Ribbink's analysis of river regime and species richness is, however, questioned by recent hydrogeological and palaeoecological studies, which indicate that climatic changes may have occurred at a much higher frequency than previously believed. In this case, the notion that fish speciation in the Great East African lakes is based on long periods of stability would be replaced by the notion of rapid allopatric speciation (Lévêque, 1994) with implications for our present understanding of riverine species richness. Species diversity estimates will certainly increase in the near future as surveys based on more sophisticated taxonomic techniques involving studies of the anatomy, genetics, biosystematics (e.g. the identification of species based on parasite infections, Paugy *et al.*, 1994) and behaviour, complete the existing records. Further surprises may be still

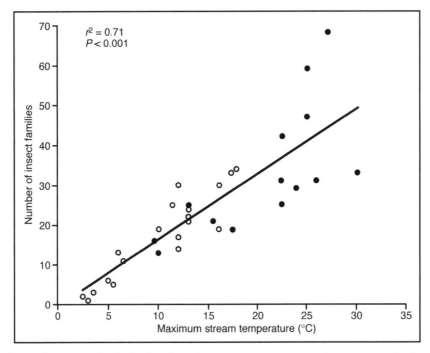

Figure 6.6   *Insect diversity at family level and maximum stream temperature in temperate Arctic and tropical streams (Jacobson* et al*., 1997). Open circles, Arctic; closed circles, tropical systems*

concealed in the great habitat diversity found in poorly investigated river fringing ecotones. Recent results from West Africa conducted in the Chari, Senegal and Niger systems as well as in coastal lagoons and secondary wetlands yielded a list of 480 freshwater fish species (Paugy *et al.*, 1994). The Congo Basin is expected to yield the greatest increase in biodiversity in the near future. At *ca* 4.0×10^6 km² it is the second largest basin in the world. The currently known number of fish species (690) is more than twice the number known for the Mississippi Basin (250 species for 3.2×10^6 km²). The current number is also far higher than the number of species that would be estimated using Hugueny's and Welcomme's relationships: respectively 462 and 512 species (see also Teugels and Guégan, 1994).

The relative abundance of species in tropical rivers has to be seen in the context of global trends of biodiversity. Analyses of species numbers along latitudinal gradients indicate that for several classes of animals and plants the number of terrestrial, marine and freshwater species increases from the higher latitudes towards the Equator (Rosenzweig, 1995; Jacobsen *et al.*, 1997). From these studies several causes – biogeographical, physical, historical – may be

invoked, and no clear-cut, single mechanism can be defined. Jacobsen *et al.* (1997) recorded latitudinal trends in riverine biodiversity by comparing selected freshwater invertebrate surveys from four continents. Their data suggest that a highly significant proportion of the variability in insect diversity may be ascribed to latitudinal temperature gradient (Figure 6.6).

The direct mechanism is unclear. Temperature may act directly by increasing mutation rates and shortening generation times (Rohde, 1992); nevertheless important indirect mechanisms can be postulated:

- The temperature gradient coincides with a gradient of higher ecosystem stability due to lower seasonal variability in stream temperature and greater continuity in life cycles; in particular, the persistence during the year of aquatic vegetation may be an important factor in the persistence of stable habitats and physical and chemical conditions.
- Higher temperature allows faster weathering rates and greater nutrient cycling efficiency (see above). This may allow more species to coexist than in temperate situations where biotic assemblages follow each other in annual succession as temperature-

Table 6.1 *Comparison between North America and tropical Africa of the families and genera of Trichoptera and Ephemeroptera*

	North America		Tropical Africa	
	Families	Genera	Families	Genera
Trichoptera	23	123	23	79
Ephemeroptera	21	70		68

inhibited periods of nutrient accumulation (winter) succeed periods of nutrient depletion (summer).

• Tropical regions were warmer during the last glaciation and represented biogeographical refugia from which animals and plants migrated back towards the colder regions of the globe; this colonization process would still be continuing (Jacobsen *et al.*, 1997).

Not all freshwater invertebrate surveys indicate a higher species diversity in the tropics; several authors, cited by Jacobsen *et al.* (1997), observed the same diversity in tropical and temperate regions. Table 6.1 compares African invertebrate species (from the Department of Entomology of the Albany Museum in Grahamstown, South Africa; A.D. Harrison, Freshwater Research Unit, University of Cape Town, personal communication) with North American freshwater invertebrates (Merritt and Cummins, 1996); the total number of taxa is comparable. Harrison suggests that taxa of North American running-water Chironomidae are far more numerous than those found in Afro-tropical streams.

However, Jacobsen *et al.* (1997) show that, notwithstanding the incomplete knowledge of tropical running waters, neotropical invertebrate families are more species-rich than North American ones, particularly in Odonata, Heteroptera, Coleoptera and, to a lesser extent, Trichoptera. In a brief review of the Trichopteran fauna of the Afro-tropical Region, Scott (1986) stressed the virtual absence of the most important Holarctic families (Phryganeidae, Rhyacophilidae, Limnephilidae) and the large number of Afro-tropical species of recent discovery. She reported that there may be as many as '50–80 probably new but undescribed species in the Albany Museum collections alone'. It is yet too soon to express judgement on tropical invertebrate diversity; nonetheless, a higher number of species in the New World tropics (Jacobsen *et al.*, 1997) and in the Far East (Bishop, 1973; Schmid, 1984) may not necessarily mean a higher diversity in the Afro-tropics.

Several limnologists working in Kenya have noticed the poor diversity of the benthic invertebrate community (Van Someren, 1952; Barnard and Biggs, 1988). Similarly Harrison and Hynes (1988) reported the impoverished invertebrate fauna of the Ethiopian highlands. Several reasons have been presented to explain the apparent low diversity of East African stream invertebrates. Among these are:

• the seasonality of flow restricting the fauna to species adapted to highly variable hydrology (Barnard and Biggs, 1988; Clark *et al.*, 1989);
• the recent geological formation of the regions investigated (Van Someren, 1952);
• local isolation and past climatic changes (Harrison and Hynes, 1988).

The East African highlands are densely populated and subject to high erosion; a possible cause of low diversity may be due to the underestimated, but extensive impacts of subsistence agriculture that typically encroaches on headwaters and stream banks. Cattle, daily water abstractions for personal needs, bathing and washing activities, as well as traditional rites and ceremonies (e.g. for some tribes, circumcision) take place at the river bank. Considering the regional scarcity of water resources, it is questionable whether or not tropical Africa may still have pristine, totally undisturbed streams. This statement may be surprising but it agrees with the conclusion of Davies *et al.* (1995) in their analysis of southern African rivers. Intensive use by humans causes bank erosion, resuspension of river-bed sediments and damage to aquatic vegetation. The siltation which follows may cause a simplification of riverine microhabitats for several kilometres downstream. Bank erosion in Ethiopian rivers and high turbidity were, indeed, described by Harrison and Hynes (1988) and Harrison (1995) who also cited similar observations carried out in the catchment of the Vaal, in South Africa (Chutter, 1969). While part of this effect may be caused by the natural irregularity of flow in tropical highlands, a significant impact may be attributed to the intensive human use of water resources.

No information is available on the macroinvertebrate fauna of the large East African floodplain rivers such as the Jubba, the Tana, the Athi and the Rufiji, that may help to place existing faunal surveys in a wider context of regional riverine biodiversity. A brief but fairly comprehensive account of the biogeography of the region, based on existing information, was recently compiled by Harrison (1995).

THE TAXONOMIC BASE

Information about species richness and habitat characteristics is vital for the identification of biodiversity foci eligible for international protection. At present, however, taxonomic knowledge is uneven. Vertebrates and flowering plants are well documented: the former because of their size and prominence and, in the case of fish, their economic importance. Knowledge of fish or plant taxonomy is by no means complete, but it does provide a good basis for understanding ecosystem functioning (e.g. Lowe-McConnell, 1975; Denny, 1985; Welcomme, 1989). The preservation of fish species and stocks can mobilize large interest groups with wider projects directed towards the study and preservation of fish prey, habitats and entire aquatic systems (Lévêque, 1997). The appreciation of the ecological requirements of endangered species such as their relationships to predators, prey and habitat quality, however, demands a sound taxonomic base.

Knowledge of invertebrates and smaller taxa is less satisfactory, with such keys that exist also being inaccessible to local researchers, as they are often published in colonial monographs and proceedings. In smaller subject areas, such as taxonomy related to agriculture and disease control, where more western aid and expertise has been involved, the picture is brighter. There is a clear need, however, for a firm invertebrate taxonomic base for the development of pollution indicators and indices beyond simple ones such as taxonomic richness, the Sequential Comparison Index, or family-based indices.

The upsurge in concern about 'biodiversity' has focused on tropical rain forests, with less emphasis on freshwater systems in the tropics (see also Benstead *et al.*, this volume; Pringle, this volume). Yet a relatively large number of endemic species in tropical African river systems, particularly fish species, are affected by dam, and other regulatory developments with, until now, little or no mitigation. This, amongst many other anthropogenic factors, places endemic mammals as well as fish at high risk of extinction.

RIVER TYPOLOGY

African watercourses comprise an estimated 13×10^6 km (Welcomme, 1989); they are all inter-tropical with the exception of the Orange system and South, and the delta of the Nile, and are dominated by four major basins: the Nile, the Congo, the Niger and the Zambezi. Most of them are poorly investigated. Nevertheless, they can be grouped into three main geographical types:

- The well-watered mountainous equatorial zone characterized by high annual precipitation (1000–2000 mm yr^{-1}), one or two well-distinguished rainy seasons, and torrential streams with waterfalls.
- The low relief sahelo-sudanian zone with large rivers (Nile, Niger, Senegal, Tana) flowing through semi-arid landscapes, with low and irregular precipitation (200–600 mm yr^{-1}). The natural ecology of this zone is driven by the extensive floods of its large rivers.
- The large deserts and marginal areas with no permanent flowing water and an annual precipitation below 200 mm yr^{-1}; these include more than one-third of the African continent (Welcomme, 1974).

This zonation is particularly well defined in West Africa where the three run in parallel from the equatorial moist highlands to the northern dry lowlands. Here, biogeographical studies demonstrated a similarity between geographic zonation and the composition of fish populations of different rivers (Paugy *et al.*, 1994). Working in West Africa, Daget and Iltis (cited in Welcomme, 1974) divided watercourses into 'Guinean' rivers running through gallery forests – low pH and high humic matter – and 'Sudanian', draining open savannah – high fish yields. Upstream sections of several 'Sudanian' rivers often include some 'Guinean' characteristics (e.g. the Senegal). Climate changes during the Quaternary have produced wide shifts of the zones along a north–south axis and have played an important role in species extinction and radiation over the sub-continent. In East Africa, relief plays a predominant role; the region has an additional high mountain zone with cool- to cold-adapted species which are unknown to West Africa where, with the notable exception of Mount Cameroon, the mountainous equatorial zone is equivalent to the East African foothills. This 'High' Africa (see also Davies and Wishart, this volume) includes the Ethiopian massifs, the Ruwenzoris, Mount Elgon, Mount Kenya with the Nyandarua range and Mount Kilimanjaro. As a unit the region has been contrasted to the 'Low' Africa of the dry low-lying savannah plains stretching west of the Rift Valley (Roberts, 1975). Localized volcanic formations, the Rift Valley and the presence of large lakes, contribute to the generation in East Africa of a more complex climatic and biogeographical pattern.

The characteristic domination, throughout tropical Africa, of relatively consistent annual temperatures in different altitude zones, led to the application of an early predictive model of faunal zonation, the River Zonation Concept, of Illies and Botosaneanu (1963), discussed by Harrison (1965, 1995) and Harrison and Hynes (1988). This created three zones – the sources zone, the torrent zone, and the foothills or plains zone. Illies and Botosaneanu further subdivided the last two into sub-zones prefixed by epi-, meta- and hypo-, but in the context of river conservation it is perhaps better to consider the plains zone more in terms of the structural features of the channel and floodplain.

Longitudinal succession concepts based on structural (geomorphology, faunistic composition) or functional gradients (organic matter processing, nutrient cycling), developed in regulated rivers of the temperate region, such as the River Continuum Concept (Vannote *et al.*, 1980) and the Serial Discontinuity Concept (Ward and Stanford, 1983) may be of little relevance for natural rivers in the tropics where discharges are often unpredictable (*sensu* Davies *et al.*, 1995) and main channel–floodplain interactions play a predominant role in the transfer of matter and energy (e.g. Junk *et al.*, 1989). Nevertheless, they form a valuable platform for a comparative limnology of downstream continua in temperate and tropical systems. Based on a detailed macroinvertebrate survey, Gibon and Statzner (1985) documented a steady increase in diversity from the source downstream with no species replacement. The trophic functional paradigms set by the River Continuum Concept remain, up to now, unsupported in Afro-tropical rivers (Lévêque, 1995).

More topically relevant to many sediment-laden Afro-tropical rivers is the zonation introduced by Chutter (1969, 1970) which associates invertebrate communities with the structure of the river bed and which distinguishes an 'upstream erosional zone' characterized by a hard river bed from an 'intermediate stable-depositing zone' and a 'lower unstable-depositing zone' controlled by floods.

In the tropics extensive floods and the natural channel migration of large rivers lead typically to the development of a wide range of extensive secondary features. These may be classified in three large groups (Welcomme, 1985):

- temporarily and permanent fringing flooded wetlands (dead arms, oxbow lakes, wetlands associated with the presence of shallow aquifers);
- inner deltas with numerous braided channels;
- coastal brackish lagoons associated with extensive estuaries.

Tropical African river ecology has also been evaluated in the light of other temperate-zone paradigms, particularly feeding guilds and downstream continua, by Harrison (1995), Lévêque (1995) and Cooper (1996). Furthermore the extensive review by Davies *et al.* (1995) of southern African streams, focusing upon the unpredictability of their discharge regimes, contains much of relevance to tropical African rivers. All four of these reviews serve to emphasize the lack of taxonomic knowledge for the continent as a whole.

THE FUNCTIONING OF AFRO-TROPICAL FLOODPLAINS

The large Afro-tropical floodplains of the Sahelian region are focal points of biodiversity. They represent relatively stable environments in dry landscapes that undergo extreme variations in temperature and moisture. Floodplains are characterized by high primary and secondary production, and have been utilized in sustainable and often complex fashion by human societies for several thousand years on all continents. Freshwater biologists tend to regard floodplains as temporary extensions of the main channel and point out their importance for the maintenance of fish stocks (Welcomme, 1989). Studies on floodplains linked to African rivers, however, also stress their role in determining the survival rates of terrestrial animals migrating from adjacent non-flooded savannas during the dry period and demonstrate their role in the ecology of the entire tropical region.

The distinctive characteristics of Afro-tropical floodplains are derived from the previously detailed hydrological cycle. In most temperate climates, where rains are scattered throughout the year, floods are of short amplitude. It is unlikely, therefore, that temperate floodplain communities would show a high degree of adaptation to these temporary events. Under tropical conditions, however, floods can persist for many weeks and, where pristine hydrological conditions are preserved, extensive shallow temporary wetlands form. With the onset of the dry season, river discharge drops dramatically and water falls back to the main channel. The floodplain, and sometimes the main channel itself, are then reduced to isolated pools. The main consequences of the tropical flood cycle are twofold:

- tropical animals and plants are adapted to lateral movement across the floodplain (see below); and

- high-amplitude variations between the aquatic and terrestrial environments, and the duration of flooding, drives higher species richness in tropical wetlands compared with their temperate counterparts (Junk *et al.*, 1989).

The lack of pronounced annual temperature changes (over most of tropical Africa the daily range is greater than the annual range) makes the advent of the rains and floods a primary phenological trigger mechanism in an otherwise unseasonal environment: plants and animals tune their reproductive cycles to correspond with the seasonal cycle (Payne, 1986; Junk *et al.*, 1989). This can be demonstrated by comparing fish reproductive behaviour within and outside floodplains. In the Sahelian region, for instance, fish spawn during the rising hydrograph (Lévêque, 1995), whilst in contrast, where river drawdown is limited in moist forest zones, the reproductive cycle may be spread across nearly the whole year (Paugy *et al.*, 1994) (Figure 6.7).

Terrestrial animals of the Sahelian region show similar adaptations to the flood cycle, of which the large mammal migrations are perhaps the most striking: the tiang, *Damaliscus korrigun tiang*; Mongalla gazelle, *Gazella thomsoni albonota*; and the Nile lechwe, *Kobus megaceros* in the Sudd. Temporary floodplain wetlands benefit mammals and birds coming from very wide areas; for example the Hadejia–Nguru Wetlands in the Chad Basin of Nigeria and the wetlands of the Djoudj National Park in the Senegal Delta are critical for palaearctic migratory birds (Schwöppe, 1994).

The high primary productivity of the flooded lands in the Sahel is the result of fertile and well-structured alluvial soils. These characteristics are tied to annual provision of silt, to accelerated nutrient cycles under the influence of a frequent wetting–drying regime, and to high moisture availability during most of the year. A diverse natural and cultivated vegetation develops along the river banks. Several cycles of the same species may coexist in succession along the spatio-temporal gradient as the floods ebb and flow providing a food source for herbivores all the year round. The flooded landscape is typically dominated by productive and nutritional annual grasses (Schwöppe, 1994).

The scale of the response to flooding, from fish to terrestrial mammals, dictates that river conservation must adopt a catchment- or basin-scale approach to be effective in conserving current patterns of biodiversity. Conservation strategies must recognize that

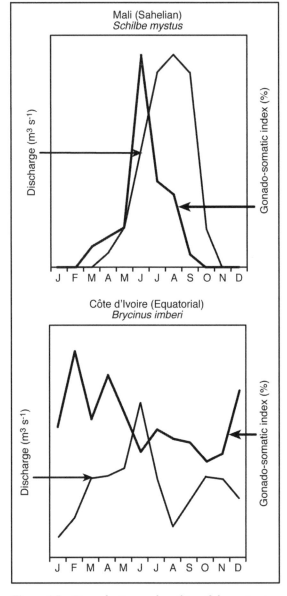

Figure 6.7   *Reproductive cycles of two fish species in relation to climatic regimes in West Africa (from Paugy* et al. *1994)*

floodplains may become grounds of fierce competition for scarce resources, but which are dominated by fragile ecological balances. Not by chance, it is these environments that have been subject to the greatest human intervention; it is here, perhaps, that humans run the greatest risk of irreversible damage to river systems.

EDUCATION

Indigenous people

Traditional African societies had a low impact on natural resources: they were widely dispersed and either nomadic pastoralists or shifting cultivators. Education and traditions were orally transmitted, and the local knowledge of the environment and its unpredictability was refined (Adams, 1992). One of the most widespread characteristics of European colonization was the imposition of European educational and cultural systems. These allocated limited resources to indigenous people, although through the influence of missionaries and medical aid, mortality was reduced enough to initiate population increases similar to those of the developed world some hundred or so years earlier. Speaking generally, at independence there was an inadequate graduate cadre and the educational system since then, in many countries, has produced an educated urban elite, but has had little impact on widespread illiteracy among the rising population in rural and urban shanty areas. Knowledge of health issues such as the basic information necessary for breaking parasitic disease cycles is still not widespread and, even where it is, poverty and a shortage of basics such as water or fuel for cooking prevent people from adequately applying knowledge. Similar problems apply in agriculture; traditional knowledge of floodplain and rain-fed agriculture has been lost or supplanted by less sustainable forms of irrigated cash-crop agriculture, or by the population growth pressures and inappropriate exploitation of marginal land. Educational programmes in health and agriculture supported by funds from international development agencies have disseminated basic information about waterborne diseases and environmental problems related to agricultural practices. These programmes have, however, failed to bring an understanding of wider natural resources such as wildlife, forests and wetlands.

'Experts'

The development of a conservation ethic and a conservation infrastructure in tropical African countries has been almost exclusively terrestrial and is primarily directed at large mammal and bird species. It is driven partly by an appreciation that humans can conserve wildlife by allocating protected areas, and partly by a recognition that it is an economically valid form of land use linked to tourism (Eltringham, 1975;

Western and Pearl, 1989). It is thus dependent upon developed-world tourists and is sustained by developed-world aid and expertise. Very little aid, expertise or tourism is directed at aquatic resources other than where they are a part of terrestrial ones, or adjacent to them in protected areas (e.g. the proposed Biosphere National Park in western Niger: Le Berre and Messan, 1995). Economic and agricultural development as currently practised are also critically dependent upon developed-world aid and expertise. Here, the ethics and practice of environmental impact assessment of development projects at least partially matches that applied in the home countries of the aid-givers (usually the USA and European nations). However, as Balek (1992) has shown, the vast scale of many projects gives rise to solutions that are inappropriate and ineffective. Adams (1992) has given several case studies where solutions imposed by international or bilateral aid projects, delivered by donor country 'experts', have both failed and caused irreversible damage to aquatic systems. Today, environmental consultants take a more humble approach. More recent large-scale programmes, such as the Onchocerciasis Control Project in West Africa, owed much of their success by applying lessons learned from previous 'failed' projects.

Educational infrastructure

The infrastructure for effective conservation – accumulated knowledge and experience in personnel, information in libraries and databases, and specimens in museums and botanical gardens and national parks – is patchily distributed in tropical Africa for the same reasons of economics and history already discussed (see also Wishart *et al.*, this volume, Part II). Most specimens, particularly type specimens, of colonial expeditions are deposited in the home museums of the participants, in cities such as London and Brussels. Universities and museums were staffed by the colonial service before independence and expatriates after independence, with slow movement towards an educated African staff only really bearing fruit in the last 5–10 years. Almost without exception, the maintenance of large unit resources such as museums and universities is only possible through the cyclical, and often fickle, support of donor countries. Successful developments are rare and are often linked to charismatic personalities, such as the maintenance of the National Museums of Kenya through the 1950s to 1970s, and the Kenya Wildlife Services through the

1980s, by two generations of Leakeys (Morell, 1995). Even in the 1990s, after the Kenyan Government gave its Wildlife Services responsibility for generating its own income from tourism, infrastructure development is still only possible with international aid. Tropical African universities and fisheries research institutes (the only aquatic research centres) are poorly resourced with books, computers and laboratory equipment unless they have specific bilateral links. Often the most modern equipment is to be found in the headquarters of international wildlife charities and United Nations facilities, staffed by expatriates: the technological and information revolution has hardly touched indigenous African conservation efforts (Wishart *et al.*, this volume, Part II).

Lack of infrastructure and availability of taxonomic data hinders the dissemination of scientific studies carried out in the tropics. No widely distributed international journal deals specifically with tropical freshwater ecology and most international journals will not publish results of surveys that are unable to provide detailed taxonomic lists, although they deal with newly investigated tropical rivers. Even in northern countries, river conservation studies are poorly represented; a review of seven mainstream international journals in applied ecology, freshwater biology and conservation showed that, during the 1980s, not more than 5.2% of articles dealt with river conservation (Boon, 1992). The combination of these factors explains why the great majority of information that could serve river conservation in tropical countries is buried in unpublished reports and local publications issued by universities and research institutes.

Besides the scarce infrastructural resources, the contents of conservation education efforts are often inappropriate. In African countries, conservation biology and wildlife management training courses are set up almost exclusively with the help of academic links with European and North American universities. Objective scientific data describing animal populations and environmental conditions constitute the main interest of wildlife managers in the northern, developed world, but are scarce in the tropics. Here instead, tropical conservation managers are daily and urgently faced with the subsistence demands of natural-resource-dependent human populations, seen by some as the single major threat to wildlife protection (Saberwal and Kothari, 1996). Particularly in the tropics, the setting up of environmental education needs a strong social sciences and humanities component; without it, such training could become an abstract exercise of natural sciences theory.

LINKING ECOLOGY, HYDROLOGY AND SOCIO-ECONOMICS

During the past two decades emphasis on the science base necessary for an optimal conservation policy has progressively moved from biological to more managerial disciplines. Autecology and wildlife population studies remain prerequisites for understanding the potential effects of habitat change. Ecological studies have progressively included stronger vegetational and hydrological components. On a continent characterized by moderate temperature fluctuations and high evaporation rates, hydrology is the main physical forcing function. This realization has led to more detailed hydrological studies linked to ecological ones (Balek, 1977; Acreman and Hollis, 1996).

Attention is slowly shifting from effects to causes of environmental damage. Several river basin development projects of the 1970s and 1980s have caused severe environmental disruption. More pervasive and dangerous are generalized poverty and lack of rural development that induce populations to exhaust their resources irrationally, jeopardizing their own survival. Conservation agencies are today involved more and more frequently in socio-economic and anthropological assessments. These approaches provide means for realistic monitoring of the effects of investment projects on rural development. At the same time they expose links between environmental degradation and natural-resource exploitation. This renewed interest in socio-economic issues is illustrated by the rural development studies carried out in the Senegal River Valley (see below) by the Institute for Development Anthropology (IDA) (Salem-Murdock, 1996). In this, as in other basins, it is apparent that seemingly free and unexploited resources (water) were in reality part of well-established complex interrelationships between wildlife and numerous human social groups. Managers also realized that social structure plays a major role in the success of any development venture, irrespective of the financial resources involved. Lack of concern for the social effects of river-basin management has had severe consequences in regions characterized by fragile economies. For instance, the building of the Diama Dam in the Lower Senegal Valley led to ethnic clashes in 1989, and to more than 1000 deaths, and more than 10 000 refugees (Verhoef, 1996).

Today, sub-Saharan Africa hosts more than 160 planned or constructed large river regulation projects, all of which have led to major environmental changes including loss of wetlands and wildlife habitat (76% in Chad, 82% in Senegal and 89% in Gambia: Schwöppe, 1994). Yet none has produced improvement to the life of resident human populations. It is clear that river conservation in tropical Africa has much ground to cover.

The framework for conservation

THE POLITICAL IMPERATIVE: SHARE THE BENEFITS OF NATURAL RESOURCES

Centralization, initiated by colonial governments, has alienated many local African communities from the management of their own resources; territories were subdivided into extensive farms and cattle ranches and distributed among European settlers. After independence, redistribution favoured officials and their clans to the detriment of more traditional groups – for example, the Masai in East Africa. As a result, once tight links between African cultural heritage and patterns of land exploitation have been affected profoundly. Village chiefs have, in many cases, lost the traditional mandate of caring for shared resources such as hunting, the grazing of commons and fishing. For example, in parts of the Logone River, traditional chiefs exerted control over fishing to allow sufficient reproduction and protected fish migration between the wetland and the main river during times critical for recruitment (Dadnadji and Van Wetten, 1993). Similarly, in the Kafue Flats of Zambia, traditional management of the terrestrial mammalian population living within the wetlands was disrupted by the strongly centralized colonial administration. Now local communities are slowly beginning again to appreciate wildlife as a manageable economic resource (Jeffery, 1993).

Today, the management of rivers, floodplains and wetlands faces the challenge of reconciling the cultural and economical divergence which pervades modern African societies:

- the interests of traditional tribes and their subsistence economy confronted by economic development and imported technologies and, often, imported labour;
- the interests of poor rural and peri-urban populations confronted by a small but highly influential, cosmopolitan, rich, urban community.

Modern economies provide easier access to natural resources and stimulate private investment; this contrasts with traditional management concerned with sustainable exploitation of shared resources. The contrast has contributed in some cases to the progressive extinction of original populations and of their land management system. In Guinea Bissau, land privatization started in 1983, but within only 10 years some 45% of the national territory had passed into the hands of foreign investment-backed private producers. Moderation of the activities of these new interest groups, oriented towards short-term revenues offered by export markets, is one of the primary concerns of the IUCN in the region (Campredon, 1993). Cases such as these advocate aggressive conservation policies based upon the establishment of exclusive protected areas in priority sites (Dugan, 1994). Modern exploitation systems are unable to integrate the complex hydrological cycles which rule ecological balances in tropical coastal wetlands and river floodplains, and conflicting economic activities including fishing, deforestation, large-scale agriculture and tourism are carried out in the coastal wetlands of Guinea Bissau without any monitoring. Only recently a rationalization programme was introduced, directed at the involvement of state authorities and decision-makers (Campredon, 1993).

Typical products of the world market economy are large structures for hydrological control which, in Africa, constitute the most substantial portion of foreign investment in basin 'management'. These developments disrupt existing land exploitation systems, force the resettlement of the resident populations and the diversion of all resources towards risky, export-oriented, cash crops, thereby alienating natural resources from the rural community to the benefit of urban managers, backed by foreign, multinational companies. Technical problems are common with such large-scale schemes: for example, the Kibimba rice scheme, in eastern Uganda, had an expected yield of 4.9 t ha^{-1} in 1986, but achieved only 1.7 t ha^{-1} by 1988, with major financial losses.

CONSERVATION OF ETHNIC DIVERSITY

Numerous examples can be cited describing local community control over hydrological unpredictability in the tropics:

- The Balanta and the Feloupe of Guinea Bissau have constructed an efficient hydraulic system for the irrigation of traditional rice plantations that is still unequalled by modern methods (Campredon, 1993).

- The traditional farming community residing between the Chari and the Logone rivers (Dadnadji and Van Wetten, 1993).
- The Sonjo, a Tanzanian irrigation-based society.
- The Bozo and Somono fishermen of the Central Delta of the Niger River in Mali, who distinguish several fishing zones (*bamo*) and manage fishing activities year-by-year according to its availability (Sissoko *et al.*, 1986).
- The complex set of floodplain activities including fishing, agriculture and pastoralism carried out in the Kona District of Mali (Skinner, 1987). A similar case study is known from the Hadejia–Nguru Wetlands of the Yobe Basin of north-western Nigeria (Hollis *et al.*, 1993).

Cases such as these demonstrate the persistence of a traditional 'submerged' subsistence economy little influenced by international funding. Small-scale inland fisheries are typically managed by local exploitation systems and most of the produce goes unaccounted. The productivity of traditional practices such as these is commonly underestimated and, consequently, economic strategies based on traditional exploitation methods have little weight when confronted by alternative proposals of resource allocation backed by international funding.

Recent development projects conducted jointly by several international agencies led by the IUCN have stimulated commitment to economic development in conjunction with wetland conservation at institutional level. In 1986 this campaign convinced the Ugandan Government to ban all large-scale drainage projects and made it the leading nation in the developing world in the matter of wetland management (Dugan, 1993, 1994).

THE SOCIO-ECONOMIC IMPERATIVE: INTEGRATION OF DEVELOPMENT AND CONSERVATION

Whilst acting as foci for the majority of human activities, river basins are also repositories of vital natural resources. No realistic conservation strategy can be proposed without tight integration of nature protection and economic development. This consideration is particularly relevant in the context of developing countries where the link between human survival and natural-resource availability is still at the base of the socio-economic structure. Industrialized

countries have 'progressed' from a direct dependence on natural resources and this lack of dependence has contributed to the establishment of an ethic of wildlife conservation *per se*. In the modern African context, however, the concept of wildlife as a legitimate resource user would not be comprehensible and cannot be applied without top-down enforcement and the restriction of public participation. An interesting exception is represented by restrictions imposed by traditional religious practices.

Top-down enforcement has managed to protect endangered species at the brink of extinction on regional scales (large felines, elephant, rhinoceros); it also provides sources of income from international tourism to the extent that tourism is of primary financial concern to many African countries. However, these achievements have been obtained at the cost of the social and economic disruption of traditional populations that have in most cases been evicted from protected areas. Managers of protected areas in developing countries are aware that competition between the rural poor and wildlife during periods of scarcity may become acute and endanger long-established equilibria. On the other hand, protection through the total exclusion of human activity is becoming less viable, as political consciousness and demands for a wider participation in decision-making, rise among modern African peoples. Only a comprehensive strategy integrating human with conservation needs, designed on wide agreement, can hope to provide effective conservation in the long term.

Conservation in practice

THE SCIENTIFIC IMPERATIVE: BEYOND THE PROTECTION OF RARE SPECIES

Lack of systematic inventories of habitats and species, of established classification schemes and of methods for the assessment of conservation value are major impediments for the implementation of effective conservation in tropical regions (Boon, 1992). Advances in taxonomic knowledge are expensive, laborious and open to the criticism that the conservation of functional ecosystems should come first. The ultimate aim of the conservation of aquatic biodiversity should be the protection of critical resources necessary for species survival, in reserves large enough to resist temporal variation (Fausch and Young, 1995). Riverine fish, for example, are known to migrate longitudinally and laterally, following regular seasonal patterns. Resource exploitation by fish is also related to the succession of development stages. Studies carried out on juvenile recruitment in North America show strong seasonal spatial variation in the distribution of fish populations (Angermeier and Schlosser, 1995), and reproduction may be concentrated in small habitats, sometimes at the margins of the main population distribution area; other habitats become critical during later stages of life.

In the case of tropical African rivers, biodiversity and natural habitat inventories have to be a considered long-term aim. While working in this direction, an effective conservation strategy has to act to preserve endangered communities and ecosystems. Priority should be given to the identification of 'centres of biodiversity' that represent genetic sources for the recolonization of perturbed areas. The relative importance of different areas can be assessed by means of conservation criteria (naturalness, representativeness, rarity) applied to different physical and biological features. Such evaluation systems are under development for UK rivers and could be applied, in a modified form, in the tropical African context (Boon *et al.*, 1997; Raven *et al.*, 1997).

Some indication of which ecosystems should be preserved comes directly from development history. Drastic changes in basin hydrology, pollution and exotic species introductions have had great impacts on large lake and river ecosystems. At the margins, unperturbed, smaller aquatic habitats have become not only rare but true refuges and repositories of original genetic pools. Future attention will no doubt focus on isolated wetlands such as floodplain oxbow lakes, mountain tarns and coastal lagoons. A recent study carried out in Lake Nabugabo within the swamps of north-eastern Lake Victoria, revealed that Nabugabo serves as refuge for several species of fish now extinct in the main lake (Chapman *et al.*, 1996).

Coastal wetlands within river deltas are well developed along the West African coastline from the Senegal Estuary down to the Equator. Several factors such as difficulty of access, presence of wildlife, waterborne diseases and difficulties in carrying out agricultural practices, due to salinity and flooding, explain why these systems remained isolated. Coastal wetlands are today vital refuges of rare and endangered species of animals and plants. For example the large estuary of the Rio Caine in Guinea Bissau shelters the last remnants of primary tropical forest (Cantanthès Forest) while the Bijagos Archipelago has an abundance of endangered terrestrial and marine aquatic species, such as marine turtle, manatee (*Trichecus manatus*), crocodile, hippopotamus and waterbirds (Campredon, 1993).

GENETIC CONSERVATION

In-depth genetic studies of riverine fish populations are few and confined mainly to northern countries. In tropical Africa much interest has been focused on the rapidly speciating fish assemblages of the East African Great Lakes and on biodiversity threats from environmental degradation (e.g. Lowe-McConnell *et al.*, 1994). Strategies for the conservation of Great Lakes species include:

- the *ex situ* conservation of rare species such as the haplochromines of Lake Victoria (Kaufman and Ochumba, 1993);
- the protection of endemic tilapines in small wetlands adjacent to the Victoria Basin (Ogutu-Ohwayo, 1993; Chapman *et al.*, 1996);
- reintroductions of fish to protected areas (Lowe-McConnell *et al.*, 1994).

Although not applied in the tropical context, genetic studies have provided conservation both with new concepts and objectives in the study of population dynamics, mechanisms of speciation, and the role of biodiversity in ecosystem functioning; they are relevant for the design of conservation strategies for tropical African rivers. A recent definition identifies four levels of biodiversity (Temple, 1991):

- genetic variation within a species
- phenotypic and morphological diversity
- species richness within an ecosystem
- community–ecosystem diversity

Management directed towards the protection of species disregards other levels of genetic variation and may lead to the failure of wildlife protection programmes. Genetic studies reveal that populations are unequal in terms of their conservation value. The survival of a given species may depend on the preservation of populations with high genetic variability or even on the preservation of assemblages.

This holistic view represents a conceptual advancement but reduces the value of the species as a practical management unit. In recognition of this, Ryder (1986) coined the term 'Evolutionarily Significant Unit' (ESU) provoking a debate among geneticists and conservation managers. A clear-cut agreement on the definition of an ESU has not been reached (Moritz, 1994). However, the debate has served to highlight both the need of genetics and spatial and temporal population dynamics for effective management of wildlife. Genetic studies can reveal much about the evolutionary potential of populations and, therefore, about the effectiveness of conservation measures, but can hardly be expected to provide decision-making tools for conservation management; a more flexible approach is needed. The protection of ESUs can be practically implemented by an ecosystem-oriented planning strategy (Angermeier and Schlosser, 1995).

These considerations are of direct practical relevance for managers of tropical river basins that typically do not possess exhaustive species lists. Under the habitat/ecosystem protection approach, lack of taxonomic information does not represent a major impediment. A strategy directed towards broad landscape units preserves communities and landscapes which provide the ecological and evolutionary context relevant for the survival of species and populations. Such units also constitute the spatial dimension of human activities and lead managers in developing countries to study human subsistence economies at the same scale of wildlife management. Programmes designed at this scale are also more likely to preserve the integrity of complex ecological processes characterizing ecosystem functioning (Angermeier and Schlosser, 1995).

Conclusions and future outlook

APPLYING THE CONCEPT OF SUSTAINABILITY

Environmental sustainability must rest on sustainable economic policies oriented towards long-term local development. The introduction of modern technologies, the sudden opening of fragile socio-economic systems to foreign investment and rapid resource privatization may, within a short time, break the balance between human activities and resource regeneration. Liberal 'aid' agencies often work *de facto* against the persistence of sustainable systems. Reports of sustainable management practices conducted before European colonization demonstrate that there is no distinction between subsistence economy and conservation. Introducing the concept of 'wise use', the Ramsar Convention on Wetlands of International Importance (1971) represented the first official move towards the promotion of conservation and natural-resources management as a means of community

development. The success of the idea is closely connected to the wide participation of the local communities which can be ensured only through the involvement of traditional forms of administration close to African cultural heritage. Village chiefs, elected by the community, and trained in collaboration with central government, offer the appropriate balance of representation and direct contact with local issues (e.g. Dugan, 1994).

On the Kafue Flats (Zambia) two Wetland Management Authorities were established to provide links between traditional and contemporary administrative authorities. Autonomy from central government allowed the authorities to develop their own income-generating projects which provided an independent source of investment for community development (Jeffery *et al.*, 1992). The Zambian experience has shown that wildlife, managed by controlled hunting and tourism, can be a major money-earner, and that poaching, by subtracting this shared resource to the benefit of people external to the wetland area, represents a limitation both to the conservation and the development potential of local people. Similar examples applied to fishery management can be found in the Senegal Valley and the Hadejia–Nguru Wetlands where traditional rules regulate the distribution of fishing and farming activities during the year to maximize sustainable resource renewal and sharing of resource benefits. The economy is managed on the traditional base of communal property. In Kano State it is forbidden for individuals to own water bodies (Thomas *et al.*, 1993). Sustainable exploitation here lies with the authority of respected traditional experts. In the Hadejia–Nguru rights of access, fishing seasons, and fishing methods are carefully assessed on a yearly basis to fit the natural variation of wet and dry years (Thomas *et al.*, 1993) (Figure 6.8).

In the context of river conservation the concept of sustainability becomes useful particularly when applied to the catchment scale. When this is done it becomes apparent that permanently flowing rivers such as the Senegal, the Chari and the Tana are themselves key centres of biodiversity within the drought stricken sub-Saharan region (Verschuren, 1996). Drought stress caused habitat fragmentation and isolation of refuge islands, several of which, in the Senegal and in the Tana basins are today protected by reserves and national parks. No provision has been made, however, for a catchment-scale strategy that would preserve downstream linkages and give river conservation its rightful place within river basin management. Under

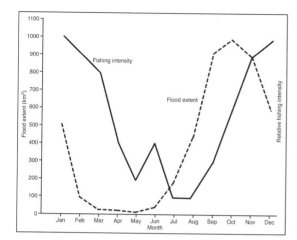

Figure 6.8 *Fishing intensity compared to floodwater levels during one year in Hadejia–Nguru Wetlands, West Africa (from Thomas* et al.*, 1993)*

present conditions, the region is not capable of allowing traditional management and is prone to rapid degradation.

The IUCN approach focuses on the ways in which natural systems work, the ways in which people have adapted to them, and the strategies necessary to enhance the productivity of the resources to maximize human benefits (Adams, 1992). This approach contrasts with the past policies of development agencies which traditionally focused upon actions directed at maintaining the income of urban populations, disregarding the needs of rural producers. One important message from the Rio Convention in 1992, repeating those of the Stockholm Conference 20 years earlier and the World Conservation Strategy in 1980, was that sustainable development must involve all 'stakeholders'. Without it, the prospects for effective river conservation in tropical Africa in the 21st century are grim.

References

Acreman, M.C. and Hollis, G.E. (eds) 1996. *Water Management and Wetlands in Sub-Saharan Africa.* International Union for the Conservation of Nature, Gland.

Adams, W.M. 1992. Sustainable agricultural development and wetland conservation in Northern Nigeria. In: Maltby, E., Dugan, P.J. and Lefeuvre, J.C. (eds) *Conservation and Development: The Sustainable Use of Wetland Resources.* International Union for the Conservation of Nature, Gland, 11–20.

Amonoo-Neizer, E.H., Nyamah, D. and Bakiamoh, S.B. 1996. Mercury and arsenic pollution and biological samples around the mining town of Obuasi, Ghana. *Water Air and Soil Pollution* 91:363–373.

Angermeier, P.L. and Schlosser, I.J. 1995. Conserving aquatic biodiversity: beyond species and populations. *American Fisheries Society Symposium* 17:402–414.

Balek, J. 1977. *Hydrology and Water Resources in Tropical Africa.* Elsevier, Amsterdam.

Balek, J. 1992. *The Environment for Sale.* Carlton, New York.

Barnard, C. and Biggs, J. 1988. Macroinvertebrates in the catchment of Lake Naivasha, Kenya. *Revue d'Hydrobiologie Tropicale* 21:127–134.

Beadle, L.C. 1981. *The Inland Waters of Tropical Africa*, 2nd edition. Longman, London.

Bishop, J.E. 1973. *Limnology of a Small Malayan River, Sungai Gombak.* Monographiae Biologicae 22. Dr W. Junk, Dordrecht.

Boon, P.J. 1992. Essential elements in the case for river conservation. In: Boon, P.J., Calow, P. and Petts, G.E. (eds) *River Conservation and Management.* John Wiley, Chichester, 11–34.

Boon, P.J., Wilkinson, J. and Martin, J. 1997. The application of SERCON (System for Evaluating Rivers for Conservation) to a selection of rivers in Britain. *Aquatic Conservation: Marine and Freshwater Ecosystems* 8:597–616.

Bowell, R.J., Warren, A., Minjera, H.A. and Kimaro, N. 1995. Environmental impact of former gold mining on the Orangi River, Serengeti National Park, Tanzania. *Biogeochemistry* 28:131–160.

Bretschko, G. 1995. Tropical River Ecology Initiative. Unpublished Workshop Proceedings, Department of Zoology, Eldoret University, Njoro, Kenya.

Bretschko, G. 1996. Tropical River Ecology Initiative. Unpublished Second Workshop Proceedings, Department of Zoology, Eldoret University, Njoro, Kenya.

Briquet, J.P., Mahe, G., Bamba, F. and Olivry, J.C. 1996. Changements climatiques récents et modification du régime hydrologique du fleuve Niger à Koulikoro (Mali). In: *L'hydrologie Tropicale: Géoscience et Outil pour le Développement.* Actes de la Conférence de Paris, Mai 1995. IAHS Publication 238, 157–166.

Cairns, J., Heath, A.G. and Parker, B.C. 1975. Temperature influence on chemical toxicity to aquatic organisms. *Journal of the Water Pollution Control Federation* 47:267–280.

Campredon, P. 1993. Coastal wetland planning and management in Guinea-Bissau. In: Davies, T.J. (ed.) *Towards the Wise Use of Wetlands.* Ramsar Convention Bureau, Gland, 44–51.

Chapman, L.J., Chapman, C.A., Ogutu-Ohwayo, R., Chandler, M., Kaufman, L. and Keiter, A.E. 1996. Refugia for endangered fishes from introduced predator in Lake Nabugabo (Uganda). *Conservation Biology* 10:554–561.

Chukwuma, C. 1997. Environmental lead exposure in Africa. *Ambio* 26:399–403.

Chutter, F.M. 1969. The effects of silt and sand on the invertebrate fauna of streams and rivers. *Hydrobiologia* 34:57–76.

Chutter, F.M. 1970. Hydrobiological studies in the catchment of the Vaal Dam, South Africa. Part 1. River zonation and benthic fauna. *Internationale Revue der gesamten Hydrobiologie* 55:445–494.

Clark, F., Beeby, A. and Kirby, P. 1989. A study of the macro-invertebrates of Lakes Naivasha, Oloidien and Sonachi, Kenya. *Revue d'Hydrobiologie Tropicale* 22:21–33.

Cohen, A.S., Kaufman, L. and Ogutu-Ohwayo, R. 1996. Anthropogenic threats, impacts and conservation strategies in the African Great Lakes: a review. In: Johnson, T.C. and Obada, E.O. (eds) *The Limnology, Climatology and Paleoclimatology of the East African Great Lakes*. Gordon & Breach Publishers, Amsterdam, 575–624.

Cooper, S.D. 1996. Rivers and streams. In: McClanahan, T.R. and Young, T.P. (eds) *East African Ecosystems and their Conservation*. Oxford University Press, New York, 133–170.

Dadnadji, K.K. and Van Wetten, J.C.J. 1993. Traditional management systems and integration of small scale interventions in the Logone floodplains of Chad. In: Davis, T.J. (ed.) *Towards the Wise Use of Wetlands*. Report to the Ramsar Convention Wise Use Project. International Union for the Conservation of Nature, Gland, 74–81.

Davies, B.R., O'Keeffe, J.H. and Snaddon, C.D. 1995. River and stream ecosystems in southern Africa: predictably unpredictable. In: Cushing, C.E., Cummins, K.W. and Minshall, G.W. (eds) *Ecosystems of the World 22. River and Stream Ecosystems*. Elsevier, Amsterdam, 537–599.

Dejoux, C. 1988. *La pollution des Eaux Continentales Africaines*. ORSTOM Travaux et Documents 213.

Denny, P. 1985. *The Ecology and Management of African Wetland Vegetation*. Dr W. Junk Publishers, Dortrecht.

Dugan, P.J. 1993. Research priorities in wetland science, wetland conservation and agricultural development. In: Maltby, E., Dugan, P.J. and Lefeuvre, J.C. (eds) *Conservation and Development: The Sustainable Use of Wetland Resources*. International Union for the Conservation of Nature, Gland, 3–10.

Dugan, P.J. 1994. The role of ecological science in addressing wetland conservation and management in the tropics. *Mitteilungen der internationalen Vereinigung für theoretische und angewandte Limnologie* 24:5–10.

Dunne, T. 1979. Sediment yield and land use in tropical catchments. *Journal of Hydrology* 42:281–300.

Ekweozor, I.K.E. 1989. A review of the effects of oil pollution in a west African environment. *Discovery and Innovation* 1:27–37.

Eltringham, S.K. 1975. *Wildlife Resources and Economic Development*. John Wiley, Chichester.

Faniran, A. and Areola, O. 1978. *Essentials of Soil Study (with Special Reference to Tropical Areas)*. Heinemann, London.

Fausch, K. and Young, M. 1995. Evolutionarily significant units and movement of resident stream fishes: a cautionary tale. *American Fishery Society Symposium* 17:360–370.

French, G.T., Awosika, L.F. and Ibe, C.E. 1994. Sea-level rise in Nigeria: potential impacts and consequences. *Coastal Research Special Issue* 14 (eds R.J. Nicholls and S.P. Letherman).

Gibon, F.M. and Statzner, B. 1985. Longitudinal zonation of lotic insects in the Bandama River system (Ivory Coast). *Hydrobiologia* 122:61–64.

Grove, A.T. 1996. African river discharges and lake levels in the twentieth century. In: Johnson, T.C. and Obada, E.O. (eds) *The Limnology, Climatology and Paleoclimatology of the East African Great Lakes*. Gordon & Breach Publishers, Amsterdam, 95–102.

Harrison, A.D. 1965. River zonation in southern Africa. *Archiv für Hydrobiologie* 61:387–394.

Harrison A.D. 1995. Northeastern Africa rivers and streams. In: Cushing, C.E., Cummins, K.W. and Minshall, G.W. (eds) *Ecosystems of the World 22. River and Stream Ecosystems*. Elsevier, Amsterdam, 507–517.

Harrison, A.D. and Hynes, H.B.N. 1988. Benthic fauna of Ethiopian streams and rivers. *Archiv für Hydrobiologie Supplement* 81:1–36.

Hollis, G.E., Adams, W.M. and Aminu-Kano, M. (eds) 1993. *The Hadejia–Nguru Wetlands: Environment, Economy and Sustainable Development of a Sahelian Floodplain Wetland*. International Union for the Conservation of Nature, Gland.

Hugueny, B. 1989. West African rivers as biogeographic islands: species richness of fish communities. *Oecologia* 79:236–243.

Ikingura, J.R. and Akagi, H. 1996. Monitoring of fish and human exposure to mercury due to gold mining in the Lake Victoria goldfields, Tanzania. *The Science of the Total Environment* 191:59–68.

Illies, J. and Botosaneanu, L. 1963. Problèmes et méthodes de la classification et de la zonation écologique des eaux courantes, considérées surtout du point de vue faunistique. *Mitteilungen der internationalen Vereinigung für theoretische und angewandte Limnologie* 12:1–57.

Jacobsen, D., Schultz, R. and Encalada, A. 1997. Structure and diversity of stream invertebrate assemblages: the influence of temperature with altitude and latitude. *Freshwater Biology* 38:247–261.

Jeffrey, R.C.V. 1993. Wise use of floodplain wetlands in the Kafue Flats of Zambia. In: Davies, T.J. (ed.) *Towards a Wise Use of Wetlands*. Ramsar Convention Bureau, Gland, 145–152.

Jeffery, R.C.V., Chabwela, H.N., Howard, G. and Dugan, P.J. (eds) 1992. *Managing the Wetlands of Kafue Flats and Bangweulu Basin*. IUCN, Gland.

Junk, W.J., Bayley, P.B. and Sparks, R.E. 1989. The flood pulse concept in river–floodplain systems. In: Dodge, D.P. (ed.) *Proceedings of the International Large Rivers Symposium. Canadian Special Publication of Fisheries and Aquatic Sciences* 106:110–127.

Kaufman, L. and Ochumba, P. 1993. Evolutionary and

conservation biology of cichlid fishes as revealed by faunal remnants in northern Lake Victoria. *Conservation Biology* 7:719–730.

Kinyua, A.M. and Pacini, N. 1991. The impact of pollution on the ecology of the Nairobi–Athi River System in Kenya. *International Journal of BioChemiPhysics (Kenya)* 1:5–7.

Koeman, J.H., Pennings, J.H., De Goeij, J.J.M., Tjioe, P.S., Olindo, P.M. and Hopcraft, J. 1972. A preliminary survey of the possible contamination of Lake Nakuru in Kenya with some metals and chlorinated hydrocarbon pesticides. *Journal of Applied Ecology* 9:411–416.

Laraque, A. and Olivry, J.C. 1996. Evolution de l'hydrologie du Congo–Zaïre et de ses affluents rive droite et dynamique des transports solides et dissous. In: *L'hydrologie Tropicale: Géoscience et Outil pour le Développement*, Actes de la conférence de Paris, Mai 1995. IAHS Publication 238, 271–288.

Le Berre, M. and Messon, L. 1995. The west region of Niger: assets and implications for sustainable development. *Nature and Resources* 31:18–30.

Lévêque, C. 1994. 'Introduction générale: biodiversité des poissons africains. In: Teugels, G.G., Guégan, J.F. and Albaret, J.J. (eds) *Biological Diversity of African Fresh- and Brackish Water Fishes*. Annales Sciences Zoologiques, 275, Musée royal d'Afrique centrale, Tervuren, Belgium, 7–16.

Lévêque, C. 1995. River and stream ecosystems of northwestern Africa. In: Cushing, C.E., Cummins, K.W. and Minshall, G.W. (eds) *Ecosystems of the World 22. River and Stream Ecosystems*. Elsevier, Amsterdam, 519–536.

Lévêque, C. 1997. *Biodiversity Dynamics and Conservation: The Freshwater Fish of Tropical Africa*. Cambridge University Press, Cambridge.

Lowe-McConnell, R.H. 1975. *Fish Communities in Tropical Freshwaters*. Longman, London.

Lowe-McConnell, R.H., Roest, F.C., Ntakimazi, G. and Risch, L. 1994. The African Great Lakes. In: Teugels, G.G., Guégan, J.F. and Albaret, J.J. (eds) *Biological Diversity of African Fresh- and Brackish Water Fishes*. Annales Sciences Zoologiques, 275, Musée royal d'Afrique centrale, Tervuren, Belgium, 87–94.

Malvestuto, S.P. and Meredith, E.K. 1989. Assessment of the Niger River fishery in Niger (1983–85) with implications for management. In: Dodge, D.P. (ed.) *Proceedings of the International Large Rivers Symposium. Canadian Special Publication of Fisheries and Aquatic Sciences* 106:533–544.

Merritt, R.W. and Cummins, K.W. 1996. *An Introduction to the Aquatic Insects of North America*, 3rd edition. Kendall/Hunt Publishers, Dubuque.

Meybeck, M. 1981. Pathways of major elements from land to ocean through rivers. In: *River Inputs to Ocean Systems*. UNEP/UNESCO, Paris, 18–30.

Meybeck, M. 1986. Composition chimique des ruisseaux non pollués de France. *Bulletin de Sciences Géologiques* 39:3–77.

Moffat, D. and Lindén, O. 1995. Perception and reality: assessing priorities for sustainable development in the Niger River Delta. *Ambio* 24:527–538.

Morell, V. 1995. *Ancestral Passions*. Collins, London.

Moritz, C. 1994. Defining 'evolutionary significant units' for conservation. *Trends in Ecology and Evolution* 9:373–375.

Nicholson, S.E. 1996. A review of climate dynamics and climate variability in Eastern Africa. In: Johnson, T.C. and Obada, E.O. (eds) *The Limnology, Climatology and Paleoclimatology of the East African Great Lakes*. Gordon & Breach Publishers, Amsterdam, 25–56.

Nicholls, R.J., Awosika, L.F., Niang-Diop, I., Dennis, K.C. and French, G.T. 1993. Vulnerability of West Africa to accelerated sea-level rise. In: *Coastlines of West Africa*. American Society of Civil Engineers, New York, 294–308.

Ogbeibu, A.E. and Victor, R. 1989. The effects of road and bridge construction on the bank-root macrobenthic invertebrates of a southern Nigerian stream. *Environmental Pollution* 56:85–100.

Ogutu-Ohwayo, R. 1993. The effects of predation by Nile perch, *Lates niloticus* L., on the fish of Lake Nabugabo, with suggestions for conservation of endangered endemic cichlids. *Conservation Biology* 7:701–711.

Olivry, J.C. 1987. Les conséquences durables de la sécheresse actuelle sur l'écoulement du fleuve Sénégal et l'hypersalinization de la Basse-Casamance. In: *The Influence of Climate Change and Climatic Variability on the Hydrologic Regime and Water Resources*. Proceedings of the Vancouver Symposium, August 1987, IAHS Publication 168, 501–512.

Pacini, N. 1989. The ecology of the Nairobi–Athi river system and the impact of polluting discharges. Unpublished MSc Thesis, University of Leicester, Leicester, UK.

Pacini, N., Harper, D.M. and Mavuti, K.M. 1999. Hydrological and ecological considerations in the management of a catchment controlled by a reservoir cascade: the Tana River, Kenya. In: Harper, D. and Brown, T. (eds) *The Sustainable Management of Tropical Catchments*. John Wiley, Chichester, 239–258.

Paugy, D., Kassoum, T. and Diouf, P.S. 1994. Faune ichtyologique des eaux douces d'Afrique de l'Ouest. In: Teugels, G.G., Guégan, J.F. and Albaret, J.J. (eds) *Biological Diversity of African Fresh- and Brackish Water Fishes. Annales Sciences Zoologiques*, 275, Musée royal d'Afrique centrale, Tervuren, Belgium, 35–66.

Payne, A.I. 1986. *The Ecology of Tropical Lakes and Rivers*. John Wiley, Chichester.

Rached, E. 1996. Introduction. In: Rached, E., Rathgeber, E. and Brooks, D. (eds) *Water Management in Africa and the Middle East, Challenges and Opportunities*. International Research Development Centre Publications, Ottawa, 3–6.

Raven, P.J., Holmes, N.T.H., Dawson, F.H. and Everard, M. 1997. Quality assessment using River Habitat Survey data. *Aquatic Conservation: Marine and Freshwater Ecosystems* 8:477–500.

Ribbink, A.J. 1994. Biodiversity and speciation of freshwater fishes with particular reference to African cichlids. In: Giller, P.S., Hildrew, A.G. and Raffaelli, D.G. (eds) *Aquatic Ecology: Scale, Pattern and Process*. Blackwell Scientific Publications, Oxford, 261–288.

Roberts, T.R. 1975. Geographical distribution of African

freshwater fishes. *Zoological Journal of the Linnean Society* 57:249–319.

Rohde, K. 1992. Latitudinal gradients in species diversity: the search for the primary cause. *Oikos* 65:514–527.

Rosenzweig, M.L. 1995. *Species Diversity in Space and Time.* Cambridge University Press, Cambridge.

Ryder, O.A. 1986. Species conservation and systematics: the dilemma of subspecies. *Trends in Ecology and Evolution* 1:9–10.

Rzóska, J. 1976. *The Nile, Biology of an Ancient River.* Dr W. Junk, The Hague.

Saberwal, V.K. and Kothari, A. 1996. The human dimension in conservation biology curricula in developing countries. *Conservation Biology* 10:1328–1331.

Salem-Murdock, M. 1996. Social science inputs to water management and wetland conservation in the Senegal River Valley. In: Acreman, M.C. and Hollis, G.E. (eds) *Water Management and Wetlands in Sub-Saharan Africa.* IUCN, Gland, 125–144.

Schmid, F. 1984. Un essai d'évaluation de la faune mondiale des Trichoptères. In: Morse, J.C. (ed.) *Proceedings of the 4th International Symposium on Trichoptera,* Clemson, South Carolina, July 1983. Dr W. Junk, The Hague, 337.

Schwöppe, W. 1994. Die Lanschaftökologischen Veränderungen im Bereich des Nationalparkes Djoudj (Senegal). Unpublished PhD Thesis, University of Hamburg.

Scott, K.M.F. 1986. A brief conspectus of the Trichoptera (caddisflies) of the Afrotropical Region. *Journal of the Entomological Society of Southern Africa* 49:231–238.

Sissoko, M.M., Malvestuto, S.P., Sullivan, G.M. and Meredith, E.K. 1986. Inland fisheries in developing countries: an opportunity for a farming systems approach to research and management. In: Flora, C.B. and Tomecek, M. (eds) *Selected Proceedings: Farming Systems Symposium.* Kansas State University, Manhattan, 297–317.

Skinner, J. 1987. Rapport d'activités, projet de création d'une 'fôret villageoise' à Bouna, arrondissement de Kona, 5ème Région, Mali. International Union for the Conservation of Nature, Gland.

Tadesse, H. 1998. The Nile river basin. *Ethioscope* 1:13–23.

Temple, S.A. 1991. Conservation biology: new goals and new partners for managers of biological resources. In: Becker, D.J., Kransy, M.E., Goff, G.R., Smith, C.R. and Gross, D.W. (eds) *Challenges in the Conservation of Biological Resources: A Practitioner's Guide.* Westview Press, Boulder, Colorado, 45–54.

Teugels, G.G. and Guégan, J.F. 1994. Diversité biologique des poissons d'eaux douces de la Basse-Guinée et de l'Afrique Centrale. In: Teugels, G.G., Guégan, J.F. and Albaret, J.J. (eds) *Biological Diversity of African Fresh- and Brackish Water Fishes. Annales Sciences Zoologiques,* 275, Musée royal d'Afrique centrale, Tervuren, Belgium, 67–85.

Thomas, D.H.L., Jinoh, M.A. and Matthes, H. 1993. Fishing.

In: Hollis, G.E., Adams, W.M. and Aminu-Kano, M. (eds) *Hadejia–Nguru Wetlands.* International Union for the Conservation of Nature, Gland, 95–115.

Thompson, J.R. 1996. Africa's floodplains: a hydrological overview. In: Acreman, M.C. and Hollis, G.E. (eds) *Water Management and Wetlands in Sub-Saharan Africa.* International Union for the Conservation of Nature, Gland, 5–20.

Timberlake, L. 1985. *Africa in Crisis.* Earthscan/IIED, London.

Twinch, A.J. 1987. Phosphate exchange characteristics of wet and dried sediment samples from a hypereutrophic reservoir: implications for the measurement of sediment phosphorus status. *Water South Africa* 21:1225–1230.

Van Someren, R. 1952. *The Biology of Trout in Kenya Colony.* Government Printer, Nairobi.

Vannote, R.L., Minshall, G.W., Cummins, K.W., Sedell, J.R. and Cushing, C.E. 1980. The river continuum concept. *Canadian Journal of Fisheries and Aquatic Sciences* 37:130–137.

Verhoef, H. 1996. Health aspects of Sahelian floodplain development. In: Acreman, M.C. and Hollis, G.E. (eds) *Water Management and Wetlands in Sub-Saharan Africa.* International Union for the Conservation of Nature, Gland, 33–50.

Verschuren, D. 1996. Comparative paleolimnology in a system of four shallow tropical lake basins. In: Johnson, T.C. and Odada, E.O. (eds) *The Limnology and Paleoclimatology of the East African Lakes.* Gordon & Breach Publishers, Amsterdam, 559–574.

Viner, A.B. 1975. The supply of minerals to tropical rivers and lakes (Uganda). In: Hasler, A.D. (ed.) *Coupling of Land and Water Systems.* Ecological Studies 10, Springer-Verlag, New York, 227–261.

Walling, D.E. and Webb, B.W. 1986. Solute transport by rivers in arid environments: an overview. *Journal of Water Resources* 5:800–822.

Ward, J.V. and Stanford, J.A. 1983. The serial discontinuity concept of lotic ecosystems. In: Fontaine, T.D. and Bartell, S.M. (eds) *Dynamics of Lotic Ecosystems.* Ann Arbor Science, Ann Arbor, Michigan, 29–42.

Welcomme, R.L. 1974. Some general theoretical considerations on the fish yield of African rivers. *Journal of Fish Biology* 8:351–364.

Welcomme, R.L. 1985. *River Fisheries.* FAO Fisheries Technical Paper No 262, FAO, Rome.

Welcomme, R.L. 1989. Review of the present state of knowledge of fish stocks and fisheries of African rivers. In: Dodge, D.P. (ed.) *Proceedings of the International Large River Symposium, Special Publication of the Canadian Journal of Fisheries and Aquatic Sciences* 106:515–532.

Western, D. and Pearl, M. (eds) 1989. *Conservation for the Twenty-First Century.* Oxford University Press, New York.

Worthington, E.B. 1958. *Science and the Development of Africa.* CCTA, London.

7

River conservation in the countries of the Southern African Development Community (SADC)

B.R. Davies and M.J. Wishart

The regional context

DEFINITION

It first made sense to confine ourselves to 'Southern Africa' within an ecological and biogeographical framework and we elected to address the area 'south of 17°S', effectively the Zambezi Valley southwards. As we progressed, however, it became clear that recent socio-political developments might have far more important consequences for river conservation in the region than, for example, purely geographical or biological considerations. Thus, we reconsidered and broadened our boundaries to cover the Southern African Development Community (SADC). Although some may disdain our socio-political, rather than biogeographical approach, we argue that throughout SADC, socio-political considerations are inextricably linked to conservation issues. The community comprises the Republics of Angola, Botswana, the Democratic Republic of Congo (DRC), Madagascar, Malawi, Mauritius, Moçambique (Mozambique), Namibia, South Africa, Tanzania, Zambia and Zimbabwe, and the Kingdoms of Lesotho and Swaziland. Even with our decision we were immediately faced with the problem of inclusivity. For instance, it made no sense to include Mauritius and Madagascar (see Benstead et al., this volume, for Madagascar) and, hence, our review excludes the Indian Ocean states, although it does, to some extent, include the recent addition to SADC, the DRC (we avoid overlap with the coverage of the DRC by Pacini and Harper (this volume)).

The uneven distribution, both of permanent surface waters and of human populations in the region has led to the construction of large water-distribution networks (interbasin transfers: IBTs), and will do so with increasing frequency. Such developments have profound implications for river conservation (see also Davies et al., this volume). Within this framework there is growing pressure to look beyond national boundaries for water. The Congo and Zambezi Rivers, for instance, are the only large perennial systems of the region (Table 7.1), and both are being investigated as potential sources of water for the southern states of the region. The abundant water of the Congo, and the increasing trend towards 'redistribution' of unevenly distributed water (Asmal, 1995; Davies and Day, 1998), makes it vitally important for SADC.

Our decision was also shaped by the recent democratic elections in South Africa, the disbanding of the so-called 'Front-Line States', South Africa's return to the international community, and the creation of a regional economic community. It is this last, we feel, that will eventually lead to rapid industrial, urban and rural development, with concomitant effects on aquatic environments. In this context, unless closer regional cooperation is strengthened (see Wishart et al., this volume), river ecosystems will be subject to even greater stresses than at present.

Global Perspectives on River Conservation: Science, Policy and Practice.
Edited by P.J. Boon, B.R. Davies and G.E. Petts. © 2000 John Wiley & Sons Ltd.

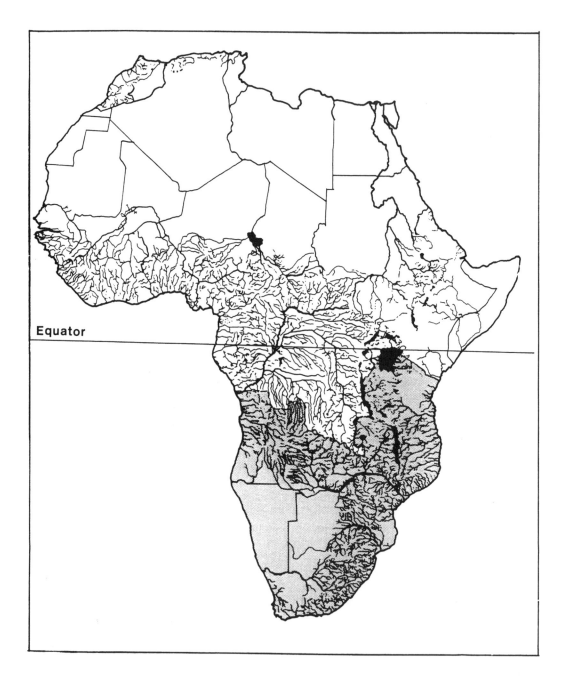

Figure 7.1   *The distribution of permanent surface water resources in Africa (from Gourou, 1970). Countries of the Southern African Development Community (SADC) are shaded*

Table 7.1 *The mean annual precipitation (MAP) and mean annual run-off (MAR) of some major rivers of the SADC countries, listed in order of decreasing catchment area. After Conley (1995): extracted from Pitman and Hudson (1994)*

River	Area (km²)	MAP (mm)	MAR (10^6 m³ yr⁻¹)
Congo[a]	3 981 000	1500[b]	1 250 000
Zambezi	1 234 000	860	110 000
Orange	973 000	330	11 860
Okavango	586 000	580	11 650
Limpopo	413 000	520	7 330
Rovuma	155 000	1100	28 000
Cunene	117 000	830	5 550
Save	104 000	680	6 200

[a] The DRC is only a recent affiliate of the SADC states; its major river basin borders several SADC countries including Angola, Zambia and Tanzania
[b] Estimated MAP
See Figure 7.3 for basin juxtaposition

REGIONAL CHARACTERIZATION

Africa comprises many varied climatic zones, from Equatorial tropical rain forest, through humid Mediterranean-type climates, to hyper-arid areas (e.g. Beadle, 1981). Examples of these extremes occur in Southern Africa. Globally, Africa is dry (e.g. Gourou, 1970; Figure 7.1), with 37% (Thomas, 1989) of the planet's 'dry-land' zones (*sensu* Davies *et al.*, 1993b), and large areas are devoid of reliable water supplies (Gourou, 1970; Alexander, 1985; Conley, 1995; Jacobson *et al.*, 1995; Figures 7.1 to 7.3; Table 7.2). This uneven distribution, coupled to rapid human population growth (Table 7.3) has exacerbated water supply. Indeed, water demand has increased to a point where most river ecosystems have become greatly stressed (see Davies *et al.*, 1993a,b, 1995; Davies and Day, 1998). In the case of hyper-arid regions such as Namibia, where assured permanent surface resources exist (Jacobson *et al.*, 1995), groundwaters are over-exploited and it is inevitable that permanent water supplies from perennial waters will be negotiated (demanded?) from neighbours. Water demand in the SADC is predicted to increase at 3% yr⁻¹ until 2020 (Table 7.3) and several countries already face considerable water-supply shortages. During the past five to 10 years, areas of Zimbabwe experienced deficits, despite the fact that projections (Heyns *et al.*, 1994) estimated that there would be sufficient water until 2020.

Many rivers form international boundaries and/or traverse more than one country (e.g. the Zambezi:

Angola, Botswana, Congo, Malawi, Moçambique, Namibia, Tanzania, Zambia and Zimbabwe: Davies, 1986; Figure 7.3) and thus are managed by more than one set of regulations. For example, virtually all rivers in Moçambique and Swaziland have sources located beyond their western borders and, in the case of southern Moçambique, major rivers (the sole water supply) have ceased to flow, partly through over-utilization by South Africa. In response, Moçambique has called for a Joint Water Commission between the two countries.

Geology and geomorphology

Continental Africa is *ca* 100 Ma old (King, 1978) and, since its separation from Gondwanaland, has undergone massive erosion, creating a remarkable uniformity (King, 1978). Almost all of the SADC countries lie on a central plateau (*ca* 1200 m above mean sea level) of ancient and impoverished soils. Northern Angola and Congo lie in the separate Congo River Basin (Figure 7.3), while nearly all of Moçambique and the northern sections of KwaZulu Natal in South Africa occupy a coastal shelf between 300 and 1000 km wide where erosion by coastal rivers forced the development of a chain of floodplains, lagoons and lakes spreading into Moçambique (e.g. Allanson *et al.*, 1990). Malawi and Tanzania, on the other hand, are situated within the southern Rift Valley of Africa, which also cuts across central Moçambique through the Zambezi Basin (e.g. Davies, 1986; Figures 7.1 and 7.3). Because of their relative rarity, coastal floodplains assume importance for regional river conservation. As well as the internal 'floodplain' of the Okavango (Figure 7.3), and the Barotse Floodplains of the Upper Zambezi, large floodplains only occur along the eastern margins – the Lower Zambezi (Marromeu and Delta: Beilfuss and Davies, 1999), and the Pongolo/Inkomati system, while a small floodplain in the southwest, the Berg River Floodplain occupies a position of extreme importance (see below; Figure 7.3).

'Older' rivers dissect the elevated '*Highveld*': the Zambezi in the north; the upper Congo in the northwest; Limpopo in the northeast; and south-central Orange–Vaal (Figure 7.3). The Orange–Vaal plays an essential role, supplying the industrial heartland of South Africa, the Province of Gauteng. Shorter, 'young' rivers of the east and south coasts are subjected to wide hydrological fluctuations, and the eastern rivers are frequently turbid. Deflections have created major aquatic environments including the

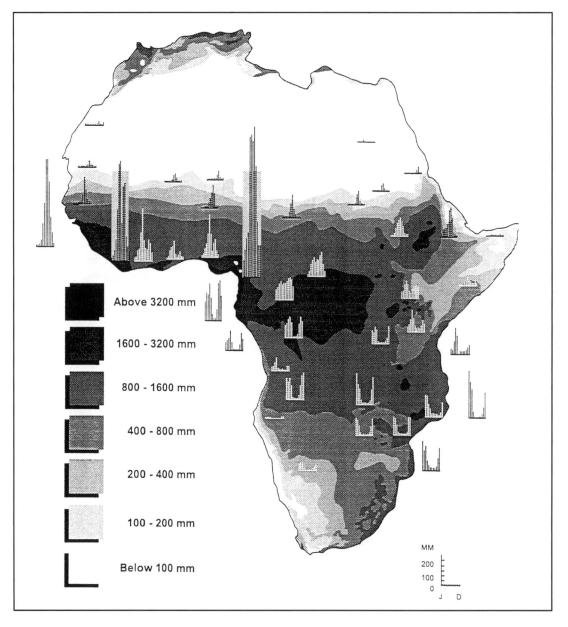

Figure 7.2 *Monthly distribution of rainfall and mean annual precipitation (MAP) in Africa (taken from Conley, 1995)*

Okavango Swamps in Botswana: linked to the Zambezi, this system was created by the deflection of the Kavango River at Makgadikgadi (King, 1978). The geomorphology of the southern coast is dominated by the Cape Fold Mountains, an ancient group lying more or less parallel to the coast. Here, rivers are short, steep and nutrient-poor, running over poorly nutrient-buffered bedrock of the Table Mountain Sandstone (TMS) series (e.g. Allanson *et al.*, 1990; see Davies *et al.*, 1993a for full description).

The geology of the region was described by Tankard *et al.* (1982) and comprises a vast sedimentary

Figure 7.3 *Major river basins, political boundaries and main rivers of the Southern African Development Community (SADC). Modified after Conley (1995)*

basin of weathered rocks of the Karoo Super Group (*ca* 265–200 Ma), topped by basalts of the Stormberg Series in Lesotho and KwaZulu Natal (*ca* 200–290 Ma). The basin is surrounded by more ancient formations.

Climate and vegetation

Aridity dramatically increases from the southeast to the northwest into the southern portions of Angola (Figures 7.1 and 7.2). Ignoring for the moment the Congo Basin and the south Cape of South Africa, aridity is certainly the overriding feature of the region. Some 38% of the area is dryland (*sensu* Davies *et al.*, 1993b), of which 7% is hyper-arid desert with a mean annual precipitation (MAP) <100 mm yr^{-1}, 15% is arid (between 100 and 400 mm yr^{-1}) and 16% is semi-arid (400–600 mm yr^{-1}), while of the remaining 62%, 19% is dry sub-humid (MAP 600–1200 mm yr^{-1}), 40% is moist sub-humid (1200–1500 mm yr^{-1}) and 3% is humid

Table 7.2   *The average rainfall and ranges in rainfall and evaporation of SADC countries (after Heyns et al., 1994)*

Country	Average rainfall (mm yr^{-1})	Range (mm yr^{-1})	Range in evaporation (mm yr^{-1})
Angola	800	25–1600	1300–2600
Botswana	400	250–650	2600–3700
DRC	1397	(1524 mm in north,	1270 mm in south)
Lesotho	700	500–2000	1800–2100
Malawi	1000	700–2800	1800–2000
Moçambique	1100	350–2000	1100–2000
Namibia	250	10–700	2600–3700
South Africa	500	50–3000	1100–3000
Swaziland	800	500–1500	2000–2200
Tanzania	750	300–1600	1100–2000
Zambia	800	700–1200	2000–2500
Zimbabwe	700	350–1000	2000–2600

(MAP >1500 mm yr^{-1}) (e.g. Conley, 1995; Table 7.1; Figure 7.2). Precipitation in much of the area occurs in summer and is highly seasonal (Tyson, 1986), yet there is a lack of pattern: this is the extreme unpredictability (*sensu* Davies *et al.*, 1995) of regional rainfall, a constraint that confounds traditional approaches to river management. Indeed, hydroclimatic stochasticity is the strongest determining feature, even in those parts with more 'assured' rainfall (e.g. winter rainfall of the western Cape, and year-round rainfall of the southern Cape of South Africa), and has given rise to what has been termed, the 'predictably unpredictable' nature of the Southern African hydroclimate (Davies *et al.*, 1995).

While it is clear that the SADC region is predominantly dryland, there are few quantitative data on temporary systems. Given the aridity of Namibia (e.g. Figures 7.1 to 7.3), all of its river systems

are temporary (e.g. Jacobson *et al.*, 1995) as are the majority of Botswanan systems. Despite the infrastructure available in South Africa, there is still great uncertainty as to the importance of temporary systems in the country. For instance, Smakhtin *et al.* (1995) have erroneously attributed Alexander (1985) as asserting that: 'only 25% of the rivers in South Africa are perennial with another 25% that flow periodically. The remaining 50% of the rivers (primarily the northern and southern interior) flow only after infrequent storms.' Careful examination of Alexander's paper reveals no such assertion. Recent estimates suggest that somewhere between 40% (J.H. O'Keeffe, Institute for Water Research, Rhodes University, South Africa, personal communication) and more than 60% of the region's rivers are 'temporary'. Figure 7.4 (adapted from Smakhtin *et al.* (1995)) shows that for South Africa, at least, there are very few areas where cessation of river flow has never been recorded.

Another contentious issue centres on whether or not there are any long-term patterns in MAP (see for instance, review in Davies *et al.*, 1993a). Although long-term cycles may be relatively predictable, short-term and annual cycles are not (e.g. Conley, 1995). For example, Alexander (1985) has summarized the major characteristic of Southern African rivers in one word – 'variability' – and as Davies *et al.* (1993a) state 'such variability is typical of Southern Hemisphere landmasses, and probably constitutes the single most important difference between Southern African lotic ecosystems and North American and European systems'. The variability is exacerbated by evaporation (e.g. Davies and Day, 1998) and by the very low percentage of MAP that is converted to mean

Table 7.3   *Population statistics and water demand estimates for southern Africa. (Based on Heyns et al. (1994) and Pitman and Hudson (1994))*

Country	Est. popn (million)	Popn growth (% yr^{-1})	Est. water demand m$^3 \times 10^6$ yr^{-1} (1993)	Est. water demand m$^3 \times 10^6$ yr^{-1} (2020)	Per capita usage (m^3 yr^{-1})
Angola	10.284	2.5	1335	2757	43
Botswana	1.245	3.1	129	336	98
Lesotho	1.891	2.6	118	268	34
Malawi	8.751	3.3	1135	2575	22
Moçambique	20.278	4.3	1967	3210	53
Namibia	1.402	3.0	265	538	n/a
South Africa	40.566	2.8	19295	30168	540
Swaziland	0.845	3.7	454	511	n/a
Tanzania	25.297	2.8	5374	12200	36
Zambia	7.948	3.7	994	2192	86
Zimbabwe	10.400	3.2	2524	5737	129

n/a = not available

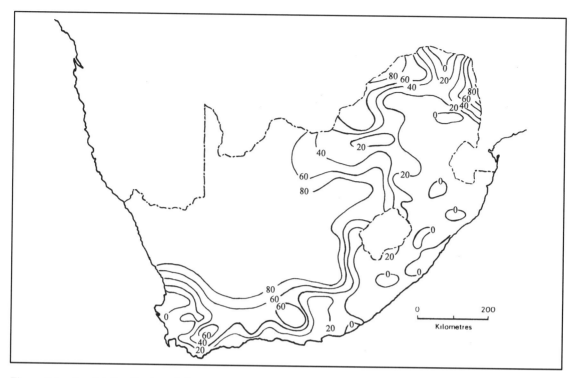

Figure 7.4 *Spatial distribution of the percentage of time for which rivers in South Africa experience zero surface stream flow (modified after Smakhtin et al., 1995)*

annual run-off (MAR) – 8.6% for South Africa (continental Australia = 9.8%; Alexander, 1985). As a direct result of these characteristics, stream flow may be extremely erratic, and the biota of many rivers will require extreme specialization and long-lived resting stages to survive. However, the systems to the north – the Congo and Zambezi – force very different conditions on the biota, and their perennial natures have allowed considerable endemism to develop; an issue important for river conservation in a totally different way.

The vegetation is complex: amongst others, Acocks (1953) and Van Zinderen Bakker (1978) have described modern vegetation patterns which fall into 11 components (Van Zinderen Bakker, 1978), several of which are rich in endemics. The most diverse is the sclerophyllous 'fynbos' of the Cape Floristic Kingdom which incorporates several endemic families and a diversity of Proteaceae, Restionaceae and Ericaceae. Fynbos leachates, coupled with the age and porosity of the bedrock (TMS: low buffering capacity), combine to produce black, acid waters that support a varied and often highly endemic fauna (e.g. Picker and Samways, 1996). The Highveld region is typically covered by 'grassveld', while savannas and woodlands cover the central-eastern and north-eastern portion of the region, into Zimbabwe, Angola, Zambia, Tanzania, and Moçambique. The Karoo, a semi-desert community, with xerophytic and dwarf succulent shrubs, is very diverse.

DEVELOPMENT PRESSURES AND HUMAN IMPACTS

Table 7.4 summarizes the major pressures on the rivers, recognizing the region's aridity/semi-aridity, and the 'developing-world' status of most SADC countries (e.g. Wishart and Davies, 1998; Wishart *et al.*, this volume, Part II). We also recognize that although there are global similarities, groups of pressures are either peculiar to dryland zones and developing countries, or are exacerbated by these conditions. For instance, destruction of riparian vegetation is common globally, but is exacerbated in developing countries by the rural and urban poor, desperate for fuel and for building materials. Riparian destruction in SADC countries is alarming, with *per capita* fuelwood consumption rates amongst the highest in the world (SADC, Environment

Table 7.4 *Pressures and human threats to rivers of the SADC region. Issues are listed in descending order of severity as a subjective guide to the pressures that we believe to be priorities for regional research and/or action*

Common pressures	Developing world/regional pressures	Dryland pressures
	Education	Hydroclimate stochasticity
Global climate change[a]		
River regulation	Present and future problem	Exacerbated
	Poverty and informal urban settlements	
	Lack of sanitation	Exacerbated
Floodplain destruction	Riparian zone destruction	Exacerbated
		Over-abstraction
		Interbasin transfers
		Salinization
Invasive organisms	Invasive plants	Exacerbated
Diffuse-source pollutants		
Point-source pollutants	Can be extreme (i.e. informal settlements)	Exacerbated
Pesticide pollution	Dumping by the developed world	
		Micro-algal (cyanobacterial) toxins
	Human disease vectors	
Canalization		
Wetland destruction	Can be extreme and rapid	Flash flooding
Mining pollution	Problematic in South Africa	
Erosion and suspensoids	Can be extreme and rapid	Exacerbated
Lack of trained personnel	Extreme	
	Poor planning	
	Taxonomic uncertainty	
Loss of endemics	Regional endemic hot spots	
Lack of funds	Survival economies	
Loss of biota		
Population growth	Rapid and redistributing	Exacerbated
Loss of habitat	Riparian removal	Exacerbated
Irrigation agriculture		
Genetic mixing		
	Political instability	
Conservation moral highground	Immediate survival *versus* conservation	

[a] We are unsure that this issue warrants priority at this level, but if current predictions (e.g. Regier and Meisner, 1990) are anywhere near reality then the implications for river conservation worldwide are enormous

and Land Management Sector – ELMS, 1990): legislation cannot solve the problem; community development and education programmes are required before control can be effected.

Such pressures also have knock-on effects that, while common elsewhere, are again exacerbated in the developing world. For example, following riparian-corridor destruction, vegetation loss, rainfall stochasticity, and dryland characteristics often lead to accelerated soil erosion, channel degradation and suspensoid production at a far greater rate than in most 'developed-world' scenarios (see also Gopal *et al.*, this volume). For instance, Starmans (1970) recorded a range of sediment transport between 420 and 586 784 t yr^{-1} (mean 62 517 t yr^{-1}) for 16 Malawian streams. The silt amounted to between 1.3 and 1202 t km^{-2} yr^{-1}, with a mean of 283 t km^{-2} yr^{-1}; deforestation around Lake Malawi threatened the

survival of several endemic species of cichlid fish for which the lake is renowned (see also Pacini and Harper, this volume).

Table 7.4 attempts to list threats to rivers that are: (1) common globally; (2) regionally specific (or 'developing-world'); and (3) problematic for dryland aquatic environments. Threats are listed in order of severity, although this has been more subjective than objective, but it indicates our priorities for resolution. To us, two of the most severe threats to regional rivers are river regulation and the redistribution of water through IBTs. Of course, the effects of regulation are well known worldwide (e.g. Ward and Stanford, 1979; Lillehammer and Saltveit, 1984; Petts, 1984; Gore and Petts, 1989). Indeed, the topic is well researched in the region (e.g. Davies, 1979; Byren and Davies, 1989; O'Keeffe *et al.*, 1990; Palmer and O'Keeffe, 1990a,b,c; Davies *et al.*, 1993a,b) but while Europe and North

America are scaling down dam construction, as water-demand management supplants water-supply management, the prevailing philosophy in the SADC tends strongly towards water supply.

Water demand will increase by 3% yr^{-1} to 2020 (e.g. Heyns *et al.*, 1994; Pitman and Hudson, 1994) and while the rivers are already greatly stressed, in South Africa alone *ca* 22 million people have no waterborne sanitation and *ca* 14 million have no potable drinking water (Asmal, 1995): *quo vadis* river conservation, when there are so many basic human 'needs'? Clearly, pressures on rivers as 'water suppliers' will continue to mount, as pleas for job creation, and provision of sanitation escalate. For example, a Namibian delegation in Botswana in 1996 discussed the proposed 250-km-long pipeline from the Okavango River, which feeds the Okavango Swamps (Figure 7.3), one of the most spectacular and least disturbed wetlands in the world, to Windhoek: this would be the first major diversion of the river's waters. Namibia's deputy permanent secretary of Water Affairs is reported to have stated to Botswanan officials 'We have to conciliate our need and your fears', and fears there must be. Namibia had just avoided a severe water deficit (all storages were 8% full on average at the time of the delegation's visit). According to *The Namibian* (18 October 1996), the pipeline would divert some 20×10^6 m^3 yr^{-1}, only 'in times of serious water shortages' (this appears to be contradictory; it can only be one or the other, not both). In addition, Windhoek has a projected growth of *ca* 6% yr^{-1}, while water use is expected to quadruple in the next 25 years. The plan, which first surfaced in 1973, is now regarded as immediately necessary in order to thwart a crisis.

Apart from the over-abstraction of east-flowing rivers from South Africa, perceptions of 'need' coupled to flimsy water-demand infrastructures have forced a culture of reliance on areas of supposed 'surplus' providing yet more water for those in 'deficit'. The Lesotho Highlands Water Project (LHWP) to Gauteng, with its industrial and mining complexes (Snaddon *et al.*, 2000), is a case in point. The LHWP is narrow in its design, with almost total reliance of economic growth in South Africa hanging on its implementation. Yet, the ecological effects of this and other schemes, and the implications for river conservation, have remained untested (e.g. Petitjean and Davies, 1988; Davies *et al.*, 1992; Davies and Day, 1998; Snaddon *et al.*, 1998, 2000). This is all the more worrying by the fact that outside South Africa, 99% of the hydroelectric potential has yet to be developed, with implications for future river regulation on a grand scale. A properly managed and shared power grid for the region would be an obvious mitigatory approach to the problem.

Of course, the greatest single threat to river conservation in Southern Africa is people. Excluding the DRC (*ca* 45 million), Southern Africa's population is currently *ca* 136 million people, and regional growth averages 3%; individual countries range between 2.2 and 3.8% (McCullum, 1994). Given the dryland constraints, the challenge facing conservation is sustainable development within a directed framework.

CATCHMENT *VERSUS* LOCAL ISSUES

The establishment of an Environment and Land Management Sector (ELMS) located in Lesotho forms the basis of a regional approach to resource management. There are three programmes: Environmental Management, Land Management, and Water-Resources Management. The last addresses integrated and efficient monitoring, planning, management and development of water resources throughout the region. Two sub-projects are directed towards ensuring appropriate, sustainable and equitable utilization of shared river basins. The first, the Zambezi River Systems Action Plan (ZACPLAN) aims to ensure maximum long-term utilization of the Zambezi for all participating states. The second is the Development of Integrated River Basin Management Planning that will implement regionally the ZACPLAN, leading to equitable and environmentally sound development and management plans. Eight major basins share two or more countries and, in an attempt to coordinate their management, ELMS has developed a Protocol on Shared Watercourse Systems. When ratified, the protocol will drive cooperative efforts at three levels: government; basin commission; and river authority, board or utility level. However, whilst they reflect a commitment to cooperative management, such agreements have no legal jurisdiction and rarely translate into local initiatives.

In the absence of such agreements, however, there is a real threat of international conflict. For example, due to poorly defined boundaries along the Congo River, a long section of the river is contentious (DRC and Congo), and both countries claim sovereignty to more than 4000 islands of the system. Other rivers, such as the Zambezi and Orange–Vaal, face potential conflict at both provincial and national levels. For instance, the Vaal in Gauteng takes water from Lesotho and from the provinces of Mpumalanga and KwaZulu Natal via

IBTs, flows through four provinces of South Africa, and is a shared watercourse with Namibia, while Botswana also has water rights to it.

In some SADC countries, governmental and municipal infrastructures are non-existent – *vide*: civil war in the DRC and Angola – these countries have neither the resources, the will, nor the infrastructures to cope. Recently, Zimbabwe formulated a catchment-based management approach to water through the secretariat of the Water Resources Management Strategy (WRMS) in the Department of Water Resources Development (DWRD). In the absence of a new water law, however, the approach has been confounded by complex access rights that stem from its colonial history. In addition, commercial farmers established River Boards, which serve to manage surface water resources through six designated hydrological zones.

In terms of environmental awareness, most of the region is thoroughly backward in its thinking. For instance, less than 5% of the landmass of South Africa is set aside for conservation, a figure well below the recommended world average (*ca* 10%) and, historically, emphasis has been misdirected towards the conservation of 'the big five' (e.g. elephant, rhinoceros, lion, buffalo and leopard), rather than of ecosystems. This still-prevailing mindset has led to the formation of large reserves, such as the Kruger National Park (KNP), which does not actually encompass an entire catchment: the rivers rise to the west of the KNP and are truncated along a north–south axis by the reserve boundary before subsequently flowing into Moçambique to the east (e.g. O'Keeffe and Davies, 1991). Over-abstraction of water to the west, coupled with multiple river regulation, has transformed the northern Luvuvhu, central Letaba and Olifants Rivers, from perennial to seasonal systems. Recently, the central Sabie River stopped flowing for the first time in recorded history (O'Keeffe and Davies, 1991; O'Keeffe *et al.*, 1996; Pollard *et al.*, 1996), whilst the Crocodile River to the south of KNP is regulated to a constant flow. These problems wrought by external influences severely limit the internal management of KNP. One of the ironies has been the development of a large, expensive and multidisciplinary research programme, the Kruger Park Rivers Research Programme (KNPRRP), with the goal of providing management with resolution to the problems associated with low, or no, flow. While the research has increased the understanding of Lowveld (lowland, sub-tropical) rivers, and has developed capacity to manage 'low flow' conditions, what is required is

education, and tough legislation to curb excessive water use outside the reserve.

MAIN AREAS AT RISK

In the absence of effective environmental protection measures, economic and social development needs have the potential further to degrade the region's rivers. Similarly, prolonged political instability in countries like Moçambique, the DRC and Angola, has created wide-scale destruction of infrastructure, and a lack of trained personnel throughout the region. While the situation in Moçambique has rapidly improved, at the time of writing Angola is again tangled in bloody civil war, and civil (almost sub-regional) war rages in the DRC: several other countries have had civil unrest (Lesotho, Zimbabwe, Kenya). These problems, and the legacy of colonialism, have retarded the training of personnel and the development of conservation-oriented infrastructure.

As we have already asserted, the natural hydrological regime is an area under great threat. Given the high degree of variability and the often unpredictable nature of river flows, conservation management often centres on flow-pattern models. These, in turn, are frequently based on inadequate data sets and are often simulated extensions of very limited information, frequently collected for catchments divorced from those in question: this often leads either to gross under- or over-estimations of natural flows. As such, many water-development projects have been incorrectly designed. A classic example is the LHWP, where full implementation would over-commit the Orange River to the tune of nearly 10^9 m^3 yr^{-1} (Snaddon *et al.*, 1998, 2000). River regulation and over-abstraction have transformed many naturally fluctuating systems either into constant flows, or into zero flows, thus disrupting both perennial and non-perennial systems. Moreover, an increasing reliance on IBTs arises from the fact that many rivers experience high levels of natural variability, the extreme manifestations of which are ephemeral streams whose natural flows are at variance with the strategies and management of water for domestic and agricultural supply (see also Schofield *et al.*, this volume).

Such manipulations create almost intractable problems for conservation (e.g. Davies *et al.*, this volume). For example, the once-seasonal Great Fish River in the Eastern Cape of South Africa is now a perennial system with a flow some 800% greater than natural flows, created by the Orange–Fish–Sundays IBT, a transfer that originates in Lake Gariep on the Orange River (e.g. O'Keeffe and De Moor, 1989). In the Western Cape, the perennial Berg River, with natural high and low flows respectively during winter and summer, now receives irrigation water from an IBT which raises summer flows by up to 4000% (Snaddon and Davies, 1998).

The rivers of SADC countries suffer from major degradation of floodplains and riparian zones. The causes are multifarious: population pressures, fuel requirements, formal and informal settlements, local channel manipulations (e.g. bulldozing river beds), mountain-catchment forestry and the invasion of riparian zones by alien vegetation. Forestry plantations have reduced the virgin MAR of the Sabie River in Mpumalanga, South Africa, by over one-third. Often, a deliberate policy of planting involves the utilization of the watershed right into the riparian zone. Elsewhere, the riparian zones of mountain catchments have been invaded by a suite of exotic plants such as Australian *Hakea* spp, European oak, *Pinus radiata* and Australian acacias and wattles (particularly *Acacia mearnsii* and *A. saligna*). Invasions are so dense as to prevent access to the channel. It is estimated that between 30 and 60% of the MAR of infested mountain catchments of the Western Cape of South Africa is lost through evapotranspiration by invasives. Furthermore, Van Wilgen *et al.* (1996) have estimated that the entire Fynbos Biome – the smallest, but most biologically diverse plant kingdom in the world – will disappear under alien vegetation by the middle of the 21st century unless urgent steps to clear them are taken (see below).

Information for conservation

SCIENTIFIC DATA

River conservation has recently received considerable attention, and this section is drawn in part from the syntheses of O'Keeffe (1986), Ferrar (1989), O'Keeffe *et al.* (1989), Allanson *et al.* (1990), Davies *et al.* (1993a, 1995) and Davies and Day (1998).

It is safe to say that the regional scientific database on rivers is patchy. Initiatives, sometimes involving the United Nations Development Programme (UNDP), the United Nations Environment Programme (UNEP) and the Food and Agriculture Organization of the United Nations (FAOUN), have aimed more at developing resources and management protocols within an ecological and economically sustainable framework than at providing a scientific platform for the long-term development of local conservation.

However, for some areas there is a relative wealth of information gathered by provincial authorities and staff of national parks on a wide variety of systems and taxa: for instance, in the historically conserved areas of South Africa (e.g. KNP, the national parks under the control of KwaZulu Natal Parks Board), some National Parks in Zimbabwe, Kenya and Zambia, and in areas of particular public interest (Okavango Wildlife Society). Although part of the 'grey literature', this information is invaluable and covers many of the rivers of the eastern seaboard and others scattered throughout the region. Private conservation authorities have also developed some knowledge of local systems, while provincial authorities have developed similar archival databases. However, the problem remains of little coordination, both in terms of information gathering and its dissemination. Notable exceptions include the Albany Museum of Natural History in Grahamstown, South Africa, which houses the Southern African Collection of Freshwater Invertebrates. This collection contains material from a very wide range of rivers and wetlands throughout the SADC region and the material is being developed to produce identification guides, while the Southern African Society of Aquatic Scientists (SASAQS), together with the South African Water Research Commission (WRC), is in the throes of publishing identification keys that should begin to appear around the year 2000. Regional specialists on invertebrates exist for the Trichoptera, Simuliidae, Ephemeroptera, Chironomidae, Diptera, Nematoda, Mollusca, Amphipoda, Decapoda atyiid prawns and other Crustacea, and the Dytiscidae.

Vertebrates have fared better and, with the clear exception of the DRC, the systematics and distribution of lotic fish, for instance, are well known (mainly through the activities of the J.L.B. Smith Institute for Ichthyology in Grahamstown; e.g. Skelton (1993)). The exception is the vast Congo Basin. Second only to the Amazon Basin in terms of fish diversity, the exact number of species of this basin is still unknown and probably constitutes *ca* 700 species, of which >500 are probably endemic (Banister, 1986; see also Pacini and Harper, this volume). Information on aquatic birds (indeed all avifauna) is overwhelming, perhaps reflecting the 'vertebro-pocentric' interests of many scientists (and lay public) of the region. Not only are there excellent field guides and general texts (e.g. Gordon MacLean's *Roberts' Birds of Southern Africa*, 1988, a work that is being updated through to 2003 by the Percy Fitzpatrick Institute for African Ornithology, University of Cape Town – UCT), while the Avian

Demography Unit (UCT) has recently produced two comprehensive volumes of Geographical Information Systems (GIS) data on bird distribution. A similar set of works to complement existing texts on the Amphibia is in the throes of production at the time of writing. Good botanical taxonomic and distribution data exist for a wide variety of taxa ranging from centric diatoms, filamentous algae, cyanobacteria and, in the higher taxa, aquatic and riparian angiosperms. However, although such data (and even field guides) are available, the expertise to identify these organisms accurately lies in the hands of very few people.

While some information on the large aquatic vertebrates, such as crocodile, hippopotamus and elephant, is available (e.g. Tinley, 1976), information on small water mammals is scarce and very little is known about their distribution or ecological requirements. It is clear that aquatic mammals have been decimated over the past centuries and there is no historical perspective of how ecologically important they may once have been; e.g. hippopotamus. This has implications for conservation planning: should mammal species be reintroduced for instance; at what densities; and in what areas? The same is true for other species, for there is no doubt that as a result of the removal of large herbivores, major vegetation, geomorphological and faunal changes have been wrought on the landscape. No doubt this type of dilemma faces conservationists elsewhere but, as pointed out earlier, it is probable that dryland aquatic ecosystems are far more vulnerable to human impacts than their more humid counterparts.

Although information for the conservation of many 'medium-sized' rivers is relatively good in places (e.g. the Berg, Buffalo and the rivers of the KNP), large rivers, and the most common form of river – ephemeral, or intermittent systems – have been overlooked (e.g. Wishart, 1998).

The Sabie River (e.g. O'Keeffe and Davies, 1991; O'Keeffe *et al.*, 1996; Pollard *et al.*, 1996) is possibly the most biologically diverse river in South Africa, both for its indigenous fish species (47; e.g. O'Keeffe *et al.*, 1989) and for its insect species (O'Keeffe *et al.*, 1989). The Sabie River also maintains dense riparian forests that arguably form the prime large-mammal habitat of the KNP. The South African Department of Water Affairs and Forestry (DWAF) has recently completed a strategic survey of the catchment, and is now trying to develop a water-management policy that will fulfil, as far as possible, the water demands in different parts of the catchment, whilst also providing sufficient water downstream into Moçambique (e.g.

Bruwer, 1991). The major conservation requirement is to estimate the in-stream flow needs of the river (e.g. Ferrar, 1989; see below), particularly in the reaches flowing through the KNP.

Since the first suggestions concerning 'water for the environment' (Roberts, 1981, 1983), the South African Department of Water Affairs (DWA, 1986) and the WRC have more formally recognized the need for quantitative estimates of water allocations (e.g. DWAF, 1992) for a variety of uses in each primary drainage. However, as Walmsley and Davies (1991) point out, not only is the allocation of conservation areas badly skewed in comparison with other parts of the world, but until very recently research funding for lotic ecology has been ludicrously inadequate.

Interestingly, the scarcity of information on the large and ephemeral/intermittent rivers of the region differs: scarcity of information for large systems (e.g. the Zambezi, Congo) is essentially spatial, while for ephemeral systems it is mainly temporal (e.g. Stanley and Boulton, this volume). Even the first large river work (e.g. the Zambezi; see Davies, 1986, for review) was management-oriented in terms of river regulation and ultimately failed to involve local personnel. With the exception of fisheries (e.g. Welcomme, 1972), pollution-related work, particularly that of W.D. Oliff and A.D. Harrison, and aspects of invertebrate ecology conducted by A.D. Harrison and F.M. Chutter (see Davies *et al.*, 1993a, 1995, for reviews), river research only got under way in the southern SADC states in the mid 1980s. Even then, it was often 'development'-directed and designed to understand the influences of river regulation by dams, rather than river research for its own sake (e.g. O'Keeffe and De Moor, 1988; Byren and Davies, 1989; O'Keeffe *et al.*, 1990; Palmer and O'Keeffe 1990a,b,c). Exceptions do occur, of course (e.g. King, 1981, 1983; King *et al.*, 1987a,b, 1988). Of those scientific databases which do exist (e.g. the water quality database of the DWAF), there is great potential for synthesis and this is now occurring (e.g. Uys, 1994; Eekhout *et al.*, 1997; Davies and Day, 1998). Elsewhere in the SADC, attempts seem to be un-coordinated and scattered. It is to be hoped that the formulation of the separate water sector, at present proposed within the SADC structure, will extend the current emphasis.

CLASSIFICATION AND EVALUATION SCHEMES

Historically, river conservation research has occurred in four stages: taxonomic; distribution; biodiversity (all are descriptive); and process studies, leading to a predictive capability for management. In South Africa, Chutter (1972) capitalized on many early taxonomic and distribution studies (not a few his own) to formulate a 'Biotic Index' of water quality for South African rivers. Chutter analysed early distribution data to identify those species that were pollution tolerant, intolerant, or ubiquitous. Scores from 0 to 10 were assigned to different groups, in which 0 indicated complete intolerance, and 10, tolerance. Some groups, such as the Simuliidae and Chironomidae, presented taxonomic problems, containing both tolerant and intolerant species. To simplify identification and quantification, groups were considered at the family level or higher, and were given sliding scores, depending on the numbers and densities of Ephemeroptera (particularly Baetidae). The total additive score for any sample was then divided by the number of individuals in the sample to give a final score:

- 0–2, clean unpolluted waters
- >2–4, slightly enriched waters
- >4–7, enriched waters
- >7–10, polluted waters

The system has since been greatly modified as the South African Scoring System 4 (SASS4: Chutter, 1994). Sensitivity-weighting scales for invertebrate taxa of 'stones-in-current', and other biotopes, have been refined with pollution tolerance scaled at 1 (most tolerant) to 15 (most sensitive to pollution), allowing the calculation of a reach-by-reach, river-by-river comparative Average Score Per Taxon (ASPT). The method is now in extensive use in the 22 drainage regions of South Africa as a rapid bioassessment tool for water quality (e.g. Dallas and Day, 1993).

An additional descriptive phase of conservation research has focused on river classification so that priority uses of rivers could be identified and allocated (e.g. King *et al.*, 1992). From this thrust, a 1:250 000 river conservation map for South Africa was developed under workshop conditions: known systems and reaches of high-, medium-, low- and severely degraded conservation value were identified and recorded.

A conservation 'Expert System' computer technique, known as the River Conservation System (RCS) has also been developed by O'Keeffe *et al.* (1987). The RCS, originally developed as a tool for understanding the conservation status of South African rivers, was designed to assess the extent to which any river has been changed from its natural state, and from its original importance for conservation. Using historical data on 'virgin' conditions, compared against proposed use, the

RCS provides a numerical assessment of the worth *for* conservation of a river by asking some 58 questions on attributes including size, degree of regulation, pollution, water use, catchment land use, vegetation condition, degree of erosion, indigenous, endemic and introduced biota, and recreational uses. The questions are weighted to indicate their relative importance and the computer programme provides an overall percentage maximum and minimum score for each system, as well as individual scores for the condition of the river channel, its catchment and the biota. Differences between the maxima and minima indicate how much is known about each river. The most important conservation attributes of each system are identified, as are those for which more information is most urgently required (see O'Keeffe and Davies, 1991). For the Sabie River in Mpumalanga, O'Keeffe and Davies (1991) assumed a conservation status of 100% under virgin conditions, and zero if all flows were intercepted by impoundment. Between these extremes, the RCS was run for prevailing conditions for two management regimes: if the Injaka Dam on the Marite tributary of the system were to be built, and for the proposed impoundment of the main stem by the Madras Dam, 40 km upstream of the KNP boundary. Predicted outcomes were that Madras would have significant negative impacts on the Sabie, reducing its conservation status to 43% (a degraded regulated river). On the other hand, slight modifications of the operating rules for Injaka Dam (from maximum water abstraction to seasonal harvesting – this last equated to a loss of 2% of the potential MAR) would lead only to a 3% loss of conservation value, versus a 15% loss for proposed management procedures (O'Keeffe and Davies, 1991).

In terms of the third and fourth stages of research, Southern Africa has relied heavily on ideas imported from elsewhere (e.g. Instream Flow Incremental Methodology – IFIM: e.g. Bovee, 1982; Gore and Petts, 1989; and PHABSIM II: e.g. Milhous *et al.*, 1989). Many have proved inappropriate: their intensive data and manpower requirements and the high costs, means that they will not be applied to many (if any) South African rivers (let alone, Southern African) (R. Tharme, Freshwater Research Unit, University of Cape Town, personal communication). Further, these methodologies operate best for single target species rather than for whole systems/river reaches. As a result, more 'user-friendly' modifications to existing methods (e.g. Gore and King, 1989; King *et al.*, 1992; King and Tharme, 1994; Tharme, 1996; Weeks *et al.*, 1996), as well as the development of new and holistic approaches to in-stream flow requirements – IFRs – are being sought (R. Tharme, personal communication), one of which is the IFR 'Building Block Methodology' developed by J.M. King and R. Tharme (1994; see also King and Louw, 1998; Tharme and King, 1998). These developments have borne fruit in the recent recognition that rivers are not just channels for water supply and effluent disposal, and the significant entrenchment of the 'water rights' of rivers *per se* in the new South African Water Act (1998; see also Palmer *et al.*, this volume). The innovation in the Act is the concept of the 'Ecological Reserve', or that amount of water required by a particular river in order to maintain essential ecological processes. The law requires IFRs for all water-development projects in order to ensure that correct volumes of water are conserved for ecological processes.

Methods for assessing the 'health' of rivers in Southern Africa are poorly developed. In South Africa, however, a number of assessment systems have been, and continue to be, developed. The Kleynhans (DWAF, Institute for Water Quality Studies) methodology for the determination of Habitat Integrity is used to assess the conservation status of rivers. Although related, there is as yet no formal method for assessing conservation importance, which still relies on the expert advice of individuals and on the dependence of rural communities along a river. However, river health viewpoints are being developed at present with Australians, and strong research and exchange ties have been forged (e.g. Uys, 1994). Perhaps this is ironic when we consider that river-health ties within SADC are not strong at all, and although some movement is beginning between Moçambique and South Africa, the same cannot be said for Botswana, Zimbabwe, Angola and the DRC.

Finally, and a priority concern, is the need to develop protocols for the classification, assessment and management of dryland rivers: to say that this important area of conservation is in its infancy would be a gross understatement.

AVAILABILITY AND DISSEMINATION

The infrastructure for the dissemination of conservation-related information is very variable. In South Africa it could be regarded as exceptionally strong with several active governmental and non-government organization (NGO) agencies. The South African DWAF and the WRC are particularly active, and although the National Research Foundation (a major funding agency in pure science) is elitist, it also

aids the process through links with DWAF and the WRC. The South African Department of Environmental Affairs and Tourism (DEAT) has recently developed strong interests in what it terms 'Wetland Conservation' into which it frequently slots river conservation issues. In contrast, most provincial nature conservation and Parks Board administrations are relatively poor, but there are one or two exceptions: the Parks Board of KwaZulu Natal and the conservation administrations of the Cape. There is great scope for improvement.

Elsewhere, the situation is poor, although Zimbabwe and Botswana have developing infrastructure, and the Namibian DWA is developing personnel and approaches to water conservation quite rapidly. In the last, the situation is gaining momentum, not least because of the enormous problems facing water supply in that country, but also because of controversial water-supply developments: the Epupa Gorge hydropower proposals on the Cunene River; the Eastern National Water Carrier (an IBT for Windhoek, a large portion of which comprises an open canal), and the proposed Okavango Pipeline (another IBT). Funding from the European Community, and from Sweden and Norway, has accelerated the production of water-conservation information (e.g. Jacobson *et al.*, 1995).

The lack of trained personnel is a major concern. Often, the lucrative salaries of the private sector, coupled to a general lack of environmental awareness of senior management in many government departments, and a concomitant lack of resources, conspire to disillusion talented staff who look elsewhere for fulfilment. Indeed, the recent 'rationalization' of many South African government departments has led to the retrenchment of hundreds of field-experienced staff who have now turned to private consultancy in order to survive.

The SADC group ELMS provides liaison and coordination at government level in the water sector, but it rarely addresses ecological considerations and is focused more towards engineering and development. Nonetheless, the fact that such groups exist is a platform from which larger things may grow and, as such, there is some degree of hope that conservation issues will gain a more representative share of resources. In this context, the role of the World Wide Web (WWW) is important, and user groups are already active with funding from the US Agency for International Development (USAID), and the Southern African Environment Programme (SAEP). Further, the WRC has, for many years, funded the

Computing Centre for Water Research, through which researchers gain access to a very wide variety of literature on water issues.

A number of NGOs are also active: the Wildlife and Environment Society of Southern Africa is one such agency that has successfully begun to move away from the conservation of the 'big five' towards ecosystem issues. This sort of activity has been enhanced by the watchdog activities of groups like 'Earthlife Africa', the South African-based Group for Environmental Monitoring (GEM), and the opening of a section of the Berkeley, California-based International Rivers Network (IRN) in Gaborone, Botswana. These initiatives (and others; see below) have recently gained momentum with the addition to the scene of the Cape-Town-based World Commission on Dams (WCD). Since its inception (1998), it has stimulated the formation of a number of WWW networks surrounding water-supply developments and river conservation. Such initiatives will only serve to strengthen professional societies in the region, SASAQS for instance.

Public awareness and education

It is sad but true that although rural communities are aware of the importance of water, if for no other reason than it daily must be fetched and carried, often over long distances, there is abysmal ignorance of it in an environmental context. Just as disturbing is the general ignorance of the double-edged sword of water – water the life giver, and the carrier of serious disease and death. Schistosomiasis (bilharzia), malaria, cholera, typhoid, dysentery, and other diarrhoeal diseases are endemic (e.g. Davies and Day, 1998; Pacini and Harper, this volume). In fact, other than the bloody and continuing civil wars in the region, waterborne diseases are the greatest child killers with some 60 children dying each day in South Africa alone.

On the other hand, most urban inhabitants do not even have the same notion or awareness of water as an issue as rural communities. Informal developments often spring up along watercourses which are already severely degraded and which subsequently degrade further, becoming positively dangerous to human health through disease and flooding. Along the Arpies River in Gauteng, many tens of thousands of people are annually threatened by floods and disease: the same is true all over the region. Indeed, severe outbreaks of cholera in Maputo (Moçambique) forced the closure of its borders with Swaziland and South Africa at least twice in 1997 and 1998. Furthermore,

just as unmanaged watercourse developments take place in urban environments, rural communities settle along natural river courses for practical purposes of water supply and stock watering, with resulting riparian and channel degradation. There is a great urgency for coordinated education campaigns aimed at all strata of society, but lack of personnel, vast distances and the present levels of education confound the problem. Often communications are difficult and language roots are many and varied. Nonetheless, a few SADC countries have begun to make inroads into the general ignorance. For example, the Namibian Environmental Education Network (NEEN) is attempting to coordinate existing, and to encourage the development of new, environmental-awareness projects. However, NEEN is a relatively recent initiative (1995) and has a long way to go to promote the proper use of water and the integration of a sustainable ethos into development needs. In Zimbabwe, recent drought and other environmental problems have forced an initiative for grassroots education in the form of fact sheets. Zimbank has a series of pamphlets on a variety of issues including water as a resource, alien invasive plants, and waterborne diseases. The pamphlets are distributed free of charge throughout Zimbabwe and other countries of the SADC (CEP: the Communicating the Environment Programme, distributed by the Southern African Research and Documentation Centre). In South Africa, a recent Government White Paper on rural and urban development is clear on the need for booklets and pamphlets.

Most exciting was the 1995 launch of the South African National Water Conservation Campaign. Directed by DWAF, it has had an immediate impact both on urban and rural communities. It is linked to schools and community initiatives: the '2020 Vision for Water' has supplied tens of thousands of water-audit and water-saving devices kits to schools; the 'Water-wise Gardening Campaign' disseminates information on garden water-conservation techniques; and the 'Working for Water Programme' provides employment and skills training to poor communities. The Programme is an extraordinary exercise – 'job creation by stealth', where clearance of alien invasive vegetation from mountain catchments provides both jobs and dignity for the unemployed, and generates water, by removing invasives that actively transpire between 30 and 60% more water than do indigenous plants. More than US$10 million have been committed and over 1500 people are employed clearing the Cape Peninsula of invasives alone. Cleared timber is now being beneficiated as fuel wood; a win–win situation for conservation and job creation.

The framework for conservation

THE SOCIO-POLITICAL/RELIGIOUS–CULTURAL BACKGROUND

Southern Africa has an intricate history, today reflected by a complex socio-political and religious–cultural climate. Some of the earliest hominid remains (if not the earliest) belong to this part of the world, with some experts intimating that the region was crucial for the evolution of modern humans (e.g. Skotnes, 1996; recent media reports indicate exceptionally old finds at Sterkfontein; 4 Ma BP). Up to 2000 BP the region was populated by hunter-gatherer groups, 'Sonqua' or 'Soaqua', now called 'San' or 'Bushmen', but from that time on an array of pastoralist peoples and cultures ebbed and flowed across the region (Skotnes, 1996). The difficulty in interpreting this pot-pourri background is confounded by a lack of written records: most early traditions appear to have been oral (Skotnes, 1996). Hence, from a modern perspective there has been a skewed view of causes and effects driven mainly by the region's recent colonial history and, of course, the scourge of apartheid.

Colonial suppression over the past three centuries has forced a number of unexpected cultural veneers, as well as disruptions of established patterns of land use, settlement and resource utilization, armed conflict, slave deportations (to the Americas and to the Middle East), and importations (mainly from India, Malaysia, Indonesia and the Far East). Overlying these switches and mixes has been the mass extermination of 'San' peoples, colonial and indigene clashes over territory and resources, mass migration, ethnic conflict, wars of independence (in SADC: Zambia, Zimbabwe, Namibia, South Africa, Angola, Moçambique, DRC; in fact it is probably easier to list the countries that have not experienced such upheavals), and civil strife, and South African apartheid. All have had major implications for the conservation and management of river systems over a variety of time frames. To reduce this history into the framework of this contribution would do it an injustice and serve little purpose. Rather, we have decided to summarize some of the cultural characteristics which have implications for the implementation of conservation initiatives.

The effects of ethnic diversity in Southern Africa are compounded by a number of languages and an array of dialects. For example, more than 80 Moçambican dialects are based on nine different roots, while in the DRC there are 435 dialects, based on four roots. Thus, perhaps one of the important results of colonialism has been nation-specific language unification, with English, French and Portuguese dominating. However, since the demise of apartheid in South Africa, there has been a shift from English and Afrikaans as official languages to the recognition of 11 official languages. While this is a welcome shift in terms of cultural recognition, its effects on conservation have still to be gauged. For instance, according to the South African Constitution, government pamphlets should be published in all 11 official languages, with obvious implications for resources.

Religious diversity and fervour has also had significant effects on, and implications for, effective conservation and management of rivers. Colonial influences have led to an extraordinary mix, ranging from the monotheistic–traditional–orthodox, to charismatic forms of Christianity, forms of Animism, monotheistic Judaism and Islam, and mixes of Hindu, Buddhist, and other eastern philosophies. Across the region, there are areas with complex mixes of such groups, in contrast to large swathes dominated by a single viewpoint. For instance, alongside calvinistic forms of Christianity in the Western Cape of South Africa is a very strong Islamic community, mainly comprising descendants of imported Malay slaves, while in KwaZulu Natal, Hinduism and other Indian philosophies imported by labourers, are strongly established. Similarly, the DRC, Angola and Moçambique exhibit 'patchiness', with firm Roman Catholic influences in some areas, and a variety of animistic beliefs in others.

Imported monotheistic religions have generally underpinned the belief systems of those in education and in governance and have had the main influence on the development and execution of conservation ordinances, policies and practices. Although this at first may appear to have been an 'advantage', it frequently has not been so, because the concept of a separate creation and the 'position' of humankind in relation to the environment (see e.g. Davies and Day, 1998), coupled to the colonial overlay of exploitation for the 'mother country', has frequently led to catastrophic environmental disruption. Furthermore, some communities have strong religious aversions to (for instance) water recycling, making water-demand management and the curtailment of water-supply developments all the more difficult. Similar political diversity prevails.

STATUTORY FRAMEWORK (NATIONAL AND INTERNATIONAL LEGISLATION)

In the more industrialized countries of SADC, water has been grossly under-valued with obvious implications for water consumption and water-supply developments. This is a direct result of colonial history and has dictated a set of Euro-centric water laws and related legislation. More often than not, such legislation has ignored the inappropriate use of water, and has overridden indigenous water rights. Furthermore, much legislation has favoured the colonists over the indigenes. The classic example, of course, is South African apartheid – five decades – where farmers of European descent were courted for votes by the Nationalist Government, partly by 'give-away' water tariffs. Similarly, the River Boards of Zimbabwe still had no black representation up until 1998, while the Water Court which decides on the granting of the right to use public water had no black representation until after independence (Derman, 1998).

The situation is further complicated by the fact that many of the region's river basins are international. While international accords and rules, such as the Helsinki Rules on the Uses of the Waters of International Rivers, the International Law Commission on the Non-Navigational Uses of International Water Courses Rules, and the SADC Protocol on Shared Watercourse Systems in the Southern African Development Community Region, provide a framework for mediation, they have no legal force and provide little more than a set of guidelines. Regional water scarcities are such that water is perceived as the most constraining factor limiting development and social improvement. Water is, therefore, a prized possession and not given up lightly and the potential for inter-regional conflict over water resources is high. For instance, in 1995, the World Bank predicted war between South Africa and Zimbabwe if South Africa pushed Zimbabwe too far over the proposed 'Zambezi Southern Transfer' to Gauteng.

At the request of the African members of the UNEP, the African Ministerial Conference on the Environment (AMCEN) was established in 1985, with rivers and lake basins forming one of the five fora (the other four being, seas, forests and woodlands, island ecosystems, and deserts and arid lands). 'Committees of Experts' deal with general issues of hydrology, fisheries, limnology, watershed management, economics, sociology, pollution control,

settlements, planning and international law. During 1994–95, priority was given to the following, relating to the priorities of Agenda 21:

- promotion of an environmentally sound management programme for inland waters;
- assistance in the development of sub-regional organizations within the intergovernmental agreements on shared water resources;
- preparation of national environmental-related plans and programmes for integrated water management;
- mobilization of international support for the implementation of basin action plans.

There are currently three types of agreements existing within SADC:

- bilateral agreements between countries sharing the same water resource;
- sub-regional agreements between three or more countries for specific water-resource sharing projects;
- a proposed SADC Protocol on Shared Watercourse Systems in the Southern African Development Community Region.

Within the sub-regional agreements there are Permanent River Basin Water Commissions (PRBWCs), agreements for specific water-resource sharing projects and for establishing technical commissions or committees, and general economic and social agreements. The commissions act as technical advisers on issues of conservation, development and the utilization of water resources. They will also advise on the long-term yield of water, the criteria for conservation, equitable allocation and sustainable utilization, development of new water resources, the prevention of pollution, and on measures to alleviate short-term difficulties caused by water shortages. Six agreements for PRBWCs have been established, while five others have nominal existence. The OKACOM (Okavango Committee), a trilateral agreement between Angola, Botswana and Namibia (1994), ensures equitable and sustainable development of the waters of the Okavango system, including the Kavango and Cuito Rivers.

Development of the SADC Protocol for Shared Watercourses will provide a foundation for the establishment of a Zambezi River Basin Authority; there is an urgent need for legal agreement between the

Table 7.5   *SADC signatories (with month/year of signing) to various international conventions*

	Ramsar (as of 30 Oct. 98)	World Heritage	Biodiversity	Bonn CMS
Angola	–	11/94	4/98	–
Botswana	4/97	11/98	10/95	
DRC	5/96	9/74	12/94	9/90
Lesotho	–	–	1/95	–
Malawi	3/97	–	2/94	–
Moçambique	–	11/82	8/95	–
Namibia	12/95	–	5/97	–
South Africa	12/75	7/97	11/95	12/91
Swaziland	–	–	11/94	–
Tanzania	–	–	3/96	–
Zambia	12/91	6/84	5/93	–
Zimbabwe	–	8/82	11/94	–

eight member states. At present the Zambezi River Authority has the sole purpose of the management of Lake Kariba. It does, however, make decisions on all operations, including downstream releases of water. Other agreements between South Africa, Namibia and Lesotho govern the development of the Orange River Basin, while Moçambique, South Africa, Zimbabwe and Botswana have established a Permanent Technical Committee for the Limpopo Basin.

These water commissions represent a commitment towards a strong conceptual framework for regional cooperation. However, due to numerous constraints – personnel, financial and political – there has been little practical implementation. Many commissions and basin authorities rarely meet, their members often do not have qualified staff or representation and, at present, the commissions have little legal authority. Furthermore, while there is seen to be a need for hydrologists, engineers and technical staff, there does not seem to be any recognition of the need for ecologists; this could have very serious implications for the future 'sustainable utilization' of SADC river systems.

Traditional approaches to water have developed within a sustainable framework and we could do well to examine some of these approaches within the local context. In many instances, the incorporation of ethnic laws may provide a feasible approach to water-resources management. In Lesotho, for instance, the rights to land and water, as stated in the Water Act of 1978, incorporates customary practices (see also Pacini and Harper, this volume). According to the Act, the local chief is guardian and trustee for the community. The chief is allowed to allocate three fields (*ca* 5–6 ha) to each married male resident, together with domestic

water-use rights: water permits for larger areas, as well as for other water uses, are issued by the Minister of Water Affairs. Obviously such laws have inherent problems, such as the perpetuation of the lowly role of women in traditional societies, whilst also opening the system to abuse of power. Nonetheless, they do recognize the important role of traditional cultures in water utilization.

Finally, the present situation in terms of the Ramsar Convention is worrying. Although South Africa is an original signatory, at least one of its major floodplains is not protected under the convention (Davies and Day, 1998). This, the Berg River Floodplain, is not only an important summer breeding ground for birds (Hockey, 1993), but is also a vital coastal-fish nursery (Bennett, 1993), with more biological diversity than any of the presently designated Ramsar systems in South Africa! Its position (Figure 7.3), to the south of a large desert expanse stretching from Angola through Namibia and Namaqualand, dictates its priority for conservation. For other SADC countries, Namibia is only a recent signatory to the convention and, whilst the Okavango in Botswana is protected, the Marromeu and the vast deltaic floodplains of the Lower Zambezi Valley are not (Davies, 1986; Beilfuss and Davies, 1999; Davies *et al.*, 2000). The fact that the Marromeu is the largest breeding ground for the endangered crane (crowned and wattled crane) in Southern Africa (R. Beilfuss, International Crane Foundation, Baraboo, Wisconsin, USA, personal communication) and is threatened by severe degradation caused by the Cahora Bassa Dam (e.g. Davies, 1998; Davies *et al.*, 2000) demands an urgent reassessment by SADC states of such protocols (see Table 7.5).

THE ROLE OF VOLUNTARY ORGANIZATIONS

Many NGOs throughout SADC have focused on large mammal conservation, rather than ecosystem conservation. This trend, however, is slowly changing, aided by international intervention in training and in identifying and implementing conservation programmes. In this context, organizations such as the World-Wide Fund for Nature (WWF) have an important role in seeding local initiatives and in generating local interest and education. In 1996, WWF identified freshwater biomes as a priority, apportioning 20% of the priority biome status and initiated a project, 'Priority Setting and Project Identification in Southern Africa'. This project will assemble spatial information on biodiversity, the threats and opportunities for conservation in Southern Africa, and the identification for intervention of fresh waters, along with marine and savanna sites and species. A WWF training programme in Malawi is also helping to train members of the Department of National Parks and Wildlife in the conservation and management of aquatic reserves.

Such international organizations also provide a function in the absence of local legislation: there is often little or no legal obligation for developers to undertake environmental/ecological impact assessments (EIAs), or to provide money for such undertakings. In the absence of such legislation it is often the lot of international organizations to act as watchdogs. In so doing many are also providing valuable services in terms of environmental education. Often, the political perception of EIAs is that they are in direct conflict with job creation and, as a result, they lack community support. Contrasting examples of this problem can be gleaned from two South African experiences: Lake St Lucia and the Berg River Floodplain. In the former, the integrity of a Ramsar river–floodplain–estuary complex was threatened by dune mining for rare minerals by Richard's Bay Minerals (RBM). Public enquiries led temporarily to the granting of a licence to mine, and it was at this stage that organizations such as the Wildlife and Environment Society of Southern Africa, and Endangered Wildlife Trust (amongst others), managed to raise the level of awareness through active engagement with the public. Ultimately, additional assessments coupled to arguments that 'job creation' would be better served by eco-tourism, stopped the further penetration of the St Lucia wetlands. On the other side of South Africa, and in contrast to the St Lucia conclusion, proposals in 1995–96 by a giant steel-producing para-state corporation for the construction of a hot-rolled-coil steel-mill at Saldanha, some 120 km north of Cape Town, eventually won the day after extensive hearings and lobbying by all sectors. The steel-mill construction site lies between an adjacent Ramsar Site, Langebaan Lagoon, a tidal wetland area exceptionally important for birds, and the Berg River Floodplain. The floodplain system has for decades failed to receive due attention from provincial conservation administrators and has not been formally recommended as a Ramsar Site. The area is depressed economically with high unemployment. Despite strong opposition from environmental groups, the problems posed by the steel-mill for the integrity of both Langebaan and the Berg River Floodplain, and the problems of water supply for human consumption and for the maintenance of ecosystem integrity (e.g. Davies and Day, 1998), the 'job creation'/industrial lobby won the day. Ironically, the area is still economically depressed for, as some cynical observers noted during the hearings, job creation was not the issue: job translocation (from Gauteng to the Western Cape) and the availability of a seaport, were.

The office of the IRN in Gaborone has been successful in highlighting the political deals of the Namibian Government surrounding the controversial Epupa Falls hydroelectric project on the Cunene River. Internally, the Epupa project is a political 'no-go': any criticism has been variously described by a succession of politicians as 'unpatriotic' and 'subversive' (see e.g. Davies and Day, 1998), so much so that there has been suppression of commentary. In 1997, the IRN organized an international review of the EIAs for the project: reviews that were highly critical of the site/s for the scheme, its economic viability in comparison with alternative sources of energy (e.g. gas), the treatment of the indigenous nomadic Himba people, and the ecological consequences for the region should the project go ahead. Criticism has been fierce to the extent that external funders of the project now have second thoughts about their involvement.

Apart from these international agencies, other more activist organizations such as Greenpeace do not have any significant support base, with the possible exception of the Okavango Pipeline controversy (see above). Many Southern African cultures are centred on a deep respect for authority, with little overt questioning of decisions made by 'the powers that be'. Such attitudes are not conducive to the almost revolutionary approaches taken by some NGOs in many Western countries. Even in instances where such organizations are involved, the subsequent dissemination of information is poor.

Conservation in practice

APPLICATION AND POLICING OF LEGISLATION

At a meeting in 1995 held in Nairobi to discuss the implementation of Chapter 18 of Agenda 21, on Freshwater Resources, 13 African nations complained of a lack of human and financial resources, and a lack of international assistance – assistance that was promised at the 1992 Earth Summit held in Rio de Janeiro. Although many National Parks are staffed by committed and highly qualified personnel, others are not. An additional and very real problem is the general lack of effective legislation and poor policing. For instance, many professionals in South Africa have long regarded the principal 'environmental agency', the DEAT, as ineffective to the point of uselessness, due to poor leadership, under-resourcing, lack of suitably trained personnel, and inadequate legislation. This is one side of the coin. On the other side are examples of excellence: policing of reserves in South African national parks is strong, as is the case in the Parks Boards of KwaZulu Natal and the Campfire Programme in Zimbabwe. An additional encouraging sign is the involvement of local communities in programmes such as Campfire, and in Moçambique the '*Tchuma Chato*' (our wealth) programme which has been enormously successful, bringing previously disenfranchised communities into active conservation by giving them a stake in eco-tourism, hunting and fishing rights. The sense of 'belonging' has greatly stimulated local conservation efforts to the point where poaching in some communities has ceased. Similar success stories may be found in South Africa with the Richtersveldt and Augrabies experiments that involve local communities in the conservation of their own resources, together with the simultaneous promotion of tourism. Such programmes are also beginning to develop in Namibia, Botswana and Tanzania, but are lacking elsewhere, most likely due to civil disorder. In this last context, it is difficult to see how field conservation and legislation enforcement can develop rapidly due to the legacy of war: parts of Angola and Moçambique still harbour hundreds of thousands of

landmines, while others are 'no-go' areas because of armed conflict.

In general, most nature conservation authorities have little power. For instance, where water pollution laws *are* enforced they can hardly be regarded as punitive, and no 'polluter-pays' principle is embodied in the constitution of any country, although South Africa has taken the first steps towards this principle.

IN SITU CONSERVATION (PROTECTED AREAS)

Where nature reserves exist, *in situ* conservation is generally excellent, but there is a lack of catchment protection outside reserves. This is a major threat to many activities associated with internal reserve regulation (see above, the KNP). But, again we make the observation that in South Africa, for instance, less than 4% of total land mass is protected in reserves – considerably below the global average. Sadly, there appears to be a similar story elsewhere in SADC. Despite possessing the second largest area of rain forest in the world, only 4% of the DRC is protected. Considering its rich biological diversity this is amazing, but perhaps understandable, given the chaos and corruption that has reigned supreme in that area for the last five decades.

Basin conservation through Integrated Catchment Management (ICM) is at last in vogue, at least in South Africa, and this bodes well for river conservation in the long term, as long as resources and personnel are available. However, with the burdens of over-population, unemployment and low levels of environmental awareness, there is a long road to walk. How, for example, does a huge country like Namibia conserve its ephemeral rivers? How, indeed, can such systems be conserved when they are so difficult to research and to understand?

EX SITU CONSERVATION (GENE BANKS)

We have noted the relatively detailed collections of aquatic invertebrates, fish and birds housed throughout South Africa, in particular. There is similarly an excellent botanical infrastructure through the Botanical Research Institute and the various National Park authorities in South Africa and Zimbabwe. However, local genetic storage elsewhere in SADC is virtually non-existent. Ironically, many resources are stored overseas, particularly in the USA and Europe (e.g. UK, Belgium, France), reflecting the 'safari' nature of scientific interest of the developed world in the region (e.g. Wishart and Davies, 1998; Denny, 2000).

RESTORATION AND REHABILITATION PROGRAMMES

The 'grassroots' concerns of the mass of people centre on survival, while political concerns centre on economic growth and job creation. There is little room for conservation, and even less for restoration and rehabilitation of degraded systems. However, there are some moves to redress this imbalance, particularly in urban environments where development, particularly of informal settlements, has led to 'eyesore' levels of riverine degradation, as well as to problems of human health (see above).

Recent collaboration between the UK and Zimbabwe has focused on urban streams, while in Cape Town the Liesbeek (reed river) has been the subject of several small rehabilitation projects (Davies, unpublished data) based both on the channel (de-canalization) and peripheral wetlands (restoration). These projects are in line with the policy of the Cape Town City Engineers Department, 'Greening the City', in which urban rivers are regarded as recreational and aesthetic corridors. The success of low-budget biotope enhancement (re-creation of weir faces, slow-run and riffles, reconnection of canalized sections with the hyporheos) has been clear, with increases in taxa from seven pollution-tolerant invertebrate species, to 35 species, and the return of waterfowl to the system (Davies, unpublished data). The level of public awareness for the plight of urban rivers has been greatly enhanced by such simple projects.

However, the major thrust of 'restoration', if it can be called such, has been the development of IFR techniques (see above). Early calculations of 'water for the environment' (Roberts, 1981, 1983) indicated that some 11% of the MAR of South Africa would be required, and that this could comprise some 13% of total water demand. These estimates were based on the water requirements of estuaries and lakes (evaporative and flooding requirements), as well as on water demand for nature conservation areas, including game watering, and the maintenance of riverine habitats. A decade and a half later, work on specific systems appears to be pushing this figure upwards. For instance, in the case of the Berg River (Western Cape) estimated IFRs fall between 20 and 40%. It is clear that under the new Water Act in South Africa, at least, restoration flows will be possible for some of the more important rivers.

FUTURE THREATS AND THEIR MITIGATION

The threats to the region's rivers have been summarized in Table 7.4. The pressure of a rapidly increasing and poorly educated population is without doubt the single greatest threat, not only to rivers, but to the Southern African environment as a whole. The population is expected to double within the next 24 years, and is already estimated at 136 million people, well over half of whom are under 18 years of age. With increasing pressures to meet the demands of market-driven economies, there is increasing tension between the cultural practices of the past and the 'modern' worlds, resulting in mono-specific cropping, over-exploitation of resources, over-abstraction of water and, with very few standing water bodies, the burden is having to be carried by the region's rivers. The demographics are in stark contrast to the current distribution and availability of water. The need for water in areas such as Gauteng in South Africa is resulting in more and more IBTs; this is combined with increasing trends towards rapid urbanization and industrialization.

The cynic might remark that pandemics such as HIV/AIDS will resolve the 'population problem'. Perhaps that is true in a way, for sub-Saharan Africa reports more than 60% of all cases of the disease, with some countries running at infection rates of between 20 and 45%, or higher. Yet a moment's thought reveals the flaw (apart from the callousness) of the argument: such diseases are a real threat. One person with full-blown AIDS (or malaria, or cholera, etc.) will, in a 'high-tech' alleopathic environment, consume orders of magnitude more water than a healthy person; the numbers of orphans are already legion, requiring additional support and resources in a region already reeling from under-supply, and the loss of people from all sectors of the work force, has already been, and will continue to be, catastrophic.

Because of the scarcity of freshwater resources, and also because the MAR, MAP and the hydrology of the region are so variable, water-storage construction has occurred on a massive scale, with almost universal river regulation. Many systems have been converted to cascades of reservoirs, and many perennial rivers have now become seasonal (e.g. Chutter, 1973; Davies, 1979, 1986; Davies *et al.*, 1993a,b, 1995; Davies and Day, 1998). However, the overriding issue is that whichever way one views the problem, most rivers are currently over-exploited, and will continue to be.

The effects of flow regulation in drylands are ecologically problematic at every level, simply because the biota is adapted to unregulated regimes. Indeed, our rivers inherit much of their character from the climate, and their flow behaviour especially mirrors the erratic patterns of rainfall, with variable periods of drought punctuated by flooding that may be equally variable in timing, duration, and in magnitude (e.g. Davies *et al.*, 1993a,b). The significance of a single *flood pulse* (Junk *et al.*, 1989) must be judged both against the history of inundation at one place, and the flood regime, describing variations in both space and time. Hydrological variability strongly influences the evolutionary character of the biota – for example, many resident species are highly tolerant of environmental extremes and are reproductively opportunistic. Where flooding is modified, as for irrigation, the character of fluvial systems is likely to change (e.g. Davies *et al.*, 2000). Moreover, the high degree of spatial and temporal variability suggests that the times for dryland river ecosystems to respond to, and to recover from, environmental changes will be prolonged (e.g. Davies *et al.*, 1993b).

Many controversial water-development projects are still going ahead, despite the growing awareness of environmental problems and the political tensions that they create. For instance, according to the South African Press Association (SAPA), the Chinese Government has pledged US$283 000 to Namibia for the Okavango Pipeline Project. Namibian President Sam Nujoma has also appealed to the German Government for soft loans on the project: tensions between Botswana and Namibia will inevitably continue to mount. Conservationists fear that the effect of the project on the ecology of the Okavango Delta, which supports a wide variety of game and more than 200 species of birds – and increasingly, tourism – will be catastrophic. 'The delta is the source of livelihood for the ordinary people and the tourism industry' (Modisa Mothoagae, head of the Hotel and Tourism Association of Botswana). Since the Botswanan Government stepped up its promotion of tourism in 1989, the sector's contribution to the economy has risen by 27.5%, and now accounts for just over 3% of the Gross Domestic Product. The Okavango Delta region accounts for 80% of the nation's tourism. A local environmentalist has written:

'This ecosystem is so sensitive that even relatively minor past dredging projects in the lower reaches of the Delta have resulted in the desiccation of stream courses, and some channels cleared over 30 years ago still show no sign of recovery. The limit at which upstream water reduction would result in loss of wetland is not known, nor does any credible model exist yet to predict the ratio of wetland loss to reduction of flow.'

Four years ago, the Botswana Government was forced to scrap a plan to extract water flowing out of the delta to supply the mining town of Orapa. Local communities and NGOs opposed the project on environmental grounds. An independent assessment of that project by the IUCN revealed that there were better alternatives, concluding that the costs had been under-estimated, and its benefits over-estimated. It is to be hoped that environmental opposition will continue to mount and that such ill-advised projects are shelved once and for all, but unless water-demand management is better implemented, similar threats will continue for the foreseeable future.

We have noted some of the damage to lotic ecosystem functioning caused by invasive species: in particular, floating plants, fish such as trout (e.g. Bruton, 1986; De Moor and Bruton, 1988; Davies and Day, 1998), terrestrial plant invasion of floodplains and riparian corridors, and the problems of population inter-mixing through IBTs (e.g. Davies *et al.*, 1992; Davies *et al.*, this volume). These are serious and constitute a major problem for river conservation. Active biological control, eradication, and research into effective screening for IBTs are urgently needed (Snaddon *et al.*, 2000). In addition, the geographical distribution of water, the continuing political instability, the major political and cultural reforms away from traditional ways to more 'Western ethics' (see also Gopal *et al.*, this volume), all pose difficult problems for conservation. Ethnic and tribal rivalries are also likely to increase as the demand for water increases and as countries begin to look beyond national borders to meet increasing demands (see Snaddon *et al.*, 1998). African one-party states are gaining an increasing stranglehold, with their governments (typically consisting of members of the same ethnic group) in partnership with large multinational corporations indulging in massive environmental exploitation.

We see little by way of mitigation of many of these problems. The population growth rates of the region will continue to exacerbate the situation unless community involvement and education are stepped up by orders of magnitude. We have already noted the Campfire and *Tchuma Chato* initiatives: in Malawi, the approach to the management of riparian vegetation also includes community involvement – indeed, we believe this to be absolutely essential for successful conservation. Here, in an attempt to protect '*Mbuna*' species (the collective colloquial name for the remarkable flock of more than 200 small, colourful or vividly marked species of cichlid in Lake Malawi, each of which has specific substratum and feeding requirements (Ribbinck *et al.*, 1983)), the Government of Malawi declared as a national park an area of 87 km^2 extending 100 m into the lake. The Park has since been listed by the United Nations Educational, Scientific and Cultural Organization (UNESCO) as a World Heritage Site. Having done so, it was realized that the fish were not threatened by over-fishing, but by increased siltation as the result of the removal of riparian vegetation. The vegetation provides an important and essential fuel source for many of the villagers and, rather than forcing the people out of the park, the Department of Parks and Wildlife has opted for an integrated educational approach to conservation. Local villagers have been educated as to the importance and significance of the fish, as well as being charged a fee for firewood collection. They have also been encouraged to use less wood. In Southern Africa, failure to recognize the need for integrated consultative approaches of this type towards effective river basin management will clearly lead to poor decisions and ineffective management.

Although research into the effective implementation of ICM is under way in South Africa, it appears at present that unless South Africa exports the technology to other SADC countries, little progress will occur elsewhere. There is a tendency for developed countries to arrogate expertise in African issues: locally, the realization that 'home grown is best' is gradually taking hold. As well as education and training programmes, the region must give urgent attention to forging international basin linkages, full-scale ICM programmes, and cost-effective water-demand strategies if the rivers of the region are to survive to the end of the 21st century: 'capacity building' is vital. It is here that institutions and individuals in the developed world can really assist.

References

Acocks, J.P.H. 1953. Veld types of South Africa. *Memoirs of the Botanical Society of South Africa* 28:1–192.

Alexander, W.J.R. 1985. Hydrology of low-latitude Southern Hemisphere land masses. In: Davies, B.R. and Walmsley, R.D. (eds) *Perspectives in Southern Hemisphere Limnology. Developments in Hydrobiology* 28. Dr W. Junk, Dordrecht, 75–83.

Allanson, B.R., Hart, R.C., O'Keeffe, J.H. and Robarts, R.D. 1990. *Inland Waters of Southern Africa: An Ecological Perspective*. Monographiae Biologicae 64. Kluwer Academic Publishers, Dordrecht.

Asmal, K. 1995. Opening address to the first national water conservation campaign conference, Kempton Park, Johannesburg, 2–3 October, 1995. Department of Water Affairs and Forestry, Private Bag X313, in manuscript.

Banister, K.E. 1986. Fish of the Zaïre system. In: Davies, B.R. and Walker, K.F. (eds) *The Ecology of River Systems*. Monographiae Biologicae 60. Dr W. Junk, The Hague, 215–224.

Beadle, L.C. 1981. *The Inland Waters of Tropical Africa: An Introduction to Tropical Limnology*, 2nd edition. Longman, New York and London.

Beilfuss, R. and Davies, B.R. 1999. Prescribed flooding and wetland rehabilitation in the Zambezi Delta, Mozambique. In: Streever, W. (ed.) *International Perspectives on Wetland Rehabilitation*. Kluwer Press, 143–158.

Bennett, B.A. 1993. An assessment of the potential effects of reduced freshwater inputs on the fish community of the Berg River Estuary. In: *Berg River Estuary Worksession, A Workshop held at Port Owen, March 1993*. Department of Water Affairs and Forestry, Pretoria.

Bovee, K.D. 1982. *A Guide to Stream Habitat Analysis using Instream Flow Incremental Methodology*. Instream Flow Information Paper 12, United States Department of Information, Fisheries and Wildlife Service, Office of Biological Surveillance, FWS/OBS-82/26.

Bruton, M.N. 1986. Life history styles of invasive fishes in Southern Africa. In: MacDonald, I.A., Kruger, W. and Ferrar, A. (eds) *The Ecology and Management of Biological Invasions in Southern Africa*. Oxford University Press, Cape Town, 201–208.

Bruwer, C. (ed.) 1991. *Flow Requirements of Kruger National Park Rivers*. Report to the Department of Water Affairs and Forestry, Pretoria, TR149.

Byren, B.A. and Davies, B.R. 1989. The effect of stream regulation on the physico-chemical properties of the Palmiet River, South Africa. *Regulated Rivers: Research & Management* 3:107–121.

Chutter, F.M. 1972. An empirical biotic index of the water quality in South African streams and rivers. *Water Research* 6:19–30.

Chutter, F.M. 1973. An ecological account of the past and future of South African rivers. *Newsletter of the Limnological Society of southern Africa* 21:22–34.

Chutter, F.M. 1994. The rapid biological assessment of stream and river water quality by means of the macroinvertebrate community in South Africa. In: Uys, M.C. (ed.) *Classification of Rivers, and Environmental Health Indicators*. Proceedings of a Joint South African–Australian Workshop, Cape Town, 7–14 February 1994. Water Research Commission, Pretoria, TT63/94, 217–234.

Conley, A.H. 1995. *A Synoptic View of Water Resources in Southern Africa*. Paper to South African Department of Water Affairs and Forestry, Pretoria.

Dallas, H.F. and Day, J.A. 1993. *The Effect of Water Quality Variables on Riverine Ecosystems: A Review.* Report to the Water Research Commission, Pretoria, TT61/93.

Davies, B.R. 1979. Stream regulation in Africa. In: Ward, J.V. and Stanford, J.A. (eds) *The Ecology of Regulated Streams.* Plenum Press, New York and London, 113–142.

Davies, B.R. 1986. The Zambezi River System. In: Davies, B.R. and Walker, K.F. (eds) *The Ecology of River Systems.* Monographiae Biologicae, 60. Dr W. Junk, The Hague, 225–267.

Davies, B.R. (ed.) 1998. *Report on the Workshop: Sustainable Utilisation of the Cahora Bassa Dam and the Zambezi Valley, Songo, 29 September through 2 October, 1997.* Arquivos Patrimonio Cultural, Maputo and the Ford Foundation, Johannesburg.

Davies, B.R. and Day, J.A. 1998. *Vanishing Waters.* University of Cape Town Press and Juta Press, Cape Town.

Davies, B.R., Thoms, M.C. and Meador, M. 1992. The ecological impacts of inter-basin water transfers and their threats to river basin integrity and conservation. *Aquatic Conservation: Marine and Freshwater Ecosystems,* 2, 325-349.

Davies, B.R., O'Keeffe, J.H. and Snaddon, C.D. 1993a. *A Synthesis of the Ecological Functioning, Conservation and Management of South African River Ecosystems.* Water Research Commission, Pretoria, Report No. WRC TT62/93.

Davies, B.R., Thoms, M.C., Walker, K.F., O'Keeffe, J.H. and Gore, J.A. 1993b. Dryland rivers: their ecology, conservation and management. In: Calow, P. and Petts, G.E. (eds) *The Rivers Handbook.* Vol. 2. Blackwell Scientific Publications, Oxford, 484–511.

Davies, B.R., O'Keeffe, J.H. and Snaddon, C.D. 1995. River and stream ecosystems in southern Africa: predictably unpredictable. In: Cushing, C.E., Cummins, K.W. and Minshall, G.W. (eds) *Ecosystems of the World 22. Rivers and Stream Ecosystems.* Elsevier, Amsterdam, 537–599.

Davies, B.R., Beilfuss, R.D. and Thoms, M.C. 2000. Cahora Bassa Retrospective, 1974–1997: Effects of flow regulation on the Lower Zambezi River. *Verhandlungen der Internationalen Vereinigung für theoretische und angewandte Limnologie* (in press).

De Moor, I.J. and Bruton, M.N. 1988. *Atlas of Alien and Translocated Indigenous Aquatic Animals in Southern Africa.* South African National Scientific Programmes Report 144, Council for Scientific and Industrial Research, Pretoria.

Denny, P. 2000. Limnological research and capacity building in tropical developing countries. *Verhandlungen der Internationalen Vereinigung für theoretische und angewandte Limnologie* (in press).

Department of Water Affairs, South Africa (DWA) 1986. *Management of the Water Resources of the Republic of South Africa,* Pretoria.

Department of Water Affairs, South Africa (DWA) 1992. *Water for Managing the Natural Environment.* Draft Policy Document, Pretoria.

Derman, B. 1998. Balancing the waters. Development and hydropolitics in contemporary Zimbabwe. In: Donahue, J.M. and Johnston, B.R. (eds) *Water, Culture, and Power. Local Struggles in a Global Context.* Island Press, Washington, DC, 73–93.

Eekhout, S., King, J.M. and Wackernagel, A. 1997. *Classification of South African Rivers.* Volume 1, *Text and* Volume 2, *Species Distribution Maps.* Report to the Department of Environmental Affairs and Tourism, South Africa, South African Wetlands Conservation Programme, Pretoria.

Ferrar, A.A. (ed.) 1989. *Ecological Flow Requirements for South African Rivers.* South African National Scientific Programmes Report, Council for Scientific and Industrial Research, Pretoria, 162.

Gore, J.A. and King, J.M. 1989. Application of the revised physical habitat simulation (PHABSIM II) to minimum flow evaluations of South African rivers. In: Kienzle, S. and Mäaren, H. (eds) *Proceedings of the Fourth South African National Hydrological Symposium,* University of Pretoria, Pretoria, 289–296.

Gore, J.A. and Petts, G.E. (eds) 1989. *Alternatives in Regulated River Management.* CRC Press, Boca Raton, Florida.

Gourou, P. 1970. *L'Afrique.* Hachette, Paris.

Heyns, P.H., Masundire, M. and Sekwale, M. 1994. Freshwater resources. In: Chenje, M. and Johnson, P. (eds) *State of the Environment in Southern Africa.* Southern African Research and Documentation Centre, IUCN, and Southern African Development Community. Penrose Press, Johannesburg, 181–206.

Hockey, P.A.R. 1993. Potential impacts of water abstraction on the birds of the Lower Berg River Wetlands. In: *Berg River Estuary Worksession, A Workshop held at Port Owen, March 1993.* Department of Water Affairs and Forestry, Pretoria.

Jacobson, P.J., Jacobson, K.M. and Seely, M.K. 1995. *Ephemeral Rivers and their Catchments: Sustaining People and Development in Western Namibia.* Desert Research Foundation of Namibia, Windhoek.

Junk, W.J., Bayley, P.B. and Sparks, R.E. 1989. The Flood–Pulse concept in river–floodplain systems. In: Dodge, D.P. (ed.) *Proceedings of the International Large River Symposium (LARS). Canadian Special Publication of Fisheries and Aquatic Sciences* 106:110–127.

King, J.M. 1981. The distribution of invertebrate communities in a small South African river. *Hydrobiologia* 83:43–65.

King, J.M. 1983. Abundance, biomass and diversity of benthic macroinvertebrates in a western Cape river, South Africa. *Transactions of the Royal Society of South Africa* 35:11–34.

King, J.M. and Louw, D. 1998. Instream flow assessments for regulated rivers in South Africa using the Building Block Methodology. *Aquatic Ecosystem Health and Managament* 1:109–124.

King, J.M. and Tharme, R. 1994. *Assessment of the Instream Flow Incremental Methodology and Initial Development of Alternative Instream Flow Methodologies for South Africa.*

Contract Research Report to the Water Research Commission, Pretoria, 295/1/94.

King, J.M., Day, J.A., Davies, B.R. and Henshall-Howard, M.-P. 1987a. Particulate organic matter in a mountain stream in the south-western Cape, South Africa. *Hydrobiologia* 154:165–187.

King, J.M., Henshall-Howard, M.-P., Day, J.A. and Davies, B.R. 1987b. Leaf-pack dynamics in a Southern African mountain stream. *Freshwater Biology* 18:325–340.

King, J.M., Day, J.A., Hurley, P.R., Henshall-Howard, M.-P. and Davies, B.R. 1988. Macroinvertebrate communities and environment in a Southern African mountain stream. *Canadian Journal of Fisheries and Aquatic Sciences* 45:2168–2181.

King, J.M., De Moor, F.C. and Chutter, F.M. 1992. Alternative ways of classifying rivers in Southern Africa. In: Boon, P.J., Calow, P. and Petts, G.E. (eds) *River Conservation and Management.* John Wiley, Chichester, 213–228.

King, L.C. 1978. The geomorphology of Central and Southern Africa. In: Werger, M.J.A. and Van Bruggen, A.C. (eds) *Biogeography and Ecology of Southern Africa.* Monographiae Biologicae 31. Dr W. Junk, The Hague, 1–18.

Lillehammer, A. and Saltveit, S.J. (eds) 1984. *Regulated Rivers.* Universitetsforlaget AS, Oslo.

McCullum, J. 1994. People and environment. In: Chenje, M. and Johnson, P. (eds) *State of the Environment in Southern Africa.* A report by the Southern African Research and Documentation Centre in collaboration with the IUCN, the World Conservation Union and the Southern African Development Community, Penrose Press, Johannesburg.

Milhous, R.T., Updike, M.A. and Schneider, D.M. 1989. *Physical Habitat Simulation System Manual – Version 2.* Instream Flow Information Paper 26, United States Department of Information, Fish and Wildlife Service, Biology Report, 89.

O'Keeffe, J.H. 1986. *The Conservation of South African Rivers.* South African National Scientific Programmes Report, 131, Council for Scientific and Industrial Research, Pretoria.

O'Keeffe, J.H. and Davies, B.R. 1991. Conservation and management of rivers of the Kruger National Park: suggested methods for calculating instream flow needs. *Aquatic Conservation: Marine and Freshwater Ecosystems* 1:55–71.

O'Keeffe, J.H. and De Moor, F.C. 1988. Changes in the physico-chemistry and benthic invertebrates of the Great Fish River, South Africa, following an inter-basin transfer of water. *Regulated Rivers: Research & Management* 2:39–55.

O'Keeffe, J.H., Danilewitz, D.B. and Bradshaw, J.A. 1987. An expert system approach to the assessment of the conservation status of rivers. *Biological Conservation* 40:69–84.

O'Keeffe, J.H., Davies, B.R., King, J.M. and Skelton, P.H. 1989. The conservation status of Southern African rivers.

In: Huntley, B.J. (ed.) *Biotic Diversity in Southern Africa: Concepts and Conservation.* Oxford University Press, Cape Town, 276–299.

O'Keeffe, J.H., Byren, B.A., Davies, B.R. and Palmer, R.W. 1990. The effects of impoundment on the physico-chemistry of two southern African river systems. *Regulated Rivers: Research & Management* 5:97–110.

O'Keeffe, J.H., Weeks, D.C., Fourie, A. and Davies, B.R. 1996. *A Pre-Impoundment Study of the Sabie–Sand River System, Mpumalanga, with Special Reference to Predicted Impacts on the Kruger National Park.* Volume 3, *The Effects of Proposed Impoundments and Management Recommendations.* Report to the Water Research Commission, 294/3/96, Pretoria.

Palmer, R.W. and O'Keeffe, J.H. 1990a. Transported material in a small river with multiple impoundments. *Freshwater Biology* 24:563–575.

Palmer, R.W. and O'Keeffe, J.H. 1990b. Downstream effects of a small impoundment on a turbid river. *Archiv für Hydrobiologie* 119:457–473.

Palmer, R.W. and O'Keeffe, J.H. 1990c. Downstream effects of impoundments on the water chemistry of the Buffalo River (Eastern Cape), South Africa. *Hydrobiologia* 202:71–83.

Petitjean, M.O.G. and Davies, B.R. 1988. Ecological impacts of inter-basin water transfers: Some case studies, research requirements and assessment procedures. *South African Journal of Science* 84:819–828.

Petts, G.E. 1984. *Impounded Rivers.* John Wiley, Chichester.

Picker, M.D. and Samways, M.J. 1996. Faunal diversity and endemicity of the Cape Peninsula, South Africa – a first assessment. *Biodiversity and Conservation* 5:591–606.

Pitman, W.V. and Hudson, J. 1994. *A Broad Overview of Present and Potential Water Resource Linkages in Southern Africa.* Report prepared by Stewart Scott Incorporated, Johannesburg, for the Development Bank of Southern Africa.

Pollard, S.R., Weeks, D.C. and Fourie, A. 1996. *A Pre-Impoundment Study of the Sabie–Sand River System, Mpumalanga, with Special Reference to Predicted Impacts on the Kruger National Park.* Volume 2, *Effects of the 1992 Drought on the Fish and Macro-invertebrate Fauna.* Report to the Water Research Commission, 294/2/96, Pretoria.

Regier, H.A. and Meisner, J.D. 1990. Anticipated effects of climate change on freshwater fishes and their habitat. *Fisheries* 15:10–15.

Ribbink, A.J., Marsh, B.A., Marsh, A.C., Ribbink, A.C. and Sharp, B.J. 1983. A preliminary survey of the cichlid fishes of rocky habitats in Lake Malawi. *South African Journal of Zoology* 18:149–310.

Roberts, C.P.R. 1981. *Environmental Considerations of Water Projects.* Department of Water Affairs and Forestry, Pretoria, Technical Report, 114.

Roberts, C.P.R. 1983. Environmental constraints on water resources development. *Proceedings of the South African Institute of Civil Engineers* 1:16–23.

Skelton, P. 1993. *A Complete Guide to the Freshwater Fishes of Southern Africa*. Southern Book Publishers, South Africa.

Skotnes, P. (ed.) 1996. *Miscast: Negotiating the Presence of the Bushmen*. University of Cape Town Press, Cape Town.

Smakhtin, V.Y., Watkins, D.A. and Hughes, D.A. 1995. Preliminary analysis of low-flow characteristics of South African rivers. *Water SA* 21:201–210.

Snaddon, C.D. and Davies, B.R. 1998. A preliminary assessment of the effects of a small inter-basin water transfer, the Riviersonderend–Berg River Transfer Scheme, Western Cape, South Africa, on discharge and invertebrate community structure. *Regulated Rivers: Research & Management* 14:421–441.

Snaddon, C.D., Wishart, M.J. and Davies, B.R. 1998. Some implications of inter-basin water transfers for river ecosystem functioning and water resources management in Southern Africa. *Aquatic Ecosystem Health and Management* 1:159–182.

Snaddon, C.D., Davies, B.R. and Wishart, M.J. 2000. *The Ecological Effects of Inter-Basin Water Transfer Schemes, with a Brief Appraisal of their Socio-Economic and Socio-Political Implications, and an Outline of Guidelines for their Management*. Contract Report to the Water Research Commission, Pretoria.

Southern African Development Community (SADC) and Environment and Land Management Sector (ELMS) 1990. *Sustaining our Common Future*. Maseru, Lesotho.

Starmans, G.A.N. 1970. *Soil Erosion of Selected African and Asian Catchments*. International Water Erosion Symposium, Prague.

Tankard, A.J., Jackson, M.P.A., Eriksson, K.A., Hobday, D.K., Hunter, D.R. and Minter, W.E.L. 1982. *Crustal Evolution of South Africa*. Springer-Verlag.

Tharme, R. 1996. *Review of International Methodologies for the Quantification of the Instream Flow Requirements of Rivers*. Contract Report to the Department of Water Affairs and Forestry, Pretoria, and the Water Research Commission, Pretoria, under the continuing Water Law Review Process.

Tharme, R. and King, J.M. 1998. *Development of the Building Block Methodology for Instream Flow Assessments and Supporting Research on the Effects of Different Magnitude Flows on Riverine Ecosystems*. Report to the Water Research Commission, 576/1/98, Pretoria.

Thomas, D.S.G. 1989. The nature of arid environments. In: Thomas, D.S.G. (ed.) *Arid Zone Geomorphology*. John Wiley, New York, 1–8.

Tinley, K.L. 1976. *The Ecology of Tongaland. Report 1. Lake Sibayi; 2. Pongolo and Mkuze Floodplains; 3. Kosi Lake System*. Natal Branch of the Wildlife Society of South Africa, Durban.

Tyson, P.D. 1986. *Climatic Change and Variability in Southern Africa*. Oxford University Press, Cape Town.

Uys, M.C. (ed.) 1994. *Classification of Rivers, and Environmental Health Indicators*. Proceedings of a joint South African–Australian Workshop, Cape Town, 7–14 February 1994. Water Research Commission, Pretoria, TT63/94.

Van Wilgen, B.W., Cowling, R.M. and Burgers, C.J. 1996. Valuation of ecosystem services: a case study from the fynbos, South Africa. *BioScience* 46:184–189.

Van Zinderen Bakker, B.M. 1978. Quaternary vegetation changes in southern Africa. In: Werger, M.J.A. and Van Bruggen, A.C. (eds) *Biogeography and Ecology of Southern Africa*. Monographiae Biologicae, 31. Dr W. Junk, The Hague, 133–143.

Walmsley, R.D. and Davies, B.R. 1991. Water for environmental management: an overview. *Water SA* 17:67–76.

Ward, J.V. and Stanford, J.A. (eds) 1979. *The Ecology of Regulated Streams*. Plenum Press, New York and London.

Weeks, D.C., O'Keeffe, J.H., Fourie, A. and Davies, B.R. 1996. *A Pre-Impoundment Study of the Sabie–Sand River System, Mpumalanga, with Special Reference to Predicted Impacts on the Kruger National Park*. Volume 1, *The Ecological Status of the Sabie–Sand River System*. Report to the Water Research Commission, 294/1/96, Pretoria.

Welcomme, R.L. 1972. *The Inland Waters of Africa*. United Nations Food and Agriculture Organization, Technical Paper 1.

Wishart, M.J. 1998. Temporal variations in a temporary sandbed stream. Unpublished MSc Thesis, Zoology Department, University of Cape Town, Cape Town.

Wishart, M.J. and Davies, B.R. 1998. The increasing divide between First and Third Worlds: science, collaboration and conservation of Third World aquatic ecosystems. *Freshwater Biology* 39:557–567.

8

River conservation in Madagascar

J.P. Benstead, M.L.J. Stiassny, P.V. Loiselle, K.J. Riseng and N. Raminosoa

Introduction

Madagascar's extraordinary biological heritage makes the island one of the highest priorities for international conservation efforts. This 'micro-continent' is one of the 12 'megadiversity countries' which together harbour 70% of the world's plant and animal species (McNeeley *et al.*, 1990). Its long isolation has resulted in the adaptive radiation of many groups and a correspondingly high level of endemism. Approximately 90% of the island's species are found nowhere else (Battistini and Richard-Vindard, 1972; Jolly *et al.*, 1984; International Union for Conservation of Nature and Natural Resources/United Nations Environment Programme/World Wildlife Fund – IUCN/UNEP/WWF, 1987; Nicoll and Langrand, 1989). Many of these taxa either belong to relic lineages that exist elsewhere only in the fossil record (Wright, 1997) or are basal representatives of their respective clades (Stiassny and DePinna, 1994; Stiassny and Raminosoa, 1994). Despite its biological riches, however, Madagascar is the 10th poorest nation in the world (Economist Intelligence Unit – EIU, 1994). Poverty, combined with a 3% population growth rate, has forced the country and its inhabitants into a familiar spiral of environmental degradation that is characterized by extensive deforestation and severe soil erosion (Jolly and Jolly, 1984; EIU, 1994). These processes are destroying Madagascar's unique environments and driving many species to extinction.

Despite their high degree of vulnerability, the island's biota and habitats remain poorly known. This is particularly true of the island's river and stream ecosystems. As a result, the island's endemic and highly threatened riverine biota are not being included in current conservation efforts. In addition, decisions relating to ecosystem protection and the design of nature reserves historically have not been made from a watershed perspective. Consequently, there is an urgent need for consideration of freshwater ecosystems and biota to be integrated into future conservation planning (Reinthal and Stiassny, 1991; Stiassny and Raminosoa, 1994).

Efforts to conserve Madagascar's river ecosystems face immediate and daunting challenges. First, many of the island's rivers have been severely affected by deforestation, subsequent erosion and sedimentation, the spread of exotic fish species, and overfishing. Restoration or rehabilitation of these damaged systems would be a monumental task. Second, those rivers on the island that are still relatively undisturbed are threatened with similar fates. The factors that are driving these changes include human population growth, a dwindling resource base, and a struggling economy (Jolly and Jolly, 1984). Third, information is a prerequisite to successful conservation. As outlined above, very little is known about Madagascar's river ecosystems.

The main goal of this chapter is to draw together the sparse and disparate literature pertaining to river conservation in Madagascar. We concentrate on the data and information sources that are available and identify major gaps in an attempt to encourage more interest and research in the region. For example, our discussion of the river biota and its conservation status is biased towards the endemic fish fauna; other taxa are

Global Perspectives on River Conservation: Science, Policy and Practice.
Edited by P.J. Boon, B.R. Davies and G.E. Petts. © 2000 John Wiley & Sons Ltd.

very poorly known. Throughout the chapter we will often discuss conservation in general as opposed to river conservation specifically. This is for two reasons. First, there are very few studies and projects focusing on river conservation in Madagascar. Second, the conservation of Madagascar's natural river communities relies heavily on efforts to protect the island's original forested habitats. We begin the chapter with a general description of the island, including brief summaries of its geology, climate, hydrology and evolutionary history.

Regional context

THE PHYSICAL ENVIRONMENT

Madagascar lies 400 km off Africa's south-eastern coast (11°57'S to 25°35'S and 43°14'E to 50°27'E), separated by the Moçambique Channel (Figure 8.1). Although dwarfed by the African continent, the island is the fourth largest in the world (587 000 km^2), comparable in size to France or Texas. The geological origin of the island is controversial; the morphological relationship of Madagascar, Africa, and the Indian sub-continent has been debated for decades. The available geological evidence indicates that before the supercontinent of Gondwana broke apart 180 million years (Ma) ago, Madagascar was a continental fragment of Greater India (Agrawal *et al.*, 1992; Storey *et al.*, 1995). This land mass started to break away from Africa during the Jurassic quiet zone (± 165 Ma) and ceased relative motion about 45 Ma later (Rabinowitz *et al.*, 1983). Recent fossil evidence supports the existence of a prolonged link between South America and Indo-Madagascar (via land bridges to Antarctica) until as late as 80 Ma ago (Sampson *et al.*, 1998). Although the timing of breakup between the island and Greater India is not known accurately, sea-floor spreading between the two land masses was under way at least 88 Ma ago (Storey *et al.*, 1995).

The island is geologically diverse with two main divisions that give rise to marked differences in soil types and relief (see Brenon, 1972, and IUCN/UNEP/WWF, 1987 for a more complete discussion). The Precambrian basement outcrops over the eastern two-thirds of the island. Faulting, uplifting and subsequent erosion of this ancient metamorphic rock is responsible for the varied and rugged topography of this part of the island (Figure 8.2). The basement rock is overlain by an extensive and thick layer of highly erodible lateritic clay. To the west of the Precambrian

basement, and in a narrow strip along the east coast, is a more recent sedimentary formation that makes up the second main geological feature of the island. This region is characterized by comparatively gentle slopes. A variety of soils are found but true lateritic clays are absent. Finally, two periods of intense volcanic activity have left widespread igneous intrusions. These form many of the major massifs, particularly in the central highlands but also in the extreme north and south of the island. These include Madagascar's highest peak, Mt Maromokotra (2876 m) in the northerly Tsaratanana Massif.

Despite the absence of peaks over 3000 m, the size and latitudinal span of Madagascar give rise to great variations in climate. Temperature varies from −15 to 35°C. Mean annual temperatures on the central plateaux normally range between 16 and 19°C, while in the lowlands mean annual temperatures vary between 27°C in the north and 23°C in the south (Figure 8.3). The annual amount of rainfall decreases from east to west and from north to south (range 310 to 3600 mm yr^{-1}). Seasonality of rainfall increases along the same gradients (Figure 8.2). There is no discernible wet season in the extreme southwest of the island. Lastly, tropical cyclones frequently hit the island (a total of 155 cyclones between 1848 and 1972: Donque, 1972). These intense storm systems occur mainly during the summer months of January to March and are capable of dropping 600–700 mm of rain in a 3–4 day period. The occurrence of cyclones is a major determinant of total annual rainfall, particularly in eastern regions, and is of major importance to the hydrology of the island's rivers.

Historically the island has been split into five natural hydrographic regions of very different sizes (Aldegheri, 1972). These are the slopes of Montagne D'Ambre, the Tsaratanana slopes, the eastern slopes, the western slopes, and the southern slopes (Figure 8.1). The Montagne D'Ambre region consists of the most northerly part of the island and covers only 1.8% of the land area. This volcanic massif is drained by narrow, torrential streams with few tributaries. The major rivers in this region are the Irodo, the Saharenana and the Besokatra. The slopes of the Tsaratanana Massif form the second region, covering 3.4% of the island's area. The major rivers are the Mahavavy, the Sambirano and the Maevarano which flow into the Moçambique Channel, and the Bemarivo which flows into the Indian Ocean. All these rivers are characterized by very steep profiles (30–40 m km^{-1}) in their upper reaches, lessening to low gradients on the coastal plain. The eastern slopes region covers 25% of

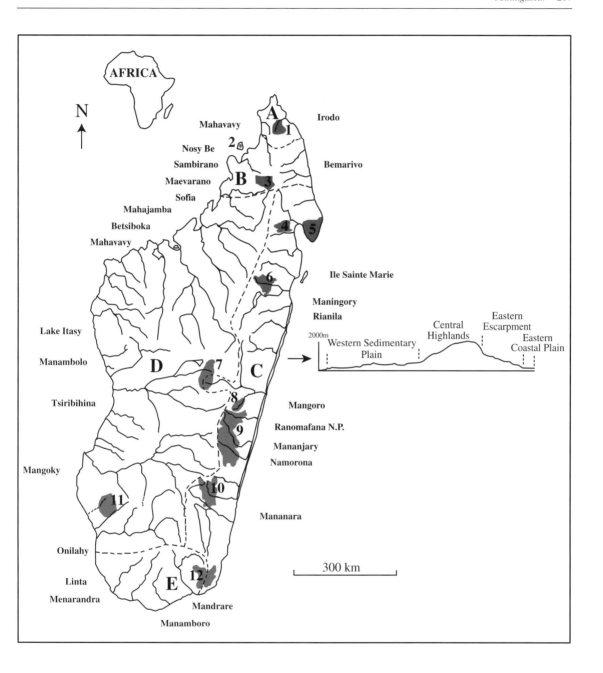

Figure 8.1 *Map of Madagascar with west–east topographic profile. Shown are the major river systems and the boundaries of the five hydrographic regions (dashed lines). A, the Montagne D'Ambre slopes; B, the Tsaratanana slopes; C, the eastern slopes; D, the western slopes; and E, the southern slopes. Shaded regions are those considered priorities for freshwater conservation efforts (by the GEF workshop; see text). 1, the Irodo River Basin; 2, the Djabala River and associated Mont Passot crater lakes on the island of Nosy Be; 3, the upper reaches of the Sambirano, Mahavavy and Bemarivo rivers; 4, the Rantabe River Basin; 5, the rivers of the Masoala Peninsula; 6, the upper reaches of the Maningory River; 7, the streams of the Ankaratra Highlands; 8, the Nosivolo River Basin; 9, the Namorona, Mananjary and Sakaleona river basins; 10, the streams of the Andringitra Massif; 11, the streams of the Analavelona Forest; and 12, the Manampanihy, Mandrare and Efaho river basins*

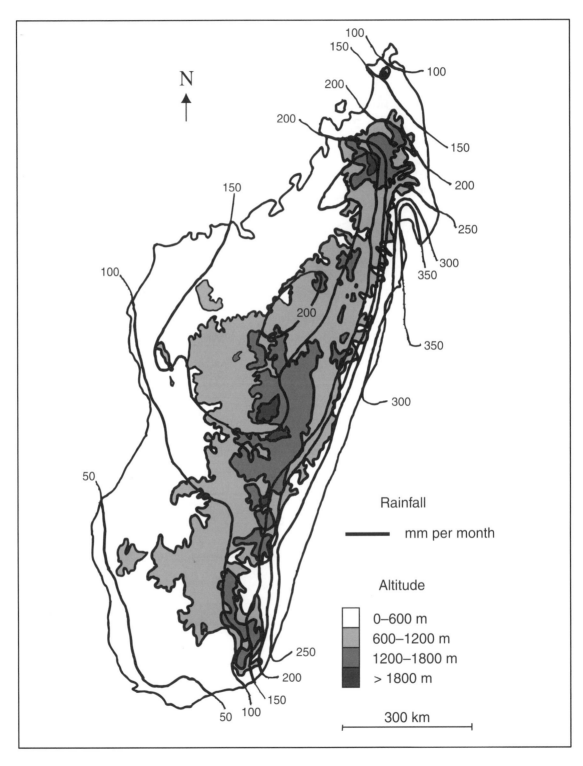

Figure 8.2   *Topography and rainfall patterns (mean monthly precipitation: Donque, 1972)*

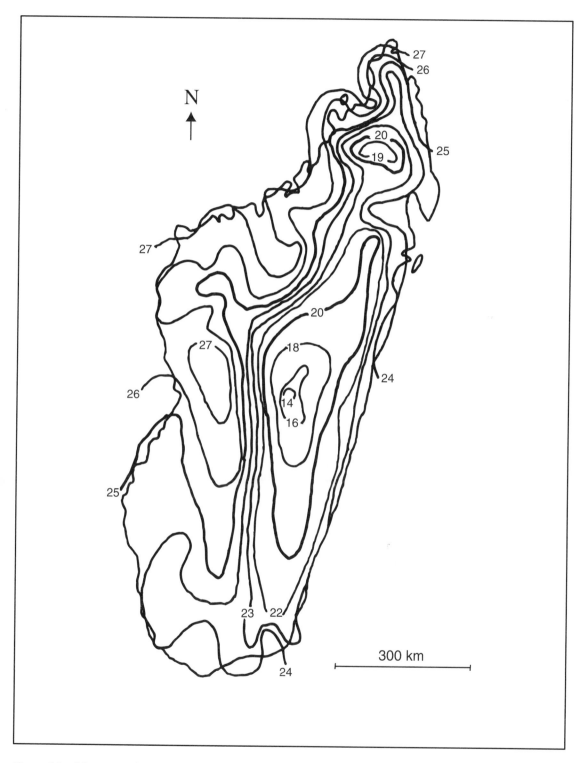

Figure 8.3   *Mean annual isotherms (°C) (Donque, 1972)*

the land area and forms a long strip 1200 km long and averaging 100 km wide. Five major rivers drain this steep escarpment: the Mananara, the Mangoro, Rianila, Maningory and Mananjary. Most are comparatively short with steep profiles although some are constrained by secondary hill chains which force them to flow parallel to the coast for long distances. The western slopes region is by far the largest, covering 60% of the island, and contains the five largest river basins. Aldegheri (1972) split the rivers of this region into two groups. In the first are the seven large rivers which drain the central plateaux. From north to south these are the Sofia, the Betsiboka–Mahajamba system (the largest basin in Madagascar at 63 450 km^2), the Mahavavy, the Manambolo, the Tsiribihina, the Mangoky (the longest river on the island at 821 km) and the Onilahy. The second group consists of small coastal rivers whose sources are on the edge of the high plateaux. All have basins of less than 8000 km^2. Lastly, the southern slopes region is divided into three parts: the Mandrare Basin in the east; the Manamboro, Menarandra and Linta basins in the extreme south; and the western Mahafely Plateau which has virtually no surface water.

Climatic differences between regions give rise to great variation in hydrology across the island (Aldegheri, 1972; IUCN/UNEP/WWF, 1987; Chaperon *et al.*, 1993). All rivers display high discharge, including violent spates, especially during the rainy season (November–April) and particularly during the January to March cyclone season. The east and north of the island receive sufficient rainfall all year round to maintain continuous river flows. River flows in the south are markedly seasonal and erratic. Few rivers in this region are perennial and many display rapid and extreme variations in discharge during the wet season. In the west, rivers can be separated into two groups: small coastal watercourses which run dry in their lower reaches between May and November; and larger perennial rivers which drain significant areas of the central plateau. Dry season flows in these larger rivers are maintained by the limited precipitation falling on the plateaux combined with the highly retentive nature of the mostly lateritic soils.

Aldegheri (1972) provides a synopsis of the island's hydrology on which the following summary is based. The island's rivers have been classified into 10 hydrological regimes based on rainfall patterns and relief (Figure 8.4). These are the northern or Montagne D'Ambre regime, the north-eastern regime, the Tsaratanana regime, the eastern coast regime, the high plateaux regime, the north-western regime, the southern-central regime, the western regime, the south-Sahelian regime, and a mixed regime that is exhibited by large rivers, the basins of which extend over more than one regime area.

The northern regime (1 on Figure 8.4) is poorly understood. High rainfall during the wet season (November to April) is largely absorbed by the fissured basalt lithology. Drainage basins are small (<100 km^2) and maximum specific discharges (i.e. m^3 s^{-1} km^{-2} catchment area) are no greater than 0.2 m^3 s^{-1} km^{-2}. The north-eastern regime (2) is distinguished on the basis of relief and rainfall patterns. No hydrological data exist for this region but less accentuated topography and lower rainfall (1200 mm as compared with 2500 mm) must give rise to lower specific discharges than those of rivers draining the northern regime.

The third regime encompasses the Tsaratanana Massif (3) which is characterized by high annual rainfall (>2500 mm) and steep topography. Low-water discharges (November) are 10–15 m^3 s^{-1} in the Sambirano (see Figure 8.1). Discharges in the wet season are typically 400–1000 m^3 s^{-1} in the same river but can approach 3000 m^3 s^{-1} during cyclones. The eastern coast regime area (4) is characterized by high annual rainfall (2500–3000 mm) and very steep slopes. Flow is maintained throughout the year and direct exposure to westward-tracking cyclones can give rise to high specific discharges (>2 m^3 s^{-1} km^{-2}).

The high plateaux regime (5) encompasses the upper catchments of Madagascar's largest rivers. Mean annual rainfall is 1200–1800 mm, mostly falling between November and April. Flows are sustained throughout the year by the high retention capacity of the region's lateritic soils. The absence of steep slopes gives rise to less violent responses to rainfall than those of the eastern coast regime (<1 m^3 s^{-1} km^{-2}). The north-western regime (6) is sharply seasonal. Annual rainfall is 1600–1800 mm, almost all of which falls between November and April. There is little flow in the dry season and discharges may decrease in the downstream direction. Rainy season discharges increase rapidly and can be extremely violent, particularly in small catchments (up to 30 m^3 s^{-1} km^{-2}).

The southern-central regime (7) is a transition regime between those of the eastern coast, high plateaux and west coast. Average annual rainfall is 800–1000 mm and low water occurs in October–November with rainy season discharges increasing rapidly from December onwards. Rainfall in the western regime area (8) is 500–800 mm yr^{-1}.

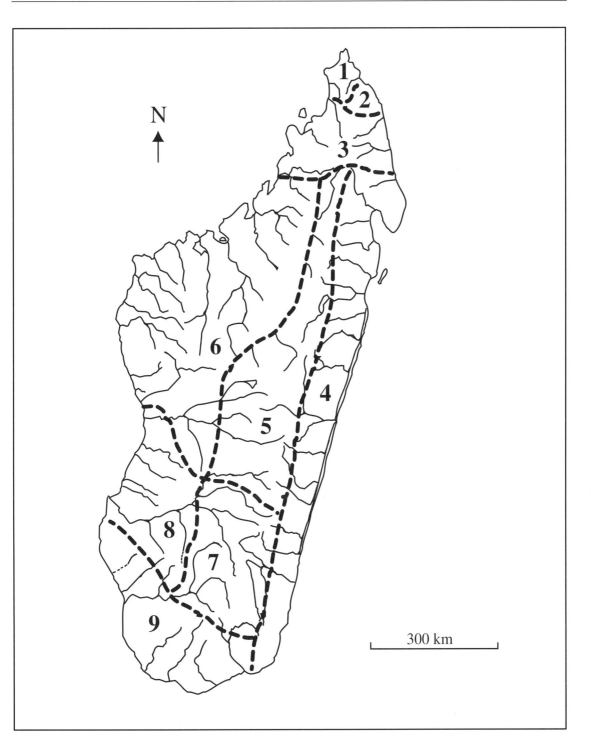

N

Figure 8.4 *Hydrological regimes of Madagascar (Aldegheri, 1972). 1, the northern or Montagne D'Ambre; 2, north-eastern; 3, Tsaratanana; 4, east coast; 5, high plateaux; 6, north-western; 7, south-central; 8, western; 9, south-Sahelian*

Seasonality is very marked and most small streams are dry from May to October. Wet season spates are extremely sudden but may last for only a few hours. Few data are available for this region. The south-Sahelian regime area (9) experiences 300–500 mm yr^{-1}. During the wet season, violent storms cause very sudden and short-lived spates (up to 1.5 m^3 s^{-1} km^{-2}) which give rise to very rapid variations in discharge. Dry season flows decrease downstream and few rivers are perennial. The Mahafely Plateau (situated between the Linta and Onilahy rivers; see Figure 8.1) is devoid of surface hydrology because of sandy soils and fissured limestone lithology.

The following rivers exhibit mixed regimes: Sofia, Betsiboka and Tsiribihina (regimes 5 and 6), Mangoky (5, 6, 7 and 8), Onilahy (7, 8 and 9), Mandrare (4, 7 and 9), Mananara (4 and 7), and Mangoro and Maningory (4 and 5; see Figure 8.1 for locations). Few data are available because of the large size of these rivers. Rainy season discharges can be extremely high and occur suddenly. For example, in 1964 rainfall associated with a tropical depression caused the Mangoky's discharge to increase from 265 m^3 s^{-1} to 4360 m^3 s^{-1} in 3 hours (Aldegheri, 1972). Even higher spates occur in exceptional rains, e.g. 14 000 m^3 s^{-1} in the Mangoky (14 year mean discharge 458 m^3 s^{-1}; Aldegheri, 1972). It is not known to what extent the hydrology of the island's rivers has been modified by the clearance of natural vegetation (see next section), but it is likely to have been considerable.

HUMAN IMPACTS ON PRE-COLONIZATION ENVIRONMENT

Madagascar was settled by palaeo-Indonesians who arrived on the island less than 2000 years ago, making the island one of the last major regions of the world to be colonized (Gade, 1996). Prior to colonization the island's central plateaux were covered with a complex and dynamic mosaic of montane rain forest, deciduous woodland, ericoid heath, and grassland (Burney, 1987, 1997; Lowry et al., 1997). Xerophyllous bush and spiny forest dominated the arid southwest. The eastern slopes were covered with dense montane and lowland rain forest (Guillamet, 1984). The island's natural vegetation assemblages have since been totally altered by burning for creation of agricultural fields and cattle pasture (Battistini and Verin, 1972). Although fire seems to have been a natural component of the central highlands' ecology (Burney, 1987, 1997; Raxworthy and Nussbaum, 1996), the extent and frequency of man-made fires have effectively arrested natural

succession and caused a complete shift to pyrogenic grassland. Almost 80% of the island is now covered by man-made prairie. This homogeneous assemblage is often described as 'pseudosteppe', as it consists of a handful of mostly exotic and highly fire-resistant grass species. Just two species of wiregrass, *Aristida rufescans* and *A. similis*, cover vast areas of the central highlands because of their tolerance to regular burning and poor soils (Gade, 1996; Lowry et al., 1997). The pattern of deforestation on the high plateaux continues to be driven by subsistence needs, such as cutting for cooking fuel and building materials, which place severe pressure on the remaining remnants of forest (Gade, 1996). The situation is exacerbated by the ritualistic burning of the pseudosteppe; at the end of each dry season approximately one-third of the island burns (Jolly and Jolly, 1984), a phenomenon clearly visible on Landsat images (Burney, 1987). Annual burning, combined with the loss of soil cover and the paucity of native tree species adapted to colonize cleared areas, means that deforestation is an essentially irreversible process (Koechlin, 1972).

Less than 11% of the island is now forested (Nelson and Horning, 1993; Faramalala, 1995). Over half of this area comprises the mountainous eastern rain forests. Much of the remainder makes up the arid southern spiny forests. The eastern rain forests are increasingly threatened with deforestation rates estimated at 111 000 ha (1.5%) yr^{-1} between 1950 and 1985; at this rate, all but the steepest slopes will be deforested by 2025 (Green and Sussman, 1990). Deforestation on the eastern slopes takes the form of slash-and-burn clearance for subsistence agriculture (*tavy* in Malagasy; see Figure 8.5). The rivers draining western deciduous forest and the spiny forest of the southwest are threatened by removal of the dense gallery forest which borders them. Sussman et al. (1994) concluded from satellite images taken in 1984 that at that time there were less than 4500 ha of gallery forest remaining in the south of the island. Gallery forest is typically cut down for creation of pasture and agricultural fields, provision of building materials for houses and cattle corrals, and charcoal production (Sussman et al., 1994; Gade, 1996). The damage to gallery forest is exacerbated by poor regeneration because of overgrazing by high densities of cattle and goats (Jolly and Jolly, 1984; Sussman et al., 1994).

These changes in vegetation cover have had catastrophic effects on the river systems of Madagascar. Removal of vegetation and destruction of humus layers by burning have given rise to some of the highest erosion rates in the world (up to 250 t ha^{-1} yr^{-1};

Figure 8.5 *Slash-and-burn subsistence agriculture (*tavy*) is the main driving force behind deforestation in eastern and north-western Madagascar. This is the Tomaro (a tributary of the Namorona) in the peripheral zone of Ranomafana National Park. (Photograph by Jonathan Benstead)*

Randrianarijaona, 1983; Wells and Andriamihaja, 1997; see Figure 8.6). Accelerated erosion causes widespread flooding and chronic sedimentation of rivers and irrigation canals downstream (Le Bourdiec, 1972; Olson, 1984; Gade, 1996). The resulting supersedimentation is clearly visible from space (Helfert and Wood, 1986), with deposition rates approaching 40×10^6 t yr^{-1} in some estuaries (Randrianarijaona, 1983). The ecological damage to river ecosystems caused by sedimentation is little understood as few data exist. Olson (1984) provides some anecdotal information relating the loss of shrimp and fish species to increased sediment loads. In addition to its direct impacts, environmental degradation seems to have facilitated the invasion of exotic fish species throughout many of the freshwater environments on the island. (The impact of introduced fish is dealt with in detail in the section 'Conservation status of the Malagasy ichthyofauna'.)

Finally, more recent and local human impacts have been identified but are also poorly understood. These include the effects of the introduced water hyacinth (*Eichhornia crassipes*), the impact of pesticide run-off from extensive cotton cultivation in the north-west of the island (Stiassny and Raminosoa, 1994), pollution of groundwater with insecticides in areas of intense cultivation (Richard and O'Connor, 1997), and contamination from artisanal mercury mining (P.V. Loiselle, unpublished data). Very little is known about industrial point-source or urban non-point-source pollution on the island. Given Madagascar's low level of industrialization, however, their impacts are likely to be insignificant compared with those of forest clearance for agriculture. The problems faced by river ecosystems in Madagascar are overwhelmingly those at the catchment scale (e.g. deforestation and introduction of exotic fish species). For this reason, and because so little information is available on more local impacts, a comparison of local *versus* catchment issues is not included here.

A framework for identifying main areas at risk was provided by Ganzhorn *et al.* (1997) in the results of the *Global Environment Facility Workshop* (see '*In situ* Conservation'). Essentially, all of the river systems considered to be of high biological importance are at risk. Of special concern are the Irodo River, the rivers draining the Tsaratanana Massif, the rivers of the Masoala Peninsula, all the rivers draining parcels of

Figure 8.6   *Accelerated erosion of lateritic soils has caused super-sedimentation in many rivers draining Madagascar's central plateaux. This photograph shows gully erosion beside the upper Mania River. (Photograph by Catherine Pringle)*

relatively intact eastern rain forest, and the rivers draining the region of Andohahela (Figure 8.1). These regions correspond to areas of high human pressure (Ganzhorn *et al.*, 1997). Threats to these river systems are, in order of importance, *tavy* subsistence agriculture, clearance for cash crops, and cutting of trees for timber and firewood.

Information for conservation

Research into the ecology of the island's rivers and streams was pioneered by researchers such as Renaud Paulian and Ferdinand Starmühlner during the period of French colonial rule (1896–1960). After the establishment of Madagascar's Second Republic in 1972 there was a hiatus in ecological research that lasted more than a decade. Few foreign scientists visited the island during this period and, due to lack of funds, little research was performed by Malagasy scientists. This situation has changed only in the last 10–15 years. Access for foreign researchers has become easier but Malagasy scientists are still chronically underfunded. Modern ecological research focuses on the ecology and conservation of the island's endemic forest species, particularly the highly threatened lemurs. Currently, freshwater research is limited to the systematics and conservation status of the native fish (e.g. Reinthal and Stiassny, 1991; Stiassny and Reinthal, 1992; Stiassny and Raminosoa, 1994; Stiassny and Harrison, 2000), the systematics of the aquatic insects (e.g. Gibon and Elouard, 1996; Lugo-Ortiz and McCafferty, 1997), and the effects of deforestation on stream communities and water quality (Riseng *et al.*, 1997; K.J. Riseng, unpublished data; J.P. Benstead, unpublished data). Freshwater ecological research has been hampered by lack of taxonomic knowledge and the paucity, location and status of museum collections that mostly date from the colonial period. For example, perhaps the largest collection of the island's freshwater insects is housed at the Muséum National d'Histoire Naturelle in Paris. In addition, many papers by early workers were written in French or German and published in obscure journals. Consequently, information on freshwater ecosystems remains relatively scarce and often difficult to find. However, this situation is likely to change in the near future (see below).

Satellite images and aerial photographs have become a rich and widely available source of information.

Figure 8.7   *The aquatic tenrec* Limnogale mergulus. *The conservation status of this endemic insectivore is unknown.* (*Photograph by Kevin H. Barnes*)

These data have been used by some workers, particularly those studying forest cover, deforestation rates, and burning of vegetation (see Green and Sussman, 1990; Nelson and Horning, 1993; Sussman *et al.*, 1994; Randriambelo *et al.*, 1998; Rasolofoharinoro *et al.*, 1998). However, to date the island's rivers have not been studied using these resources (but see Kramer *et al.*, 1997). Remotely sensed data may offer potentially useful tools for the study of erosion and sedimentation as these processes are clearly visible on aerial images (Helfert and Wood, 1986). In addition, the setting of conservation priorities among the island's watersheds (and subsequent monitoring) would be greatly facilitated by the use of remote sensing and GIS technology.

The scarcity of scientific information concerning Madagascar's river ecosystems represents a considerable challenge to conservation efforts. Beyond the separation of rivers into hydrographic regions and hydrological regimes, no evaluation or classification schemes have been applied to the island's river networks. Some hydrological data are available (Aldegheri, 1972; Chaperon *et al.*, 1993), but this information is patchy in its coverage. No physical information exists for large areas of the country. The

situation regarding the biota of the island's rivers is little better. What follows is intended as a summary of current knowledge. Although ecological understanding of the island's native fish fauna is 'rudimentary' (Stiassny and Raminosoa, 1994), the fish represent the best known taxa inhabiting Madagascar's rivers. Our summary is therefore biased towards this taxonomic group. However, the status of the fish serves to illustrate the threats faced by river biota throughout the island.

RIVER COMMUNITIES IN MADAGASCAR

Madagascar's rivers support some fascinating animals. One of the island's tenrecs (an endemic family of insectivores), *Limnogale mergulus*, is adapted to river and stream habitats (Figure 8.7). This small mammal (80–100 g) swims with webbed feet and feeds on tadpoles, aquatic insects and crustaceans (Malzy, 1965; Eisenberg and Gould, 1984; J.P. Benstead, unpublished data). It is completely nocturnal, emerging from its burrow at dusk to forage exclusively within the stream along ranges that can extend to 1500 m of stream channel in smaller (3–4 m wide) streams (J.P. Benstead, unpublished data). An important finding of recent research is that the aquatic tenrec is not an obligate

Table 8.1   *List of endemic mammals and birds that are associated with river and stream habitats in Madagascar*

Species		Notes
Mammals		
Limnogale mergulus	Aquatic tenrec	Semi-aquatic; found in small rivers in eastern region[a]
Fossa fossana	Striped civet	Sometimes feeds in streams (frogs, fish, crustaceans)[b,c]
Galidea elegans	Ring-tailed mongoose	As above[b,c]
Mungotictis decemlineata	Narrow-striped mongoose	As above[b]
Birds[d]		
Tachybaptus pelzelnii	Madagascar little grebe	Frequents lakes and running water
Ardea humbloti	Humblot's heron	Found in all aquatic habitats
Lophotibis cristata	Madagascar crested ibis	Prefers forested habitats, near small streams
Anas melleri	Meller's duck	Frequents lakes, rivers and streams
Haliaeetus vociferoides	Madagascar fish eagle	Rare; found near large rivers, mangroves, lakes[e]
Corythornis vintsioides	Malagasy kingfisher	Frequents all aquatic habitats
Acrocephalus newtoni	Madagascar swamp-warbler	Found in aquatic vegetation near lakes and rivers

[a] Malzy, 1965; Eisenberg and Gould, 1984
[b] Haltenorth and Diller, 1980
[c] Glaw and Vences, 1994
[d] Langrand, 1990
[e] Rabarisoa *et al.*, 1997

forest species; as long as the community that makes up its prey is not affected by sedimentation, it appears able to maintain healthy populations in streams draining deforested catchments (e.g. peripheral zone of Ranomafana National Park; J.P. Benstead, personal observation). Unlike some of the terrestrial tenrec species, *Limnogale* is not hunted for food. However, it sometimes drowns in the traditional bamboo crayfish traps used by villagers. Three other mammals, all endemic viverrids, exhibit a partial reliance on river and stream habitats (see Table 8.1).

Madagascar's avifauna is not particularly diverse (256 species) but, as in other groups, many species are endemic (105 species; mostly restricted to forested habitats). Approximately 30 species of birds are to be found in or near river habitats. Seven of these are endemic (see Table 8.1) though none is restricted to riverine environments (Langrand, 1990). Of special concern is the Madagascar fish eagle *Haliaeetus vociferoides* which is restricted to western and north-western Madagascar. This species may have been reduced to as few as 100 breeding pairs and is considered to be one of the world's rarest birds of prey (Langrand, 1990; Rabarisoa *et al.*, 1997).

At least three species of large, aquatic skinks inhabit streams in the east and northwest of the island. These are *Amphiglossus astrolabi*, *A. waterloti* and *A. reticulatus* (Millot, 1951; Blanc, 1984; Glaw and Vences, 1994). *Amphiglossus* is one of the least understood genera of Malagasy lizards (R.A. Nussbaum, University of Michigan, personal communication) and there is likely to be at least one

other undescribed aquatic member. These aquatic skinks may reach large sizes (up to 0.5 m) but their ecology and conservation status are not known. One species, *Amphiglossus astrolabi*, can be found in the streams draining the deforested peripheral zone of Ranomafana National Park (J.P. Benstead, personal observation), which suggests that at least this species is tolerant of some degree of habitat degradation. However, like the aquatic tenrec, this species sometimes drowns in traps set for crayfish and eels.

Four species of freshwater turtle are found in suitable habitat, typically in lowland rivers, backwaters and floodplain lakes. These are the endemic *Erymnochelys madagascariensis*, and three species also found in Africa, *Pelusios castanoides*, *Pelusios subniger* and *Pelomedusa subrufa* (Blanc, 1984; Glaw and Vences, 1994). All are hunted for food, including *Erymnochelys* which is fully protected by Malagasy law, and appear to be overexploited as a resource (IUCN/UNEP/WWF, 1987; Kuchling, 1988; Kuchling and Mittermeier, 1993). Lastly, with respect to reptiles, the Nile crocodile *Crocodylus niloticus* (sometimes regarded as a distinct sub-species *C. n. madagascariensis*) is still found in some rivers, despite hunting pressure. Large numbers exist in the northwest and in the subterranean rivers of the northern limestone region. Fortunately, commercial farming has recently reduced pressure on these wild populations (Glaw and Vences, 1994).

The streams and rivers of the island's forests are also the larval and adult habitat of many species of frog (Glaw and Vences, 1994). These include most species in the large genus *Boophis* (ca 40 spp) and some of the

Mantidactylus species, many of which, being tree frogs, require forested habitat as adults.

The island's undisturbed rivers support a rich community of endemic decapod crustaceans. Six species of parastacid crayfish (*Astacoides* spp) inhabit the rivers of the eastern rain forest (Hobbs, 1987). These species are currently threatened by deforestation and over-harvesting for food (Dixon, 2000). At least 20 species of atyid shrimp (*Caridina* spp), three species of palaemonid shrimp (*Macrobrachium* spp) and nine species of potamonid crab are also found in the island's rivers (Bott, 1965; Holthuis, 1965; IUCN/UNEP/ WWF, 1987). All are harvested for food.

The lotic insect fauna is poorly known but endemism is again very high at the generic and species level. Some groups are better known than others. For example, about 144 species of Odonata have been described from Madagascar (Schmidt, 1951; Fraser, 1956), but, until recently, there were descriptions for only 24 Ephemeroptera, 22 Trichoptera and 12 simuliids (Gibon *et al.*, 1996). Over 90% of species in major aquatic insect orders (e.g. Trichoptera) may be undescribed. The freshwater invertebrates (with the aquatic macrophytes and fish) are the subject of an ambitious, long-term project (Biodiversity and Biotypology of Malagasy Continental Waters) headed by Jean-Marc Elouard and François Gibon of the Institut Française de Recherche Scientifique pour le Développement en Coopération, the Centre National de Recherches sur l'Environnement (CNRE) and the Laboratoire de Recherche sur les Systèmes Aquatiques et leur Environnement (LRSAE) in Madagascar and France. To date, more than 700 streams and rivers, mostly in the south of the country, have been sampled. Sampling in the north is in progress. Some results of this research have already been published (e.g. Gibon and Elouard, 1996; Gibon *et al.*, 1996; Oliarinony and Elouard, 1997; Elouard *et al.*, 1998) with many more descriptions in progress. As a result of this research effort, it is becoming clear that Madagascar's streams and rivers support a highly diverse insect fauna, e.g. approximately 600 species of Trichoptera and 180 species of Ephemeroptera (J.-M. Elouard, LRSAE, personal communication). Presumably the island's long isolation and complex hydrography have given rise to a high level of speciation within these groups. Other patterns are evident: forest and open savanna rivers support different communities; many groups exhibit elevational zonation; savanna river species generally have wider distributions than those of humid forest environments; and groups presenting the greatest level of micro-endemism (e.g. the *Paulianodes* caddisflies,

Megaloptera, Plecoptera) are restricted to rivers draining primary and high-altitude forest (Gibon *et al.*, 1996). These patterns highlight the importance of maintaining original forest habitats in the conservation of the island's river communities.

All data from the Biodiversity and Biotypology of Malagasy Continental Waters project is managed by sophisticated database management software and will soon be made available on CD-ROM for use by other researchers. A definitive volume on the freshwater biodiversity of Madagascar is also planned.

Conservation status of the Malagasy ichthyofauna

In addition to their extraordinary level of endemism and their status as the basal representatives of their respective clades, the freshwater fish of Madagascar are also characterized by an extreme level of vulnerability. Four of the 64 freshwater fish endemic to Madagascar have not been collected within the last quarter of a century and are feared extinct. If the criteria of the United States Endangered Species Act are applied to the remainder, 12 species must be considered critically endangered, and 11 threatened. We believe that a further 15 species should be considered vulnerable, while insufficient data preclude accurate determination of the status of 16 of the remaining species (Table 8.2). The freshwater fish of Madagascar are therefore considered the island's most endangered vertebrates (Stiassny and Raminosoa, 1994).

Three factors have been proposed to account for the current status of the Malagasy ichthyofauna. The first of these is the degradation of aquatic habitats that follows deforestation. The removal of riparian forest has two immediate consequences. The first is to increase significantly insolation upon a given body of water. The second is to reduce the quantity of canopy-derived plant matter and terrestrial insects reaching its surface. Higher water temperatures and accelerated algal growth are the most immediate effects of increased insolation. Combined with reductions in inputs of allochthonous food items, these changes favour physiologically plastic species able to exploit increased *in situ* productivity and work to the disadvantage of those that depend upon infall of food items from the forest canopy. Such a change appears to have taken place in the streams draining Perinet–Andasibe Reserve where Cyclone Giralda significantly thinned out riparian forest cover. According to park staff, numbers of the green swordtail (*Xiphophorus helleri*), an exotic poeciliid that feeds upon periphyton, increased markedly after

Table 8.2   *Conservation status of the endemic freshwater fish of Madagascar*
Footnotes refer to information sources for undescribed species. Code: X, probably extinct; E, critically endangered; T, threatened; V, vulnerable; S, secure; ?, conservation status unknown

Family: Clupeidae – herrings
S   *Sauvagella madagascarensis* (Sauvage 1883)
?   *Sauvagella* sp./Ambomboa River[a]

Family: Ariidae – sea catfish
S   *Arius madagascarensis* Vaillant 1894

Family: Anchariidae – Malagasy catfish
?   *Ancharius brevibarbus* Boulenger 1911
T   *Ancharius fuscus* Steindachner 1881

Family: Cyprinodontidae – killifish
V   *Pachypanchax omalonotus* (Dumeril 1861)
V   *Pachypanchax* sp./Anjingo River[b]
V   *Pachypanchax* sp./Betsiboka River[c]
T   *Pachypanchax sakaramyi* (Holly 1928)
X   *Pantanodon madagascarensis* (Arnoult 1963)
E   *Pantanodon* sp. nov./Manombo[d]

Family: Atherinidae – silversides
?   *Teramulus waterloti* (Pellegrin 1932)
?   *Teramulus keineri* Smith 1965

Family: Bedotiidae – Madagascar rainbowfish
V   *Bedotia longianalis* Pellegrin 1914
V   *Bedotia madagascariensis* Regan 1903
V   *Bedotia* sp. nov./Lazana River[e]
V   *Bedotia* sp. nov./Nosivolo River[f]
V   *Bedotia* sp. nov./Namorona River[g]
V   *Bedotia* sp. nov./Masoala[d]
V   *Bedotia marojejy* sp. nov./Marojejy[h]
E   *Rheocles alaotrensis* (Pellegrin 1904)
?   *Rheocles lateralis* Stiassny & Reinthal 1992
?   *Rheocles pellegrini* (Nichols & Lamonte 1931)
E   *Rheocles sikorae* (Sauvage 1891)
E   *Rheocles wrightae* Stiassny 1990
?   *Rheocles* sp. nov./Masoala[i]
?   *Rheocles* sp./Ambomboa River[j]
?   *Rheocles* sp. nov./Rianila River[k]

Family: Terapontidae – target perch
V   *Mesopristes elongatus* (Guichenot 1866)

Family: Ambassidae – glassfish
?   *Ambassis fontoynoti* Pellegrin 1932

Family: Cichlidae – cichlids
T   *Paratilapia bleekeri* Sauvage 1882
T   *Paratilapia polleni* Bleeker 1868
X   *Paratilapia* sp. nov./Lac Ihotry[l]
T   *Ptychochromis oligacanthus* Bleeker 1868
V   *Ptychochromis nossibeensis* Bleeker 1868
S   *Ptychochromis* sp. nov./Saroy mainty[m]
S   *Ptychochromis* sp. nov./Saroy mavou[n]
X   *Ptychochromis* sp. nov./Kotro[o]
E   *Ptychochromoides betsileanus* (Boulenger 1899)
E   *Ptychochromoides* sp./Fia Potsy[p]
T   *Ptychochromoides katria* Reinthal and Stiassny 1997
T   *Paretroplus dami* Bleeker 1878
T   *Paretroplus keineri* Arnoult 1960
E   *Paretroplus maculatus* Keiner and Mauge 1966
E   *Paretroplus menarambo* Allegayer 1996
E   *Paretroplus petiti* Pellegrin 1929
S   *Paretroplus polyactis* Bleeker 1878
E   *Paretroplus* sp. nov./Lac Andropongy[q]
V   *Paretroplus* sp. nov./Lamena[r]
E   *Paretroplus* sp. nov./Maevatanana[s]
V   *Paretroplus* sp. nov./Tsimoly[s]
T   *Oxylapia polli* Keiner & Mauge 1966

Family: Eleotridae – sleeper gobies
?   *Eleotris pellegrini* Mauge 1984
?   *Eleotris vomerodentata* Mauge 1984
S   *Hypseleotris tohizonae* (Steindachner 1881)
V   *Ophiocara macrolepidota* (Bloch 1792)
E   *Ratsirakia legendrei* (Pellegrin 1919)
X   *Ratsirakia* sp. nov./Tohobe[t]
T   *Typhleotris madagascarensis* Petit 1933
T   *Typhleotris pauliani* Arnoult 1959

Family: Gobiidae – true gobies
?   *Acentrogobius therezieni* Kiener 1963
?   *Chonophorus macrorhynchus* (Bleeker 1867)
?   *Glossogobius ankaranensis* Bannister 1994

Family: Trichinotidae
?   *Gobitichinotus arnoulti* Kiener 1963

[a] P. de Rham (personal communication); [b] J. Sparks and P.N. Reinthal (in preparation); [c] Schaller (1991); [d] P.N. Reinthal and J. Sparks (in preparation); [e] P.V. Loiselle and R. Haeffner (unpublished data); [f] Lucanus (1996); [g] Reinthal and Stiassny (1991); [h] Stiassny and Harrison (2000); [i] P.N. Reinthal, J. Sparks and K.J. Riseng (in preparation); [j] P.N. Reinthal, J. Sparks and K.J. Riseng (unpublished data); [k] P.V. Loiselle, R. Haeffner and J. Robinson (unpublished data); [l] Loiselle and Stiassny (1993); [m] Nourissat (1992); [n] O. Lucanus (personal communication); [o] Kiener (1963); [p] De Rham (1996); [q] Nourissat (1993); [r] Nourissat (1992); [s] J.-C. Nourissat (personal communication); [t] Kiener (1963)

the passage of Giralda. Abundance of the endemic bedotiid *Rheocles alaotrensis*, which feeds largely upon stranded terrestrial insects, decreased.

Such a pattern of local decline, frequently dramatic in the near term, need not always be irreversible. Substantial populations of *Bedotia madagascariensis* and *B. longianalis* persist in lowland streams of the eastern versant of Madagascar whose watersheds have

long been devoid of primary forest. The colonization of the riparian zones of such streams by such exotics as torch ginger, banana, mango, and Chinese bamboo apparently provides sufficient shade and an adequate source of terrestrial insect prey to support these endemic atherinoids. A similar situation exists on the island of Nosy Be, whose cover of native semideciduous forest has been almost completely replaced

Figure 8.8 *Madagascar's endemic fish are its most threatened taxa. This is* Pachypanchax omalonotus *from the Djabala River on the offshore island of Nosy Be. (Photograph by Paul Loiselle)*

by an assemblage of exotic plants. Notwithstanding the degraded nature of its riparian flora, its streams still support robust populations of the endemic cyprinodont *Pachypanchax omalonotus*, another species whose diet includes terrestrial insects (Figure 8.8).

Other effects of deforestation are less readily reversed. In tropical regions characterized by a sharply seasonal pattern of rainfall, such as the western versant of Madagascar, the presence or absence of forest cover in the upper reaches of drainages means the difference between rivers that flow perennially and those that either flow on an intermittent basis or, in the case of small streams, disappear altogether during the dry season. Small streams often serve as refugia for species whose small adult size puts them at risk of predation in the main channel habitat of larger rivers. Thus, while emigration downstream may appear to offer some hope of dry season survival for mobile species, the reality typically proves otherwise. This pattern of habitat loss is particularly evident on the slopes of the Massif d'Ambre, to the detriment of *Pachypanchax sakaramyi*, a cyprinodont endemic to the streams draining its eastern slopes (Loiselle and Ferdenzi, 1997).

Greatly accelerated erosion is the third long-term consequence of deforestation. Subsequent sedimentation negatively affects river communities in several ways (Waters, 1995). First, sediment imposes a direct physiological burden upon fish (and other branchially respiring organisms) by fouling gill surfaces. Second, siltation poses an obvious threat to species that deposit demersal eggs. Suitable substrata may be so burdened with sediment that they can no longer serve as spawning grounds. Alternatively, deposition of silt after demersally spawning fish have bred may suffocate developing zygotes, resulting in large-scale reproductive failure. There is considerable evidence that such loss of spawning sites played a role in the extirpation of Madagascar's largest endemic cichlid, *Ptychochromoides betsileanus* from Lake Itasy (Kiener, 1959). Finally, through its effects on benthic invertebrate assemblages, sedimentation profoundly disturbs aquatic food webs, with predictable results upon organisms at higher trophic levels. It appears probable that this process accounts in part for the continuing decline of the seven *Paretroplus* species native to north-western Madagascar (Kiener, 1963; Nourissat, 1992). Selective feeders on aquatic invertebrates (Kiener and Thérezien, 1963), these large cichlids are particularly vulnerable to any disruption of the benthos.

The second factor that has negatively affected the freshwater fish of Madagascar is overfishing. The failure of recent reviews (Reinthal and Stiassny, 1991; Stiassny and Raminosoa, 1994) to take this most direct of human impacts into consideration is in all probability explained by the fact that representatives of the family Cichlidae, the most commercially important of the island's endemic fish, are at present so rare that they can no longer support dedicated fisheries. However, it is quite clear that this was not formerly the case (Grandidier, 1886; Louvel, 1930; Petit, 1930; Kiener, 1963). While overexploitation is only specifically mentioned as playing a role in the decline of *Ptychochromoides betsileanus* in Lake Itasy and the rivers of the southern highlands (Kiener, 1959), the problem of declining yields from traditional fisheries was clearly recognized in the 1950s (Kiener, 1963). Indeed, the introduction of tilapias to Madagascar was largely undertaken in response to this situation.

While environmental degradation was certainly implicated in this decline, the well-documented collapse of Lake Victoria's aboriginal tilapia-based fishery between 1910 and 1950 (Khudongania and Chitamwebwa, 1995) strongly suggests that overfishing also played a role. The operative factor in Lake Victoria was the introduction of beach seines and gill nets into a trap-based artisanal fishery. Combined with the demands of an ever growing market and a total lack of effective regulation, the end result was the commercial extinction of Lake Victoria's two endemic tilapias by the early 1950s. Data presented by Kiener (1963) strongly suggest that a similar process of gear substitution also took place in Madagascar between the turn of the century and the early 1950s. Given the increased demand for fish made by an expanding population and the extraordinary logistical difficulties involved in enforcing any sort of environmental regulations, a similar outcome in Madagascar would seem to have been inevitable.

Interaction with exotic species is the third factor that threatens the survival of Madagascar's native fish (Table 8.3). Twenty-one of the 81 freshwater fish known to occur in Madagascar, some 26% of the total ichthyofauna, are exotics (Stiassny and Raminosoa, 1994). The impact of this assemblage of aquaculture and ornamental species upon aquatic ecosystems in Madagascar has been profound. Naturalized exotics have completely replaced native fish in the central highlands of Madagascar (Reinthal and Stiassny, 1991).

Table 8.3 *Representation of native Malagasy and exotic fish species at localities sampled from 1989 to 1994*

Region	Number of localities with		
	Natives only	Natives and exotics	Exotics only
East Coast	10	26	8
Plateau	0	0	9
West Coast	2	20	21
Total	12	46	38

Except where physical or physiological barriers prevent their dispersal, exotic fish are widespread in other parts of the island (Stiassny and Raminosoa, 1994; Loiselle and Ferdenzi, 1997), where they typically greatly outnumber native species (Nourissat, 1992; Loiselle, 1995, 1996).

While the dominant position of exotic species in the fresh waters of Madagascar is beyond doubt, the nature of the interactions between introduced and native species is not always obvious. The replacement of native cichlids of the genera *Paratilapia* and *Ptychochromoides* by the largemouth bass in Lake Itasy is well documented (Kiener, 1963; Moreau, 1979; Reinthal and Stiassny, 1991). However, *Ptychochromoides betsileanus* was in decline well before the introduction of the bass (Kiener, 1959) and the extent to which interactions between the two species contributed to the extirpation of the native species from its type locality is unclear. Direct trophic competition between these two top carnivores may well have been a factor, as might predation on smaller size classes of *Ptychochromoides betsileanus* by adult bass (see Contreras and Escalante, 1984; Williams and Deacon, 1986; Minckley *et al.*, 1991, for examples of impacts of bass introductions on endemic fish). Both processes also seem to have played a role in the dramatic decline reported by local residents (Loiselle, 1994, 1995) in the numbers of *Paratilapia bleekeri* in the lower Betsiboka Basin following the establishment of the Asian snakehead, *Ophiocephalus striatus*, in 1975. However, there is no way retrospectively to ascertain the relative importance of each factor.

Six tilapia species were introduced between 1950 and 1956 for aquacultural purposes by the Service des Eaux et Forêts and the Service des Recherches Agricoles. The relationship between the establishment of these herbivorous and microphagous tilapias of the genera *Tilapia* and *Oreochromis* and the virtual disappearance of five of the seven *damba* (*Paretroplus*) species native to north-western Madagascar is even more obscure.

The nature of their buccal and pharyngeal dentition (Kiener and Maugé, 1966), taken with what is known of their trophic ecology (Kiener and Thérezien, 1963), suggests that *dambas* feed primarily upon macroinvertebrates picked from the substratum. The *Tilapia* species naturalized in this part of Madagascar are omnivores with a strong preference for aquatic macrophytes, while *Oreochromis* feed upon algae and detritus. While trophic competition between juveniles of *Paretroplus* species and those of either genus of *Tilapia* may well occur, any such interactions between adults appear unlikely. Competition for another limiting resource – spawning sites – is effectively precluded by the very different reproductive modalities of these cichlids. *Oreochromis* are arena-breeding maternal mouthbrooders (Trewavas, 1983), whose spawning site preferences do not overlap with those of biparentally custodial substratum spawners like *Paretroplus* or *Tilapia*. While these two genera have the same reproductive modality, they differ markedly in their choice of spawning sites. *Paretroplus* deposit their eggs upon vertical surfaces, such as plant stems or pieces of waterlogged wood (Kiener and Thérezien, 1963; Catala, 1977) or in caves (Nourissat, 1992; De Rham, 1995). In contrast, *Tilapia rendalli* and *T. zillii* prefer flat surfaces as oviposition sites and typically place their nests in the open (Loiselle, 1977).

A similar problem is posed by the situation of the endemic bedotiid genera *Bedotia* and *Rheocles* and that of the naturalized poeciliids *Xiphophorus helleri* and *X. maculatus*. The two exotic species have become virtually ubiquitous inhabitants of the rivers draining the eastern versant of Madagascar, where they typically outnumber the indigenous bedotiids. While adults of both bedotiids and poeciliids are quite capable of opportunistically preying upon one another's young, the three genera in question are not specialized piscivores. As previously noted, the diets of adults do not show extensive overlap. Neither do their reproductive modalities: bedotiids are oviparous, while poeciliids are viviparous. It thus seems unlikely that predation, trophic competition or spawning site competition accounts for this state of affairs.

It is worth noting that the exotic fish most successfully established in Madagascar are characterized by a shared suite of characteristics. All are physiologically plastic species. All are characterized by elevated fecundity and most have reproductive features, such as viviparity or parental care, that maximize fry survival. Relative to Malagasy fish, they are also far less seasonal in their reproductive patterns. Female poeciliids deliver a brood of young every 28 to 32 days, while female *Oreochromis* can spawn on a monthly cycle if adequately nourished. *Tilapia* species and *Ophiocephalus* can also produce several broods annually when food is abundant. This stands in contrast to the markedly seasonal reproductive pattern reported for some bedotiids (Kiener, 1963) and all Malagasy cichlids (Kiener, 1963; Kiener and Thérezien, 1963; Catala, 1977; Loiselle, 1996).

Apart from the few instances in which habitat degradation has resulted in the disappearance of aquatic habitat, no single factor can account for the inexorable decline of the indigenous Malagasy ichthyofauna. It is rather the interaction of two or more of the three factors previously elaborated that account for the disappearance of native fish and their replacement by naturalized exotics. What has occurred is in essence a sequential process, initiated by habitat degradation. This places indigenous fish under stress, resulting in declines in population size and range contraction. In many regions, these negative trends were exacerbated by the impact of introduced predators and, in the case of the native cichlids and possibly the larger *Rheocles* species, by heavy fishing pressure. The final stage in the process, the replacement of native by exotic species, is in the majority of cases due less to any direct interaction between representatives of the two faunas than to the fact that the exotics are simply better adapted to severely modified habitats and, by virtue of their superior reproductive potential, can more rapidly replace losses caused by predation and natural mortality.

The Watershed Project in Ranomafana National Park

While there is a paucity of information on aquatic organisms, there is even less known about the chemical and physical characteristics of the rivers and streams (see Aldegheri, 1972, large river discharge; Weninger, 1985, lake chemistry). In June 1994, the first comprehensive water monitoring programme in Madagascar was established at Ranomafana National Park. The Ranomafana Watershed Project was designed to collect long-term data in order to compare the chemical, physical and biological attributes of rivers draining disturbed *versus* non-disturbed forest catchments (Riseng *et al.*, 1997). The goals of this project are (1) to increase the understanding of water chemistry in Malagasy streams by identifying the patterns and controls of chemical variation, (2) to study the effects of disturbance on these streams, and (3) to relate water chemistry and disturbance to patterns of biological

Figure 8.9 *The Namorona River near Vohiparara in Ranomafana National Park. This high-gradient river is typical of those draining remaining forested areas on the island's eastern slopes. (Photograph by Jonathan Benstead)*

diversity. To study the effect of watershed disturbance, two sites on the Menarano (a tributary of the Namorona; Figure 8.1) are being monitored: one upstream site that drains pristine forest and one disturbed downstream site. These sites are both on granite bedrock at approximately 600 m elevation. In addition, the Namorona, the largest river that drains the park (Figure 8.9) has been monitored. All sites are located in the eastern slopes region and drain into the Indian Ocean approximately 90 km to the east. The sites have been surveyed for catchment vegetation and are sampled monthly or twice monthly for pH, temperature, oxygen, alkalinity, plankton,

phosphorus, carbon and nitrogen. Quantitative fish sampling was conducted in 1994 and 1996.

Preliminary results show that all sites have extremely low alkalinity and nutrient levels. Strong seasonal trends are apparent. Seasonal patterns in water chemistry follow seasonal changes in water temperature, and are controlled in part by the distinct rainy season which has a dilution effect on nutrient levels. At the Menarano sites, the average water temperature is significantly greater at the disturbed site than at the forested site. The increase in water temperature at the disturbed site is due to 80% less canopy cover than the forested site (the sites have comparable widths). Oxygen concentration ranges between 7.4 and 13.0 mg L^{-1} and is significantly lower at the disturbed site than the forested site. The change in canopy cover is a direct result of deforestation. After the trees are burned or cleared, the area is used for crops, such as rice, coffee, banana or cassava depending on the productivity of the soil. Due to the presence of these crops, the percentage cover of bank vegetation is not greatly different at the two sites. However, the vegetation height and species composition have changed drastically. The effects of these changes on water quality will remain unclear until this project is completed. It is hoped that the final results will guide the design of conservation programmes that minimize the negative impacts of deforestation on aquatic communities (K.J. Riseng, unpublished data).

PUBLIC AWARENESS AND ENVIRONMENTAL EDUCATION

The island's 12 million residents are 85% rural; most are subsistence farmers. Access to information is very difficult for all levels of Malagasy society. As a consequence, many Malagasy are unaware of their country's unique biodiversity. However, many people have an innate understanding of their local area's environmental problems (Kull, 1996). For example, many subsistence farmers appreciate the causal link between deforestation and subsequent erosion and flooding, but cannot afford to change their agricultural practices without assistance. Lack of awareness is not restricted to rural areas. The urban population, which holds all the political and economic power, has little understanding of rural people's problems and needs. In addition, Madagascar's education system historically did not contain an environmental component. There is still little teaching of environmental sciences at university level. Environmental education is an integral component of the National Environmental Action Plan (see below) and has been promoted by conservation organizations for many years. The programme established by the World Wide Fund for Nature (WWF) once reached most school districts (Kull, 1996), but has been reduced in scope in recent years. Although environmental education is a compulsory component of curricula, teachers typically lack the resources and training to teach this subject effectively (S. Sheldon, United States Peace Corps, personal communication).

Recently USAID has funded a linkage programme to create educational exchange between Malagasy and US universities (Université d'Antananarivo, and Université de Fianarantsoa; State University of New York at Stony Brook, Eastern Michigan University, and Duke University). The linkage programme provides opportunities for students and staff from the associated universities to participate in an annual field course at Ranomafana National Park in the southeast of the island. The field course comprises three weeks of intensive classwork followed by the design and implementation of individual projects. The classwork covers topics ranging from primatology to ichthyology. This programme has been extremely successful in promoting student enthusiasm, providing the Malagasy with the means and competitive ability to earn graduate degrees at US institutions, and increasing collaboration between Malagasy and US scientists. A related watershed project designed by P.N. Reinthal and K.J. Riseng (and funded by USAID and the US National Science Foundation) was described in more detail above.

The framework for conservation

ROLE OF NATIONAL AND INTERNATIONAL NGOs

Madagascar's economy is a major constraint to conservation efforts. The country faces a debt of US$2.5 billion and average *per capita* income is only US$210 yr^{-1} (EIU, 1994). In 1987, the total domestic budget for the entire protected areas system was less than US$1000 (Kull, 1996). Consequently, the country is increasingly reliant on foreign institutions to fund the protection of its biological heritage. The global importance of Madagascar's biodiversity is reflected in the high number of outside organizations involved in conservation and development work. Kull (1996) examined the extraordinary boom in international conservation that has taken place since the early 1980s. For example, spending by WWF increased more than 10 times between 1983 and 1993. This boom has involved development of a National Environmental Action Plan (NEAP; see below), establishment of at least 14 integrated conservation and development projects, and coordination of several large debt-for-nature swaps. The timing of this drastic increase in conservation activity can be attributed to a number of interacting factors, including the combination of 'megadiversity' and severe environmental degradation, the growth of the global environmental movement, and Madagascar's political and economic situation (Kull,

1996; Gezon, 1998). By the late 1980s, annual average bilateral aid had reached US$331 million, or 14% of Madagascar's GNP. A large proportion of this foreign funding was pledged to the World Bank-sponsored NEAP. The NEAP is primarily financed by the United Nations Development Programme, the United Nations Educational, Scientific, and Cultural Organization, USAID, the World Bank, the African Development Bank, and the WWF, as well as French (CIRAD), German (GTZ and KfW), Swiss (Coopération Suisse) and Norwegian (NORAD) overseas development agencies (Kull, 1996).

In the last decade, many small NGOs have sprung up within Madagascar. Most are involved with development efforts. Some Malagasy conservation organizations exist, however, including the Association pour la Sauvegarde de l'Environnement (Kull, 1996). Religious organizations involved in conservation include the Agricultural Development Department of the Malagasy Lutheran church, SAF-FJKM (of the Eglise de Jesus Christ à Madagascar), and Catholic Relief Services (Hough, 1994; Kull, 1996).

We know of no NGO projects that explicitly involve river conservation. However, many NGOs have contributed towards river conservation indirectly through their involvement with various protected areas. For example, the Institute for the Conservation of Tropical Environments, based at the State University of New York at Stony Brook, was instrumental in the establishment of Ranomafana National Park. This park protects some of the catchment of the Namorona River. Similarly, WWF has long been involved with the region now protected by Andohahela National Park, which includes the upper catchments of several south-eastern rivers.

STATUTORY FRAMEWORK AND POLICING OF ENVIRONMENTAL LEGISLATION

Currently, no laws exist that relate explicitly to river conservation. However, Madagascar has a long history of environmental legislation. Laws banning the cutting of live wood were introduced by King Andrianampoinimerina (1782–1810) and the 1881 Code of 305 Articles prohibited the burning of forest (Andriamampianina, 1984; Kull, 1996). The first reserves were established in 1927. In addition, Madagascar was one of the first developing countries to implement a National Environmental Action Plan (NEAP). The process began with the National Strategy for Conservation and Development which was drafted in 1984. This was followed by a Conference on

Conservation for Sustainable Development in 1985 which appointed a special commission to carry out the goals of the strategy. During 1987–88, the more detailed NEAP was drafted by the government in collaboration with donor agencies (Gezon, 1998). The NEAP was passed into law as the Charte de l'Environnement in 1990, began in 1991, and is to be carried out in three 5-year phases. It is hoped that the 15-year NEAP 'will help Madagascar to put in place the legal and institutional framework and the skills and tools needed to manage its environmental heritage' (World Bank, 1990). Its primary goals are: protection and management of biodiversity through establishment of a network of protected areas (adding 400 000 ha and increasing the number of reserves from 36 to 50); soil conservation and re-afforestation; mapping; establishment of reserve boundaries, and improvement of land security through titling; environmental training, education, and awareness; ecological research; and strengthening of environmental institutions, protective procedures and databases (World Resources Institute – WRI, 1994). Improving watershed protection and strengthening the environmental legislative framework are stated goals of the NEAP. The NEAP is now in its second 5-year phase (PE2) in which emphasis has shifted from building institutional capacity to assuring Madagascar's ability to continue conservation activities in the long term (Gezon, 1998). Under PE2, an additional US$155 million in donor support has been made available for environmental conservation (Hannah *et al.*, 1998).

Development and coordination of the NEAP is the responsibility of the National Office of the Environment (French acronym: ONE), a small public agency attached to the Ministry of the Environment. Responsibilities for managing Madagascar's reserves and parks have recently been transferred from the Department of Water and Forests (French acronym: DEF) to a newly created management agency, the National Association for Management of Protected Areas (French acronym: ANGAP), a private association with representation from the public and private sectors, and national/international environmental NGOs. Coordination of mini-projects for watershed protection, re-afforestation and other areas indirectly related to river conservation is the responsibility of the National Association for Environmental Actions (French acronym: ANAE). In excess of 460 mini-projects for soil conservation were coordinated by ANAE up to 1996 (Kull, 1996).

The DEF and ANGAP are responsible for enforcing protective statutes with regard to cutting of trees and burning, as well as those pertaining to the hunting or possession of protected species such as lemurs. However, environmental legislation in Madagascar historically has not been backed up by adequate policing. The earliest laws banning deforestation were flaunted through a lack of enforcement (Andriamampianina, 1984). Encroachment of agriculture, hunting, fishing, grazing and collection of timber within protected areas continues to be widespread (Jolly and Jolly, 1984; IUCN/UNEP/WWF, 1987). This situation is unlikely to change in the near future. Problems with enforcement are not restricted to habitat protection. Enforcement of species protection has also been meagre (Kull, 1996). For example, the Madagascan big-headed turtle *Erymnochelys madagascariensis* is the only freshwater organism fully protected by Malagasy law and cannot be hunted, killed, captured, or collected without authorization. However, the species is often caught incidentally by fishermen and invariably killed for local consumption (Kuchling and Mittermeier, 1993). Policing of this activity is virtually impossible. Environmental education programmes are urgently required if such species are to be protected from subsistence hunting pressure.

Conservation in practice

IN SITU CONSERVATION

Priority areas for aquatic conservation were identified by a 1995 Global Environmental Facility working group meeting, organized by Conservation International and funded by the World Bank (Ganzhorn *et al.*, 1997; Hannah *et al.*, 1998). The group's findings represent a preliminary blueprint for freshwater conservation in Madagascar. The rivers included as priority areas by the working group are: (1) the Irodo River Basin; (2) the Djabala River and associated Mont Passot crater lakes on the island of Nosy Be; (3) the upper reaches of the Sambirano, Mahavavy and Bemarivo rivers; (4) the Rantabe Basin; (5) the rivers of the Masoala Peninsula; (6) the upper reaches of the Maningory River; (7) the streams of the Ankaratra Highlands; (8) the Nosivolo Basin; (9) the Namorona, Mananjary and Sakaleona basins; (10) the streams of the Andringitra Massif; (11) the streams of the Analavelona Forest; and (12) the Manampanihy, Mandrare and Efaho basins (Figure 8.1). These were selected largely on the basis of two criteria: species richness and uniqueness. Information on fish, invertebrates and aquatic macrophytes was factored

into the decision-making process. Other criteria included degree of human pressure (e.g. population density and growth, road access, etc.) and amount and quality of data available for the area (Ganzhorn *et al.*, 1997). Financial support for this conservation initiative is required if these areas are to receive adequate protection from human-induced change.

Currently the only river reaches accorded any formal protection are those draining the various categories of reserves. These reserves include nine National Parks, 11 Strict Nature Reserves, 23 Special Reserves, and 158 Classified Forests. However, none of these reserves was designed to protect whole catchments (but see Figure 8.10). Historically, reserve locations have been chosen because of the protection afforded by their topography and inaccessibility (Andriamampianina, 1984; Sussman *et al.*, 1994). This natural protection is likely to disappear as pressure increases for agricultural land.

Difficulties relating to the conservation of the native fish are exacerbated by the presence of naturalized exotic species, among them two voracious predators and the impact of fishing pressure on declining or severely depleted fish stocks. The offshore island of Nosy Be still supports significant populations of native fish, and there are reports of *Paratilapia* on Ile Sainte Marie and *Bedotia* on the island of Nosy Mangabe. Poeciliids, tilapias and *Ospronemus goramy* are present on Nosy Be but, to date, neither black bass nor snakeheads have been introduced. Excluding these two predators from these offshore islands is the single most efficacious *in situ* conservation measure that can presently be undertaken on behalf of Madagascar's endemic fish fauna.

Restoration of Madagascar's degraded river ecosystems presents a considerable challenge because of the scale and ongoing nature of the threats they face. No restoration projects are currently in progress. Any successful restoration project would have to include large-scale re-afforestation, with replacement of vegetated riparian zones as a minimum measure. A necessary second step would entail the elimination of exotic fish and the reintroduction of native species. Given the challenges such a project would face we do not consider restoration an option at this time.

The role of captive breeding programmes in the conservation of Malagasy fish

It is a truism that the *in situ* conservation of endangered organisms is always preferable to efforts undertaken *ex situ*. In Madagascar, such efforts are synonymous with the preservation of forested watersheds. Given the

Figure 8.10   *The upper reaches of the Onive River have been placed within the new Masoala National Park because of their unique assemblage of fish. (Photograph by John Sparks)*

present extent and rate of deforestation in Madagascar, the severity of the pressures driving the assault on the forest that remains, and the scope of the difficulties confronting any conservation effort on the ground, it is difficult to be optimistic about the viability of this approach even in so far as it applies to representatives of the island's 'charismatic megafauna'. In addition, there are no means of selectively removing naturalized exotic species from a river system (Rinne and Turner, 1991). Short of suspending all fishing in a given water body, there is no way to stop the incidental capture of endangered species. In the light of these facts, captive breeding represents the most effective approach to saving Madagascar's endemic fish from extinction in the short term. Such an endeavour might seem overly ambitious. In reality, such is not the case. First, the Malagasy ichthyofauna is neither highly speciose nor are all of its approximately 65 species equally endangered. Using the most liberal selection criteria, 30 species would warrant inclusion in such a programme. The North American member institutions of the Association of Zoos and Aquariums have the resources needed to manage a captive breeding programme for this number of species. Second, the most highly endangered of Madagascar's freshwater fish are representatives of the families Cyprinodontidae, Bedotiidae and Cichlidae, all three of which are easily maintained and bred in captivity.

Under ideal circumstances, captive breeding programmes for endangered organisms should be undertaken within the country of origin. In practice the availability of institutional resources and technical expertise usually dictates that initial efforts along these lines are undertaken by foreign institutions. This situation applies in the case of the Malagasy fish. The initiative in this area has been taken by amateur aquarists in France and Switzerland and by a coalition of amateur aquarists, ornamental fish farmers and public aquaria in North America. Since 1990, founders of 20 species of Malagasy fish have been placed in the hands of qualified breeders on both sides of the Atlantic (Table 8.4). Seventeen have been successfully bred, among them both species of *Paratilapia* and the four most endangered *Paretroplus* species. Immediate future efforts need to focus upon securing breeding stock of species that are to date unrepresented in the programme and to secure its formal recognition from the Association of Zoos and Aquariums as a Species Survival Programme.

In the long run, suitable captive breeding facilities must be established in Madagascar. While traditional conservation strategies may not be applicable to

Table 8.4 *Malagasy freshwater fish presently in culture*

Species	Status
Pachypanchax omalonotus (Dumeril 1861)	A
Pachypanchax sp./Anjingo River	B
Pachypanchax sp./Betsiboka River	B
Pachypanchax sakaramyi (Holly 1928)	A
Bedotia madagascariensis Regan 1903	C
Bedotia sp. nov./Lazana River	D
Rheocles pellegrini (Nichols & Lamonte 1931)	D
Paratilapia bleekeri Sauvage 1882	E
Paratilapia polleni Bleeker 1868	E
Ptychochromis oligacanthus Bleeker 1868	D
Ptychochromis nossibeensis Bleeker 1868	B
Ptychochromis sp./Saroy mainty	E
Ptychochromis sp./Saroy mavou	B
Ptychochromis katria Reinthal & Stiassny 1997	D
Paretroplus dami Bleeker 1878	B
Paretroplus kieneri Arnoult 1960	A
Paretroplus maculatus Kiener & Maugé 1966	B
Paretroplus menarambo Allegayer 1996	B
Paretroplus petiti Pellegrin 1929	B
Paretroplus polyactis Bleeker 1878	D
Paretroplus sp. nov./Lac Andropongy	D
Paretroplus sp. nov./Lamena	A

A: Successfully bred. F_2 present generation in captivity
B: Successfully bred. F_1 present generation in captivity
C: Established as an ornamental aquarium fish since 1958
D: Not bred in captivity
E: Successfully bred. F_3 present generation in captivity

Malagasy fish, both aquaculture and managed fisheries offer some hope for the persistence of native cichlids in their country of origin. All of these species were formerly important food fish whose disappearance is deeply felt by the Malagasy people. The aquacultural potential of the *dambas* of the genus *Paretroplus* clearly warrants serious field evaluation. Another possible approach to maintaining populations of native cichlids in Madagascar is to manage small impoundments such as Grand Lac on the grounds of Parc Ivoloina, or Lac Tsimbazaza in Antananarivo, in a manner that favours their survival through a programme of selective fishing for exotic species. A dedicated breeding facility within the country would represent the most cost-effective means of supplying the fingerlings necessary to implement these management options.

FUTURE THREATS AND PRIORITIES FOR CONSERVATION EFFORTS

Current threats to the river ecosystems of Madagascar are set to continue. The future is likely to bring human encroachment into the remaining 'pristine' watersheds, further degradation of riparian areas, overfishing, and the increased spread of exotic fish species. Mitigation of these continuing impacts on the island's rivers is inextricably linked to development of the country's economy, changing agricultural practices, success of biodiversity conservation measures, and increased environmental education (i.e. successful implementation of the NEAP).

All river ecosystems on the island are threatened by the spread of exotic species. In the short term, the impacts of introductions are likely to depend on the presence of geomorphological barriers to dispersal, human population density, and cultural attitudes towards the consumption of fish. Undisturbed river systems which are most at risk from the effects of deforestation (and its consequences) are those draining areas of low relief and close to areas of high population density (Green and Sussman, 1990). These include large areas of remaining eastern and north-western rain forest (Sussman *et al.*, 1994). Preserving the integrity of these rivers and streams depends heavily on the success of current and proposed conservation and development projects.

Our recommendations for future river conservation efforts are as follows:

- Physical and biological surveys of the island's rivers will provide sorely needed information for further conservation planning. Access to some areas of the island is particularly difficult (e.g. the upper Nosivolo Basin). However, these are the areas most likely to have escaped degradation until now. Such surveys will require concerted and reliably funded programmes.

- Aquatic biodiversity studies should be expanded. Research on the taxonomy and ecology of freshwater species is restricted to the activity of a handful of workers. More researchers are needed if Madagascar's freshwater biodiversity is to be documented and conserved. The Biodiversity and Biotypology of Malagasy Continental Waters project is training several students in the systematics and taxonomy of river invertebrates and macrophytes and is an excellent example for future efforts in this direction.

- With the possible exception of hydrology, long-term data on the river ecosystems of Madagascar are conspicuously lacking. We therefore recommend the establishment of long-term monitoring projects to assess levels of human-induced change at key river sites on the island (e.g. a representative site in each of the hydrological regime areas: Figure 8.4). Some monitoring programmes should be associated with a proposed system of regional research stations (Wright, 1997). The selection and subsequent monitoring of chosen sites would be greatly strengthened by the application of remote sensing and GIS technologies.

- The introduction of exotic species into previously unaffected river systems should be strongly discouraged through public awareness campaigns. This is particularly true of the offshore islands of Madagascar, which have to date remained relatively free of exotic fish introductions. The Asian snakehead *Ophiocephalus striatus* has recently been found on the island of Île Sainte Marie for the first time (K.H. Barnes and J.P. Benstead, personal observations). This exotic predator is an important factor in native fish extinctions on the mainland and its introduction could possibly have been avoided by increased awareness of the problem and effective policing at points of entry to the island.

- Consideration of aquatic habitats should be included in future conservation planning and nature reserve design (i.e. implementing the recommendations of the GEF workshop; see Ganzhorn *et al.*, 1997).

- A regional approach to prioritization was not used by the GEF workshop (Ganzhorn *et al.*, 1997). Given Madagascar's large size, complex hydrography, levels of endemism and wide variety of river types, we recommend a regional perspective for future conservation efforts. In addition, representative river sites in each region should be given some form of legal protection.

- Establishment of protected areas and captive breeding programmes for known, threatened species, e.g. *Erymnochelys madagascariensis* and many of the endemic fish, are essential. These captive breeding programmes would benefit from secure, long-term funding and from formal recognition from the Association of Zoos and Aquariums as Species Survival Programmes.

While the state of many of Madagascar's rivers is depressing, there are reasons for hope. These include the recently strengthened national institutions for environmental conservation, the growing system of protected areas and a small but active biodiversity research programme. Adoption of the above recommendations would further reinforce efforts to conserve the river ecosystems of this fascinating island.

Acknowledgements

Duncan Elkins, Jamie March and Alex Worden are thanked for helping prepare the maps. We are grateful to Alison Shaw, Cathy Pringle, Amy Rosemond and Alex Worden for comments on the manuscript. J.P.B. is grateful to the Douroucouli Foundation for supporting his research in Madagascar. K.J.R. thanks Peter Reinthal and George Kling for advice and financial support (NSF-DEB-95-53064 and NSF-DEB-9300996) and Richard Randriampionona for assistance with the Ranomafana Watershed Project. Finally, the authors wish to thank the Association Nationale pour la Gestion des Aires Protégées, the Direction des Eaux et Forêts, the Service des Pêche, and the Malgache Institut pour la Conservation des Environnementes Tropicaux for facilitation of their research in Madagascar.

References

Agrawal, P.K., Pandey, O.P. and Negi, J.G. 1992. Madagascar: A continental fragment of the paleo-super Dharwar craton of India. *Geology* 20:543–546.

Aldegheri, M. 1972. Rivers and streams on Madagascar. In: Battistini, R. and Richard-Vindard, G. (eds) *Biogeography and Ecology of Madagascar*. Dr W. Junk, The Hague, 261–310.

Andriamampianina, J. 1984. Nature reserves and nature conservation in Madagascar. In: Jolly, A., Oberlé, P. and Albignac, R. (eds) *Madagascar*. Pergamon Press, Oxford, 219–227.

Battistini, R. and Richard-Vindard, G.. 1972. *Biogeography and Ecology of Madagascar*. Dr W. Junk, The Hague.

Battistini, R. and Verin, P. 1972. Man and the environment in Madagascar. In: Battistini, R. and Richard-Vindard, G. (eds) *Biogeography and Ecology of Madagascar*. Dr W. Junk, The Hague, 311–337.

Blanc, C.P. 1984. The reptiles. In Jolly, A., Oberlé, P. and Albignac, R. (eds) *Madagascar*. Pergamon Press, Oxford, 105–114.

Bott, R. 1965. Die Süsswasserkraben von Madagascar (Crustacea, Decapoda). *Bulletin Muséum National d'Histoire Naturelle, 2nd Sér* 37:335–350.

Brenon, P. 1972. The geology of Madagascar. In: Battistini, R. and Richard-Vindard, G. (eds) *Biogeography and Ecology of Madagascar*. Dr W. Junk, The Hague, 27–86.

Burney, D.A. 1987. Late Quaternary stratigraphic charcoal records from Madagascar. *Quaternary Research* 28:274–280.

Burney, D.A. 1997. Theories and facts regarding Holocene environmental change before and after human colonization. In: Patterson, B.D. and Goodman, S.M.

(eds) *Natural Change and Human Impacts in Madagascar*. Smithsonian Press, Washington, DC, 75–89.

Catala, R. 1977. Poissons d'eau douce de Madagascar. *Revue Française d'Aquariologie* 1:27–32.

Chaperon, P.J., Danloux, J. and Ferry, L. 1993. *Fleuves et rivières de Madagascar*. Office de la Recherche Scientifique et Technique Outre-Mer (ORSTOM), Paris.

Contreras, S. and Escalante, M.A. 1984. Distribution and known impacts of exotic fishes in Mexico. In: Courtenay, W.R. and Stauffer, J.R. (eds) *Distribution, Biology and Management of Exotic Fishes*. Johns Hopkins University Press, Baltimore, 102–129.

De Rham, P. 1995. Reproduction du 'Lamena'. *Revue Française des Cichlidophiles* 145:21–33.

De Rham, P. 1996. Die endemischen Cichliden Madagaskars – kaum bekannt abergefährdet. In: Stawikowski, R. (ed.) *Cichliden Festschrift zum 25 jährigen Jubiläum der DCG*. Deutsche Cichliden-Gesellschaft, Frankfurt, 162–178.

Dixon, H.M. 2000. Species identification and described habitats of the crayfish genus *Astacoides* (Decapoda: Parastacidae) in the Ranomafana National Park regions of Madagascar. In: Wright, P.C. (ed.) *Biodiversity in Ranomafana National Park, Madagascar*. Island Press, Washington, DC (in press).

Donque, G. 1972. The climatology of Madagascar. In: Battistini, R. and Richard-Vindard, G. (eds) *Biogeography and Ecology of Madagascar*. Dr W. Junk, The Hague, 87–144.

Eisenberg, J.F. and Gould, E. 1984. The insectivores. In: Jolly, A., Oberlé, P. and Albignac, R. (eds) *Madagascar*. Pergamon Press, Oxford, 155–165.

Economist Intelligence Unit (EIU) (eds) 1994. *Madagascar. Country Profile: Madagascar 1994–95*. London.

Elouard, J.-M., Oliarinony, R. and Sartori, M. 1998. Biodiversité aquatique de Madagascar. 9. Le genre *Eatonica* Navás (Ephemeroptera, Ephemeridae). *Bulletin de la Société Entomologique Suisse* 71:1–9.

Faramalala, M.H. 1995. *Formations végétales et domaine forestier national de Madagascar*. 1:1000 000 colour map. Conservation International, Washington, DC.

Fraser, F.C. 1956. Odonates Anisoptères. *Faune de Madagascar* 1:1–125.

Gade, D.W. 1996. Deforestation and its effects in highland Madagascar. *Mountain Research and Development* 16:101–116.

Ganzhorn, J.G., Rakotosamimanana, B., Hannah, L., Hough, J., Iyer, L., Olivieri, S., Rajaobelina, S., Rodstrom, C. and Tilkin, G. 1997. Priorities for biodiversity conservation in Madagascar. *Primate Report* 48–1 Special Issue.

Gezon, L. 1998. Institutional structure and the effectiveness of integrated conservation and development projects: case study from Madagascar. *Human Organization* 56:462–470.

Gibon, F.-M. and Elouard, J.-M. 1996. Étude préliminaire de la distribution des insectes lotiques à Madagascar (exémples des trichoptères philopotamidae et diptères simuliidae). In: Lourenço, W.R. (ed.) *Biogéographie de Madagascar*. Office de la Recherche Scientifique et Technique Outre-Mer, Paris, 507–516.

Gibon, F.-M., Elouard, J.-M. and Sartori, M. 1996. Spatial distribution of some aquatic insects in the Réserve Naturelle Intégrale d'Andringitra, Madagascar. In: Goodman, S.M. (ed.) *A Floral and Faunal Inventory of the Eastern Slopes of the Réserve Naturelle Intégrale d'Andrinitra, Madagascar: With Reference to Elevational Variation*. Fieldiana, New Series No. 85, 109–120.

Glaw, F. and Vences, M. 1994. *A Field Guide to the Amphibians and Reptiles of Madagascar*, 2nd edition. Zoologisches Forschunginstitut und Museum Alexander Koenig, Bonn.

Grandidier, A. 1886. Les canaux et les lagunes de la côte orientale de Madagascar. *Bulletin de la Société Géographique* 1:132–140.

Green, G.M. and Sussman, R.W. 1990. Deforestation history of the eastern rain forests of Madagascar from satellite images. *Science* 248:212–215.

Guillamet, J.-L. 1984. The vegetation: an extraordinary diversity. In: Jolly, A., Oberlé, P. and Albignac, R. (eds) *Madagascar*. Pergamon Press, Oxford, 27–54.

Haltenorth, T. and Diller, H. 1980. *The Collins Field Guide to the Mammals of Africa Including Madagascar*. The Stephen Greene Press, Lexington.

Hannah, L., Rakotosamimanana, B., Ganzhorn, J., Mittermeier, R.A., Olivieri, S., Iyer, L., Rajaobelina, S., Hough, J., Andriamialisoa, F., Bowles, I. and Tilkin, G. 1998. Participatory planning, scientific priorities, and landscape conservation in Madagascar. *Environmental Conservation* 25:30–36.

Helfert, M.R. and Wood, C.A. 1986. Shuttle photos show Madagascar erosion. *Geotimes* 31:4–5.

Hobbs, H.H. 1987. A review of the crayfish genus *Astacoides* (Decapoda: Parastacidae). *Smithsonian Contributions to Zoology* 443:1–50.

Holthuis, L.B. 1965. The Atyidae of Madagascar. *Mémoires du Muséum National d'Histoire Naturelle, Nouvelle Séries. Séries A, Zoologie* 23:1–48.

Hough, J.L. 1994. Institutional constraints to the integration of conservation and development: a case study from Madagascar. *Society and Natural Resources* 7:119–124.

International Union for Conservation of Nature and Natural Resources, United Nations Environment Programme, and World Wildlife Fund (IUCN/UNEP/WWF) 1987. *Madagascar: An Environmental Profile*. Jenkins, M.D. (ed.). IUCN, Gland.

Jolly, A. and Jolly, R. 1984. Malagasy economics and conservation: a tragedy without villains. In: Jolly, A., Oberlé, P. and Albignac, R. (eds) *Madagascar*. Pergamon Press, Oxford, 211–217.

Jolly, A., Oberlé, P. and Albignac, R. (eds) 1984. *Madagascar*. Pergamon Press, Oxford.

Khudongania, A.W. and Chitamwebwa, D.B.R. 1995. Impact of environmental change, species introductions and ecological interactions on the fish stocks of Lake Victoria. In: Pitcher, T.J. and Hart, P.J.B. (eds) *The Impact of Species*

Change in African Lakes. Chapman & Hall, London, 19–32.

Kiener, A. 1959. Le Marakely à bosse de Madagascar. *Bulletin Malgache* 157:501–512.

Kiener, A. 1963. *Poissons Pêche et Pisciculture à Madagascar.* Centre Technique Forestier Tropical, Nogent-sur-Marne.

Kiener, A. and Maugé, M. 1966. Contributions à l'étude systématique et écologique des poissons Cichlidae endémiques de Madagascar. *Mémoires du Muséum National d'Histoire Naturelle Séries A, Zoologie* 40:4–99.

Kiener, A. and Thérezien, Y. 1963. Principaux poissons du Lac Kinkony: Leur biologie et leur pêche. *Bulletin Malgache* 204:395–440.

Koechlin, J. 1972. Flora and vegetation of Madagascar. In: Battistini, R. and Richard-Vindard, G. (eds) *Biogeography and Ecology of Madagascar.* Dr W. Junk, The Hague, 145–190.

Kramer, R.A., Richter, D.D., Pattanayak, S. and Sharma, N.P. 1997. Ecological and economic analysis of watershed protection in eastern Madagascar. *Journal of Environmental Management* 49:277–295.

Kuchling, G. 1988. Population structure, reproductive potential, and increasing exploitation of the freshwater turtle *Erymnochelys madagascariensis. Biological Conservation* 43:107–113.

Kuchling, G. and Mittermeier, R.A. 1993. Status and exploitation of the Madagascan big-headed turtle *Erymnochelys madagascariensis. Chelonian Conservation and Biology* 1:13–18.

Kull, C.A. 1996. The evolution of conservation efforts in Madagascar. *International Environmental Affairs* 8:50–86.

Langrand, O. 1990. *Guide to the Birds of Madagascar.* Yale University Press, New Haven, Connecticut.

Le Bourdiec, P. 1972. Accelerated erosion and soil degradation. In: Battistini, R. and Richard-Vindard, G. (eds) *Biogeography and Ecology of Madagascar.* Dr W. Junk, The Hague, 227–259.

Loiselle, P.V. 1977. Colonial breeding by an African substratum-spawning cichlid fish *Tilapia zillii* (Gervais). *Biology of Behavior* 2:129–143.

Loiselle, P.V. 1994. The cichlids of Madagascar – Going, going, gone? *Journal of the American Cichlid Association* 161:1–8.

Loiselle, P.V. 1995. The cichlids of Jurassic Park Part I. *Cichlid News* 4:18–23.

Loiselle, P.V. 1996. The cichlids of Jurassic Park Part III. Review of the genus *Paretroplus. Cichlid News* 5:21–25.

Loiselle, P.V. and Ferdenzi, J. 1997. The natural history and aquarium husbandry of *Pachypanchax sakaramyi* (Holly 1928), the 'lost killifish' of Madagascar. *Journal of the American Killifish Association* 30:29–41.

Loiselle, P.V. and Stiassny, M.L.J. 1993. How many Marakely? *Journal of the American Cichlid Association* 157:2–8.

Louvel, M. 1930. *L'Exploitation des Eaux Douces de Madagascar.* Imprimerie Pitot, Tananarive.

Lowry, P.P., Schatz, G.E. and Phillipson, P.B. 1997. The classification of natural and anthropogenic vegetation in Madagascar. In: Patterson, B.D. and Goodman, S.M. (eds) *Natural Change and Human Impacts in Madagascar.* Smithsonian Press, Washington, DC, 93–123.

Lucanus, O. 1996. Field notes on behavior and ecology of Malagasy cichlids. *Cichlid News* 5:24–27.

Lugo-Ortiz, C.R. and McCafferty, W.P. 1997. *Edmulmeatus grandis* – an extraordinary new genus and species of Baetidae (Insecta, Ephemeroptera) from Madagascar. *Annales de Limnologie* 33:191–195.

Malzy, P. 1965. Un mammifère aquatique de Madagascar: le *Limnogale. Mammalia* 29:399–411.

McNeeley, J.A., Miller, K.R., Reid, W.V., Mittermeier, R.A. and Werner, T.B. 1990. *Conserving the World's Biological Diversity.* International Union for the Conservation of Nature, Gland.

Millot, J. 1951. Un lézard d'eau à Madagascar. *Le Naturalist Malgache* 3:87–90.

Minckley, W.L., Meffe, G.K. and Soltz, D.L. 1991. Conservation and management of short-lived species: the cyprinodontoids. In: Minckley, W.L. and Deacon, J.E. (eds) *Battle Against Extinction.* University of Arizona Press, Tucson, 247–282.

Moreau, J. 1979. Biologie et évolution des peuplements des cichlides (Pisces) introduits dans les lacs malgaches d'altitude. Thesis Institut National Polytechnique, Toulouse.

Nelson, R. and Horning, N. 1993. AVHRR–LAC estimates of forest area in Madagascar, 1990. *International Journal of Remote Sensing* 14:1463–1475.

Nicoll, M.E. and Langrand, O. 1989. *Madagascar: revue de la conservation et des aires protégées.* World Wildlife Fund, Gland.

Nourissat, J.-C. 1992. Madagascar. *Revue française des Cichlidophiles* 118:9–29.

Nourissat, J.-C. 1993. Madagascar – 1992. *Revue française des Cichlidophiles* 129:8–36.

Oliarinony, R. and Elouard, J.-M. 1997. Biodiversité aquatique de Madagascar: 7 – *Ranorythus*, un nouveau genre de Tricorythidae définissant la nouvelle sous-famille des Ranorythinae (Ephemeroptera, Pannota). *Bulletin de la Société Entomologique de France* 102:439–447.

Olson, S.H. 1984. The robe of the ancestors. *Journal of Forest History* 28:174–186.

Petit, G. 1930. *L'Industrie des Pêches à Madagascar.* Éditions Géographiques, Maritimes et Coloniales, Paris.

Rabarisoa, R., Watson, R.T., Thorstrom, R. and Berkelman, J. 1997. Status of the Madagascar fish eagle *Haliaeetus vociferoides* in 1995. *Ostrich* 68:8–12.

Rabinowitz, P.D., Coffin, M.F. and Falvey, D. 1983. The separation of Madagascar and Africa. *Science* 220:67–69.

Randriambelo, T., Baldy, S. and Bessafi, M. 1998. An improved detection and characterization of active fires and smoke plumes in south-eastern Africa and Madagascar. *International Journal of Remote Sensing* 19:2623–2638.

Randrianarijaona, P. 1983. The erosion of Madagascar. *Ambio* 12:308–311.

Rasolofoharinoro, M., Blasco, F., Bellan, M.F., Aizpuru, M., Gauquelin, T. and Denis, J. 1998. A remote sensing based methodology for mangrove studies in Madagascar. *International Journal of Remote Sensing* 19:1873–1886.

Raxworthy, C.J. and Nussbaum, R.A. 1996. Montane amphibian and reptile communities in Madagascar. *Conservation Biology* 10:750–756.

Reinthal, P.N. and Stiassny, M.L.J. 1991. The freshwater fishes of Madagascar: A study of an endangered fauna with recommendations for a conservation strategy. *Conservation Biology* 5:231–243.

Richard, A.F. and O'Connor, S. 1997. Degradation, transformation, and conservation: the past, present, and possible future of Madagascar's environment. In: Patterson, B.D. and Goodman, S.M. (eds) *Natural Change and Human Impacts in Madagascar*. Smithsonian Press, Washington, DC, 406–418.

Rinne, J.N. and Turner, P.R. 1991. Reclamation and alteration as management techniques and a review of methodology in stream renovation. In: Minckley, W.L. and Deacon, J.E. (eds) *Battle Against Extinction*. University of Arizona Press, Tucson, 219–244.

Riseng, K.J., Reinthal, P.N., Kling, G.W. and Randrapioni, R. 1997. Effect of deforestation on the water chemistry of a rainforest stream in Madagascar. *Abstracts of the American Society of Limnology and Oceanography Annual Meeting, Santa Fe, New Mexico, USA*, 283.

Sampson, S.D., Witmer, L.M., Forster, C.A., Krause, D.W., O'Connor, P.M., Dodson, P. and Ravoavy, F. 1998. Predatory dinosaur remains from Madagascar: implications for the Cretaceous biogeography of Gondwana. *Science* 280:1048–1051.

Schaller, D. 1991. Anmerkungen zum Madagaskar-Hechtling *Pachypanchax omalonotus*. *DATZ Aquarien Terrarien* 44:692–693.

Schmidt, E. 1951. The Odonata of Madagascar (Zygoptera). *Mémoires Institut de Recherche Scientifique de Madagascar* 6:115–279.

Stiassny, M.L.J. and DePinna M.C.C. 1994. Basal taxa and the role of cladistic patterns in the evaluation of conservation priorities: a view from freshwater. In: Forey, P., Humphreys, C.J. and Vane-Wright, R.I. (eds) *Systematics and Conservation Evaluation*. Systematics Association Special Volume, Oxford University Press, Oxford, 235–249.

Stiassny, M.L.J. and Harrison, I.J. 2000. Notes on a small collection of fishes from the Réserve Naturelle Intégrale de Marojejy, northeastern Madagascar, with a description of a new species of the endemic genus *Bedotia* (Atherinomorpha: Bedotiidae). *Fieldiana* (in press).

Stiassny, M.L.J. and Raminosoa, N. 1994. The fishes of the inland waters of Madagascar. *Annales du Musée Royal de l'Afrique Centrale Zoologie* 275:133–149.

Stiassny, M.L.J. and Reinthal, P.N. 1992. Description of a new species of *Rheocles* (Atherinomorpha, Bedotiidae) from the Nosivolo Tributary, Mangoro River, Eastern Malagasy Republic. *American Museum Novitates* No. 3031.

Storey, M., Mahoney, J.J., Saunders, A.D., Duncan, R.A., Kelley, S.P. and Coffin, M.F. 1995. Timing of hot spot-related volcanism and the breakup of Madagascar and India. *Science* 267:852–855.

Sussman, R.W., Green, G.M. and Sussman, L.K. 1994. Satellite imagery, human ecology, anthropology and deforestation in Madagascar. *Human Ecology* 22:333–354.

Trewavas, E. 1983. *Tilapiine Fishes of the Genera Sarotherodon, Oreochromis and Danakilia*. British Museum (Natural History), London.

Waters, T.F. 1995. Sediment in streams: sources, biological effects and control. *American Fisheries Society Monograph* 7.

Wells, N.A. and Andriamihaja, B.R. 1997. Extreme gully erosion in Madagascar and its natural and anthropogenic causes. In: Patterson, B.D. and Goodman, S.M. (eds) *Natural Change and Human Impacts in Madagascar*. Smithsonian Press, Washington, DC, 44–74.

Weninger, G. 1985. Principal freshwater types and comparative hydrochemistry of tropical running water systems. *Revue d'Hydrobiologie Tropicale* 18:79–110.

Williams, J.E. and Deacon, J.E. 1986. Subspecific identity of the Amargosa pupfish, *Cyprinodon nevadensis*, from Crystal Spring, Ash Meadows, Nevada. *Great Basin Naturalist* 46:220–223.

World Bank 1990. *Staff Appraisal Report: Democratic Republic of Madagascar Environment Program*. The World Bank, Washington, DC.

World Resources Institute (WRI) 1994. *World Resources 1994–95*. Oxford University Press, Oxford.

Wright, P.C. 1997. The future of biodiversity in Madagascar: A view from Ranomafana National Park. In: Patterson, B.D. and Goodman, S.M. (eds) *Natural Change and Human Impacts in Madagascar*. Smithsonian Press, Washington, DC, 381–405.

9

River conservation in the Indian sub-continent

B. Gopal
with contributions from B. Bose and A.B. Goswami

Introduction

The Indian sub-continent, lying between 0° and 37°6'N, and 61° and 97°25'E, is a diamond-shaped land mass, bounded to the north by the Great Himalayan arc, and to the south, east and west by deep sea. The Great Himalaya is contiguous on its western side with the Hindukush Ranges that extend in the southwest as the Suleiman and Kirthar Ranges. On the east, the Himalayan arc is represented by the Arakan Yoma Ranges in Myanmar, and the Naga and Garo hills in eastern India. The sub-continent comprises six sovereign nations – India, Pakistan, Nepal, Bhutan, Bangladesh and Sri Lanka – and is unique in the diversity of its geology, climate, vegetation and fauna, as well as human cultures.

The rivers of the region have played a major role in shaping the history of human civilization in the sub-continent. The early agrarian civilizations of Harappa and Mohenjodaro that flourished in the Indus Basin depended upon intensive irrigation which required diversion of river water through an extensive system of canals. Human settlements on the banks of the River Ganga (Ganges) and its tributaries have continued to exist for more than 5000 years. The rivers were extensively used for irrigation, potable water supply, recreation, fishing and transport. Interestingly, rivers were revered as mothers and worshipped as goddesses. During the past few decades, an exponential increase in human population, rapid urbanization and industrialization, intensive agriculture, and growing demands for energy, have all severely affected the rivers of the region. The regulation of river flows by channelization and dams, and the discharge of domestic and wastewater effluents, have led to a sharp decline in riverine water quality and biological resources. However, while river conservation and management have recently begun to receive some attention, scientific studies are few and are generally confined to hydrology and water quality. This chapter provides only a bird's eye view of the river systems in the sub-continent, their scientific understanding, the state of their degradation and its causes, and the conservation and management policies and actions in different countries of the region.

Almost all the rivers originating in the Himalayan belt pass through India in their upper and/or middle reaches. They share drainage basins between two or more countries and, hence, also their problems. Like other areas of science, rivers have also received relatively more attention in India than in other countries of the sub-continent. Therefore, this chapter focuses primarily on the Indian rivers.

The regional context

Extensive information on the geology, geomorphology, climate, soils, water resources, flora and fauna of the Indian sub-continent is available in many publications (e.g. Ahmad, 1951, 1969; Raychaudhuri et al., 1963; Mani, 1974; Robinson, 1976; Fernando, 1985; Cooray 1984; Erb, 1984; Costa and De Silva, 1995); thus only a few salient features are mentioned below.

Global Perspectives on River Conservation: Science, Policy and Practice.
Edited by P.J. Boon, B.R. Davies and G.E. Petts. © 2000 John Wiley & Sons Ltd.

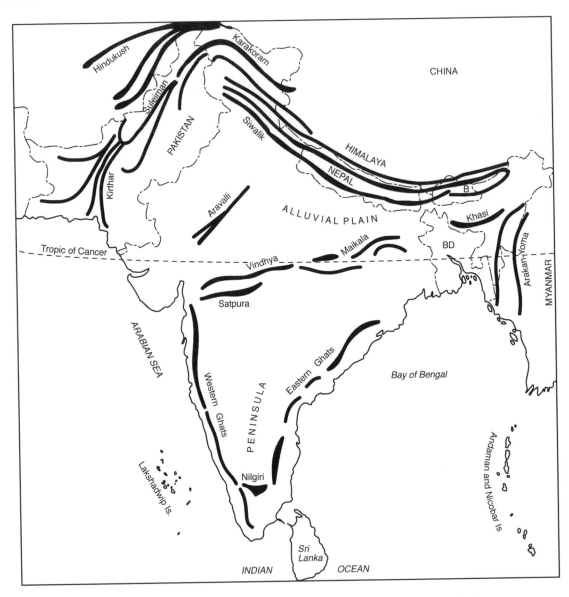

Figure 9.1　*The Indian sub-continent and the mountain ranges. B = Bhutan; BD = Bangladesh*

GEOMORPHOLOGY AND GEOLOGY

The sub-continent is the direct result of the split of Gondwanaland during the Jurassic, followed by the subsequent north-northeastward drift of the part now represented by peninsular India. The northerly thrust of the drifting Indian land mass under the Asian land mass obliterated the Tethys Sea giving rise to the Himalayas. Today, the sub-continent, covering more than 4.57×10^6 km², is physiographically divided into

two, roughly along the Tropic of Cancer: the vast northern alluvial plain and the southern peninsula (Figure 9.1). The peninsula includes the geologically oldest Precambrian rocks, while the Himalayas are the youngest (and yet highest of the world's mountain ranges). The Himalayas have, to a great extent, shaped the geomorphology of the sub-continent. The vast plain between the Himalayan Ranges and the peninsula has been formed by pre-Tertiary alluvial debris from the peninsula, and Upper and post-

Tertiary alluvium from the Himalayas. An aero-magnetic survey has revealed that the thickness of these deposits in the Ganga Plain varies from 1300 to more than 8000 m (Hari Narain, 1965). The Himalayas rise almost abruptly from the plains – from less than 500 m AMSL, to over 8000 m. The northern plain is traversed on its western side by relatively low hill ranges, the Aravallis, extending northeast to southwest (Figure 9.1).

The inverted pyramid-shaped peninsula is a plateau, varying in altitude from 300 to 900 m AMSL, which slopes towards the east. It is bordered on all sides by a series of hilly ranges. Along the western seaboard these ranges rise somewhat abruptly – the Western Ghats – whereas on the eastern side, they are rather low and occur in a broken chain – the Eastern Ghats (Figure 9.1). To the north, the plateau is bounded by the Satpura and Vindhya Ranges, which run along the Tropic of Cancer. In the southern corner, the Western and Eastern Ghats meet in the Nilgiri Ranges.

Much of the north-western part of Pakistan is covered by mountain ranges – the Hindukush and Suleiman Ranges – whereas the remainder comprises alluvial plain. Nepal and Bhutan are covered by high mountains in the Central Himalaya, whereas Bangladesh (East Pakistan before 1971) comprises mostly the deltaic plain formed by the Ganga and Brahmaputra rivers. Sri Lanka, a pear-shaped island (65 525 km^2), lies to the south of the Indian mainland from which it is separated by an isthmus. The latter is interrupted by a chain of rocky outcrops. The island has a central core of hills, which rise to 2520 m AMSL (Pidurutalagala Peak) and which is flanked by an alluvial plain. Another chain of numerous islands which constitute Andaman and Nicobar in the Bay of Bengal, represent a continuation of the Great Himalayan arch, with hills rising to 732 m.

CLIMATE AND HYDROLOGY

Climatically, the Indian sub-continent includes areas with some of the greatest contrasts of anywhere in the world between temperature and rainfall regimes: the Himalayas play a dominant role in these contrasts. The world's most rainy place (Cherrapunji in north-eastern India), as well as the hottest and driest (Quetta in Pakistan), lie respectively at the eastern and western ends of the Himalayan arch.

The climate of the sub-continent is described as tropical to sub-tropical although there is great spatial variability. The western parts are arid to semi-arid, whereas at higher altitudes in the western Himalaya

and also on the southern plateau the climate is better classified as warm temperate or montane temperate. Summer (April–June) temperatures in most parts of the sub-continent rise above 40°C, with several areas on the plains experiencing temperatures of 50°C or even higher. Winters are mild (temperature >20°C) on the peninsula while, northwards in the plains, temperatures fall with increasing latitude, reaching near freezing in the foothills. In the mountains, temperatures may drop to below −30°C in many areas for many weeks. The sub-continent is, however, well protected against the cold winds from the north. Lying closer to the Equator, Sri Lanka experiences a more equitable tropical climate, with relatively small annual variations in the temperature (Costa and De Silva, 1995).

The climate, and consequently the water resources and biota of the region, are primarily governed by monsoons. The term 'monsoon' refers to winds that seasonally change their direction along the shores of the Indian Ocean, especially in the Arabian Sea. Monsoon winds blow from southwest to northeast during the summer of the Northern Hemisphere, and from northeast to southwest during winter (Das, 1968; Fein and Stephens, 1987). This reversal of winds is caused by differential heating and cooling of the land and ocean. Typical monsoon winds are experienced in the tropical and sub-tropical belt of the Eastern Hemisphere (Ramage, 1971; Webster, 1987). The monsoon has been described as 'one of the most dramatic of all weather events, tantalizingly complex, rich in variations from place to place and year to year, day to day, and difficult to predict' (Fein and Stephens, 1987). The activity and influence of the monsoons are most spectacular on the Indian sub-continent because of its specific geomorphological characteristics. Towards the end of April, the monsoon winds originating over the Indian Ocean split into two streams as they touch the peninsula. One stream moves from the southwest over the west coast of India towards the north-northeast, bringing heavy rain over the Western Ghats. The areas in the immediate shadow of the Western Ghats, however, receive no rain. The other stream passes over the Bay of Bengal towards Bangladesh and north-east India. This stream is trapped between the high hill ranges, and is deflected westwards. Thus, it results in heavy rains over Bangladesh and in the north-eastern parts of India, and is gradually depleted of its moisture as it moves westwards (Figure 9.2). The Aravallis stop its further westward progression. By early October, the south-west monsoon starts to retreat. The north-east monsoon is gentle as the Himalayas prevent its

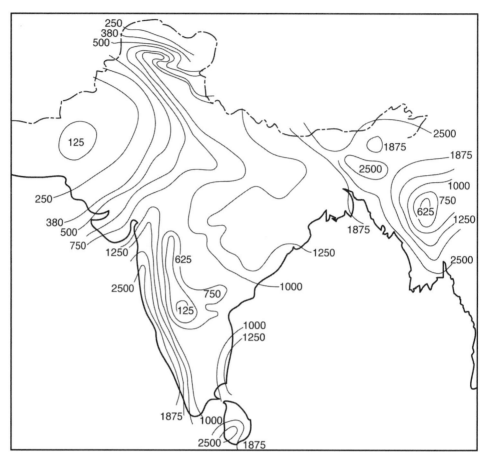

Figure 9.2   *Average annual precipitation (mm) over the Indian sub-continent*

southward movement. However, it brings sufficient rain during winter (November to January) over southern India and Sri Lanka. The physiographic features of Sri Lanka also influence the precipitation that decreases from west to east (Costa and De Silva, 1995). Besides its physiography and the monsoon, the climate of Sri Lanka is strongly influenced by its proximity to the Equator, and maritime influences (low diurnal and annual temperature variations). Climatically, Sri Lanka is usually divided into a south-western wet zone with heavy rainfall throughout the year, and a larger dry zone that experiences a summer drought.

Thus, there is large spatial and temporal variability in the total annual precipitation over the sub-continent (Figure 9.3), and this variability increases towards the west. In recent years, there has been growing evidence that El Niño also influences the movement of monsoons, causing further year-to-year variability.

SOILS

The most widespread soil types on the sub-continent are alluvial, black and lateritic soils (Raychaudhuri *et al.*, 1963). Bangladesh is almost wholly covered by the alluvial soils of the Ganga and Brahmaputra. Alluvial soils often contain large amounts of calcium in the form of calcium carbonate granules (kankar) which form a hard pan leading to waterlogging. Most alluvial silts also have a high clay content. Black soils are derived chiefly from the Deccan Trap in peninsular India. These are generally fertile but are also prone to waterlogging. Red soils are derived from ancient metamorphic rocks and are poor in nitrogen, phosphorus and organic matter. Lateritic soils develop in the intermittently moist climate of peninsular India and are rich in the hydrated oxides of aluminium and iron. Saline soils are common in the northern and western parts of the sub-continent.

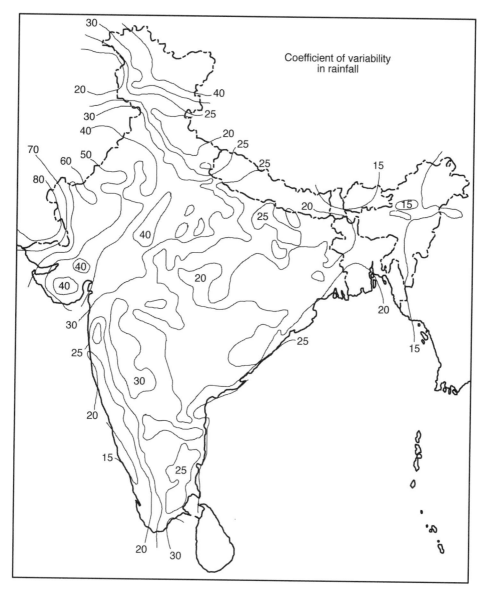

Figure 9.3   *Percentage inter-annual variability in precipitation over the Indian sub-continent*

Organic soils are nearly non-existent: the organic content is high only in forest soils and the soils of hilly areas. There are no significant peat deposits. Typical desert soils occur in the semi-arid and arid parts of the sub-continent, mostly west of the Aravalli Ranges (western Rajasthan and Pakistan).

Panabokke (1984) has recognized 14 soil groups in Sri Lanka. Of these, the most common are alluvial soils, various latosols, rugosols, low humid grey soils and non-calcic brown soils.

VEGETATION

The sub-continent, with its distinct geological history, is also a well recognized biogeographic region which has a characteristic flora and fauna, although there is also a clear influence from neighbouring regions as well as from its Gondwanan affinities (for detailed accounts see Mani, 1974; Fernando, 1985).

The vegetation of the sub-continent varies from tropical rain forest, through moist and dry evergreen and deciduous forests, to desert scrub (Champion and Seth, 1968). Mangrove forest occurs in all river deltas along the coast, of which the Sunderban Mangroves of the Ganga–Brahmaputra Delta cover an area of more than $10\,000\,km^2$, the single largest mangrove formation in the world (Gopal and Krishnamurthy, 1993). The flora and fauna of the Indian sub-continent exhibit high species diversity together with a high degree of endemism. Indeed, the Western Ghats and the north-eastern Himalaya are recognized 'hot spots' of biodiversity (e.g. Myers, 1988, 1990): about 34% of the flowering plants of the Indian mainland are endemic (Chatterjee, 1940; Nayar, 1980), while on Sri Lanka, the degree of endemism is still higher.

WATER RESOURCES

The characteristic physiographic and climatic features of the sub-continent determine the type and extent of water resources and their management. With the exception of a crater lake (Lake Lonar) in western India, and oxbows, as well as seasonally flooded depressions, natural lakes are few, relatively small and are exclusively confined to the Himalayan Belt. However, there is a dense network of several hundred rivers and streams that flow practically in every direction across the region. In Sri Lanka alone there are 103 perennial rivers (Costa and De Silva, 1995). Thus, the rivers constitute the most important water resources of the sub-continent. The majority of the rivers, except for those in the arid western region, are perennial, although they do exhibit very large seasonal variability in discharge. The rivers originating in the Himalayas are fed both by snowmelt (during the summer), and rainfall during the rest of the year. Peninsular rivers are entirely rain fed, receiving run-off both during the south-west and north-east monsoons.

THE RIVER SYSTEMS

The river systems are grouped on the basis of their drainage-basin area into major (more than $20\,000\,km^2$), medium ($2000–20\,000\,km^2$) and minor ($<2000\,km^2$) rivers. Accordingly, the Indian mainland is drained by 15 major, 45 medium and over 120 minor systems, besides numerous ephemeral streams in the western arid region (Rao, 1975; Khondker, 1995; Nazneen, 1995). Among more than 100 rivers in Sri Lanka, only seven can be categorized as medium rivers. The major river systems of the mainland are shown in Figure 9.4. Data on their length, discharge, basin area, and sediment yields are listed in Table 9.1.

The river systems of the Indian mainland can be grouped, according to their origin, into Himalayan and Peninsular rivers. The rivers Indus, Ganga and Brahmaputra are the major rivers originating in the Himalaya. However, several tributaries of the Ganga arise on the northern flanks of the peninsular plateau and flow northwards. The River Meghna arises in the Baraila Ranges in Assam on the eastern side of the Himalayan arch. The remaining rivers arise on the peninsula and flow eastwards or westwards (Figure 9.4).

Similarly, Sri Lankan rivers rise in the central highlands and flow radially through the peripheral lowland before meeting the Indian Ocean (Costa and De Silva, 1995). These rivers have been grouped into four categories (Erb, 1984):

- rivers with headwaters on the mountain plateaux and plains of the inner region of the Central Massif, flowing outwards to the coastal plain;
- rivers with headwaters in the marginal hills of the plateaux of the Central Massif, flowing down to the coastal plain;
- rivers with headwaters in the mountain hills and plateaux of the Sabaragamuwa Hills and Elahera Ridges, traversing to the coastal plain;
- rivers with headwaters on the circum-island peneplains, traversing the coastal plain.

Himalayan rivers

The Indus (2880 km length) rises in Lake Mansarovar in Tibet at an altitude of *ca* 5180 m AMSL and flows westwards through India before turning south-southwest into Pakistan and finally discharging into the Arabian Sea, forming a delta near Karachi. The Brahmaputra (2900 km length) rises in Lake Kanggyu Tsho in the Kailash Ranges (east of Lake Mansarovar) at 5150 m AMSL, and flows eastwards through Tibet and China for 1600 km before taking a 'U'-turn, and then flows south, entering India. It flows southwest and west through Arunachal Pradesh and Assam for about 800 km, and then bends southwards near Dhubri to enter Bangladesh, where it joins the Ganga at Golaundo. It is variously known as the River Tsangpo in Tibet, the Dihang or Siang along its southward bend towards India, and the Jamuna in Bangladesh. Along its course through the Assam Valley, the river often spreads laterally up to 10 km on either side. Interestingly, the Brahmaputra descends steeply from *ca* 3600 m in Tibet to 150 m at Sadiya in India, shortly after it enters India. The River Ganga (2525 km length) rises on the Gangotri Glacier

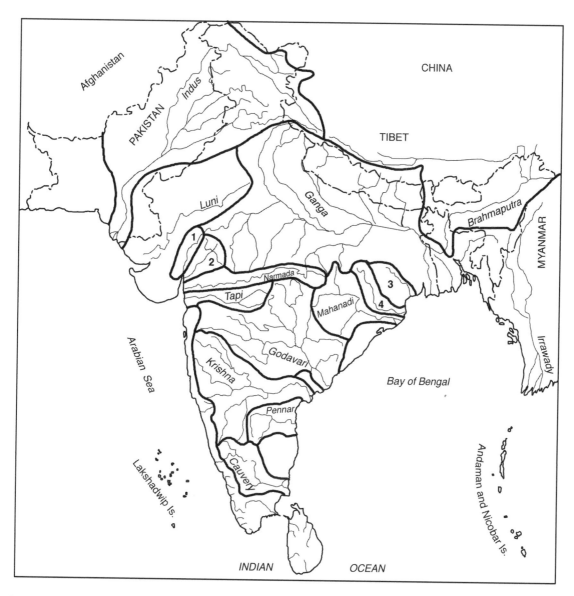

Figure 9.4 *Drainage basins of major rivers in the Indian subcontinent. 1, R. Sabarmati; 2, R. Mahi; 3, R. Subarnarekha; 4, R. Brahmani*

(7010 m) in the Himalaya within India. It flows southwest before descending to the plains at Rishikesh. It then flows south and turns eastwards, meandering across the plain to Farakka, where it turns south and divides into two main channels one of which (now called the Padma) flows through Bangladesh and meets the Jamuna before forming an extensive delta. The Padma finally joins the Meghna in Bangladesh in its last stretch just above the delta.

All three rivers have many major perennial tributaries most of which arise in the Himalaya. The River Yamuna, the largest tributary of the Ganga (1376 km long; see Figure 9.4), also rises in the Himalaya on the Yamunotri Glacier (6320 m AMSL) and flows almost parallel to the Ganga until their confluence at Allahabad. The Yamuna itself is joined by another major tributary, the Chambal, that drains a large area to the south and west of the Yamuna. The

Table 9.1 Data on the length, basin areas, annual sediment yields, gross average annual flows and percentage utilization of the major river systems of the Indian sub-continent and Sri Lanka. (B. Bose and A.B. Goswami)

River	Provenance	Outfall	Length (km)	Drainage area (km²)	Gross average annual flow (×10⁶ m³)	Water Potential Index (WPI) (×10⁶ m³ km⁻²)	Annual Sediment yield (×10⁶ t yr⁻¹)	Status of utilization (% of total)
A Rivers of the Northern Plain (between Himalayas and Vindhya Range)								
Indus	Himalayan Range	Arabian Sea	2800	1 117 570	209 780	0.188	133 200	87
Luni	Aravalli Range	Arabian Sea	482	193 392	12 278	0.063	—	—
Sabarmati	Aravalli Range	Arabian Sea	371	21 674	3 663	0.169	0.05	93
Mahi	Aravalli Range	Arabian Sea	583	34 842	11 812	0.339	6.0	81
Ganga	Himalayan Range	Bay of Bengal	2525	914 400	567 201	0.620	500	—
(i) Brahmaputra	Himalayan Range	Bay of Bengal	2900	935 508	563 900	0.603	900	—
(ii) Barak	Naga Patkoi Range	Bay of Bengal	900	78 150	41 198	0.527	—	—
Manipur (within India)	Naga Patkoi Range	Irrawaddy river (Myanmar)	320	19 000	1 900	0.100	—	—
Subarnarekha	Chotanagpur Plateau	Bay of Bengal	395	19 296	7 941	0.412	3.6	54.8
Coastal drainages	Alluvial Uplands	Arabian Sea and Bay of Bengal	—	42 400	17 190	0.405	—	—
Total of A				3 376 232	1 436 863	0.426[1]		
B. Rivers of Peninsular India (south of the Vindhya Range)								
Narmada	Vindhya Range	Arabian Sea	1 312	99 171	54 600	0.551	42.4	23
Tapi	Vindhya Range	Arabian Sea	724	65 145	19 736	0.303	56.2	—
Brahmani–Baitarani	Chotanagpur Plateau	Bay of Bengal	1 355	42 822	23 762	0.555	11.8	—
Mahanadi	Deccan Plateau	Bay of Bengal	890	141 589	66 644	0.471	28.4	34
Godavari	Deccan Plateau	Bay of Bengal	1 465	312 812	118 000	0.377	83	50
Krishna	Deccan Plateau	Bay of Bengal	1 400	257 923	67 675	0.262	8.2	81
Pennar	Deccan Plateau	Bay of Bengal	597	55 213	3 238	0.059	1.3	75
Cauvery	Deccan Plateau	Bay of Bengal	800	87 900	20 950	0.238	0.9	95
Coastal drainages	Western and Eastern Ghat Ranges	Arabian Sea and Bay of Bengal	—	320 046	250 704	0.783	—	—
Total of B				1 382 621	625 309	0.452[2]		
Total A and B				4 758 853	2 062 172	0.433[3]		
C. Sri Lanka								
Mahaweli Ganga	Central hilly tracks	Bay of Bengal	329	10 320	21 314	2.065		
Other streams			—	55 680	127 884	2.300		

[1] mean of Area A
[2] mean of Area B
[3] mean of Areas A + B

Chambal rises in the Vindhan Ranges and delivers more water at their confluence than the Yamuna. In this context, the discharge of the Yamuna at Allahabad exceeds that of the Ganga. Among other tributaries are many rivers which originate in Nepal (Sharma, 1997): the Mahakali, Karnali, Sapt Gandaki and Sapta Kosi rivers in Nepal are, in India, known respectively as the Sarda, Ghagara, Gandak and Kosi. These systems contribute more than 40% of the annual flow of the Ganga at Farakka (Khan, 1994). The Ramganga, Gomti and Burhi Gandak rivers also rise in Nepal, while the Mahananda rises in the Darjeeling Hills, flows south, and meets the Ganga inside Bangladesh. Several tributaries, such as the Tons, Son and Punpun, which flow northwards from the hills east of the Vindhyan Ranges, join the River Ganga on its right.

Five major rivers, the Chenab, Sutlej, Ravi, Beas and Jhelum, originating in the Indian Himalaya join the Indus from the east, while the Kunnar, Kabul and several others which originate in the Hindukush and Suleiman Ranges in Afghanistan, join the Indus from the west (Nazneen, 1995).

The Brahmaputra is joined by several tributaries in its mid-reaches. Of these, several originate in Bhutan, Sikkim and even in Tibet. The more important tributaries are the Subansiri, Jia Bhareli, Manas, Sankosh (= Mo in Bhutan), Raidak, Torsa, Jaldhaka and Tista (all joining from the right), and the Lohit, Dibang, Buri Dihing and Dhansiri (joining from the left).

Among other rivers that originate in the Himalayan Belt, mention must be made of the Meghna and Imphal. The River Meghna rises in the hills of Assam, flows southwards into Bangladesh, and is joined by several tributaries that drain parts of north-eastern India. The Imphal, one of the medium rivers, and known as the Manipur in its southern reaches, originates in the Manipur Hills, and flows south through the valley into Myanmar where it joins the Chindwin, a tributary of the Irrawady.

Peninsular rivers

As well as some tributaries of the Ganga, 11 major rivers originate in, and drain, Peninsular India. These systems meander through incised valleys in the plateau where narrow, deep gorges result in many waterfalls. The Tons, a tributary of the Ganga, passes through a 113-m-deep gorge in the Kaimur Hills in central India. The Jog Falls on the Sharavati River, a tributary of the River Krishna, has a drop of 253 m; the Cauvery River falls by 101 m at Sivasamundram; the Subarnarekha

has a 74-m-high fall near Ranchi on the Chotonagpur Plateau; and the Narmada River drops 15 m at Marble Falls near Jabalpur.

The Narmada River originates in the Amarkantak Hills (1057 m altitude) and flows westwards through a narrow basin lying between the Vindhyan and Satpura ranges. It is joined by 18 main tributaries and after flowing through several gorges for 1312 km it meets the Arabian Sea near Broach (Bharuch). The River Tapi (also called the Tapti) flows westwards almost parallel to the Narmada, from its source at the eastern end of the Satpura Ranges (730 m altitude) to the Arabian Sea.

The Mahanadi River rises in the Chattisgarh region of south-eastern Madhya Pradesh and flows east to the Bay of Bengal where it forms a large delta (7000 km^2) to the north of Lake Chilka. The largest peninsular river, the Godavari, rises in the northern part of the Western Ghats (near Nasik), flows east-southeast and forms a delta (2500 km^2) near Rajamundhry in Andhra Pradesh. Among its several tributaries, the major ones are the Manjira, Waingunga and Indrawati. The River Krishna and its major tributaries, the Bhima and Tungabhadra, rise in the central parts of the Western Ghats. The Tungabhadra carries more water than the Krishna at their confluence, while the Bhima nearly dries during summer. The Krishna also flows eastwards and forms another delta (*ca* 2500 km^2) just south of the Godavari Delta, before discharging into the Bay of Bengal.

The River Cauvery, originating in the lower parts of the Western Ghats, is a relatively small system (800 km), but is joined by many tributaries. In its lower reaches, the river bifurcates first into the Coleroon and the Cauvery which rejoin 18 km downstream, only to split again into the Vennar and the Cauvery. Finally, the system forms a large delta (*ca* 5000 km^2).

Of the other major rivers, the Subarnarekha and Brahmani drain most of the eastern area between the basins of the Ganga and Mahanadi, flowing eastwards into the Bay of Bengal. The River Pennar, another east-flowing river, drains an area of 55 200 km^2 between the Krishna and Cauvery basins. Two west-southwest-flowing rivers, the Mahi and the Sabarmati, are much smaller (respectively 34 800 and 21 700 km^2) and drain the semi-arid parts of Rajasthan and Gujarat, west of the Yamuna Basin.

Almost all medium and minor rivers lie within peninsular India. Most of them are small, and flow westwards from the Western Ghats into the Arabian Sea, or rise on the Eastern Ghats and flow into the Bay of Bengal. A few are seasonal or ephemeral streams in the western arid and semi-arid regions of the sub-continent.

Rivers in Sri Lanka

In Sri Lanka, the River Mahaweli Ganga is the longest (329 km) and has the largest basin: catchment area of 10 300 km², total discharge >11 000 × 10⁶ m³. Although next in terms of length and catchment area, the Aruvi Aru and Kala Oya have much smaller discharges (<600 × 10⁶ m³), while somewhat perversely, the shorter and smaller basins of the Kalu Ganga and Kelani Ganga respectively have second and third largest discharges (7862 and 5474 × 10⁶ m³).

DEVELOPMENT PRESSURES AND HUMAN IMPACTS

With continuous human habitation for more than 5000 years, the Indian sub-continent accounts for more than 28% of the world's total human population and, with an area far less than that of China, it is also the most densely populated region of the world (Table 9.2). At the turn of the millennium, the total population will be about 1.35 billion, with an average density of *ca* 300 persons km⁻². Bangladesh, with an estimated population of 132 million, has the highest density: more than 900 persons km⁻². Within the countries of the region there is great variation in population density. For instance, in India, densities range between <40 persons km⁻² in Jammu and Kashmir, to 600 in Bihar and Uttar Pradesh, 850 in Kerala and West Bengal, and >8000 persons km⁻² in Delhi, the National Capital Territory.

As pointed out earlier, the countries of the sub-continent became independent sovereign states only 50 years ago, after prolonged British colonial rule. Political independence aroused the aspirations of the people for economic independence through rapid agricultural, industrial and technological development. Consequently, there has been unplanned urbanization, industrialization, and intensification of agriculture. During the 1950s, the region had to import food grains. The use of high-yielding varieties and intensive use of agro-chemicals, and irrigation, led to a 'green revolution' bringing countries to self-sufficiency in food production. The human population, however, increased at a greater rate, often negating the benefits of other developmental activities.

The natural environment has been the victim of this vicious cycle. In India, for instance, deforestation has

Table 9.2 Land area and population of the countries of the Indian sub-continent. The values in parentheses are densities (individuals km⁻²)

Country	Area (× 10⁶ km²)	Population (× 10⁶)		
		1951	1991	2001
India	3.288	361 (110)	844 (257)	1025 (312)
Pakistan	0.88	40 (45)	118 (134)	155 (176)
Bangladesh	0.144	45.5 (316)	110 (764)	132 (917)
Sri Lanka	0.066	7.7 (117)	17 (258)	19 (288)
Nepal	0.148	8 (55)	18.5 (126)	22 (150)
Bhutan	0.047	0.73 (16)	1.44 (31)	1.74 (37)

reduced forest cover from more than 23% before independence to a present 13%; land degradation and soil erosion have increased, and both air and water are severely polluted, particularly near major urban and industrial centres. The all-round environmental degradation has obviously affected the rivers that have been over-exploited and developmentally abused.

Agriculture is the major consumer of water. It also requires an assured supply in the wake of unpredictable precipitation. The construction of reservoirs and irrigation tanks for storing surface run-off, and the diversion of river water through canals for irrigation, has been practised throughout the sub-continent since pre-historic times. The Grand Anicut on the Coleroon River (a branch of the River Cauvery) was built in the 1st century AD. In Sri Lanka, the Parakrama Samudra Reservoir was constructed in AD 386 (Brohier, 1934), while in Rajasthan, two dams – Rajasamand and Jaisamand (= Dhebar Lake) – were constructed in 1671 and 1730, respectively.

The waters of the Ganga and Yamuna rivers were diverted in the late 18th and 19th centuries by the Upper Yamuna Canal (1789), the Western and Eastern Yamuna Canals (1810), the Upper Ganga Canal (1854), Lower Ganga Canal (1868) and the Sone Canal (1879). Water from the eastern tributaries of the Indus was diverted between 1872 and 1922. In addition, the construction of large dams had commenced at the beginning of the 20th century, with Krishnarajasagar (1932), Chamrajasagar (1934) and Mettur Dam (1934) on the River Cauvery. Many rivers were also dammed to create reservoirs for drinking water supply. The Upper Lake in Bhopal was created on the Kolans as early as the 11th century.

During the past 50 years, river regulation has proceeded rapidly. Besides barrages for diverting water, thousands of multipurpose reservoirs have been created by building high dams for water supply, irrigation, hydro-power production and fisheries. Long

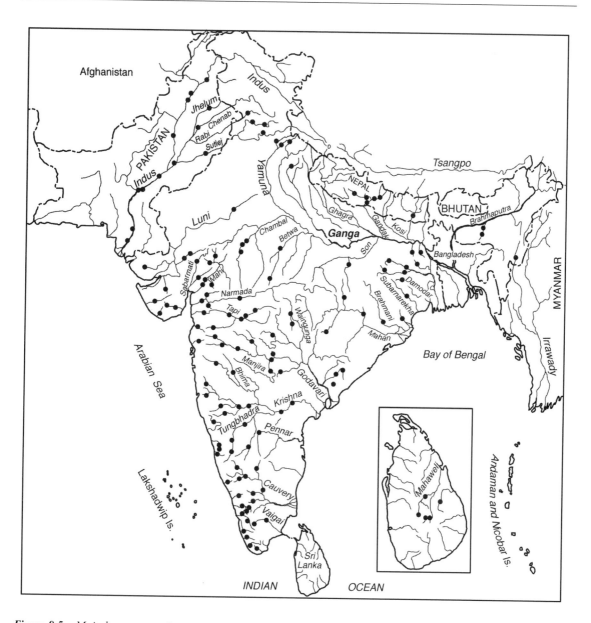

Figure 9.5 *Main large reservoirs on various rivers in the Indian sub-continent. Inset: Sri Lanka (magnified)*
(B. Bose and A.B. Goswami)

stretches of rivers, particularly those passing through urban areas, have been channelled by levee construction and floodplain 'reclamation', both for flood control and for urban development. In India alone, there are now more than 1500 large- and medium-sized dams and 100 barrages (Figure 9.5) on all the major rivers (except the Brahmaputra) and their tributaries. In Sri Lanka, there are about 250 large and medium reservoirs, and over 10 000 smaller reservoirs (2–2000 ha surface area). Recently, a series of large multipurpose reservoirs has been built on the Mahaweli Ganga and its tributaries (Hewavisenthe, 1992), while Nazneen (1995) reports some 55 reservoirs in Pakistan. More than 75% of the discharge of the Indus is withdrawn for domestic and agricultural uses. Although Bangladesh, with its flat deltaic topography, has avoided the construction of major water diversions and storage systems, two reservoirs, the Kaptai and Feni, have been constructed in the eastern Chittagong Hill Tracts for hydropower production and flood control. Embankments have also been constructed along many rivers in Bangladesh for flood mitigation, drainage improvements, river transport and salinity control (Ahmad *et al.*, 1994). In Nepal and Bhutan, dams and barrages have also been constructed for irrigation, hydropower generation and for flood control, downstream in India.

Another human activity as important as the construction of large reservoirs has been 'water harvesting', using small check dams or tanks for collecting run-off in the name of 'watershed development'. Such water harvesting invariably results in considerable flow reduction in the streams and rivers downstream.

Urban and industrial growth has also had a major impact on river water quality through the discharge of untreated domestic sewage and industrial effluents. These impacts have been aggravated by greatly reduced flows and channelization (e.g. see Pacini and Harper, this volume; Davies and Wishart, this volume). In several reaches, many rivers have turned into open sewers. Furthermore, human settlements, deforestation, mining, quarrying, grazing and other activities right into the headwaters of many systems have extensively degraded catchments and have increased sediment loads of all rivers in the region. The increasing use of fertilizers and pesticides in agriculture has also contributed to the degradation of water quality. These impacts are quite similar throughout the sub-continent, including Sri Lanka and, consequently, the river biota has been seriously affected. Riverine fisheries have declined considerably, and many species have nearly disappeared.

The rivers of the sub-continent are also used for navigation and recreation. Inland river transport historically has been very important for communication, trade and commerce and, even on a small scale, navigation also contributes to water pollution from motorized boats. The River Ganga has been used for regular ferry services between Calcutta and Allahabad since 1842, while the Yamuna was also extensively used between Allahabad and Agra until 1940. Similarly, the Brahmaputra and Meghna were regularly used until the 1950s, and the Indus and all major peninsular rivers were also navigable for considerable distances. However, except for some lower reaches (and the use of shallow-draft boats), water extraction, dams and siltation have made them unfit for navigation. Only Bangladesh regularly maintains its rivers for transport. In India, the River Hoogly, the estuarine section of the Ganga, requires regular dredging for navigation, and cargo ships have to be guided from the sea to the port.

Recreation in Indian sub-continental rivers is largely in the form of mass bathing and religious offerings to the river. In India, on several occasions every year, hundreds of thousands of people congregate for bathing in rivers, irrespective of the flow and water quality.

CATCHMENT AND LOCAL ISSUES FOR CONSERVATION

The human impacts on rivers throughout the region extend well beyond the direct use of water to all activities in the floodplain and the entire catchment. Floodplains have been traditional grazing grounds and have also been under cultivation for millennia, but during the past few decades the natural riparian vegetation has been almost completely removed. Tree species such as *Tamarix dioica*, *Anthocephalus kadamba* and *Mitragyna parviflora*, and reeds (*Phragmites* and *Arundo*), which once dominated the banks of the Yamuna have either disappeared, or occur only rarely. The riparian forests, dominated by species of *Barringtonia*, *Syzygium* and *Calamus*, which were common in many parts of India (e.g. Champion and Seth, 1968) occur now rarely, and in small patches. Floodplain cultivation makes extensive use of agrochemicals. After the floods recede, the river-bed is ploughed, terraced, or transformed into furrows and ridges right to the water margin, and vegetables or early-ripening winter season crops are grown. These practices have intensified the erosional forces as well as pollution in the main channels, while the natural breeding and feeding habitats of the aquatic biota have almost disappeared.

In the hills, where industrial development is relatively slow, deforestation and agricultural activities on the steep slopes causes heavy erosion, and silt and nutrient loads in the rivers are high, with severe consequences for riverine biota. Elsewhere throughout the catchments, there is intense pressure from the huge population of grazing animals. More than 45 million head of cattle, goats, sheep and camels decimate the vegetation in the Yamuna catchment alone. Quarrying and mining, particularly in the upper catchment areas, have also affected run-off amounts and rates, as well as water quality.

Table 9.3   *Grossly polluted stretches of Indian rivers where water is fit only for irrigation or waste disposal (from Ministry of Environment and Forests – MOEF, 1994)*

River	Polluted stretch
Sutlej (tributary of the Indus)	Below Ludhiana to Harike and below Nangal
Yamuna (tributary of the Ganga)	Delhi to Agra, and partly up to the confluence with the Chambal
Chambal (tributary of the Yamuna)	Downstream of the Nagda and below Kota
Gomti (tributary of the Ganga)	Lucknow to the confluence with the Ganga
Khan (tributary of the Chambal)	In and downstream of Indore
Kshipra (tributary of the Chambal)	In and downstream of Ujjain
Sabarmati	Ahmedabad to Veutha
Damodar	Below Dhanbad to Haldia
Godavari	In and below Nasik to Nanded
Krishna	Karad to Sangli; below Nagarjunasagar to Repella
Narmada	Near Jabalpur
Cauvery	Most of the lower stretches (Pugalur to Kumbhakonam)
Tapi	In and below Nepanagar to Burhanpur

MAIN AREAS AT RISK

The withdrawal of water, to the extent of the complete cessation of natural river flow on one hand, and the uncontrolled discharge of untreated (or at best partly treated) municipal sewage and industrial effluents on the other hand, have reduced many stretches of the rivers of the sub-continent to little more than waste-water drains: 'Class E' (Table 9.3; Ministry of Environment and Forests – MOEF, 1994). Among the best known examples are the Yamuna between Delhi and Agra, the Sabarmati from Ahmedabad to Ventha, the Gomti near Lucknow, the Khan near Indore and downstream, and the Kshipra near Ujjain. At present, plans are being discussed to channelize the Yamuna at Delhi and to reclaim the remaining riparian fringes for commercial and recreational use. If implemented, these plans will lead to further deterioration of downstream areas. Similarly, the estuarine section of the Narmada is endangered by the Sardar Sarovar Dam. Proposals for interbasin water transfers (IBTs; see below) have also been under consideration for some time, and may cause serious ecological damage if implemented (see also Davies *et al.*, this volume).

Agreements and protocols

The three Himalayan rivers – the Indus, Ganga and Brahmaputra – together with their tributaries, are shared by two or more countries. Shared watercourses were not recognized as a major problem during the British colonial period largely because of relatively low demand. Indeed, the Sarada Barrage on the River Mahakali in Nepal was constructed in 1920 after an agreement concluded by a simple exchange of letters between India and Nepal (Sharma, 1997). The sub-continent, however, has had a large number of princely states and the construction of reservoirs often led to disputes that were only resolved through intervention by the British. After the emergence of India and Pakistan as independent nations, shared rivers became a matter of prolonged discussion. The Indus River waters were readily shared under the Indus River Treaty of 1960 between India and Pakistan (Gulhati, 1973), but the Ganga became a matter of dispute between India and Bangladesh after India decided to construct a barrage at Farakka to divert water in order to ensure sufficient water for the port at Calcutta. The barrage was located only a few kilometres above the border between the two countries. This dispute has recently been resolved, with a long-term agreement between the two countries in 1996 to share the discharge at Farakka on a 50:50 basis. Various tributaries of the Ganga, originating in Nepal, bring huge amounts of water and sediments to the plains, causing recurrent floods. In order to alleviate some of these problems, India and Nepal entered into a treaty (1996) for the construction of several reservoirs within Nepal, particularly for hydropower generation and water sharing. A similar arrangement exists between India and Bhutan (the Chukha Hydel Project).

Within India, several tributaries of the River Ganga, as well as all of the peninsular rivers, are shared by several states and some are in dispute. Under the National Water Policy of India, mechanisms have been provided for the resolution of disputes through tribunals and inter-state coordination councils. Regular disputes occur between Himachal Pradesh, Haryana, Uttar Pradesh, Delhi and Rajasthan over the Yamuna River; the River Narmada is a regular source of conflict between Madhya Pradesh and Gujarat; and the Cauvery, between Karnataka and Tamil Nadu. However, most of the disputes are resolved through mutual dialogue, or judicial intervention at the highest level.

Current understanding of rivers: information for conservation

Rivers are not just channels with running water. They are complex ecological systems that interact with their drainage basins, collecting from them water, nutrients and organic matter, and as such they support large biological diversity besides humans and their activities. These systems are (i) driven by their hydrology and fluvial geomorphology; (ii) structured by food webs; (iii) characterized by spiralling processes; and (iv) are dependent upon changes of flows, sediment movements and channel shifts (e.g. Calow and Petts, 1992, 1994). While considerable advances have been made in many countries in the study of rivers as dynamic living systems (e.g. Boon *et al.*, 1992; Calow and Petts, 1992, 1994; Petts and Amoros, 1996), studies of rivers in the Indian sub-continent are at best only fragmentary. Considerable emphasis has been laid on hydrology, fluvial geomorphology and sediment transport. Comprehensive treatments of these studies are available in many books and reports (e.g. Rao, 1975; Fernando, 1985; Elahi *et al.*, 1991; Shroder, 1993; Sharma, 1997).

RIVER DISCHARGE, SEDIMENTS AND CHANNEL DYNAMICS

Like other Asian rivers, the rivers of the sub-continent are characterized by large seasonal variations in discharge, a direct result of seasonal and inter-annual variations in precipitation. Himalayan rivers are fed during the dry summer months by snow-melt which, however, provides only a small proportion of total annual flow. Indeed, more than 70% of total discharge occurs during the short rainy season, which temporally varies from east (starting in May) to the west (end of June) with the onset of the south-west monsoon. Spatial and temporal variation in rainfall intensity also leads to great variation in the timing and magnitude of peak discharges across the region. For instance, the Sri Lankan and peninsular rivers derive their flow from both the south-west and the north-east monsoons. As such, their discharges are seasonally moderated. At the other extreme, the rivers of the western, arid regions of India and Pakistan exhibit large fluctuations in

discharge, from absence of flow to very high flows in some years. Hydrographs of the Ganga and Brahmaputra rivers are illustrated in Figure 9.6. The major rivers of the Indian sub-continent discharge some $2\,162\,172 \times 10^6\,\text{m}^3\,\text{yr}^{-1}$, derived from *ca* 476×10^6 ha of land. Of this, the Indus, Ganga and Brahmaputra together account for about two-thirds, from approximately the same proportion of land area.

The drainage basins of the Himalayan rivers are the most densely populated areas in the world, let alone the sub-continent. Furthermore, the rivers originate in mountain ranges that, being geologically young, comprise unconsolidated sedimentary rocks and, hence, are highly prone to erosion. It is, therefore, hardly surprising that these rivers also carry high silt loads – amongst the highest of the world's rivers – as well as high solute loads (Subramanian, 1979; Subramanian *et al.*, 1987). The overall silt load in these systems amounts to more than 50% of the world's total. In this context, the Ganga–Brahmaputra–Meghna systems together form the world's largest delta (about 6×10^6 ha – Mha) which is well known for its mangroves. The delta of the Indus covers some 2.95 Mha. These same rivers also account for about 5% of the total solutes carried by the world's rivers to the oceans. The peninsular rivers carry relatively smaller sediment and solute loads (Vaithiyanathan *et al.*, 1988).

Of course, the channel dynamics of these systems (meandering/braiding) are closely related to the high sediment loads; their dynamics are also influenced by the tectonics of the region. Extensive sediment deposits and abrupt breaks in slope have caused shifting of channels over long distances. In some cases, channel shifts have occurred over centuries, while for others, channels shift far more frequently. For example, the Son has moved back and forth during the past few centuries (Figure 9.7), while the lower reaches of the Brahmaputra have moved some 80–100 km westwards since 1830. The River Tista, a Himalayan tributary of the Brahmaputra, has also moved course to a similar extent and the Kosi, a tributary of the Ganga, is notorious for changing its course across the plains of north Bihar: it has moved about 120 km westwards during the past 220 years (Gole and Chitale, 1966; Wells and Dorr, 1987; Agarwal and Chak, 1991; Figure 9.8). Tectonic activity and morphogenetic uplift in the west are both causing an eastward shift of the River Ganga in its lower reaches (Gupta, 1957; Chowdhury, 1966). Interestingly, the channel dynamics of the Indus in its middle and lower reaches have been investigated in detail in search of an explanation for the

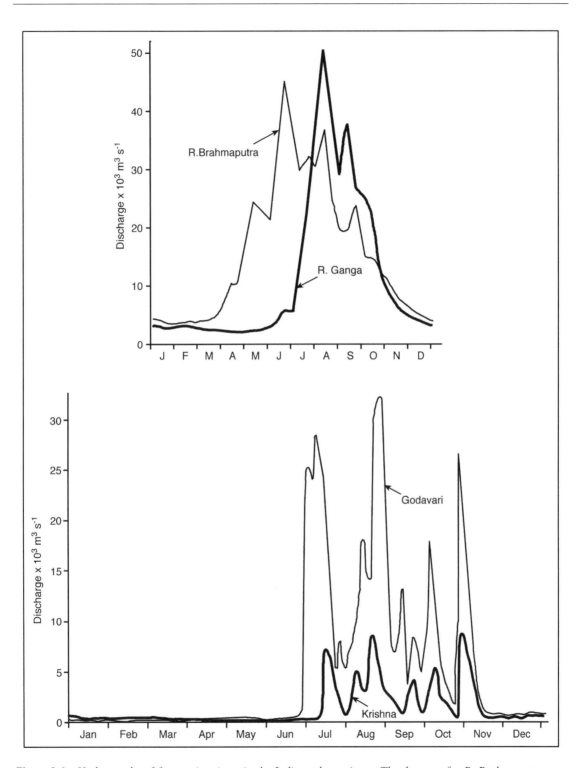

Figure 9.6 *Hydrographs of four major rivers in the Indian sub-continent. The data are for R. Brahmaputra at Pandu, R. Ganga at Farakka, R. Godavari at Polavarim, and R. Krishna at Deosugur (data from Rao, 1975)*

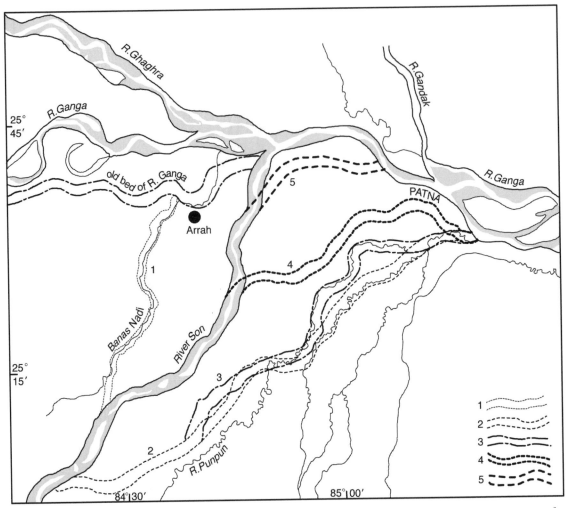

Figure 9.7 *Shifting courses of the River Son, a right-bank tributary of the Ganga. 1 to 5 mark the positions of channels over the past 3000 years. 1, 2, 3, ca 1000–600 BC; 4, 400–300 BC; 5, 1200–1500 AD*

abandonment of the ancient city of Mohenjodaro (*ca* 2500 BC). Jorgensen *et al.* (1993) observe that increased flooding and sediment deposition, followed by avulsion of the river away from the city reasonably explains the abandonment. Tectonic disturbances result in variations in channel planform, channel dimensions and frequencies of flooding and, consequently, in the great temporal and spatial variability in patterns of floodplain sedimentation.

ECOLOGY AND WATER QUALITY

There has been considerable interest in the riverine fisheries of the region for more than a century. In his book on the fish of the Ganga, Hamilton (1822)

described 269 species from the river. Day (1878) published a comprehensive monograph of fish of the Indian sub-continent, together with an account of their habitats. Comprehensive studies of fish were made by Hora, who published extensively on all aspects of their biology and ecology, and their responses to human activities (Hora, 1921, 1923, 1935, 1937a,b). The riverine fisheries have been extensively investigated by the researchers at the Central Inland Capture Fisheries Research Institute and other fisheries institutes in different countries of the sub-continent (Shreshtha, 1990; Jhingran, 1991).

However, ecological studies on the rivers, their biotas, as well as on water quality were only initiated early in the 1950s in India. The rivers in other countries

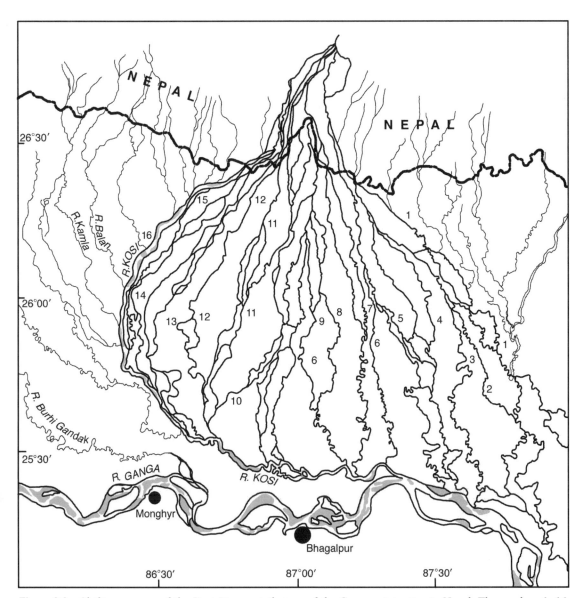

Figure 9.8 *Shifting courses of the Kosi River, a tributary of the Ganga originating in Nepal. The numbers 1–16 represent various channels in the past. The approximate period of some of these channels are: 5, 1731; 6, 1770–79; 7, 1807–39; 8, 1840–73; 9, 1873–93; 10, 1893–1921; 11, 1921–26; 12, 1926–30; 13, 1930–35; 14, 1936–42; 15, 1942–48; 16, present*

of the sub-continent started to receive attention much later. The early Indian studies focused mostly on certain sections of the Ganga and Yamuna, where universities or other research institutes were located close by. These studies include those of Roy (1949, 1955), Dutta *et al.* (1954), Chakrabarty *et al.* (1959), Rai (1962, 1974a,b), Lakshminarayana (1965) and Ray *et al.* (1966). A beginning was also made in south India with some

minor rivers (e.g. Chacko and Ganapati, 1949; Iyengar and Venkataraman, 1951; Chacko *et al.*, 1953). During the 1960s and 1970s, the physical and chemical characteristics and the phyto- and zooplankton communities of several smaller rivers were investigated (e.g. Venkateswarlu and Jayanti, 1968; Venkateswarlu, 1969, 1970, 1986), but river systems in general remained ignored by biologists and limnologists. Similarly, in Sri Lanka and Pakistan, some preliminary descriptive studies on rivers were started in the 1960s but relatively little progress has been made so far (see Costa and De Silva, 1995; Khondker, 1995; Nazneen, 1995).

Soon after the Stockholm Conference, the Indian Government took the lead in the region by enacting a Water (Prevention and Control of Pollution) Act in 1974 and steps were taken to reduce the pollution of surface waters. In 1982, the newly created Department of Environment of the Government of India initiated a coordinated project on the River Ganga, involving many universities and research institutes along its course. This stimulated investigations of water quality and the biological communities of rivers throughout India. At the same time, a national water quality monitoring programme in major and medium rivers commenced through the Central Pollution Control Board (CPCB, 1996). This programme focuses on several physical and chemical variables and faecal coliforms. Thus, during the past 20 years or so, considerable information has been generated for a few limnological components of several Indian rivers (see Trivedy, 1988, 1990). Of the larger rivers, only the Cauvery (Sivaramakrishnan *et al.*, 1996) and Narmada (Unni, 1997) have been investigated to some extent. Two other large international rivers, the Indus and Brahmaputra, have also been sparingly investigated. Most of these studies have been summarized in an extensive review of Indian limnology by Gopal and Zutshi (2000).

The River Ganga, along with its tributaries, has a drainage basin covering 1.06×10^6 km^2 of which the Indian component (8.6×10^5 km^2) accounts for 26.2% of the country. The river's flow (468.7×10^9 m^3 yr^{-1}) represents 25.2% of India's total water resources and the basin sustains about 300 million people in India, Nepal and Bangladesh. It has been the most investigated river of the sub-continent and a brief account of our knowledge of the Ganga and its major tributaries can also be taken to represent the current state of our knowledge of river systems in the region.

The River Ganga

The river and its tributaries descend very rapidly from the high altitude Himalayan sources to Rishikesh, after which the gradient is very shallow – less than 10 cm km^{-1} in the lower reaches. Thus the subsequent low gradient and huge sediment loads result in a vast floodplain, stretching over 100 km to either side of the main channel in some places. Less than 10% of the

alluvial plain is under natural forest, and the forest cover in the hills is fast disappearing. The most important land use in the basin, and even on the hills, is agriculture sustained by extensive irrigation systems. The Upper Ganga Canal diverts more than 60% of the annual flow, and almost 100% of the dry season flow, at Hardwar only 30 km downstream from Rishikesh. The Lower Ganga Canal diverts more water some 240 km downstream. The shallow gradient of 36 m between Hardwar and Aligarh (*ca* 200 km) has been exploited for 13 small hydroelectric projects. Now a dam is being constructed in the hills at Tehri on the major tributary, the River Bhagirathi, to exploit its hydroelectric potential, despite the fact that the area lies in a seismically active belt. There are more than 100 important urban settlements along the banks of the Ganga alone. Numerous industries are also located along the river, particularly in its middle and lower reaches.

Geographers and geomorphologists have extensively studied the fluvial geomorphology of the river basin for land-use planning and flood control. In the upper reaches, where the river flows southwards, the floodplain (locally known as Khadar) is characterized by a combination of extensive braiding and meandering channels (Figure 9.9). The course oscillates laterally from northwest to southeast, resulting in the formation of terraces, alluvial fans, marshes, point-bar deposits and shallow water bodies. Six micro-geomorphological belts have been identified in accordance with a depositional succession from west to east (Dasgupta, 1975). Braiding is relatively less common in the lower reaches, but several tributaries from Nepal develop extensive braiding, and change their course very frequently, mainly by avulsion. Recently, Sinha and Friend (1994) have discussed the sediment fluxes of these tributaries in detail. They observed that the mountain-fed rivers are characterized by very high discharges and low suspended sediment loads; rivers originating in foothills have moderate discharges and sediment loads, while plain-fed systems have low discharges and high sediment loads. More than 90% of the sediments are retained in the basin and the remainder are carried to the delta. Floods are a regular feature of every year and, hence, long reaches have already been regulated by embankments. However, these works have failed to reduce flood hazards.

The Ganga is one of the most polluted rivers of the sub-continent. Studies conducted by 14 universities between 1983 and 1989 along the river (part of the coordinated   programme   of   Ganga   Basin

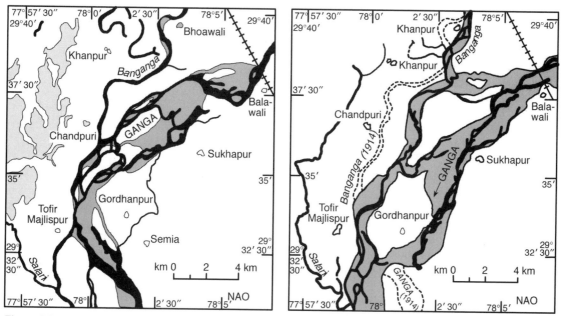

Figure 9.9 *Changes in the course of the Ganga in the upper Gangetic plain, from 1914–15 (left) to 1964 (right) (from Dasgupta, 1975)*

Ecodevelopment), included analyses of water quality, biotic inventories, the identification of pollution bio-indicators, assessments of heavy metals, and aquatic microbiology (Krishnamurti *et al.*, 1991). However, the work was uncoordinated and failed to investigate the same variables with similar methodologies. Furthermore, no efforts were made to identify trends along the river, nor were any efforts made to observe the influences of various tributaries. The studies did, however, reveal increasing levels of pollution in the river on the plains.

The Himalayan stretch of the Ganga (represented by the Bhagirathi and Alaknanda tributaries) was relatively unpolluted, while the middle reaches at Kanpur were heavily polluted by heavy metals from industrial effluents. Inventories of the phyto- and zooplankton, benthic invertebrates, molluscs, amphibians, reptiles, fish and other fauna, as well as bacteria and fungi, have also provided some information on seasonal changes in densities and on biological diversity. Although information on changes in species composition and total numbers along the river was generated, no clear horizontal trends, or direct correlations with recorded pollution levels were established. Information on food webs, nutrient dynamics, ecosystem functioning and riparian (floodplain) interactions is unavailable and work on the effects of dams and barrages on water quality and

biology are rare. Chakraborty and Chattopadhyay (1989) have reported shifts in species composition, and increases in abundance of the plankton community in the estuarine section of the Ganga (the Hoogly) after the construction of Farakka Barrage, owing to the greater availability of fresh water and lowered salinities.

The only long-term studies that are available concern the fisheries resources of the system. These show a clear, sharp decline in catches, coupled to significant changes in species composition. Dams have long been considered to be problematic for fish communities (e.g. Jhingran, 1991) and, in the Himalayan reaches, habitat alteration, including dams, has had drastic effects. Stocks of the characteristic cold-water schizothoraciine fish, such as mahseer (*Tor putitora* and *T. tor*). and the snow trout (*Schizothorax richardsonii* and *S. plagiostomus*), have all declined due to lack of recruitment; the dams prevent migration (Sehgal, 1994; Dhanze and Dhanze, 1996). In the lower reaches, the Farakka Barrage has frequently been considered to be a primary candidate for the decline of hilsa (*Hilsa* (=*Tenualosa*) *ilisha*) and other species. *Hilsa* is a typical long-distance anadromous species that migrates beyond the middle reaches of the Ganga: since construction of the barrage, its catch has declined from 19.3 to 1.04 t at Allahabad, and from 32 to 0.6 t at Buxar (Chandra,

Table 9.4 *Changes in water quality of the River Yamuna between 1958 and 1989 (adapted from Gopal and Sah, 1993)*

Variable	Wazirabad Upstream		Okhla Barrage	
	1958–59	1989	1958–59	1989
pH	7.8–8.4	7.5–9.0	7.7–8.3	7.2–10.0
Dissolved O_2 (mg L^{-1})	7.1	4.0–11.2	6.1	0.6–18.0
BOD (mg L^{-1})	2.1	2.6–2.3	3.6	4.3–98
PO_4-P (mg L^{-1})	0.1	0.04–0.64	0–0.2	0.08–2.35
Total Hardness (mg L^{-1})	74–182	76–192	82–200	76–320

1989). In the plains, carp (*Cirrhinus mrigala, Catla catla* and *Labeo* species), and catfish (*Mystus* sp., *Wallago attu, Ompok* sp. and *Pangasius pangasius*) constitute the major fisheries. Freshwater prawns (*Macrobrachium* spp) are also abundant. Fish landings have been steadily declining over the past 40 to 50 years, from 961 kg km^{-1} in 1956–1960, to 630 kg km^{-1} in 1981–1987 (Chandra, 1989; De *et al.*, 1989). The loss of feeding and breeding habitats in the floodplain lakes due to embankment construction, and to increased silt loads and macrophytic growth which has reduced the deep perennial pools in the river channel, have been identified as major causes for declining fish catches along these stretches of river (Jhingran, 1991). Other projects on tributaries such as the Kosi and Gandak have also adversely affected the fisheries in northern Bihar.

Of the other fauna that have been seriously affected by changes to habitat and water quality, the Ganges River dolphin (*Platanista gangetica*) is probably the most important. The dolphin was once abundant and widely distributed throughout the Ganga–Brahmaputra; its penetration upstream was limited by lack of water and rocky shorelines, and downstream by salinity. In recent years, however, the species has declined rapidly from about 2000 individuals in 1988–1989 to only 152 (Sinha, 1997), and great concern has been raised about the threats of pollution, water abstractions and dams to the dolphin (e.g. Perrin *et al.*, 1989; Shreshtha, 1989; Reeves *et al.*, 1991; Ahmed, 1992; Smith *et al.*, 1994; Biswas *et al.*, 1997). Similar concerns have also been expressed about the threats to otter (*Lutra* spp), which are widely distributed throughout the sub-continent (De Silva, 1993).

The River Yamuna, the largest tributary of the Ganga, has a somewhat better database than the main stem, probably because three important cities, Delhi, Mathura and Agra, are situated on its banks. Several tributaries of the Yamuna, such as the Khan and the Kshipra, have also been investigated. These studies

have been reviewed in some detail by Gopal and Sah (1993). The extraction of water via a series of barrages, extensive channelization, and heavy discharges of untreated domestic and industrial effluent, have turned the river into a sewer, at least between Delhi and Agra. It partly regains its riverine characteristics only below its confluence with the Chambal, which has a greater flow. The water quality and the biota have changed greatly over the past four decades (Table 9.4; Gopal and Sah, 1993). Turtle and crocodile, once abundant in the river, have almost disappeared. Even in the Chambal, the crocodile has declined, necessitating the establishment of a National Chambal Sanctuary for the Gharial.

EVALUATION OF WATER QUALITY AND CLASSIFICATION OF RIVERS

Rivers are generally evaluated for their water quality and can be classified into zones based on progressive changes in benthos and fish which reflect the totality of physical and chemical characteristics (e.g. Hawkes, 1975). Over recent years, all countries of the sub-continent have invested considerable effort to evaluate river water quality. However, such evaluations are based almost entirely on oxygen-related (biological oxygen demand (BOD), chemical oxygen demand (COD) and dissolved oxygen concentrations (DO)), and microbial parameters (coliforms and streptococci). Other physical and chemical components are also often considered, but pesticide and heavy metal concentrations are usually examined only along industrially-affected reaches. Biological criteria have yet to be developed largely due to problems of identification, a situation common to many developing communities (see e.g. Davies and Wishart, this volume; Wishart *et al.*, this volume, Part II). Many studies have identified organisms that are tolerant to pollution or that occur in waters with known pollutants, and experimental eco-toxicological studies are also common, but biological indices for regular monitoring have not yet been developed (see De Zwart, 1991; Krishnamurti *et al.*, 1991). A detailed study of the Yamuna by the CPCB in India, in collaboration with Dutch scientists, has shown that no single biological index could fully represent water quality correctly and reliably (De Kruijf *et al.*, 1992). Another study of the benthic fauna along the entire course of the Cauvery similarly noted discrepancies between different assessment methods and indices, although Biological Monitoring Working Party (BMWP) and Rapid Bioassessment Protocol III (RBP III) (respectively, Armitage *et al.* (1983); Plafkin *et al.* (1989)) showed the best correlation to habitat quality, and were least influenced by naturally occurring physical habitat gradients (Sivaramakrishnan *et al.*, 1996).

Regular (quarterly) monitoring of water quality only started in India in 1977. The CPCB has set up a network of *ca* 400 stations on all major rivers and their tributaries, as well as on a few medium and minor rivers. The data are periodically published and are used to classify and map stretches according to the Designated Best Use Classification (DBU) that recognizes five categories (A–E) in increasing order of pollution:

- A signifies water suitable for drinking without conventional treatment but after disinfection;
- B is suitable for outdoor bathing;
- C is suitable for drinking after conventional treatment and disinfection;
- D includes water appropriate for fisheries and wildlife;
- E signifies waters for irrigation, controlled waste disposal and industrial cooling.

According to the latest available data (for 1993–1994; CPCB, 1996), water pollution has escalated, and rivers that in the 1980s were polluted over a few short stretches are now polluted up to their headwaters. The coliform and the BOD levels exceed minimum desired levels often near headwaters, and increase steadily downstream. Several sections of most rivers are now classified as 'E' (Table 9.3). The rivers in all other countries of the sub-continent are similarly polluted (e.g. Costa and De Silva, 1995; Khondker, 1995; Nazneen, 1995; Sharma, 1997). In Sri Lanka, industrial effluents, agro-chemicals and domestic wastes have been identified as major sources of river pollution. The National Aquatic Resources Agency (NARA), established in 1981, is the major organization responsible for monitoring water quality and the

management of aquatic resources in Sri Lanka. Specific monitoring programmes in other countries of the region are not known.

AVAILABILITY AND DISSEMINATION OF INFORMATION

There are several sources of information relating to Indian rivers, including Government agencies such as the Central Water Commission (CWC), the Central Board of Irrigation and Power (CBIP) and the CPCB. The CWC is directly responsible for the collection of hydrological data and is also interested in water quality surveys (CWC, 1987); the CPCB activities have already been noted. Fisheries research institutes and Departments of Fisheries in various states collect data on fisheries, other biotic communities, and also on water quality, whilst numerous universities are engaged in river research. The data generated by Government agencies and departments are not freely disseminated and are difficult to obtain; much of it remains in internal reports. Furthermore, hydrological data on the international rivers (Indus, Ganga, Brahmaputra and Barak) are classified and are not readily available even to other Government agencies. On the other hand, data generated by the universities and research institutes are generally published in journals and conference proceedings, mostly at national level. However, even these publications are not readily accessible to everyone interested. The national abstracting services, such as those provided by the Indian National Scientific Documentation Centre, New Delhi, or the Environmental Information Service (ENVIS) of the Ministry of Environment and Forests, New Delhi, are of very little help. In addition, substantial amounts of information remain unavailable in the form of dissertations and project reports, as there is no system available for publicising, abstracting or collating the material. The situation is similar in all other countries of the region and, although efforts are made by individuals to compile some information in reviews and books, the picture remains sketchy and incomplete.

PUBLIC AWARENESS OF CONSERVATION

Until the late 1960s, the environment in general was not an issue of public concern, although water pollution and waterborne diseases were common. After the United Nations Conference on Environment and Development in Stockholm (1972), India was the first country in the region to adopt a Water (Prevention and Control of Pollution) Act (1974). The legislation was aimed primarily at the regulation of industrial effluents in surface waters. During the past 25 years, many efforts have been made throughout the sub-continent to create environmental awareness and to educate people about various environmental issues. In 1984, the Government of India initiated the Ganga Action Plan (GAP), aimed at reducing pollution loads in the system by providing sewerage and sewage treatment facilities in all major urban settlements along its course (Department of Environment, Government of India, 1985). This brought into focus the state of pollution of all Indian rivers. The Plan also highlighted the decline of fisheries and other biota in the rivers. The programme has now been extended countrywide as the National River Action Plan (MOEF, 1994). These official activities, supported by extensive television coverage, seminars and workshops, have resulted in considerable enhancement of public awareness about the pollution of rivers by domestic sewage and industrial effluents, and its consequences for humans and other biota. However, this provides only a partial, skewed picture of reality, because the importance of catchment-based human activities, and the importance of river flows for the biota and the self-purification potential of rivers, is neither communicated nor appreciated.

The framework for conservation

SOCIO-POLITICAL AND RELIGIOUS-CULTURAL BACKGROUND

The conservation of natural resources and biota has been an integral part of the socio-cultural ethos of the sub-continent. All major rivers have been considered sacred and numerous mythological events are associated with practically every river. The Ganga is believed to have been brought to the Earth from the heavens after the rigorous penance of King Bhagirath. When Ganga descended in a torrent, Lord Shiva held her in his locks, tamed her and allowed her to flow slowly to the Earth. The Yamuna is considered to be the daughter of the Sun and sister of Lord Yama – the god of death, while the Narmada is considered to be the daughter of Lord Shiva. Many aquatic animals have also been treated as gods and often worshipped: fish, turtle and crocodile were considered as incarnations of gods, and were protected. Thus, most water-polluting activities and waste disposal of any kind were prohibited, although the self-purification ability of rivers was recognized. Mass bathing is still practised on certain auspicious days, and even the

disposal of the dead was allowed. Conservation extended to the floodplains, including their biota, as they were important grazing lands; people also recognized their role in flood regulation.

During the past few decades, however, socio-cultural values have dramatically changed, influenced by rapid industrialization, the influences of Western cultures and, more recently, by rapid globalization. The result is that although rivers are still symbolically worshipped by the religious-minded, they are treated more with contempt. Rapid population increase and urbanization have led to larger demands for food and water, while waste-disposal systems have not been developed. Policy- and decision-makers treat rivers simply as water sources. Within this changing scenario, conservation has become a buzzword which is used more in speech than in practice, and conservation efforts are driven primarily by considerations other than national or regional requirements. Political actions and priorities are often directed by the availability of international financial assistance. Thus, policy-makers invariably follow sectoral, short-term approaches instead of adopting holistic long-term planning strategies; in this scenario effective conservation policies cannot be anticipated.

NATIONAL AND INTERNATIONAL LEGISLATION

There is no specific statutory provision for the conservation of rivers in any country of the region. The Indian Constitution provides for protection of nature and the environment, and various other laws provide for the prevention and control of water pollution and the conservation of wildlife.

The first formal comprehensive document on river water utilization in India was the 1972 Report of the Irrigation Commission. It led to the formulation of an Irrigation Policy that aimed to promote maximum crop production per unit area of arable land, and the greatest possible use of river water, to bring the maximum possible area of agricultural land under a single irrigation system. The policy, however, did recognize that unlimited withdrawal of water from rivers would damage their natural hydraulic regimes and ecosystems, and it stipulated the withdrawal of 37% of the annual flow of any river as the safe limit.

Ever increasing demands for water exposed the limitations of the prevailing practices of using both surface waters and groundwaters. Accordingly, the Government of India set up the National Water Resources Council (NWRC) in 1985 to develop a policy for the rational management of water. The resulting National Water Policy (NWP), finalized in 1987 (MOWR, 1987), recognized the need for the development, utilization, management and conservation of water resources, guided by national perspectives. It listed water-use priorities, with potable supply as top priority, followed by irrigation, hydropower production, navigation, and industrial and other uses.

To implement the NWP, a National Water Board was formed in 1990. This body has since formulated:

- the broad guidelines for an Irrigation Management Policy
- a Water Information Bill
- guidelines for the rehabilitation of individuals or communities displaced by reservoir projects
- National Policy Guidelines for inter-state river water allocations
- Policy Guidelines for river basin organizations
- an approach to organizational procedural changes in the irrigation sector (Indian Water Resources Society – IWRS, 1995)

It has further recognized the need to modify prevailing patterns of agriculture, and the introduction of accountability and transparency in water use, and regulatory measures and codes. The work plan prepared so far has yet to be implemented.

In 1986, Bangladesh also formulated its own Water Plan, giving prominence to the recurrent menace of floods. Again, water uses have been prioritized, with potable supply and industrial uses at the top, followed by navigation and salinity control. The plan specifies that 40% of the annual flow must be retained in any river, to maintain navigation routes, fisheries and salinities in lower reaches. Water in excess of this figure can be used for irrigation which otherwise depends mainly on groundwater.

Pakistan is heavily dependent on the Indus River water, from which water availability has been ensured by the Indus Water Treaty of 1960 with India (Gulhati, 1973). While drinking water and irrigation receive priority use, in 1991 Pakistan formulated a policy on inter-provincial water allocation of the Indus in order to avoid conflict.

The Water Resources Act, 2049 (=1992 AD) of Nepal, closely follows the Indian Water Policy and focuses on the use of water for (in order of priority) drinking and domestic use, irrigation, agriculture and fisheries, hydroelectricity production, cottage industries, navigation and recreation (Sharma, 1997).

Article 19 of the Act states that 'No one shall pollute [a] water resource by way of using or putting any litter, industrial waste, poison, chemical or toxicant to the effect that the pollution is to tolerance limit of the water resource as prescribed' [*sic*]. The Act also insists that 'while utilising water resources, it shall be done so in such a manner that no substantial adverse effect be made on [the] environment by way of soil erosion, flood, landslide or similar other cause'.

THE ROLE OF NGOs

For the past 20 years, non-governmental organizations (NGOs) have been playing an important role in creating public environmental and conservation awareness. Some NGOs have brought pressure to bear on government policies. Water resources in general, and water pollution in particular, have been important areas of interest. The growing pollution of the Ganga, Narmada and Cauvery systems, and declining fisheries, have been major issues taken up by Indian NGOs. The Sankat Mochan Foundation, headed by V.B. Misra – a professor of Civil Engineering at Varanasi – has been actively running the 'Clean Ganga Campaign' to raise public consciousness about the pollution of the system, and pollution is regularly monitored. The most active groups protesting against the construction of large dams are the Narmada Bachao Andolan (Save the Narmada Movement) against the Sardar Sarovar and Maheshwar Dams on the Narmada, and the Committee against the Tehri Dam on the Ganga. The Centre for Science and Environment has focused on the futility of dams and embankments in flood control (Agarwal and Chak, 1991). Similar NGOs have been active in other countries of the region.

It must be pointed out, however, that NGOs have rarely been successful in achieving their goals. There are so many water-resources stakeholders that they often find themselves in conflict with each other, and the necessary scientific background for sustainable management is so poor that NGOs are mainly unable to convince decision-makers of their viewpoints. Furthermore, the relationships between the catchment and the river are conveniently ignored both by the policy-makers and NGOs.

Conservation in practice

Rivers play a major role in economies by sustaining agriculture, industry, energy generation and biological resources. The impacts of extensive flow regulation and high pollution loads are already appearing in the form of fisheries decline, increased flooding, falling water tables and an increase in waterborne diseases. Thus, management requires a holistic approach combining water quantity and quality and the conservation of biological diversity. Unfortunately, management emphasis on the sub-continent continues to focus on water extraction which often exceeds 80% of total discharge. Indeed efforts are being directed to use the remaining water by further flow regulation. This focus has led to inter-state and international disputes. Several proposed IBTs include a Ganga–Cauvery link to transfer water from the Ganga to areas of water deficiency in central, western and southern India, while in the early 1970s United Nations Development Programme (UNDP) endorsed the concept of a National Water Grid. Other proposals include a link between the Brahmaputra and Ganga, passing through Bangladesh, as well as between the rivers of the Western Ghats to the east (Rao, 1975), and a Jamuna–Padma Canal within Bangladesh (Crow *et al.*, 1995) has been proposed.

There is also an increase in emphasis on hydropower generation with no consideration of the environmental impacts of reservoirs, while the reservoirs themselves are silting more rapidly than envisaged. Multipurpose projects have raised many environmental and socio-economic issues; riverine fisheries are being substituted by culture fisheries in reservoirs; and although the impacts of dams on river fisheries were pointed out long ago (e.g. Hora, 1940, 1942), relatively little attention has been paid to them.

On the other hand, in India, considerable attention has been paid to water pollution control. The Water (Prevention and Control of Pollution) Act of 1974 and its subsequent amendments require that industrial effluents are treated to meet prescribed standards before discharge into the receiving waters. However, implementation has been difficult and remains far from satisfactory. Further, municipal bodies responsible for the discharge of domestic sewage are not covered by the Act and have failed, therefore, to provide sewage treatment plants. These facilities are inadequate even in large urban centres. The Act applies only to point discharges from specific units, while numerous small and large drains continue to receive wastes from smaller industries, and domestic wastes from slums. Thus, the legislative measures have remained virtually ineffective, and the need for stringent measures to ensure wastewater diversion and treatment has been voiced time and again (Central Board for the

Prevention and Control of Water Pollution – CBPCWP, 1982; Bhargava, 1985a). The effluent standards for wastewater disposal disregard the total amount of effluents discharged at any point, or upstream, the assimilative capacity of the river, as well as changes in both the effluent quantity and assimilative capacity of the river with time. Since water quality varies greatly in different zones, it is necessary to develop standards that will determine effluent quality according to flow, assimilative capacity and local conditions. Such standards will require constant monitoring, regular updating and efficient effluent treatment.

The catchment impacts of human activities are generally ignored and are aggravated by reduced availability of water and its poor quality. Bhargava (1983, 1985a,b), in his studies of the Yamuna, suggested improved agricultural practices through economical use of chemicals, erosion control and the recycling of organic wastes to control non-point source pollution. However, the Yamuna River Action Plan aimed at improving the water quality in the river does not include these measures.

Floodplains generally have a high nutrient recycling capacity, usually accumulate nitrogen, phosphorus and other nutrients and also sequester heavy metals and toxic compounds in anaerobic organic sediments. Thus, they have the capacity to process the waste waters flowing through them and to regulate nutrient input (e.g. Lowrance *et al.*, 1984; Richardson, 1990). Floodplain vegetation also reduces run-off velocity and traps sediments, thereby reducing river siltation (Boto and Patrick, 1979). Another important role of floodplain and riparian vegetation is soil erosion reduction (Dagar, 1987; Dhingra, 1989; Prajapati, 1989). The importance of the interactions between the river channel and its floodplain for groundwater recharge, river water quality, fisheries and biodiversity (e.g. Naiman and Décamps, 1990; Jensen, 1994) have not been appreciated in the region.

Reduced flow (and often complete absence) only aggravates the problem of pollution. The conflict between the maintenance of river flow to minimize pollution, and water extraction, to meet increasing demands for irrigation and domestic supply, is indeed difficult to resolve. Often suggestions are made for flow augmentation during low-flow periods in critical reaches, after storing the monsoon run-off (e.g. CBPCWP, 1982; Bhargava, 1985a), but such steps have not been taken.

The myopic view of rivers as carriers of water ignores their ecological functioning and biological diversity. Many studies have shown that fish diversity is directly related to habitat diversity, which includes substratum diversity, diversity of food organisms, flow regimes and riparian vegetation, as well as to disturbance and pollution (Welcomme, 1979; Lowe-McConnell, 1987; Ward and Stanford, 1989; Arunachalam *et al.*, 1997). Thus, flow regulation, channelization and pollution have led to the reduction or extinction of populations of many species. Dehadrai and Poniah (1997) report that in India at least 69 species of fish are threatened or endangered.

APPLICATION AND POLICING OF LEGISLATION

Although conservation is the declared policy of the governments of the region, and although several legislative measures have been taken, the situation has only been deteriorating instead of showing signs of improvement. The report of the CPCB (1996) admits that most rivers are polluted right to their headwaters – a condition that did not exist 20 years ago (CBPCWP, 1982, 1983, 1984). This is the combined effect of several factors, of which non-implementation of legislative measures is one. In India, implementation of the water pollution control legislation of 1974 has been slow because state governments took several years to adopt the federal law. Economic factors are generally cited for the slow implementation; industries are not installing treatment plants. However, the law was not enforced on the municipal authorities responsible for the collection and treatment of domestic wastes. Even today, wastes are discharged without treatment in Delhi itself. The GAP was directed towards the provision of sewage treatment facilities in major towns, but treatment plants are still not operational in many places, 15 years after GAP. Official corruption is an important factor that makes policing almost meaningless. Effluents are discharged without meeting standards, and/or reports are manipulated. Some industries try to escape penalties by diluting their wastes to meet the effluent standards. To compound the problem, the judicial process is often cumbersome and time-consuming and, hence, hampers the rapid implementation of the law. Finally, there are two additional aspects: the laws refer only to effluent standards with no regard for stream flow or water quality, and there is no limit on the quantity of effluents discharged at specific points; and non-point sources have not been taken into account.

PROTECTED AREAS

India has a large network of protected areas – 54 national parks and 372 wildlife sanctuaries – covering about 0.11×10^6 km^2, or 3.3% of the total land area (Rodgers and Panwar, 1988). Some include small sections of rivers and streams passing through them (e.g. the Manas in Manas National Park, the Periyar in Periyar Wildlife Sanctuary and the Ramganga in the Corbett National Park), or floodplain patches, such as those in Kaziranga National Park (floodplain of the Brahmaputra). However, except for a part of the Chambal (below Kota, to its confluence with the Yamuna) which is a declared crocodile sanctuary, no other stretch of river has been considered for protection. It is interesting that two Ramsar wetlands are associated with the rivers. Lake Harike is a reservoir immediately below the confluence of the Sutlej and Beas in Punjab. The reservoir supports, among others, seven species of turtle, of which four are endangered, and a large number of migratory wildfowl. Lake Wular in Kashmir is a shallow lake through which the Jhelum flows.

In Pakistan, the 135 km stretch of the Indus from the Sukkar Barrage to the Giddu Barrage has been declared a sanctuary for the Indus dolphin (*Platanista indica*). In Sri Lanka, all of the 11 national parks (466 000 ha) and five nature reserves (64 124 ha) lie in the dry zone, and no part of any river has been considered for conservation (Costa and De Silva, 1995).

Gene banks

Efforts are now being made to conserve some river fauna such as the Ganges river dolphin, *Hilsa*, snow trout, and the gharial in the Ganga Basin. *Ex situ* conservation measures for fish, such as cryopreservation of gametes and embryos and live gene banks, have also been initiated (Dehadrai and Poniah, 1997).

Restoration and rehabilitation programmes

The only programme related to river rehabilitation is the effort of the Indian Government to divert and to treat domestic sewage in all major towns in the Ganga Basin (Department of Environment, Government of India, 1985). Similar plans have been made for other rivers under the National River Action Plan but they have not been implemented because of financial constraints. We are not aware of any other river restoration programme in other countries of the region. River restoration will remain meaningless without the provision of minimum flow regimes, and the control of point and diffuse sources of pollution.

FURTHER THREATS TO RIVERS AND THEIR MITIGATION

The rivers of the sub-continent continue to be threatened by an ever-increasing withdrawal of water, and by channelization. There seems to be no possibility in the near future of dismantling dams and levees. The growing human population is the single major threat and, even if the population stabilizes in the next 25 years or so, the rivers will have been reduced to wastewater drains by then. Mitigation of the problem requires a check on population growth, and evolving strategies to minimize the wasteful use of water in agriculture. Efficient wastewater treatment systems and recycling are required to reduce the pressure on the rivers of the region.

References

Agarwal, A. and Chak, A. 1991. *Floods, Flood Plains and Environmental Myths.* State of India's Environment – A Citizens' Report 3. Centre for Science and Environment, New Delhi.

Ahmad, K.S. 1969. *A Geography of Pakistan.* Oxford University Press, Karachi.

Ahmad, Q.K., Ahmad, N. and Sajjadar, R. 1994. *Environment and Development in Bangladesh.* Academic Press, Dhaka.

Ahmad, S. 1951. *Climatic Regions of West Pakistan.* Oxford University Press, Karachi.

Ahmed, B. 1992. Impact of Kaptai Dam and industries on Karnaphuli River in Bangladesh, with special reference to loss of habitat for the River Dolphin, *Platanista gangetica. Proceedings of the Seminar on the Conservation of River Dolphin in the Indian Subcontinent*, 18–19 August 1992, New Delhi. Conservation of Nature Trust, Calicut.

Armitage, P.D., Moss, D., Wright, J.F. and Furse, M.T. 1983. The performance of a new biological water quality score system based on macroinvertebrates over a wide range of unpolluted running water sites. *Water Research* 17:333–347.

Arunachalam, M., Madhusoodanan Nair, K.C., Vijverberg, J. and Kotmulder, K. 1997. Food and habitat partitioning among fishes in stream pools of a south Indian river. *International Journal of Ecology and Environmental Sciences* 23:271–295.

Bhargava, D.S. 1983. Effect of subsurface water on Yamuna water quality at Delhi. *Journal of Environmental*
Engineering Division, Institution of Engineers of India 64:33–34.

Bhargava, D.S. 1985a. Water quality variations and control technology of Yamuna River. *Environmental Pollution A* 37:355–376.

Bhargava, D.S. 1985b. Matching river quality to uses. Role of monitoring, zoning and classification. *The Environment Professional* 7:240–247.

Biswas, S.P., Baruah, A. and Mohan, R.S.L. 1997. Current status of the River Dolphin (*Platanista gangetica*) in the River Brahmaputra. *International Journal of Ecology and Environmental Sciences* 23:357–361.

Boon, P.J., Calow, P. and Petts, G.E. (eds) 1992. *River Conservation and Management.* John Wiley, Chichester.

Boto, K.G. and Patrick, W.H. 1979. Role of wetlands in the removal of suspended sediments. In: Greeson, P.E., Clark, J.R. and Clark, J.E. (eds) *Wetland Functions and Values: The State of our Understanding.* American Water Resources Association, Minneapolis, 479–489.

Brohier, R.L. 1934. *Ancient Irrigation Water in Ceylon.* Ceylon Government Press, Colombo.

Calow, P. and Petts, G.E. (eds) 1992. *Rivers Handbook*, Vol. 1. Blackwell, Oxford.

Calow, P. and Petts, G.E. (eds) 1994. *Rivers Handbook*, Vol. 2. Blackwell, Oxford.

Central Board for the Prevention and Control of Water Pollution (CBPCWP) 1982. *The Ganga Basin. Part I: The Yamuna Sub-basin.* Assessment and Development Study of River Basin Series ADSORBS/2/1980–81. New Delhi.

Central Board for the Prevention and Control of Water Pollution (CBPCWP) 1983. *Quality and Trend of River Yamuna (1977–82).* Assessment and Development Study of River Basin Series ADSORBS/10/1982–83. New Delhi.

Central Board for the Prevention and Control of Water Pollution (CBPCWP) 1984. *The Ganga Basin. Part II.* Assessment and Development Study of River Basin Series ADSORBS/7/1982–83. New Delhi.

Central Pollution Control Board (CPCB) 1996. *Water Quality Status and Statistics (1993 & 1994).* Monitoring of Indian Aquatic Resources (MINARS/10/1995–96). New Delhi.

Central Water Commission (CWC) 1987. *Water Quality Studies: Ganga System – Status report (1978–85).* Ministry of Water Resources, New Delhi.

Chacko, P.I. and Ganapati, S.V. 1949. Some observations on the Adyar River with special reference to hydrobiological conditions. *Indian Geographical Journal* 24: 21–28.

Chacko, P.I., Abraham, J.G. and Andal, R. 1953. Report on a survey of flora, fauna and fisheries of Pulicat Lake, Madras State, India, 1951–52. *Contributions of the Fisheries Biology Station, Madras* 8:1–20.

Chakrabarty, R.D., Roy, P. and Singh, S.B. 1959. A quantitative study of the plankton and physicochemical condition of the River Yamuna at Allahabad in 1954–55. *Indian Journal of Fisheries* 6:186–203.

Chakraborty, P.K. and Chattopadhyay, G.N. 1989. Impact of Farakka Barrage on the estuarine ecology of the Hooghly–Matlah system. In: Jhingran, A.G. and Sugunan, V.V. (eds)

Conservation and Management of Inland Capture Fisheries Resources of India. Inland Fisheries Society of India, Barrackpore, 189–196.

Champion, H.G. and Seth, S.K. 1968. *Revised Forest Types of India.* Manager of Publications, New Delhi.

Chandra, R. 1989. Riverine fishery resources of the Ganga and the Brahmaputra. In: Jhingran, A.G. and Sugunan, V.V. (eds) *Conservation and Management of Inland Capture Fisheries Resources of India.* Inland Fisheries Society of India, Barrackpore, 52–60.

Chatterjee, D. 1940. Studies on the endemic flora of India and Burma. *Journal of Asiatic Society of Bengal, Science* 5:19–68.

Chowdhury, M.I. 1966. On the gradual shifting of the Ganges from west to east in delta building operations. In: *Humid Tropics Research: Scientific Problems of the Humid Tropical Zone Deltas and their Implications.* Proceedings of the Dacca Symposium. UNESCO, Paris, France, 35–40.

Cooray, P.G. 1984. *An Introduction to the Geology of Ceylon.* Ceylon National Museum Publication, Colombo.

Costa, H.H. and De Silva, P.K. 1995. Limnological research and training in Sri Lanka: state of the art and future needs. In: Gopal, B. and Wetzel, R.G. (eds) *Limnology in Developing Countries*, Vol. 1. International Association of Limnology, and International Scientific Publications, New Delhi, 63–103.

Crow, B., Lindquist, A. and Wilson, D. 1995. *Sharing the Ganges.* Sage Publications, New Delhi.

Dagar, J.C. 1987. Studies on reclamation of Kshipra ravines. *Indian Journal of Forestry* 10:83–89.

Das, P.K. 1968. *Monsoon.* National Book Trust, New Delhi.

Dasgupta, S.P. (ed.) 1975. *Upper Gangetic Flood-plain.* National Atlas Organization, Calcutta.

Day, F. 1878. *The Fishes of India: Being a Natural History of the Fishes Known to Inhabit Seas and Freshwaters of India, Burma and Ceylon.* William Dawson and Sons, London.

De, D.K., Ghosh, A. and Unnithan, V.K. 1989. Biology and migration of Hooghly *Hilsa* in the context of Farakka barrage. In: Jhingran, A.G. and Sugunan, V.V. (eds) *Conservation and Management of Inland Capture Fisheries Resources of India.* Inland Fisheries Society of India, Barrackpore, 197–202.

Dehadrai, P.V. and Poniah, A.G. 1997. Conserving India's fish biodiversity. *International Journal of Ecology and Environmental Sciences* 23:315–326.

De Kruijf, H.A.M., De Zwart, D. and Trivedi, R.C. 1992. *Proceedings of the Indo-Dutch Workshop on Water Quality Yardstick Development.* Report No. 768602009. National Institute of Public Health and Environmental Protection, Bilthoven, and Central Pollution Control Board, New Delhi.

Department of Environment, Government of India. 1985. *An Action Plan for Prevention of Pollution of the Ganga.* New Delhi.

De Silva, P.K. 1993. The otter in the wetland ecosystems of south and southeast Asia and the impact of human activities on its survival. In: Gopal, B., Hillbricht-Ilkowska,

A. and Wetzel, R.G. (eds) *Wetlands and Ecotones: Studies on Land Water Interactions.* National Institute of Ecology and International Scientific Publications, New Delhi, 217–225.

De Zwart, D. 1991. *Report on an Expert Mission for the Evaluation of Yamuna River Biomonitoring Data.* Report No. 768602008. National Institute of Public Health and Environmental Protection, Bilthoven.

Dhanze, J.R. and Dhanze, R. 1996. Impact of habitat shrinkage on the indigenous fish genetic resources of Bas drainage system. In: *Proceedings of the Symposium on Fish Genetics and Biodiversity Conservation for Sustainable Production.* National Bureau of Fish Genetic Resources, Lucknow, 9–10.

Dhingra, R.K. 1989. Studies on the vegetation of Yamuna ravines at Chhalesar with special reference to soil and water conservation. Unpublished PhD Thesis, Agra University, Agra.

Dutta, N., Malhotra, J.C. and Bose, B.B. 1954. Hydrology and seasonal fluctuations of the plankton in the Hooghly Estuary. *Fifth Symposium on Marine and Freshwater Plankton in the Indo-Pacific*, Indo-Pacific Fisheries Council and UNESCO, Bangkok, 35–47.

Elahi, K.M., Ahmed, K.S. and Mafizuddin, M. (eds) 1991. *Riverbank Erosion, Flood and Population Displacement in Bangladesh.* Riverbank Erosion Study. Jahangirnagar University, Dhaka.

Erb, D.K. 1984. Land forms and drainage. In: Fernando, C.H. (ed.) *Ecology and Biogeography in Sri Lanka.* Dr W. Junk, The Hague, 35–63.

Fein, J.S. and Stephens, P.L. (eds) 1987. *Monsoon.* John Wiley, New York.

Fernando, C.H. (ed.) 1985. *Ecology and Biogeography in Sri Lanka.* Dr W. Junk, The Hague.

Gole, C.V. and Chitale, S.U. 1966. Inland delta building activity of Kosi River. *American Society of Civil Engineers Journal, Hydraulic Division* 92 HY2:111–126.

Gopal, B. and Krishnamurthy, K. 1993. Wetlands of south Asia. In: Whigham, D.F., Dykyjova, D. and Hejny, S. (eds) *Wetlands of the World. I. Inventory, Ecology and Management.* Handbook of Vegetation Science 15/2. Kluwer Academic, Dordrecht, 345–414.

Gopal, B. and Sah, M. 1993. Conservation and management of rivers in India: case study of the River Yamuna. *Environmental Conservation* 20:243–254.

Gopal, B. and Zutshi, D.P. 2000. Past, present and future of limnology in India. In: Gopal, B. and Wetzel, R.G. (eds) *Limnology in Developing Countries*, Vol. 3. International Association for Theoretical and Applied Limnology, New Delhi (in press).

Gulhati, N.D. 1973. *Indus Water Treaty, 1960. An Exercise in International Mediation.* Allied Publishers, Calcutta.

Gupta, A.C. 1957. The Sunderbans: its problems, its possibilities. *Indian Forester* 83:481–487.

Hamilton, B. 1822. *An Account of Fishes Found in the River Ganges and its Branches.* A. Constable and Co., Edinburgh.

Hari Narain 1965. Airborne magnetic surveys. *Proceedings of*

Seminar on Earth Sciences, Part I. Geophysics. Indian Geophysical Union, Hyderabad, 119–129.

Hawkes, H.A. 1975. River zonation and classification. In: Whitton, B.A. (ed.) *River Ecology.* Studies in Ecology 2. Blackwell, Oxford, and University of California Press, Berkeley, 312–374.

Hewavisenthe, A.C. de S. 1992. Mahaweli water resources project (Sri Lanka). *Water International* 17.

Hora, S.L. 1921. Fish and fisheries of Manipur with some observations on those of the Naga hills. *Records of the Indian Museum* 27:165–214.

Hora, S.L. 1923. Observations on the fauna of certain torrential streams in the Khasi hills. *Records of the Indian Museum* 25:579–600.

Hora, S.L. 1935. Ecology and bionomics of the gobioid fishes of the Gangetic delta. *Proceedings of the International Congress of Zoology* 12:841–863.

Hora, S.L. 1937a. Comparison of fish faunas of the northern and southern faces of the great Himalayan range. *Records of the Indian Museum* 39:241–250.

Hora, S.L. 1937b. Geographical distribution of Indian freshwater fishes and its bearing on the probable land connection between India and the adjacent countries. *Current Science* 7:351–356.

Hora, S.L. 1940. Dams and the problem of migratory fishes. *Current Science* 9:406–407.

Hora, S.L. 1942. The effect of dams on the migration of the hilsa fish in Indian waters. *Current Science* 11:470–471.

Indian Water Resources Society (IWRS) 1995. Water management – need for public awareness. Mimeographed Report, Calcutta.

Iyengar, M.O.P. and Venkataraman, G. 1951. The ecology and seasonal succession of the algal flora of the river Cooum at Madras with special reference to the Diatomaceae. *Journal of Madras University* 21 B:140–192.

Jensen, B.L. 1994. Fish refugia and captive propagation: A viable aid to conservation and restoration. In: Dehadrai, P.V., Das, P. and Verma, L.R. (eds) *Threatened Fishes of India.* Natcon Publication 4. Society of Nature Conservators, Muzaffarnagar, India, 311–320.

Jhingran, V.G. 1991. *Fish and Fisheries of India,* 3rd edition. Hindustan Publishing Corporation, New Delhi.

Jorgensen, D.W., Harvey, M.D., Schumm, S.A. and Flam, L. 1993. Morphology and dynamics of the Indus River: Implications for the Mohen-jo-daro site. In: Shroder, J.F. (ed.) *Himalaya to the Sea: Geology, Geomorphology and the Quaternary.* Routledge, London, 288–326.

Krishnamurti, C.R., Bilgrami, K.S., Das, T.M. and Mathur, R.P. (eds) (1991). *The Ganga: A Scientific Study.* Northern Book Centre, New Delhi.

Khan, T.A. 1994. Challenges facing the management and sharing of the Ganges. *Transboundary Research Report* 8 (1), University of New Mexico, Albuquerque.

Khondker, M. 1995. Limnological research in Bangladesh. In: Gopal, B. and Wetzel, R.G. (eds) *Limnology in Developing Countries,* Vol. 1. International Association of Limnology,

and International Scientific Publications, New Delhi, 105–120.

Lakshminarayana, J.S.S. 1965. Studies on the phytoplankton of the River Ganges, Varanasi, India. I–IV. *Hydrobiologia* 25:119–175.

Lowe-McConnell, R.H. 1987. *Ecological Studies of Tropical Fish Communities.* Cambridge University Press, Cambridge.

Lowrance, R., Todd, T., Fail, J., Hendrickson, O., Leonard, R. and Asmussen, L. 1984. Riparian forests as nutrient filters in agricultural watersheds. *Bioscience* 34:374–377.

Mani, M.S. 1974. *Ecology and Biogeography of India.* Monographiae Biologicae 23. Dr W.Junk, The Hague.

Ministry of Environment and Forests (MOEF) 1994. *National River Action Plan.* Government of India, New Delhi.

Ministry of Water Resources (MOWR) 1987. *National Water Policy.* Government of India, New Delhi.

Myers, N. 1988.Threatened biotas: 'hot spots' in tropical forests. *The Environmentalist* 8:187–208.

Myers, N. 1990. The biodiversity challenge: expanded hot spots analysis. *The Environmentalist* 10:243–256.

Naiman, R.J. and Décamps, H. 1990. *Ecology and Management of Aquatic–Terrestrial Ecotones.* UNESCO, Paris, and Parthenon Publishing, Carnforth.

Nayar, M.P. 1980. Endemism and pattern of distribution of endemic genera (angiosperms). *Journal of Economic and Taxonomic Botany* 1:99–110.

Nazneen, S. 1995. State of limnology in Pakistan. In: Gopal, B. and Wetzel, R.G. (eds) *Limnology in Developing Countries,* Vol. 1. International Association of Limnology, and International Scientific Publications, New Delhi, 191–230.

Panabokke, C.R. 1984. *Sri Lanka, Showing the Approximate Distribution of Great Soil Groups.* Survey Department, Colombo.

Perrin, W.F., Brownell, R.L., Zhou, K. and Jiankang Liu (eds) 1989. *Biology and Conservation of River Dolphins.* Occasional Papers of the Species Survival Commission 3. World Conservation Union (IUCN), Gland.

Petts, G.E. and Amoros, C. 1996. *Fluvial Hydrosystems.* Chapman & Hall, London.

Plafkin, J.L., Barbour, M.T., Porter, K.D., Gross, S.K. and Hughes, R.M. 1989. *Rapid Bioassessment Protocols for Use in Streams and Rivers. Benthic Macro-invertebrates and Fish.* EPA/444/4-89/001. Office of Water Regulations and Standards, US Environmental Protection Agency, Washington, DC.

Prajapati, M.C. 1989. Studies in conservation ecology of a vegetatively stabilized Yamuna River flat at Chalesar, Agra. Unpublished PhD Thesis, Agra University, Agra.

Rai, H. 1962. Hydrobiology of River Yamuna at Okhla, Delhi, India. Scientific papers from the Institute of Chemical Technology, Prague. *Technology of Water* 6:77–98.

Rai, R. 1974a. Limnological studies on the River Yamuna at Delhi, India. I. Relation between the chemistry and the state of pollution in the River Yamuna. *Archiv für Hydrobiologie* 73:369–393.

Rai, R. 1974b. Limnological studies on the River Yamuna at Delhi, India. II. The dynamics of potamoplankton in the River Yamuna. *Archiv für Hydrobiologie* 73:492–517.

Ramage, C.S. 1971. *Monsoon Meteorology*. Academic Press, New York.

Rao, K.L. 1975. *India's Water Wealth: Its Assessment, Uses and Projections*. Orient Longman, New Delhi.

Ray, P., Singh, S.B. and Sehgal, K.L. 1966. A study of some aspects of the ecology of rivers Ganga and Jamuna at Allahabad (U.P.) in 1958–59. *Proceedings of the National Academy of Science, India* 36:235–272.

Raychaudhuri, S.P., Aggarwal, R.R., Datta Biswas, N.R., Gupta, S.P. and Thomas, P.K. 1963. *Soils of India*. Indian Council of Agricultural Research, New Delhi.

Reeves, R.R., Chaudhry, A.A. and Khalid, U. 1991. Competing for water on the Indus Plain: is there a future for Pakistan's River Dolphins? *Environmental Conservation* 18:341–350.

Richardson, C.J. 1990. Freshwater wetlands: transformers, filters or sinks? In: Sharitz, R.R. and Gibbons, J.W. (eds) *Freshwater Wetlands and Wildlife*. US Department of Energy, Office of Health and Environmental Research, Washington, DC, 25–46.

Robinson, H. 1976. *Monsoon Asia*, 3rd edition. McDonald and Evans, Estover.

Rodgers, W.A. and Panwar, H.S. 1988. *Planning a Wildlife Protected Area Network in India*. Vols 1 and 2. Wildlife Institute of India, Dehradun.

Roy, H.K. 1949. Some potamological aspects of the River Hooghly in relation to Calcutta water supply. *Science and Culture*. 14:318–324.

Roy, H.K. 1955. Plankton ecology of the River Hooghly at Palta, West Bengal. *Ecology* 36:169–175.

Sehgal, K.L. 1994. State-of-art of endangered, vulnerable and rare coldwater fishes of India. In: Dehadrai, P.V., Das, P. and Verma, L.R. (eds) *Threatened Fishes of India*. Natcon Publication 4. Society of Nature Conservators, Muzaffarnagar, India, 127–135.

Sharma, C.K. 1997. *A Treatise on Water Resources of Nepal*. Sangeeta Sharma, Kathmandu.

Shreshtha, T.K. 1989. Biology, status and conservation of the Ganges River Dolphin, *Platanista gangetica*, in Nepal. In: Perrin, W.F., Brownell, R.L., Zhou, K. and Jiankang Liu (eds) *Biology and Conservation of River Dolphins*. Occasional Papers of the Species Survival Commission 3. World Conservation Union (IUCN), Gland, 173.

Shreshtha, T.K. 1990. *Resource Ecology of the Himalayan Waters: A Study of Ecology, Biology and Management Strategy of Fresh Waters*. Curriculum Development Centre, Tribhuvan University, Kathmandu.

Shroder, J.F. (ed.) 1993. *Himalaya to the Sea: Geology, Geomorphology and the Quaternary*. Routledge, London.

Sinha, R.K. 1997. Status and conservation of Ganges River Dolphin in Bhagirathi–Hooghly river system in India. *International Journal of Ecology and Environmental Sciences* 23:343–355.

Sinha, R. and Friend, P.F. 1994. River systems and their sediment flux, Indo-Gangetic plains, Northern Bihar, India. *Sedimentology* 41:825–845.

Sivaramakrishnan, K.G., Hannaford, M.J. and Resh, V.H. 1996. Biological assessment of the Kaveri River catchment, South India, using benthic macroinvertebrates. Applicability of water quality monitoring approaches developed in other countries. *International Journal of Ecology and Environmental Sciences* 22:113–132.

Smith, B.D., Sinha, R.K., Regmi, U. and Sapkota, K. 1994. Status of Ganges River Dolphins (*Platanista gangetica*) in the Mahakali, Karnali, Narayani and Saptakosi rivers in Nepal and India. *Marine Mammal Science* 10:368–375.

Subramanian, V. 1979. Chemical and suspended sediment characteristics of rivers of India. *Journal of Hydrology* 44:37–55.

Subramanian, V., Sitasawad, R., Abbas, N. and Jha, P.K. 1987. Environmental geology of the Ganga River basin. *Journal of the Geological Society of India* 30:335–355.

Trivedy, R.K. (ed.) 1988. *Ecology and Pollution of Indian Rivers*. Ashish Publishing House, New Delhi.

Trivedy, R.K. (ed.) 1990. *River Pollution in India*. Ashish Publishing House, New Delhi.

Unni, K.S. 1997. *Ecology of River Narmada*. Ashish Publishing House, New Delhi.

Vaithiyanathan, P., Ramanthan, A.L. and Subramanian, V. 1988. Erosion, transport and deposition of sediments by the tropical rivers of India. In: *Sediment Budgets* (*Proceedings of the Porto Alegre Symposium*), IAHS Publication No. 174, 561–574.

Venkateswarlu, V. 1969. An ecological study of the algae of the River Moosi, Hyderabad (India) with special reference to water pollution. II. Factors influencing the distribution of algae. *Hydrobiologia* 33:352–378.

Venkateswarlu, V. 1970. An ecological study of the algae of the River Moosi, Hyderabad (India) with special reference to water pollution. IV. Periodicity of some common species of algae. *Hydrobiologia* 35:45–64.

Venkateswarlu, V. 1986. Ecological studies on the rivers of Andhra Pradesh with special reference to water quality and pollution. *Proceedings of the Indian Academy of Science (Plant Science)* 96:495–508.

Venkateswarlu, V. and Jayanti, T.V. 1968. Hydrobiological studies of the River Sabarmati to evaluate water quality. *Hydrobiologia* 31:442–448.

Ward, J.V. and Stanford, J.A. 1989. Riverine ecosystems: the influence of man on catchment dynamics and fish ecology. *Canadian Journal of Fisheries and Aquatic Sciences, Special Publication* 106:56–64.

Webster, P.J. 1987. The elementary monsoon. In: Fein, J.S. and Stephens, P.L. (eds) *Monsoon*. John Wiley, New York, 3–32.

Welcomme, R.L. 1979. *Fisheries Ecology of Floodplain Rivers*. Longman, London.

Wells, N.A. and Dorr, J.A. 1987. Shifting of the Kosi River, northern India. *Geology* 15:204–207.

10

River conservation in central and eastern Asia

L. Li, C. Liu and H. Mou

The regional context

This chapter reviews policy and practice of river conservation in central and eastern Asia (Figure 10.1). The area includes 10 countries but, geographically, is dominated by China which has 60% of the area and over 80% of the population. The other countries are Japan, the Democratic People's Republic of Korea (North Korea), the Republic of Korea (South Korea), Mongolia, Kazakhstan, Uzbekistan, Kirghizia, Tajikistan and Turkmenistan. The total area of the region is 15 748 500 km^2 with an overall population density of 93.5 persons km^{-2} (Table 10.1).

It is a region of political, demographic and environmental contrasts. In terms of topography, Mongolia, Kirghizia and Tajikistan are mountainous countries. Mongolia has an average elevation of 1600 m a.s.l. and 50% of Tajikistan and 30% of Kirghizia are higher than 3000 m. In contrast, eastern China is dominated by the Great Plain of China comprising the North China Plain, the Songlen Plain, the Yangtze River delta, and the Pearl River delta – in total, an area of about 3 × 10^6 km^2. In terms of climate, the extensive Gobi Desert which forms over 30% of Mongolia contrasts with the East Asia Monsoon area where, for example, annual rainfall of more than 3000 mm characterizes southern Japan. With regard to population density, Japan is a highly urbanized country with half of the population concentrated in three large cities: Tokyo, Osaka and Nagoya. Kazakhstan and Turkmenistan have sparse rural populations, with densities of less than 7.5 persons km^{-2}. Within China the majority of the population is in the plain and delta areas in the eastern part of the country. There are also different political and economic systems: central

government and socialist political systems in Mongolia and North Korea, capitalist systems with market economies in Japan and South Korea.

With regard to biodiversity, China alone has more than 10% of the global species. The reasons for this include its large area, marked variations in climate and geographical conditions, and geologically stable land mass. The major ecosystems in China are divided into six groups (Mackinnon and Wang, 1996): (1) temperate and tropical forests (10%); (2) mountain grasslands (60%); (3) pasture and desert (15%); (4) main river corridors, lakes and wetlands; (5) wide seashore area and (6) agricultural land (11%). River corridors are particularly rich in biodiversity. The Yangtze River is the richest one in China; the number of fish species increases with catchment area and with river run-off depth and river gradient, and decreases with latitude (Shoukun, 1997).

Yet many species are under threat. For example, the Yangtze sturgeon (*Acipenser dabryanus*), Chinese river dolphin (*Lipotes vexillifer*) and Chinese alligator (*Alligator sinensis*) have become endangered because of habitat degradation, fishing pressure and accidents caused by river traffic and hydropower generation (see Dudgeon, 1992, for review). Habitats under threat include the huge lateral lakes of the Yangtze floodplain which are important wintering grounds for the rare white and black storks (*Ciconia ciconia* and *C. nigro*) and 98% of the world population of the endangered Siberian white crane (*Grus leucogeranus*) (Zhao et al., 1990). Mountain brooks at altitudes above 1000 m in central and south-west China are also under threat and the giant salamander (*Andrias davidianus*) is now endangered because of intensive hunting (Dudgeon, 1992).

Global Perspectives on River Conservation: Science, Policy and Practice.
Edited by P.J. Boon, B.R. Davies and G.E. Petts. © 2000 John Wiley & Sons Ltd.

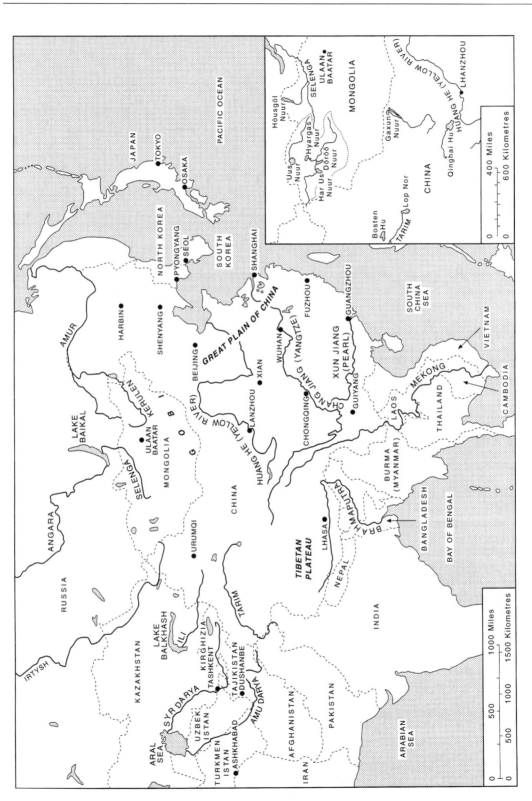

Figure 10.1 *The central and eastern Asia region showing the major rivers and (inset) the main interior drainages of western Mongolia and north-west China*

Table 10.1 *General information on China, central and eastern Asia*

Countries	Population	Territory (km^2)	Popn density (persons km^{-2})
China	1 227 000 000	9 600 000	128
Japan	125 251 000	377 800	332
Kazakhstan	16 690 000	2 717 300	6.1
Kirghizia	4 370 000	198 500	22.0
Mongolia	2 410 000	1 556 500	1.5
North Korea	23 917 000	120 410	199
South Korea	44 851 000	99 390	451
Tajikistan	5 510 000	143 100	38.5
Turkmenistan	3 620 000	488 100	7.4
Uzbekistan	20 320 000	447 400	45.4
Total	1 471 939 000	15 748 500	93.5

Table 10.2 *Main rivers of China, and central and eastern Asia*

Country/ Region	River	Length (km)	Basin area (km^2)	Mean annual run-off (km^3)
Central Asia	Amu	1415	227 000	63.1
Central Asia	Ili	1439	140 000	14.8
Central Asia	Syr	2212	462 000	14.8
China	Hai	1090	263 631	22.60
China	Heilongjiang	3420	1 620 170	207.90
China	Huai	1000	269 283	35.10
China	Lanchangjiang	1826	167 486	74.25
China	Liao	1390	228 960	9.53
China	Nujiang	1659	137 818	70.09
China	Pearl	2214	453 690	349.20
China	Songhuajiang	2308	557 180	79.85
China	Tarim	2046	194 210	4.94
China	Yaluzanbujiang	2057	240 480	116.70
China	Yangtze	6300	1 808 500	979.35
China	Yellow	5464	752 443	57.45
Japan	Shino	367	11 900	
Japan	Tone	322	16 840	
Mongolia	Kerulen	–	120 000	
Mongolia	Selenga	745	4 469 000	
North Korea	Taedong	439	17 000	
South Korea	Han	510	34 000	
South Korea	Nakdong	525	24 000	

THE MAIN RIVERS IN THE AREA

There are a large number of important rivers in the region (Table 10.2), most of them flowing through China which includes more than 1500 river basins with areas in excess of 1000 km^2. The Yangtze (Chang Jiang), also known as the 'golden waterway' (Changming and Dakang, 1987), is the largest river in China, with a total annual run-off of 980×10^9 m^3. However, the Lanchangjiang is also important as it is an international river, becoming the Mekong River after it leaves China. The Yellow River (Huang He) is famous for its huge annual sediment load of 1.6×10^9 t. In central Asia, the Selenga River flows northwards into the Arctic Ocean via Lake Baikal. The Kerulen River (Figure 10.1) flows in an easterly direction into the Hulun Lake in China and during the flood period it can flow into the upper reaches of the Amur River.

In contrast to these major basins, the rivers of the Korean peninsula and Japanese islands are short and steep with small catchment areas. The Japanese Archipelago consists of four main islands (Kyushu, Shikoku, Honshu and Hokkaido) and some 39 000 small ones. Only four rivers have catchment areas greater than 10 000 km^2. The longest rivers are the Shino River and the Tone River in Japan; the Taedong River in North Korea; and the Han and Nakdong rivers in South Korea. The Taedong River flows through the capital city of North Korea (Pyongyang), while the Han River flows through the capital city of South Korea (Seoul).

The region is also characterized by large interior drainage systems (Figure 10.1). In the arid northwest, most rivers, such as the Ili and the Tarim, have interior drainages. The Ili River flows into Lake Balkhash. In

Mongolia, inland drainages are associated with lakes Uvs and Hŏusgöl. The former, an interior salty lake, is the largest lake in Mongolia, with an area of 3350 km^2. The latter is the second largest lake and the largest freshwater lake in Mongolia with an area of 2620 km^2. Central Asia includes the Aral–Caspian interior river basin where the Amu and Syr rivers are fed by snowmelt from the mountains, and run-off is lost and evaporated in the plains. However, in most of the plains of central Asia, there are no notable rivers. The Kzyl Kum and Kara Kum Deserts cover most of Kazakhstan (60%), Turkmenistan (90%) and Uzbekistan.

HYDRO-CLIMATIC VARIATIONS

Monsoon and arid climates dominate the region. Japan is within the East Asia monsoon area. Annual precipitation is 1800 mm, with extremes of more than 3000 mm, and there are three wet seasons in a year and an autumn dry season. In winter (January to March) heavy snowfalls are experienced in the northern and north-west areas. North Korea has a temperate monsoon climate. It is cold and dry in winter and hot and humid in summer. The mean annual precipitation is more than 1000 mm. The climate of South Korea belongs to the temperate oceanic type. There are four distinct seasons: summer is hot and humid, and winter is cold and dry. The mean annual precipitation is 1274 mm (850–1800 mm), with rainfall (66%) concentrated in June to September. The distribution of annual precipitation varies widely from 745 mm in a dry year to 1683 mm in a wet year.

The location of China in relation to the Indian Ocean in the south, the Pacific Ocean to the east, and the continent of Asia to the northwest, leads to a particularly steep gradient of rainfall which increases from north to south, and from northwest to southeast (Dakang, 1983). Thus, across China mean annual precipitation varies from more than 1600 mm in the south to less than 50 mm in the northwest. In the catchment of the Yangtze River, rainfall averages a reliable 1100 mm yr^{-1}. However, precipitation is concentrated in the summer season (June to September) when severe floods often occur, and large-scale droughts are very frequent in winter, spring and early summer. In spring and winter, the rivers on the North China Plain receive only 10 and 8% of the total annual run-off, respectively, and most of the summer run-off flows directly to the sea.

The continental climate of central Asia is dominated by drought but rainfall can reach 1000 mm in the eastern mountain region, such as in Uzbekistan. Seasonal contrasts are extreme with, in some years, 100% of the annual rainfall in the summer season. Most water is needed during spring, which is the growing season for wheat, and the sowing season for autumn-harvested crops. However, throughout the region spring rainfall is scarce. During spring, air temperatures rise fairly rapidly during the day, and dry winds lead to increased evaporation rates, which often exceed rainfall. Thus, agricultural development on the North China Plain, for example, is severely limited by the climate, and the area has been known to experience drought in nine years out of 10 (Dakang, 1983).

Annual run-off has great variation too. The ratio of maximum to minimum annual run-off is higher than 10 in the north of the Yangtze Basin, but lower than five in the south of the basin. The coefficient of variation (maximum discharge *versus* minimum discharge) of the Han and Nakdong Rivers in Japan is 300–500. Inter-annual variations are also extreme. For example, run-off of the Yellow River in 1922–1932 was 30% lower than the mean annual value while it was 8% higher than the mean value in the wet period of 1943–1951. In central Asia, the long-term variation in annual run-off shows a strong periodicity with run-off in a dry year

being 30% of the mean value, and run-off in wet years five to six times the mean value. For example, run-off in Kazakhstan in an 11-year dry period (1929–1939) was 30% of the mean value, while in a two-year wet period (1941–1942) run-off was 600% of the mean value.

SOCIO-ECONOMIC CONTEXT

Population growth and poverty are the main driving forces behind water resources development throughout most of the region. Only the east and south-east rim, notably Japan, has seen significant economic growth over the past 40 years. In China, Shenzhen, adjacent to Hong Kong, is a 'special economic zone' and southern China has benefited from agricultural reforms and the 'open door' policy towards the West. In many ways, 'modernization' has accentuated regional differences. Furthermore, with development have come problems not only of water resource exploitation but also of water pollution and land degradation.

Urbanization, industrialization and intensification of irrigation agriculture are having dramatic impacts upon the rivers of the region. Irrigation agriculture is causing the decline of lake and river levels as well as desertification and soil salinization (e.g. Mengxiong, 1995); dams for hydropower and flood storage, and navigation works are having major impacts on rivers such as the Yangtze (e.g. Rushu, 1996); and in urban areas 'environmental problems' focus on waste-water control (e.g. Jun and Baoqing, 1997). Many of the issues affecting central and eastern Asia are paralleled by those of south-east Asia (Dudgeon *et al.*, this volume) and the Indian sub-continent (Gopal *et al.*, this volume). The central Asian countries of Kazakhstan, Uzbekistan, Kirghizia, Tajikistan and Turkmenistan became independent from the former Soviet Union in 1991 and Khaiter *et al.* (this volume) discuss the environmental effects of such a political change.

DEVELOPMENT PRESSURES AND HUMAN IMPACTS

The distribution of water resources is uneven both across the region (Table 10.3) and within individual countries, especially China. There is a long history of water resources development. In an attempt to harness the flood waters of the Yellow River, and to transport water to agricultural areas, dikes and irrigation canals had been completed by 4000 years BP (Snaddon *et al.*,

Table 10.3 *The status of water resources and level of water use in the area*

Country	Water resource (km³)	Water resources per capita (m³)	Water resources per unit agricultural land (m³ ha⁻¹)	Water use (km³)	Water use: water resources (%)
China	2812	2 292	292 917	522.4	19
Japan	547	4 373	137 783	90.8	17
Kazakhstan	125.4	7 328.6	3 933	37.9	30
Kirghizia	48.7	10 263.4	59 246	11.7	24
Mongolia	24.6	10 207	18 650	0.55	2
North Korea	67	2 801	3 937	14.2	21
South Korea	66.1	1 469	36 989	27.6	42
Tajikistan	95.3	15 620.3	116 219	12.6	13
Turkmenistan	70.9	17 296.9	50 643	22.8	32
Uzbekistan	107.6	4 710.4	26 244	82.2	76

2000). Changming (1989) records the works of the Great Yu, about 2200 BC, in controlling floods along the lower Yellow River. One of the earliest large dams was the 27 m-high Dujiangyan Dam on the Min Jiang, completed in AD 833, which is still being used for irrigation today. By the end of the 13th century, the 1780 km long Beijing–Hangzhou Grand Canal had been built to link five river basins, including the Hai He, Yellow (Huang He), Huai He, Yangtze (Chang Jiang) and Qiantang Jiang Rivers, and to transfer water from the Yangtze at Jiangdu to the dry North China Plain (Changming *et al.*, 1985; Changming and Dakang, 1987). The Grand Canal was designed as a shipping channel that flowed from the Yangtze in the south to the Yellow River in the north (Hangzhou to Beijing). Reservoirs were built along its length in order to feed the upper sections. The first lock-gates were built on the Grand Canal to allow boats to enter it from the Huai River. The purpose of this ambitious water project was to allow the transport of grain from the warm south to the cooler and drier regions in the north. Its annual transport capacity was 400 000 t of grain.

The total water resources in China are 2812 km³, the sixth richest in the world. Water abstractions amount to about one-fifth of the total water resources, and 87% of the water withdrawn is consumed by irrigation agriculture. However, water resources are concentrated in the humid southern areas while large areas in the north and west are deficient in water, especially in the North China Plain. Run-off in the south is, on average, 41 700 m³ ha⁻¹, which is provided by the Yangtze, and represents some 38% of the total surface run-off of the country, while in the north, the Huai and Huang rivers provide respectively 4230 and 4290 m³ ha⁻¹ (Dakang, 1983). More than 95% of the total population lives in the eastern part of China, where there are more than

300 persons km⁻². Adding further to the uneven distribution of water in China is the fact that most of the cultivated and productive land lies in the north. There is thus enough water in the south, but limited agricultural land, while in the north the opposite is the case. The huge population and continuing rapid population growth have made water resources a limiting factor for economic development. With rapid urbanization and industrial development, water demands for economic developments and wastewater discharge from domestic and industrial sectors have increased dramatically.

Human impacts on riverine ecosystems often relate to complex combinations of factors. Thus, in a study of fisheries on the Pearl River – a sub-tropical river in southern China (Figure 10.1) – between 1982 and 1984 (Liao *et al.*, 1989) the decline of fish catches was attributed to: (i) overfishing and use of inappropriate fishing techniques; (ii) pollution; and (iii) dam construction. The survey reported 381 species including 262 freshwater fish dominated by cyprinids. Mean annual catches in the 1950s of 10 367 t yr⁻¹ had declined to only 6464 t yr⁻¹ in the early 1980s. Overfishing had eliminated older fish, notably at spawning grounds of the guangdong bream (*Megalobrama hoffmanni*). The use of gill nets with a mesh of only 1.0 to 1.5 cm had led to the loss of fingerlings. More than 3000 dams had been built without fish passes and these blocked migrations of the Chinese shad (*Macrura reevesii*) and gizzard shad (*Clupanodon thrissa*) which had virtually disappeared from the eastern river by 1970. Pollution from mining, smelters and factories added 1.2×10^6 t of untreated industrial wastewater to the middle river each year and pesticides from paddy fields added levels of benzene hexachloride above the maximum allowed by state legislation.

Water resources development

Human activities have significant impacts on rivers, not least through changes in the natural hydrological regime. There were more than 80 000 dams in China by the end of 1993. Within the Hai River Basin, for example, there are 30 large reservoirs ($>100 \times 10^6$ m^3) which can control about 83% of annual run-off; the lack of compensation flows means that there is no flow in the lower reaches of many rivers throughout the dry season. Studies of impacts of dams and reservoirs include changes in channel form and processes (Chien, 1985) and changes in fish communities (Zhong and Power, 1996). Impacts are not restricted to surface water developments but include groundwater abstraction and land-use changes. There are many examples where over-abstraction of groundwater has caused a lowering of the water table and land subsidence has appeared in Shanghai, Tianjin, Beijing and other large cities (Li, 1998).

Rapid urbanization and industrial development in Japan and South Korea since the mid 1950s led to a great increase in water demands. In South Korea, for example, 19% of water use in 1992 was for domestic water use, 35% for industrial use and 46% for agricultural use. Dams were built mainly for irrigation and power generation in the 1950s and 1960s, but subsequently, large integrated reservoirs were built mainly to support domestic and industrial water supplies and river regulation. Nine large multipurpose reservoirs were built between 1962 and 1992. Two more integrated reservoirs will be built by 2001 adding 0.33 km^3 water storage.

In Japan, many dams and reservoirs have been constructed, including 88 high dams (>100 m high) for flood control, irrigation, urban water supply and power generation. Studies of impacts have focused on water quality: changes following diversion (Kagawa, 1992), the accelerated algal production in a river after the construction of a river mouth barrage (Murakami *et al.*, 1998), the development of 'red tides' of *Peridinium* spp. and *Ceratium hirundinella* in reservoirs (Kagawa, 1989; Kitamura and Onishi, 1995), and the influence of irrigation and drainage water from paddy fields on water quality (Nagasawa *et al.*, 1997). The protection of cultural heritage within river projects has also attracted attention (Nakagawa and Miyae, 1990; Nakagawa, 1988).

The main areas at risk from water resource developments are (i) the lower reaches of rivers and (ii) urban rivers. Rivers in the former category are affected mainly by water abstraction and dam construction, and in the latter by the return of untreated wastewater. Other confounding impacts include accelerated soil erosion and overfishing. The Yangtze and Yellow Rivers have experienced, and are continuing to experience, a wide range of impacts; however, it is the inland drainages that are experiencing particularly intense problems.

The Yangtze

In the middle and lower reaches of the Yangtze River many economically important fish species, and also endangered and endemic species, are at risk because of the construction of large dams and from overfishing. Zhong and Power (1996) reviewed the impacts on fish of four hydroelectric power dams, two in the Yangtze Basin and two in a coastal river, the Qiantang near Hangzhou. The major dams on the coastal river changed the fish community of the estuary, reducing the number of freshwater species from 96 to 85 and the changed flow regime of the Qiantang caused the extinction of Chinese shad (*Macrura reevesii*), a highly valued fish. All dams blocked fish migrations because of the lack of fish passage facilities. Spawning runs of some rare species, such as Chinese sturgeon (*Acipenser sinensis*), Chinese sucker (*Myxocyrinus asiaticus*) and white sturgeon (*Psephurus gladius*) were obstructed by the Gezhouba Dam on the Yangtze River, the reach below the dam became a target for a local fishery, and the populations became endangered by overfishing. Athough some species, including Chinese sturgeon and Chinese sucker, found new spawning grounds below the dams, regulated flows with low velocities caused some spawning grounds to be abandoned and lower water temperatures delayed spawning by 20–60 days. Today, the populations of both species are certainly augmented, and possibly sustained, by artificial breeding programmes. Below the Xinanjiang Dam on the Qiantang River, low water temperatures eliminated warm-water fish from a 15-km reach, but introduced rainbow trout, *Onchorhynchus mykiss*, have established a strong population.

Particular interest has focused on the Baiji dolphin (*Lipotes vexillifer*), an endangered freshwater mammal found only in the middle and lower reaches of the Yangtze River (Chen and Hua, 1989). The numbers of dolphin in one 150-km reach declined from nine groups and 43 individuals in 1986 to three groups and 11 individuals in 1991 (Zhong and Power, 1996). Causes of the decline were attributed to channel degradation below the Gezhouba Dam and associated habitat loss and disturbance, changes in the prey fish community, and indirect injuries from contact with boats and fishing gear.

The Yellow River

Problems for river management are particularly severe within the Yellow River Basin. There are seven major hydropower dams; the largest, the 175 m high Longyangxia Dam, has a storage of 24.7 × 10⁹ m³ and an installed capacity of 1280 MW (Changming, 1989). In the lower reaches, the river bed has become dry for extended periods since 1972. Reservoir storage and irrigation abstractions along the middle and lower reaches caused the river to dry up for 226 days at Lijin Station in 1997. In the delta of the Yellow River, declining freshwater run-off from upstream has endangered the wetland habitats, which are important for migratory birds from Australia to northern Asia.

Extreme flood peaks exceed 30 000 m³ s⁻¹ but the worst flood problems occur in February, associated with ice runs and hyper-concentrations of sediment (Changming, 1989). Intensive cultivation is causing severe soil erosion on the loess plateau, producing sediment loads of 1.6 × 10⁹ t at the Sanmenxia Station on the Yellow River and leading to serious channel sedimentation at rates of 10 cm yr⁻¹ (Changming, 1989). The river-bed has risen by 7 m near Kaifeng City, increasing the flood hazard. Much of the 750 000 km² basin is mantled by wind-blown silt (loess or *huang tu* in Chinese) reaching thicknesses of over 300 m. In the semi-arid areas of the catchment, rates of soil erosion exceed 50 000 t km⁻² yr⁻¹ (Derbyshire and Wang, 1994).

Degradation of the loessic areas over a long period of human utilization, driven by the growing population and inappropriate land-use policies, has reduced grain yields and the carrying capacity for sheep. Along the river's lower reaches where gradients are as shallow as 0.1% the river flows through levees up to 10 m above the floodplain. In flood the river can inundate the entire 250 000 km² alluvial plain – the breadbasket of China – where 100 million people live including many towns forming the industrial base of northern China (Derbyshire and Wang, 1994). Typically, rivers such as the Yellow River shift across their entire floodplain surface and 26 major shifts of the channel have been recorded over the past 2000 years (Changming, 1989).

Over the last 50 years problems of the Yellow River have been tackled by increasingly impressive and expensive engineering schemes designed mainly to reduce the sediment influx to the lower river. By 1985 there were also more than 150 hydroelectric power reservoirs with a total storage capacity of over

53 × 10⁶ m³. The larger dams, such as Sanmenxia, play a key role in flood mitigation, and over the 20-year period to 1985 nine floods with peak discharges in excess of 10 000 m³ s⁻¹ were controlled by reservoir manipulation and emergency works on levees (Derbyshire and Wang, 1994). Derbyshire and Wang (1994) conclude that the key remediation measure must be to minimize soil loss from hillslopes but this 'gargantuan task' will remain a severe drain on the gross national product well into this century. In such circumstances, sustainable development incorporating sound river conservation practices will be no more than a long-term dream.

Interior basins

Impacts of human activities on the hydrological regime and ecosystems in closed interior drainage systems in arid, north-west China have been dramatic. The impact of surface-water storage and diversions, and groundwater abstractions for irrigation agriculture have been: (i) to reduce groundwater levels, reduce lake levels, reduce spring flows, degrade water quality and cause extensive desertification; and (ii) in irrigated areas to cause the extension of soil salinization as a result of raising water levels. Problems have been accentuated by declining run-off from headwater mountain catchments (Zuming, 1995). The Ejina Oasis in the arid region of Inner Mongolia is an extreme example. The oasis has shrunk dramatically as a result of abstractions from the upstream Hei River in the Hexi corridor in Gansu Province: 350.9 × 10³ ha of water, forest and grassland have become salty land and desert since the 1960s. Annual run-off at the Zhengyixia Station in the lower reaches of the river has declined by 40% in the last five decades following construction of 95 reservoirs with a storage capacity of 360 × 10⁶ m³. All the tributaries of the river have dried up. Many floodplain forest plants such as *Populus diversifolia* and *Elaeagnus angustifolia* have declined or even disappeared.

Impacts in north-west China have been reviewed by Mengxiong (1995). During the past 20 years the water table has declined by 0.1–0.9 m yr⁻¹ and springflows have been reduced by 16–60%. Lake East Juyanhai has shrunk from over 120 km² in the 1940s to less than 50 km² since 1960; salt content has increased from less than 1 mg L⁻¹ to over 30 mg L⁻¹ in the same period. Lake Luobupo, once covering an area of 1900 km², is now dry. Soil salinization has affected more than 1 × 10⁶ ha. On the Tarim River, flow diversions in the upper basin have caused the annual flow in the middle river to be reduced from 49.8 × 10⁸ m³ to 9.5 × 10⁸ m³;

the lower river is now dry and 300 km of green corridor has been lost – the poplar/willow forest corridor died, 20 000 ha of grassland disappeared, and 6700 ha of cultivated land was abandoned because of desertification.

The Aral Sea crisis

Serious water shortage and river pollution caused by agricultural production in central parts of the former Soviet Union are major problems. In central Asia irrigation agriculture is the basis of economic development. Voropaev and Velikanov (1985) predicted that by the end of the 20th century the annual water deficit in the region would amount to 20 km^3. Already, huge volumes of river water have been used for irrigated water-consuming crops such as cotton and rice. In 1990 a total of 7×10^6 ha in the Aral Sea region were irrigated, with annual water withdrawals of 60 and 45 km^3 from the Amu and Syr Rivers, respectively.

In the 1950s, the Aral Sea was the world's fourth largest lake, a brackish water body with a volume of some 1000 km^3 and an area of more than 68 000 km^2. Water abstractions have led to the severe degradation of the Aral Sea with a decline of water level and increase in salinity (Williams and Aladin, 1991): water levels have fallen by 15 m, the lake volume has declined to less than 350 km^3, and salinity has increased from 10 mg L^{-1} in 1960 to 30 mg L^{-1} in 1990. The economic, social and environmental effects of the decline in the Aral Sea are numerous and severe. The sea formerly contained more than 20 species of fish, most of which have died out as shallow spawning grounds have dried up and food resources disappeared in the increasingly saline waters. Of the once healthy fishery, only four species remain and these now face extinction (Williams and Aladin, 1991). The number of nesting bird species has been reduced by nearly 50% and mammals have declined from 70 to 30 species (Kotlyakov, 1991).

The exposed bed of the Aral Sea has become the source of large-scale dust storms, blowing up to 75 000 t of dust annually from the saline soils. Much of this salty material is deposited on the irrigated cropland of the delta, and adversely affects soils and crop yields. Furthermore, pesticide abuses have degraded water quality of the main rivers and the Aral Sea (Levintanus, 1992). The heavy use of toxic chemical pesticides in irrigation areas has influenced local drinking water supplies, which have also deteriorated through inadequate purification and sewage treatment plants.

Urban expansion

In China, urbanization and industrialization, together with rapid population growth, are seen as major problems (Lian, 1995) and 90% of water bodies adjacent to urban areas are subject to pollution, especially eutrophication (Zhiqing *et al.*, 1995). Environmental problems remain focused on wastewater control in the face of increasing water supply demands (Jun and Baoqing, 1997). Water shortage is a particular problem in the Hai River Basin (North China Plain), which has rapid population growth and economic development, especially within and around Beijing and Tianjin. Water resource *per capita* in the Hai River Basin was 351 m^3 in 1996, which is the lowest in China. The ratio of water consumption to total water resources is as high as 83% – the highest value in China. However, water pollution is the most serious issue requiring attention in the North China Plain. Huge volumes of wastewater (80% of the total volume) have been discharged into the river networks without any treatment; 69.4% of the Hai River drainage network was polluted in 1996. The major pollutants were COD, non-ionic ammoniacal nitrogen and volatile phenol (Shen, 1998). Similarly, in the middle Yellow River, conflicts exist between city water supply and irrigation demands on the one hand, and severe pollution from industrial and domestic sources on the other, with projections for a 300% increase in COD load from 14.8 to 42.9×10^4 t yr^{-1} by 2010 (Jun and Baoqing, 1997).

The local water resources for metropolitan areas are also a concern in Japan and Korea, even though the annual precipitation of these countries is high. The population concentration in urban areas has caused water supply deficits and serious river water pollution, especially in dry years. In South Korea, there is particular concern for the Han River, the main freshwater source for Seoul in the 1960s, because it is no longer a safe source of drinking water supply. One million tonnes of domestic wastewater and 450 000 t of industrial wastewater are being discharged into rivers every day – 72% of the total wastewater is discharged without any treatment.

In Japan, three metropolises – Tokyo, Osaka and Nagoya – are at risk. The overall population density of Japan is 332 km^{-2}, but 76% of the land area is covered by mountains and forests, where the population density is much lower than the mean value, so the population density is extremely high in urban areas. By the early 20th century, the Tama River Water Supply System produced 50×10^6 m^3 yr^{-1} to meet the water needs of Tokyo (Snaddon *et al.*, 2000). By the mid 1930s, however, two storage tanks had to be built in Tokyo to store some of the 300×10^6 m^3 yr^{-1} which was required from the Tama River. In order to supply water to its 7 million inhabitants, the total transfer of water to Tokyo in the 1960s exceeded 1×10^9 m^3 yr^{-1}. An impoundment on the Tama River became a necessity, but to ensure a continued supply while this dam was constructed, water was diverted from the Edo and Sagami rivers to the Tama River. This supply soon proved inadequate, and the city looked further afield to the Tone River, the largest river in Japan. Subsequently, tributaries of the Tone River were impounded to augment water supply, and an additional interbasin transfer from the Sakawa to the Sagami River and on to Tokyo is now operating.

Tokyo City is currently considering diverting water from rivers even further away, such as the Naka, Agamo, Shinano and Fuji Rivers. However, water diversion adds to the pollution problems caused by the discharge of untreated wastewaters into watercourses. Of growing concern are the levels of endocrine-disrupting chemicals and organic micropollutants including alkyphenols, phthalates, polychlorinated biphenyls and dioxins, which are the subjects of current research. High levels of nonylphenols have been reported for river waters and wastewater effluents from Tokyo (Isobe *et al.*, 1999).

Information for river conservation

Data on water resources (i.e. water conservation in a hydrological sense) are available for most areas but there is little information for river conservation, i.e. on the environmental and ecological character and status of river corridors. There has been little collaboration between disciplines, e.g. between hydrologists and ecologists, in order to evaluate options for sustainable development. 'Sustainability' has been addressed with regard to human systems but little attention has been given to ecological systems.

SCIENTIFIC DATA

There are many observation stations throughout the region for meteorological, hydrological and water quality monitoring, not least in central Asia where countries in the former Soviet Union were part of a standardized system (see Khaiter *et al.*, this volume). Since independence, data have been difficult to find.

In China, observations have been made since the 1950s. Water Yearbooks have been published by the Ministry of Water Resources, including data on catchment areas, discharges, water levels, precipitation, sediment loads and the main ionic components (Ca, Mg, Na, K, CO_3, HCO_3, SO_4 and PO_4). Water quality and ecological data have been collected and published by the Bureau of Environmental Conservation of China since the 1980s. By 1990 there were 2547 water quality monitoring stations with sampling frequencies ranging from less than once per month to at least once per day where stations are designed for the dynamic detection monitoring of pollution (Zhiqing *et al.*, 1995). Most water quantity data have been stored on computer databases since about 1990, but the data are not freely available to the public. More recently, the Chinese Academy of Sciences has investigated biodiversity data.

CLASSIFICATION AND EVALUATION SCHEMES

There is considerable variation in the application of classification and evaluation schemes. However, where they exist, economic values are particularly important. In China, river basins are classified according to ecological and economic functions, into (i) protected areas for drinking water supply, (ii) special wildlife habitat and (iii) natural protected areas. There are more than 700 natural protected areas in China. The seven main rivers in China are important at the national level. The Yangtze River is one of them, containing more than 300 species of fish and the original habitat of four economic fish species. Furthermore, there are many endemic and endangered animals and plants. As mentioned above, the wetland delta of the Yellow River is also a very important habitat for migratory birds.

In Japan, under the River Law, all watercourses have been placed into one of four classes according to their importance for the national economy: first class, second class, secondary and normal. There are 109 first-class rivers, 2636 second-class rivers, 11 890 secondary rivers and 112 900 normal ones.

PUBLIC AWARENESS AND EDUCATION

Public awareness is improving through the media as well as through education systems. In the past, the common perception of water resources was of an inexhaustible natural resource, which could be used without any limitation. The ecological roles and functions of rivers were ignored. This commonly held perception is now being changed through public education. Tangible evidence is found in controls on deforestation, and incentives for reforestation to prevent soil erosion. However, there is little evidence of planning for river conservation as explored in this book.

Within the region, public understanding of environmental issues is most advanced in Japan, although most attention has been given to waste minimization, effluent treatment and control, flood control and protection of cultural heritage. The River Council's *Recommendation on Future Policy for Improvement of River Environments* provides the framework for advancing river restoration and conservation practices (Table 10.4). The Council's publication *Inviting Rivers* (River Council, 1997) emphasizes the value of rivers for recreation, amentity and conservation, the relationship between rivers and communities, and the need to involve local residents in restoration, enhancement and protection schemes. The dissemination of information is through the annual National Census on River Environments, published as yearbooks and CD-ROMs and a twice-yearly magazine *Rivers and Japan* published by the Foundation of River and Basin Integrated Communications. Although the latter focuses on hazard mitigation (floods, pollution, etc.) and cultural heritage, the former includes information from biological surveys, such as fish and plant surveys, and on river habitats and use of river space.

A good example of the limited anthropocentric view of water resources is found in central Asia where the discussion of river conservation is based around water charges. Until 1982, water resources were free of charge, which did not encourage '*water* conservation'. Since 1982 there has been a very low charge. A solution to the Aral Sea crisis proposes to establish water charges for water abstractions for irrigation taken from the Amu and Syr Rivers, as well as penalties for their pollution. It is considered that an average price for water of 0.5 copecks m^{-3} would provide enough money for modernizing the water management system without adversely affecting incomes of the rural population.

The framework for conservation

National, regional and local governments are in charge of river conservation. In China, several ministries are involved in river conservation at the national level. For example, the Ministry of National Resources manages ecological conservation throughout the country. The Ministry of Water Resources manages water supply to rural areas and abstractions of groundwaters. The Ministry of Construction administers water supply to urban areas, including domestic and industrial water use. The Bureau of State Environmental Protection is in charge of water quality monitoring and wastewater discharge control. All organizations have their provincial and local branches, and there are also some organizations with responsibilities for river management at the basin level. Basin Administration Committees for the major river basins, such as the Yangtze River, Yellow River, Pearl River, Songhuajiang and Liao River, Hai River, and Huai River are responsible for monitoring river water quantity and quality.

In central Asia, the integrated development of national water resources and conservation policy is approved by a Council of Ministers. The context for policy development comprises the quantity of water resources, water consumption in national economic departments, measures to increase water resources, and solutions for water shortages and flood control. Environmental and ecological objectives receive little attention.

THE SOCIO-POLITICAL AND RELIGIOUS–CULTURAL BACKGROUND

In China, the introduction of some principles of a market economy system within the central government and socialist political systems has reinforced the anthropocentric view of natural resources. The public has recognized that water resources are valuable natural resources, especially in the places facing water deficits. The consumer-pays principle has been introduced so that every user must pay for water resource consumption whatever the purpose. Water charging is seen as an economic measure to encourage people to save water.

In central Asia, the socialist political system, and Islamic religion and culture affect river conservation. Inadequate population policies drive rapid population growth. Population in the area increased three times in the period 1917–1990. A sharp deterioration in the living conditions of the population has led to an increase in sickness rate and mortality up to the highest levels ever registered. The public health problem of the Aral region is closely interrelated with a complex set of ecological, social and economic issues having an extremely adverse effect on the human environment. The solution to the Aral Sea crisis will determine the fate of the whole area and it has become a subject of awareness and concern to the world community.

Table 10.4 *Summary of 'Recommendations on Future Policy for Improvements of River Environments' in Japan (River Council, 1997)*

Context
'As river environments draw increasing public attention, there is a growing need for comprehensive, basin-wide efforts to realize the ideal of a river environment where life and property are safe and which is biologically diverse and scenic, by using three basic approaches:

- protecting diverse habitats,
- protecting the hydrological cycle,
- re-establishing the river–(human) community relationship.'

I Basic policy for protection and creation of river environments

Basic policies
(i)   Protect diverse habitats recognizing the important role of rivers in maintaining biodiversity.
(ii) Protect the hydrological cycle by addressing issues including: urban flood run-off, overpumping of groundwater and land subsidence, contamination of lakes and salinization of groundwater, depletion of spring flows and decreases of normal stream-flow.
(iii) Re-establish the relationships between rivers and local communities with concern for river-based industries and culture (including rice cultivation, fisheries, transport, recreation and amenity).

Important considerations
(i)   Long-term trends in lifestyles, industrial structure, land use and global environmental problems.
(ii) Closer cooperation with local residents, local governments and administrative agencies.
(iii) Need for interdisciplinary and basin-scale approaches.

II Implementation measures to improve river environments

Basic measures
(i)   To protect diverse habitats: formulate a Master Plan for Improvement of River Environments to include programmes aimed at creating nature-rich scenic rivers and fish-friendly rivers. Implement measures to: secure the required amount of space within river corridors to maintain/restore natural channel forms and associated diverse habitats; increase green spaces along rivers; maintain the longitudinal continuity of rivers to enable free movement of fish and wildlife; protect rare plant and animal species; upgrade the National Census on River Environments; and restrict activities which adversely affect wildlife habitats.
(ii) To protect the hydrological cycle by: improving communication of information on impacts of hydrological change (quality and quantity); implement restoration measures and improved monitoring; and integrate basin measures and river administration.
(iii) To re-establish river–community relationship by: regenerating river environments and riverscapes; improving access; promoting riverfront developments; improving disaster-preventing functions; and involving the local community.
(iv) To further technological cooperation in advancing measures to tackle global environmental problems.

Actions to be taken
(i)   Expand river environment plans to attach more importance to protecting habitats and hydrological cycles and to involve local communities.
(ii) Support the Council on Management of River Improvement for the formulation of a Master Plan for Management of River Environments and other schemes promoting the roles of local residents and voluntary organizations, etc.; establish a River Environment and Basin Council composed of community representatives, local governments, river administrators, academic experts and non-governmental organizations.
(iii) Improve communication with local communities and foster environmental education.
(iv) Advance R&D programmes including cooperation between ecology and river engineering; undertake not only scientific and technical research but also studies on human health and consensus building.
(v)   Improve standards, guidelines and institutional frameworks.

STATUTORY FRAMEWORK (NATIONAL AND INTERNATIONAL LEGISLATION)

Much legislation related to river management has been promulgated in China in the last three decades. Rushu (1996) identified three key components for the advancement of river conservation in China. First, the Constitution of the People's Republic of China (4 December 1982) Article 26 states that: 'The state protects and improves people's environment and the ecological environment, prevents and controls pollution and other public hazards.' Second, the Environmental Protection Law of the PRC (26 December 1989) Article 19 states that: 'Measures must be taken to protect the ecological environment while natural resources are being developed or utilized.' Third, the Agenda 21 White Paper on *China's Population, Environment and Development in the 21st Century* is the Chinese Government's guiding document designed to balance protection of the environment and natural resources with sustained economic growth. It focuses on population control, energy savings and pollution reduction. Population control policy is reducing population pressures on water resources and on river ecosystems. Environmental protection policy is beginning to control pollution sources from industrial and domestic sectors. A 'Water Law' and 'Wildlife Conservation Law' have also been enacted. Conventions on Biological Diversity and Wetland Conservation have also been adopted in recent years and legislation on natural reserves has been established. The Chinese Government also cooperates on international river development, such as with east Asian countries on the estuary of the Tumenjiang, and on the Lanchangjiang–Mekong River Basin.

Some legislation on water use and water pollution prevention has been enacted in all countries in East Asia. For example, in Japan, this includes: Water Course Law (1957), Legislation on Water Resources Development Improvement (1961), River Law (1964), Water Pollution Protection Law (1970), Special Legislation on Water Source Area Strategy (1973) and Special Measures on Water Quality Protection in Lakes (1984). The River Law (JIDI, 1997) covers all aspects of river administration: flood control and damage mitigation; water resources utilization and development; land and river water administration; and conservation of the fluvial environment. However, the last of these aspects refers only to the 'conservation of the river-banks or river administration facilities'.

River management in Japan is carried out at different levels. The Ministry of Construction manages the first-class rivers: provincial, regional and county government manages the second-class ones, and secondary ones are managed by governors of cities, towns or villages. Furthermore, in terms of water rights and according to Japanese law, all surface run-off is regarded as public property. Water rights must be granted by the government before a river is impounded or diverted but owing to the long history of agricultural water use, most agricultural water rights were entrenched long before the institution of the 'public water' law (Okamoto, 1983). The practical outcome of this is that farmers have automatic and undisputed entitlement to water rights, which has led to many disputes, especially in cases where water is diverted from downstream users.

In South Korea, water resources management is carried out by the Ministry of Construction and the regional Land Management Bureau in five regions. South Korea Water Resources Company, a comprehensive water management institute which carried out an integrated survey for four main rivers in 1967, is in charge of multipurpose reservoirs, water supply and water treatment.

ROLE OF VOLUNTARY ORGANIZATIONS

The role of voluntary organizations is limited because all matters related to water are controlled by governments at different levels. There are some non-governmental academic organizations to exchange and disseminate scientific information on river conservation in China, including the Society of Water Conservancy of China, the Geographic Society of China and the Society of Environmental Sciences. There are some voluntary organizations in Japan but they focus on the effective use of water resources, water quality protection and the development of new water sources.

Conservation in practice

IN SITU CONSERVATION (PROTECTED AREAS)

There are more than 900 natural protected areas in China for wildlife habitat conservation. Special protected areas have been established for saving endangered and rare plants and animals, such as *Lipotes vexillifer*, *Alligator sinensis*, *Acipenser sinensis*, *Metasequoia glyptostroboides*, *Cathaya argyrophylla* and *Davidia involucrata* in the lower reaches of the Yangtze River. There are many forest reserves for endemic plant species in the Lanchangjiang Basin.

EX SITU CONSERVATION (GENE BANKS)

In China, a national gene bank has been established for saving and storing genetic material for all kinds of natural and agricultural species from throughout the country. Artificial reproduction and propagation have been carried out for endangered and rare species including *Alligator sinensis*, a famous endangered and endemic species in the Yangtze River. The centre for *A. sinensis* reproduction and research was established in Anhui Province in 1982. Successful artificial breeding techniques were developed by 1994, and in 1996 there were 4376 *A. sinensis* in the centre, comprising 248 breeding adults, 1542 juveniles and 2586 others (Chengyuan, 1996). There are also many zoos and botanical gardens in China for *ex situ* conservation as well as for public education. There are about 120 botanical gardens growing more than 18 000 plants, including 65% of the protected plant species in China (Mackinnon and Wang, 1996).

RESTORATION AND REHABILITATION PROGRAMMES

Limited progress has been made in this area and the full nature of the environmental impacts in many areas remains to be documented. However, as noted above, reforestation has been carried out in the headwaters of some rivers, and a water and soil conservation programme has been implemented in the loess plateau, and these initiatives have reduced sedimentation problems. Soil conservation measures including reforestation have reduced sediment yields in north-east China by up to 80% in some catchments (Hsueh-Chun, 1995). Special water pollution control programmes in the Huai, Liao and Hai River Basins have been put in place and all small-scale factories producing heavy pollution loads were closed by the end of September 1995.

Some water pollution control has been carried out in Japan. Most notable are the plans to rehabilitate Lake Biwa: *Special Measures Regulation on Integrated Development of Biwa Lake* and *Integrated Development Planning of Biwa Lake* made in 1972 and revised in 1982. The measures included: (a) water quality conservation; (b) flood control; and (c) water conservation for domestic, agricultural and industrial use, construction of fish ponds, pearl ponds and fish harbours, and for tourism development. The total investment of these measures has reached 1524.8×10^9 Japanese yuan. The Japan National Committee for Hydrological Sciences (JNCHS, 1999) reports that the integration of traditional water resource management approaches with human, social and ecological systems is growing in importance. Methods are being developed to 'value' environmental quality including recreational uses and 'nature-friendly' river works which have been intensively executed since 1990 (Tamai *et al.*, 1998). Contingent valuation procedures have been used to assess the value of nature conservation measures including augmentation of river flows and use of self-purification facilities (reed-bed technologies), and preliminary results show that the willingness to pay is greater than the cost of the restoration works.

The major breakthrough in this field was made in 1989 when the USSR Supreme Soviet, by the decree of 27 November 1989, constituted the government commission for the elaboration of measures for restoring the ecological balance in the Aral Sea region. In its turn the commission announced in June 1990 a competition to work out a concept for the conservation and restoration of the Aral Sea, and normalization of the ecological, sanitary, medical, biological and socio-economic situation in the region. After reviewing the results of the competition, the commission formed an *ad hoc* working group in November 1990 made up of prominent Soviet scientists. This group was entrusted with summarizing the best proposals submitted to the competition and using them to formulate a plan to combat the Aral Sea crisis that could become a foundation for official government policy. The first draft of this document was produced early in 1991 and concluded that the restoration of the Aral Sea should be part of a broader strategy for overcoming the ecological, economic and social crisis of the whole Aral Sea region.

FUTURE THREATS AND THEIR MITIGATION

Throughout the region, the main threats are water shortage and water pollution, and flood and drought hazards. Rapid population growth and urbanization have made the water crisis acute both in cities and in rural areas. Sustainable development in the 21st century is seen to be dependent upon interbasin water transfers.

The imbalances in land, water and human distribution throughout the region are perceived as major limitations to economic growth and development. In China, to address these imbalances the Ministry of Water Resources and Electric Power has investigated the feasibility of transferring water from the south to the north, primarily to allow further agricultural and urban development in recipient areas (Dakang, 1983; Stone, 1983; Harland, 1988). The proposed transfers would divert water from the Yangtze to the Yellow River, and other smaller rivers in the north. The schemes are collectively referred to as the South-to-North Water Project (SNWP) (Yiqiu, 1981). Forecasts (Rushu, 1996) suggest that the central feature of the scheme, the Three Gorges Dam and reservoir on the Yangtze, will benefit 75 million people along the lower river, protect 6×10^6 ha of agricultural land from floods, and create a 70 000 km inland waterway allowing river shipping capacity to increase five-fold.

The proposals and their impacts have been reviewed by Snaddon *et al.* (2000). Three routes were proposed. First, the Western Route comprises three transfers: (i) from the Tongtian River to the Qaidam Basin through the Kunlun Mountains; (ii) from Jinsha Jiang, a tributary in the upper reaches of the Yangtze, through the Jishi Mountains to the Huang River; and (iii) from the Nu Jiang through various rivers to Dingxi County. Each transfer included in this scheme cuts through mountain ranges, considerably increasing construction costs, and consideration of this route has been postponed (Changming *et al.*, 1985; Changming and Dakang, 1987). Second, the Middle Route is a two-phase scheme transferring water initially from the extant Dangjiangkou Reservoir (on the Han Jiang, a tributary of the Yangtze) to the Beijing district, and later drawing water from the Three Gorges Dam (under construction) on the middle reaches of the Yangtze, and transferring it to the Dangjiangkou Reservoir. Third, the East Route would use part of the extant Beijing–Hangzhou Grand Canal to transfer water from the lower reaches of the Yangtze at Jiangdu to the North China Plain, and is thus a cheaper and easier alternative. The Grand Canal is, at present, silted up, but would be dredged, pumps installed, and water pumped from the Yangtze to the Yellow River.

The East Route was approved by the State Council in 1983, and construction has already begun; hence this route has been described in detail by Changming *et al.* (1985). It comprises two stages. The first will abstract water at the Jiangdu Pump Station on the Yangtze transferring it via Hongze Lake (on the Huai River), Luoma Lake (on the Yi River) and Nansi Lake, to Dongping Lake on the Dawen River, along a 646 km long canal. This first stage coincides with an existing interbasin transfer in the north of Jiangsu Province, on which construction began in the early 1960s. This multipurpose scheme involves the transfer of water from the Yangtze River to the Huai River and then northwards to the Yi-Shu-Si Basin. The second stage will continue under the Yellow River into Beidagang Reservoir, which is situated near, and supplies water to, Tianjin City. The average volume of water that will be transferred is 1000 m^3 s^{-1}, which represents a total transfer of 14×10^9 m^3 yr^{-1}, rising to 30×10^9 m^3 yr^{-1} in drought years with the addition of water from the Yellow River. The transfer passes under the Yellow River in order to avoid the inevitable massive increase

in suspended solids load that an overland route would entail. The Yellow is the most turbid river in the world, exporting some 2×10^9 t yr^{-1} of sediment to the sea (Harland, 1988). Thus, Yellow River water will only be used when absolutely necessary. The East Route will also harness the run-off of the Huai Basin. The three schemes together would eventually transfer an annual average flow of 1200 m^3 s^{-1}.

With regards to the Middle Route, the Three Gorges Dam will form the major storage unit for the transfer. Its main purpose is to provide flood control in the catchment and hydroelectric power generation, and to improve navigation along the Yangtze River (Kwai-Cheong, 1995). Construction began in December 1992, despite pressure from within China and from other countries to reconsider the project (Burton, 1994; Pearce, 1995). The dam will take approximately 15 years to complete, at a cost of between US$22 and US$34 billion, although, taking interest and inflation into account, some estimates reach US$70 billion (Snaddon *et al.*, 2000). The dam wall will be approximately 175 m high and nearly 2 km long, impounding approximately 500 km of river behind it. Hydroelectric power generating capacity is estimated at 18 000 MW, which is substantially greater than that generated by the world's largest hydroelectric power station at Itaipú in Paraguay (Pearce, 1995; McCully, 1996). It is argued (Rushu, 1996) that the environmental benefits of the scheme are: (i) to replace the need for 10 nuclear power stations; (ii) to reduce coal consumption by 40–50×10^6 t yr^{-1}; (iii) to reduce CO_2 emissions by 100×10^6 t; (iv) reduce SO_2 emissions by 2×10^6 t; and (v) to reduce NO_x emissions by 370 000 t yr^{-1}.

To reduce the adverse environmental effects of the Three Gorges Project an environmental impact assessment has been completed and countermeasures have been proposed (Wenxuan, 1987). As a result of this study, a fish ladder and navigation locks have been added to the project. Some protected areas for endangered and endemic plant and animal species have also been established. Furthermore, the operation of the reservoir will consider the needs of ecosystems and the environment. However, voiced concerns about the impacts of the proposed interbasin transfers on river conservation are non-existent, while a handful of authors have raised a few disparate issues. In the lower reaches of the Yangtze, siltation is anticipated as a result of decreased river flows resulting from abstraction for the East Route diversion, especially during low-flow months (Changming and Dakang, 1987). This is likely to have detrimental effects on the Yangtze estuary, as well as on estuarine and brackish-water fish (Xuefang, 1983). Concerns have been voiced about the seepage of water from unlined transfer canals, which would be a likely consequence of the proposed East Route (Changming *et al.*, 1985), while salinization has been, and still is, a potentially serious threat to agriculture in the areas to which water will be transferred by this route, as well as in areas along the route. More recently there has been a reduction in salinization, but care must be taken to prevent the problem recurring (Huanting *et al.*, 1983). The one anticipated area of great concern for the East Route is the impact on the recipient region of the transfer of pollutants including phenols, cyanides and mercury (Jinghua and Yongke, 1983).

In addition, the proposed East Route will transfer water through a series of shallow lakes – the Hongze, Luoma, Dongping and the Nansi – which serve as important freshwater fisheries, and it is feared that high water levels resulting from the use of these lakes as a transfer route could reduce their productivity (Xuefang, 1983). The lakes are also rich growing areas for various economically important plant species, such as *Phragmites* reeds, lotus root, wild rice, and water chestnuts. Furthermore, reduced dry-weather flows, as a possible consequence of the diversion of water along the East Route, have raised fears of a substantial reduction in freshwater supplies to the industrial, agricultural and domestic water for Shanghai Municipality (Yuexian and Jialian, 1983). Altered flow regimes, sediment transport and tidal gradients in the Yangtze estuary caused by transfers away from the system will also adversely affect navigation in these parts of the river and estuary (Snaddon *et al.*, 2000). This is contrary to the predictions of Yuexian and Jialian (1983), who, in their review of the probable effects of the East Route water transfer, anticipated that the use of the Yangtze as China's main navigation channel would not be affected by this transfer, even when all three proposed routes are operational.

The future

Water shortage and pollution have become serious problems in urban areas. More measures have to be introduced to limit water consumption, such as water recycling in industrial systems, water charging, and so on. At least in Japan, concerns about the environmental and social consequences of large water projects have

arisen following the proliferation of diversion schemes. Many local people are beginning to doubt the necessity of an unlimited search for further water supplies, when reductions in water consumption and more efficient use of the present resource could prolong the adequacy of current supply (Okamoto, 1983). However, throughout the region, a view remains that the challenge is to capture and exploit natural resources. In Tibet, the Lhasa River, a major tributary of the Yaluzangbu River or Brahmaputra, is the political, economic and cultural centre of the Tibet Autonomous Region. With forecasts of rapid growth of irrigation agriculture and hydroelectric power development, and the need for flood control around expanding urban areas, river regulation is seen as the key to economic development (Dajun, 1995). With regard to China, Rushu (1996) stated: 'The annual run-off of water to the sea is one third of the country's total . . . what a pity so much water power is going to waste!' The continuing ecological degradation of river corridors throughout the region is likely to be the inevitable outcome of current demographic, economic and political trends.

References

Burton, S. 1994. Taming the River Wild. *Time* 19 December, 42–44.

Changming, L. 1989. Problems in management of the Yellow River basin. *Regulated Rivers: Research & Management* 3:361–370.

Changming, L. and Dakang, Z. 1987. Environmental issues of the Three Gorges Project, China. *Regulated Rivers: Research & Management* 1:267–273.

Changming, L., Dakang, Z. and Yuexian, X. 1985. Water transfer in China: the East Route Project. In: Golubev, G.N. and Biswas, A.K. (eds) *Large-Scale Water Transfers: Emerging Environmental and Social Experiences*. United Nations Environmental Programmes, Water Resources Series, 7. Tycooly Publishing Company, Dublin, 103–118.

Chen, P. and Hua, Y. 1989. Distribution, population size and protection of *Lipotes vexillifer*. *Occasional paper of IUCN/SSC* 3:81–85, Gland.

Chengyuan, L. 1996. Present status of *Alligator sinensis* species resources. *Chinese Biodiversity*, 4:83–86.

Chien, N. 1985. Changes in river regime after the construction of upstream reservoirs. *Earth Surface Processes* 10:143–160.

Dajun, S. 1995. Research on the rational use of water resources on the Lhasa River, Tibet. In: Simonovic, S.P., Kundzewicz, Z., Rosbjerg, D. and Takeuchi, K. (eds) *Modelling and Management of Sustainable Basin-scale Water Resource Systems*. IAHS Publlication 231, Wallingford, 151–158.

Dakang, Z. 1983. China's south-to-north water transfer proposals. In: Biswas, A.K., Dakang, Z., Nickum, J.E. and Changming, L. (eds) *Long Distance Water Transfer, A Chinese Case Study and International Experiences*. Water Resources Series, 3. Tycooly International Publishing Limited, Dublin, 91–96.

Derbyshire, E. and Jingtai Wang 1994. China's Yellow River Basin. In: Roberts, N. (ed.) *The Changing Global Environment*. Blackwell, Oxford, 417–439.

Dudgeon D. 1992. Endangered ecosystems: a review of the conservation status of Tropical Asian rivers. *Hydrobiologia* 248:167–191.

Harland, D. 1988. Once more unto the breach. *New Scientist*, 21 July, 31–32.

Hsueh-Chun, Y. 1995. Impact of human activity on hydrology: Liaoning Province, China. In: Petts, G. (ed.) *Man's Influence on Freshwater Ecosystems and Water Use*. IAHS Publication 230, Wallingford, 41–45.

Huanting, S., Zhichang, M., Guochuan, G. and Pengling, X. 1983. The effect of south-to-north transfer on saltwater intrusion in the Chang Jiang estuary. In: Biswas, A.K., Dakang, Z., Nickum, J.E. and Changming, L. (eds) *Long Distance Water Transfer, A Chinese Case Study and International Experiences*. Water Resources Series, 3. Tycooly International Publishing Limited, Dublin, 315–360.

Isobe, T., Satoh, M., Ogura, N. and Takada, H. 1999. Nonylphenols in river waters and wastewater effluents collected in Tokyo. *Journal of the Japanese Society of Water Environment*, 22:118–126.

JIDI 1997. The River Law. Infrastructure Development Institute Water Series No. 2, Tokyo.

Jinghua, W. and Yongke, L. 1983. An investigation of the water quality and population in the rivers of the proposed water transfer region. In: Biswas, A.K., Dakang, Z., Nickum, J.E. and Changming, L. (eds) *Long Distance Water Transfer, A Chinese Case Study and International Experiences*. Water Resources Series, 3. Tycooly International Publishing Limited, Dublin, 361–372.

Japan National Committee for Hydrological Sciences (JNCHS) 1999. *Japan National Report on Hydrological Sciences 1995–1998*. Science Council of Japan, Tokyo.

Jun, X. and Baoqing, H. 1997. Water-related environmental problems and sustainability of water resources: a case study of the Sabn–Hua region, China. In: Rosbjerg, D., Boutayeb, N.-E., Gustard, A. and Kundzewicz, Z.W. (eds) *Sustainability of Water Resources under Increasing Uncertainty*. IAHS Publication 240. Wallingford, 459–468.

Kagawa, H. 1989. Proposals for inhibiting abundant phytoplankton growth at the head of a river reservoir. *Regulated Rivers: Research & Management* 3:123–132.

Kagawa, H. 1992. Effects of diversion on the chemistry of a stream in Japan. *Regulated Rivers: Research & Management* 7:291–302.

Kitamura, Y. and Onishi, S. 1995. Study on environmental conditions and freshwater red tide in man-made lake. *Journal of the Japan Society of Hydrology and Water Resources* 8:297–308.

Kotlyakov, V.M. 1991. The Aral Sea basin: a critical environmental zone. *Environment* 4–9 and 36–38.

Kwai-Cheong, C. 1995. The Three Gorges Project of China: resettlement prospects and problems. *Ambio* 24:98–102.

Levintanus, A. 1992. Saving the Aral Sea. *International Journal of Water Resources Development* 8:60–64.

Li, L. 1998. Impacts of human activities on water cycle in Haihe basin. *The Journal of Chinese Geography* 8: 306–310.

Lian, G.S. 1995. Impact of climate change on hydrological balance and water resource systems in the Dongjiang Basin, China. In: Simonovic, S.P., Kundzewicz, Z., Rosbjerg, D. and Takeuchi, K. (eds) *Modelling and Management of Sustainable Basin-Scale Water Resource Systems*. IAHS Publication 231. Wallingord, 141–150.

Liao, G.Z., Lu, K.X. and Xiao, X.Z. 1989. Fisheries resources of the Pearl River and their exploitation. *Proceedings of the International Large River Symposium, Canadian Special Publication of Fisheries and Aquatic Sciences* 106:561–568.

Mackinnon, J. and Wang Song 1996. Working report of the Biodiversity Working group of CCICED in the third year (1994/5). *Chinese Biodiversity* 4:54–62.

McCully, P. 1996. *Silenced Rivers: the Ecology and Politics of Large Dams*. ZED Books, London.

Mengxiong, C. 1995. Impacts of human activities on the hydrological regime and ecosystems in an arid area of northwest China. In: Petts, G. (ed.) *Man's Influence on Freshwater Ecosystems and Water Use*. IAHS Publication 230, Wallingford, 131–139.

Murakami, T., Kuroda, N. and Tanaka, T. 1998. Accelerated phytoplankton growth in a Japanese river after the construction of a rivermouth barrage. *Japan Journal of Limnology* 59:251–262.

Nagasawa, T., Umeda, Y., Muneoka, T. and Yamamoto, T. 1997. Influence of irrigation and drainage on water quality environment of paddy field area – a case study on the Ishikari River Basin in Hokkaido, Japan. *Journal of the Japan Society of Hydrology and Water Resources* 10:477–484.

Nakagawa, T. 1988. A flood alleviation of the Kakehashi basin, Japan. *Regulated Rivers: Research & Management* 2:187–194.

Nakagawa, T. and Miyae, S.-I. 1990. Management of the Sai River and the Tatsumi Canal, Japan. *Regulated Rivers: Research & Management* 5:183–188.

Okamoto, M. 1983. Japanese water transfer: a review. In: Biswas, A.K., Dakang, Z., Nickum, J.E. and Changming, L. (eds) *Long Distance Water Transfer, A Chinese Case Study and International Experiences*. Water Resources Series, 3. Tycooly International Publishing Limited, Dublin, 65–75.

Pearce, F. 1995. The biggest dam in the world. *New Scientist*, 28 January, 25–29.

River Council 1997. *Inviting Rivers*. River Bureau, Ministry of Construction, Technology Research Centre for Riverfront Development, Tokyo, Japan.

Rushu, W. 1996. Three Gorges Project and the environmental impacts. *Proceedings RIVERTECH '96*. International Water Resources Association, Urbana, Illinois, 146–168.

Shen, Z. 1998. Ten crucial challenges to water security for agricultural use and the expansion of irrigation area in China. In: *Research on Countermeasures to Water Shortage for Agriculture in China*. Press of Agricultural Scientific Techniques, Beijing.

Shoukun, W. 1997. The relationship between fish species distribution and biodiversity and river characteristics of main rivers in China. *Chinese Biodiversity* 5:197–201.

Snaddon, C.D., Davies, B.R. and Wishart, M.J. 2000. *The Ecological Effects of Inter-Basin Water Transfer Schemes, with a Brief Appraisal of their Socio-Economic and Socio-Political Implications, and an Outline of Guidelines for their Management*. Report to the South African Water Research Commission, Pretoria.

Stone, B. 1983. The Chang Jiang Diversion Project: an overview of economic and environmental issues. In: Biswas, A.K., Dakang, Z. Nickum, J.E. and Changming, L. (eds) *Long Distance Water Transfer, A Chinese Case Study and International Experiences*. Water Resources Series, 3. Tycooly International Publishing Limited, Dublin, 193–214.

Tamai, N., Shirakawa, N. and Matsuzaki, H. 1998. Cost and benefit in economic evaluation of river restoration works. *Annual Journal of Hydraulic Engineering, Japan Society of Civil Engineers*:271–276.

Voropaev, G.V. and Velikanov A.L. 1985. Partial southward diversion of Northern and Siberian rivers. In: Golubev, G.N. and Biswas, A.K. (eds) *Large-Scale Water Transfers: Emerging Environmental and Social Experiences*. Tycooly International Publishing Limited, Dublin, 67–83.

Wenxuan, C. 1987. Preliminary assessment of impacts of the Three Gorge Project on fish resources of the Changjiang (Yangtze) River and approaches to the resources proliferation. In: *Proceedings of Impacts of the Three Gorge Project on Ecosystem and Environment and Countermeasures Study*. Science Press, 2–20.

Williams, W.D. and Aladin, N.V. 1991. The Aral Sea: recent limnological changes and their conservation significance. *Aquatic Conservation: Marine and Freshwater Ecosystems* 1:3–23.

Xuefang, Y. 1983. Possible effects of the proposed Eastern Transfer Route on the fish stock of the principal water bodies along the course. In: Biswas, A.K., Dakang, Z., Nickum, J.E. and Changming, L. (eds) *Long-Distance Water Transfer: A Chinese Case Study and International Experiences*. Water Resources Series, 3. Tycooly International Publishing Ltd, Dublin, 373–388.

Yiqiu, C. 1981. Environmental impact assessment of China's Water Transfer Project. *Water Supply and Management* 5:253–260.

Yuexian, X. and Jialian, H. 1983. Impact of water transfer on the natural environment. In: Biswas, A.K., Dakang, Z., Nickum, J.E. and Changming, L. (eds) *Long-Distance Water Transfer: A Chinese Case Study and International Experiences*. Water Resources Series, 3. Tycooly International Publishing Ltd, Dublin, 159–168.

Zhao, J., Zheng, G., Wang, H. and Xu, J. 1990. *The Natural History of China*. McGraw-Hill, New York.

Zhiqing, G., Weijun, C. and Chuanliang, J. 1995. On the optimization of the water quality monitoring network in China in view of scale effects. In: Simonovic, S.P., Kundzewicz, Z., Rosbjerg, D. and Takeuchi, K. (eds) *Modelling and Management of Sustainable Basin-scale Water Resource Systems*. IAHS Publication 231, Wallingford, 353–358.

Zhong, Y. and Power, G. 1996. Environmental impacts of hydroelectric projects on fish resources in China. *Regulated Rivers: Research & Management* 12:81–98.

Zuming, L. 1995. Variation of water resources in the inland regions of northwest China: a review. In: Petts, G. (ed.) *Man's Influence on Freshwater Ecosystems and Water Use*. IAHS Publication 230, Wallingford, 89–94.

11

River conservation in south-east Asia

D. Dudgeon, S. Choowaew and S.-C. Ho

The regional context

INTRODUCTION

Asia is the most populous region of the world, with 13% of the land area and around 50% of the people. It includes some countries with highly developed economies (e.g. Singapore), but more people live in poverty in Asia than in Africa and Latin America combined (Braatz *et al.*, 1992). Six of the longest rivers in the world are situated in Asia south of latitude 30°N: the Yangtze (Chang Jiang), Mekong, Indus, Brahmaputra, Ganges and Irrawaddy (Figure 11.1). In addition, Bangladesh has more than 50 'important' rivers; India, 400; Indonesia, 200; and Thailand, 10 (Van Der Leeden, 1975; Jalal, 1987; see also Gopal *et al.*, this volume). One estimate is that these rivers (together with the Palaearctic Huang Ho) transport over 80% of the sediment carried by all rivers worldwide (Jalal, 1987; but see Degens *et al.*, 1991). The ecology of tropical Asian rivers has been reviewed recently by Dudgeon (1992, 1995a,b, 1999) who emphasizes the importance of discharge seasonality on the biota. The predominant influence of monsoons gives rise to a pattern whereby predictable periods of drought and water scarcity in the dry season alternate with times of increased discharge, spates and floodplain inundation during the wet season.

Asia, and especially the south-east Asian countries of Malaysia, the Philippines, Vietnam and Indonesia, is extremely biodiverse, with tremendous species richness and high levels of endemism. Indonesia, for example, is among the world's top 10 countries for numbers of species of flowering plants, birds, reptiles and amphibians, with more species of plants and birds than the African continent (Braatz *et al.*, 1992). It is estimated to support an overall total of at least 15% of the species in the world and, in terms of aquatic biodiversity, has 900 species of amphibians and more dragonflies (666 species) than any other country (Caldecott, 1996). Because many of the Indonesian islands have been isolated for long periods, there are high rates of endemism among dragonflies and freshwater fish. Fish are highly diverse elsewhere in the region also, and the Indochinese Peninsula has 930 species in 87 families (Kottelat, 1989). Despite such richness, it is apparent that existing inventories of the fish fauna are far from complete (Zakaria-Ismail, 1994).

Much of Asia is characterized by large and rapidly growing urban complexes with associated problems of potable water supply, sanitation, water pollution, flooding and depletion of groundwater aquifers. As a result, 'the tropics, environmentally and economically, are in trouble' (Baker, 1993). It might be argued that 'the tropics' has no more homogeneity than 'the non-tropics', but tropical south-east Asia provides many examples that illustrate Baker's view. While substantial economic progress has been made by some countries in the region, this has almost invariably been associated with environmental degradation and ever-increasing demands on the resource base (Baker, 1993). A compilation of figures by Niacin (1992: Fig. 6.14) shows that total water consumption in south-east Asia has risen from 82 km^3 yr^{-1} in 1900 to 187 in 1950, 609 in 1990 and a projected 741 km^3 yr^{-1} in 2000. Irretrievable water losses (due to consumption in agriculture, especially on irrigated land) have increased from 65 (1990) to 142 (1950) to 399 (1990) to a projected 435 km^3 yr^{-1} in 2000. This loss denotes a

Global Perspectives on River Conservation: Science, Policy and Practice.
Edited by P.J. Boon, B.R. Davies and G.E. Petts. © 2000 John Wiley & Sons Ltd.

Figure 11.1   *The south-east Asia region showing the major rivers*

steady decline from 79 to 58% of total water consumption. These data illustrate increasing pressures for the development of water resources and conversion of wildlands to 'productive' use.

Human impacts throughout the region have been, and are, all-pervasive. South-east Asia has lost about 67% of the original wildlife habitat, including a great deal of the forest and more than half of the freshwater wetlands (Braatz *et al.*, 1992). There are now probably no large Asian rivers in pristine condition (Hynes,

1989; Dudgeon, 1992). Riverine wetlands are especially threatened in Malaysia and the Philippines, but more than 50% of all wetlands in Cambodia (Kampuchea), Indonesia, Laos and Burma (Myanmar) are at risk, while few sites in Vietnam have protected status (Scott and Poole, 1989). Forest clearance is particularly relevant to river ecology – especially when coupled with the misuse and over-exploitation of drainage basins – because of the concomitant downstream effects of increased run-off, sedimentation and flash floods (Dudgeon, 1992; Tejwani, 1993 and references therein). For example, rates of sedimentation in Philippine rivers are among the highest in the world: rivers in central Luzon receive from 11 t sediment ha^{-1} yr^{-1} (Pampanga River) up to 45 t ha^{-1} yr^{-1} (Agno River) compared with 0.5 and 1.5 t ha^{-1} yr^{-1} for the Mississippi and Mekong Rivers respectively (Villavicencio, 1987). Some action has been taken to reverse the effects of drainage-basin degradation, but innovative management programmes at the appropriate scale are needed urgently (Hufschmidt, 1993; Low, 1993; Burbridge, 1994; Choowaew, 1995). After devastating floods claimed several hundred lives in 1988, the Thai Government banned logging and revoked all logging concessions in January 1989 (Charoenphong, 1991). Unfortunately, this decision has had detrimental effects on Burma, Laos and Vietnam because it caused logging companies to shift their activities into these countries. Clearly, threats to biodiversity must be addressed at a regional or sub-regional scale, and conservation policies must reflect this reality. Plans for mainstream dams on the Mekong River (which will be considered in detail below) provide further evidence of the need for conservation initiatives to extend beyond local scales.

It is unclear whether poverty – with its pressures to survive – or affluence – with its pressures to consume – drive environmental degradation in Asia. It is likely that both causes are influential. Regardless of their relative importance, it is obvious that indigent people cannot conserve natural resources if this conflicts with their survival needs. The challenge is to reconcile the needs of the Asian populace with global interests in biodiversity conservation (see Wishart *et al.*, this volume, Part II). In this chapter we review river conservation in Asia in the context of planned and ongoing water-resource development projects and environmental degradation. In particular we use case studies to uncover some of the threats to aquatic biodiversity – most especially river fish which have a direct relevance to human welfare. In addition, we examine the effectiveness of existing legislation that deals with pollution control, environmental impact assessment, and protected-area systems in the region. We pay particular attention to the conflicts that can arise between the authorities who initiate or plan projects which change river ecology and the local communities that have to bear the consequences of those changes. One underlying theme in this account is the requirement to match the pressures for rapid economic growth with what the environment requires for persistence and what it can supply at a sustainable

rate. We have focused mainly on Indonesia, Malaysia, Thailand and the Philippines since these countries make up the bulk of the land area and population of south-east Asia, but some information has been drawn from other countries within the region.

DEVELOPMENT PRESSURES AND LOCAL IMPACTS

The Mekong River: a case study

In order to illustrate the extent and range of threats to the ecology of tropical Asian rivers, we will discuss planned and ongoing water-resource developments in the Mekong River Basin. This case demonstrates the conflict between the pressures for economic development, which are usually driven by governmental imperatives, and the ecological consequences of such development which are most often felt by local communities some distance from centres of political power and policy.

Plans for the Mekong Basin date back to the 1950s when Raymond Wheeler, a retired general of the United States Army Corps of Engineers, headed a mission to study the hydropower potential of the Mekong. The annual discharge of the Mekong is more than 475×10^6 m^3, yielding a potential energy capacity of 58 000 MW and the possibility of irrigating some 6 000 000 ha (Chomchai, 1987). In 1957, Wheeler's report to ECAFE, an Economic Commission for Asia and the Far East created by the United Nations (renamed ESCAP – the Economic and Social Commission for Asia and the Pacific – in 1974), identified seven sites on the river that were suitable for the multipurpose development of water resources. Also in 1957, the riparian states formed a coordinating committee to function under the auspices of the United Nations as represented by ECAFE. Thus the Committee for the Coordination of Investigations of the Lower Mekong Basin came into being.

The Mekong Committee (serviced by the Mekong Secretariat comprising an array of water-resource specialists) assigned first priority to dams at Pa Mong and Sambor (Thailand), and to a barrage which would control the movement of waters into and out of Tonlé Sap River and Le Grand Lac in Cambodia. Work on these was envisaged to proceed in parallel with projects on smaller tributaries of the Mekong in Cambodia, Laos, Vietnam and Thailand. By 1967, however, only two small tributary projects (both in Thailand) had been completed. War and political instability in the region made the construction of large infrastructural projects and work of an international nature impossible, and for nearly three decades the Mekong Committee sat more or less idle, unable to fulfil its mandate. More recently, the political situation stabilized, and by 1986 there were 14 dams on tributaries in the lower basin (Pantulu, 1986a). Following completion of a major dam on a tributary of the Mekong at Nam Ngum in Laos, further feasibility studies were launched (Chomchai, 1987). Despite the potential for detrimental impacts on river biota, the reservations of former staff of the Mekong Committee regarding the acceptability of impacts on fish and fisheries (Usher, 1996), and some criticism by the World Bank, in December 1994 the Mekong Secretariat published the *Mekong Mainstream Run-of-River Hydropower* study document which identified 12 potential dam sites on the mainstream, 11 of them with generating capacities of more than 1000 MW. Five months later, the Mekong Committee was replaced by the Mekong Commission formed by the four lower riparian states. At the same time, the Swedish International Development Authority strengthened its previous support for the authority (Usher, 1996). The Commission was set up as an autonomous, inter-governmental organization to develop policies on the Mekong Basin, one of the first being that member states were allowed to use the river waters without seeking the approval of other members – except during the dry season (*South China Morning Post* – SCMP, 1995). The Mekong Commission has stressed its commitment to dam construction at 12 sites along the mainstream of the river in Laos, Thailand and Cambodia (Roberts, 1995). The environmental impacts of these dams on, for example, fish spawning and migrations, have yet to be assessed fully, but are hardly likely to be positive (Pantulu, 1986b; Lohmann, 1990; Roberts, 1992, 1993a,b, 1995; Baird, 1994). This has important socio-economic implications given that the annual fisheries yield from the lower Mekong Basin is approximately 0.5×10^6 t (Lohmann, 1990), and estimates for the Vietnamese portion alone are 100 000–160 000 t (Tran Thanh Xuan, Ministry of Fishery, Vietnam, personal communication). Adequate data for sound management decisions are still lacking (Pantulu, 1986b; Roberts, 1993a), insufficient or unreliable (Hori, 1993), although the Mekong Committee has been attempting to rectify this deficiency (Petersen and Sköglund, 1990; Choowaew, 1993).

It should be stressed here that although plans for mainstream dams in the lower Mekong Basin have been stalled for many years, this is not the case in the upper basin. For example, China is not a member of the Mekong Committee, but nevertheless has its own schemes for the Mekong: the 1500 MW Manwan Dam in Yunnan Province (completed in 1995) is only one of at least 14 dams that China intends to build on the river (SCMP, 1993, 1995). China and Burma have recently agreed to participate in a sub-committee of the Mekong Commission which is designed to facilitate information sharing and to study upstream hydropower development (*The Nation* (*Bankok*), 1996). Elsewhere in the Mekong Basin, more than 60 dam projects were in various stages of consideration by the Laotian Ministry of Industry and Handicrafts during 1995 (Usher, 1996). Although only two projects (Nam Ngum and Xeset) had reached the construction stage by 1995, several others were in the planning stage. If even a fraction of these projects materialize, there are expected to be serious implications for the Laotian environment and for the people whose livelihoods are dependent on riverine ecosystems (Usher, 1996). There is particular concern over the potentially damaging effects of these projects because the electricity generated will be sold abroad, mostly to Thailand, and thus the financial benefits are likely to accrue to those far from the sites where the impacts of dam construction are felt. Second, foreign private companies – especially from Nordic countries – are investing in Laotian hydropower development, and it seems that there is an inverse relationship between the costs of avoiding or mitigating impacts in Laos and the profits to be gained abroad by these firms (Usher, 1996). There has been concern also over the role of private consultancy firms in the hydropower development in Laos, since they are hired not only to assess social and economic interests, but also to design the projects. This dual role 'practically ensures that the negative effects of projects are glossed systematically over' (Usher, 1996) thereby leading to 'appraisal optimism'. Despite this apprehension and the recent economic depression in south-east Asia, the development of hydropower and irrigation seems inevitable given the potential benefits: Laos is one of the world's poorest countries (annual per capita income <US$300), and depends largely upon rain-fed agriculture; average life expectancy is only 49 years (Jacobs, 1994).

Among other dams to be completed on Mekong tributaries in Laos is the Nam Theun Hinboun, a 210 MW hydropower project on the Theun River. Construction – funded by companies in Sweden and Norway – began in 1994 and was completed in late 1997. In another example of 'appraisal optimism', the interests of the rural villagers, who depend on fish stocks and river transport routes, were given scant consideration by foreign consultants who concluded that there would be no negative effects on fish and significant beneficial environmental impacts (Usher, 1996). This is despite a projection that the river bed will be reduced to a series of pools below the dam during the three-month dry season. It is evident that the dam will block the migration routes and destroy the downstream habitat of some of the 140 species of fish that inhabit the Theun Basin and must inevitably decrease this source of food and income for local people.

There is a persistent view that 'water resources development may create positive effects especially on fisheries through the integration of freshwater aquaculture in reservoir projects' (ESCAP, 1992; see also Djuangsih, 1993), and that 'the nutritional status of rural people is further improved by the availability of animal protein through . . . the development of inland fisheries in the newly created reservoirs' (Biswas and El-Habr, 1993). The project consultant's view in the Nam Theun Hinboun case was that increasing the water level behind the dam would provide an ideal environment for rearing fish, and would improve the situation for local people (Usher, 1996). This statement is, however, based upon a lack of understanding of fish ecology (see Roberts, 1993b) and reflects the conflict of interest that (in this case) foreign consultants face if they can benefit from one outcome over another (Usher, 1996). An increased fisheries yield at Nam Theun Hinboun might have to depend on exotic species, and would certainly be accompanied by a loss of indigenous biodiversity.

Hydropower development in Vietnam includes the Yali Falls Dam which is part of a six-dam scheme proposed by the Mekong Committee for the Se San River. Although the environmental impact assessment of the project states that the general pressure on wildlife habitat will be increased by the project, the prevailing government view seems to be that hydropower will decrease the use of wood and charcoal for fuel thereby reducing deforestation and resulting in a net ecological benefit (Lang, 1996). This ecological equation seems rather simplistic. After completion of the Hao Binh Dam in northern Vietnam – which is the largest dam in south-east Asia – displaced people moved to the upstream part of the drainage basin where they cleared forest to plant crops, thereby increasing run-off and sedimentation (Lang, 1996). This situation could occur again around the Yali Falls Dam, unless there is an increase in the

government will and wherewithal to preserve forested catchments upstream (and so, incidentally, prolonging the useful life of the dam). Even if such pollution does occur, downstream impacts of the dam on the river biota are likely to be such that the projected net ecological gain of the scheme disappears.

Important environmental consequences of dam construction along the Mekong will include changes in flow regime, especially inundation patterns, which have, for example, affected aquatic productivity of waters at Nam Ngum (De Bont and Kleijn, 1984; Beeckman and De Bont, 1985). Roberts (1993b, 1995) reports that the recently completed Pak Mun Dam on the Mekong in Thailand has been extremely damaging to fish populations, which declined from 1991 (when dam construction began) until 1994 (when it was finished). Affected fishers received a one-off compensation payment of US$3600 per family. Dam building on the Mekong mainstream will lead to changes in flow and temperature regime and may remove important directive factors which stimulate fish migratory and breeding behaviour. Such dams will obstruct the passage of several long- and medium-range whitefish migrants. These include Pangasidae catfish (especially *Pangasius* spp which contribute 10–20% of total landings from the river; Pantulu, 1970), cyprinids such as *Cirrhinus auratus*, *P. jullieni* and other fisheries species. Of these, the giant 3 m Mekong catfish, *Pangasias* (=*Pangasianodon*) *gigas* is believed to undergo spectacular long-range migrations for spawning, travelling as far upstream as Yunnan Province (Smith, 1945; Pantulu, 1970; but see Bhukaswan, 1983). Attention was drawn to declining populations of this poorly known species in the mid 20th century (Smith, 1945), when it was the basis of an important fishery in Cambodia wherein the fish was caught for its flesh and the extraction of oil due to its high fat content. It has declined greatly in recent years (Nandeesha, 1994). *Pangasius siamensis* may be extinct already. *Pangasius krempfi* is an anadromous species, with a life history resembling that of salmon; it appears to comprise a single population in the Mekong River (Roberts and Baird, 1995). These large fish, reaching up to 15 kg and over 1 m in length, are of particular economic importance to fishers along the river (Baird, 1994).

Roberts (1995) is of the opinion that the series of mainstream dams proposed by the Mekong Committee in 1994 will cause 'extirpation or extinction of many fish species, including strongly migratory species that are the main basis for Mekong wild-capture fisheries'. Some of these species are confined to the Mekong: for example, *Pangasius gigas* and *P. krempfi*, the carp *Aaptosyax grypus* and *Probarbus labeamajor*, and the freshwater herring *Tenualosa thibaudeaui* which is already endangered (see also Roberts, 1992; Roberts and Baird, 1995). They could be driven to extinction by a single mainstream dam (Roberts, 1995). Fish ladders – which have been proposed as a means of mitigating the impacts of dams on migratory fishes – are unlikely to be successful in the Mekong, because the majority of species in the river do not jump (Roberts, 1993b, 1995). Moreover, ladders are designed for species (such as the north-temperate salmon) which swim upstream only; little consideration has been given to the fact that potamodromous Asian fish make downstream return migrations. As Hill (1995) points out, research and development of passage facilities for tropical fish species has rarely been addressed and there is virtually no relevant information upon which to draw. Meisner and Shuter (1992) raise the additional issue of the effects of global climate change on the annual flow regimes and breeding migrations of floodplain fish in the Mekong. The extent to which possible changes might magnify or ameliorate the effects of mainstream dams is unknown.

The impacts will extend beyond fish to other taxa. The Irrawaddy dolphin, *Orcaella brevirostris*, enters rivers throughout south-east Asia from India to Vietnam and south through the Indonesian archipelago (Sigurdsson and Yang, 1990), and is in need of protection in some rivers (Wirawan, 1986; Baird, 1994) but has already disappeared from others (e.g. the Chao Phraya in Thailand: Baird and Mounsouphom, 1994). Dolphin numbers in the Mekong have fallen because of declines in fish prey and death of animals trapped in gill nets, but the large-scale dams proposed for the river basin will pose greater threats to *O. brevirostris* in Laos and Cambodia (Baird and Mounsouphom, 1994). Considering possible impacts on invertebrate taxa, we do not know what effect the proposed dams along the Mekong will have on the endemic species-flocks of gastropods (at least 119 species: Davis, 1979) in the river. Attwood (1995) considers that modified flow regimes caused by dams may lead to changes in food availability and substrate characteristics which will favour some snails – in particular, *Neotricula aperta* which is a host of the human parasite *Schistosoma mekongi* (Trematoda: Schistosomatidae). Such changes, together with the translocation of riparian peoples and the influx of casual labour from endemic areas in Laos, may result in an epidemic of human schistosomiasis in north-east Thailand (Attwood,

1995). Srivardhana (1987) reports that there has been a great increase in flukes and intestinal parasites among residents of the irrigated area associated with the Nam Pong Dam in north-east Thailand.

In addition to the impacts that may arise from implementation of the Mekong Secretariat's proposals for the lower basin, and various hydropower plants in Laos and Vietnam, there may be cumulative impacts on river ecology from other projects. The Thai Government has been formulating a master plan for dam building and drainage-basin development covering the north and north-east parts of the country where rainfall is low and unreliable (Office of the National Economic and Social Development Board, 1994). It is likely that the latter project will go ahead because of the urgent need to improve soil conditions, agriculture and forest management, and to enhance the economic development of this, the poorest region of Thailand (Krishna, 1983; Hori, 1993; Jacobs, 1994). Significantly, the World Bank has been critical of the scale of the developments proposed by the Mekong Secretariat for the lower basin, and considers that a more appropriate approach to development in the riparian states should, in the first instance, be centred on improvements in agricultural productivity through small-scale irrigation projects, rural development and related hydropower projects (Kirmani, 1990). The Thai Government plans seem more in line with World Bank thinking in this regard and, because of their smaller scale, they are likely to be less damaging – and more 'sustainable' – than the large mainstream dams that have been proposed for the Mekong. Indeed, the Mekong Committee and Secretariat have recently broadened their programme of work to take greater account of environmental and social concerns (Jacobs, 1994).

Significantly, Usher (1996) reports that 'some 30 Thai NGOs and local water basin groups issued a statement . . . opposing the influence of the dam-building industry in the creation of the new Commission'. Similarly, Kirmani (1990) considers that the lower Mekong Basin project is a 'classic example of external effort, external management and external planning with little involvement of the beneficiaries' and 'the assistance of donor countries was utilized mostly to finance their own experts and consulting firms' while the 'Mekong Secretariat was managed by foreign experts'. While such comment and criticism are well founded (see also Lohmann, 1990; Roberts, 1993b) and, in combination with the economic downturn of 1997–98, will probably delay the construction of mainstream dams on the lower Mekong for the foreseeable future, Jacobs (1994) points out that without dam construction for hydropower (and the concomitant impacts on river ecology) to fuel economic development, Thailand will burn more lignite while Cambodia and Laos will continue to clear their forests for fuelwood. 'To reject outright the prospect of hydropower project construction denies these countries the opportunity for development and keeps them upon a path of environmental degradation and poverty' (Jacobs, 1994). McNeely (1987) suggests that drainage-basin protection could have been used to justify reserves and preservation of riparian habitats along the Mekong, while the economic impetus arising from water-resource development might stimulate government control and enforcement of environmental protection. It is not yet clear whether such hopes for the integration of conservation with development along the Mekong will bear fruit, but there is little room for optimism over the fate of migratory river fish.

Tropical deforestation and river fisheries

Reference has been made above to the rate and extent of deforestation in south-east Asia. Most of this loss has been recent: forest cover in Thailand decreased from 53% of the total area in 1961 to 26% in 1993 (Royal Forest Department, 1994). Likewise, 90% of lowland forest in the Philippines has disappeared during the last three decades, and only 5% of the land area remains under natural forest. Rates of loss there and in Vietnam are more than 1.5% yr^{-1}, while forest clearance in Indonesia proceeds at a rate of $10\,000$ km^2 yr^{-1} (Braatz *et al.*, 1992). The general pattern over the last three decades has 'been for forests to be degraded and destroyed, for much native biodiversity to be eroded and lost, and for any distinctive local cultures to be erased' (Caldecott, 1996). This has important implications for river conservation since a significant proportion of the rich fish fauna of the region exploits inundated forest and allochthonous foods (see Dudgeon, 1995a and references therein). Moreover, the economic pressures for logging which cause deforestation often have national or international origins whereas the effects are felt by local communities, especially fishers. There is thus considerable potential for conflict among different interest groups.

As an example, we consider an issue that came to the fore in Sarawak during the late 1980s and involved the side-effects of logging on the lifestyles of indigenous people who were mainly shifting cultivators. Shifting cultivation requires access not only to land, but to a wide range of materials for use or sale which are derived from the surrounding forests and rivers: fish were of particular importance and accounted for up to one-third of the food intake of forest-inhabiting Kenyah and Iban communities (Caldecott, 1996; Parnwell and Taylor, 1996). Prawns and turtles were harvested also. Fish are especially significant as a source of protein because hunting land animals carries a greater risk of failure than fishing (Parnwell and Taylor, 1996). In the Baram River drainage basin, mechanized logging at high intensities was detrimental to fish populations in the river through muddying, increased sedimentation and diesel pollution. This disturbance led to a decline in average ration harvested from 54 to 18 kg per person per year during the first decade of logging, and a local perception that logging was primarily to blame for a significant increase in hardship in rural areas (Caldecott, 1996). The Sarawak Forest Department was sympathetic to the aim of preserving biodiversity and drainage-basin integrity of the Baram, but the department's position was compromised because one of its roles was to supervise a forestry system which functioned by awarding and exploiting industrial logging concessions in order to support the Sarawak economy (Caldecott, 1996). Conflicts thus arose between the demands of this system on natural resources, and the needs of wildlife and local communities in the affected areas.

One possible solution to conflicts such as these is the maintenance of protected areas or drainage basins in which logging is not permitted. Sustainable harvests of fish from protected rivers are feasible as long as the ecological requirements of the target population continue to be met and exploitation is monitored so that overharvesting is prevented. However, a particular danger arises when a management strategy devised for one set of harvesting techniques is upset by the introduction of a new, more effective means of capture. The impact of fishing by hooks, harpoons, traps and plant toxins is very different from rates of exploitation that can be achieved when they are replaced by more efficient nets, explosives, electricity and artificial toxins. For example, more damaging fishing practices were adopted by the Iban in Sarawak because logging led to a scarcity of fish and prawns, and spurred an increased use of pesticides (replacing natural phytochemicals) and electro-fishing (using power generators) to increase catches (Parnwell and Taylor, 1996). It also initiated a change in community practices. Traditional fish poisoning by the Iban, for example, was a labour-intensive social activity involving extraction of poison from plants, damming rivers, and harvesting fish (Caldecott, 1996; Parnwell and Taylor, 1996). Such poisons (e.g. rotenone from *Derris*) are usually biodegradable and non-persistent. The use of agricultural pesticides and artificial toxins can have disastrous effects, since the concentrations of commercial insecticide needed to disable fish will be extremely damaging to aquatic communities. By contrast, rotenone is efficacious against fish, degrades rapidly in bright sunlight, and has relatively minor effects on benthic invertebrates (Dudgeon, 1990). Commercial markets for fish encourage more frequent use of poisons, and the intensity of fishing can also be affected by transport and trading patterns. During the 1970s, for example, unharvested fish in remote locations in interior Borneo were exploited by urban markets because of the increased availability of speedboats and cold-storage facilities (Caldecott, 1996). Obviously, continued monitoring and flexibility must be built into systems for managing river fisheries if they are to be responsive to changes such as these.

In addition to logging for timber, plantation agriculture for the growing pulp and paper industry in Thailand and Indonesia poses a significant threat to the river biota. Natural forests have been transformed to monocultures of acacia and eucalyptus and huge pulp-mills have been constructed along major waterways. Plantations in Indonesia 'are a direct cause of legal and illegal deforestation, devastating biodiversity, agricultural watersheds, soils and fisheries' (Lohmann, 1996); pollution from pulp-mills has resulted in 'poisoning of drinking and bathing water and in large-scale fish kills which in some areas have devastated one of the rural villagers' most important protein sources and marketable products'. Lohmann (1996) also reports that rivers in Kalimantan became sediment filled and 'useless for transport to market or fishing' as a result of logging and plantation establishment. Interestingly, Tejwani (1993) states that there is 'no direct evidence of an adverse impact of heavy load of sediment in water on inland fish and fish breeding grounds . . . however, it is reasonable to conclude that siltation of . . . rivers and streams . . . adversely influences the breeding and multiplication of fish'.

BIOTA AT RISK

The biota of many rivers in south-east Asia are at risk of impoverishment or – in extreme cases – extinction.

Reference has been made above to the possible effects of dams on Mekong River fish, and the impacts of deforestation or logging upon fisheries. The overall pattern throughout the region is one of decline and species loss, reflecting widespread urbanization, pollution, river regulation and land-use changes (Dudgeon, 1992; Zakaria-Ismail, 1994; Kottelat and Whitten, 1996). This loss of fish biodiversity has parallels in streams and rivers in all countries (Maitland, 1995), and 20% of the world's freshwater fish fauna is already extinct or in danger of extinction in the foreseeable future (Moyle and Leidy, 1992). A particular problem that is obvious from three recent reviews of threatened fish of the world (Moyle and Leidy, 1992; Bruton, 1995; Maitland, 1995) is the critical shortage of information on the conservation status of Asian freshwater fish. A report on the biodiversity of the region (Kottelat and Whitten, 1996) underscores this point. Indeed, it is obvious that fish biodiversity has been declining throughout the region for more than 30 years. Human impacts on biodiversity are often rather conspicuous, and threats to fish biodiversity – resulting from overfishing and pollution by tin-mine effluents – were highlighted in Malaysia over three decades ago (Alfred, 1968; Johnson, 1968; Prowse, 1968), but have persisted and been exacerbated in recent years (Zakaria-Ismail, 1987, 1994; Ho, 1994). Massive fish kills have been reported in rivers throughout the regions, and recolonization of defaunated areas is often prevented by dams which block fish movements (Hunt, 1992).

Many Asian fish species are poorly known and their conservation status cannot be determined accurately. Among these are several species of freshwater dasyatid stingrays (Monkolprasit and Roberts, 1990; Compagno and Cook, 1995); one of them, *Dasyatis* (= *Himantura*) *chaophraya*, has declined due to pollution of the Chao Phraya River and may soon be included on the protected species list under the 1992 Wildlife Conservation Act. Like other species of obligate freshwater elasmobranchs, these stingrays are vulnerable to extinction because they occupy a restricted range of habitats which 'strictly limit their opportunities to evade pollutants, habitat modifications, or directed and incidental capture in local fisheries' (Compagno and Cook, 1995). Their vulnerability may be increased by low fecundity: *D. chaophraya* give birth to only a single offspring at a time. Pethiyagoda (1994) points out that few Asian countries have made systematic assessments of their native fish resources (or faunal diversity in general), and emphasizes that recently published faunal inventories are based more on dated literature or old museum collections than on recent surveys. In such circumstances, it is unlikely that extinctions will become known until long after they have occurred and 'the state of taxonomic knowledge of the fishes of the Asian region in general is so poor that many extinctions may never be known' (Pethiyagoda, 1994). Problems with fish taxonomy and correct identifications have been highlighted recently by Kottelat and Whitten (1996), and can give rise to situations where protected status is sometimes granted to species which do not exist in the country or are misidentified.

Reptiles and amphibians, as well as fish, are at risk from human modification of south-east Asian rivers. Hunting and habitat changes have caused a general decline in crocodilian populations throughout tropical Asia, and a variety of river turtles and terrapins are also in need of protection (Belsare, 1994; Thirakhupt and Van Dijk, 1994) – particularly in Thailand which hosts an exceptionally diverse turtle community (Thirakhupt and Van Dijk, 1994). Dussart (1974) urged that more research be undertaken on Asian amphibians to inform conservation measures, and the situation must now be considered as grave given the increase in the extent of habitat destruction, human exploitation and pollution over the last two decades. Human modification of south-east Asian rivers and floodplains has also placed riparian vegetation and gallery forest at risk (Cox, 1987; Le, 1994). Their loss would affect terrestrial mammals and birds which depend upon the riverine habitat during the dry season (Cox, 1987), and would cause the demise of plant and animal species which are associated exclusively with the riparian zone or floodplains. This matter lies beyond the scope of the present chapter, but some relevant issues have been reviewed by Dudgeon (1992), Belsare (1994) and Kottelat and Whitten (1996).

Information for conservation

SCIENTIFIC DATA

South-east Asian rivers and their associated wetlands have long been subject to modification by humans and we do not know what these habitats were like in their original state (Dudgeon, 1995a). Indeed, 'virtually all reviews [of biodiversity in Asia] indicate a dearth of knowledge about inland wetlands . . . this information gap needs to be corrected as a matter of urgency, since these areas are facing increasing threat' (Braatz *et al.*,

1992). Among studies that have attempted to identify priority sites for biodiversity conservation in south-east Asia, only one has concentrated on inland waters. Scott and Poole (1989) give a status overview of 947 wetland sites in Asia (including coastal sites and many within the Palaearctic realm), providing a description of each one, their legal protection, threats and effectiveness of conservation measures. The authors say little about rivers, although many of the wetlands included are on seasonally inundated floodplains. A global review of river basin management (Niacin, 1992) underscores the general paucity of information: Asian examples are confined to the Indian sub-continent (mainly the Ganges) and China (albeit in little detail), and next to nothing is said about south-east Asia. According to Lim and Valencia (1990), for 'most countries in the region, the present pool of knowledge with regard to the geographical . . . and limnological characteristics affecting water masses and aquatic resources, and fish resources themselves, continues to be extremely limited.' Sustainable development of water resources is practical only where there is adequate knowledge of the resources (including biodiversity) and the constraints within which they must be managed (Barbier, 1993). Without such data it is, for example, impossible to calculate potential yields and sustainable levels of fishing activity, or to design policies which will ensure sustainable returns to small-scale fishers. In effect, development decisions are often made on the basis of biological information which is poor, inadequate or almost non-existent.

To summarize, river conservation in south-east Asia faces a number of obstacles arising from the inadequacy of relevant data (modified from Barrow, 1983, 1987):

- An established body of knowledge is lacking. Tropical limnology still draws heavily upon temperate latitude research with research largely in the hands of expatriates and few indigenous experts.
- Transfer of ideas from temperate to tropical water-resource development has not been particularly successful. In addition, exchange of data among countries in the same region is not common and, where it does occur, there can be a lack of standardization so comparison is difficult.
- Biogeochemical processes in tropical latitudes are complex and rapid, and are not moderated by cool seasonal conditions.
- Water resources needs are so pressing that development may be hurried and take place

without consideration of the information that is available. As a result, they do not take account of long-term environmental costs.

• Some countries are unable (because of funding constraints) or unwilling to invest in adequate environmental surveys and continued monitoring, and face political and administrative difficulties which hinder such research.

• No large south-east Asian rivers have been studied in enough detail over sufficient time to yield insights that would enable the synthesis of general planning guidelines. Moreover, each river basin is unique, and to generalize the results derived from one or a few cases is unwise.

• Where data are available they may not be understood by planners or decision-makers, perhaps because the research is too sectoral. The application of existing information may be impractical because it is irrelevant to the particular problem requiring a solution, or too site-specific so that it cannot be transferred to the project at hand.

For the reasons set out above, environmental problems are often unforeseen and manifest only after the completion of a project when consultants or funds are no longer available. Although awareness of increasing rates of resource exploitation and associated environmental degradation has grown, it has not been matched by appropriate regional efforts to quantify and map the progress and process of change (McGregor *et al.*, 1996). Effective drainage-basin management requires the incorporation of data at different scales from a range of sources, including information on topography, climate, soils, vegetation cover, population distribution and density, ecological characteristics, and so on. This information can be captured and stored in a Geographic Information System (GIS), and such systems have emerged as important tools for handling spatial data in a variety of hydrological studies and attempts at water-resource management (e.g. Kovar and Nachtnebel, 1993). Although GIS technology identifies rather than solves environmental problems, it can provide information which may serve as a basis for appropriate ecological practices (McGregor *et al.*, 1996). The Mekong Secretariat has been developing a 'Mekong GIS', to include data from remote-sensing technologies (Choowaew, 1993; Michelson, 1993). Although the situation is changing quickly, most GIS applications to Asian river basins (e.g. Meijerink *et al.*, 1993) have addressed patterns of erosion and sedimentation, and linked them to agriculture within

a catchment. Issues of biological conservation have yet to be confronted.

CLASSIFICATION AND EVALUATION SCHEMES

Widespread and serious pollution is a major contributing factor to the degradation of many south-east Asian rivers, and is a matter which must be addressed in the conservation context. Most streams and rivers adjacent to tropical cities are grossly polluted because centralized collection and disposal of urban-generated sewage is inadequate and continues to lag seriously behind urban growth (Hufschmidt, 1993). Extensive networks of monitoring stations covering many river basins have been established in Peninsular Malaysia, Thailand, the Philippines and Indonesia (Chia, 1987, and references therein; ESCAP, 1987, 1992; Ho, 1996). The general impression is that pollution is bad and that, in many areas, river water quality is poor and may be worsening (Low, 1993; see also 'Implementation of Legislation').

As part of an initial attempt to tackle pollution, the quality of rivers in Thailand has been monitored since 1976. The National Environment Board (established in 1975) uses a river classification scheme which establishes goals and objectives for the use of river water, and hence determines the level of water quality to be maintained. The aim has been to develop an acceptable trade-off between river water quality and the need for waste treatment 'so that they will not add an unnecessary burden on public and private enterprises' (Setamanit, 1987). The river water quality standards which have been developed involve a five-scale classification from water which is suitable for drinking without conventional treatment (first grade) down to water which is suitable for navigation only (fifth grade). Only the first two grades are of sufficient quality to ensure the 'conservation and protection of aquatic animals' (Setamanit, 1987).

In order to deal with river pollution the Malaysian Government carried out regular monitoring of 94 rivers (49 in Peninsular Malaysia, 21 in Sarawak and 24 in Sabah). This revealed that, over a two-year period in the early 1980s, water quality (based on NH_3, Biochemical Oxygen Demand (BOD), pH and suspended solids) remained constant at 48% of sampling sites and declined at 14%, with an improvement in 39% (Ong *et al.*, 1987). In 1985, the Division of Environment (DOE) commissioned a follow-up study to develop water quality criteria and standards for Malaysia (Ho, 1996). This organization was established in 1976 to give form and substance to the various regulations included under the umbrella of the 1974 Environmental Quality Act (plus subsequent additions and amendments; see Ho, 1996) which was designed to address pollution-related activities. In Phase I (1985–86), the DOE established a set of interim national water quality standards (INWQS), while Phase II (1988–89) dealt with the development of a river classification system based on the INWQS for six river basins. A third phase (1990) reviewed the status of discharges and compliance with existing Effluent Discharge Standards and, under Phase IV (1992–93) another 10 basins were classified. One outcome has been the development of a computerized integrated river basin information system (IRBS), which includes a GIS interface and will enable the DOE to develop strategies and set priorities for river water quality management. However, as Ho (1996) points out, the monitoring and law enforcement arms of DOE must be strengthened if the well-established legislation to regulate pollution under the Malaysian Environmental Quality Act is to be employed effectively (see 'Implementation of Legislation').

The framework for conservation

THE SOCIO-POLITICAL BACKGROUND

Threats to biodiversity can often be traced back to one or more underlying problems. They may be linked to failure of planning, markets, policies or institutions, to distorted distributions of wealth and power or poverty and weakness, to excessive numbers of people, or open-access to renewable resources (Caldecott, 1996). It is possible to find examples from south-east Asia that epitomize all of these problems. For instance, reports of ESCAP (Water Resources Series: ESCAP, 1987, 1992) provide summaries of the ongoing situation in countries of the region, describe trends in water use and supply, and set out concerns over the extent of pollution, salinization, catchment degradation and sedimentation, as well as the development of monitoring systems and legislation to protect water resources. Unfortunately, while 'the need for properly managed water-resource development was recognized . . . policies for environmentally sound development and sustainable use of freshwater resources had largely failed to materialise in many countries' (ESCAP, 1992). The concept of 'sustainability' is widely used – indeed, it is ubiquitous in such documents – yet it is not always particularly clear what this term means, and a precise

definition remains elusive (Jacobs, 1994). As Biswas (1994) points out 'sustainable development' means different things to different people and operationalization of this concept is not yet possible. (Similar remarks might also be applied to the term 'integrated river-basin management': Downs *et al.*, 1991). Nevertheless, the concept of sustainable development is useful since it has raised awareness of ecological and social trade-offs associated with river-basin development (Jacobs, 1994).

In most instances, the concept of sustainability seems to relate to the potential for continued (long-term) use of water resources by humans, and hence goes in tandem with the need to avoid pollution and salinization. In other words, sustainability is defined broadly in terms of direct consumption or use by humans, and ecological considerations do not figure (e.g. Bonell *et al.*, 1993). Even if this definition is accepted, in some developing countries the reality for many subsistence fishers and farmers may be the need to ensure short-term benefits and even survival, and thus societal or governmental goals of sustainability will remain out of reach. Leaving aside the difficulty of determining whether a particular approach to water-resource management is actually sustainable, it is striking that the ESCAP (1987, 1992) reports give so little attention to the maintenance of river fisheries or biodiversity conservation. This is despite their own report that the construction of the Hoa Binh hydropower plant and its associated 100 m-high dam on the Da River in Vietnam blocked fish migration routes and reduced downstream fish catches by 50% (ESCAP, 1992). Only one recommendation within these reports deals with ecological issues, and relates to impoundments: 'reservoir operation should be developed to guarantee a controlled flooding of the floodplains downstream of the dam in order to maintain the subsistence functions of the river basin, including agriculture and fishing' (ESCAP, 1992). Precisely how it will be possible to determine whether or not the subsistence functions have been maintained is far from obvious, but the report recognizes that technical assistance regarding downstream release of water to maintain ecosystem functioning (and reduce pollution loads) is required (ESCAP, 1992), and training is needed to make water-resource managers and planners more aware of environmental issues (e.g. in Vietnam: ESCAP, 1992).

Low (1993) gives two examples of the direct conflict between the 'consumptive use' of rivers for development and maintenance of their ecological integrity or 'sustainability'. The city of Kuala

Lumpur, in the Sungai Kelang Basin (1200 km^2), extracts so much potable water from two dams located upstream of the city that downstream flows are greatly reduced. The river fails to flush pollutants originating in urban areas with the effect that, during the dry season, the river becomes an open sewer. The situation is exacerbated by road-building and urbanization of foothills in the valley, and sediment load into the rivers is approximately 500 t km^{-2} yr^{-1}. Transfer of water from the nearby Sungai Semenyih (which, presumably, has impacts on flows downstream of the offtake) is needed to meet demands for potable water, and it is estimated that this supply will have to be augmented by transfer from other rivers around the year 2000. In this scenario, the demands of a growing urban population override concerns about river ecology. The impacts of industrialization on Malaban City, north of Manila, on the Tinajeros River are also instructive. Factory development proceeded apace in the 1970s, and the river began to receive large amounts of untreated effluents. This devastated fish catches, and destroyed an important river fishery which had been operating for over 200 years (Low, 1993). Conflict and legal actions followed, involving fishers, industrialists and the government, and culminated in the prosecution of several factory owners. The situation could have been avoided with appropriate and effective planning and pollution-control legislation.

History cautions that water-resource development involves a complex interrelationship between social and environmental processes that can, unless carefully managed, result in ruined soils and lost water supplies (Barrow, 1983). Reduced biodiversity should be added to this list. In densely populated south-east Asia, water-resource development affects significant numbers of people even in relatively sparsely populated areas. Their interests must be taken into account, and forced resettlement (often in marginal areas) may lead them into environmentally damaging practices (Barrow, 1983). National policies supporting agricultural development also represent direct or indirect threats to biodiversity because they lead to degradation of wildlands and the rivers draining them (Braatz *et al.*, 1992). For example, the Indonesian Government actively promotes clearing and agricultural settlement on more remote and less densely settled islands. Between 1980 and 1986, this transmigration programme moved over two million people from Bali and Java to other islands, reducing forest cover (logged primary forest and secondary regrowth) by at least 6000 km^2. Other environmental consequences of transmigration settlements are

discussed by Sage (1996). While some Asian governments provide subsidies (including grants, low-interest loans and tax deductions) for the conversion of forest lands to agriculture, in parts of the Philippines, Indonesia and Thailand insecure land tenure has acted as a disincentive to land improvements and encouraged expansion of agriculture onto upland drainage basins and marginal lands (Braatz *et al.*, 1992). Programmes in the Philippines and Thailand are now providing land titles on the assumption that it might slow the conversion of land, but this can exacerbate the situation if titles are provided to smallholders only for forest lands which have been 'improved'. In Sabah, for example, any indigenous inhabitant can obtain title to forest land by clearing and cultivating it. Remaining forest areas in Thailand may be threatened also by a government plan to give land certificates to villagers living in conservation areas (SCMP, 1997) – a scheme which, NGOs believe, will lead to the loss of all primary forest in the country within a decade.

The export-led growth strategy of south-east Asian governments has focused attention away from the need to maintain the integrity of the natural environment. Instead, it has resulted in misguided infrastructural development, swamp draining, and the establishment of polluting industry along river courses or next to traditional fishing grounds, thereby destroying spawning areas and denying fishing communities easy access to rivers (Lim, 1990). Water pollution arising as a consequence of economic expansion is also being caused by the excessive use of fertilizers, insecticides and herbicides. In consequence, conflicts over natural resources occurs at the local level among competing groups of users (e.g. tribal communities, peasants, fishermen, miners, loggers and corporations) and threaten to worsen in future. These conflicts are the immediate result of the dramatic increase in population and the corresponding increase in use of natural resources, the polarization of rival claimants, and the failure of many governments and authorities in the region to mediate effectively (Lim, 1990).

The 205 m Bakun Dam on the Balui River in the upper Rajang Basin, Sarawak, provides an outstanding example of the conflicts that can arise among different interest groups over water-resource development, and the important influence of government policy in this regard. The Bakun Dam is intended to provide cheap power to fuel development and to enhance opportunities for investment. Although the project is backed strongly by the Malaysian Government there is opposition to it from Malaysian NGOs and residents along the Balui River. An environmental impact assessment prepared for the developers notes that (among other things) fish habitat will be degraded and fishery resources will be lost both downstream and within the inundation area. However, the consequences of these changes for biodiversity and subsistence fishery resources, and the effects of reduced frequency and duration of inundation of the floodplain, are not addressed in detail. In June 1996, the Kuala Lumpur High Court ruled that residents of the area to be submerged had been deprived of their right under federal environmental law to be consulted before official approval was given for the dam. In mid-July, however, project proponents persuaded the High Court to suspend the ruling pending an appeal, and work at the site by a Swiss–Swedish engineering multinational company began in March 1997. Press statements by Malaysian Prime Minister Dr Mahathir Mohamad, who visited Sarawak in August 1996, and his ministers have made it abundantly clear that the Federal Government is in favour of the dam (*Guardian Weekly*, 1996; SCMP, 1996a), asserting also that the government would ensure that construction would have no adverse effect on the environment. Interestingly, this statement follows a 1990 decision to cancel dam construction on environmental grounds that was reversed when the government revived the project in 1993. Flow of information on the status of the project was effectively curtailed in 1996 when access to the area around the dam site was prevented without permission from the government or state police (SCMP, 1996a). Financial constraints led to the indefinite postponement of work on the Bakun Dam in April 1998 and a scaled-down version of the scheme is now under construction.

Conflicts between local and national interest groups in south-east Asia are heightened further where international players become involved. As Usher (1996) shows, northern interests and agencies, especially from the Nordic countries, have provided both the means and incentive to harness the hydropower potential of Laotian rivers (see 'The Mekong River: a case study'). Though the 'pervasive appraisal optimism' which exudes from the proponents and evaluators of dam schemes (as well as aid donors and financiers), the negative environmental and social consequences are down-played, leaving an impression of a 'win–win' solution that will generate hard currency and a 'green' environmentally benign energy (Usher, 1996). Where NGOs are underdeveloped or lack the ability to act effectively, there are very limited opportunities for the opinions of local communities to be heard. In addition to Nordic influences, Japanese interests in south-east Asia have significant effects on river ecology (Cameron, 1996). Slag outflows from copper mines operated (at least in part) by Japanese companies have caused serious downstream damage in rivers in a number of countries. The many Japanese-funded golf-course developments in Thailand deplete scarce water reserves during the dry season, and pollute rivers with pesticides; water theft is reported to be increasing (Cameron, 1996). In cases where development of hydropower potential, forest exploitation, and so on are driven by foreign businesses and international agencies, the national government must make decisions 'on the run', without the opportunity to develop a coherent policy (Bryant and Parnwell, 1996). This situation does not contribute to sustainable development or conservation.

STATUTORY FRAMEWORK

Examples of the existing statutory framework for pollution control have been described above under the context of river monitoring and classification schemes; the effectiveness of this legislation will be considered below (see 'Implementation of Legislation'). It remains to consider the statutory framework for protected areas and potential impacts of water-resource developments.

Protected areas

Protected areas are of considerable importance for river conservation, since they maintain the integrity of the drainage basin and thus serve to maintain the ecosystem as a whole. Many Asian countries have made substantial efforts to establish or enhance well-managed and representative protected-areas systems, although the protection afforded to rivers is generally an incidental consequence of the selection of an area of terrestrial habitat. Indonesia and Thailand have (respectively) 10 and 11% of their land gazetted as protected areas. By contrast, Cambodia and Laos have no legally protected areas, while Burma has gazetted <1% of the total land area (Braatz *et al.*, 1992). Even where areas are designated as 'protected' however, there may be inadequate management, trained staff, legislation and/or infrastructure. In many cases the flora and fauna have yet to be surveyed adequately and a viable conservation plan is lacking. While they may be protected by law, such 'paper parks' continue to be degraded (Braatz *et al.*, 1992). An additional problem is that while most countries (except Cambodia and Laos) have basic legislation concerning conservation and environmental protection, it is weak and focused on species rather than habitats. Very few of these species are aquatic. In essence, it may be illegal to kill or to possess a particular species, but acceptable to pollute or modify its habitat.

Environmental impact assessment

The World Bank, Asian Development Bank, United Nations Development Programme and various bilateral aid agencies now insist that, before a project can be approved for funding, it must undergo some kind of environmental impact assessment (EIA). This, according to Biswas (1992) is the single most important factor enhancing the cause of the environment in many developing countries. However, because the EIAs are carried out primarily to satisfy the internal requirements of external organizations, they do not necessarily ensure that the projects become environmentally sound. Indeed, the EIA is often considered to be an end in itself, and there is no follow-up monitoring by the project authorities, government or international funding agencies (Biswas, 1992). As of 1996, four countries in south-east Asia – the Philippines, Indonesia, Thailand and Malaysia – had specific laws and acts requiring EIAs (ESCAP, 1992; Briffett, 1996). Some countries lack a formal requirement for an EIA (e.g. Laos, Cambodia, Brunei, Vietnam), although they may have administrative measures through which EIAs for specific types of projects may be required. Where there are no formal requirements at all (as in Burma), there may be informal procedures to incorporate environmental considerations into planning some water resource projects. All of these systems are being improved and upgraded gradually (Chia, 1987; ESCAP, 1987, 1992; Briffett, 1996). A recent survey of EIA legislation in south-east Asia is given by Briffett (1996), and only issues relevant to river conservation will be discussed herein.

The introduction of EIA systems can be seen as a tool to integrate environmental issues into development planning: for example, the Act of the Republic of Indonesia No. 11 of 1974 concerning water resource development; the Malaysian Environmental Quality Act of 1974 which was followed by the creation of the Division of Environment in 1975 (including a Water Pollution Control Unit which deals with river monitoring: Ong *et al.*, 1987); and the creation of the National Environmental Protection Council and promulgation of the Philippine Environmental Policy which took effect in 1977. The general pattern is for countries to have environmental legislation which empowers a government agency to require an EIA for particular projects. In Indonesia, for example, EIAs are required for large dams and related infrastructure, as well as for the conversion of natural forest to transmigration settlements and agriculture (Partoatmodjo, 1987). Indonesian Government Regulation 51 of 1993 clarified procedures for analysing and managing the environmental and social impacts of development projects (collectively referred to as the AMDAL process). All projects which might impact upon a protected area must undergo AMDAL before a decision is made to proceed with them (Caldecott, 1996). This pertains to Nature Conservation Areas and Sanctuary Reserves defined under Act 5 of 1990 (which sets out the nature and management objectives of the various kinds of nature reserve). In addition, it applies to Protection Forests and a range of sensitive terrestrial and aquatic ecosystems including water catchments, riverine edges, areas surrounding lakes and springs, aquatic ecosystems with high biodiversity, and other areas with special attributes (Caldecott, 1996).

The type of water resource development project which might require an EIA varies among countries and, in most cases, is a matter for ongoing discussion (see, for example, Chia, 1987). This means that many decisions are made on an *ad hoc* basis. In Malaysia, EIAs must generally be undertaken on projects with potentially high pollution impact (Ong *et al.*, 1987). In Thailand, only large-scale water resource development projects (e.g. where storage capacity $>100 \times 10^6$ m^3 or where >15 km^2 is inundated) require an EIA (Setamanit, 1987); small- and medium-scale ones ($<100 \times 10^6$ m^3) do not, although they may be requested by the National Environmental Board (ESCAP, 1987) which was established in 1977 following promulgation of Thailand's first Environmental Act in 1975. In the Philippines, all developmental activities undertaken by government which relate to fresh water (and the associated drainage basin) must be carried out in accordance with the current five-year Philippines Development Plan and the provisions of the Water Code (ESCAP, 1987). The goals of these plans include improvement in the quality and levels of water supplies, pollution control and hydropower development to minimize reliance on oil imports. Conservation and protection of water resources is also seen as important and (presumably) includes concern for biodiversity. The problem, as elsewhere in Asia, is that there is a need to establish criteria for sustainable resource management and protection of biodiversity (Ong *et al.*, 1987; see also Burbridge, 1994; Choowaew, 1995).

In Thailand and Malaysia, the EIA is undertaken by a committee of experts and representatives from interested parties; they report to government agencies (i.e. the relevant department(s) responsible for the environment: e.g. the EIA Unit of the Division of Environment in Malaysia) who decide on the acceptance of the EIA. In some countries (e.g. Thailand) an Initial Environmental Examination (IEE), involving a small budget and existing information, is undertaken to assess the potential environmental effects of a proposed project. In effect, the IEE is a screening and scoping procedure used to determine whether a full EIA is needed. The IEE does not usually deal with ecological impacts, because only certain topics are considered at this stage of the project. However, in Thailand, at least, the full EIA usually includes some consideration of 'fisheries and aquatic biology' (ESCAP, 1992). Indeed, as early as 1952, the Thai Government consulted United Nations' experts regarding the need for a fish ladder on the Chao Phraya Barrage.

One view is that the EIA methodology must be 'flexible and can be carried out within the constraints of limited funding, time and expertise available in developing countries' (ESCAP, 1992). The question is whether or not these perceived constraints will lead to an EIA system that allows a project to proceed in the absence of clear evidence of detrimental impacts or if, when data and expertise are limited, a more cautious approach is adopted: i.e. the uncertainty over impacts means that development is not allowed to proceed. Biswas and El-Habr (1993) question EIA protocols from a different perspective, and cite evidence to support their view that the indirect beneficial impacts of water resource developments on health, transportation and education are often underestimated. Significantly, however, they do not consider the repeated failure of EIAs to address the potential long-term, cumulative and detrimental impacts of dams and irrigation projects on biodiversity. This omission is the rule rather than the exception. For example, in a 'multi-objective analysis' of the allocation of water to storage and release in the Bili-Bili Reservoir, south Sulawesi (Santosa and Goulter, 1991), there is no consideration of ecological matters or the impacts of release on the integrity of the downstream habitat. Full utilization of the irrigable areas and regional economic development, which are objectives of the Indonesian Government, take priority over ecological concerns in this and many other

instances, and so constrain the analysis. Likewise, a feasibility study by Nielsen and Strom (1984) involving simulations of water availability that would result from various combinations of project developments in the Nam Kam Basin, a Mekong tributary in north-east Thailand, does not take account of the volume of flow needed to maintain river ecology. Neither does a discussion of water allocation and resources planning in the Chao Phraya River basin (Vadhanaphui *et al.*, 1992).

ROLE OF NON-GOVERNMENTAL ORGANIZATIONS

There are many NGOs with interests pertaining to conservation and 'the environment' in south-east Asia (Eccleston and Potter, 1996). Vietnam has three small NGOs based in Hanoi: CNRES (Centre for Natural Resources Management and Environmental Studies, formed in 1985); ECO-ECO (Institute of Ecological Economy, formed in 1990); and CERED (Centre for Environmental Research, Education and Development, formed in 1991). In addition, the World Wide Fund for Nature (WWF) has a branch in Vietnam. Malaysia has more groups and they have been established for considerably longer. The three largest and most active are WWF (Malaysia), the Malaysian Nature Society (MNS) and Sahabat Alam Malaysia (SAM). Active Thai NGOs include the Thailand Environment Institute, the Natural History branch of the Siam Society (= the Thailand Research Society), the Wildlife Fund of Thailand, and the Thai Association for the Conservation of Wildlife, as well as TERRA (Towards Ecological Recovery and Regional Alliance) which specializes in ecological matters in Burma and Indochina. There are more than 300 NGOs in Indonesia. Most are linked together in WAHLI (Indonesian Forum for the Environment) but a number of others are included under the umbrella of SKEPHI (NGO Network for Forest Conservation). Still others are not affiliated with either umbrella group (Eccleston and Potter, 1996). In the Philippines, the Community-Based Forest Management Programme (CBFMP) of the Department of Natural Environment and Resources (DENR) – the state agency responsible for management and protection of upland resources – is a major driving force for sustainable development. Although associated with the government, it has a community focus which is similar to that typical of NGOs. The United States-based International Rivers Network also has interests in river conservation in the region.

Local community groups and NGOs with membership drawn from local communities have a critical role to play in conservation. Projects directed at implementing conservation goals should employ local people, either directly or through NGOs, and should especially train them to work within their own communities (Caldecott, 1996; see also Choowaew, 1995). Successful conservation projects with local support are those which improve local inhabitants' security of tenure, because there is a close link between security of tenure and confidence in the future, and hence willingness to invest in conservation (Hufschmidt, 1993; Caldecott, 1996). Thus, land ownership by farmers, exclusive fishing rights, and so on are essential institutional arrangements needed for effective implementation of drainage-basin management programmes. In essence, conservation is most likely to happen when the people most likely to benefit from it are the ones that decide whether or not to do it (Caldecott, 1996) and it is for this reason that the involvement of NGOs and community groups in river conservation is essential.

There is a critical need to develop project strategies which integrate conservation and development and so serve the ends of preserving biodiversity with minimal detriment to local social and economic development (see also McNeely, 1987). Such projects must increase the incomes of people living around protected areas or sensitive habitats, so that they have a stake in ensuring that the protection is maintained; the pressure for further exploitation of the habitat is thereby reduced (Braatz *et al.*, 1992). The Dumoga-Bone National Park in north Sulawesi (Indonesia) is such a project: the 3000 km^2 reserve was established in 1982 to protect the upland drainage basin of rivers supplying two irrigation projects used by 8000 farmers (many translocated from Bali) to grow rice. Funding was provided by the World Bank and, of a total of US$60 million, about $1 million was used to establish the reserve. This seems to be one of the few successful models linking ecology, conservation and development because the maintenance of the protected drainage basin was clearly tied to the success of an irrigation system that was fundamental to human welfare (McNeely, 1987; Braatz *et al.*, 1992).

The ability of NGOs and community organizations to change resource practices that lead to environmental degradation is contingent on the state's willingness to allow such protest in the first place. A detailed review of the relationship and effectiveness of environmental NGOs in the different political contexts of Malaysia, Indonesia and Vietnam is given by Eccleston and

Potter (1996). They believe that NGOs may have more opportunity for effective action in Malaysia, but the difference is one of degree rather than kind because all three nations have limits to the openness of party competition and the power of parliament. A more extreme contrast in NGO effectiveness is provided by the examples of Burma and the Philippines (Bryant and Parnwell, 1996). The ruling Burmese junta has opposed all forms of popular discontent over the last decade, while the demise of the Marcos regime as a result of popular action in the Philippines during the mid-1980s has given rise to a regime where popular protest (people's power) is the norm. Abracosa and Ortolano (1988) give an example of an instance where the National Power Corporation was forced to cancel the Binongan hydropower project in the Abra River basin in northern Philippines because of local resistance to proposed and suspected resettlements. In the Philippines also, environmental damage such as deforestation has led to calamitous flash floods (almost 2000 dead in Omroc City in 1991) which have strengthened calls for community action and the need to involve the public in water-resource planning (Abracosa and Ortolano, 1988).

NGOs in south-east Asia have, in general, received greater tolerance from governments following the 1992 Rio Earth Summit, which has led a closer dialogue and (in some cases) the narrowing of formerly diametrically opposed positions (Bryant and Parnwell, 1996). Nevertheless, the prevalence of large-scale logging and dam-building in the face of local and international protest indicates that, despite their importance in some countries, NGOs and community organizations are not the main players determining the outcome of events. Instead, such outcomes reflect the relative influences of a diverse and competing array of groups representing interests both within and beyond national boundaries (see 'The Socio-Political Background').

Conservation in practice

IMPLEMENTATION OF LEGISLATION

The weakest link in water resource management in most developing countries is implementation, and the literature is replete with examples demonstrating the failed implementation of project or programme plans (Lim and Valencia, 1990; Hufschmit, 1993). Likewise, enforcement of existing conservation legislation in south-east Asia is generally poor, with – for example – illegal hunting, logging and agricultural encroachment within protected areas (Braatz *et al.*, 1992). One difficulty facing nature reserves and protected areas is that boundaries may be unmarked or unclear to local people so that the 'protected' areas receive little or no effective protection and derive no benefit from their special status. The main problem, however, is a shortage of funds for operation, enforcement, education, research and monitoring. There is little correlation between a country's ability and its willingness to pay for conservation. For example, Indonesia with a per capita GNP of US$430 spends 0.06% of its total annual national budget on conservation – the same percentage as Malaysia which has a GNP of $1870 (Braatz *et al.*, 1992). In Vietnam – estimated GNP US$ <500 – the percentage expenditure is similar (0.05%). Among south-east Asian countries, Thailand (GNP US$1000) devotes the greatest proportion of its annual budget (0.19%) to conservation. When expenditure on environmental protection is added the percentage is somewhat higher, but only a tiny proportion of this additional expenditure is aimed at reducing human impacts on riverine ecosystems.

Abdul Rahim (1985) and Braatz *et al.* (1992) have highlighted the critical problem of a shortage of relevant technical expertise and trained staff for conservation agencies in Asia, including field staff, mid- and top-level conservation planners and administrators. This problem affects countries with relatively well-developed protected-area systems that need improved management (Indonesia, the Philippines), and is a serious bottleneck for those wishing to establish such systems (Laos, Vietnam). Programmes must be developed to give training in the more traditional conservation subjects, and to enable staff to work effectively with local communities.

Limited political support is a further obstacle to conservation because, typically, the relevant government agencies lack political influence and operational capabilities. They may not have the necessary authority to carry out their responsibilities effectively, especially in cases where over-centralization hampers communication and relationships with local authorities and community leaders (Braatz *et al.*, 1992). Even where decentralization of the responsibilities from the seat of government has occurred, there can be major difficulties in coordination, especially where several departments play important roles in a conservation project (see 'Conflicts and coordination'). Regardless of management plans and organizational structures, the relevant agencies must have the legal authority and administrative capacity to perform the tasks within their spheres of responsibility. This effectiveness is particularly important during the implementation phases of management. For example, government organizations must provide fair and consistent administration of private rights and obligations to water, land, fisheries and forest resources. Management of open-access resources, such as fisheries, requires rigorous and equitable enforcement of restrictions on use, while rights to irrigation water must be administered and policed evenhandedly (Hufschmidt, 1993). Regulatory failure can lead to inequalities and conflicts among resource users, widespread evasion, over-exploitation and eventual deterioration of the resource.

The limited success of protected-areas systems in conserving the environment in south-east Asia, is paralleled by the EIA process – even in those countries (Thailand and the Philippines) which have been implementing EIA legislation of some kind for 20 years or so (Briffett, 1996). Clearly, it is not necessarily the legislation which counts but rather its effective implementation. Although a description of mitigating measures and post-project monitoring will typically form part of an EIA, the actual implementation of these activities is problematic (Briffett, 1996). While the problems and reasons for failure may be well understood, attempts to deal with the problems have met with limited success. Implementation of mitigation measures which are not the direct responsibility of the project proponent – such as may be needed at the scale of the drainage basin (Burbridge, 1994) – are not addressed in existing EIA legislation in the region (ESCAP, 1992). Moreover, the lack of an operational definition of 'sustainability' makes it impossible to apply this concept in the EIA process (Briffett, 1996). Improvements might come from incorporating implementation strategies and tools into development schemes from the start of planning, examining institutional capabilities for implementation and – where necessary – involving the public in implementation (Hufschmidt, 1993). In addition, it will be necessary to define ecologically acceptable flow regimes for river sections downstream of dams, and water abstractions or diversions (Petts, 1996), and to develop indices of 'biotic integrity' (e.g. Simon and Emery, 1995) that are relevant for the biota of south-east Asian rivers.

While it may be possible to predict the general impacts of dams and other water-development projects by means of an EIA, this does not mean that environmental impacts will be avoided. According to

Barrow (1987), there are three sources of disruption whereby successful achievement of the primary project goals will be overshadowed by hydrological, biological or socio-economic difficulties:

- Inadvertence: i.e. a lack of knowledge and expertise, or inadequate application of expertise.
- Convenience, whereby expert advice is ignored or not sought for political or economic reasons. This is an unpredictable and persistent category of problems since, in many cases, the development scenario is to achieve the primary goal and then deal subsequently with the unforeseen problems as they arise in the hope that they will not be too intractable or expensive.
- Adoption of a narrow outlook that considers only technical feasibility and economic gains. Difficulties of this type should decline in importance with improved, integrated planning.

In addition to deficiencies in the implementation of EIA laws in the region, there are significant weaknesses in the ability or willingness of south-east Asian governments to enforce pollution-control legislation. Although measures have now been taken in many Asian countries to contain pollution, including enactment of laws to control point-source pollution in Indonesia and Malaysia (Y.C. Ho, 1987; Jalal, 1987; Lim, 1987; Ho, 1994, 1996; Nontji, 1994), pollution from non-point sources has been neglected and is in urgent need of attention. Despite vigorous efforts, however, pollution levels in Asian rivers have not been reduced substantially. In some cases they have actually risen (Pantulu, 1986a; Jalal, 1987; Lim, 1987; Welch and Lim, 1987; Baconguis *et al.*, 1990; Djuangsih, 1993; Low, 1993; Bukit, 1995), and may continue to do so. For example, despite enforcement efforts, 40 of the 119 principal rivers of the Philippines are biologically dead because of pollution by industrial, domestic and mining wastes (Villavicencio, 1987). In a recent incident, toxic tailings from a copper mine caused mass mortality of fish and shrimp and poisoned livestock along 27 km of the Boac River on Marinduque Island (SCMP, 1996b; see also Low, 1993). Saeni *et al.* (1980) concluded that almost all of Indonesia's rivers and streams were polluted by domestic wastes, but industrial effluents have become increasingly important (Djuangsih, 1993; Nontji, 1994; Bukit, 1995; Palupi *et al.*, 1995). Conditions in the Citarum River in West Java have been described as 'supercritical' because of pollution by industrial wastes, organochlorines, pesticides and fertilizers (Djuangsih,

1993). In parts of north-east Thailand, as well as in the Chao Phraya River and the lower Mekong (predominantly in the delta), pollution by domestic wastes, agricultural run-off, and recent industrial developments is becoming acute (Srivardhana, 1984; Pantulu, 1986a: Hunt, 1992; Vadhanaphuti *et al.*, 1992; Roberts, 1993b; Muttamara and Sales, 1994). Likewise, many of Malaysia's major rivers have been contaminated with industrial and agricultural wastes (including oil palm and rubber processing residues), and over 40 of them have been classified as 'biologically dead' (Khoo *et al.*, 1987; Phang, 1987; Y.C. Ho, 1987; Lim, 1990; Ho, 1994). This contrasts with the situation 30–40 years ago when inorganic sediments from tin mines were the major pollutant in most of the country's rivers (Johnson, 1957, 1968; Alfred, 1968; Prowse, 1968). More recently, fertilizers, pesticides, herbicides and a range of industrial wastes have become important (Lim, 1990).

Common problems with pollution control include inadequate enforcement of laws and regulations, ineffectiveness of water quality standards, regulations and penalties, and a general lack of treatment of domestic and industrial effluents (e.g. Srivardhana, 1984; Y.C. Ho, 1987; Vadhanaphuti *et al.*, 1992; Low, 1993; Muttamara and Sales, 1994; Choowaew, 1995; Afsah *et al.*, 1996; Ho, 1996). Centralized collection and disposal of urban-generated sewage is inadequate over most of the region, and much of what is collected continues to be discharged into streams and rivers without treatment (Hufschmidt, 1993). Even in the few cases (e.g. Singapore) where point-source pollution is treated, non-point sources contribute to gross water pollution during periods of high rainfall and associated run-off. The capital investment required for sewage collection and treatment is far beyond the immediate financial capacity of many cities – especially fast-growing ones. With few exceptions (again, Singapore), wastewater treatment to the secondary level is not a realistic option, at least over the next decade or two (Hufschmidt, 1993; see also Hjorth and Nguyen, 1993). According to Low (1993), 'water pollution monitoring and enforcement are particularly weak as they depend on adequate manpower and funding, as well as the will to act . . . which many countries lack. The financial and economic gains from unfettered development are too attractive to be hampered by enforcement of anti-pollution legislation'. Likewise, Setamanit (1987) writes with regard to river pollution in Thailand that 'the difficulty appears to lie more with the enforcement of laws rather than with the lack of them', and 'there appears to be a lack of urgency in formulating clear policies and developing strategies to deal with the problem'. Similarly, while the Philippines has some of the most comprehensive environmental legislation in Asia, 'regulation and enforcement of anti-pollution laws remain weak . . . [and] purely regulatory solutions to environmental management have not produced the expected results' (Villavicencio, 1987).

Because monitoring of the regulated community and the enforcement of environmental standards are often extremely weak, the incentives to comply with standards and to control emissions is small. Innovative approaches are needed. The Indonesian Ministry for Population and Environment introduced a 'Clean River Programme' (PROKASIH) in 1989 and, upon its establishment in 1990, the Environmental Impact Management Agency (BAPEDAL) used PROKASIH to require the largest industrial polluters along certain degraded rivers to reduce pollution outputs (Nontji, 1994; Afsah *et al.*, 1996). Although participation in the programme was not voluntary, a particular characteristic of the agreement that a polluter signed with the authorities was that compliance with the terms of the agreement was voluntary, largely because resources available for monitoring were limited. Preliminary data indicate that PROKASIH resulted in a decline in total BOD discharge, although this effect was due to reductions by only 25% of polluters (Afsah *et al.*, 1996). BAPEDAL intends to develop a compliance management system but in the meantime has set in place PROPER PROKASIH which will announce publicly the environmental performance of individual polluters. Nevertheless, it is instructive to note that a system of voluntary compliance serves as a feasible and cost-effective strategy in the initial stages of the development of a comprehensive framework of public intervention to improve river water quality.

Conflicts and coordination

Effective environmental management is hindered and conflicts can arise where the responsibility for protected areas, conservation projects and drainage basins lies with – as is often the case – forestry ministries or departments of governments where production – and not protection – may be seen as a priority (Braatz *et al.*, 1992). Additional difficulties arise in (for example) Malaysia where the responsibility for conservation is divided among two or more national agencies (such as the Department of Wildlife and National Parks, the Forestry Department, and the Federal Fisheries Department), and where conflicts can arise between federal and state governments (Abdul Rahim, 1985; Braatz *et al.*, 1992). For instance, the state governments of Sabah and Sarawak are responsible for natural resources management (agriculture, forestry and water resources) which also falls within the purview of the national agencies. In essence, as far as drainage basin conservation and management are concerned, there is no single government agency in Malaysia which has overall authority or responsibility (Abdul Rahim, 1985). The National Water Code for Malaysia, which was promulgated in the early 1980s, proposes that – for purposes of planning and development – river basins be designated for either systematic development, management, or conservation. However, these categorizations might be set aside if there is a greater demand for water supply, flood mitigation or hydropower generation in a particular basin (Abdul Rahim, 1985). In Malaysia, also, NGO work in the local state context faces different constraints and compromises when compared with work at the federal level (Eccleston and Potter, 1996). Increased political awareness may allow NGOs to exploit differences between federal and state governments to the advantage of the environment. International influences (as sources of finance for hydropower or consumers of timber) will provide further difficulties for national NGOs, who face the problem that policy-makers often give priority to the short-term economic benefits of resource exploitation over environmental and social costs.

Difficulties in coordination among authorities responsible for drainage basin management are by no means confined to Malaysia, and an example is given by Srivardhana (1987) in connection with the Nam Pong Reservoir on a tributary of the Mekong in north-east Thailand. A productive fishery developed after the impoundment was completed. However, this fishery depended upon annual stocking by the Department of Fisheries, and control of the number of fishermen (or their opportunities for access) was considered to prevent declining yields. Deforestation of the upstream catchment and increased turbidity in the reservoir due to high suspended sediment loads may have affected the fishery also (Hufschmidt and McCauley, 1988), but management of the upstream catchment was not considered during the planning phase of the dam, and hence no effective action has been taken (Srivardhana, 1987). This neglect reflects the division of responsibility between the Electricity Generating Authority of Thailand (who were responsible for managing the dam and reservoir but not the areas upstream), and the Royal Forestry Department which had responsibility for the drainage basin but lacked a management plan (Srivardhana, 1987). Thus some of the environmental problems that have arisen in the Nam Pong Basin are a result of failures in 'human resource management' (Jacobs, 1994). Another instance of complications arising from the need to coordinate the efforts of various government departments has been described by Srivardhana (1987) in connection with the Nam Pong Reservoir in Thailand. By contrast, coordination among various government agencies with responsibility for management of the Chao Phraya River basin has worked well, although this has not solved the problems of pollution from large-scale irrigation and industry along the river (Srivardhana, 1984; Vadhanaphuti *et al.*, 1992; Muttamara and Sales, 1994).

Hufschmidt and McCauley (1988) and Hufschmidt (1993) describe the case of the Wonogiri Dam (and reservoir) on the Solo River in central Java where differences over priority for management of the upstream drainage basin arose among government departments. The Forestry Department proposed that priority be given to areas with highest erosion rates; the Public Works Ministry (responsible for reservoir management) favoured giving priority to erosion control in the areas closest to the reservoir, while the Agriculture Ministry prioritized the areas of catchment with the greatest prospects for increased productivity. The Home Affairs Ministry and local government emphasized improving the livelihood of inhabitants in the poorest areas of the basin. Hufschmidt and McCauley (1988) contrast this situation with the case of the successful management of the Angat River Basin in the Philippines where the National Power Corporation, which manages the dam and the reservoir, had responsibility also for the upstream catchment. Other reasons for success in this instance include the relatively low population density, control of immigration, and the development of a fishery in the reservoir based on exotic *Tilapia* (*sensu lato*) and Chinese major carp (Briones and Castro, 1986). Note that while this case may be treated as 'successful' with respect to management of the catchment, the effects of the dam on the biodiversity of native fish would not have been positive.

Integrated approaches

Effective management of water resources must achieve integrated planning, given the complex reciprocal links between physical, economic, ecological and social factors in drainage basins and the associated agricultural and urban systems (Hufschmidt, 1993). Cases of the successful application of integrated basin-wide management are few. Most water resource developments are undertaken on a sectoral basis – for irrigation, hydropower or domestic consumption – and the individual project forms the dominant element (and hence defines the management unit). Basin-wide planning, integration and implementation of projects is rare, arising – in some cases, at least – from problems of scale. The Mekong, for example, has many large tributary basins, each of which represents a sizeable management unit. Extra complexity is added because the Mekong crosses international boundaries. The Mekong Commission represents a special case of management of an international drainage basin but, to date, its activities have been limited mainly to collecting and analysing basic data and to framework planning. While the size of the Mekong constrains integrated management of the entire basin, much of south-east Asia comprises islands and peninsulas with rather small – often densely populated – drainage basins. Only six basins in Peninsular Malaysia, Indonesia (excluding Kalimantan) and the Philippines exceed 20 000 km^2, and most urban centres are situated within basins of 2000 km^2 or less (Low, 1993). It is desirable and should be feasible to plan water resource management at the basin scale for these relatively small rivers. Unfortunately, political and administrative boundaries rarely coincide with those of river basins. Implementation of programmes is usually the responsibility of provincial or district government, or regional offices of national sectoral agencies. Even when water resource management plans are prepared using river drainage boundaries, their implementation is not likely to proceed on this basis. Exactly how the basin can be used most effectively as a spatial unit for management remains an unresolved issue.

As an example of integrated river basin management, the Cisadane–Cimanuk Integrated Water Resources Development Study (which started in 1985, with completion projected for 2015) is instructive (ESCAP, 1992). The plan was to develop a series of reservoirs on six rivers, centred around the Citarum River in western Java, with the aim of flood control and providing water for agricultural and urban areas. The project objectives included an exhortation to 'prevent or minimize negative impacts on flora and fauna and on wildlife habitats' (ESCAP, 1992). Of 32 potential reservoir sites, 11 were selected for detailed consideration.

Downstream impacts from the selected sites were eventually given little weight in the final project assessment because of the condition of the rivers, especially the Citarum River which is severely polluted (see Djuangsih, 1993; Bukit, 1995). To quote: 'Because little natural flora and fauna remains in the lower and middle catchment areas and in the coastal plain, the impacts were thought to be of minor importance' (ESCAP, 1992). These predictions were 'based on expert judgement by the members of the multidisciplinary project team' (ESCAP, 1992), but the extent of available pre-impact data on lotic biodiversity is unknown and seems to have been limited. As this example demonstrates, there seems to be some way to go before conservation and ecological considerations will become a fully integrated component of river basin management plans.

Hufschmidt (1993) has identified three elements that should be incorporated in plans for integrated management of river basins and water resource developments. First, it is essential to include 'multiple objectives': i.e. ecological and social objectives as well as economic development objectives. The main issue is just how these objectives can be embodied in a given project. While much has been made of the application of multiple objective approaches to river management, a major problem is incommensurability: measurements of achievement of various objectives cannot be made in a common currency.

A second element is 'multiple purposes': an integrated water-resource plan requires the balanced consideration of a wide range of water uses and management purposes, including withdrawal uses (irrigation, domestic and industrial supply), in-stream uses (navigation, hydropower, fish and wildlife), as well as other problems such as pollution and flooding. Environmental purposes – especially the maintenance of wildlife habitat and biodiversity – are most often neglected under this approach. A balance may be difficult to achieve where water development agencies take a narrow sectoral approach to planning, and when different purposes are the responsibility of different management agencies (see 'Conflicts and coordination'). The problems will be exacerbated when the funding agency for a project is international while the impacts of planning failures are local.

The third element is 'multiple means'. In most situations, such means are narrowly defined – e.g. physical facilities to solve problems (dams, canals, pumping stations, hydropower plants). Management options which involve human adjustments to floods or periods of water scarcity are given little consideration, and approaches which give priority to biodiversity needs are unheard of in south-east Asia.

IN SITU AND EX SITU CONSERVATION

Given the severity of threats to the river biota in south-east Asia, including many species of fish which have direct socio-economic relevance to humans, it is disappointing that institutional conservation efforts are poorly developed. Nevertheless, the Cambodian Government passed a freshwater fisheries law in 1989, which included regulation of fishing activities, protection of areas of inundated floodplain forest, and the establishment of fish sanctuaries. Regulation of activities is accomplished by dividing the fishery into three sectors: large-scale fishing based on stationary gear (including vast fenced fishing lots in Tonlé Sap) is regulated through an auction system of site licences which set out permissible activities; medium-scale fishing using large-scale mobile gear is subject to a gear licence; and small-scale or family fishing is open access (Maclean, 1994). The regulations have not been formulated on the concept of sustainability because the necessary data are lacking, but ongoing monitoring by the Cambodian Department of Fishery (initiated by the Mekong Secretariat in 1994) is directed at collecting primary data (Maclean, 1994). Concern over declining fish catches in Laos led to the establishment of the Laotian Community Fisheries and Dolphin Protection Project supported by NGOs and the Laotian Government. The project has focused on the Mekong close to the Laotian border with Cambodia, and addresses resource management problems and conservation of threatened Irrawaddy dolphins (Baird, 1994). Much of the initial effort has been devoted to collecting primary data and, during the first year of the project, more than 150 species (of the 400 or more species in the Mekong) were identified from fishers' catches. Investigations of the importance of seasonally inundated forest were also initiated. In accordance with Laotian Government policy to decentralize government, 'village-based conservation strategies' have been established involving the local stakeholders and interest groups. They have begun to identify unsustainable harvest techniques (e.g. the use of explosives, chemicals and electricity for fishing, inappropriate and excessive deployment of gill nets and fish traps) which are now banned (Baird, 1994). Locally managed 'fish conservation zones' have been set up to protect deep pools in the Mekong which, during the dry season, are key habitats for many important fishery species. This process of empowering local people seems to be effective, in that villages in other districts are setting up their own 'fish conservation zones' and, moreover, villagers have identified problems in need of attention that government officials were unaware of. However, such community-based conservation efforts may be derailed by dam-building on the Mekong mainstream which would significantly affect inundation patterns and fishery stocks along the river (Baird, 1994).

Concern has been expressed in Thailand over the decline of some spectacular long-distance migratory fish due to overfishing, blockage of migration routes, and destruction of feeding or breeding grounds. They include *Catlocarpio siamensis* (which can exceed 120 kg), three *Probarbus* spp (including *P. jullieni*, and *P. labeamajor* which reaches 80 kg) and *Pangasius gigas* (300 kg: Nandeesha, 1994). Two of them (*P. jullieni* and *Pangasius* [= *Pangasianodon*] *gigas*) are listed in the appendices of the Convention on International Trade in Endangered Species of Wild Fauna and Flora (CITES) and the Migratory Species Convention (Kottelat and Whitten, 1996). The Thailand Fisheries Department has collected mature individuals of all three species from the wild in order to induce spawning. The hatchlings were reared and resulting juveniles released into the wild. Success with induced breeding of captive *Pangasius gigas* (Pholprasith, 1983, 1993; Chang, 1992) has averted, at least temporarily, the listing of the giant catfish as an endangered species in Thailand (Nandeesha, 1994), although the effort that been devoted to this species is as much an attempt to exploit its potential for aquaculture as it is a conservation imperative. More recently, the Wildlife Fund of Thailand (WFT) has been working to increase awareness of the need to protect *P. gigas* and, in 1996, began a high-profile programme to buy live specimens and return them to the wild. WFT figures show a decline in commercial catches of *P. gigas* from 69 fish in 1990, to 48 in 1993, and 18 in 1995. In Cambodia, too, attention has been paid to *Probarbus* spp and *P. gigas*, which disappeared from commercial catches in the Mekong during the early 1990s. As a result, the government outlawed fishing for these species. Collection of *Pangasius* spp eggs/spawn from the Mekong supports a major aquaculture industry (especially in Vietnam), but is believed by fishers to be damaging to natural populations (Nandeesha, 1994). Accordingly the Cambodian Government banned *Pangasius* eggs/spawn collection in 1994, and emphasized the importance of re-establishing natural stocks by a means of well-publicized release back to the Mekong of *Pangasius* hatchlings resulting from harvest of eggs from the wild.

The aquarium fish trade has been accused of driving species to extinction because of selective overfishing (especially of stenotopic blackwater fish). However, there is little direct evidence of this since declines in important trade species have always been associated with other causes (Kottelat and Whitten, 1996). An exception to this may be the endangered golden dragon fish (*Scleropages formosus*: Osteoglossidae), as there has been little effective regulation of its capture in Malaysia (Zakaria-Ismail, 1994). It is protected by law in Thailand, as is another aquarium fish, *Botia sidthimunki*, although in this case habitat destruction rather than exploitation has caused its decline in the wild (Kottelat and Whitten, 1996). *Scleropages formosus* is protected in Indonesia also (Kottelat and Whitten, 1996), but these animals are nevertheless fished intensively in Sumatra during the wet season when they breed in shallow floodwaters adjacent to rivers (Claridge, 1994). Recent success with the propagation of *S. formosus* in fish farms in Indonesia and Singapore should reduce pressures on wild stocks (Andrews, 1990; Ng and Tan, 1997) and, in theory, a proportion of the animals bred under artificial conditions in West Kalimantan have to be returned to the Kapuas River to restore a much exploited population (Kottelat and Whitten, 1996). While the aquarium trade has been largely responsible for the harvest of *S. formosus* in the wild, it is possible that the trade may help save species from extinction. For instance, the popular aquarium fish *Epalzeorhynchos* (= *Labeo*) *bicolor* may no longer exist in the wild in Thailand, but captive-bred specimens are in abundant supply within the trade (Kottelat and Whitten, 1996). Other examples are given by Ng and Tan (1997). Writing of Sri Lankan stream fish in deteriorating environments, Pethiyagoda (1994) states that 'given existing trends, no means of protection is likely to ensure their survival in the wild in the medium term' and concludes that 'the strategy most likely to succeed is the maintenance of . . . captive populations with a view to reintroduction once the pressures on wild populations have been controlled'. *Ex situ* protection of this type can effectively conserve only a small – usually critically endangered – proportion of total biodiversity. These efforts usually involve 'charismatic' species – probably the most ornamental or unusual small fish. The number of species which can be maintained in living collections is limited and, for large river fish, will be constrained by the size of the facilities and the high maintenance cost per species.

One approach to river conservation, which has yet to be adopted widely in south-east Asia, is to focus on education, and this strategy has achieved some initial success in Malaysia (Ho, 1996). A year-long, nation-wide 'Love Your River' community awareness campaign, with a budget of US$200 000, was launched in 1994 by the Drainage and Irrigation Department with the involvement of other government agencies and NGOs. The project was targeted particularly at schools and riverine towns and villages, and gave rise to a series of symposia and activities ('Adopt a River', and 'River Watch' which will monitor the pollution status of selected rivers). There is evidence (most notably in Malacca State) that local authorities are adjusting planning policies to include 'river beautification' schemes (Ho, 1996). The extent to which this campaign will reverse the decline in biodiversity of Malaysian rivers is limited. Ecological restoration measures will be required to restore channelized and degraded rivers, but these have yet to be attempted in Malaysia (Ho, 1996) or, indeed, anywhere else in the region.

Prospects and prognosis

Many people no longer accept that economic growth and environmental degradation must go hand in hand. Tolerance thresholds have been surpassed, or have been reduced by a growing awareness of the externalities of environmental exploitation. NGOs are protesting against environmental abuse, the media are championing environmental causes, yet the problem continues and, by most counts, appears to be growing (Parnwell and Bryant, 1996). Why? Perhaps people are aware but do not care enough to act, or are aware and do care but are unable to act. The latter group may include those who are forced by poverty to exploit resources beyond sustainable limits. Inaction by the former group may reflect the fuzziness or vagueness of the sustainable development concept, coupled with the difficulty of separating 'good' from 'bad' ecological practices. Thus the harmful environmental impacts of a particular action can be rationalized by emphasizing economic benefits which may reduce poverty and which might, through this 'trickle-down' effect, slow the pace of environmental degradation. For example, Chia and Chionh (1987) state that environmental protection has been facilitated by Singapore's 'rapid economic growth which in itself may have brought environmental ruin but in so far as it provided the wherewithal and the acquisition of technical know-how to deal with environmental problems, it has been an important favourable factor'. The governmental view is frequently along the lines that 'growth is essential to create the means with which to ameliorate environmental degradation'. As a result the prevailing paradigm is 'grow now, clean up later'. There is, as yet, no strong evidence of a shift in priorities towards conservation instead of growth, nor clear signs of a change in outlook by politicians and businessmen. Moreover, supposedly 'green' developments such as hydropower, which bring economic benefits (and a 'feel-good' factor) to urbanites, are ecologically damaging and undermine the sustainable livelihoods of those who must be resettled, leading to further degradation. The economic slump of 1997–98 may well create 'breathing space' for conservation while large projects are delayed, but the pressures for development will not disappear and are unlikely to be ameliorated in the long term. Effective river conservation may well be hugely expensive but the necessary investment is needed now, extending over the next decade and beyond, rather than at some point in the future when economic development has taken place but there is far less to save (Caldecott, 1996). So what are the prospects?

A summary of the current situation with regard to legislation, measures and practices related to river conservation in south-east Asia is given in Table 11.1. While recent economic changes in the region make any extrapolation difficult, improvements in nature conservation and management of biodiversity can be anticipated in Malaysia because, based on GNP, this country has the greatest *potential* availability of financial resources (Braatz *et al.*, 1992). Elsewhere, past and current performance of national governments suggests that improvements are likely to occur in Laos, the Philippines and Vietnam – albeit rather slowly – because of institutional weakness and political or social constraints and funding shortages. At the time of writing, the parlous economic situation and civil strife in Indonesia threatens to roll back past conservation gains, and prospects for the immediate future are not good. There seems little chance of improvement in Cambodia and Burma where there is low commitment and poor institutional structure. Most south-east Asian countries have only small areas of remaining natural forests and wildlands, and work to save these habitats must continue against the ongoing social and economic pressures which caused the habitat loss in the first place (Caldecott, 1996). Thus conservation projects have no option but to try to reconcile human needs with the constraints of ecology and biodiversity preservation. The situation is exacerbated by the fact that some

Table 11.1 *National status or implementation of measures relevant to river conservation in major south-east Asian countries. Further details and references are given in the text*

	Protected areas system	Protected species ordinance	Pollution control legislation	Statutory requirement for EIAs	Established river monitoring scheme	Classification scheme for river water quality	Trends in water pollution	Percentage of GNP spent on conservation	Presence of environmental NGOs
Burma	Limited	No (?)	No	No	No	No	?	?	No
Cambodia	No	Limited	No	No	No (?)	No	?	?	Limited
Indonesia	Yes	Yes	Yes	Yes	Yes	Limited	Worsening	0.06	Yes
Laos	Yes	Limited	No	No	No	No	?	?	Limited
Malaysia	Yes	Yes	Yes	Yes	Yes	Yes	Worsening	0.06	Yes
Philippines	Yes	Yes	Yes	Yes	Yes	Limited	Worsening	?	Yes
Thailand	Yes	Yes	Yes	Yes	Yes	Yes	Worsening	0.19	Yes
Vietnam	Yes	Yes	Yes	No	Limited	No	Worsening (?)	0.05	Yes

governments have committed themselves to a range of major development projects without taking biodiversity or any other aspect of the environment into account. What options remain?

Large dams tend to have adverse consequences for biodiversity, but may also provide a source of funds for conservation. Enlightened self-interest can result in the diversion of such money to drainage basin conservation and management since protection of natural vegetation upstream will reduce sedimentation and prolong the usefulness of a dam. Braatz *et al.* (1992) suggest that one way that this could come about is if land inundated or affected by dam construction were assigned a scarcity value, costed appropriately, and the revenue from power generation used to protect surrounding drainage basins or other areas of natural habitat. This option needs to be explored by conservationists since, at present, government agencies (such as the Electricity Generating Authority of Thailand) favour dam-building in areas of high conservation value (near-pristine forest) because the land is virtually free. Moreover, hydropower is a renewable energy source where costs are not subject to oil-price fluctuations or associated with the polluting effects of burning lignite. Some indirect environmental problems of dams in Thailand are highlighted by Roberts (1993b): dams are typically followed by the establishment of riverside industries, based on cheap hydroelectric power, which use rivers for the disposal of a variety of toxic wastes. Such pollution has had devastating impacts on fish and fisheries along the Nam Pong, Chee, Mekong, Tachin, Chao Phraya and Tapi rivers (Roberts, 1993b) but is not factored into the costs of hydropower projects. Similarly, rain-forest logging has a direct economic benefit (the timber yield) and a variety of indirect costs which include soil erosion and related hydrological changes which affect rivers and fish stocks by sedimentation, and damage infrastructure and farmland by flooding. Many of these can be costed, although their economic impact will be specific to each drainage basin. One estimate for the cost of drainage basin degradation in Java suggested an annual indirect use value for intact forests in the basin of about US$3000 km^{-2} $year^{-1}$ (Caldecott, 1996).

Other development decisions, such as diverting water away from river floodplains, are likewise frequently taken without considering the loss in wetland benefits arising from damages and conversion. The underlying assumption is often that the net benefits to society of any development option must be greater, although there is much evidence to suggest that this assumption is not always correct (Barbier, 1993; Burbridge, 1994; Choowaew, 1995). The values of riverine wetlands are not easy to assess directly but fall into three categories:

- Direct-use values: i.e. 'informal' economic activity that supports human livelihoods such as fishing and harvesting other material (see Choowaew *et al.*, 1994; Choowaew, 1995).
- Indirect-use values: i.e. ecological functions such as flood control and groundwater recharge, which also provide wide benefits through support of economic activity and property (Aylward and Barbier, 1992; Burbridge, 1994; Choowaew, 1995).
- Non-use values: inherent value as potentially unique environments or habitats (Barbier, 1993).

Clearly, there is a need for a basic methodology for assessing the value and economics of benefits of tropical wetlands, a process which – according to Barbier (1993) – is relatively straightforward although difficult to apply because of data and resource constraints. Some applications of a valuing methodology are given by Barbier (1993) – including one example of a tropical floodplain – who discusses appraisal of development options that involve habitat alteration and conversion (see also Aylward and Barbier, 1992).

The view taken by many south-east Asian politicians (e.g. Malaysia's Prime Minister Dr Mahathir Mohamad) is that global environmental problems are primarily due to over-consumption in the North, and that the South should not be prevented from mobilizing its resources by becoming a scapegoat for these ills. However, this view has little merit in relation to biodiversity conservation where the problems are local or national – reflecting species' distributions in river systems – rather than global. A realistic foundation for sustainable development in south-east Asia depends upon a gradual modification of directions and practices, where levels of environmental awareness steadily increase through education and the media; where legislation steadily improves and its implementation becomes more effective; where government, NGOs and business work in concert rather than in opposition; and where there is a genuine fusion of top-down and bottom-up initiatives (Parnwell and Bryant, 1996). While such changes may be taking place slowly, on a modest scale within the region, without increased momentum it is doubtful whether the degradation of Asian rivers can be halted. We require (but lack) a pragmatic means of matching the pressures for rapid economic growth with the requirements and potential of the natural environment. Obviously, some compromise must be reached. A major obstacle is that it is not in the interests of those who perpetrate environmental damage to change their attitudes and behaviour. The challenge is therefore to effect such changes.

References

Abdul Rahim, N. 1985. Watershed management in Malaysia: a perspective. *Wallaceana* 42:3–8.

Abracosa, R. and Ortolano, L. 1988. Public involvement and EIA. The case of the Binongan hydroelectric project in the Philippines. *Water Resources Development* 4:176–183.

Afsah, S., Laplante, B. and Makarim, N. 1996. Programme-based pollution control management: the Indonesian PROKASIH programme. *Asian Journal of Environmental Management* 4:75–93.

Alfred, E.R. 1968. Rare and endangered fresh-water fishes of Malaya and Singapore. In: Talbot L.M. and Talbot, M.H. (eds) *Conservation in Tropical South East Asia*. IUCN Publications New Series No. 10, IUCN, Morges, Switzerland, 325–331.

Andrews, C. 1990. The ornamental fish trade and fish conservation. *Journal of Fish Biology* 37 (Suppl. A):53–59.

Attwood, S.W. 1995. A demographic analysis of *y-Neotricula aperta* (Gastropoda: Pomatiopsidae) populations in Thailand and southern Laos, in relation to the transmission of schistosomiasis. *Journal of Molluscan Studies* 61:29–42.

Aylward, B.A. and Barbier, E.B. 1992. Valuing environmental functions in developing countries. *Biodiversity and Conservation* 1:34–50.

Baconguis, S.R., Cabahug, D.M. and Alonzo-Pasicolan, S.N. 1990. Identification and inventory of Philippine forested-wetland resources. *Forest Ecology and Management* 33/34:21–44.

Baird, I.G. 1994. Community management of Mekong River resources in Laos. *Naga, the ICLARM Quarterly* 17:10–12.

Baird, I.G. and Mounsouphom, B. 1994. Irrawaddy dolphins (*Orcaella brevirostris*) in southern Lao PDR and northeastern Cambodia. *Natural History Bulletin of the Siam Society* 42:159–175.

Baker, R. 1993. *Environmental Management in the Tropics: An Environmental Perspective*. Lewis Publishers, Boca Raton, Florida.

Barbier, E.B. 1993. Sustainable use of wetlands. Valuing tropical wetland benefits: economic methodologies and applications. *Geographical Journal* 159:22–32.

Barrow, C.J. 1983. The environmental consequences of water resource development in the tropics. In: Ooi, J.B. (ed.) *Natural Resources in Tropical Countries*. Singapore University Press, Singapore, 439–476.

Barrow, C.J. 1987. *Water Resources and Agricultural Development in the Tropics*. Longman, Harlow.

Beeckman, W. and De Bont, A.F. 1985. Characteristics of the Nam Ngum reservoir eco-system as deduced from the food of the most important fish-species. *Verhandlungen der Internationalen Vereinigung für theoretische und angewandte Limnologie* 22:2643–2649.

Belsare, D.K. 1994. Inventory and status of vanishing wetland wildlife of South-east Asia and an operational management plan for their conservation. In: Mitsch, W.J. (ed.) *Global Wetlands Old World and New*. Elsevier, Amsterdam, 841–856.

Bhukaswan, T. 1983. *Pla buk* (*Pangasianodon gigas*) in Chiang Khong. *Thai Fisheries Gazette* 36:339–346.

Biswas, A.K. 1992. Water for Third World development. A perspective from the South. *Water Resources Development* 8:3–9.

Biswas, A.K. 1994. Sustainable water resources development: some personal thoughts. *Water Resources Development* 10:109–116.

Biswas, A.K. and El-Habr, H.N. 1993. Environment and water resources management: the need for a holistic approach. *Water Resources Development* 9:117–125.

Bonell, M., Hufschmidt, M.M. and Gladwell, J.S. 1993. *Hydrology and Water Management in the Humid Tropics*. UNESCO/Cambridge University Press, New York.

Braatz, S., Davis, G., Shen, S. and Rees, C. 1992. Conserving biological diversity. A strategy for protected areas in the Asia–Pacific Region. *World Bank Technical Paper* 193:1–66.

Briffett, C. 1996. Monitoring the effectiveness of environmental impact assessment in South-east Asia. *Asian Journal of Environmental Management* 4:53–63.

Briones, N.D. and Castro, J.P. 1986. Effective management of a tropical watershed: the case of the Angat River watershed in the Philippines. *Water International* 11:157–161.

Bruton, M.N. 1995. Have fishes had their chips? The dilemma of threatened fishes. *Environmental Biology of Fishes* 43:1–27.

Bryant, R.F. and Parnwell, M.J.G. 1996. Politics, sustainable development and environmental change in South-east Asia. In: Parnwell, M.J.G. and Bryant, R.L. (eds) *Environmental Change in South-east Asia. People, Politics and Sustainable Development*. Routledge, London, 1–20.

Bukit, N.T. 1995. Water quality conservation for the Citarum River in West Java. *Water Science and Technology* 31:1–10.

Burbridge, P.R. 1994. Integrated planning and management of freshwater habitats, including wetlands. *Hydrobiologia* 285:311–322.

Caldecott, J. 1996. *Designing Conservation Projects*. Cambridge University Press, Cambridge.

Cameron, O. 1996. Japan and South-east Asia's environment. In: Parnwell, M.J.G. and Bryant, R.L. (eds) *Environmental Change in South-east Asia. People, Politics and Sustainable Development*. Routledge, London, 67–93.

Chang, W.Y.B. 1992. Giant catfish (*pla beuk*) culture in Thailand. *Aquaculturists' Magazine* 18:54–58.

Charoenphong, S. 1991. Environmental calamity in southern Thailand's headwaters. Causes and remedies. *Land Use Policy* 1991:185–189.

Chia, L.S. 1987. *Environmental Management in South-east Asia. Directions and Current Status.* Faculty of Science, National University of Singapore.

Chia, L.S. and Chionh, Y.H. 1987. Singapore. In: Chia, L.S. (ed.) *Environmental Management in South-east Asia. Directions and Current Status.* Faculty of Science, National University of Singapore, 108–168.

Chomchai, P. 1987. The Mekong project: an exercise in regional cooperation to develop the lower Mekong basin. In: Ali, M., Radosevich, G.E. and Ali Khan, A. (eds) *Water Resources Policy for Asia.* A.A. Balkema Publishers, Boston, 497–508.

Choowaew, S. 1993. Inventory and management of wetlands in the lower Mekong Basin. *Asian Journal of Environmental Management* 1:1–10.

Choowaew, S. 1995. Sustainable agricultural development in Thailand's wetlands. *TEI Quarterly Environmental Journal* 3:2–13.

Choowaew, S., Chandrachai, W. and Petersen, R.C. 1994. The socio-economic conditions in the vicinity of Huai Nam Un wetland, lower Mekong Basin. *Mitteilungen der Internationalen Vereinigung für theoretische und angewandte Limnologie* 24:41–46.

Claridge, G. 1994. Management of coastal ecosystems in eastern Sumatra: the case of Berbak Wildlife Reserve, Jambi Province. *Hydrobiologia* 285:287–302.

Compagno, L.J.V. and Cook, S.F. 1995. The exploitation and conservation of freshwater elasmobranchs: status of taxa and prospects for the future. *Journal of Aquariculture and Aquatic Sciences* 7:62–90.

Cox, B.S. 1987. Thailand's Nam Chaon dam: a disaster in the making. *The Ecologist* 17:212–219.

Davis, G.M. 1979. The origin and evolution of the gastropod family Pomatiopsidae with emphasis on the Mekong River Triculinae. *Monographs of the Academy of Natural Sciences, Philadelphia* 20:1–120.

De Bont, A.F. and Kleijn, L.J.K. 1984. Limnological evolution of Lake Nam Ngum (Lao P.D.R.). *Verhandlungen der Internationalen Vereinigung für theoretische und angewandte Limnologie* 22:1562–1566.

Degens, E.T., Kempe, S. and Richey, J.E. 1991. Summary. In: Degens, E.T., Kempe S. and Richey, J.E. (eds) *Biogeochemistry of Major World Rivers.* SCOPE/John Wiley, Chichester, 323–347.

Djuangsih, N. 1993. Understanding the state of river basin management from an environmental toxicology perspective: an example from water pollution at Citarum River Basin, West Java, Indonesia. *The Science of the Total Environment, Supplement* 1:283–292.

Downs, P.W., Gregory, K.J. and Brookes, A. 1991. How integrated is river basin management? *Environmental Management* 15:299–309.

Dudgeon, D. 1990. Benthic community structure and the effect of rotenone piscicide on invertebrate drift and standing stocks in two Papua New Guinea streams. *Archiv für Hydrobiologie* 119:35–53.

Dudgeon, D. 1992. Endangered ecosystems: a review of the conservation status of tropical Asian rivers. *Hydrobiologia* 248:167–191.

Dudgeon, D. 1995a. The ecology of rivers and streams in tropical Asia. In: Cushing, C.E., Cummins, K.W. and Minshall, G.E. (eds) *Ecosystems of the World 22: River and Stream Ecosystems.* Elsevier, Amsterdam, 615–657.

Dudgeon, D. 1995b. River regulation in southern China: ecological implications, conservation and environmental management. *Regulated Rivers: Research & Management* 11:35–54.

Dudgeon, D. 1999. *Tropical Asian Streams: Zoobenthos, Ecology and Conservation.* Hong Kong University Press, Hong Kong.

Dussart, B.H. 1974. Biology of inland waters in humid tropical Asia. In: *Natural Resources of Humid Tropical Asia.* Natural Resources Research, XII, UNESCO, Paris, 331–353.

Eccleston, B. and Potter, D. 1996. Environmental NGOs and different political contexts in South-east Asia. Malaysia, Indonesia and Vietnam. In: Parnwell, M.J.G. and Bryant, R.L. (eds) *Environmental Change in South-east Asia. People, Politics and Sustainable Development.* Routledge, London, 49–66.

Economic and Social Commission for Asia and the Pacific (ESCAP) 1987. *Water Resources Development in Asia and the Pacific: Some Issues and Concerns.* Water Resources Series No. 62. Bangkok, Thailand, and United Nations, New York.

Economic and Social Commission for Asia and the Pacific (ESCAP) 1992. *Towards an Environmentally Sound and Sustainable Development of Water Resources in Asia and the Pacific.* Water Resources Series No. 71. Bangkok, Thailand, and United Nations, New York.

Guardian Weekly 1996. Lies, statistics and dam lies. 30 June, 1996:13.

Hill, M.T. 1995. Fisheries ecology of the lower Mekong River: Myanmar to Tonlé Sap River. *Natural History Bulletin of the Siam Society* 43:263–288.

Hjorth, P. and Nguyen, T.D. 1993. Environmentally sound urban water management in developing countries: a case study of Hanoi. *Water Resources Development* 9:453–465.

Ho, S.C. 1994. Status of limnological research and training in Malaysia. *Mitteilungen der Internationalen Vereinigung für theoretische und angewandte Limnologie* 24:129–145.

Ho, S.C. 1996. Vision 2020: towards an environmentally sound and sustainable development of freshwater resources in Malaysia. *GeoJournal* 40:73–84.

Ho, Y.C. 1987. Control and management of pollution of inland waters in Malaysia. *Archiv für Hydrobiologie Beiheft, Ergebnisse Limnologie* 28:547–556.

Hori, H. 1993. Development of the Mekong River Basin, its problems and future prospects. *Water International* 18:110–115.

Hufschmidt, M.M. 1993. Water resource management. In:

Bonell, M., Hufschmidt, M.M. and Gladwell, J.S. (eds) *Hydrology and Water Management in the Humid Tropics.* UNESCO/Cambridge University Press, New York, 471–495.

Hufschmidt, M.M. and McCauley, D.S. 1988. Water resources management in a river/lake basin context. A conceptual framework with examples from developing countries. *Water Resources Development* 4:224–231.

Hunt, P. 1992. Sweet smell of death on Thailand's rivers. *New Scientist* 9 May 1992: 7.

Hynes, H.B.N. 1989. Keynote address. *Canadian Special Publications in Fisheries and Aquatic Sciences* 106:5–10.

Jacobs, J.W. 1994. Toward sustainability in Lower Mekong River Basin development. *Water International* 19:43–51.

Jalal, K.F. 1987. Regional water resources situation: quantitative and qualitative aspects. In: Ali, M., Radosevich, G.E. and Ali Khan, A. (eds) *Water Resources Policy for Asia.* A.A. Balkema Publishers, Boston, 13–34.

Johnson, D.S. 1957. A survey of Malayan freshwater life. *Malayan Nature Journal* 12:57–65.

Johnson, D.S. 1968. Water pollution in Malaysia and Singapore: some comments. *Malayan Nature Journal* 21:221–222.

Khoo, K.H., Leong, T.S., Soon, F.L., Tan, S.P. and Wong, S.Y. 1987. Riverine fisheries in Malaysia. *Archiv für Hydrobiologie Beiheft, Ergebnisse Limnologie* 28:261–268.

Kirmani, S.S. 1990. Water, peace and conflict management: the experience of the Indus and Mekong River basins. *Water International* 15:200–205.

Kottelat, M. 1989. Zoogeography of the fishes from Indochinese inland waters with an annotated checklist. *Bulletin of the Zoological Museum, University of Amsterdam* 12:1–56.

Kottelat, M. and Whitten, T. 1996. Freshwater biodiversity in Asia with special reference to fish. *World Bank Technical Paper* 343:1–59.

Kovar, K. and Nachtnebel, H.P. 1993. *Application of Geographic Information Systems in Hydrology and Water Resources Management.* IAHS (International Association of Hydrological Sciences) Publication No. 211, IAHS Press, Wallingford.

Krishna, J.H. 1983. Water resources development in Thailand. *Water International* 8:154–157.

Lang, C.R. 1996. Problems in the making. A critique of Vietnam's tropical forestry action plan. In: Parnwell, M.J.G. and Bryant, R.L. (eds) *Environmental Change in South-east Asia. People, Politics and Sustainable Development.* Routledge, London, 225–234.

Le, C.K. 1994. Native freshwater vegetation communities in the Mekong Delta. *International Journal of Ecology and Environmental Science* 20:55–71.

Lim, R.P. 1987. Water quality and faunal composition in the streams and rivers of the Ulu Endau area, Johore, Malaysia. *Malayan Nature Journal* 41:337–347.

Lim, T.G. 1990. Conflict over natural resources in Malaysia: the struggle of small-scale fishermen. In: Lim, T.G. and Valencia, M.J. (eds) *Conflict over Natural Resources in South-East Asia and the Pacific.* Oxford University Press, Oxford, 145–181.

Lim, T.G. and Valencia, M.J. 1990. *Conflict over Natural Resources in South-East Asia and the Pacific.* Oxford University Press, Oxford.

Lohmann, L. 1990. Remaking the Mekong. *The Ecologist* 20:61–66.

Lohmann, L. 1996. Freedom to plant. Indonesia and Thailand in a globalizing pulp and paper industry. In: Parnwell, M.J.G. and Bryant, R.L. (eds) *Environmental Change in South-east Asia. People, Politics and Sustainable Development.* Routledge, London, 23–48.

Low, K.S. 1993. Urban water resources in the humid tropics: an overview of the ASEAN region. In: Bonell, M., Hufschmidt, M.M. and Gladwell, J.S. (eds) *Hydrology and Water Management in the Humid Tropics.* UNESCO/Cambridge University Press, New York, 526–534.

Maclean, J. 1994. Management of Cambodian freshwater fisheries. *Naga, the ICLARM Quarterly* 17:4–6.

Maitland, P.S. 1995. The conservation of freshwater fish: past and present experience. *Biological Conservation* 72:259–270.

McGregor, D., McMorrow, J., Wills, J., Lawes, H. and Lloyd, M. 1996. Mapping the environment of South-east Asia. The use of remote sensing and geographical information systems. In: Parnwell, M.J.G. and Bryant, R.L. (eds) *Environmental Change in South-east Asia. People, Politics and Sustainable Development.* Routledge, London, 190–224.

McNeely, J.A. 1987. How dams and wildlife can coexist: natural habitats, agriculture, and major water resource development projects in tropical Asia. *Conservation Biology* 1:228–238.

Meijerink, A.M.J., Mannaerts, H.A., De Brouwer, H.A. and Valenzuela, C.R. 1993. Application of ILWIS to decision support in watershed management: case study of the Komering river basin, Indonesia. In: Kovar, K. and Nachtnebel, H.P. (eds) *Application of Geographic Information Systems in Hydrology and Water Resources Management.* IAHS (International Association of Hydrological Sciences) Publication No. 211, IAHS Press, Wallingford, 35–44.

Meisner, J.D. and Shuter, B.J. 1992. Assessing potential effects of global climate change on tropical freshwater fishes. *GeoJournal* 28:21–27.

Michelson, D.B. 1993. GIS supports wetlands land use analysis. *GIS World* 6:56–59.

Monkolprasit, S. and Roberts, T.R. 1990. *Himantura chaophraya*, a new freshwater stingray from Thailand. *Japanese Journal of Ichthyology* 37:203–208.

Moyle, P.B. and Leidy, R.A. 1992. Loss of biodiversity in aquatic ecosystems: evidence from fish faunas. In: Fielder, P.L. and Jain, S.A. (ed.) *Conservation Biology: The Theory and Practice of Nature Conservation, Preservation and Management.* Chapman & Hall, New York, 128–169.

Muttamara, S. and Sales, C.L. 1994. Water quality

management of the Chao Phraya River (a case study). *Environmental Technology* 15:501–516.

Nandeesha, M.C. 1994. Fishes of the Mekong River – conservation and need for aquaculture. *Naga, the ICLARM Quarterly* 17:17–18.

Ng, P.K.L. and Tan, H.H. 1997. Freshwater fishes of southeast Asia: potential for the aquarium fish trade and conservation issues. *Aquarium Sciences and Conservation* 1:79–90.

Niacin, M. 1992. *Land, Water and Development. River Basin Systems and their Sustainable Management.* Routledge, London.

Nielsen, M.D. and Strom, B. 1984. Simulation of water resources development projects in the Nam Kam basin, Thailand. *Nordic Hydrology* 15:297–306.

Nontji, A. 1994. The status of limnology in Indonesia. *Mitteilungen der Internationalen Vereinigung für theoretische und angewandte Limnologie* 24:95–113.

Office of the National Economic and Social Development Board, 1994. *Study of Potential Development of Water Resources in the Mae Khong River Basin (Executive Summary).* Water Resources Engineering Program, Asian Institute of Technology, Bangkok, Thailand.

Ong, A.S.H., Maheswaran, A. and Ngan, M.A. 1987. Malaysia. In: Chia, L.S. (ed.) *Environmental Management in South-east Asia. Directions and Current Status.* Faculty of Science, National University of Singapore, 14–76.

Palupi, K., Sumengen, S., Inswiasri, S., Augustina, L., Nunik, S.A., Sunarya, W. and Quraisyn, A. 1995. River water quality study in the vicinity of Jakarta. *Water Science and Technology* 31:17–25.

Pantulu, V.R. 1970. Some biological considerations related to the lower Mekong development. In: *Regional Meeting of Inland Water Biologists in Southeast Asia. Proceedings.* Unesco Field Science Office for Southeast Asia, Djakarta, 113–119.

Pantulu, V.R. 1986a. The Mekong River system. In: Davies, B.R. and Walker, K.F. (eds) *The Ecology of River Systems.* Dr W. Junk, The Hague, 695–719.

Pantulu, V.R. 1986b. Fish of the lower Mekong basin. In: Davies, B.R. and Walker, K.F. (eds) *The Ecology of River Systems.* Dr W. Junk, The Hague, 721–741.

Parnwell, M.J.G. and Bryant, R.L. 1996. Towards sustainable development in South-east Asia? In: Parnwell, M.J.G. and Bryant, R.L. (eds) *Environmental Change in South-east Asia. People, Politics and Sustainable Development.* Routledge, London, 330–343.

Parnwell, M.J.G. and Taylor, D.M. 1996. Environmental degradation, non-timber forest products and Iban communities in Sarawak. Impact, response and future prospects. In: Parnwell, M.J.G. and Bryant, R.L. (eds) *Environmental Change in South-east Asia. People, Politics and Sustainable Development.* Routledge, London, 269–300.

Partoatmodjo, S. 1987. Indonesia. In: Chia, L.S. (ed.) *Environmental Management in Southeast Asia. Directions and Current Status.* Faculty of Science, National University of Singapore, 1–13.

Petersen, R.C. and Sköglund, E. 1990. *Wetlands Management Programme. Study to Formulate Plans for the Management of the Wetlands in the Lower Mekong Basin (1.3.13/88/SWE).* The Mekong Secretariat, Bangkok, Thailand.

Pethiyagoda, R. 1994. Threats to the indigenous freshwater fishes of Sri Lanka and remarks on their conservation. *Hydrobiologia* 285:189–201.

Petts, G.E. 1996. Water allocation to protect river ecosystems. *Regulated Rivers: Research & Management* 12:353–365.

Phang, S.M. 1987. Agro-industrial wastewater reclamation in Peninsular Malaysia. *Archiv für Hydrobiologie Beiheft, Ergebnisse Limnologie* 28:77–94.

Pholprasith, S. 1983. Induced breeding of *pla buk* (*Pangasianodon gigas*). *Thai Fisheries Gazette* 36:347–360 (in Thai).

Pholprasith, S. 1993. Development techniques for induced spawning of the Mekong giant catfish. *Thai Fisheries Gazette* 46:399–415 (in Thai).

Prowse, G.A. 1968. Pollution in Malayan waters. *Malayan Nature Journal* 21:149–158.

Roberts, T.R. 1992. Revision of the Southeast Asian cyprinid fish genus *Probarbus*, with two new species threatened by proposed construction of dams on the Mekong River. *Ichthyological Exploration of Freshwaters* 3:37–48.

Roberts, T.R. 1993a. Artisanal fisheries and fish ecology below the great waterfalls in the Mekong River in southern Laos. *Natural History Bulletin of the Siam Society* 41:31–62.

Roberts, T.R. 1993b. Just another dammed river? Negative impacts of Pak Mun Dam on the fishes of the Mekong basin. *Natural History Bulletin of the Siam Society* 41:105–133.

Roberts, T.R. 1995. Mekong mainstream hydropower dams: run-of-the-river or ruin-of-the-river? *Natural History Bulletin of the Siam Society* 43:9–19.

Roberts, T.R. and Baird, I.G. 1995. Traditional fisheries and fish ecology on the Mekong River at Khone Waterfalls in southern Laos. *Natural History Bulletin of the Siam Society* 43:219–262.

Royal Forest Department 1994. *Forestry Statistics of Thailand.* Data Centre, Information Office, Royal Forest Department, Bangkok.

Saeni, M.S., Sutamihardja, R.T.M. and Sukra, J. 1980. Water quality of the Musi River in the city area of Palembang. In: Furtado, J.I. (ed.) *Tropical Ecology and Development. Proceedings of the Vth International Symposium of Tropical Ecology.* International Society of Tropical Ecology, Kuala Lumpur, 717–724.

Sage, C.L. 1996. The search for sustainable livelihoods in Indonesian transmigration settlements. In: Parnwell, M.J.G. and Bryant, R.L. (eds) *Environmental Change in South-east Asia. People, Politics and Sustainable Development.* Routledge, London, 97–122.

Santosa, D. and Goulter, I. 1991. Application of multi-objective analysis to water storage. The Bili-Bili multipurpose reservoir, Indonesia. *Water Resources Development* 7:82–91.

Scott, D.A. and Poole, C.M. 1989. *A Status Overview of Asian Wetlands.* Asian Wetland Bureau, Kuala Lumpur.

Setamanit, S. 1987. Thailand. In: Chia, L.S. (ed.) *Environmental Management in Southeast Asia. Directions and Current Status.* Faculty of Science, National University of Singapore, 169–211.

Sigurdsson, J.B. and Yang, C.M. 1990. Marine mammals of Singapore. In: Chou, L.M. and Ng, P.K.L. (eds) *Essays in Zoology.* Department of Zoology, National University of Singapore, 25–37.

Simon, T.P. and Emery, E.B. 1995. Modification and assessment of an index of biotic integrity to quantify water resource quality in great rivers. *Regulated Rivers: Research & Management* 11:283–298.

Smith, H.M. 1945. The freshwater fishes of Siam, or Thailand. *Bulletin of the United States National Museum* 188:1–622.

South China Morning Post (SCMP) 1993. Power-hungry Thais eye mighty Mekong. 6 November 1993: 10.

South China Morning Post (SCMP) 1995. Pact signed to protect Mekong. 6 April 1995: 10.

South China Morning Post (SCMP) 1996a. Electors 'don't give a dam'. 24 August 1996: 13.

South China Morning Post (SCMP) 1996b. Island faces disaster as mine waste destroys river. 30 March 1996: 13.

South China Morning Post (SCMP) 1997. Greens prepare for forest battle. 10 May 1997: 11.

Srivardhana, R. 1984. No easy management: irrigation development in the Chao Phraya Basin, Thailand. *Natural Resources Forum* 8:135–145.

Srivardhana, R. 1987. The Nam Pong case study. Some lessons to be learned. *Water Resources Development* 3:238–246.

Tejwani, K.G. 1993. Water management issues: population, agriculture and forests – a focus on watershed management. In: Bonell, M., Hufschmidt, M.M. and Gladwell, J.S. (eds) *Hydrology and Water Management in the Humid Tropics.* UNESCO/Cambridge University Press, New York, 496–525.

The Nation (*Bangkok*) 1996. Mekong nations to share information. 27 July, 1996: 5.

Thirakhupt, K. and Van Dijk, P.P. 1994. Species diversity and conservation of turtles in western Thailand. *Natural History Bulletin of the Siam Society* 42:207–259.

Usher, A.D. 1996. The race for power in Laos. The Nordic connections. In: Parnwell, M.J.G. and Bryant, R.L. (eds) *Environmental Change in South-east Asia. People, Politics and Sustainable Development.* Routledge, London, 123–144.

Vadhanaphuti, B., Klaikayai, T., Thanopanuwat, S. and Hungspreug, N. 1992. Water resources planning and management of Thailand's Chao Phraya River basin. *World Bank Technical Paper* 175:197–202.

Van Der Leeden, F. 1975. *Water Resources of the World.* Water Information Centre Inc., New York.

Villavicencio, V. 1987. Philippines. In: Chia, L.S. (ed.) *Environmental Management in Southeast Asia. Directions and Current Status.* Faculty of Science, National University of Singapore, 77–107.

Welch, D.N. and Lim, T.K. 1987. Water resources development and management in Malaysia. In: Ali, M., Radosevich, G.E. and Ali Khan, A. (eds) *Water Resources Policy for Asia.* A.A. Balkema Publishers, Boston, 71–81.

Wirawan, N. 1986. Protecting the *Pesut* (freshwater dolphin) in the Mahakam River of Kalimantan, Borneo. *Wallaceana* 44:3–6.

Zakaria-Ismail, M. 1987. The fish fauna of the Ulu Endau River system, Johore, Malaysia. *Malayan Nature Journal* 41:403–411.

Zakaria-Ismail, M. 1994. Zoogeography and biodiversity of the freshwater fishes of Southeast Asia. *Hydrobiologia* 285:41–48.

12

River conservation in Australia and New Zealand

N.J. Schofield, K.J. Collier, J. Quinn, F. Sheldon and M.C. Thoms

Regional overview

THE SETTING

Australia and New Zealand have many cultural and historical similarities, but the interplay of contrasting physiographic characteristics, as well as politics, result in variations in river conservation research, policy and practice between these neighbours.

Australia is a large and varied island continent (State of the Environment Australia – SOE, 1996). Its land mass covers 7 682 300 km^2 and its coastline about 37 000 km. The land is ancient and flat, with an average altitude of only 300 m AMSL. The country can be broadly divided into three main physiographic units (Jennings and Mabbutt, 1986):

- the Eastern Highland Belt consisting of a series of low mountain ranges, including the Great Dividing Range and Australia's highest peak, Mt Kosciusko (2228 m);
- the Central Eastern Lowlands with an altitude mostly below 150 m, including Lake Eyre at 14 m below sea level;
- the Great Western Plateau covering most of the north and west with a relatively flat, uniform surface, giving way to incised valleys near the coast.

New Zealand's climate is largely marine-temperate, whereas Australia spans a range of climatic zones. The northern part of the continent is in the wet–dry tropics, the centre is semi-arid to arid, giving way to temperate and mediterranean climates in the south. The mean annual rainfall for the country is 465 mm, making it the driest of all of the permanently inhabited continents. Some 80% of the land has an annual rainfall <600 mm, whilst only 4% receives above 1200 mm yr^{-1}. This low rainfall is exacerbated by high evaporative potential, with annual pan evaporation rates >2000 mm for most of the continent. The combination of low rainfall and high evaporation results in a meagre mean annual run-off of only 52 mm. Water is, therefore, a key limiting resource for human, economic and social development. Australia's small human population of 18 million (*ca* 2 km^{-2}; see Gopal *et al.*, this volume, for contrast) is mostly concentrated in major urban areas around the relatively well-watered eastern coast.

New Zealand is much wetter and more mountainous than Australia. Mean annual rainfall ranges from 350 mm in central Otago, on the south-east of the South Island, to more than 12 000 mm yr^{-1} in the Southern Alps (Statistics New Zealand – SNZ, 1998). New Zealand is also much smaller in area (270 543 km^2) and approximately half lies above 300 m AMSL, with slopes often greater than 28° (SNZ, 1998). Over 70% of the population (3.5 million – density 12.9 km^{-2}) reside in the North Island.

RIVERS OF AUSTRALIA AND NEW ZEALAND

The distinct climatic and geological regions of Australia give rise to a diversity of river types. The rivers draining east from the Eastern Highlands tend to be short, relatively high energy systems with steep headwaters in the range section, and meandering sections across the coastal plain. Those of the Central Eastern Lowlands are

Global Perspectives on River Conservation: Science, Policy and Practice.
Edited by P.J. Boon, B.R. Davies and G.E. Petts. © 2000 John Wiley & Sons Ltd.

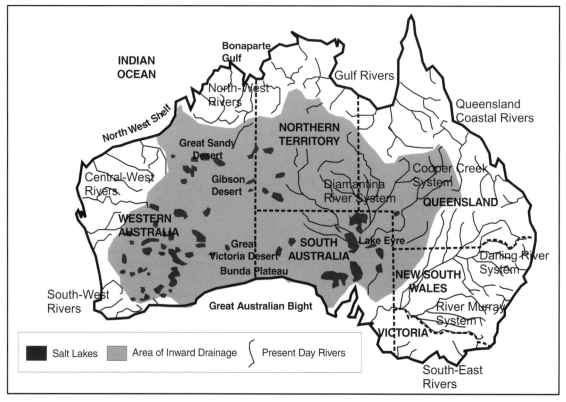

Figure 12.1   *The main drainage divisions and river systems of Australia*

typically low gradient, slow moving and hydrologically highly variable (Thoms and Walker, 1993) such as the rivers of the Murray–Darling Basin, and Cooper Creek and the Diamantina River within the Lake Eyre Basin (Figure 12.1). Many are ephemeral, particularly the lowland reaches of these systems. The Great Western Plateau is mostly defined by ephemeral, discontinuous surface drainage. There are only two main river systems in this region, the Fitzroy and Ord Rivers in the northwest. The continent's drainage system has been divided into 12 main divisions (Australian Water Resources Council, 1987) which can be further subdivided into 245 large river basins. A number of the larger rivers have substantial floodplains, especially in the lowland areas, with high ecological values. Despite the diversity of river types within the country there is no national classification or even broad description of its rivers.

Because of its elongate shape (the main islands span 13° of latitude), mountainous topography and high rainfall–run-off ratio, there is an abundance of river systems in New Zealand. The majority are short (only four exceed 250 km in length), high-energy systems

(Collier and McColl, 1992); the largest river (Waikato) drains only 5% of New Zealand's land area: the Murray–Darling drains 14% of Australia. Main river types include: braided gravel-bed rivers on the east coasts of both main islands; turbid glacial-fed rivers and acid brown-water streams in the western regions of the South Island; spring-fed, sandy-bottomed streams (some of which are geothermally influenced) draining extensive volcanic ash deposits in the central North Island; turbulent mountain rivers draining the main ranges; and low-energy systems of the coastal plains. The New Zealand dominion also includes more than 500 offshore islands, which are characterized by short or ephemeral lotic habitats, that sometimes have endemic aquatic invertebrate species. The diverse and geologically active landscape is reflected in the 79 ecological regions identified by McEwen (1987).

A history of development

Australia has a relatively short history (200 years) of European occupation. Land, vegetation and rivers have been accorded little protection from broad-scale agricultural (and, to a lesser degree, urban and industrial) development. Although New Zealand has a similar development history to Australia, water shortage is much less of an issue due to abundant rainfall in most areas. However, the location of the largest metropolitan area (Auckland) on a very narrow part of the North Island, with limited catchment areas for water supply, has caused serious water shortages. In the drought of 1994 moves were made to invoke emergency legislation to permit an interbasin transfer (IBT) from the Waikato River to supply the city. So, even in well-watered areas, development can lead to marked impacts on the river systems. This section provides a brief summary of the impact that development has had on the river systems of both countries.

LAND-USE IMPACTS

Many rivers in Australia and New Zealand have been adversely affected by agricultural development. In 1788 forest covered about 10% and woodlands about 23% of the Australian continent (Carnahan, 1986). By 1990 this had been reduced respectively to 5 and 15% with an average loss of more than 500 000 ha of woody vegetation yr^{-1} (Commonwealth Scientific and Industrial Research Organization – CSIRO, 1990). In the south, native vegetation has been extensively cleared and replaced principally with annual pastures and crops. In the central and northern regions, the land has been given over to large pastoral leases for beef production. This broad-scale vegetation clearance and subsequent agricultural development has led to severe degradation of riparian zones, increased sediment loading and turbidity in rivers, and to surface water nutrient enrichment. Land and river salinization is also emerging as a major problem across large areas of Australia (Warner, 1995; Prime Minister's Science, Engineering and Innovation Council – PMSEIC, 1999).

Even though more than half of Australia's native vegetation has been cleared, clearing of the remainder is still occurring at a significant pace, although the worst affected states have introduced clearing embargoes. Overall, riparian forest is poorly protected in Australian agricultural areas. Riparian zones are predominantly privately owned, with little active management for the protection of riparian values (Bunn *et al.*, 1993). A number of states are developing, or have developed, policies that advocate appropriate management of riparian lands, but little

active management is occurring on anything other than a demonstration scale.

Forest covered 80–90% of New Zealand before the arrival of Polynesians about 1000 years ago, and much of this remained when European colonization began 160 years ago. Following the onset of European colonization in the 1840s, large tracts of lowland and hill-land forest were cleared for pastoral farming. There are now over 13×10^6 ha, approximately 50% of New Zealand's land surface, converted to grazing land compared with 1.7×10^6 ha in production forest and 91 000 ha in horticulture (SNZ, 1998). Agriculture is now the dominant land use in the middle and lower catchment areas of most of New Zealand's streams and rivers.

Recent nationwide surveys of resource managers (Sinner, 1992; Smith, 1993) have highlighted widespread concern about the effects of agricultural land-use on water quality. Sedimentation, nutrient contamination, alteration of physical characteristics, and faecal contamination of surface water were identified as the most significant impacts. Comparative studies of stream ecosystems under pasture and native forest throughout New Zealand (Smith *et al.*, 1993; Harding and Winterbourn, 1995; Quinn *et al.*, 1997; Townsend *et al.*, 1997) have indicated fairly consistent patterns of stream habitat change, and biotic response, to the conversion of native forest or grassland to pasture. Commonly observed changes in physical and chemical habitat include:

- increased water yield (e.g. by 60% in a central North Island basin (Dons, 1987)) and storm flows (Fahey and Rowe, 1992);
- reduced shade in low-order streams, with consequent increases in stream temperature and in-stream algae and macrophyte growth (Quinn *et al.*, 1997);
- narrower stream channels in the small streams (<30 km^2 catchment area) (Davies-Colley, 1997);
- lower in-stream leaf litter and woody debris, and higher in-stream plant production and algal and macrophyte biomass (particularly adventive species), due to less riparian vegetation (Howard-Williams and Pickmere, 1994; Harding and Winterbourn, 1995; Quinn *et al.*, 1997; Townsend *et al.*, 1997);
- often higher concentrations of suspended sediment, faecal contamination, nitrogen and phosphorus (e.g. Smith *et al.*, 1993; Quinn *et al.*, 1997);
- loss of habitat diversity due to channelization to improve drainage (Quinn *et al.*, 1992b; Williamson *et al.*, 1992) in lowland areas in New Zealand,

where excessive soil moisture has been a major limiting factor on the agricultural productivity of 2×10^6 ha of land (Bowler, 1980).

Dairy farming is widespread in lowland areas of New Zealand and the effluents from milking sheds are by far the most numerous point-sources of effluent to streams. Although most effluents receive treatment in waste stabilization lagoons, they still contain significant loads of organic matter, ammonia, nutrients and faecal indicators, and require large in-stream dilution to meet water quality criteria (Hickey *et al.*, 1997).

Irrigated agriculture in Australia covers about 2×10^6 ha, mostly concentrated in the southeast (80% in the Murray–Darling Basin). Many irrigated areas are increasingly suffering raised water tables and land salinization. The drainage of irrigated land to rivers often contributes significant loads of salt, nutrients and pesticides. Irrigation accounts for about 70% of the country's harnessed water resources and has a major impact on river hydrology, both through direct abstraction and storage operations (Schofield, 1996a).

Irrigated agriculture covers 0.23×10^6 ha in New Zealand, mostly on the east coast of the South Island (SNZ, 1995). Salinization has not been a major issue in New Zealand, although seawater intrusion into groundwater aquifers has occurred near Nelson, where irrigation of coastal orchards was interrupted (Thorpe, 1992). The impact of forestry operations on rivers has generally been minor in the long term, when compared with agriculture. Examples of increased river sediment and salinity have been cited where extensive clear-felling has taken place, but these problems diminish rapidly with regeneration (Schofield, 1996b). Forest management impacts on rivers are often reduced by the inclusion of riparian buffer zones, and the retention of riparian forest and understorey, although use of this management technique is recent and inconsistent between states. Some transient ecological impacts of forestry pesticide spraying have been recorded (Davies *et al.*, 1994).

Forestry in New Zealand is based on exotic conifers (96% of wood volume in 1992 and increasing: Maclaren, 1996) of which the short-rotation (average 28 years at harvest) *Pinus radiata* is strongly dominant (91%) (SNZ, 1998). Less than 2% of New Zealand's timber production is from indigenous forest, and sustainable harvest management is required. Exotic afforestation of pasture land, particularly on hill-land that is marginal for agriculture, has increased in recent

years (SNZ, 1998). Studies to date indicate that forestry has less impact on stream water quality and biota than pastoral agriculture, at least during the growth phase before harvest (Harding, 1995; Quinn *et al.*, 1997; Rowe *et al.*, 1999). Harvest impacts on river ecosystems are variable and are only now receiving detailed research (Collier *et al.*, 1997). It is now common practice in many areas to leave the area adjacent to larger streams unplanted to provide a buffer of naturally regenerating riparian vegetation.

Mineral resources are a mainstay of the Australian economy but mining has generally had little impact on rivers because of the small area developed (<1%). Some important mineral deposits do occur, however, within very high-value ecosystems (e.g. Kakadu National Park, Northern Territory; Jarrah Forest, Western Australia) and potential riverine impacts are of concern (Schofield and Bartle, 1984; East *et al.*, 1988). Extraction of sand and gravel deposits from river channels does occur and effects of these activities are well documented (e.g. Warner, 1995). Perhaps Australia's most notable and ongoing river-related mining disaster is in the Queen and King rivers, Tasmania (Knighton, 1989; Davies *et al.*, 1996). Most of Australia's other industrial activity is concentrated within or close to urban and coastal areas. Recent studies (Thoms and Thiel, 1995; Thoms *et al.*, 1999) have shown that significant lengths of downstream river and floodplain reaches can be influenced by these activities. Large-scale industrial point-source discharges to rivers are rare.

Mining plays a much less significant role in the economy of New Zealand, employing <0.3% of the workforce (SNZ, 1998). Alluvial (placer) mining in floodplains of gravel-bed rivers in Otago, Southland and Westland can cause increased turbidity downstream, with impacts on riverine plants and invertebrates (Davies-Colley *et al.*, 1992; Quinn *et al.*, 1992a). Effluents from current hardrock mining are usually well controlled through resource-consent processes, although occasional problems have occurred with tailings-dam failures (Taylor *et al.*, 1997). However, the legacy of tailings and mine drainage from historic mining operations persists at local hot spots (Penny, 1987; Hickey and Clements, 1998).

Australia's population is realtively small and concentrated in a few large, coastal cities. Hence, urbanization has had only local impacts on rivers (Thoms *et al.*, 1999). A number of small to medium inland towns do, however, contribute significant sewage effluent loads to rivers, particularly in the

Murray–Darling Basin. Here sewage outfalls are the principal nutrient input during low-flow periods and have been implicated in blue-green algal blooms (Gutteridge, Haskins and Davey – GHD, 1992; Thoms and Flett, 1993).

Much of New Zealand's population is also concentrated in coastal cities. Nevertheless, inland cities, and food and fibre processing plants do discharge significant loads of organic matter and nutrients to some major rivers, such as the Waikato and Manawatu (Figure 12.2). In general, however, improved wastewater treatment has led to reduced organic matter loads over the last 20 years (e.g. Quinn and Gilliland, 1989). Most smaller towns treat their sewage using waste stabilization lagoons prior to discharge to rivers, and ecological impacts are generally acceptable provided the dilution is greater than 30- to 50-fold (Quinn and Hickey, 1993). However, nutrients in these discharges sometimes result in nuisance growths of benthic algae (Quinn and Gilliland, 1989; Welch *et al.*, 1992; Huser *et al.*, 1994).

IN-STREAM IMPACTS

Flow regulation, through impoundments, water abstractions and other regulatory structures, has had a major impact on the structure and functioning of Australian riverine environments. More than 447 large dams have been built for a variety of purposes, including urban water supply, hydroelectric power generation, irrigation and flood mitigation (Australian National Committee on Large Dams – ANCOLD, 1990). Until very recently these structures were designed on engineering and water-supply criteria alone, with little thought to riverine ecological impacts. As a result most dams have been designed without fish ladders and consequently impede upstream migration (Harris and Gehrke, 1997). Many dams frequently experience eutrophication, with the build-up of nutrients in sediments and in the water column, which may stimulate the proliferation of green or toxic blue-green algae and aquatic weeds (Mitchell and Rogers, 1985; Harris, 1994). Severe eutrophication and the release of algal toxins have repeatedly prevented the utilization of water storages in several areas (Cooperative Research Centre for Water Quality and Treatment – CRCWQT, 1998). Owing to high evaporation rates and flow variability, storages within Australia must be larger per unit of supply than other inhabited continents (Munro, 1974; Table 12.1; see also Davies and Wishart, this volume).

Table 12.1 *Approximate water storage km^{-2} for irrigation in selected countries (source: Munro, 1974)*

Country	Storage ($\times 10^3$ m^3 km^{-2})
India	150
Egypt	380
USA	760
Australia	1520

There are 24 IBTs in Australia that transfer water from 'surplus' to 'deficit' areas, the best known being the Snowy Mountains Hydro-Electric Scheme (Davies *et al.*, 1992). As with dam releases, water transfers may increase bank erosion and sediment transport, adversely affect the source river through reduced flows and effect biotic transfers, thereby disrupting river basin biotic integrity (Davies *et al.*, 1992; see also Davies *et al.*, this volume).

Australia has now ceased its large dam-building phase, with the exception of Queensland (Department of Natural Resources – DNR, 1997) and, perhaps, Western Australia. However, in addition to large dams, the Australian rural landscape is typified by many small 'farm dams' designed for local use, particularly for stock water. Approximately 400 000 farm dams have been estimated to occur in each of south-western West Australia and Victoria, and the nation probably supports more than several million farm dams. These dams collectively cause modification of river flows (particularly low flows), reduce river water quality, but act as refuges both for wildlife and native and exotic aquatic biota.

FORMS OF DEGRADATION

Salinization

There are many forms of river degradation in Australia and it is difficult to generalize as to which are the most severe. One of the earliest observed was increasing river salinity. The first detailed account comes from a railway engineer (Wood, 1924) who observed local railway water supplies going saline a few years after the clearing of native vegetation. His concept of rising water tables, brought about by the change in the hydrological balance following vegetation clearing, was validated much later through detailed research in the 1970s, by which time stream salinities were rising rapidly in many areas (Schofield and Ruprecht, 1989).

Salinity in Australia is a problem for both dryland farming and irrigation farming (Ghassemi *et al.*, 1995).

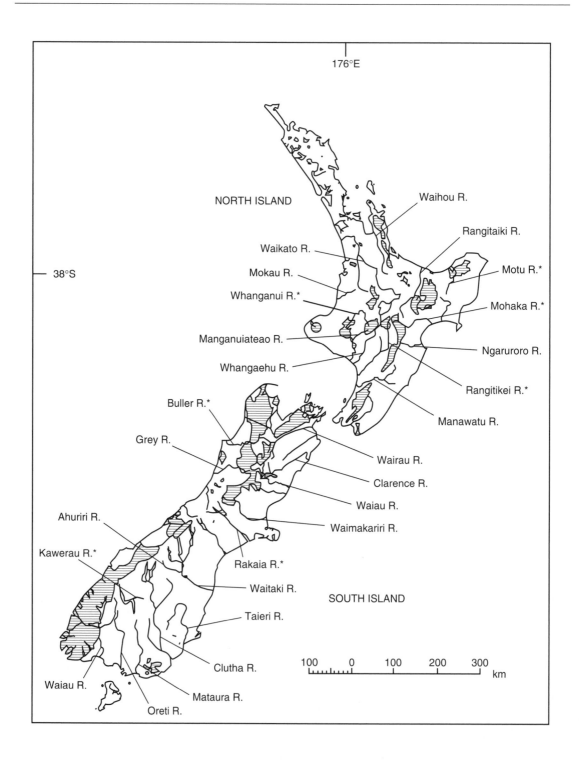

Figure 12.2   *Map of the main islands of New Zealand showing main parks and reserves (shaded), major rivers (>150 km in length (Statistics New Zealand, 1996), and rivers with passed or pending Water Conservation Orders (*)*

Steadily rising salinities in the Murray–Darling Basin have been temporarily abated by expensive groundwater interception schemes adjacent to the river in which saline water is pumped away to on-land disposal sites (PMSEIC, 1999). The ecological impact of increasing stream salinity is not well understood but is likely to be severe. Macroinvertebrate community structure is altered and diversity decreased when salinities rise to the 1000–2000 mg L^{-1} range, while the reproductive capacity of macrophytes is impaired as salinities rise to 6000 mg L^{-1} (Hart *et al.*, 1991; Bailey and Warwick, 1998).

Eutrophication

The frequency and severity of algal blooms in Australia's rivers, lakes, reservoirs and estuaries are an indicator of the severity of aquatic degradation. In 1991/92 Australia set a dubious world record for the longest riverine algal bloom, covering some 1100 km in the Darling River, western New South Wales (Thoms and Flett, 1993; Bowling and Baker, 1996). The principal nuisance algae are the blue-greens (cyanophytes) *Microcystis*, *Anabaena* and *Nodularia*. The former two produce toxins, which frequently lead to closure of drinking water supplies. In rivers, the common sites for algal blooms are river pools created by weirs and barrages. Increased frequency and intensity of blooms is associated with nutrient enrichment and modified river flows due to water storage and extraction (Harris, 1994; Johnstone *et al.*, 1996). Despite some 20 years of active research, practical solutions to this problem are few.

In the case of nutrients, there is still lively debate as to the most important sources, e.g. point sources such as sewage effluent or diffuse sources such as fertilizers, bank and gully erosion or sub-surface colloidal movement (Land and Water Resources, Research and Development Corporation – LWRRDC, 1998). A number of management programmes have focused on reducing nutrient input to rivers but their success and effectiveness have not been demonstrated. On the in-stream side, there is still limited understanding of the initiation and development of algal blooms and the relative contributions of such factors as flow regime, stratification, turbidity, light, temperature and nutrients (Harris, 1994; Banens and Davis, 1998). The release of nutrients from river sediments is an area of current research (LWRRDC, 1996). Within reservoirs and other impoundments, however, there is an increasing understanding of such factors as

stratification, algal buoyancy, hydrodynamics and grazing activity. Good data and understanding are available for one weir pool where manipulation of the flow regime to reduce stratification proved to be the most effective means of algal bloom control (Webster *et al.*, 2000).

The much shorter residence times in New Zealand's relatively short rivers help to prevent occurrence of the nuisance phytoplankton blooms that have plagued Australian rivers in recent years. Phytoplankton tend to build up in the chain of eight hydroelectric reservoirs along the Waikato River and can exceed desirable levels for avoiding clogged water intakes (Huser *et al.*, 1994). However, with appropriate filtration technology (e.g. back-flushing filters), phytoplankton have not caused severe problems. However, many of New Zealand's shallow gravel-bed rivers are prone to proliferation of filamentous periphyton that have nuisance effects on human uses and in-stream values (Biggs and Price, 1987; Biggs, 1990). Guidelines for managing these and other nuisance growths in rivers have been developed by the Ministry for the Environment (1992).

The first five years' data from the 77 river sites in New Zealand's National Rivers Monitoring Network showed encouraging national trends of declining concentrations in total phosphorus, nitrate and ammonia between 1989 and 1994 (Smith *et al.*, 1996).

Pesticides

Pesticides have been detected in a number of Australian rivers at concentrations of concern to aquatic ecological health and drinking water potability (Cooper *et al.*, 1996; Schofield and Simpson, 1996). The use of chemicals to control insect pests and weeds has grown about five-fold over the last 15 years. Herbicide use has shown the fastest rate of increase. Atrazine is now widely detected in rural Australian rivers, typically in the low parts-per-billion concentration range. Most public attention, however, has focused on the cotton industry, a heavy user of pesticides. At present, the principal cotton chemical of concern is endosulfan, which often occurs in rivers at concentrations above the national guidelines for the protection of aquatic ecosystems (Muschal and Cooper, 1999). Atmospheric transport of endosulfan (spray drift, volatilization, dust) are important pathways for river contamination (Raupach and Briggs, 1999; Woods *et al.*, 1999). Reductions in the abundance of common macroinvertebrate taxa have been correlated with the presence of pesticides during the cotton growing season (Leonard *et al.*, 2000) and accidental spills or off-target spray events have been associated with local fish kills.

Storm events that transport water and sediment into rivers are the most significant episodic pesticide contamination events (Cooper and Riley, 1996). Even irrigation farms with substantial on-farm water storage capacity and recycling systems cannot withstand major storms. Such events are associated with peak loading and peak pesticide concentrations in rivers. Whilst much attention has been focused on cotton, many other rural industries are substantial users of pesticides, some of which are known to be contaminating rivers (Schofield and Simpson, 1996).

Although river biomonitoring programmes have not yet clearly identified the in-stream impacts of pesticides, laboratory toxicity testing and mesocosm studies are being used to set more appropriate aquatic protection guideline values for Australian ecosystems (Chapman, 1999). At the same time, a precautionary approach to industrial practices is being fostered through the development and implementation of 'best practices' to minimize the off-farm transport of pesticides (Williams, 1999).

Pesticide use is considered a vital part of conventional agricultural practice in New Zealand, with about 4000 t of active ingredient applied annually in the late 1980s (Wilcock, 1989; Wilcock and Close, 1990) and higher levels are probably currently applied. Most water monitoring has focused on groundwaters, rather than on surface waters, and has shown these to be generally free from pesticide contamination (Close, 1996; Taylor *et al.*, 1997). Pentachlorophenol (PCP) was extensively used in New Zealand by the timber sawmilling industry from the 1950s to the mid 1980s, before being deregistered in 1991. Despite widespread use, studies indicate only localized contamination of water and aquatic biota near areas of high use, rather than widespread contamination.

The agricultural sector widely used DDT in New Zealand until the 1970s (Boul, 1995). The extent of DDT contamination in New Zealand rivers is not well documented, but Hickey *et al.* (1997) found generally increasing levels of total DDT in freshwater mussels with distance along the Waikato River. The greatest levels were associated with agricultural drainage. However, the levels were considered low enough for safe human consumption and did not affect shellfish condition. These findings suggest that pollution from this persistent agrochemical is unlikely to be a major issue in New Zealand rivers.

Sediment

Another form of degradation of considerable concern in both Australia and New Zealand is soil erosion leading to increased river sediment concentrations and turbidity. Accelerated erosion is common in developed agricultural areas, particularly where riparian vegetation has been removed or substantially degraded. Most Australian research has been conducted in eastern states where stream bank collapse and gully erosion have been considered to be the principal sources of sediment. The fine clay component of this eroded material is moderately high in phosphorus and may be the main cause of river eutrophication in these areas (Donnelly *et al.*, 1996). The rate of gully erosion is now thought to be slowing, following the high rates initiated by the early clearance of native vegetation in rural Australia. Other sources of increased sediment loads to rivers include river-bank slumping; in regulated rivers this has been associated with short-duration flow releases, while in unregulated systems it may reflect increased rates of water level fall associated with irrigation diversions (Thoms and Sheldon, 2000). In the Adelaide Hills, subsurface movement of colloidal material in macropores, such as root channels and fissures, has been implicated in increased sediment and chemical transport (Kirkby *et al.*, 1997). In tropical areas, large sediment loads are often generated from farmed land during cyclonic events. For example, 890 t (95% of the annual load) were exported from the Johnstone River catchment in north-east Queensland by Cyclone Sadie (Hunter *et al.*, 1996) with serious implications for the ecology of the Great Barrier Reef (Devlin and Schofield, 1999).

Whatever the mechanism, the majority of Australian rivers are highly turbid. This was frequently the case before European settlement but agricultural development has substantially increased sediment loads and turbidity (Olive *et al.*, 1995). Combined with dramatic changes to the flow regime, this has resulted in modifications to riverine habitats, including infilling of river pools, and radical changes to channel morphology, resulting in loss of biodiversity.

New Zealand is situated at the junction of two colliding (Indo-Australian and Pacific) continental plates where tectonic forces cause uplift and high rates of erosion. The rate of erosion, however, has been exacerbated by forest clearance. Erosion rates range from 20 000 t km^{-2} yr^{-1} in the soft sedimentary rocks of the North Island's East Cape, to less than 50 t km^{-2} yr^{-1} in areas of low relief (Griffiths, 1981, 1982; Jowett, 1984; Griffiths and Glasby, 1985). Black disk visibility showed a correspondingly wide variation between rivers (e.g. 0.25–10.75 m) in a national survey under summer base-flow conditions (Davies-Colley and Quinn, 1998). Monthly monitoring of 77 river sites in the National Rivers Monitoring Network gave an overall median of 1.3 m in 1989–90 (Smith and Maasdam, 1994) and there was a national trend of increasing water clarity (and reduced turbidity) between 1989 and 1994 (Smith *et al.*, 1996). New Zealand rivers are estimated annually to carry 916 t km^{-2} to the sea in the North Island, and 1856 t km^{-2} in the South Island (Griffiths and Glasby, 1985).

Regulation

As stated earlier, Australia has more than 447 large dams, a number of IBTs, vast numbers of weirs, 124 barrages and other barriers to flow, and possibly several million small farm dams capturing low flows. These structures can cause radical changes to flow regimes (Thoms and Walker, 1993). The most explicit examples of the impacts of water impoundment, flow regulation and extraction occur in the Murray–Darling Basin. Water extraction from the system is approaching 90% of mean annual run-off (MAR), with discharge to the ocean becoming a rare event (Murray–Darling Basin Commission – MDBC, 1995). Flood magnitudes and frequencies have been reduced with serious consequences for floodplain vegetation, billabongs and wetlands. A review of the effects of regulation on the ecology of the River Murray can be found in Walker (1985).

Alien invasive organisms

Regulation not only changes the nature of flows in a river, but can have impacts that often extend right through the food chain. In the Murray–Darling Basin the abundance and diversity of native fish is declining, whereas exotic species such as carp (*Cyprinus carpio*) and mosquito fish (*Gambusia holbrooki*) are proliferating. In the Murrumbidgee and Murray rivers, carp now account for over 90% of the biomass (Harris, 1995). The introduced brown trout predominate in cooler streams of south-eastern Australia. Australian rivers are not without their plant invaders. Infestations by exotic plants also lead to modified habitats and changes in food-web structures. The major nuisance species are para grass (*Mimosa pigra*), alligator weed (*Lippia*) and Kariba weed (*Salvinia molesta*).

Exotic 'oxygen weeds' (*Egeria*, *Elodea* and *Lagarosiphon*) are well established in many lowland streams and lakes in New Zealand, often crowding out native plants and causing night-time dissolved oxygen depletion (Wilcock *et al.*, 1998). In many cases, clearance of riparian shade has assisted aquatic weed invasions (Howard-Williams *et al.*, 1987). Several introduced fish are listed as 'noxious' including rudd, *Scardinius erythrophthalmus*, and European or koi carp, *Cyprinus carpio* (McDowall, 1990). Catfish, *Amerinus nebulosus*, are also considered a nuisance (Taylor *et al.*, 1997). Brown and rainbow trout (*Salmo trutta* and *Oncorhynchus mykiss*) are probably the most widespread fish invaders but, along with salmon (*Oncorhynchus* spp), which are largely restricted to rivers of the South Island's east coast, are protected because they provide a significant recreational resource. However, trout have displaced, or become predators of, some of the native galaxiid fishes (Townsend and Crowl, 1991).

The need for change

The picture outlined so far suggests that the health of Australia's aquatic habitats is in decline. The Wild Rivers survey of the Australian Heritage Commission showed that very few systems can be classed as truly 'Wild Rivers' (Stein *et al.*, 1997). There are, however, a number of systems that have so far escaped the major ravages of development. These include a few river stretches (impacted upstream) in the west of the country, and a few intact rivers and catchments in the tropics and Tasmania. The large river systems of central Australia (Cooper Creek, Diamantina River and the Neales River complex) have been impacted by land-use changes through grazing but have so far escaped large-scale water-resource development.

In recognizing this degradation Australia is now at a critical point in managing its aquatic resources. Socio-political and economic influences will now determine the treatment of rivers in Australia. Until the last two decades the support of Western economic development, based on primary resources, was the dominant ethos, whether through land clearing, water use or mining. For example, land release policies, right up to the late 1970s, required minimum clearing levels, often as high as 95%.

The growing environmental movement has initiated radical changes in natural resource management in Australia. Unfortunately this change has come too late to retain many of the unique natural environmental values of the country, including river ecosystems. The radical changes in resource management occurring in Australia are exciting and will be detailed below. The success of these changes, however, will depend on how easy it is to transcend the existing institutional obstacles and vested interests.

Despite the apparent community acceptance of an environmental ethic in Australia, the dominant socio-political and economic institutional arrangements continue to hamper progress towards truly sustainable land and water management. The recent trend toward privatization of natural resource management, particularly water, may be an added barrier to improved environmental performance. Fundamental to this problem, in relation to rivers, is private land ownership, poor riparian management and a lack of clear administrative responsibilities for rivers or water resources (at regional, state and national levels) (Schofield and Price, 1999).

The majority of land in Australia is in private ownership and improved land-use or land-management controls or incentives have been, and continue to be, all but absent. Whilst landowners have riparian rights and/or water allocations, rivers are generally 'managed' as a public entity, by default. However, until recently this management has largely been in the context of resource utilization, principally for water supply and effluent disposal. Whilst dams were sited and constructed generally on engineering criteria, in a number of cases their catchments were fortuitously protected for urban water supply (e.g. Perth, Melbourne, Sydney) with clearing of native vegetation largely prohibited. However, rivers with no water resource value were not managed in any respect and have often suffered massive degradation with little sense of loss by the general community. The administrative situation today is far from ideal, with no national water or river management agency, and few coherent, consistent water laws at either national or state levels (Schofield and Price, 1999). With responsibility fragmented between agencies and levels of government, sustained action on river management is rare. Despite this, frameworks are being established at national, state and local levels to address land- and water-management issues.

The state of New Zealand's environment was comprehensively reviewed by Taylor *et al.* (1997). They noted the fragmentation of aquatic habitat and declining water quality brought about by land-use change, the increasing threat that pests and weeds posed for aquatic ecosystems, and catalogued a series of issues that require attention. Much of the pressure to address these issues comes from consumer demand in export markets for goods that are produced in an environmentally sustainable manner. In future, product certification requirements are likely to lead both to improvements in land-management techniques and to the health of streams flowing through developed catchments. In addition, there is a groundswell of support at local government and community levels for initiatives to restore the health of rural waterways.

The framework for conservation

AUSTRALIAN POLICIES AND INITIATIVES

Partly in frustration at its own ineffectiveness and lack of resources, the Australian Federal Government has established a number of recent policies and initiatives to assist the sustainable management of the country's water resources. These include the Council of Australian Governments (COAG) Water Reforms, the National Landcare Programme and Integrated Catchment Management (ICM).

For a number of reasons, the statutory framework in Australia is complex, duplicative and ineffective in relation to river management and conservation. First, Australia has three levels of government – federal, state, local (some call ICM the fourth) – resulting in diffusion and confusion of responsibility. Second, most of the powers for 'managing' rivers devolve to the states, resulting in a wide variety of activities and protocols that often cannot be nationally aggregated or coordinated. Third, Australia has no national water or rivers agency, and both nationally, and in the states, responsibility for rivers is dispersed. Fourth, Australia is currently breaking up its large regional water bodies in two ways: (i) separating water utilities from water resource managers from water regulators; and (ii) dividing rural water management (especially irrigation) into small local boards. Fifth, river management is often a small responsibility of a large number of agencies and, in some cases, entirely 'falls through the cracks'!

National River Health Programme

The National River Health Programme (NRHP, 1993) is the most significant federal action on rivers in Australia in the last decade. The programme comprises two initiatives, one on river health assessment (Schofield and Davies, 1996) and the other on environmental flow management. In excess of A\$20 million has been allocated over the period 1993–2001.

To date, the key outcomes have been the development of a national rapid bioassessment protocol for the measurement of river health (known as AUSRIVAS - the Australian River Assessment Scheme) (Norris, 1999; Simpson and Norris, 2000); application of the protocol in all states and territories as the First National Assessment of River Health; and commissioning a range of research projects on environmental flows as they relate to allocation techniques, management of storages, impacts on biota, relationships to algae, and requirements for wetlands. Numerous outputs are already available, including state-by-state river health assessments (Read and Franski, 1998; Choy, 1999), a comprehensive evaluation of environmental flow assessment techniques (Arthington, 1998; Arthington and Zaluki, 1998; Arthington *et al.*, 1998), and an environmental flows decision support system (Davis and Young, 1998).

COAG water reforms

In February 1994, COAG outlined a national agenda for micro-economic reform (COAG, 1994). A key part was a national water resource policy aimed at supporting higher sustainable economic and employment growth. The Council endorsed a comprehensive strategic framework for efficient and sustainable reform of the Australian water industry. Elements of the framework include (from Russell, 1996):

- water pricing based on consumption-based pricing and full cost recovery
- reduction or elimination of cross-subsidies
- clarification of property rights
- allocation of water for the environment
- adoption of trading arrangements for water
- institutional reform
- public consultation and participation

The Water Resource Policy endorsed the 'National Principles for Provision of Water for Ecosystems'. It stated that 'environmental requirements, wherever possible, should be determined on the best scientific information available and have regard to the inter-temporal and inter-spatial water needs required to maintain the health and viability of river systems and groundwater basins' (COAG, 1994).

In response to these requirements a series of federal and state agency initiatives have been developed to facilitate decision-making on land- and water-resource management. For example, the National Land and Water Resources Audit is federally funded through the Natural Heritage Trust programme and has five broad objectives:

- to provide a clear understanding of the status of, and changes in, Australia's land and water resources and the implications for their sustainable use;
- to interpret the costs and benefits of changes to these resources;
- to construct a compatible and readily accessible national database;
- to encourage collaboration between all relevant parties and other initiatives;
- to provide a framework for monitoring of Australia's land and water resources in a structured way.

Moreover, the Murray–Darling Basin Ministerial Council, the inter-government council that coordinates activities for the Murray–Darling Basin (MDB), agreed to place a cap on further increases in water extractions within the basin (MDBC, 1995) shortly after the announcement of the COAG Water Resource Policy. This process will be discussed later but, in brief, each state in the MDB has developed an environmental flow management strategy which will aid in meeting the Cap. The approaches taken at a state level to the MDBC Cap differ according to the level of existing development, approach to property rights for water, and potential for future development in each state and territory.

Catchment management

The concept of Integrated (or Total) Catchment Management has been well accepted in Australia (e.g. see Hooper and Margerum, this volume). A large number of catchment management groups have been established and catchment planning is well advanced in some areas. An active programme to establish catchment-by-catchment water quality objectives has been initiated in New South Wales (NSW). New South Wales first implemented its 'Total Catchment Management Policy' (TCM) in 1994. Its aim is the sustainable use and management of the state's natural resources – particularly soil, water and vegetation. With its passage in 1989, the Catchment Management Act is now part of NSW legislation and provides institutional mechanisms for putting policy into effect. The Act enabled the establishment of a state-level coordinating committee, the Catchment Management Co-ordinating Committee, and two kinds of catchment-level committee – Catchment Management Committees and Catchment Management Trusts.

The other states have all implemented variations of a TCM approach in managing a large number of their waterways. Both Queensland and Western Australia have state-level integrated catchment management policies (Wasson *et al.*, 1996) with Queensland employing a system similar to NSW termed Integrated Catchment Management (ICM). Western Australia has adopted a series of Land Conservation District Committees (LCDCs) comprising a number of farmers or community representatives. These groups tend to be community driven. Higher levels of ICM, where either the community or state government takes a proactive role, have been formed in this state based on need (e.g. Blackwood, Avon and Harvey groups). South Australia and Victoria are also using similar principles to bring about whole-catchment management, with South Australia using a system of Local Action Planning Groups and Catchment Management Boards. In Victoria the ICM model has evolved rapidly and currently comprises relatively large Catchment Management Authorities (CMAs) which have a strong legislative basis and the ability to raise levies for river management.

Overall, there is also a national effort to develop farm and catchment-based indicators of catchment management effectiveness, which requires substantial investment. The principal issues facing 'catchment management' initially relate to resolving local questions, including integration with local government planning schemes. This is usually dealt with during the development of catchment strategies or plans, a process that can take three or more years to achieve. Once plans have been agreed, the next issue is implementation, which involves resourcing as well as the assimilation and application of 'best knowledge'. Clearly, ICM is at a critical phase in Australia and requires strong government support if it is to yield the expected benefits.

Natural Heritage Trust

One of the most successful government initiatives has been the National Landcare Programme (NLP), funded under the Federal Government's Natural Heritage Trust initiative. This is an attempt to inform and mobilize the rural community to take sustainable land and river care into their own hands. Growth of the movement has been phenomenal, with over 2400 community-based Landcare groups now established. The NLP promotes partnerships between the community, industry and government in the management of natural resources. It provides funds either directly to community groups for joint Commonwealth–State initiatives, or for combined government and community efforts (Russell, 1996).

Much of the initial focus of these groups has been in rural demonstration projects as well as in the preparation of local and catchment-based plans to meet environmental as well as other objectives. The critical issue now is large-scale, effective implementation and 'delivery'. This will require substantial financial resources that need to be efficiently and effectively allocated, with a sound understanding of the investment needs, as well as effective national coordination. There is currently a strong national political will to 'fix things', i.e. to solve some of the salient land and water environmental problems. This will require careful consideration of appropriate investment in remediation and management practices if effective, long-lasting environmental goals are to be achieved.

AUSTRALIAN EXAMPLE: THE MURRAY–DARLING BASIN

The MDBC is a major inter-governmental agency, with representatives of five state governments (NSW, Queensland, Victoria, South Australia, Australian Capital Territory), the Commonwealth, and local government. It has established a Natural Resource Management Strategy for the basin including the Sustainable Rivers Program. The Murray–Darling is Australia's largest river system (>3600 km), covering one-seventh of the continent, and yielding in excess of A\$10 billion of the nation's agricultural produce, of which A\$3 billion is from irrigation activities. The MDBC is a unique experiment in large-scale ICM and, to date, has a sound record in addressing its very substantial riverine management and conservation issues.

Water diversions and flow regulation in the Murray–Darling Basin have increased significantly since the 1950s with significant changes to its rivers' flow regimes (e.g. Thoms and Sheldon, 2000). Median annual flows from the basin to the sea are only 21% of those prior to development. Studies have highlighted that many rivers in the basin are displaying signs of stress because of changes to flow regimes (e.g. Thoms and

Walker, 1992; Walker *et al.*, 1992; Sheldon, 1994; Harris and Gerhrke, 1997). In response to these studies and the outcomes of the Murray–Darling Basin Ministerial Council's audit of water use in the basin a Cap on water diversions was implemented in 1997. The Cap was set to the volume of water that would have been diverted under 1993/94 levels of development. It is seen as the first step towards a balance between consumptive and non-consumptive users. From its inception, the Cap has had the support of the governments of each of the Basin states and of the Commonwealth. Each state has dedicated resources to implement the Cap within their jurisdictions. However, individual states have responded differently in terms of water management strategies. For example, in NSW, compliance is undertaken on a valley by valley basis and the aim is to keep long-term average diversions under the Cap with monitoring to be approached from a long-term modelling perspective. Environmental water allocation rules have not been specifically designed as a Cap management strategy in this state. In Queensland the development of a Cap on diversions is based on a comprehensive Water Allocation Management Plan (WAMP). Despite the Cap being in place for three years no WAMPs have been completed for rivers draining the Queensland section of the basin.

In order to monitor compliance to the Cap an Independent Audit Group (IAG) was established. This group reported a variable compliance by individual states during the 1997/98 water year. Some states have complied with the Cap; others have failed totally. However, in one state, some individual valleys were shown to be Cap compliant whilst a number had diversions well above 1993/94 levels. As a result of their investigations the IAG has suggested a number of changes to Cap implementation rules. These include supplementing the reliance on computer models with other measures in order to assist in determining whether or not there has been an increase in diversions; the creation of an independent regulator to assess all information on compliance issues; and the continued development of an open and transparent process for reporting on water diversions.

NEW ZEALAND POLICIES AND INITIATIVES

The framework for water management in New Zealand received detailed attention during the mid 1980s when the government embarked on a review of the plethora of laws relating to air, land and water management. This produced the Resource Management Act (RMA)

1990 under which air, water, and land resources (excluding minerals) are managed by regional and district councils whose areas are based on catchment boundaries. The role of central government is greatly reduced, with a high level of devolution to the regional and district councils. The Minister for the Environment has the ability to set national standards and to develop national policy statements, but to date none has been developed, although a number are being worked on. This framework has evolved from a system of catchment boards set up in the early 1940s primarily for flood control. In the 1950s and 1960s, these were broadened to include water quality, and the concept of managing waters for multiple use was introduced in the Water and Soil Conservation Act of 1967. The 1970s saw increasing public concern about damming of 'icon rivers' under this scheme, and the Wild and Scenic Rivers Act of 1981 represented a move away from the multiple-use doctrine. The RMA attempts to integrate natural resource management, to incorporate the values of the indigenous Maori people, and to use an effects-based approach, rather than a proscriptive approach, for sustainable management. One of the first tasks of regional councils under the RMA was to develop regional policy statements, setting a broad policy for the development of regional and district plans that contain the more detailed framework of policy and rules for resource management. The first round of regional-plan development is still in progress.

The RMA promotes sustainable management of natural and physical resources: it provides for 'safeguarding the life-supporting capacity of air, water, soil and ecosystems [as well as] avoiding, remedying, or mitigating any adverse effects of activities on the environment'. The RMA stresses an integrated approach to the management of all natural resources and recognizes the preservation of the natural character of rivers and their margins as a matter of national importance. Discharges to rivers are not permitted if they are likely to cause the production of surface films, scums, a conspicuous change in colour or clarity, an objectionable smell, or significant adverse effects on aquatic life, and 11 national water quality classes are established, based on use. The Department of Conservation administers 23 Acts of Parliament that provide it with responsibility for, amongst others, indigenous and recreational freshwater fisheries and their habitats, advocacy for the conservation of aquatic life in general, management of the diadromous 'whitebait' (Galaxiidae) fishery, fish passage and the management of noxious fish (Department of Conservation, 1996).

Information database

AUSTRALIAN AQUATIC RESEARCH

The research and development scene for rivers in Australia is a little better than the management scenario as there is a national body covering water and river research, the Land and Water Resources Research and Development Corporation (LWRRDC). The portfolio for this body is broad, and it administers and funds (with partner agencies) a number of national programmes focused on rivers, covering river health, eutrophication, riparian management, pesticides and wetlands. Each programme has broad-based objectives and may contain 20–50 individual projects, each managed and coordinated in a rigorous programme management framework. The LWRRDC, whilst demonstrating national leadership, is only a small provider of funds overall, with other key sources being the CSIRO, the Cooperative Research Centres (CRCs), the Australian Research Council (ARC), the MDBC and individual federal and state agencies. The private sector (e.g. mining, farmers) generally only plays a role in their own local environments, often to meet regulatory requirements.

AQUATIC RESEARCH IN NEW ZEALAND

New Zealand has not had a national body specifically overseeing aquatic research since the National Water and Soil Conservation Authority was disbanded in the 1980s. Since the government science reforms of the early 1990s, the Ministry for Research Science and Technology has set the broad research agenda for water research that is funded through the Public Good Science Fund administered by the Foundation for Research, Science and Technology. The main river research providers are universities and Crown Research Institutes, notably the National Institute of Water and Atmospheric Research. Funding is also available from the Ministry for the Environment for developing management techniques and monitoring tools for river systems (e.g. for biomonitoring, flow management, river classification), and there has been some input to river research programmes from industry (e.g. forestry and fertilizer companies), local bodies and the Department of Conservation (e.g. for development of riparian management guidelines, Collier *et al.*, 1995).

Conservation of aquatic systems

APPLICATION

The regulatory side of river management in Australia falls principally to the state environmental protection agencies. Nationally, principles and guidelines are espoused, but key activities such as land-use zoning, water allocation, industrial-discharge licensing and pollution prosecution, are undertaken at the state or local government level. The state agencies often use a combination of approaches, from prosecution of offences committed under their Acts, to encouragement of self-management, such as the adoption of best practices by industry. Some 'powers' are devolved to local government, but they frequently do not have the resources to implement major river initiatives.

Similarly, in New Zealand, responsibility for the direct management of water rests mainly on Regional Councils. Regional Plans set out the circumstances in which a resource-use consent is required to dam, abstract or discharge to a river, or to extract gravels from the bed, and councils are able to stipulate provisions associated with any consent. Minor uses (e.g. small farm dams) are usually managed as 'permitted uses', with a generic set of controls set out in Regional Plans.

IN SITU CONSERVATION (PROTECTED AREAS)

There is little direct activity in reserving river conservation areas at the national level in Australia. The Register of the National Estate is an inventory of natural and cultural heritage places but has no control over the activities of states, local governments, or private landowners. Similarly, the National Wilderness Inventory is simply a database to assist managers to monitor wilderness loss or impacts. The Federal Government can, however, intervene in river development issues when they are deemed to be of national significance (e.g. following nomination for World Heritage Listing) as it did to prevent the damming of the Franklin River in Tasmania. Reservation of National Parks and Reserves is principally the responsibility of state governments and is based on a wide range of natural features. In some cases the rivers in these reserves are severely degraded upstream of the reserve boundaries, through land, or other development.

In New Zealand, there is also little direct activity at the national level in preserving river conservation areas, and most protection is provided by national parks and reserves whose status also confers protection to the rivers passing through them. This network encompasses 13 National Parks covering almost 3×10^6 ha (including Whanganui National Park which was established mainly to protect the unique features of a large river), 20 Forest Parks (1.8×10^6 ha), 3500 reserves and 61 750 ha of private land protected by covenants (Department of Conservation, 1996). Water Conservation Orders, which are administered by the Ministry for the Environment, have previously been a main means of conserving rivers, that have outstanding characteristics either regionally or nationally. The process has proved cumbersome, however, and to date only six have been gazetted. The main tool for river protection in the future will probably be regional freshwater plans prepared under the RMA. In New Zealand, most National Parks are in headwater areas and the lower parts of river systems are poorly represented in the protected areas network (Collier and McColl, 1992).

EX SITU CONSERVATION

Internationally, Australia is party to a number of treaties and agreements relevant to the management and conservation of rivers and wetlands. For example, Australia is a Contracting Party to the Ramsar Convention, the broad aims of which are to halt the worldwide loss of wetlands and to conserve, through wise use and management, those that remain. Australia has 49 listed Ramsar Sites, mostly scattered around the coast. There are, however, a few significant sites on inland river systems such as the Coongie Lakes on Cooper Creek in northern South Australia; the Riverland region of the lower River Murray; Currawinya Lakes on the Paroo River, western MDB; the large red gum forests of the middle River Murray; the Ord River Floodplain in Western Australia; and the Kakadu Wetlands in the Northern Territory.

Australia is also party to a number of international agreements relating to migratory birds. Most notable of these are JAMBA, the Japan–Australia Migratory Bird Agreement, and CAMBA, the China–Australia Migratory Bird Agreement. These agreements recognize that migratory birds require protection for all three phases of the annual cycle: breeding, migration and non-breeding. In Australia, the birds covered by JAMBA and CAMBA agreements include many species that feed in the shallow waters of inland wetlands and floodplains. Thus, the management of Australia's rivers and wetlands has direct implications for these international agreements.

In 1974, Australia became one of the first countries to ratify the International Convention for the Protection of the World Cultural and Natural Heritage (The World Heritage Convention). The Convention aims to ensure international cooperation to safeguard places of 'outstanding universal value'. At present Australia has 13 listed World Heritage Areas (WHAs). Of these, only four would contain representatives of streams or river systems within Australia: the Wet Tropics of Queensland WHA, the Tasmanian Wilderness WHA, the Central Eastern Rainforest WHA and the Kakadu WHA. Although these contain a range of stream types, all are representatives of wet-temperate or tropical systems. The lower floodplain regions of the Diamantina River and Cooper Creek as well as Lake Eyre itself were nominated for World Heritage listing but the nomination was never put forward to the Convention owing to local community opposition. If successful, this listing would have added to the World Heritage List a unique and relatively untouched arid floodplain river-wetland system, the ecology of which is unparalleled elsewhere in the world (Morton *et al.*, 1995).

RESTORATION AND REHABILITATION PROGRAMMES

Australia is now increasingly focusing its river activities on the task of rehabilitation, following an emphasis in the early to mid 1990s on river condition assessment. The key pursuits in river rehabilitation are to:

- improve water quality
- protect and rehabilitate riparian areas
- manipulate flow regimes for environmental benefits
- control erosion
- enhance in-stream habitat

In a recent national review of the research and development needs for river restoration (Rutherford *et al.*, 1998), 12 general principles were identified:

- Stream rehabilitation is about attempting to restore the physical, hydrological, hydraulic and biological complexity of stream systems in an effort to develop sustainable ecosystems – it is not just about having stable, attractive streams.

- It is unlikely that any Australian streams can be restored to their original condition.
- Protecting the remaining pieces of stream that are in good condition should always take precedence over attempts to rehabilitate damaged reaches. This principle applies at a national scale meaning that resources may need to be expended in remote regions where there is no apparent human benefit.
- Stream rehabilitation is a sub-set of catchment management. Streams cannot ultimately be rehabilitated in isolation from improvement in the condition of the catchment.
- It has taken nearly two centuries to reduce streams to their present parlous state – it will take at least as long to rehabilitate them: stream rehabilitation will take decades, rather than years, to accomplish.
- Australia's comprehensive stream rehabilitation effort should concentrate on smaller streams in which there is a good chance of success and in which variables can be controlled and measured. Rehabilitation of large streams is likely to be expensive and problematic such that efforts should be restricted to flow and to water quality issues. Obviously there will be exceptions to this principle, where simple changes to a large stream can produce great ecological benefits, or where the publicity value of the project may outweigh the technical challenges.
- The main reasons for the failure of stream rehabilitation projects are poor definition of objectives, the wrong diagnosis of the real problem, or the failure to consider the catchment. The easiest part of a stream rehabilitation project is the selection of the tools to fix the problems in the stream (e.g. particular structures), yet this is often the issue that attracts most research and development attention. Again, often the main impediment to stream rehabilitation can be social and political factors, rather than poor understanding of what to do.
- Stream rehabilitation efforts should first target the 'limiting variable' for recovery. In most cases variables should be dealt with in the following order: water quality; hydrology; hydraulics; interaction with the floodplain; and habitat condition.
- All stream rehabilitation projects must have some evaluation of success, and a few projects must be rigorously examined in order to achieve a high level of confidence in the outcomes. Few projects have been evaluated in the past.
- One of the key advances that can be made in stream rehabilitation practice in Australia is the development of well-evaluated 'flagship' projects that attempt to rehabilitate long reaches of stream. These projects must be a focus of research.
- Australian streams and their biotas are different in many ways from Northern Hemisphere streams. Physical differences, as well as political and resource differences, mean that international experience must be critically evaluated.
- Stream rehabilitation requires input from many previously disparate disciplines and groups. Because stream rehabilitation is a young and evolving activity, communication and interaction between these groups is poor.

To initiate a nationally coordinated effort in river restoration, a National Rivers Consortium of key federal agencies has been formed, involving LWRRDC, CSIRO, MDBC, CRCs for Catchment Hydrology and Freshwater Ecology, and Agriculture, Fisheries and Forestry – Australia.

The Consortium's vision is for there to be continuous improvement in the condition of rivers throughout Australia not constrained by lack of understanding of physical, ecological or social processes. The Consortium will achieve its vision by leading the acquisition, marshalling and uptake of knowledge to protect, restore and to enhance Australia's rivers. A number of preliminary activities include a review of river legislation, the development of a river restoration framework, a review of training and education needs, and the development of methods to identify and to protect high value rivers or river reaches. In the past large sums of money have been expended on river improvements but these have not been conducted in an ecological context, often increasing the degradation of riverine habitat. A major challenge of the new initiative is to turn these old practices around.

Improving river water quality has been a major objective in Australia for the past two decades. Whilst more than 1800 water quality programmes exist at a cost of some A\$100 million yr^{-1}, there has been little national evaluation of improvements or benefits (Aquatec, 1995).

Addressing increasing land and stream salinization in southern Australia has been a preoccupation since the late 1970s. Salinization is the result of broad-scale land-use change and requires an equally broad-scale response. The largest rehabilitation example is the establishment of 6000 ha of riparian buffer strips in the Collie Catchment, Western Australia, at a cost of A$10 million, to control increasing river salinity (Schofield, 1989). This, and related work, has led to a national ethos of tree planting for environmental and riverine enhancement (Schofield *et al.*, 1989; Schofield, 1992). More recently, LWRRDC has led a national programme on riparian land management and has produced national guidelines suitable for professionals and practitioners.

Strategies for controlling point-source nutrient discharges to streams have frequently been implemented (e.g. from sewage treatment plants and storm waters), but less frequently for diffuse-catchment sources. The former is predicated on achievement and/ or compliance with national or state-based water quality guidelines for ecosystem protection, or for recreational values. In-stream rehabilitation measures have focused on providing fish ladders for migration, the provision of environmental flows and, recently, investigating hydraulic manipulation for improved fish habitat and stream stability. To date, such programmes have been limited.

Interest in riparian management as a restoration tool has recently increased in New Zealand due to the statutory importance of river margins provided for in the RMA. Guidelines have been produced to assist with planning and implementing riparian management schemes (e.g. Collier *et al.*, 1995). The potential for riparian management to mitigate some adverse impacts, particularly of agricultural development, has also been demonstrated in some studies (Ngongotaha, Whangamata Stream: Howard-Williams and Pickmere, 1994; Williamson *et al.*, 1992, 1996). Some private landholders and streamcare groups, often facilitated by Regional Council staff, have begun to take the initiative in fencing and replanting river margins. Other activities have focused on restoring the passage for diadromous native fish over culverts and dams, often using wetted plastic pipes with brush or gravel linings (Mitchell *et al.*, 1984), and large-scale rehabilitation of floodplains and river-beds in some southern South Island rivers, to restore habitat for some threatened native wading birds (Rawlings, 1993).

The future

The inland rivers of Australia are principally at risk from agriculture, including dryland, irrigation and pastoral. Since most of the high potential farming land (and a good deal of low potential farming land!) has already been developed, and as most of Australia's rivers are already seriously affected, major future risks are associated with further farming development in pristine areas. Examples include proposed cotton-growing developments in the catchments of the Cooper Creek (Walker *et al.*, 1995) and Paroo and Warrego rivers. Inappropriate developments in such areas pose significant environmental risks to ephemeral aquatic environments of great national and international significance, such as Lake Eyre.

Future land 'development' is mostly targeted in Queensland where, for example, sugar cane (Australia is already the world's largest exporter) is expected to increase by a further 40%, partly at the expense of floodplains and wetlands, and associated with plans for a number of new dams that will affect in-stream flows. Slow but significant rural and irrigation development is also commencing in the tropical regions of the Northern Territory.

Urbanization, tourism and recreation may also have some riverine impacts, particularly in coastal NSW and Queensland. The coastal emphasis of these activities, however, does mitigate the riverine impacts. Mining, particularly uranium and bauxite, can be expected to have localized impacts in high conservation areas (e.g. Kakadu, Jarrah Forest), although these are closely scrutinized by environmental protection agencies, and high quality rehabilitation is required.

Some of the current threats to river conservation in Australia, such as salinity, have proved intractable and will worsen. Increasing stream salinity has principally been considered as a water-supply problem and very little is known about its ecological impact. However, improved flow management may provide some improvements in river condition. Control of sediment erosion and diffuse nutrient sources remains difficult and expensive in both countries. In Australia, any benefits will be long-term due to nutrient storage in sediments of the long, low-gradient rivers, whereas New Zealand's shorter, steeper rivers are expected to show more rapid responses to improved diffuse-source management.

One of the primary risks posed to Australia's rivers is the recent trend towards privatization of water resource management. Economic imperatives following privatization may reduce investment in appropriate environmental management (including remediation, restoration and monitoring) of aquatic ecosystems, unless state and national policies are strong, coordinated and are effectively enacted.

Sediment input to streams and rivers continues to be a problem in many areas of New Zealand, particularly on the east coast of the North Island, where clearance of forest from catchments with soft sedimentary parent material has accelerated large-scale erosion and sedimentation (Hicks, 1991). Efforts to control erosion and, hence, sediment input, and downstream water quality are currently focusing on land-management techniques, principally converting pastoral land to production forestry. Although this will reduce erosion risk for much of the growing cycle (Hicks, 1991), this risk will increase during crop harvesting.

The current expansion of dairy farming into traditional sheep farming areas, particularly in the South Island, represents a threat to streams due (for example) to increased point-source discharge of milking-shed effluent and greater stream-bank damage by cows than sheep (Williamson *et al.*, 1992). In addition, increased urban-fringe development and land-use intensification is likely further to degrade agricultural streams and rivers. Greater demand for water, to increase pastoral production in areas with low summer rainfall (e.g. Northland), also represents a future threat if not appropriately managed.

In both Australia and New Zealand, exotic pests are likely to assume greater importance and provide enormous challenges for their control and possible eradication. Although the destructive effect of the spread of European carp (*Cyprinus carpio*) in Australia has been well documented, in New Zealand the spread of koi carp (*Cyprinus carpio*) also has the potential to become a significant future threat to river systems, especially in the North Island where these species are already well established (McDowall, 1990). In New Zealand, the recent Biosecurities Act (1993) has reduced the potential for intentional introductions of exotic species, as evidenced by the decision to destroy stocks of marron (*Cherax quadricarinatus*) and channel catfish (*Ictalurus punctatus*) that had been imported for aquaculture trials under quarantine (e.g. Townsend and Winterbourn, 1992).

Other impacts on rivers, such as acid drainage and climate change, may come to the fore and yet unforeseen impacts will arise. Protection of remnant pristine rivers must be a high priority, whilst moves to reduce the rate of degradation, or to provide some improvement in rivers in developed areas, will require sustained, expensive, well-informed and diligent efforts. Increased emphasis will be given to the protection of high-value rivers or river reaches; rehabilitation of degraded but recoverable rivers; management of catchment-to-ocean interactions; address of critical areas of past failure, including legislation, institutional, market and policy failures;

and support for community efforts, through provision of information, training and implementation frameworks, and tools (Schofield and Price, 1999).

References

Aquatec 1995. *Water Quality Monitoring in Australia*. Report to the Environment Protection Agency, Canberra.

Arthington, A.H. 1998. *Comparative Evaluation of Environmental Flow Assessment Techniques: Review of Holistic Methodologies*. Land and Water Resources Research and Development Corporation. Occasional Paper 26/98.

Arthington, A.H. and Zalucki, J.M. (eds) 1998. *Comparative Evaluation of Environmental Flow Assessment Techniques: Review of Methods*. Land and Water Resources Research and Development Corporation. Occasional Paper 27/98.

Arthington, A.H., Brizga, S.O. and Kennard, M.J. 1998. *Comparative Evaluation of Environmental Flow Assessment Techniques: Best Practice Framework*. Land and Water Resources Research and Development Corporation. Occasional Paper 25/98.

Australian National Committee on Large Dams (ANCOLD) 1990. *Register of Large Dams in Australia*. Hobart.

Australian Water Resources Council 1987. *1985 Review of Australia's Resources and Water Use*. AGPS, Canberra.

Bailey, P. and Warwick, N. 1998. Salinity – a threat to our streams and wetlands. *Water Journal*, May/June. Australian Waste Water Association: 8–12.

Banens, R.J. and Davis, J.R. 1998. Comprehensive approaches to eutrophication management: the Australian example. *Water Science and Technology* 37:217–225.

Biggs, B.J.F. 1990. Periphyton communities and their environments in New Zealand rivers. *New Zealand Journal of Marine and Freshwater Research* 24:367–386.

Biggs, B.J.F. and Price, G.M. 1987. A survey of filamentous algal proliferations in New Zealand rivers. *New Zealand Journal of Marine and Freshwater Research* 21:175–191.

Boul, H.L. 1995. DDT residues in the environment – a review with a New Zealand perspective. *New Zealand Journal of Agricultural Research* 38:257–277.

Bowler, D.G. 1980. *The Drainage of Wet Soils*. Hodder & Stoughton, Auckland.

Bowling, L.C. and Baker, P.D. 1996. Major cyanobacterial blooms in the Barwon–Darling River, Australia, in 1991 and underlying limnological conditions. *Marine and Freshwater Research* 47:643–657.

Bunn, S.E., Pusey, B.J. and Price, P. 1993. *Ecology and Management of Riparian Zones in Australia*. Land and Water Resources Research and Development Corporation, Occasional Paper Series 05/93.

Carnahan, J.A. 1986. Vegetation. In: Jeans, D.N. (ed.) *The Natural Environment – Australia A Geography*. Sydney University Press, Sydney, 223–259.

Chapman, J. 1999. Laboratory ecotoxicology studies and implications for key pesticides. In: Schofield, N.J. and Edge, V.E. (eds) *Proceedings of Conference on Minimising the Impact of Pesticides on the Riverine Environment: Key Findings from Research with the Cotton Industry*. Land and Water Resources Research and Development Corporation, Occasional Paper 23/98, 62–67.

Choy, S. 1999. Assessing river health in Queensland. *Rivers for the Future Magazine* 8:15–17.

Close, M.E. 1996. Survey of pesticides in New Zealand groundwaters, 1994. *New Zealand Journal of Marine and Freshwater Research* 30:455–461.

Collier, K.J. and McColl, R.H.S. 1992. Assessing the natural value of New Zealand rivers. In: Boon., P.J., Calow, P. and Petts, G.E. (eds) *River Conservation and Management*. John Wiley, Chichester, 195–211.

Collier, K.J., Cooper, B., Davies-Colley, R., Rutherford, J.C., Smith, C.M. and Williamson, R.B. 1995. *Managing Riparian Zones: A Contribution to Protecting New Zealand's Rivers and Streams* (Vol. 1 *Concepts*; Vol. 2 *Guidelines*). Department of Conservation, Wellington.

Collier, K.J., Baillie, B., Bowman, E., Halliday, J., Quinn, J. and Smith, B. 1997. Is wood in streams a dammed nuisance? *Water and Atmosphere* 5:17–21.

Commonwealth Scientific and Industrial Research Organization (CSIRO) 1990. *Australia's Environment and its Natural Resources, An Outlook*. Canberra.

Cooper, B. and Riley, G.T. 1996. Storm transport of pollutants from dryland agriculture. In: Hunter, H.M., Eyles, A.G. and Rayment, G.E. (eds) *Downstream Effects of Land Use*. Department of Natural Resources, Queensland, 173–175.

Cooper, B., Jones, G., Burch, M. and Davis, R. 1996. Agricultural chemicals and algal toxins – emerging water quality issues. In: *Managing Australia's Inland Waters*. Prime Minister's Science and Engineering Council, Canberra, 89–110.

Cooperative Research Centre for Water Quality and Treatment (CRCWQT) 1998. *Algal Toxins Research Directions*. Proceedings of a LWRRDC/MDBC/WSAA workshop, 23–24 March 1998. Occasional Paper No. 2.

Council of Austrialian Goverments (COAG) 1994. *Communique from the COAG Meeting of 25 February 1994*. Canberra.

Davies, B.R., Thoms, M.C. and Meador, M. 1992. An assessment of the ecological impacts of inter-basin water transfers and their hidden threats to river basin integrity and conservation. *Aquatic Conservation: Marine and Freshwater Ecosystems* 2:325–349.

Davies, P.E., Cook, L.S.J. and Barton, J.L. 1994. Atrazine herbicide contamination of Tasmanian streams: sources, concentrations and effects on biota. *Australian Journal of Marine and Freshwater Research* 45:209–226.

Davies, P.E., Mitchell, N. and Barmuta, L. 1996. *The Impact of Historical Mining Operations at Mount Lyell on the Water Quality and Biological Health of the King and Queen River Catchments, Western Tasmania*. Mount Lyell Remediation Research and Demonstration Programme. Supervising Scientist, Report 118. Supervising Scientist, Environment Australia, Canberra.

Davies-Colley, R.J. 1997. Stream channels are narrower in pasture than in forest. *New Zealand Journal of Marine and Freshwater Research* 31:599–608.

Davies-Colley, R.J. and Quinn, J.M. 1998. Stream lighting in five regions of the North Island, New Zealand: control by channel size and riparian vegetation. *New Zealand Journal of Marine and Freshwater Research* 32:591–605.

Davies-Colley, R.J., Hickey, C.W., Quinn, J.M. and Ryan, P.A. 1992. Effects of clay discharges on streams. 1. Optical properties and epilithon. *Hydrobiologia* 248:215–234.

Davis, J.R. and Young, W.J. 1998. A decision support system for planning environmental flows. In: Loucks, D.P. (ed.) *Restoration of Degraded River Systems: Challenges, Issues and Experiences*. Kluwer Academic Publishers, Dordrecht, 357–376.

Department of Conservation (DoC) 1996. *Conservation in New Zealand – A Guide to the Department*. Wellington.

Department of Natural Resources (DNR) 1997. *Water Infrastructure Planning and Development 1997–98 to 2001–02 Implementation Plan, July 1997*. Queensland Government, Brisbane, 18–26.

Devlin, M. and Schofield, N.J. 1999. River management to protect the Great Barrier Reef. In: *Proceedings of the Second International Conference on Natural Channel Systems*, 1–4 March, Niagara Falls, Canada.

Donnelly, T.H., Olley, J.M., Murray, A.S., Caitcheon, G.G., Olive, L.J. and Wallbrink, P.J. 1996. Phosphate sources and algal blooms in Australian catchments. In: Hunter, H.M., Eyles, A.G. and Rayment, G.E. (eds) *Downstream Effects of Land Use*. Department of Natural Resources, Queensland, 189–194.

Dons, A. 1987. Hydrology and sediment regime of a pasture, native forest, and pine forest catchment in the Central North Island, New Zealand. *New Zealand Journal of Forestry Science* 17:161–178.

East, T.J., Cull, R.F., Murray, A.S. and Duggan, K. 1988. Fluvial dispersion of radioactive mill tailings in the seasonally-wet tropics, Northern Australia. In: Warner, R.F. (ed.) *Fluvial Geomorphology of Australia*. Academic Press, Sydney, 303–322.

Fahey, B.D. and Rowe, L.K. 1992. Land-use impacts. In: Mosley, M.P. (ed.) *Waters of New Zealand*. New Zealand Hydrological Society Inc., Wellington, 265–284.

Ghassemi, F., Jakeman, A.J. and Nix, H.A. 1995. Australia. In: *Salinisation of Land and Water Resources – Human Causes, Extent, Management and Case Studies*. University of New South Wales/CAB International Publishers, Sydney, 143–212.

Griffiths, G.A. 1981. Some suspended sediment yields from South Island catchments, New Zealand. *Water Resources Bulletin* 17:662–671.

Griffiths, G.A. 1982. Spatial and temporal variability in suspended sediment yields of North Island basins, New Zealand. *Water Resources Bulletin* 18:575–584.

Griffiths, G.A. and Glasby, G.P. 1985. Input of river-derived sediment to the New Zealand continental shelf: I. Mass. *Estuarine Coastal and Shelf Science* 21:773–787.

Gutteridge, Haskins and Davey (GHD) 1992. *An Investigation of Nutrient Pollution in the Murray–Darling River System*. Report for the Murray–Darling Basin Commission.

Harding, J.S. 1995. Lotic aquatic ecoregions and land use influences on benthic stream communities. Unpublished PhD Thesis, University of Canterbury, Canterbury, New Zealand.

Harding, J.S. and Winterbourn, M.J. 1995. Effects of contrasting land use on physico-chemical conditions and benthic assemblages of streams in a Canterbury (South Island, New Zealand) river system. *New Zealand Journal of Marine and Freshwater Research* 29:479–492.

Harris, G.P. 1994. *Nutrient Loadings and Algal Blooms in Australian Waters – A Discussion Paper*. Land and Water Resources Research and Development Corporation. Occasional Paper No. 12/94.

Harris, J.H. 1995. Carp: the prospects for control? *Water*, May/June, 25–28.

Harris, J. and Gehrke, P.C. (eds) 1997. *Fish and Rivers in Stress: The NSW Rivers Survey*. New South Wales Fisheries Office of Conservation and the Cooperative Research Centre for Freshwater Ecology, Sydney.

Hart, B.T., Bailey, P., Edwards, R., Hortle, K., James, K., McMahon, A., Meredith, C. and Swadling, K. 1991. A review of the salt sensitivity of the Australian freshwater biota. *Hydrobiologia* 210:105–144.

Hicks, D.L. 1991. Erosion under pasture, pine plantations, scrub and indigenous forest: a comparison from Cyclone Bola. *New Zealand Journal of Forestry* 36:21–22.

Hickey, C.W. and Clements, W.H. 1998. Effects of heavy metals on benthic macroinvertebrate communities in New Zealand streams. *Environmental Toxicology and Chemistry* 17:2338–2346.

Hickey, C.W., Buckland, S.J., Hannah, D.J., Roper, D.S. and Stuben, K. 1997. Polychlorinated buphenyls and organochlorine pesticides in the freshwater mussel *Hyridella menziesi* from the Waikato River, New Zealand. *Bulletin of Environmental Contamination and Toxicology* 59:106–112.

Howard-Williams, C. and Pickmere, S. 1994. Long-term vegetation and water quality changes associated with the restoration of a pasture stream. In: Collier, K.J. (ed.) *Restoration of Aquatic Habitats*. New Zealand Limnological Society, Wellington, 93–109.

Howard-Williams, C., Clayton, J.S., Coffey, B.T. and Johnstone, I.M. 1987. Macrophyte invasions. In: Viner, A.B. (ed.) *Inland Waters of New Zealand*. Department of Scientific and Industrial Research, Wellington, 307–331.

Hunter, H.M., Walton, R.S. and Russell, D.J. 1996. Contemporary water quality in the Johnstone River Catchment. In: Hunter, H.M., Eyles, A.G. and Rayment, G.E. (eds) *Downstream Effects of Land Use*. Department of Natural Resources, Queensland, 339–345.

Huser, B.A., Prowse, H.J., McBride, G.B. and Timperley, M.H. 1994. Water quality monitoring in the Waikato River, North Island, New Zealand. In: *Water Environment Federation 67th Annual Conference and Exposition*. Chicago, Illinois, 457–468.

Jennings, J.N. and Mabbutt, J.A. 1986. Physiographic outlines and regions. In: Jeans, D.N. (ed.) *The Natural Environment – Australia A Geography*. Sydney University Press, Sydney, 48–96.

Johnstone, P., Fabbro, L.D., Blake, T., Robinson, D., Noble, R.M., Rummenie, S.K. and Duivenvoorden, L.J. 1996. Land use, water use and blue-green algae in Australia. In: Hunter, H.M., Eyles, A.G. and Rayment, G.E. (eds) *Downstream Effects of Land Use*. Department of Natural Resources, Queensland, 141–149.

Jowett, I. 1984. Sedimentation in New Zealand hydroelectric schemes. *Water International* 9:172–176.

Kirkby, C.A.K., Smythe, L.J., Cox, J.W. and Chittleborough, D.J. 1997. Phosphorus movement down a toposequence from a landscape with texture-contrast soils. *Australian Journal of Soil Research* 35:399–417.

Knighton, A.D. 1989. River adjustment to changes in sediment load: The effects of tin mining on the Ringarooma River, Tasmania 1875–1984. *Earth Surface Processes and Landforms* 14:333–359.

Land and Water Resources Research and Development Corporation (LWRRDC) 1996. *National Eutrophication Management Program, Program Plan 1995–2000*. Canberra.

Land and Water Resources Research and Development Corporation (LWRRDC) 1998. *Phosphorus in the Landscape: Diffuse Sources to Surface Waters*. Occasional Paper No. 16/98.

Leonard, A.W., Hyne, R.V., Lim, R.P. and Chapman, J.C. 2000. Relationship between endosulfan and common macroinvertebrate populations in the cotton growing region of the Namoi River, Australia. *Ecotoxicology and the Environment* (in press).

Maclaren, J.P. 1996. *Environmental Effects of Planted Forests in New Zealand: The Implications of Continued Afforestation of Pasture*. New Zealand Forest Research Institute. Rotorua.

McDowall, R.M. 1990. *New Zealand Freshwater Fishes, a Natural History Guide*. Heinemann Reed MAF, Wellington.

McEwen, M. (ed.) 1987. *Ecological Regions and Districts of New Zealand*. Biological Resources Centre publication No. 5. Department of Conservation, Wellington.

Ministry for the Environment 1992. *Guidelines for the Control of Undesirable Biological Growths*. Resource Management Water Quality Guidelines No. 1. Wellington.

Mitchell, C.P., Huggard, G. and Davenport, M. 1984. Home grown fish pass helps eels get by. *Freshwater Catch (N.Z.)* 24:21–22.

Mitchell, D.S. and Rogers, K.H. 1985. Seasonality of aquatic macrophytes in Southern Hemisphere inland waters. *Hydrobiologia* 125:137–150.

Morton, S.R., Doherty, M.D. and Barker, R.D. 1995. *Natural*

Heritage Values of the Lake Eyre Basin in South Australia: World Heritage Assessment. Report to the World Heritage Unit, Commonwealth Department of the Environment, Sport and Territories.

Munro, C.H. 1974. *Australian Water Resources and their Development.* Angus & Robertson, Sydney.

Murray–Darling Basin Commission (MDBC) 1995. *Water Use and Healthy Rivers – Working Towards a Balance. An Audit of Water Use in the Murray–Darling Basin.* Canberra.

Muschal, M. and Cooper, B. 1999. Regional level monitoring of pesticides and their behaviour in rivers. In: Schofield, N.J. and Edge, V.E. (eds) *Proceedings of a Conference on Minimising the Impact of Pesticides on the Riverine Environment: Key Findings from Research with the Cotton Industry.* Land and Water Resources Research and Development Corporation. Occasional Paper 23/98, 65–74.

National River Health Program (NRHP) 1993. The National River Health Program. Brochure available from Land and Water Resources Research and Development Corporation, Canberra.

Norris, R. 1999. Bugs and computers – used to piece together a picture of river health. *Rivers for the Future Magazine* 8:12–13.

Olive, L.J., Olley, J.M., Murray, A.S. and Wallbrink, P.J. 1995. Variations in sediment transport at a variety of temporal scales in the Murrumbidgee River N.S.W., Australia. International Association of Hydrological Sciences, 226:275–284.

Penny, S.F. 1987. Stream biology of three Coromandel catchments containing past mines. In: Livingstone, M.E. (ed.) *Preliminary Studies on the Effects of Past Mining on the Aquatic Environment, Coromandel Peninsula.* National Water and Soil Conservation Authority, Wellington, 49–117.

Prime Minister's Science, Engineering and Innovation Council (PMSEIC) 1999. *Dryland Salinity and its Impact on Rural Industries and the Landscape.* Occasional Paper, No. 1.

Quinn, J.M. and Gilliland, B.W. 1989. The Manawatu River cleanup: has it worked? *Transactions of the Institute of Professional Engineers of New Zealand* 16:22–26.

Quinn, J.M. and Hickey, C.W. 1993. Effects of sewage waste stabilization lagoon effluent on stream invertebrates. *Journal of Aquatic Ecosystem Health* 2:205–219.

Quinn, J.M., Davies-Colley, R.J., Hickey, C.W., Vickers, M.L. and Ryan, P.A. 1992a. Effects of clay discharges on streams. 2. Benthic invertebrates. *Hydrobiologia* 248:235–247.

Quinn, J.M., Williamson, R.B., Smith, R.K. and Vickers, M.L. 1992b. Effects of riparian grazing and channelization on streams in Southland, New Zealand. 2. Benthic invertebrates. *New Zealand Journal of Marine and Freshwater Research* 26:259–269.

Quinn, J.M., Cooper, A.B., Davies-Colley, R. J., Rutherford, J.C. and Williamson, R.B. 1997. Land use effects on habitat, water quality, periphyton, and benthic invertebrates in Waikato, New Zealand, hill-country streams. *New Zealand Journal of Marine and Freshwater Research* 31:579–597.

Raupach, M.R. and Briggs, P.R. 1999. Integrative assessment of endosulfan transport from farm to river. In: Schofield, N.J. and Edge, V.E. (eds) *Proceedings of a Conference on Minimising the Impact of Pesticides on the Riverine Environment: Key Findings from Research with the Cotton Industry.* Land and Water Resources Research and Development Corporation, Occasional Paper 23/98, 51–57.

Rawlings, M. 1993. Lupins, willows, and river recovery: restoration in the upper Waitaki. *Forest and Bird* 10:15.

Read, M. and Franski, T. 1998. Assessing river health – progress in Tasmania. *Rivers for the Future Magazine* 6:30–33.

Rowe, D.K., Chisnall, B.J., Dean, T.L. and Richardson, J. 1999. Effects of land use on native fish communities in east coast streams of the North Island of New Zealand. *New Zealand Journal of Marine and Freshwater Research* 33:141–151.

Russell, L. 1996. Current commonwealth and state organisations, policies and programs. In: *Managing Australia's Inland Waters.* Prime Minister's Science and Engineering Council, Department of Industry, Science and Tourism, 111–121.

Rutherford, I.D., Ladson, A., Tilleard, J., Stewardson, M., Ewing, S., Brierley, G. and Fryirs, K. 1998. *Research and Development Needs for River Restoration in Australia.* Land and Water Resources Research and Development Corporation. Occasional Paper No. 15/98.

Schofield, N.J. 1989. *Stream Salinity and its Reclamation in South-west Western Australia.* Steering Committee for Research on Land Use and Water Supply, Water Authority of Western Australia Report No. WS 52.

Schofield, N.J. 1992. Tree planting for dryland salinity control in Australia. *Agroforestry Systems* 20:1–23.

Schofield, N.J. 1996a. Irrigation – on the brink? *Proceedings of the 23rd Hydrology and Water Resources Symposium,* Hobart, 363–369.

Schofield, N.J. 1996b. Forest management impacts on water values. Recent research developments in hydrology. *Research Signpost* 1:1–20.

Schofield, N.J. and Bartle, J.R. 1984. *Bauxite Mining in the Jarrah Forest: Impact and Rehabilitation.* Steering Committee for Research on Land Use and Water Supply, Department of Conservation and Environment Western Australia, Bulletin 169.

Schofield, N.J. and Davies, P.E. 1996. Measuring the health of our rivers. *Water,* May/June:39–43.

Schofield, N.J. and Price, P. 1999. River management and restoration in Australia. In: *Proceedings of the Second International Conference on Natural Channel Systems,* 1–4 March 1999, Niagara Falls.

Schofield, N.J. and Ruprecht, J.K. 1989. Regional analysis of stream salinisation in southwest Western Australia. *Journal of Hydrology* 112:19–39.

Schofield, N.J. and Simpson, B.W. 1996. Pesticides – an emerging issue for water and the environment. In:

Proceedings of the 23rd Hydrology and Water Resources Symposium, Hobart, 229–236.

Schofield, N.J., Loh, L.C., Scott, P.R., Bartle, J.R., Ritson, P.C., Bell, R.W., Borg, H., Anson, B. and Moore, R. 1989. Vegetation strategies to reduce stream salinities of water resource catchments in south-east Western Australia. Steering Committee for Research on Land Use and Water Supply. Water Authority of Western Australia, WS 33.

Sheldon, F., 1994. Littoral ecology of a regulated dryland river (River Murray, South Australia) with reference to the Gastropoda. Unpublished PhD Thesis, University of Adelaide, Adelaide.

Simpson, J. and Norris, R.H. 2000. Biological Assessment of Water Quality: Development of AUSRIVAS Model and Outputs. In: Wright, J.F., Sutcliffe, D.W. and Furse, M.T. (eds) *Assessing the Biological Quality of Fresh Waters: RIVPACS and Similar Techniques*. Special Publication of the Freshwater Biological Association, Ambleside (in press).

Sinner, J. 1992. *Agriculture and Water Quality in New Zealand*. Ministry of Agriculture and Fisheries. MAF Policy Technical Paper 92/17.

Smith, C.M. 1993. *Perceived Riverine Problems in New Zealand, Impediments to Environmentally Sound Riparian Zone Management, and the Information Needs of Managers*. Water Quality Centre, Ecosystems, NIWAR, Hamilton.

Smith, C.M., Wilcock, R.J., Vant, W.N., Smith, D.G. and Cooper, A.B. 1993. *Towards Sustainable Agriculture: Freshwater Quality in New Zealand and the Influence of Agriculture*. Ministry of Agriculture and Fisheries, Wellington.

Smith, D.G. and Maasdam, R. 1994. New Zealand's National River Water Quality Network 1. Design and physico-chemical characterisation. *New Zealand Journal of Marine and Freshwater Research* 28:19–25.

Smith, D.G., McBride, G.B., Bryers, G.G., Wisse, J. and Mink, D.F.J. 1996. Trends in New Zealand's National River Water Quality Network. *New Zealand Journal of Marine and Freshwater Research* 30:485-500.

State of the Environment Australia (SOE) 1996. *State of the Environment Australia 1996*. Department of Environment Sports and Territories, Canberra.

Statistics New Zealand (SNZ) 1995. *New Zealand Official Yearbook 94*. Wellington.

Statistics New Zealand (SNZ) 1996. *New Zealand Official Yearbook 95*. Wellington.

Statistics New Zealand (SNZ) 1998. *New Zealand Official Yearbook. 101*. GP Publications, Wellington.

Stein, J.L., Stein, J.A., Nix, H.A. and Hutchinson, M.S. 1997. *Wild Rivers of Australia*. Centre for Resource and Environmental Studies. Annual Report, Australian National University, 37–38.

Taylor, R.W., Smith, I., Cochrane, P., Stephenson, B. and Gibbs, N. 1997. *The State of the NZ Environment 1997*. Ministry for the Environment and GP Publications, Wellington.

Thoms, M.C. and Flett, D.J. 1993. Blue green algae and our degraded waterways. *Geography Bulletin* 25:172–176.

Thoms, M.C. and Sheldon, F. 2000. Water resource development and hydrological change in a large dryland river: the Barwon–Darling River, Australia. *Journal of Hydrology* (in press).

Thoms, M.C. and Thiel, P. 1995. The impact of urbanisation on the bed sediments of South Creek. *Australian Geographical Studies* 33:31–43.

Thoms, M.C. and Walker, K.F. 1992. Morphological changes along the River Murray, South Australia. In: Carling, P.A. and Petts, G.E. (eds) *Lowland Floodplain Rivers: Geomorphological Perspectives*. John Wiley, Chichester, 235–249.

Thoms, M.C. and Walker, K.F. 1993. Channel changes associated with two adjacent weirs on the River Murray, South Australia. *Regulated Rivers: Research & Management* 8:271–284.

Thoms, M.C., Parker, C.R. and Simons, M. 1999. The dispersal and storage of trace metals in the Hawkesbury River valley. In: Finlayson, B.R. and Brizga, S.A. (eds) *River Management in Australia*. John Wiley, Chichester, 197–219.

Thorpe, H. 1992. Groundwater – the hidden resource. In: Mosley, M.P. (ed.) *Waters of New Zealand*. Caxton Press, Wellington, 167–186.

Townsend, C.R. and Crowl, T.A. 1991. Fragmented population structure in a native New Zealand fish: An effect of introduced brown trout? *Oikos* 61:347–354.

Townsend, C.R. and Winterbourn, M.J. 1992. Assessment of the environmental risk posed by an exotic fish: the proposed introduction of channel catfish (*Ictalurus punctatus*) to New Zealand. *Conservation Biology* 6:273–282.

Townsend, C.R., Arbuckle, C.J., Crowl, T.A. and Scarsbrook, M.R. 1997. The relationship between land use and physico-chemistry, food resources and macroinvertebrate communities in tributaries of the Taieri River, New Zealand: A hierarchically scaled approach. *Freshwater Biology* 37:177–191.

Walker, K.F. 1985. A review of the ecological effects of river regulation in Australia. *Hydrobiologia* 125:111–130.

Walker, K.F., Thoms, M.C. and Sheldon, F. 1992. Effects of weirs on the littoral environment of the River Murray, South Australia. In: Boon, P.J., Calow, P. and Petts, G.E. (eds) *River Conservation and Management*. John Wiley, Chichester, 271–292.

Walker, K.F., Sheldon, F. and Puckridge, J.T. 1995. A perspective on dryland river ecosystems. *Regulated Rivers: Research & Management* 11:85–104.

Warner, R.F. 1995. Human impacts on Australian rivers. *Australian Geographical Studies* 33:3–15.

Wasson, B., Banens, B., Davies, P., Maher, W., Robinson, S., Volker, R., Tait, D. and Watson-Brown, S. 1996. Inland rivers. In: *State of the Environment – Australia 1996*. CSIRO Publishing, Collingwood.

Webster, I.T., Sherman, B.S., Bormans, M. and Jones, G. 2000. Management strategies for cyanobacterial blooms in an impounded lowland river. *Regulated Rivers: Research & Management* (in press).

Welch, E.B., Quinn, J.M. and Hickey, C.W. 1992. Periphyton biomass related to point-source nutrient enrichment in seven New Zealand streams. *Water Research* 26:669–675.

Wilcock, R.J. 1989. *Patterns of Pesticide Use in New Zealand. Part 1. North Island 1985–1988.* Directorate of Scientific and Industrial Research, Water Quality Centre, Wellington.

Wilcock, R.J. and Close, M.E. 1990. *Patterns of Pesticide Use in New Zealand. Part 2. South Island 1986–1989.* Directorate of Scientific and Industrial Research, Water Quality Centre, Wellington.

Wilcock, R.J., Nagels, J.W., McBride, G.B., Collier, K.J., Wilson, B.T. and Huser, B.A. 1998. Characterisation of lowland streams using a single-station diurnal curve analysis model with continuous monitoring data for dissolved oxygen and temperature. *New Zealand Journal of Marine and Freshwater Research* 32:67–79.

Williams, A. 1999. The Australian cotton industry's Best Management Practices Manual. In: Schofield, N.J. and Edge, V.E. (eds) *Proceedings of a Conference on Minimising the Impact of Pesticides on the Riverine Environment: Key Findings from Research with the Cotton Industry.* Land and Water Resources Research and Development Corporation. Occasional Paper 23/98, 109–115.

Williamson, R.B., Smith, R.K. and Quinn, J.M. 1992. Effects of riparian grazing and channelisation on streams in Southland, New Zealand. 1. Channel form and stability. *New Zealand Journal of Marine and Freshwater Research* 26:241–258.

Williamson, R.B., Smith, C.M. and Cooper, A.B. 1996. Watershed riparian management and its benefits to a eutrophic lake. *Journal of Water Resources Planning and Management* 122:24–32.

Wood, P.J. 1924. Increase of salt in soil and streams following the destruction of native vegetation. *Journal of the Royal Society of Western Australia* 10:35–47.

Woods, N., Craig, I.P. and Dorr, G. 1999. Aerial transport: spray application and drift. In: Schofield, N.J. and Edge, V.E. (eds) *Proceedings of a Conference on Minimising the Impact of Pesticides on the Riverine Environment: Key Findings from Research with the Cotton Industry.* Land and Water Resources Research and Development Corporation. Occasional Paper 23/98, 19–22.

CONSERVING NATURAL RIVERINE FEATURES

Plate 1a ↑   Dawn mist on the lowland River Murray in South Australia below its confluence with the Darling. This reach supports typical stands of *Eucalyptus camaldulatus*, the river red gum or 'canoe' tree. Flooding is now a rare occurrence owing to over-abstraction of water and exceptionally heavy regulation by dams and weirs. (Photograph: B. R. Davies.)

Plate 1b →   A lowland reach of the blackwater Suwanee River in the south-eastern United States. The diverse riparian vegetation and unconstrained river corridor are important features of high conservation value. (Photograph: B. R. Davies.)

Plate 1c ↑   Orinoco – riparian habitat. Riparian forests of tropical rivers such as the Orinoco, driven by the predictable annual flow, present a diversity of habitats supporting a rich flora, typically with a high diversity of Myrtaceae and Euphorbiaceae. (Photograph: G. E. Petts.)

CONSERVING NATURAL RIVERINE FEATURES

Plate 2a ➔ Tagliamento. The Fiume Tagliamento in north-east Italy, one of the few remaining near-natural rivers in Europe, displays a longitudinal sequence of channel forms including steep, single-channel headwaters, braided middle reaches, sometimes with wooded islands, and meanders along the lower river, all connected to a semi-natural riparian/floodplain corridor. Multi-thread, braided channels once characterized the piedmont zones of rivers across Europe but most were regulated by engineering works during the 18th, 19th and 20th centuries.
(Photograph: G. E. Petts.)

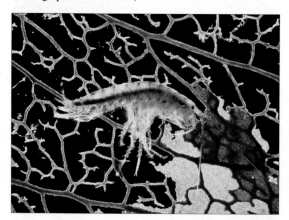

Plate 2b ⬅ Waterfalls, on the River Hvita, Iceland. The Hvita, photographed here in autumn, flows through arctic tundra. The catchment is largely natural, with few pressures other than low-level grazing.
(Photograph: P. S. Maitland.)

Plate 2c ➔ Typical billabong along the lower River Murray in South Australia. These oxbows provide refuge and maintenance habitat for floodplain species.
(Photograph: B. R. Davies.)

Plate 2d ⬆ A shredding stream amphipod, *Paramelita nigroculus* from the Liesbeek, Table Mountain, South Africa. Recent gel-electrophoresis studies have illustrated the high degree of endemism in this group, trebling the number of headwater amphipod species known from streams in the mountains of the Western Cape.
(Photograph: N. Eden.)

GEOGRAPHICAL SETTINGS

Plate 3a ← The Liesbeek (reed river), originates on the eastern flank of Table Mountain, Cape Town. Within 4 km of its source it becomes a typical urban system, degraded by trapezoidal channelization for stormwater control, and by industrial and domestic pollution, litter, loss of riparian wetland habitat, and invasion by alien plants and animals. (Photograph: B. R. Davies.)

Plate 3b → This landscape photographed near the town of Renmark, South Australia, epitomizes the consequences of over-abstraction and irrigation farming in arid environments. The apparent winter 'snowscape' is in fact a summer 'saltscape' – several centimetres of salt crust cover the soil surface, and the river redgum trees, *Eucalyptus camaldulatus*, have died. (Photograph: B. R. Davies.)

The challenges of conserving rivers in tropical latitudes are often quite different from those in temperate regions.

Plate 3c ← The River Legi, central Java, lies within a densely populated part of the world, and experiences not only human impacts from intensive rice cultivation and its associated irrigation, but also the natural hydrological extremes of a monsoon climate. (Photograph: P. J. Boon)

Plate 3d ↑ The catchment of the River Dee, in temperate north-east Scotland, is sparsely populated, and is subject to relatively limited impacts. (Photograph: L. Gill/Scottish Natural Heritage.)

KEY CONSTRAINTS AND RIVERINE IMPACTS

These two photographs illustrate the immense differences in scale between large and small rivers that impinge on all aspects of 'river' conservation – including classification and evaluation, research, monitoring, protection, the amelioration of impacts, and river restoration.

Plate 4a ↑ Caura River. The high-energy, bed-rock channel of the blackwater, Caura River drains the ancient Guyana Shield of Venezuela. (Photograph: G. E. Petts.)

Plate 4b ↑ Allt na Bogair, a small woodland stream in the central Highlands of Scotland. (Photograph: L. Gill/Scottish Natural Heritage.)

KEY CONSTRAINTS AND RIVERINE IMPACTS

Plate 5a ← The Augeigas River, Namibia. The majority of river systems on the planet are intermittent. This photograph shows a typical flood-front generated by a thunder storm many kilometres upstream bringing organic debris to the lower parts of the river. (Photograph: C. Boix-Hinzen.)

Plate 5b → Outfall to the Berg River of the Riviersonderend-Berg-Eerste Government Water Scheme, an inter-basin transfer (IBT) in the Western Cape, South Africa. The water is required during low-flow conditions in summer for irrigation farming and elevates the river flow by 4000%. The transfer (as in all other IBTs) injects organisms from the donor to the recipient catchment thereby breaching basin integrity, mixing gene pools, and confounding conservation efforts. (Photograph: B. R. Davies.)

Plate 5c ↑ Guri Dam, completed in 1986, has a hydropower rating of 10 GW and is one of the largest hydropower dams in the world. The dam impounds the Caroni River, Venezuela creating a reservoir capacity of 138×10^6 m³. Guri is one of a chain of reservoirs on the Caroni which, when completed, will inundate a combined area of more than 9,000 km². Projects of this magnitude create a wide range of impacts downstream on features such as flow regimes, temperature patterns, and migratory routes. (Photograph: G. E. Petts.)

KEY CONSTRAINTS AND RIVERINE IMPACTS

Plate 6a ➜ Swartboschkloof sub-catchment of the Eerste River, Jonkershoek Forestry Reserve, near Cape Town, South Africa. The 'fynbos' vegetation of the region is fire-adapted, and controlled burning is one of the major conservation tools in catchment management in the region. This late-summer burn led to rapid regeneration and had virtually no effects on stream chemistry or the invertebrate fauna. (Photograph: J. A. Day.)

Plate 6b ⬅ Lake Rawapening, in central Java, and its outflow the River Tuntang (seen here) are heavily infested with water hyacinth *Eichhornia crassipes*. This species is one of a suite of aquatic plants that invade tropical and sub-tropical fresh waters, often with serious ecological consequences. (Photograph: P. J. Boon.)

Plate 6c ➜ Downstream from the dam at the lake outlet, the River Tuntang is subjected to the dual impacts of gross fluctuations in flow coupled with scouring from large quantities of suspended sediment and water hyacinth. (Photograph: P. J. Boon.)

Plate 6d ⬅ The presence of non-native species of animals pose a serious environmental threat to many river systems. American signal crayfish *Pacifastacus leniusculus* has recently become established in several streams in southern Scotland. This species damages aquatic habitats, preys on other invertebrates and on fish eggs, and carries a disease that elsewhere in Britain has decimated populations of native crayfish. (Photograph: P. J. Boon.)

CONSERVATION IN PRACTICE

Plate 7a ↓ The European Union (EU) Habitats Directive lists a range of riverine species (e.g. Atlantic salmon) for which EU Member States are to designate statutorily protected sites known as Special Areas of Conservation (SACs). The River Avon in northern Scotland is a tributary of the Spey, and provides important spawning grounds for salmon. The main stem of the Spey has been proposed as an SAC for the presence of this and other listed species. (Photograph: J. M. Baxter.)

Plate 7b ↓ Streams are protected from development in Riding Mountain National Park, located on the Manitoba escarpment in central Canada. As they cross the park boundary they are channelized to follow roadways and public rights-of-way, dramatically simplifying the natural drainage pattern and destroying in-stream habitat. (Photograph: R. Newbury.)

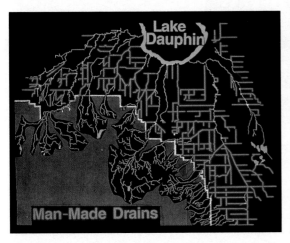

Plate 7c → Jack Stanford (University of Montana) and James Ward (then of the University of Colorado, Fort Collins) drilled augers into the glacial gravels of the Flathead River in Montana. Their work demonstrated the connectivity of the river with a large and diverse hyporheic community stretching laterally many kilometres from the main river channel. (Photograph: B. R. Davies.)

CONSERVATION IN PRACTICE

Plate 8a ➔ The physical removal and translocation of endangered species threatened by developments is a conservation tool of last resort. This picture shows an individual freshwater pearl mussel *Margaritifera margaritifera* being marked before being moved to another location in a small Scottish stream, prior to the construction of a 'run-of river' hydropower scheme. (Photograph: P. J. Boon.)

Plate 8b ⬅ The River Skerne, running through Darlington in north-east England, has recently been used as a demonstration site for river restoration. Environmentally sensitive techniques have been used to re-create meanders, improve water quality, and enrich the bankside vegetation. The result is a transformation in the ecological, visual and recreational quality of a previously degraded watercourse in a confined urban setting. (Photograph: Northumbria Water/Air Photos.)

Plate 8c ➔ Landsat Thematic Mapper (TM) image of the River Helmsdale in north-east Scotland. The image is displayed using bands 4, 3, 2 draped over a Digital Elevation model of the area. The picture has been included to emphasize that river conservation activities can only be truly effective when they encompass a catchment perspective. (Photograph: Scottish Natural Heritage.)

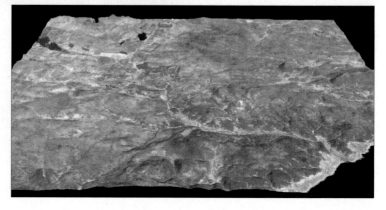

PART II

Constraints and Opportunities: Problems and Solutions

A   Introduction

13

Conservation, ecosystem use and sustainability

J.L. Gardiner and N.C. Perala-Gardiner

Man has lost the capacity to foresee and to forestall. He will end by destroying the earth.
Albert Schweitzer

Introduction

This chapter attempts a holistic examination of the factors influencing river conservation globally, drawing on examples from the United Kingdom (UK) and the United States of America (USA). Subsequent chapters focus on specific issues in different countries. John Muir reflected that one could not examine anything without finding that it is attached to the rest of the universe, and Noel Hynes confirmed the need for holism in river conservation, because the river corridor reflects, sooner or later, the health of the catchment (Hynes, 1985). It is land use which determines the quality and quantity of the water environment. So, to put river conservation into a practical context, we may ask, 'How well are we doing with our husbandry of the land and its wildlife, with nature conservation as a whole?'

What are we losing?

The fossil record suggests that 95% of all the species that ever lived are extinct, with the average life span of a mammal species being about one million years (Pettifer, 1997). By contrast, although there may be some 13 million species on earth, with extraordinary diversity such as 473 tree species in a single hectare (Ecuador), and no less than 1200 species of beetle on a single tree (Panama), extinction rates in the past century show a species life span now averaging only 10 000 years. A reduction to between 100 and 1000 years is currently threatened, so that extinction rates are 40 times greater than the past average 'natural' level for mammals and a staggering 1000 times for birds (Pettifer, 1997); an estimated 40 to 100 species become extinct every day (Owen and Chiras, 1995).

By contrast to this loss of species, the world's human population doubled to two billion in the 100 years from 1830 to 1930, doubled again to four billion in the 44 years to 1974 and passed six billion in the next 25 years (during October 1999 – see Davies *et al.*, Introduction to this volume). The present global erosion of 24 billion tonnes of topsoil per year will amount over a decade to

Global Perspectives on River Conservation: Science, Policy and Practice.
Edited by P.J. Boon, B.R. Davies and G.E. Petts. © 2000 John Wiley & Sons Ltd.

a loss equal to half of America's cropland soils (Owen and Chiras, 1995). This has a devastating effect on water quality, the morphology of rivers and their carrying capacity for wildlife. By 2016, a further 169 000 ha of England will have been converted from rural to urban use, making the land 12% urbanized; reaches of river conserved in their truly natural state are increasingly hard to find.

The devastating impacts of post-war farming policy on land and rivers, the loss of the small family farm and the rural exodus, are graphically portrayed by Harvey (1997) in *The Killing of the Countryside*. His message is that it is not too late, that scrapping subsidies and imposing taxes in accordance with the 'polluter-pays' principle, with a return to organic farming, can deliver 'safe, wholesome food and a safe, living countryside', feeding the British nation sustainably. The UK Soil Association has calculated that organic farming is approximately 80% as productive in yields as intensive farming methods, but organically farmed topsoils increase in fertility over time (Geier, 1998).

UK taxpayers are spending £3 billion per annum to subsidize agriculture, while native birds, mammals, reptiles, amphibians, insects and plants are being decimated by industrial agribusiness (Campbell and Cooke, 1997), not least from physical and chemical impacts on water and river corridors (Page, 1997). Compared with the billions of currency units being spent on agricultural subsidies, how much is being spent by governments on river conservation, including support for organic farming or the purposeful creation of buffer zones to protect fluvial and riparian ecosystems from the impact of industrial farming? Once fully exposed in the 21st century, will the true costs of industrial agriculture (from wildlife loss by pesticides to land ruined by irrigation, etc.) become a devastating indictment of the technological 'green revolution'?

If rural nature is eroding, our food supply has become suspect, and nature near to our cities is fast disappearing. Water shortages are more frequent, and promise to increase. In countries lacking a strong planning system, urban sprawl continues unabated to replace good agricultural land. This offers an illusion of economic growth without increasing either real human wealth or happiness (Kinsley and Lovins, 1995), but it does make further demands on our services, especially water and power.

With so many blatant examples of uncaring land use, it is all the more surprising that in 1990 the National Rivers Authority (England and Wales) chose to refer to ecosystem 'use' (Boon and Howell, 1997) in its Catchment Management Plans. This was not appropriate language for an authority charged with environmental protection, and it encouraged the idea that natural attributes could be traded for some development use of comparable 'value', as though human artifice could in any way substitute for the ecosystem services that Nature provides.

Some implications for environmental policy and legislation

Environmental legislation lacks a historical basis, being relatively new, and has largely taken the form of throwing a *cordon sanitaire* around development, attempting piecemeal protection for parts of natural, open systems. The resulting inadequate protection for catchments or ecosystems becomes more evident daily, as legislation fails to protect even such features as the UK Sites of Special Scientific Interest (SSSIs), when 84% (85 sites) of wetland showed signs of eutrophication, changing their nature conservation interest in 69 sites (Cavalho and Moss, 1995). Until such special areas are given 'inalienable' status (as for National Trust lands), national policy will fail to secure these ecological assets for future generations. The precautionary and prevention principles from the Maastricht Treaty should be fully embedded in European general legislation, prior to the application of the subsidiarity principle, to address concerns over the effectiveness of environmental protection in the face of international agreements over free trade.

While the continuing net loss of hedgerows in the UK is sufficiently visible and quantifiable to generate national concern and government action, the more insidious loss of riverine habitat has yet to attract controlling legislation of a general nature. Few rivers or floodplains in the UK are partly or wholly designated as SSSIs (Boon, 1991), and few mechanisms exist to control unsympathetic activities around existing SSSIs (Everard, 1998). Some river corridors have been declared Environmentally Sensitive Areas which, together with Countryside Stewardship and Long-Term Set-Aside (partly aimed at riparian zones), give farmers financial incentives to adopt environmentally more benign land-use practices. The EU Water Framework Directive may offer some hope for river corridor conservation, if catchment or river-basin management plans become statutory. In the UK, apart from the few SSSIs, no large riparian corridors or floodplains have received substantial protection.

For protection of specific areas to be fully effective, environmental protection must also be rooted in the local domain. Public awareness and activism made the difference for California's Mono Lake case, where diversion of feeder streams by the Los Angeles Department of Water and Power had progressively shrunk the huge lake. A few committed individuals, the Mono Lake Committee, invoked legislation known as the Public Trust Doctrine, which imposed a physical solution to conserve and restore the lake. Without the means to protect the community's interests through recourse to such powerful legislation, as it stands in the UK and elsewhere, natural areas of vital importance have little defence when development threatens. Increasingly, there is a reliance on estimating the economic value of natural areas to justify their protection.

Ecosystem 'services' and economic valuation

At present, unless a 'value' can be ascribed to ecosystems which is greater than the value of the alternative land use, any ecosystem seems expendable – a predicament that will only be exacerbated by population growth. Hence the interest in environmental economics, much applied to the UK river environment in recent years, and the growing interest in applying concepts such as 'environmental capacity', maintaining a constant 'stock of environmental assets' and 'critical natural capital'.

In an effort to quantify the benefits to humans from the functioning ecosystem, Ehrlich and Ehrlich (1981) identified several functions which underpin human civilization, but which have rarely been identified on the balance sheet:

- maintaining atmospheric quality, regulating gas ratios and filtering dust and pollutants;
- controlling and ameliorating climate through the carbon cycle and effects of vegetation in stimulating local and regional rainfall;
- regulating freshwater supplies and controlling flooding (wetlands, for example, can act as giant sponges to soak up moisture during rainy periods and release water slowly during dry periods);
- generating and maintaining soils through the decomposition of organic matter and the relationships between plant roots and mycorrhizal fungi and other soil micro-organisms;

- disposing of wastes, including domestic sewage and wastes produced by industry and agriculture, and cycling of nutrients;
- controlling pests and diseases, for example through predation and parasitism on herbivorous insects;
- pollinating crops and useful wild plant species by insects, bats, hummingbirds and other pollinators.

There have been notable efforts to extend this understanding in the river context, such as Maltby's *Waterlogged Wealth* (Maltby, 1986) and the seminal study of the hydrology and economics of the Sahelian Hadejia–Nguru floodplain wetlands (Hollis *et al.*, 1993). In 1972, the US Army Corps of Engineers acquired and protected >3200 ha of wetlands within the Charles River (Massachusetts) floodplain to absorb flood flows and then slowly release them, estimating that this system of 'natural valley storage' is equivalent to a new dam, for only 10% of the cost (US Army Corps of Engineers – US ACoE, 1972). A recent study of the floodplain of the Willamette River in Oregon reported that restoration of >20 000 ha of woodlands and wetlands within the floodplain could reduce the peak flows of a major flood by up to 18%, which would have saved many millions of dollars in the 1996 flood (Philip Williams and Associates – PWA, 1996). A summary of wetlands issues, including threats to their existence, valuation of benefits from them and approaches to conservation, can be found in Gardiner (1994).

Most recently, a group of scientists has estimated the annual value of global ecosystem 'services' at 33 trillion US dollars, just over twice the global annual GDP (Costanza *et al.*, 1997). Such valuation is therefore not impossible, but should it ever be formalized for use in determining the environmental cost of development, the phrase 'costing the earth' could take on a new and significant meaning. The concern over such valuations, which are unlikely ever to encompass 'total economic value' (including both 'use' and 'non-use' values), is that they should be seen as a target 'cost' for development 'benefit' to overcome – which is why *strong sustainability* (see below) is a vital principle to uphold.

Conservation and biodiversity

Perhaps the earliest pioneers of biodiversity were the naturalists of the era of Linnaeus (1709–1778), who went out to 'primitive' places in search of new species to classify. Conservation became the Great Collections, ranging from preserving lifeless flora and fauna to importing exotic species to create the great gardens and zoos of the world. Today, the role of gardens, zoos and seed banks in gene-pool maintenance has provided contemporary justification for their existence (Pain, 1997), although this conservation strategy *ex situ*, out-of-context, poses great potential dangers to the species being conserved, such as the losses to plant species of their pollinators and mycorrhizal associates. The dangers of introducing exotic species have been demonstrated many times. European willows grow as aggressive weeds in Australia; kudzu covers the entire landscape in parts of the Lower Mississippi; Japanese knotweed is a ubiquitous pest in the UK; and arundo has so choked rivers in California that eradication programmes have targeted thousands of hectares for removal, enough to produce fuel for power generation.

Estimates of tropical species extinction rates are reckoned at between 10 000 and 150 000 per year over the next few decades (Wilson, 1988; Diamond, 1990), as natural ecosystems are devoured to meet the demands of societal development. In the UK, some habitats cover less than 10% of their post-war area, and some species have lost 70% of their numbers in just two decades (Young, 1997). In response to the Biodiversity Agreement signed at the 'Earth Summit' (UN Conference on Sustainable Development, Rio de Janeiro, 1992) and the EU Habitats Directive, the UK Government has adopted a target-based approach; initially, with action plans for 116 species and 14 habitats. By the end of 1999, plans for 391 species and 40 habitats had been produced.

There have already been successful efforts, for example to encourage the return of the otter (*Lutra lutra*) to its former haunts in England; the timid water vole (*Arvicola terrestris*) is more challenging! Experiences such as the efforts to restore the River Kissimmee in Florida from the impacts of the 1960s channelization (Maltby, 1986; Karr *et al.*, this volume) show that river conservation is far preferable to rehabilitation. Species abundance and genetic diversity depend on conservation at the ecosystem level, so landscape-scale efforts must include habitat and ecosystem biodiversity in both conservation and rehabilitation.

The legal protection of natural areas in the USA (such as Yellowstone National Park, established in 1872) does not guarantee protection of biodiversity

and was not intended so to do (Noss and Cooperider, 1994). There are not enough parks in the right areas to represent potential natural vegetation types (Cooperider, 1993); they are not large enough to support viable populations of many species (Clark and Zaunbrecher, 1987) either in the USA or Africa. They suffer from incompatible uses, often being designated for their recreational rather than habitat value, and are typically under threat from surrounding land uses, if not actually invaded by poachers as well as exotic species. Following Hynes' dictum, rivers in such stressed areas will reflect that stress; river conservation therefore must match the carrying capacities for wildlife and human activities well beyond the river corridor.

Early efforts in river restoration have typically focused on a river site or reach with scant consideration of the river as a hydrological, geomorphological and ecological continuum in the catchment context. Ensuring that meandering migration will not outflank bank protection works in a major flood event may be as important as keeping grazing animals from the new shoots of bank revegetation projects. Typically such projects do not address biodiversity issues, and may risk becoming monocultures when only willows are used. Soil bioengineering and the new discipline of biogeomorphology have emerged to address the gap between hard river engineering and ecological approaches to river management.

Biodiversity seems to be set as a major focus for UK countryside planning, with the success of agri-environment support schemes such as the Tîr Cymen (Countryside Council for Wales, 1992), which has helped arrest biodiversity decline. There is a challenge for landscape planners to understand biodiversity and the UK Action Plan (Department of the Environment – DoE, 1994), and adopt approaches which integrate ecology, landscape heritage and local rural economics (Hesketh, 1997). In England, however, the apparent success of the Countryside Stewardship and Hedgerow Incentive schemes has been tempered by the predominance of *laissez faire* conservation over the more involved 'restoration' of habitats and landscapes, a tendency also observed with Environmentally Sensitive Areas (Morris, 1997). Do we expect substantial river conservation, enhancement or restoration from such economic incentives? How far can we expect these institutional initiatives to help farmers restore the biodiversity of the lost heritage of river corridors? And what of urban rivers?

Managing development

The institutional context to river conservation has its operational side, where the environment (and human communities) win or lose on a daily basis. Many countries have a planning system in which development proposals are received by local authorities and copied to statutory consultees like the Environment Agency in England and Wales, who return comments which may include objections and conditions. This process is not mainstream business for the statutory consultee (typically, the environmental bodies and public services), and has historically often been poorly regarded and under-resourced. The result is that comments can be made in a very superficial way, as those responsible for this planning liaison mechanically churn out their standard comments in an effort to meet their operational targets for the month.

It is not just mitigation of environmental damage that is missed, but opportunities for conservation, enhancement, rehabilitation and creation of channel, riparian and floodplain habitat go unregarded. A synergy can be induced, however, by wedding the planning liaison function of the environmental regulator(s) to their consenting and licensing functions. This was shown in the UK's National Rivers Authority (NRA) Thames Region (TR) to bring spectacular enhancements in river landscape ecology through development planning and construction.

Incorporating a three-year team training schedule, the approach was tried between 1988 and 1993 in the NRATR, and tested by measuring outcomes in terms of negotiated agreements with developers, either written into planning conditions or separately contracted. In 1992, the following results were achieved (counting only those 178 major proposals in which enhancements over £1000 were agreed):

- a high level of environmental asset damage averted;
- total asset creation (a majority of them flood defence) of over £32.5 million, including environmental enhancements to a total cost of £11.6 million; 50 km of river enhancements; and 3 km of river saved from culverting, typically by a few metres per site.

It should be acknowledged that most developers were pleased with the outcome, which invariably enhanced the value of their development, even when it slightly reduced the number of housing units.

Floodplain developments were accommodated when the overall conveyance and essential storage were maintained; the property was elevated above the appropriate flood level and a package of ecological, landscape and recreation/amenity measures were agreed.

From project to strategic appraisal

Such operational activities must be supported by an institutional framework in which strategic planning can create a complementary context of policy at central/regional government, county and district levels. Strategic planning can not only influence planning at the macro-level, but can also provide advanced warning of areas of development, allowing the statutory consultee to check and remedy any shortfalls in knowledge or predictive modelling. The NRATR pioneered this activity, employing professional land-use planners to meet the challenge of some 12 million people plus tourists using the Thames water environment in a catchment of just 13 000 km². Development pressures range from the fastest-growing urban area, Swindon, in its headwaters, to the demands of London downstream.

In the last days of the NRA, *Thames 21: A Planning Perspective and a Sustainable Strategy for the TR* was published (NRATR, 1995), providing a bridge at policy level between the NRA and external organizations and setting a regional context for catchment plans. The sequel *Thames Environment 21 – the Environment Agency Strategy for Land Use Planning in the TR* was produced in 1997. This applied the Agency's set of six Sustainability Principles to the provision of guidance, for example, on environmental assessment (EA) of projects and environmental appraisal of development plans.

In the UK, there is renewed institutional support (cf. section 39 of the Environment Act 1995) for applying benefit/cost (B/C) assessment (BCA) to decision-making. This is a powerful way of prioritizing preferred options for expenditure in the absence of alternative policy or political direction. Yet attempts to have BCA as the decision-making framework for project appraisal, relegating EA to an obstacle course separated from the main decision-making process, is inimical to achieving sustainable development (SD). The fact that ecosystems enable economic growth, and not the reverse, must be reflected in any decision-support system purporting to identify SD.

Ironically, commercial greed has been so blind to this truth that the current perception looks to economic success as the means to conserve and recover what we can of our ecosystems – undermining our foundation for life in order to increase standards of living and hope for regaining our quality of life 'in the future'. In retrospect, it seems that there can never be enough money to spare for looking after the environment; when should 'the future' begin? Are we not now far more prosperous than our grandparents ever dreamed possible for themselves?

FALLACIES IN NEO-CLASSICAL ECONOMIC THEORY

The fundamental flaw in the paradigm that encourages us to spend natural resources at the expense of present and future generations is continually glossed over by the intransigence of discounting the costs falling on future generations, without investing now to cover these costs. Misuse of BCA as the framework to rank options within a project usually leads to a situation where environmentally insensitive options have a high B/C ratio owing to exclusion of environmental costs because natural resources are effectively treated as free, exploitable goods in neo-classical economic theory. Thus insensitive options may survive to the last stages of the decision-making process. The only way then to reduce the B/C ratio is to 'value' the intangible benefits of the environment that will be lost or damaged, which can be an expensive process if contingent valuation methods, such as testing the public's 'willingness to pay', are employed.

In the recent UK Axford Inquiry, an attempt was made by the Environment Agency to justify reduction in the licensed abstraction of water on account of environmental impact. An attempt to transfer the values obtained from similar cases to avoid the estimated £40 000 (minimum) site-specific study, failed to find favour with the Inquiry Inspector. The Agency's estimate of about £14 million for the present value of reducing abstractions from the River Kennet was hotly contested by Thames Water (TW), which put forward £300 000 as the more probable value. Which is correct? The argument showed how difficult it can be to build a robust case for environmental benefits, especially when seeking a change in existing conditions. TW won the case; the river environment was held to be of little value compared with the alternative use of the aquifer for supplying potable water resource, despite

the fact that neither demand management nor strategic recycling had been fully implemented.

None of us is ever likely to know the intrinsic value of any natural 'resource'. It seems that 'free-market accounting', while locked in a marital embrace with BCA and unable to account for ecological value, will hasten the end of our race. This self-evident truth remains, despite valiant efforts to identify constant natural assets, critical natural capital, strong sustainability, etc., which all appear to fail when tested against the real political and other pressures of the global market-place. Initiatives such as the Earth Summit, the Montreal Protocol and their derivatives are hailed as global breakthroughs, only to be rendered more or less ineffectual by political intransigence in the face of pressure from national treasuries or multinational corporate interests.

Environmental Assessment

To avoid the artificial and difficult situation in which valuation of the environment is needed to reduce the apparent B/C ratio of an environmentally insensitive option, EA is used as the framework for the decision-making process, as in the Lower Colne Study (Gardiner, 1988, 1991). Environmentally insensitive options are eliminated at an early stage, dispensing with the need for so-called environmental valuation, which is anthropocentric at best and can never identify the intrinsic values of ecosystems. Valuation of all the identifiable environmental benefits of the riverine and wetland system should of course be calculated (Barbier *et al.*, 1993), in order to calculate the net benefits forgone after the scheme is completed.

This holistic EA approach, emphasizing the primacy of ecological and social factors in decision-making and using BCA wisely, supports the definition of SD in *Caring for the Earth* as that which 'improves the quality of human life while living within the carrying capacity of supporting ecosystems' (International Union for Conservation of Nature and Natural Resources, United Nations Environment Programme and World Wildlife Fund – IUCN/UNEP/WWF, 1991). This definition implies that SD is not a synonym for 'sustainable growth', an oxymoron even in economic terms (Daly and Townsend, 1993); 'growth is quantitative increase in physical dimensions; development is qualitative improvement in nonphysical characteristics' (Daly, 1991). Practical application of EA as a means of holistic appreciation of ecological needs and societal aspirations has shown

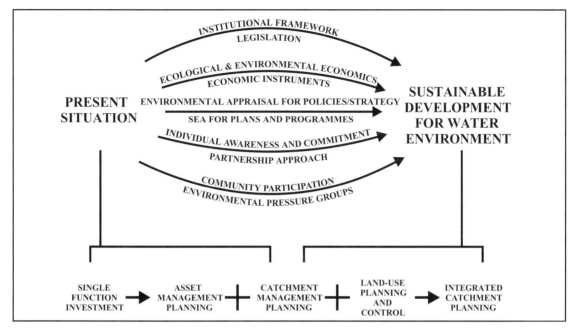

Figure 13.1 *Pathways to sustainable development for the water environment (SEA, Strategic Environmental Assessment)*

the EA process to be very effective in delivering value for money and social approval (Gardiner, 1991). However, the misuse of EA and Environmental Impact Assessment (EIA) has led to distrust and alienation of the community both from developers and from the system of decision-making (Sherman, 1997), including the failure of appeal to the European Commission (Sheate, 1994).

Nevertheless, contingent valuation methods (CVM) in the form of 'public willingness to pay', has been made legitimate by the UK Government as the justification for the alleviation of low flows and water quality improvements not included in the current Asset Management Plan of the privatized water utilities. This is a plan of proposed investment which seeks agreement from the regulators to pass the costs on to the users through the water rates. A new focus by the Environment Agency on multi-functional issues in catchment plans, together with the decentralization of each Region into multi-functional Areas, brings hope that the more holistic approach to river conservation will extend to the currently fragmented budget arrangements. Single-function funding (see Figure 13.1), which severely distorts decision-making in river management, as in any field, is clearly anathema to SD for the water environment, even when subject to EA.

Strategic Environmental Assessment (SEA) and the sustainable cycle

Referring to Figure 13.1, functional investment regimes built on single functions (e.g. flood defence) within separate sectors (e.g. the water industry) seldom acknowledge the regional issues to which they are irrevocably connected. So ingenious flood gates, wonders of engineering but hugely expensive to build and maintain, are proposed for Venice *before* a regional catchment-cum-coastal management plan (CCMP) is produced. The CCMP would link trends, data and processes in industry, navigation, agricultural and urban pollution, sedimentation of Venice Lagoon, etc., all of which are related and should affect decision-making on sustainable flood defence measures for Venice (Penning-Rowsell *et al.*, 1998). Such holistic assessment of what development should take place where, related to the carrying capacities of the social, ecological and economic environment, needs the support of institutionalized structures and processes – called institutional carrying capacity by the World Bank. Before the EA for the proposed Venice flood gates answered the question 'How should it be done?', the question of 'What should be done and where?' should have been addressed in production of the CCMP.

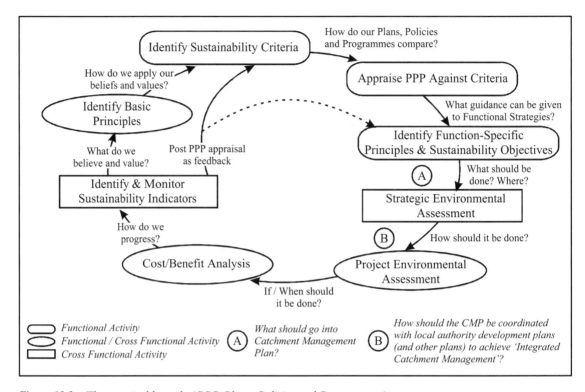

Figure 13.2   *The sustainable cycle (PPP-Plans, Policies and Programmes)*

In the UK, questions of what and where development should take place should be addressed by Strategic Environmental Assessment (SEA), involving all sectors of society (e.g. water, energy or transport) and structured on the land-use planning system. At the time of writing, the European Commission's much-diluted proposals for SEA are again being considered by Member States. When first introduced in 1991, SEA for plans, policies and programmes was torpedoed by the UK, Germany and Denmark.

SEA needs to be supported by sophisticated communication between sectors, stakeholders and, indeed, among functions within any one sector (e.g. water resources, flood defence, and water quality functions within the water sector). 'The Sustainable Cycle' (Figure 13.2) has been proposed as such a framework (NRATR, 1995) for consensus-oriented decision-making, providing an iterative process building on agreement among stakeholders over basic beliefs and values. This cycle connects SEA, EA, BCA, monitoring and post-project appraisal using sustainability indicators (Gardiner, 1996). Gradually, the importance of focusing on the process (as compared with the product, traditionally the only thing that mattered) is being acknowledged, despite the persistent lack of adequate monitoring and post-project appraisal. SD will elude us until we have some effective means to question together and affirm as a society what our needs are, and what our land will support.

Population growth and demand management

The context for sustainable river conservation would be incomplete without reference to human population growth, the resource limits to that growth, and the current priorities for global investment. Notwithstanding the threatened impacts of climate change (Department of the Environment, Transport and the Regions and The Meteorological Office, 1997), the single greatest threat to the future of human well-being and ecosystem stability worldwide is the exponential growth in human population. Without addressing this factor in nature conservation, all efforts to protect natural areas, ecosystems or native plant and animal species will soon or ultimately be overrun by the pressure to sustain ever-increasing numbers of humans. From January 1996 to January 1997, the global number of humans increased by 80 million (US Census Bureau, 1997), a rate of increase of over 10 times the population of London in a single year, with frightening implications for water, energy and other global resource use. Between 1900 and 1990, there was a 700% increase in water use worldwide, and a 566% increase in water withdrawal from the world's freshwater resources (Newson, 1992). Water storage, coupled with hydroelectric power production, has been the universal response to increased demand, until recently.

The scope of dam impacts on rivers since World War II has been difficult to appreciate. In 1950, there were 5270 dams worldwide of 15 m or more; today there are about 40 000, with 1118 under construction in 1995 and some 300 started each year (Pottinger, 1997). In his seminal paper 'Energy Strategy: The Road Not Taken', published in the influential journal *Foreign Affairs* in October 1976, Lovins insisted that American energy policy should change its focus from the 'hard' path of costly fossil/nuclear fuels and demand-led production, onto the 'soft' path of safer, cheaper renewables (solar and wind power) and demand management, especially efficiency of use (Lovins, 1977). He articulated one of the most important changes in growth trends, and despite being initially ridiculed by the energy industries, he has since been amply vindicated.

Similar issues eddy around the energy, water and transport debates in the UK. It has met with the familiar fierce opposition from those whose knowledge, experience and livelihoods are chained to meeting (and effectively encouraging increases in) demand, while treating natural resources as 'free goods'. Thus river conservation in the UK is again threatened by any number of structural schemes for water resources, while there is still much slack to be taken up through demand management and strategic planning measures. Most water resource managers seem far from grasping the longer-term importance of taking a systems approach to supporting the hydrological cycle, where source control and recycling schemes at local scale have much more to offer for water conservation than demand management. It seems not to be in any one organization's interests to promote such sustainable solutions, which depend greatly on inter-organizational communication and common purpose.

The publication *Limits to Growth* (Meadows *et al.*, 1972), commissioned by The Club of Rome, concluded that unchanged growth trends in population, industrialization, pollution, food production and resource depletion would bring a sudden and uncontrollable decline in both population and industrial capacity by 2072. However, these growth trends could be altered to establish a sustainable ecological and economic stability, and the sooner this truth is recognized and acted on, the greater our chances of success (Meadows *et al.*, 1972).

In the sequel *Beyond the Limits*, Meadows *et al.* (1992) concluded that the challenges are still present, although some sustainability limits have already been exceeded and there is a real prospect of global collapse. The achievement of a sustainable society will require adoption of radical new directions away from policy and practice supporting continuing expansion (cf. the oxymoron 'sustainable growth') in population and material consumption, and towards maximizing efficiency in using materials and energy. Implied are emphases on reliance on local community resources, and equity and quality of life rather than quantity of production. In *Factor Four* (Von Weizsäcker *et al.*, 1997), Lovins' early misgivings about the globalization of resource use have matured into the concept of resource (rather than labour) productivity – doing more with less, rather than doing less or doing without; this is as true for water as it is for energy.

'Megaprojects' and appropriate technology

Riverine megaprojects have had huge impacts on the human-scale landscape, the river corridor and its inhabitants, often while generating small amounts of hydropower relative to the costs of dam construction (Palmer, 1994). Numerous studies have documented that big dams have often increased poverty locally, eliminated fisheries, introduced disease and effected genocide or disastrous 'resettlement' of indigenous peoples in developing countries (Goldsmith and Hildyard, 1984). Huge loans from agencies such as the World Bank have funded megaprojects in developing countries, which not only squandered both fiscal and natural capital, but also represented financial losses to the Bank because the country could not afford to repay these loans or even maintain the capital assets in working order (Caufield, 1997). There can be an alarming difference between the reassuring technical reports of irrigation schemes, for example, and the perception of native professionals required to implement, operate and maintain these schemes.

In the face of such close inspection, leading to the advertisement of so many investments proven uneconomic, institutions such as the World Bank have pulled out from involvement with controversial dam projects such as Sardar Sarovar in India and Three Gorges Dam in China. Unfortunately, the perceptions of the World Bank are not shared by the commercial concerns and export credit agencies in Germany, Japan and Switzerland, who are funding the dam on the River Yangtze. The International Rivers Network believes that domestic and international concerns over Three Gorges Dam led to the unusual step by Jexim (Japan's credit agency), of stating that the situation would be monitored and that the agency would 'reconsider this internal decision if environmental and human rights problems arise' (McCully, 1997). Navigation is an important part of the dam; the tonnage is predicted to increase from 10×10^6 t at present to 50×10^6 t, with dire consequences for conservation.

This parallels the phenomena observed on the Upper Mississippi River, where biodiversity and size of native species populations have crashed as the river geomorphology responded to regulation (Upper Mississippi River Conservation Committee – UMRCC, 1993). By 1900, navigation had altered some 8750 km of the world's rivers; by 1980, this figure had increased to 498 000 km (Pottinger, 1997). Extensive navigation severely limits the scope of river conservation, and global transport must be considered in rethinking river conservation.

The seductiveness of megaprojects is not hard to understand. As well as the project 'benefits', the lure of local employment and rise in local standards of living (neither guaranteed), together with the 'heroic' idea of 'taming nature', has resulted in approval of the construction of thousands of dams, overcoming arguments for river conservation and more appropriate smaller-scale works. In Europe, the River Loire seems constantly under threat of losing its status as the last wild river in France; river conservationists must time and again rally in its defence. It will prove far more difficult to mount effective opposition to Hidrovia, a huge plan to regulate the River Paraná in South America, a fate which also threatens the River Mekong from China to South Vietnam, following the formation of the Mekong Commission comprising all the riparian countries through which it passes (see Dudgeon *et al.*, this volume).

Exporting technology is likewise far more dangerous than the International Trade Agencies of the Developed Countries reflect in current policies on lending. The hunger relief promised in the Green Revolution has backfired in many countries (Shiva, 1991), leaving rusting tractors and saline soils, overtapped aquifers, pest problems, local economies in chaos and often widespread hunger (and elevated population numbers) as farmers either no longer grow traditional crops or have been forced to become farm workers or migrants to the city's shanty town, thus increasing rather than decreasing inequality (Levins, 1995). In such conditions, river conservation cannot be a priority.

Water-based public health engineering is clearly inappropriate as an export from temperate industrial Europe to the arid regions of Africa and India, yet is now held to be a highly desirable symbol of wealth. Such transfers have suffered from the common syndrome of being a capital gift without adequate training or resourcing of its operation and maintenance – as in Zambia, where a new sewage treatment plant has had no maintenance for the past 15 years, and sewage overflows threaten local water supplies (Majura and Banda, 1996). Again, river conservation is regarded as an ideal unlikely to be realized until far into the future, rather than a top national priority for today.

Sustainability

The concept of sustainability is based on the idea of inter-generational equity, conveyed by the Brundtland Report definition of SD (World Commission on Environment and Development – WCED, 1987). The economist's view, as expressed in *Blueprint One* (Pearce *et al.*, 1989), is strictly utilitarian (harking back to the start of this chapter): future generations will need a 'stock' of assets no less than the present, and that we should therefore aim at a constant asset resource to achieve sustainability. A widely held perception is that 'we do not inherit the world from our parents, we borrow it from our children' (Jacobs, 1991).

Yet every day, development converts natural capital, including river corridors, into human capital such as buildings and machinery; this is the ruling paradigm of *weak sustainability*, allowing for the mix of natural and human assets to 'remain constant'. Surely, only an ivory-towered, neo-classical economist with total ignorance of how the earth's ecosystem works would cling to this dangerously naive belief? Our knowledge of what we need to survive is adequate to embrace *strong sustainability*, keeping natural assets constant, so that loss of a natural asset would need to be mitigated by creation of a natural asset of *equal* value to the ecosystem. Deep ecologists believe that natural assets should not be regarded as a collection of goods and services for human use at all – a *very strong sustainability* position (Turner, 1993).

The difficulties of achieving *constant natural assets* in practice are exemplified by the USA's 'no net loss' rule applied to wetlands. American wetland scientists found that the functional value of a replacement wetland could not be guaranteed as replacing the lost asset, when a change was made in its geographical position within the catchment. River corridors have suffered so much degradation in most countries that there are virtually limitless opportunities for conservation and enhancement, if not rehabilitation, to mitigate the unavoidable impacts of development. For reasons of rarity, irreplaceability or importance to human life-support, some natural assets cannot be substituted, and have to be protected as '*critical natural capital*'.

What is the threshold between constant and critical natural capital? What constitutes true substitution to achieve constant natural assets? These two questions should be debated daily in decision-making over SD (Figure 13.2). They rely so heavily on regional, national and even international considerations that the process of decision-making itself is rapidly assuming prime importance. The confusion over what development should go where, and how it should be achieved, has to be clarified in a new approach to decision-making which separates the two questions, controlling both of them with processes that assess the environment (ecological, socio-political, etc.) at strategic and project levels respectively. In preparation for the strategic phase, stakeholders need to have identified the sustainability objectives (and indicators) for their functions and sectors, in a process of consensus-oriented debate (Gardiner, 1996).

O'Riordan defined sustainability as a 'mediating term' between developers and environmentalists which has stayed the course; according to him, SD is with us for all time (O'Riordan, 1993). It has enabled consumer perception to dictate policy to major companies – witness the successful campaigns against Burger King for allegedly buying meat from former Brazilian rain-forest farms, SunKist tuna of the Kellogg Corporation for failing to control the deaths of dolphin caught in its tuna nets, and against Shell for not consulting widely enough over the disposal of the Brent Spar accommodation platform in deep waters (O'Riordan, 1993).

Global progress in sustainability

The publication of the World Conservation Strategy (IUCN/UNEP/WWF, 1980) led the first change in global perceptions of the nature of development and the need for conservation of nature and natural resources. It originated the phrase 'sustainable development', led to the establishment of the World Commission on Environment and Development (WCED) and the subsequent publication of *Our Common Future*, commonly known as the Brundtland Report (WCED, 1987). This provides the most commonly quoted definition of SD, and called for the marriage of ecology and economy, together with institutional change to tackle the challenges of population, food security, species and ecosystems, energy, industry, urbanization, the role of women and managing the commons. Sadly this report did not treat management of the water environment as one of the key considerations, reflecting what surely will become one of the great historical paradoxes – that such a fundamental element of existence could be downplayed so consistently or ignored in a century which has seen such dramatic increases in the use of this natural resource.

The role of holistic management of the hydrological cycle as the basis for SD was clearly identified in the seminal publication *Caring for the Earth* (IUCN/UNEP/WWF, 1991). However, this vital topic was poorly represented at the Earth Summit at Rio de Janeiro in 1992, the failure being attributed to the rather low-key efforts by the prior Dublin Conference to improve its global profile. This lack of serious international attention, evidenced in the content of Agenda 21 from the Earth Summit, has impoverished national policy, and made progress towards SD of the water environment via control of land use much slower than it could have been.

One of the powerful messages of river conservation is that change should be incremental, small-scale and appropriate, facilitating adaptive management. This is at the core of the Intermediate Technology Development Group (ITDG), a charity which addresses the root causes of both environmental degradation and poverty with appropriate technology, as promoted by Schumacher (1973) in *Small is Beautiful: Economics as if People Mattered*. The charity he founded matches the scale of assistance with the scale of need. A hundred thousand efficient cooking stoves, wind-up lights and radios can do far more for environmental and social sustainability than one large hydropower plant. For the river and its catchment to be conserved, appropriate solutions must be applied (Swallow *et al.*, 1997).

Surface water management

Progressive river management has benefited from the endorsement in Agenda 21 of coordination of catchment management plans with asset management plans (in the UK) and local authority development plans, leading (Figure 13.1) to integrated catchment management as a corner-stone of SD (Newson, 1992; Gardiner, 1997). The lack of appreciation for the holistic approach surfaced in the European Community's draft Water Framework Directive, which seeks to recover costs from abstractors and users, but fails to address land-use impacts, one of the prime causes of poor water management. Increasing the resilience of the catchment to change in both land use and climate means moving from the traditional paradigm of surface water *disposal* to *management*, enabling conservation of the hydrological cycle.

A number of measures generically known as 'source control' address the root causes of urban degradation of the water environment (Urbonas and Stahre, 1993), particularly when using vegetative treatment of the run-off. Source control techniques can be implemented everywhere to the benefit of people, wildlife, water resources and the water environment generally (Construction Industry Research and Information Association – CIRIA, 1992), and are actively being promoted within the UK and elsewhere (Scottish Environment Protection Agency – SEPA and Environment Agency, 1997). Effective source control, in both the rural and urban context, should be seen as a necessary concomitant to river conservation, but needs economic justification as well as technical credibility.

Industrial agribusiness and sustainable farming

River quality is largely determined by catchment run-off, especially in first- and second-order streams, so river conservation is hugely dependent on farming methods and policies. Since the publication of Carson's *Silent Spring* in 1962, total volumes of chemicals may have decreased, but the toxicity and diversity of agricultural chemicals has increased (Campbell and Cooke, 1997). In the UK, steep declines in all forms of wildlife have been documented as indirectly attributable to pesticide use (Campbell and Cooke, 1997). However, a small but growing public movement is working to reverse the trend of dependence on

chemicals in food production. State and national standards now exist in many countries, notably the California Organic Agriculture Code, the UK Soil Association Organic Standard Code, and the International Federation of the Organic Agriculture Movement (IFOAM). These set standards for organic farm production and consumer recognition of the products for many neighbouring states and countries. IFOAM is working with the United Nations to set international standards for food and fibre production and processing, to improve access to information, to assist farmers in making the transition from chemical dependence to sustainable agriculture, and to create more stable markets for sustainably produced goods (IFOAM, 1997).

Sustainable agriculture encourages buffer zones along streams, and guides farmers to focus on enhancing soil fertility and structure by increasing soil organic carbon or humus, protecting biological/microbial biomass (earthworms, bacteria, fungi, etc.), increasing predator/prey interactions and reducing chemical inputs. Soils with higher humus content hold water better, reducing water demand and also the rate of run-off, which determines soil erosion, river baseflow and the subsequent health of the river. By eliminating chemicals designed to poison pests, and increasing farmer/farm contact, the net balance of predators and pests (prey) is enhanced. For example, a farmer can rotate legume crops with grains to increase soil nitrogen, carbon and microbial activity to increase crop yields without use of nitrogen fertilizers which harm the soil micro-flora and fauna.

There has been considerable development recently in the USA about the application of buffer zones within farming landscapes. In addition to the more than US$200 million funds identified for the application of buffer zones to Chesapeake Bay, there have been further initiatives by the US Department of Agriculture (USDA) 'to establish 2 million miles of conservation buffers by the year 2002'. This programme is being directed by the USDA and managed by the National Resources Conservation Service (NRCS).

Linking economics with ecology

The links between equity, our ecological life-support systems and river conservation are becoming clearer. For example, trading essential elements of ecosystems such as forest preserves for short-term profit is not equitable, not good for our collective well-being, not good for business and is disastrous for rivers. Single-function interests such as forestry need to expand their horizons to diversify the forest economic base; it is a nonsense to claim that loggers have no other livelihood to make from the forest than cutting it down. In the coming decades, we will see more examples of communities who are investing in protecting their environment, and gaining economically by so doing. Examples can be found in the Australian Landcare initiative (which supports integrated catchment management), and the California Urban Streams Restoration Program. Good ecological thinking is good economics in the short, medium and long term, when all the costs of doing business are included in the calculus. Altruistic behaviour and building sustainable human communities, in harmony with nature, are matters of enlightened self-interest.

Thus accurate public perception of ecological functions is vital to river conservation. How important is it perceived to be? We face a huge problem of rapid and excessive natural resource exploitation, leading to loss of the physical habitat, loss of species diversity and production of mostly polluting waste. While there is enough food to feed the current population, distribution and economic issues mean that millions continue to starve despite the best efforts of aid agencies. Far from the US$10 billion of extra aid which Maurice Strong hoped would come from the rich countries as a result of the Earth Summit, there has actually been a substantial fall behind the agreed target of 0.7% of developed countries' GDP going to overseas aid (Pearce, 1997).

We have made many mistakes, and have learned something from them; we now have a shrewd idea of the decision-making processes needed to achieve SD. We have made progress in identifying ecosystem 'services' and their valuation, albeit anthropocentric. There is healthy debate over the process of decision-making, with increasing public involvement. Genuine consensus-oriented application of (socio-) environmental assessment will generally survive the test of project implementation better than theoretical edifices based on 'valuing' and 'weighting' environmental impacts. Sustainability objectives and indicators are being identified, with carrying capacities in view. Children are learning the importance of the water environment and clean, healthy rivers in supportive catchments, by participating in educational programmes such as RiverWatch, Adopt-A-Watershed, and RiverKeepers. These are significant factors for conservation in the developed world.

Conservation in developing countries probably depends first on bringing the 'peace dividend' to bear on the reduction of poverty, unemployment, malnutrition, illiteracy, low status of women, exposure to environmental risks, family planning, and limited access to social and health services. Incremental success also depends on the export of appropriate technology – not old technologies with their pollution, nor poorly conceived genetic engineering products – from developed to developing countries (e.g. see Wishart *et al.*, this volume, Part II).

Aid must be tied firmly to the needs of the local communities affected, not the lending agencies nor the politicians of the day. Care for the supporting ecosystem is unlikely to precede care for families, neighbours and employees. Global river conservation will follow acts of global compassion, courage and common sense within supportive institutional structures; economics must evolve to support the ecosystems on which all economic activity ultimately depends.

References

Barbier, E.B., Adams, W.M. and Kimmage, K. 1993. An economic valuation of wetland benefits. In: Hollis, G.E., Adams, W.M. and Amin-Kano, M. (eds) *The Hadejia-Nguru Wetlands: Environment, Economy and Sustainable Development of a Sahelian Floodplain Wetland.* IUCN, Gland and Cambridge, 191–208.

Boon, P.J. 1991. The role of Sites of Special Scientific Interest (SSSIs) in the conservation of British rivers. *Freshwater Forum* 1:95–108.

Boon, P.J. and Howell, D.L. 1997. Defining the quality of fresh waters: theme and variations. In: Boon, P.J. and Howell, D.L. (eds) *Freshwater Quality: Defining the Indefinable?* The Stationery Office, Edinburgh, 522–533.

Campbell, L.H. and Cooke, A.S. (eds) 1997. *The Indirect Effects of Pesticides on Birds.* Joint Nature Conservation Committee, Peterborough.

Carson, R. 1962. *Silent Spring.* Penguin Books, London.

Caufield, C. 1997. *Masters of Illusion.* HarperCollins, New York.

Cavalho, L. and Moss, B. 1995. The current status of a sample of English sites of Special Scientific Interest subject to eutrophication. *Aquatic Conservation: Marine and Freshwater Ecosystems* 5:191–204.

Clark, T.W. and Zaunbrecher, D. 1987. The Greater Yellowstone Ecosystem: the ecosystem concept in natural resource policy and management. *Renewable Resources Journal* 5:8–16.

Construction Industry Research and Information Association (CIRIA) 1992. *Scope for Control of Urban Runoff.* Report 123, London.

Cooperider, A.Y. 1993. Ecological quality in the national parks. Paper presented at *Diamonds or Dust? – A Forum on the Quality of our National Parks*. San Francisco, California.

Costanza, R., D'Arge, R., De Groot, R., Farber, S., Grasso, M., Hannon, B., Limburg, K., Haeem, S., O'Neill, R.V., Paruelo, J., Raskin, R.G., Sutton, P. and Van Den Belt, M. 1997. The value of the world's ecosystem services and natural capital. *Nature (London)* 387:253–260.

Countryside Council for Wales (CCW) 1992. *Tir Cymen – a Farmland Stewardship Scheme*. Bangor.

Daly, H.E. 1991. *Steady-State Economics*. Island Press, Washington, DC.

Daly, H.E. and Townsend, K.N. 1993. *Valuing the Earth: Economics, Ecology, Ethics*. MIT Press, London.

Department of the Environment (DoE) 1994. *Biodiversity: The UK Action Plan*. HMSO, London.

Department of the Environment, Transport and the Regions and The Meteorological Office, 1997. *Climate Change and its Impacts: A Global Perspective*. The Meteorological Office, Bracknell, UK.

Diamond, J.M. 1990. Playing Dice with MegaDeath. *Discover* (April):55–59.

Ehrlich, P.R. and Ehrlich, A.H. 1981. *Extinction: The Causes and Consequences of the Disappearance of Species*. Random House, New York.

Everard, M. 1998. Floodplain protection: challenges for the next millennium. In: Bailey, R.G. José, P.V. and Sherwood, B.R. (eds) *United Kingdom Floodplains*. Westbury Academic and Scientific Publishing, Otley, UK, 117–143.

Gardiner, J.L. 1988. Environmentally sensitive river engineering: examples from the Thames catchment. In: *Regulated Rivers: Research & Management* 2:445–470.

Gardiner, J.L. 1991. *River Projects and Conservation: A Manual for Holistic Appraisal*. John Wiley, Chichester.

Gardiner, J.L. 1994. Pressures on wetlands. In: Falconer, R.A. and Goodwin, P. (eds) *Wetland Management*. Thomas Telford, London, 47–75.

Gardiner, J.L. 1996. The use of EIA in delivering sustainable development through integrated water management. *European Water Pollution Control* 6:50–60.

Gardiner, J.L. 1997. River landscapes and sustainable development: a framework for project appraisal and catchment management. *Landscape Research* 22:95–115.

Geier, B. 1998. Yes, organic farming can feed a growing world. *Living Earth* 198:22–23.

Goldsmith, E. and Hildyard, N. 1984. *The Social and Environmental Effects of Large Dams*. Wadebridge Ecological Centre, Camelford, UK.

Harvey, G. 1997. *The Killing of the Countryside*. Jonathan Cape, London.

Hesketh, F. 1997. A brief history of diversity. *Landscape Design* 260:25–34.

Hollis, G.E., Adams, W.M. and Aminu-Kano, M. 1993. *The Hadejia–Nguru Wetlands; Environment, Economy and Sustainable Development of a Sahelian Floodplain Wetland*. IUCN, Gland and Cambridge.

Hynes, H.B.N. 1985. The stream and its valley. *Verhandlungen*

der Internationalen Vereinigung für theoretische und angewandte Limnologie 19:1–5.

International Federation for Organic Agriculture Movement (IFOAM) 1997. Internet address http://www.ecoweb.dk/ ifoam.index.html

International Union for Conservation of Nature and Natural Resources, United Nations Environment Programme and World Wildlife Fund (IUCN/UNEP/WWF) 1980. *World Conservation Strategy: Living Resource Conservation for Sustainable Development*. Gland.

International Union for Conservation of Nature and Natural Resources, United Nations Environment Programme and World Wildlife Fund (IUCN/UNEP/WWF) 1991. *Caring for the Earth: a Strategy for Sustainable Living*. Gland.

Jacobs, 1991. *The Green Economy*. Pluto Press, London.

Kinsley, M.J. and Lovins, L.H. 1995. *Paying for Growth, Prospering from Development*. Rocky Mountain Institute, Snowmass, Colorado.

Levins, R. 1995. Preparing for uncertainty. *Ecosystem Health* 1:150–169.

Lovins, A.B. 1977. *Soft Energy Paths*. Ballinger, Cambridge, Massachusetts.

Majura, P.B. and Banda, A.F. 1996. Sustainability of Lusuka Sewage Works, Zambia. In: *Sustainability of Water and Sanitation Systems*. Intermediate Technology Publications, London, 147–149.

Maltby, E. 1986. *Waterlogged Wealth*. Earthscan, London.

Meadows, D.H., Meadows, D.L., Randers, J. and Behrens, C.W. 1972. *Limits to Growth*. Pan, London.

Meadows, D.H., Meadows, D.L. and Randers, J. 1992. *Beyond the Limits: Confronting Global Collapse, Envisioning a Sustainable Future*. Earthscan, London.

McCully, P. 1997. Trio of nations supports Three Gorges Dam. *World Rivers Review* 12:4.

Morris, C. 1997. Effete or effective stewardship? *Landscape Design* 260:19–23.

National Rivers Authority, Thames Region (NRATR) 1995. *Thames 21: a Planning Perspective and a Sustainable Strategy for the Thames Region*. Reading.

National Rivers Authority, Thames Region (NRATR) 1997. *Thames Environment 21 – The Environment Agency Strategy for Land Use Planning in the TR*. Environment Agency Thames Region, Reading.

Newson, M. 1992. *Land, Water and Development: River Basin Systems and their Sustainable Development*. Routledge, London.

Noss, R.F. and Cooperider, A.Y. 1994. *Saving Nature's Legacy*. Island Press, Washington, DC.

O'Riordan, T. 1993. The politics of sustainability. In: Turner, R.K. (ed.) *Sustainable Environmental Economics and Management: Principles and Practice*. Belhaven Press, London.

Owen, O.S. and Chiras, D.D. 1995. *Natural Resource Conservation; Management for a Sustainable Future*. Prentice-Hall, New Jersey.

Page, R, 1997. Factory farm or countryside: time to put the culture back into our agriculture? *Birds* 16:31–32.

Pain, S. 1997. Conservation comes home. *Kew Journal* Spring 1997:21.

Palmer, T. 1994. *Lifelines: the Case for River Conservation.* Island Press, Covelo, California.

Pearce D.W., Markyanda, A. and Barbier, E.B. 1989. *Blueprint for a Green Economy.* Earthscan, London.

Pearce, F. 1997. The state we're in. *The Royal Geographical Society Magazine* LXIX:75–80.

Penning-Rowsell, E., Winchester, P. and Gardiner, J.L. 1998. New approaches to sustainable management for Venice. *Geographical Journal* 164(1).

Pettifer, J. 1997. The world – a stunning place. *Birds* (RSPB, Sandy) (Summer): 20–27.

Pottinger, L. 1997. Learn to live with rivers. In: *Raise the Stakes*, No. 27, Planet Drum Foundation, San Francisco.

Philip Williams and Associates (PWA) 1996. *An Evaluation of Flood Management Benefits through Floodplain Restoration on the Willamette River, Oregon, USA.* Prepared for International Rivers Network, Portland, Oregon. Philip Williams and Associates, San Francisco.

Schumacher, E.F. 1973. *Small is Beautiful: Economics as if People Mattered.* Penguin Books, London.

Scottish Environment Protection Agency and Environment Agency 1997. *A Guide to Sustainable Urban Drainage.* Stirling.

Sheate, W. 1994. *Making an Impact: A Guide to EIA Law and Policy.* Cameron May, London.

Sherman, J. 1997. Some consequences of cheap trees and cheap talk – pulp mills and logging in Northern Alberta. *Ecologist* 27:64–68.

Shiva, V. 1991. The failure of the Green Revolution: a case study of the Punjab. *Ecologist* 21:57–61.

Swallow, B., Prugh, T., Costanza, R., Cumberland, J., Daly, H., Goodland, R. and Norgaard, R. 1997. Review of natural capital and human economic survival. *Ecosystem Health* 3:142–164.

Turner, R.K. 1993. Sustainability: principles and practice. In: Turner, R.K. (ed.) *Sustainable Environmental Economics and Management: Principles and Practice.* Belhaven Press, London, 12–54.

United States Army Corps of Engineers (USACOE) 1972. *Charles River Watershed*, Massachusetts, New England Division, Waltham, Massachusetts.

Upper Mississippi River Conservation Committee (UMRCC) 1993. *Facing the Threat: an Ecosystem Management Strategy for the Upper Mississippi River.* A Call for Action from the Upper Mississippi River Conservation Committee. UMRCC Co-ordinator, Rock Island, Illinois.

Urbonas, B. and Stahre, P. 1993. *Stormwater: Best Management Practices and Detention for Water Quality, Drainage and CSO Management.* Prentice-Hall, New Jersey.

Von Weizsäcker, E., Lovins, A.B and Lovins, L.H. 1997. *Factor Four: Doubling Wealth, Halving Resource Use.* Earthscan, London.

Wilson, E.O. 1988. The current state of biological diversity. In: Wilson, E.O. (ed.) *Biodiversity*. National Academy Press, Washington, DC, 3–18.

World Commission on Environment and Development (WCED) 1987. *Our Common Future.* Oxford University Press, Oxford.

Young, B.S. 1997. Dear Prime Minister: an open letter from the RSPB. *Birds* (Summer):2–3.

B   Geographical Settings

14

Global disparities in river conservation: 'First World' values and 'Third World' realities

M.J. Wishart, B.R. Davies, P.J. Boon and C.M. Pringle

Introduction

The divide between the North and the South, encompassing the developed countries of the First World and those developing or non-developing countries of the Third World, now extends well beyond the socio-economic and political grounds within which it was first conceived. Given that there is often an inextricable link between economic and environmental concerns, and with an increasing global awareness as to the state of the natural environment, the divide between the North and South provides an interesting template upon which to examine the issues and challenges facing the conservation of lotic environments.

In modern terms, conservation has become a rather nebulous and vague concept, reflecting as much about the cultural beliefs of the individual, or the society, as it does the state of the natural environment. Definitions such as those proposed by the World Conservation Union are centred around concepts of utilization of natural resources within a sustainable framework to ensure the prevention of species extinction and the maintenance of ecosystem viability (International Union for Conservation of Nature and Natural Resources, United Nations Environment Programme, and World Wide Fund for Nature – IUCN/UNEP/WWF, 1991). Irrespective of definition, however, it could be argued that, until recently, the dominant conservation paradigm has been one of preservation, advocating the maintenance of ecosystems in isolation from human populations. With the global population at 5.84 billion in mid-1997, and predicted to increase to 8.04 billion by 2025 (United Nations, 1997), such exclusionary philosophies are becoming increasingly difficult to justify when pitched against the immediate needs of human populations. This is particularly true in the Third World, where the majority of the world's people without access to waterborne sanitation (>2.5 billion) or safe potable water (1.2 billion in 1990) reside, and where more than 10 million people die annually because of poor sanitation.

Conservation efforts have been directed historically toward terrestrial and, more recently, marine environments, despite the fact that inland aquatic ecosystems represent one of the most diverse and as yet largely undescribed group of environments. The neglect of lotic ecosystems ignores the fact that these

Global Perspectives on River Conservation: Science, Policy and Practice.
Edited by P.J. Boon, B.R. Davies and G.E. Petts. © 2000 John Wiley & Sons Ltd.

environments include some of the most threatened ecosystems on earth. For example, it has been estimated that the rates of imperilment within major aquatic taxa such as fish, crayfish and mussels are three to eight times those for birds and mammals in North America (Masters, 1990; in Angermeier, 1995). Lotic conservation efforts are often confounded because of the longitudinal nature of lotic ecosystems, coupled to the intrinsic linkages between rivers and their catchments. As such, the implementation to lotic ecosystems of a 'preservation' philosophy would necessitate complete control over an entire catchment. While this may be possible on smaller spatial scales, preservation of international rivers such as southern Africa's Zambezi River would require detailed and coordinated efforts between eight different countries, covering an area of 1 234 000 km^2. Other constraints to lotic conservation include a general lack of taxonomic information and the continuing taxonomic confusion surrounding many aquatic taxa. This is confounded by the historical development of conservation efforts which have arisen not from recognition of the need to conserve rivers as functioning ecosystems, but more from the perspective of their maintenance as systems for safe, potable water supply for human consumption and for transport and other human utilities.

Despite the altruistic global intent of science, based primarily upon the premise of advancing human understanding through increased knowledge, without respect for, or recognition of, artificial boundaries, it has failed to be truly global in its extent and understanding (e.g. Gibbs, 1995; Wishart and Davies, 1998). For example, although the vast majority of the world's tropical rain forest is found within the developing world, these areas are home to only 7% of the world's trained ecologists (Colvin, 1992). As a result of this scientific maldistribution, the framework for developing alternatives, and for identifying sustainable economic activities and conservation priority areas, does not exist in many Third World countries. Overcoming these impediments to conservation has traditionally been reliant on scientific tourism and the implementation of technologies developed elsewhere in the world. In this last case, however, the fundamental problem centres upon the appropriateness of the extrapolation from one region to another of scientific constructs developed for what are often very disparate areas (e.g. Winterbourn *et al.*, 1981; Lake *et al.*, 1986; Junk *et al.*, 1989), a process in which biogeographic boundaries and biomes are often poorly considered. For example, the northern temperate environments, from which

much of our current understanding has been derived, can be considered representative of fewer than 20% of the world's aquatic biomes (Williams, 1988). Crossing such boundaries has many intrinsic and inherent problems, and the problems experienced, and solutions required, by the countries of the Third World are often very different and far removed from those provided by the First World.

While typically less is known about the biodiversity and biology of developing regions, this situation is amplified in freshwater ecosystems, where even in industrialized nations the question of aquatic diversity is poorly recognized (e.g. Lydeard and Mayden, 1995). Faced with harsh economic realities, many developing countries view science as an ill-affordable luxury, and thus it is that our scientific understanding, irrespective of discipline, is largely confined to those nations wealthy enough to support its pursuit. The collection and dissemination of information, essential for development of a conservation framework is at present heavily biased towards those countries of the North, and with increasing reliance on technologies such as the Internet for dissemination there is potential for even greater divisions.

Definitions

Precise meaning of the terms 'First World' and 'Third World' has been distorted over time (e.g. Thomas *et al.*, 1994). Modern reference to the Third World has come to mean *countries of Asia, Africa and Latin America not politically aligned with Communist or with Western nations* (*Concise Oxford Dictionary*), with the western and communist nations in this definition comprising the nations of the First and Second Worlds. Such definitions are centred around social, economic and political considerations, invoking connotations of poverty and a lack of development. However, the term was originally coined by the French economist and demographer Alfred Sauvy (1952), in reference to the exclusion and aspirations of the Third World, in which he saw the Third World ('Tiers Monde') as a modern parallel to the Third Estate ('Tiers Etat') of the French Revolution – the class of commoners. Sauvy's term carried not only the connotation of exclusion from power but also the idea of revolutionary potential, with the Third World 'excluded from its proper role in the world by two other worlds', the East and the West, whose conflict monopolized the spotlight of history. In essence the Third World came to represent a 'third' way of doing things. Modern reference to the Third World, however, has become more synonymous with the 'North/South' divide proposed by the Brandt Report of the Independent Commission on International Development Issues (Brandt, 1982; Figure 14.1). In essence, the Brandt Commission delineated the globe according to the developed and non-developed nations of the world, a collective definition that threw together a disparate group of countries. These included developing countries, such as many of the rapidly industrializing nations of south-east Asia, along with the poorest of African nations, which are at present unable, or unwilling, to alter the current *status quo*.

While definitions of any kind invariably tend to be value laden, it is not our aim to become embroiled in a discussion as to the relative merits or demerits of various delineations. Reference to the developed/ developing world has become largely euphemistic, with little more than a descriptive meaning, while reference to the North–South divide risks imposing a geographical divide, or dictate, upon what are essentially social, economic and political problems. We have therefore chosen to use each of the terms interchangeably which, despite the broad differences, ambiguities and problems, reflect countries that have something in common. Our reference to the Third World arises more from a historical analysis, stressing the colonial experience and separation from power shared by these countries in the modern world. We do not, nor could we, attempt to identify or review the individual problems facing these nations. The issues affecting countries of the Third World will invariably include many in tropical Asia, where problems are primarily focused on issues pertaining to drainage-basin degradation through, for example, deforestation (Dudgeon, 1992; Pethiyagoda, 1994; Dudgeon *et al.*, this volume). These problems, while common to other regions of the world, will undoubtedly be different from those experienced (for instance) in the dryland regions of southern Africa, where problems are centred on over-abstraction, pollution and hydroclimatic stochasticity (Davies *et al.*, 1993, 1995; Davies and Wishart, this volume). Similarly, the problems threatening lotic ecosystems in South America are different from those of developing island states, such as Madagascar (see, respectively, Pringle *et al.* and Benstead *et al.*, this volume).

Such social, political and economic dichotomies also exist at different scales within the same country. For example, development in the south-eastern states of America has lagged behind that of the northern and western states (Lydeard and Mayden, 1995), while in

most Third World countries the problems faced by the rural poor are typically far removed from their urban counterparts. Given these disparities, we have attempted to identify commonalties in the general perceptions of river conservation between countries of different economic, political and social standing, defined along the lines of Brandt's North/South divide, the implications of which are important with respect to the science, conservation and management of river ecosystems. Our discussion, therefore, is limited to the problems and threats facing the developed First World nations, essentially equivalent to Brandt's 'North' divide, including North America, Europe, Australasia, and the developing Third World countries of Brandt's 'South' divide, which includes many 'non-developing' countries. Developing countries are defined here as those with transitional economies, or those that are attempting to improve their social and economic position within the global economy. In contrast, non-developing nations are those which, for whatever reasons (usually economic or political) are not at present striving toward improved living conditions and the quality of life within the global context, and are trapped in a cycle of perpetuating poverty.

Issues facing river conservation

DEVELOPED WORLD

Habitat loss and degradation, the spread of exotic species, over-exploitation, secondary extinctions, chemical and organic pollution, and climate change have all been identified as constituting major threats to the conservation of biodiversity in lotic environments worldwide (Allan and Flecker, 1993). Of interest, however, is the undeniable fact that a misperception resides in the private and public psyche of First World communities: namely, that the Third World is 'bad' (essentially 'worse') in terms of its pollution, problems and the general degradation of its aquatic environments. The reality is that although only accounting for less than one-fifth of the global population, First World communities are responsible for production of two-thirds of all greenhouse gases, 85% of the worlds CFCs, and most of the sulphur dioxide and nitrogen oxide emissions that produce acid rain, through over-consumption and often poorly regulated industries.

Indeed, most rivers of the developed world are already largely degraded. For instance, aquatic degradation in North America through habitat loss,

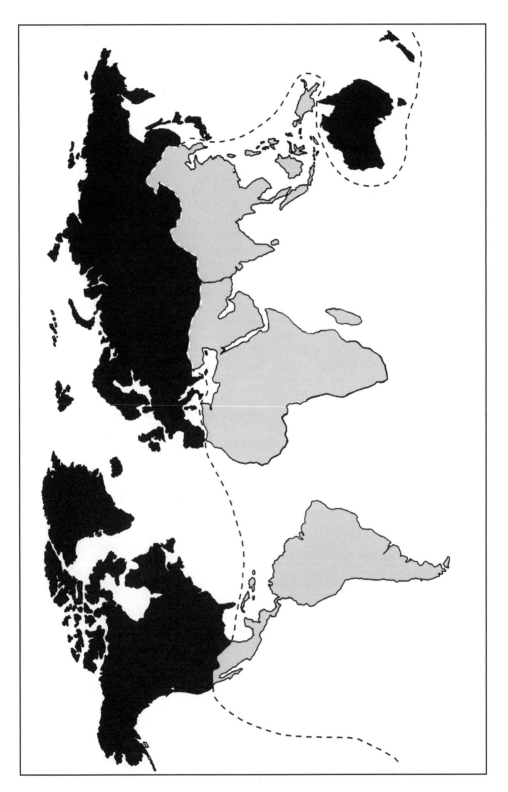

Figure 14.1 *Delineation of 'First' (black) and 'Third' (grey) worlds as defined by the Brandt Report (1982)*

introduction of exotic species and pollution, continues to result in high rates of endangerment and extinction among aquatic species (Miller *et al.*, 1989; Williams *et al.*, 1989, 1993; Karr *et al.*, this volume). However, the greatest threat to rivers in the First World undoubtedly comes from the regulation of natural flows for power production, flood control, improvement of navigation, potable water supply and large irrigation schemes for industrialized agriculture (e.g. Ward and Stanford, 1979; McCully, 1996). For example, in 48 contiguous states of the USA less than 2% (<100 000 km) of the total 5 200 000 km of natural stream length was regarded as being of sufficient quality to warrant federal protection status, with only 42 high-quality, free-flowing rivers longer than 200 km still in existence (Benke, 1990). Furthermore, river regulation has seriously damaged anadromous fisheries stocks, with a drastic decline along the rivers of the US Pacific coast, from 400 or so salmon and steelhead stocks to 214, of which 169 vary between moderate to high risk of extinction (e.g. McCully, 1996). Similar reports come from Europe with Billen *et al.* (1995) noting the elimination of Atlantic salmon from the French rivers – the Dordogne, Meuse and Moselle – by dams around the start of the 20th century, whilst more recently, regulation has added the Garonne and the Seine to the list. Only the Loire and one of its tributaries, the Allier, still retain wild salmon stocks.

As a result of widespread degradation of this nature, and also stemming from growing public concern, the past decade has seen an increased focus on the plight of many rivers of the First World. As a direct result a groundswell in research has documented and developed protocols for the assessment of status and 'health' of aquatic ecosystems. Indeed, there have even been recent moves to remove regulating dams from a number of rivers flowing through areas of prime conservation value in the USA, in order to restore their 'wild' nature (McCully, 1996). The river conservation groundswell has been assisted by public awareness and a general availability of trained personnel (albeit patchy). Research efforts into rehabilitation and mitigation technologies, in particular, have grown rapidly. As such, status assessments of rivers in many developed nations, the issues and pressures facing them, and their conservation are relatively well recognized and have been the subject of numerous discussions and publications (e.g. Benke, 1990; Barmuta *et al.*, 1992; Boon *et al.*, 1992; Allan and Flecker, 1993; Pringle *et al.*, 1993; Arthington and Welcomme, 1996).

As a direct result of the long-term and sustained environmental degradation of rivers in the developed world, coupled with the 'luxury' of little or no poverty spiral and its attendant struggles, as well as a relative wealth in trained research personnel, the rather patronizing view of the developed world that 'pristine' systems or segments of systems should be protected at all costs, has been foisted on many Third World communities as the 'only route' for conservation efforts (Bonner, 1993). This view overlooks the very real fact that many developing nations have yet to go through the process of industrialization responsible, in part, for the economic entrenchment of the developed nations, as well as for the massive environmental degradation of their own systems. It also conveniently overlooks the fact that many companies from First World countries, through their presence and exploitation of Third World resources, cheap labour markets and often rudimentary environmental legislation, have been partly responsible for much of the present degradation in those countries. For example, the strengthening of regulations governing disposal of toxic wastes in First World countries has also resulted in an increased cost of disposal, so much so that the burgeoning international trade in toxics currently represents a rapidly emerging and major threat to freshwater ecosystems. In the United States costs of disposal rose 12-fold between 1980 and 1992, creating a market in Africa, Latin America, and now Eastern Europe and the former Soviet Union. When Poland opened its borders in 1989 it was inundated with toxics from many First World countries and by late 1990 it was estimated that at least 22×10^6 t of toxics had been offered to Poland, 46 000 t of which is known to have crossed the border (Bochniarz, 1993; Kruszewska, 1993; Athanasiou, 1996). The improper disposal of First World toxic wastes is certain to exacerbate the pre-existing problems facing freshwater ecosystems in most Third World countries as they make their way into ground- and surface-waters. At the same time, countries of the developed world are investing in rehabilitation and mitigation technologies for their own use and for export, while most developing countries have neither sufficient trained personnel to mount restoration programmes, nor the will or energy to achieve anything other than social upliftment and survival, irrespective of the consequences for medium- to long-term sustainability of the natural environment.

Concomitant with the efforts invested in restoration technologies, a large amount of effort in North America and Europe is also being directed towards developing and amending legislation, both through government departments and agencies such as the

Environmental Protection Agency in the United States, the Environment Agency and the Scottish Environment Protection Agency in Britain, and the European Environment Agency. These agencies often have wide-ranging legal jurisdiction and public support, and their influence is enhanced by relatively healthy budgets and well-trained personnel. In the United States, the EPA had sufficient powers and support to stop a water-supply development which would have threatened the survival of the snail darter within the Colorado River Basin. In the United States federal and state agencies have been working with conservation-based organizations, such as American Rivers, National Wildlife Federation and the Sierra Club, to lobby and implement legislation aimed at the preservation of rivers. In 1968 the United States Government passed the Wild and Scenic Rivers Act which, once applied to a stream segment, prohibits any federal agency from approving or assisting any land or water development project that has the potential to affect adversely the designated segment. Unfortunately, such legislation has little jurisdiction over private developments at this time. In 1988, 119 river segments were protected under the Act, a total river length of 14 900 km (Benke, 1990), which, with major commitments made recently by some of the largest federal land and reserve holders, is likely to be increased dramatically with hundreds more segments proposed for inclusion (see also Karr *et al.*, this volume). Over and above this, many states have their own wild and scenic rivers system which affords protection to more than 22 000 km of river (Palmer, 1986).

Despite the advanced recognition of the need for development of more stringent and comprehensive efforts for the conservation of lotic ecosystems, the developed nations of the world are struggling to ensure full protection. For example, in the United States, while the Federal Clean Water Act of 1972 has succeeded in reducing annual discharges of toxic chemicals and raw sewage into US lakes and rivers by about 3.8×10^5 t and 900×10^6 t, respectively (Adler *et al.*, 1993), the Act has many weaknesses which include lack of emphasis on toxics, non-point-source pollution, and problems with regulatory enforcement. Also, some political forces threaten to weaken the Clean Water Act rather than strengthen it. Revisions of the Act that were proposed in 1995 threatened to reverse hard-won progress in environmental clean-up of fresh waters in the United States (Pringle, 1996). It remains to be seen whether US citizens will remain vigilant enough to prevent future attempts at weakening the Act.

DEVELOPING WORLD

If lotic ecosystems in the First World are already largely degraded then the rivers of the Third World, which are generally in much better condition, are rapidly being degraded. This degradation is due primarily to the easing of social and economic constraints in order to promote economic development, without the implementation of appropriate protective infrastructures. Often the dominant paradigm is that the advantages derived from water development projects, such as improved human health, social and/or economic conditions, negate the consideration of possible environmental impacts. This uncritical approach has been reinforced and perpetuated by inadequate policies, ineffectual legislation and a lack of power or resources for adequate implementation. In the United States, for example, the National Environmental Policy Act of 1969 made it a requirement for Environmental Impact Assessments to be prepared by any federal agency planning a project that would significantly affect the quality of the human environment. Nearly 30 years later, such policies simply do not exist in most developing countries. Thus, the very development of Third World countries probably represents the greatest potential threat to the conservation of their rivers. In many instances where policies have been prepared and legislation enacted Third World countries are unable, or unwilling, to enforce laws effectively and so they become little more than words.

Where poverty drives survival scenarios, conservation is largely viewed as an unaffordable luxury, leading to the unbridled development of catchments. Furthermore, rapid economic development inevitably leads to rapid urbanization with associated stresses placed on rivers in terms of water supplies, catchment hardening, land hunger and pollution. All of these stresses mount at a rate that outstrips the development of appropriate infrastructure and protective legislation, even if these last are perceived to be necessary. In this context, it should be noted that during the 1990s approximately 42 million people entered the global labour force annually. Of these, 39 million live in the developing world (International Labour Organization, unpublished data; cited in Livernash, 1994). With mounting pressures from the unemployed, it increasingly becomes politically difficult to constrain development, or to sacrifice job creation for environmental considerations. For instance, the water protection agency in South Africa – the Department of

Water Affairs and Forestry (DWAF) – also acts as the major dam-construction agency and licenses all water development projects. Thus, the agency is forced simultaneously to wear the hats of both 'gamekeeper' and 'poacher'. In such cases, ever-increasing development is coupled with the belief that water supply must also always increase to meet ever-growing demands (e.g. DWAF, 1986; Basson *et al.*, 1997). The realization that there are alternatives to continuous development of water supply projects and many other ways for managing and moderating demand, and that aquatic ecosystems also have their limits, is slow to develop, if at all.

One of the ironies of the past few years is that while the pace of river regulation has slowed in the developed countries of the world, and they have begun to realize the potentially deleterious effects of water development projects (e.g. McCully, 1996), countries of the developing world are entering a phase of rapid construction of large-scale water projects. These projects, aimed at providing electricity and/or potable water for rapidly expanding and urbanizing populations, will prove to be the greatest threat to the rivers of the developing world over the next few decades. For example, in Brazil alone there are at present about 80 large dam projects under consideration, with the potential to flood 100 000 km^2 of Amazonian rain forest. The list is almost endless, and includes potential mega-projects to harness the Congo River for hydropower production for a southern African electricity grid; vast projects under construction in China with potential eviction threats for nearly two million people (Three Gorges Dam on the Yangtze); the controversial Epupa Falls development in Namibia; the Katse and Mohale components of the giant Lesotho Highlands Water Project in Lesotho, supplying water to South Africa; a myriad projects in India, including the highly controversial Sardar Sarovar Project on the Narmada River; projects in Thailand, Vietnam, Pakistan; the Pangue and Ralco projects in Chile on the Bíobío River (hydropower); and the Bakun Dam in Malaysia.

The Paraguay–Paraná Hidrovia Project in South America involves a proposal to convert 3400 km of the Paraguay and Paraná River system into an industrial shipping channel (see case study in Pringle *et al.*, this volume, Part I). The project, which has been called 'the backbone' of Mercosur, the Southern Cone Common Market, is part of a broader plan to expand agribusiness and mining activities in the area which could have irreversible impacts on the world's largest wetlands, the Brazilian Pantanal, and other valuable ecosystems in Bolivia, Paraguay and Argentina. On the other side of the world the Mekong, the lifeline of Indochina, which rises on the Tibetan Plateau and runs south through China, Burma, Thailand, Cambodia, Laos and Vietnam, supporting the world's second most diverse riverine fishery, is now the subject of more than 100 dam proposals. The centrepiece of these proposals is a scheme of 10 dams proposed for the mainstem. Sixty dams are proposed for Laos alone. The Nam Theun 2 Dam in Laos will inundate part of the Nakai Plateau, an area of extensive biological diversity which has been recommended for protection by conservation groups worldwide, including the World Bank's own Global Environment Facility. Although there is a newly established agreement for the sustainable development of the Mekong, and international interests in assisting the Mekong countries in implementing sound river management practices, dam construction and project proposals continue to proliferate at a feverish rate.

Aside from the issues surrounding large-scale water development projects, many rivers in developing countries (urban rivers in particular) are threatened through a general lack of education. Rivers and other water bodies are used as canals, for rubbish removal and waste disposal, as well as for domestic purposes and supplies of water. While it is often recognized that such uses are conflicting and fraught with dangers (waterborne diseases, for instance), in urban developments without rubbish removal and domestic water supply infrastructure there are few options other than to abstract from the river, whilst simultaneously using it as a method of waste disposal. In such areas immediate action needs to be taken to improve waste removal and domestic supplies, but such social development programmes need to grow in association with appropriate education programmes.

Although over-population places huge pressures on the natural environment, over-exploitation of freshwater biota is generally not considered to be a major threat. While there are few examples of over-exploitation leading to the demise of any temperate stream-dwelling species (Miller *et al.*, 1989) there are examples where overfishing has resulted in the local extinction of some fish species in developing countries. This is a function of a far greater reliance by poor people on the exploitation of fish species both for domestic consumption and for providing income from sales, for instance through the international pet trade. While it is unlikely that exploitation for domestic or commercial consumption could be solely responsible for the entire extinction of a species there are examples of local extinctions, such as that of the barramundi in the Oueme River, Benin, Africa (Arthington and Welcomme, 1996). Secondary threats, however, through shifts in community composition, structure and trophic dynamics could culminate in extinction, and the aquarium fish trade currently threatens some species with local extinction and significantly reduced abundances (Welcomme, 1979; McLarney, 1988).

Many of the issues facing non-developing countries are similar to those facing the poor of the developing world. The difference lies in the pressures caused by the *development* of developing world economies. Similarities on the other hand, include human-effluent pollution, unregulated and informal use of floodplains and riparian zones, more often than not within frequent-return floodlines, and similar shortages of trained personnel and the absence of appropriate legislation and its enforcement. Unregulated floodplain development perhaps reflects the biggest threat, both to river integrity and to human safety. It is difficult to reconcile the health and safety problems associated with living on the floodplain, or in areas of heavy mosquito infestation, with the need for water. Many hours are spent daily collecting and transporting water – the closer one is to the water supply the less the energy requirements needed to gather the resource (e.g. Elmendorf and Buckles, 1980; Dankelman and Davidson, 1993), but the greater the potential risks.

Whereas in the rest of the world there have been attempts to integrate the human use of rivers with their broader ecological functions (Boon, 1992a) in these non-developing countries the struggle is for survival, with little regard, awareness or available time for conservation. However, many indigenous peoples have a natural understanding of ethical and behavioural conservation considerations. Too frequently they are ignored, marginalized, 'assimilated' or distanced from the dominant paradigms of modern Western thought, which are essentially exploitative. Whilst floodplains are often utilized extensively, this is done in a manner which recognizes that the river–floodplain is a dynamic system, and that lateral exchanges between the two enhance biological productivity, providing refuges and resources both for riverine and floodplain species (e.g. Junk *et al.*, 1989) as well as for migratory human societies.

THIRD WORLD GROWTH VERSUS DEVELOPMENT

In developing countries the need for economic 'development' to improve social conditions is often

carried out at the expense of the environment, under inappropriate and ineffectual legislation. Many development models are environmentally unsustainable, confusing concepts of growth and development. The worrying trend is that 'growth' is often considered synonymous with 'development'. Growth, as designed by most economies, is an exponential function, ever expanding upon previous growth and as such is simply unsustainable. Unless it is sustainable, development cannot take place (see Gardiner and Perala-Gardiner, this volume). For many, there is a pervading assumption that in the Third World the environment must be compromised for short-term economic gain, which, in the long term will allow for *economically* viable, intensive rehabilitation. For example, Conley (1995) states that:

'While demand management needs due attention, appropriate dam building must also be considered. It is only when countries achieve economic strength that they can support the sophisticated environmental management approaches which are made incumbent by population pressures. Before they have achieved economic security, people struggling for survival will wreck their environment out of sheer necessity. A choice is possible only after greater security and financial strength is attained through development.'

On the contrary, the realities are such that much of the (short-term) economic empowerment involves *over-exploitation* of the environment. For example, in the developing countries of Asia and South America, logging of natural-growth forests is being used for short-term economic benefits. Indeed, such exploitation is often conducted by a few individuals and organizations for their own profit, rather than for the benefit of the general population. Most of the so-called economic gains of such activities, including large-scale water-supply and hydropower projects, disappear into First World, rather than local Third World, economies (e.g. McCully, 1996). Such practices have massive effects on rivers, let alone on forests, and the idea that economic strength is required to ensure effective environmental management and to facilitate restoration and rehabilitation, neglects several fundamental facets of ecosystem functioning.

The environmental situation has been referred to as a vicious circle, since in many countries environmental degradation has progressed to the extent of actually limiting development, and the lack of development is correspondingly limiting environmental protection (Marek and Kassenberg, 1990). This trend is perhaps best illustrated in regions where the low quality and quantity of freshwater resources result in both a lack of drinking water for human populations and an unsuitable supply for industry. For example, according to data from the Polish Inspectorate for Environmental Protection (Srodowiska, 1992), 82% of Polish rivers are not suitable for industrial use. Waters suitable for industrial purposes comprise only 14.5%, and those suitable for agricultural use a mere 3.3%. With respect to water quantity, drainage projects have led to lowered groundwater tables in Poland and excessive drying of considerable areas of land. Increasing needs for water have, in turn, led to further stresses on water supply; the area of excessively dried land in Poland amounts to *ca* 4×10^6 ha and increased drying of Poland's central region is also associated with deforestation (Ryzkowski, 1990).

As a word of caution, and in the context of rehabilitation, it is interesting to note that only 4% of the threatened and endangered aquatic species in the United States for which there are plans for rehabilitation or preservation (published by the US Fish and Wildlife Service) have recovered significantly (Williams and Neves, 1992: in Angermeier, 1995). Indeed, between 1979 and 1989, and despite many efforts, no species have been removed from the American Fisheries Society's list of endangered fish as a result of the implementation of recovery activities. In fact, the rate at which fish and mussel species are becoming threatened continues to increase (Williams *et al.*, 1989; Allan and Flecker, 1993; Williams *et al.*, 1993; Warren and Burr, 1994). Several species have been removed from the list, however, as a result of more detailed information on their distribution (five species) coupled to taxonomic revisions (11 species), while 10 species have been removed as a result of their extinction. Since the initial assessment, more than three times as many fish taxa have declined compared with those that have improved in status (24 versus 7: Williams *et al.*, 1989). Such attempts at restoration and rehabilitation ignore the fact that biological stocks are required from which to restore and rehabilitate natural systems. Rivers by their very nature are dynamic and resilient, with considerable scope to recover from disturbances, but rehabilitation after major disturbance requires a robust biological stock.

The framework for conservation

SCIENTIFIC

As a result of the historical development of limnology most of the existing data and theoretical templates are derived from northern temperate biomes. These comprise less than 20% of those biomes found in the remaining arid, semi-arid, tropical and sub-tropical parts of the world (Williams, 1988), regions comprising predominantly Third World countries. With a solid scientific framework necessary for effective conservation, initiatives in developing and non-developing nations are often hampered by the lack of availability and poor dissemination of scientific information. Given the paucity of data, coupled to rapid population growth, poor education and poverty, attempts are often made to extrapolate from the classic templates of northern-temperate limnology in an effort to meet the requirements of the Third World. The validity of such extrapolation is highly questionable and to a large extent remains unresolved (Winterbourn *et al.*, 1981; Lake *et al.*, 1986; Williams, 1988, 1994; Dudgeon *et al.*, 1994).

The under-representation of Third World scientists in leading international journals has long been recognized. This is reflected in the proportion of scientific contributions from the developing world, which has not increased over the past two decades, as well as the decline in the number of journals from developing countries listed by the Science Citation Index, falling from 80 to 50 between 1981 and 1993 (Gibbs, 1995). As a discipline, however, freshwater ecology exhibits one of the highest levels of collaboration in the sciences, approximating that of physics, but higher than medicine, biology, chemistry and engineering (Resh and Yamamoto, 1994). In the light of the increasing pressures being placed on river ecosystems in developing nations, Wishart and Davies (1998) assessed the scientific foundations for conservation of river systems in the Third World, examining 10 of the leading journals in the field of freshwater ecology pertaining to lotic ecosystems.

Not surprisingly, yet despite a willingness by several journals to publish in other languages, English was the dominant language of publication, reinforcing the biases of the Science Citation Index Journal Citation Reports (JCR) and indicating the dominance of English as the international medium of science. Only 21 of the 8960 papers surveyed were published in French and 19 in German. The majority of papers examined by Wishart and Davies were multi-authored and published

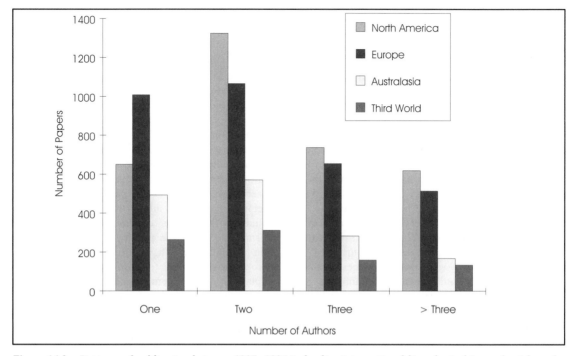

Figure 14.2   *Patterns of publication between 1987–1996 in leading international limnological journals. Adapted from Wishart and Davies (1998)*

by authors from First World countries (Figure 14.2). International collaboration, involving authors from different countries, accounted for only 12.61% of the 6545 multi-authored papers, with most of this between authors from First World countries (72.36%). Research carried out in Third World countries accounted for 11.04% of the 8960 papers examined. While most of these were either single (23.15%) or multi-authored (42.37%) papers from Third World authors, First World authors published 18.30% of research papers from the Third World without the inclusion of any local Third World co-authors; this category was the third most common type of publication (Table 14.1). In contrast, only four papers reflected the reciprocal situation with publications by Third World authors documenting work carried out in a First World country.

Boon *et al.* (1992) found that nearly half (42%) the citations of papers presented at a conference on river conservation came from unpublished reports, theses, conference proceedings, or journals with very restricted distribution. This could well be true of conservation efforts in countries of the Third World. However, the problem is that if not published in an international forum, such information is not open to peer review and cannot be properly assessed. Furthermore, while much

Table 14.1   *Publication patterns of Third World research. From Wishart and Davies (1998). (Single authored papers are denoted by an absence ('–') of junior authorship)*

Region of research	Authorship		Number	%
	Senior	Junior		
3rd World	3rd World	3rd World	419	40.4
3rd World	3rd World	–	229	22.1
3rd World	1st World	–	181	17.4
3rd World	1st World	3rd World	91	8.8
3rd World	3rd World	1st World	69	6.6
1st World	1st World	3rd World	45	4.3
1st World	3rd World	–	4	0.4
Totals			1038	100

science has conservation relevance, there appears to be a reluctance for scientists to apply this to issues of river conservation. The incidence of 'conservation' oriented papers has not increased over the past decade, and in a survey of seven journals less than 2% of 5249 papers examined concerned the conservation or management of rivers (Boon, 1992b). Conservation measures are therefore either being drawn from the 'grey' literature of unpublished reports, theses, conference proceedings, etc. or being abstracted from outwardly more theoretical pursuits published in international journals.

DISSEMINATION

Publication of information in an international journal cannot guarantee that it is widely disseminated in the country within which the work was carried out. In many instances the primary reason a host country does not conduct its own research is the lack of finance, so it is unlikely that it will be able to afford the luxury of a subscription to one of the leading journals. Resh (1985) concluded that subscription to the core journals examined in his review of aquatic entomology and benthic biology in freshwater ecology (36 journals) would cost, at 1984 prices, US$5041. He went on to make the point that this does not include the cost of back issues and that these core journals only contained from 24.1 to 65.9% of the total journal citations in each of the bibliographic sources examined. A decade later, Wishart and Davies (1998) estimated that subscription to their list of 10 journals examined would cost US$7801 at 1994 prices (*Urlich's International Periodicals Directory*, 1994–95). Such costs are prohibitive for many libraries, even those in the developed world.

Many Third World countries do not have the economic resources to invest in scientific research. This is highlighted by the fact that while Third World scientists account for 24.1% of the world total, they spend only 5.3% of the total global research budget (Gibbs, 1995). As a result one of the greatest threats to extant and future conservation efforts is the absence of any solid scientific information base to help guide effective conservation and management strategies. Examination of the Southern African Bibliographic Network (SABINET), a database documenting the holdings of over 1929 institutions throughout southern Africa, shows that 9 of the 10 journals surveyed by Wishart and Davies (1998) can be found within the region. On average, journals were subscribed to by 14 different institutions (range: two institutions subscribe to the *Journal of the North American Benthological Society* – 29 to the *Australian Journal of Marine and Freshwater Research*), with only six journals received by institutions beyond the borders of South Africa and many states, such as Moçambique and Botswana, not subscribing to any.

Disseminating information from the world's developing nations is one of the main challenges facing science in the decades to come. With increasing dependence upon computer technology both for dissemination and retrieval of information, journal indexes, such as the Science Citation Index (SCI) are assuming greater importance in providing workers with the latest developments in their field. The number of journals from developing nations listed on the SCI fell between 1981 and 1993 from 80 to 50 (Gibbs, 1995). Inclusion of prospective journals in such databases is based on the journals' citation rates. Without proper and sufficient exposure, it is impossible to obtain the citation rates required and thus the circle of exclusion is perpetuated. In Latin America, it has been estimated that 70% of the journals are not included in any index (see Gibbs, 1995). There are problems though. For example, of the 1500 scientific journals published in India, it is estimated that only 20% are refereed and appear regularly (Gibbs, 1995).

In response to the problems associated with disseminating scientific information from the Third World, the United Nations has sponsored three commercial indexes of Third World journals. The International Association of Theoretical and Applied Limnology (SIL) has also embarked on a programme of promoting limnology in developing countries through a series of publications (Mori and Ikusima, 1980; Gopal and Wetzel, 1995). Despite the problems associated with telecommunications in many Third World countries there are also a number of initiatives to connect Third World countries to the Internet, thus providing immediate access to a rapidly expanding system containing a vast amount of information. While this addresses the problems associated with the availability and dissemination of information, it is largely restricted to those in technologically advanced urban areas. It is estimated that fewer than one library in 10 in the Third World has a computer, let alone access to CD-ROM drives and the Internet. Failure to recognize such constraints and to address the problems surrounding the provision and reliability of telecommunications in these areas will ultimately limit the viability of any initiatives.

It has been emphasized that science and its dissemination is essential for providing the theoretical framework for conservation and management. While science is essential in providing the foundations for successful conservation, the key to conservation in all instances is education. Strong internal public support must be developed to advocate environmental remediation and environmentally sound protection. As in First World countries, such as the USA and Britain (Pringle and Aumen, 1993), collaboration between scientists and non-governmental

Table 14.2   *Demographic breakdown and geographical representation of membership (numbers, N, and %) to professional societies and associations in freshwater ecology. From Wishart and Davies (1998). (SIL: International Association of Theoretical and Applied Limnology; NABS: North American Benthological Society; FBA: Freshwater Biological Association; ASL: Australian Society for Limnology; SASAQS: Southern African Society of Aquatic Scientists). Numbers in parentheses after the organization abbreviations indicate the number of countries represented in each*

	SIL (44*)		NABS (40)		FBA (61)		ASL (9)		SASAQS (9)	
	N	%	N	%	N	%	N	%	N	%
Europe and North America	2458	81.77	2210	95.51	1328	91.34	7	1.43	8	4
Middle East and Asia	242	8.05	23	1.00	29	2.00	1	0.21	–	
C&S America	150	4.99	26	1.12	7	0.48	–		–	
Australasia	74	2.46	47	2.03	47	3.23	481	98.36	2	1
Africa	82	2.73	7	0.30	38	2.61	–		190	95
Other	–		1	0.04	5	0.34	–		–	
Total	3006	100	2314	100	1454	100	489	100	200	100

* Note that this is actually an underestimate as several countries were grouped into broader categories, these being: Middle East (various), Far East (various), Africa (various) and C&S America (various), which contain a total of 81 members.

organizations involved in conservation and management is beginning to emerge as a force in the development of environmental reforms in Third World countries. However, society's participation in the process of decision-making constitutes a challenge for those with no experience in these forms of governance and state functioning (e.g. National Foundation for Environmental Protection of Poland, 1993). It is here that well-organized and well-staffed organizations, such as American Rivers, the International Rivers Network and the Hydropower Reform Coalition, are increasingly assuming importance at all levels, including the international scene. Many such organizations, while involved in establishing and coordinating river conservation efforts at a local level, operate in collaboration with programmes in many Third World countries. Through capacity building and increased community education, these organizations have an important role to play in putting pressure on development aid agencies from the industrialized nations, to ensure that a properly costed EIA is part of any development project.

IMPLEMENTATION – TRAINING AND PERSONNEL

The documented increase in international collaboration in science throughout the past century (Storer, 1970; Luukkonen *et al.*, 1992) has been viewed as 'representing a response to increasing professionalization of science' (Beaver and Rosen, 1978, 1979), while presumably also reflecting technological advances in communication, such as the

Table 14.3   *Membership of the International Association of Theoretical and Applied Limnology (SIL) Working Group on the Conservation and Management of Running Waters*

	Number	%
First World		
EC/Western Europe	117	63
North America	30	16
Eastern Europe	10	5
Australia/New Zealand	8	4
Third World		
South America/Carribbean	7	4
Indian sub-continent	6	3
Africa/Middle East	4	2
South Africa	3	2
Tropical SE Asia	1	1

Internet. Resh and Yamamoto (1994) suggested that the increased level of collaboration is perhaps 'an inevitable outcome of the formation of scientific societies and associations'. In assessing the contribution of these to increased collaboration and internationalization of science, Wishart and Davies (1998) found that the demographic distribution of members to the world's leading professional societies still reflected their European and North American origins (Table 14.2). The SIL Working Group on the Conservation and Management of Running Waters shows similar patterns with a domination of First World representation (Table 14.3).

The implications of these biases are many and varied. The absence of any collaboration with local personnel in research initiatives in Third World regions represents a failure to recognize the need for training and financial support in Third World research. Failure to ensure proper representation within societies, associations and working groups limits the potential application of research, and the implementation of conservation initiatives. It must be recognized, however, that many efforts and contributions of visiting researchers are not reflected in the international literature. International publications are often the final product in a progression which begins with a locally published internal report. Whilst such reports are not always readily accessible to the broader community, they do contribute to the scientific infrastructure of a country. However, without any feedback into local communities invaluable opportunities for training are often missed, and politicians and planners will have an inadequate basis for making decisions on developments affecting aquatic environments.

In most Third World scenarios, community involvement and education must be central to conservation. Women, in particular, are often seen as the instrument to success. In most rural communities, women are largely responsible for the tasks essential for survival, including the daily collection of water (Dankelman and Davidson, 1993). In this context, the implications of basic water provision to such communities – a simple tap – are enormous. Women would be freed from onerous daily chores and be able to devote more time to family care and education. In East Africa, women can use 12% of the calorie intake (and in drier regions as much as 27% – Dankelman and Davidson, 1993), in collecting water. This can take more than one hour per day on average, with each load weighing as much as 25 kg. Such scenarios are common all across the countries of the developing world. For example, before the installation of water supply, the only source of water for the people of Chijtinimit and Chontala villages (India) was the river and several small water holes. People walked an average of 712 m (46 to 1236 m) in Chontala and 733 m (131 to 1477 m) in Chijtinimit (Elmendorf and Buckles, 1980) every day, just to collect water.

In an attempt to protect *Mbuna* – a flock of more than 200 small, colourful or vividly marked species of cichlid in Lake Malawi – the government declared an area of 87 km^2 extending 100 m into the lake as a national park, which was also listed by UNESCO as a World Heritage Site. Having done so, however, it was realized that these species were not threatened from overfishing by local communities, but from the increased siltation resulting from the removal of riparian vegetation from streams feeding into, and surrounding, the lake. This vegetation is an important and essential fuel source for many of the villagers, so rather than forcing the people out of the park the Department of Parks and Wildlife opted for an integrated educational approach to conservation. Local villagers were educated on the importance and significance of the fish as well as being charged a small fee for firewood collection in the park and encouraged to use less wood. The promotion of integrated consultative approaches for effective river basin management in most Third World countries is ultimately dependent on education and disseminating information to those using the resource.

Quo vadis?

In the First World, many river systems have suffered major or partial degradation and resources are now being directed towards rehabilitation and mitigatory measures. Although degradation of rivers in the countries of the Third World is commonly accepted as a fact of life, many systems remain in good condition compared with those in the First World. Thus, there is an opportunity for these countries to achieve economic and social development within a framework of environmental protection and appropriate management, but major obstacles lie ahead. For example, Third World countries account for 98% of the world's population growth, a population in which 35%, on average, is under 15 years of age (as of mid-1996). This fact alone will lead to enormous increases in future pressures on land and resources, and raises the question of whether or not the present scientific framework provides an adequate foundation for river conservation.

An international course in limnology, coordinated by the Limnology Institute of the Austrian Academy of Sciences in Vienna, and supported through UNESCO, has contributed to training more than 250 limnologists from the developing world, yet it has failed in terms of true capacity-building with few graduates from the courses being able to continue in active research (Gopal and Wetzel, 1995). Unfortunately, training programmes such as these often do not address the political framework within which their graduates will work when they return to their own countries. There is a real need to develop or consolidate regional cooperative frameworks to achieve a long-term sustainable critical mass (Denny, 2000). Such frameworks, drawing on the expertise and experience of Northern institutions, would result in education and research better able to address regional needs within their own socio-cultural and political environments (Denny, 2000). Furthermore, many Third World countries need to develop and secure their own human resources. For example, in the 1980s the United Nations estimated that there were about 70 000 professionals from Africa working overseas, while at the same time there were 80 000 expatriates working in Africa.

Approaches to education and conservation are changing. A conference held at Yale University in 1997 brought together African, American and European scholars and conservation professionals specializing in the densely forested Sangha region of the north-western Congo River Basin. This region typifies many of the problems associated with the Third World. A site of contact among myriad African language and culture groups for millennia, the colonial relationships was within this region were historically contested by French and German governments and concessionary companies around 1900. Relationships are no less complex today, with American NGOs, World Wildlife Fund (WWF–US) and Wildlife Conservation Society (WCS) having established several contiguous parks and reserves since 1990, and the European organizations, die Gesellschaft für Technische Zusammenarbeit (GTZ) and Ecosystèmes Forestiers en Afrique Centrale (ECOFAC), maintaining a number of forest management projects. Meanwhile, private concessions for timber and tourism companies also carry out activities within the region. All of this is superimposed upon local communities and cultures operating within a post-colonial governance. With plans to host a second conference in a more local setting, the project aims to bring together relevant information, personnel and organizations to explore the region's past ecological and social context, and to consider the relevance of the region as a unit of future analysis and action. Similarly, a training workshop held in Moçambique in October 1997 to address the sustainable utilization of the Cahora Bassa Dam and the design of rehabilitation flood flows from the dam provides another template for future action. One of the primary features of this workshop was to facilitate exchange between scientists from numerous disciplines and nationalities, and to involve senior politicians and important policy analysts in the decision-making process.

There is also an increasing awareness of the need for cooperation between development and conservation agencies. River conservation cannot be solely directed to water supply and sanitation at the exclusion of ecosystem functioning, and recognition of this fact is leading to greater cooperation between organizations such as the WWF and the United Nations Children's Fund (UNICEF). Such cooperative efforts serve not only to provide safe water supplies and protection of natural resources, but also contribute to environmental education, which is equally important to the long-term sustainable utilization of rivers.

As well as improvements in education, training and awareness, there is a need for increased recognition that the import of inappropriate First World technologies aimed at improving conditions within Third World communities more often results in environmental degradation and water shortages. In Asia, it is not uncommon for 70–80% of the water drawn from a river for irrigation never to reach its intended destination (World Resources Institute, 1986). Many of the early efforts during the Decade of Water Supply and Sanitation failed because they relied on high-technology pumps, for which was provided no support, servicing or technical training. Furthermore, many of these programmes failed due to the inability of many women and young girls to operate pumps that were too heavy (Dankelman and Davidson, 1993). More recent developments have acknowledged the importance of using locally derived materials and relying on locally derived technologies, many of which involve communities in making decisions. For example, AFOTEC, the International Service for the Support of Training and Technologies in West Africa/Sahel, based in Senegal, aims to strengthen the abilities of community groups to organize and complete their own development projects through the use of indigenous techniques and methods which have been perfected in Africa. In doing so they put communities wanting water in touch with the technology institute in Mali who have developed a simple, sturdy pump. They also advise groups on appropriate containers which are safer for transporting and storing water, reducing the risk of infections and contamination, as well as encouraging the use of taps on containers rather than dipping cups into water and therefore spreading disease (Dankelman and Davidson, 1993).

The question will always remain one of responsibility. The countries of the Third World are not necessarily looking for, nor do they necessarily need, solutions from the First World. There is potential for change, and for effective conservation, appropriate legislation and utilization and such changes have taken place, and are still continuing. The immediate pressure being experienced in many Third World countries has created a sense of urgency which has fostered the development of a number of affordable and innovative ideas and solutions. South Africa, for example, facilitated by the transition towards a new nation state, continues to develop and implement solutions to problems, many of which are common to the Third World. Many of these solutions are aided by, or in collaboration with, institutions and agencies from First World countries. Similarly, in Madagascar the development of a new environmental policy document is attempting to provide new and innovative techniques to reconcile the needs of the people within the framework of a finite and fragile environment. Nevertheless there are many in the First World, along with many in the Third World, who retain the mythological mind-set that 'West is best'. This mythology needs to be replaced. New and innovative techniques and solutions are required that arise from, and are subsequently used in, a local context situated within a broader regional framework.

Acknowledgements

Thanks to Patrick Denny, Sean Eekhout, Onno Huyser, Warona Seane and Silindiwe Sibanda for discussions and comment on this and an earlier, related publication. Thanks also to all those who have academically and otherwise contributed to the formation not only of the chapter but many of the ideas presented therein. The meeting of the developing country limnology working group at the 1998 SIL Congress provided lively and thought-provoking debate; the convenors and contributors to this working group are acknowledged and thanked, as are the congress organizers. R.N. Gwynne is thanked for reviewing and providing comment on the manuscript.

References

Adler, R.W., Landman, J.C. and Cameron, D.M. 1993. *The Clean Water Act 20 Years Later*. Island Press, Washington, DC.

Allan, J.D. and Flecker, A.S. 1993. Biodiversity conservation in running waters. *Bioscience* 43:32–43.

Angermeier, P.L. 1995. Ecological attributes of extinction-prone species: loss of freshwater fishes of Virginia. *Conservation Biology* 9:143–158.

Arthington, A.H. and Welcomme, R.L. 1996. The condition of large river systems of the world. In: Armantrout, N.B. and Wolotira, R.J. (eds) *Condition of the World's Aquatic Habitats; Proceedings of the World's Fisheries Congress,*

Theme 1. Oxford & IBH Publishing Co. Pvt. Ltd., New Delhi, 44–75.

Athanasiou, T. 1996. *The Divided Planet: The Ecology of Rich and Poor*. Little, Brown and Company, New York.

Barmuta, L.A., Marchant, R. and Lake, P.S. 1992. Degradation of Australian streams and progress towards conservation and management in Victoria. In: Boon, P.J., Calow, P. and Petts, G.E. (eds) *River Conservation and Management*. John Wiley, Chichester, 65–79.

Basson, M.S., Van Niekerk. P.H. and Van Rooyen, J.A. 1997. *Overview of Water Resources, Availability and Utilization in South Africa*. Department of Water Affairs and Forestry, Pretoria. Report PRSA/00/0197.

Beaver, D. de B. and Rosen, R. 1978. Studies in scientific collaboration. Part 1. The professional origins of scientific co-authorship. *Scientometrics* 1:65–84.

Beaver, D. de B. and Rosen, R. 1979. Studies in scientific collaboration. Part 2. Scientific co-authorship, research productivity and visibility in the French scientific elite, 1799–1830. *Scientometrics* 1:133–149.

Benke, A.C. 1990. A perspective on America's vanishing streams. *Journal of the North American Benthological Society* 9:77–88.

Billen, G., Décamps, H., Garnier, J., Boet, P., Meybeck, M. and Servais, P. 1995. Atlantic river systems of Europe (France, Belgium, The Netherlands. In: Cushing, C.E., Cummins, K.W. and Minshall, G.W. (eds) *Ecosystems of the World 22. River and Stream Ecosystems*. Elsevier, Amsterdam, 389–418.

Bochniarz, Z. 1993. East–West aspects of sustainable development: market access and environmental responsibilities. In: *Proceedings of International Symposium on Trade and the Environment*, 10–12 November 1993. Minnesota State Bar Association, St Paul, Minnesota.

Bonner, R. 1993. *At the Hand of Man: Peril and Hope for Africa's Wildlife*. Simon & Schuster, London.

Boon, P.J. 1992a. Channelling scientific information for the conservation and management of rivers. *Aquatic Conservation: Marine and Freshwater Ecosystems* 2:115–123.

Boon, P.J. 1992b. Essential elements in the case for river conservation. In: Boon, P.J., Calow, P. and Petts, G.E. (eds) *River Conservation and Management*. John Wiley, Chichester, 11–33.

Boon, P.J., Calow, P. and Petts, G.E. (eds) 1992. *River Conservation and Management*. John Wiley, Chichester.

Brandt, W. 1982. *North–South: A Programme for Survival*. Report of the Independent Commission on International Development Issues (1980). Pan Books, London.

Colvin, J.G. 1992. A code of ethics for research in the Third World. *Conservation Biology* 6:309–313.

Conley, A.H. 1995. *A Synoptic View of Water Resources in Southern Africa*. Paper to South African Department of Water Affairs and Forestry, Pretoria.

Dankelman, I. and Davidson, J. 1993. *Women and*

Environment in the Third World: Alliance for the Future (reprinted after 1988). Earthscan, London.

Davies, B.R., O'Keeffe, J.H. and Snaddon, C.D. 1993. *A Synthesis of the Ecological Functioning, Conservation and Management of South African River Ecosystems.* Water Research Commission, Pretoria, South Africa, Report No. WRC TT62/93.

Davies, B.R., O'Keeffe, J.H. and Snaddon, C.D. 1995. River and stream ecosystems in southern Africa: predictably unpredictable. In: Cushing, C.E., Cummins, K.W. and Minshall, G.W. (eds) *Ecosystems of the World 22. Rivers and Stream Ecosystems.* Elsevier, Amsterdam, 537–599.

Denny, P. 2000. Limnological research and capacity building in tropical developing countries. *Verhandlungen der Internationalen Vereinigung für theoretische und angewandte Limnologie* (in press).

Department of Water Affairs and Forestry (DWAF) 1986. *Management of the Water Resources of the Republic of South Africa.* Pretoria.

Dudgeon, D. 1992. Endangered ecosystems: a review of the conservation status of tropical Asian rivers. *Hydrobiologia* 248:167–191.

Dudgeon, D., Arthington, A.H., Chang, W.Y.B., Davies, J., Humphrey, C.L., Pearson, R.G. and Lam, P.K.S. 1994. Conservation and management of tropical Asian and Australian inland waters: problems, solutions and prospects. In: Dudgeon, D. and Lam, P.K.S. (eds) *Inland Waters of Tropical Asia and Australia: Conservation and Management. Mitteilungen der Internationalen Vereinigung für theoretische und angewandte Limnologie* 24:369–386.

Elmendorf, M. and Buckles, P. 1980. *Appropriate Technology for Water Supply and Sanitation: Sociocultural Aspects of Water Supply and Excretal Disposal.* The World Bank, Washington, DC.

Gibbs, W.W. 1995. Lost science in the Third World. *Scientific American*, August:76–83

Gopal, B. and Wetzel, R.G. 1995. *Limnology in Developing Countries*, Volume 1. Special Publication of the International Association of Theoretical and Applied Limnology (SIL), by International Scientific Publications, New Delhi.

International Union for Conservation of Nature and Natural Resources, United Nations Environment Programme, and World Wide Fund for Nature (IUCN/UNEP/WWF) 1991. *Caring for the Earth: A Strategy for Sustainable Living.* The World Conservation Union, Gland.

Junk, W.J., Bayley, P.B. and Sparks, R.E. 1989. The flood–pulse concept in river–floodplain systems. In: Dodge, D.P. (ed.) *Proceedings of the International Large River Symposium (LARS). Canadian Special Publication of Fisheries and Aquatic Sciences*, 106, 110–127.

Kruszewska, I. 1993. *Open Borders, Broken Promises: Privatization and Foreign Investment: Protecting the Environment through Contractual Clauses.* Special report of Greenpeace International, Amsterdam.

Lake, P.S., Barmutta, L.A., Boulton, A.J., Campbell, S. and Claire, R.M.St. 1986. Australian streams and Northern

Hemisphere stream ecology: comparisons and problems. *Proceedings of the Ecological Society of Australia* 14:61–82.

Livernash, R. 1994. Population and the environment. In: *World Resources 1994–1995: A Guide to the Global Environment.* World Resources Institute/Oxford University Press, London.

Luukkonen, T., Persson, O. and Sivertsen, G. 1992. Understanding patterns of international scientific collaboration. *Science, Technology and Human Values* 17:101–126.

Lydeard, C. and Mayden, R.L. 1995. A diverse and endangered aquatic ecosystem of the southeastern United States. *Conservation Biology* 9:800–805.

Marek, M.J., and Kassenberg, A.T. 1990. The relationship between strategies of social development and environmental protection. In: Grodzinski, W., Cowling, E.B. and Breymeyer, A.I. (eds) *Ecological Risks: Perspectives from Poland and the United States.* National Academy Press, Washington, DC, 41–59.

Masters, L. 1990. The imperiled status of North American aquatic animals. *Biodiversity Network News* 3:1–2, 7–8.

McCully, P. 1996. *Silenced Rivers: The Ecology and Politics of Large Dams.* Zed Books, London.

McLarney, W.O. 1988. Still a dark side to the aquarium trade. *International Wildlife* 18:46–51.

Miller, R.R., Williams, J.D. and Williams, J.E. 1989. Extinctions of North American fishes during the past century. *Fisheries* 14:22–38.

Mori, S. and Ikusima, I. 1980. *Proceedings of the First Workshop on the Promotion of Limnology in the Developing Countries.* Organising Committee of XXI SIL Congress, Kyoto, Japan.

National Foundation for Environmental Protection of Poland 1993. *The Green Lungs of Poland.* Special Publication of the National Foundation for Environmental Protection, Bialystok, Poland.

Palmer, T. 1986. *Endangered Rivers and the Conservation Movement.* University of California Press, Berkeley.

Pethiyagoda, R. 1994. Threats to the indigenous freshwater fishes of Sri Lanka and remarks on their conservation. *Hydrobiologia* 285:189–201.

Pringle, C.M. 1996. Expanding scientific research programs to address conservation challenges in freshwater ecosystems. In: Pickett, S.T.A., Ostfeld, R.S., Shachak, M. and Likens, G.E. (eds) *Enhancing the Ecological Basis of Conservation: Heterogeneity, Ecosystem Function and Biodiversity.* Proceedings of the Sixth Cary Conference, Institute of Ecosystem Studies. Chapman and Hall, New York, 305–319.

Pringle, C.M. and Aumen, N.G. 1993. Current efforts in freshwater conservation. *Journal of the North American Benthological Society* 12:174–176.

Pringle, C.M., Vellidis, G., Heliotis, F., Bandacu, D. and Cristofor, S. 1993. Environmental problems in the Danube Delta. *American Scientist* 81:350–361.

Resh, V.H. 1985. Periodical citations in aquatic entomology

and freshwater benthic ecology. *Freshwater Biology* 15:757–766.

Resh, V.H. and Yamamoto, D. 1994. International collaboration in freshwater ecology. *Freshwater Biology* 32:613–624.

Ryzkowski, L. 1990. Ecological guidelines for management of rural areas in Poland. In: Grodzinski, W., Cowling, E.B. and Breymeyer, E.I. (eds) *Ecological Risks: Perspectives from Poland and the United States.* National Academy Press, Washington, DC, 249–264.

Sauvy A. 1952. *Théorie générale de la population. Vol. I: Economie et population.* University Press of France, Paris.

Srodowiska, U. 1992. *Environmental Protection 1992.* Glowny Urzad Statystyczay (Main Statistical Office, Warsaw).

Storer, N.W. 1970. The internationality of science and the nationality of scientists. *International Social Science Journal* 22:89–104.

Thomas, A., Crow, B., Frenz, P., Hewitt, T., Kassam, S. and Treagust, S. 1994. *Third World Atlas*, 2nd edition. Open University Press, Buckingham.

United Nations 1997. *Comprehensive Assessment of the Freshwater Resources of the World.* Report of the Secretary-General. UN Commission on Sustainable Development, Fifth Session, 5–25 April 1997, New York.

Ward, J.V. and Stanford, J.A. (eds) 1979. *The Ecology of Regulated Streams.* Plenum Press, New York.

Warren, M.L. and Burr, B.M. 1994. Status of freshwater fishes of the United States: overview of an imperiled fauna. *Fisheries* 19:6–18.

Welcomme, R.L. 1979. *Fisheries Ecology of Floodplain Rivers.* Longman, London.

Williams, J.D., Warren, M.L., Cummings, K.S., Harris, J.L. and Neves, R.J. 1993. Conservation status of freshwater mussels of the United States and Canada. *Fisheries* 18:6–22.

Williams, J.E. and Neves, R.J. 1992. Introducing the elements of biological diversity in the aquatic environment. *Transactions of the 57th North American Wildlife and Natural Resources Conference* 57:345–354.

Williams, J.E., Johnson, J.E., Hendrickson, D.A., Contreras-Balderas, S., Williams, J.D., Navarro-Mendoza, M., McAllister, D.E. and Deacon, J.E. 1989. Fishes of North America endangered, threatened, or of special concern: 1989. *Fisheries* 14:2–20.

Williams, W.D. 1988. Limnological imbalances: an antipodean viewpoint. *Freshwater Biology* 20:407–420.

Williams, W.D. 1994. Constraints to the conservation and management of tropical inland waters. In: Dudgeon, D. and Lam, P.K.S. (eds) *Inland Waters of Tropical Asia and Australia: Conservation and Management. Mitteilungen der Internationalen Vereinigung für theoretische und angewandte Limnologie* 24:356–362.

Winterbourn, M.J., Rounick, J.S. and Cowie, B. 1981. Are New Zealand stream ecosystems really different? *New Zealand Journal of Marine and Freshwater Research* 15:321–328.

Wishart, M.J. and Davies, B.R. 1998. The increasing divide between First and Third Worlds: science, collaboration and conservation of Third World aquatic ecosystems. *Freshwater Biology* 39:557–567.

World Resources Institute (WRI) 1986. *World Resources 1986.* Basic Books, New York.

15

River conservation in tropical versus temperate latitudes

C.M. Pringle

Introduction

The conservation of tropical rivers presents special challenges given their evolutionary history, hydrological and ecological characteristics, and our limited understanding of how these systems function. Moreover, differences in socio-economic conditions, rates of human population growth, and related effects of globalization in tropical versus temperate countries often play an overriding role in determining the status of conservation efforts. Global disparities in river conservation resulting from such socio-economic differences are discussed in detail in Wishart *et al.* (this volume, Part II) and have great relevance to the issues discussed here. The approach of this chapter, however, is to focus on those aspects of tropical riverine systems which make them particularly vulnerable to human disturbance (i.e. hydrological and ecological characteristics) and ultimately to stress priority research and management needs for conservation. At the same time it is recognized that many tropical developing countries lack the financial resources for research and conservation. Moreover, research itself is inadequate if not backed by political will, and administrative and managerial infrastructure; this is an acute problem in developed and developing countries alike.

The evolutionary history of tropical riverine ecosystems is extremely complex and has contributed to high levels of biodiversity and endemism. Tropical regions have been characterized by long periods of climatic stability relative to temperate zones. The term 'tropical' is used here to refer to the broad and irregular equatorial area (approximately 30° north and south of the Equator) between the northern and southern subtropical zones of the drylands (i.e. as used here tropical areas are not confined to the latitude limits of Capricorn and Cancer; Pereira 1989). While the ice ages extirpated fish populations from many temperate freshwater environments during glacial periods, climatic fluctuations and associated changes in sea level throughout the Quaternary may actually have contributed to speciation in the tropics by aiding freshwater fish dispersal, particularly in South America (Weitzman and Weitzman, 1982; Lowe-McConnell, 1987). More than 2000 species of fish occur in the Amazon alone, with approximately 90% endemic. Of the 700 species in the Congo River of Africa about 70% are endemic. In contrast, there are about 250 species in the Mississippi River of North America (30% endemic) and fewer than 70 species (10% endemic) in the Danube River of Europe (World Conservation Monitoring Centre – WCMC, 1992). Lowe-McConnell (1987) summarizes general differences between 'classical' temperate rivers and their complex counterparts in tropical regions.

Given the differences between tropical and temperate riverine systems, theories and conservation strategies that have been developed based on studies of temperate streams may not be applicable or effective for tropical streams. For example, an understanding of the fate and transport of environmental contaminants in northern latitudes has little applicability to the tropics (Bordeau *et al.*, 1989). Likewise, the applicability of hydropower technology (developed for rivers in the temperate zone) to tropical regions must be evaluated with respect to differing hydrological and biological

Global Perspectives on River Conservation: Science, Policy and Practice.
Edited by P.J. Boon, B.R. Davies and G.E. Petts. © 2000 John Wiley & Sons Ltd.

features. Tropical rivers are often characterized by higher water temperatures (with little or no seasonality), heavy precipitation during at least part of the year, high variation in discharge, floods, and high biodiversity of migratory fauna. Warm tropical waters which have become eutrophic also foster a high diversity of microbial agents deleterious to humans: about 80% of tropical diseases are water related and can be attributed to poor sewage treatment and/or general lack of sanitation.

Tropical versus temperate differences notwithstanding, the growing human demand for water worldwide (Postel, 1998) is a major determining force in the evolution of inland water systems in general, along with socio-economic pressures on future development. Humans already appropriate over 50% of the accessible surface water, and this is expected to increase to 70% by the year 2025 (Postel *et al.*, 1996). This human dominance of fresh water is not only degrading valuable ecological elements such as fish and other wildlife but also the ecosystem services upon which human populations depend. The fate of tropical watersheds should be a matter of concern for the entire international community since it is estimated that the tropics will produce 80% of the world's population increase in the next two decades (Pereira, 1989). Correspondingly, the world food problem is concentrated in the broad belt of tropical and sub-tropical developing countries. By the time this book is published (2000) almost 50% of the world's total population will live within the humid tropics alone (Gladwell and Bonell, 1990). Rapid increases in population accompanied by changes in land use, including urban and industrial development, are resulting in concomitant increases in toxic and organic compounds in aquatic systems with little time for adjustment (Meybeck *et al.*, 1989).

Many tropical countries with developing economies do not have adequate resources to invest in water quality – let alone riverine conservation efforts which relate to 'non-human' ecology. For example, poor to non-existent sewage treatment has resulted in a lack of safe drinking water in many tropical regions. Consequently, the maintenance of water quality is being driven primarily by immediate human health concerns (Pan American Health Organization – PAHO, 1992), in much the same way as it occurred in developed countries such as the United States before environmental legislation led to sewage treatment programmes and water clean-up. This is evidenced by the fact that surface water quality problems in tropical developing countries (e.g. high organic and toxic

loading, faecal coliforms) often resemble those that were common in developed countries before massive water clean-up programmes.

Generalized history of river conservation in tropical versus temperate regions

While it is difficult to generalize regarding the ecological status of riverine ecosystems and related conservation efforts in temperate versus tropical countries, some general patterns do emerge. Many tropical rivers are currently being altered at an unprecedented rate, experiencing rapid increases in levels of pollutants on a time frame of years and decades. By contrast, these changes often occurred over centuries in temperate riverine systems, with corresponding declines in both physical and biotic integrity. For instance, in the United States, over half of the wetlands that existed 200 years ago have been lost; less than a quarter of bottomland hardwood forests remain in mid-western and southern portions and virtually none of it virgin; more than 85% of the nation's inland surface waters are artificially controlled; there are more than 5500 dams greater than 15 m high; of 800 fish species, 34% are rare, endangered, critically endangered or extinct; of native unionid mussel and crayfish species respectively, 73 and 65% fall into these aforementioned categories (see also Karr *et al.*, this volume).

In contrast to the information that exists for temperate streams there is little on the basic ecology of tropical rivers – let alone the effects of human activities on riverine ecology (but see Goulding *et al.*, 1996; Lowe-McConnell, 1987; Mohd, 1994). The ecology of tropical waters is quantitatively less well understood than that of temperate waters (Golterman *et al.*, 1993; Boon, 1996). For many tropical waters it is not even known what nutrients limit primary production, and communities of aquatic animals are poorly understood. History does seem to repeat itself, however: just as very little baseline information existed on ecological conditions in temperate streams before they were dramatically altered, there is also a paucity of baseline/pre-disturbance information for tropical streams. Perhaps the major difference is that, for some tropical rivers, it is not too late to collect this information (e.g. Araujo-Lima and Goulding, 1997; Barthem and Goulding, 1997).

Many tropical countries are experiencing annual population growth rates of more than 3% (particularly in Africa e.g. 3.4% in South Africa), which means that their populations will double in about 23 years. Thus, there is great pressure to increase all aspects of national infrastructure to support burgeoning populations, and this has serious implications for the protection of water resources and riverine conservation. Tropical landscapes are being altered at accelerating rates: the percentage of land used for agriculture has consistently increased throughout the tropics (in part in response to relatively high population growth rates), yet it has remained constant or declined slightly in developed countries (World Resources Institute, 1994). Agribusinesses are also expanding in tropical countries as a result of foreign investment. This has resulted in the emigration of peasants from rural areas to urban areas, and thus many tropical countries are now among the most urbanized nations on earth (Lacher and Goldstein, 1997). Mining has been expanded both for internal consumption and for export, and numerous hydropower projects are under way. The rate of landscape alteration is exacerbated by the acceptance of foreign development projects which often have large-scale environmental impacts with negative impacts on water quality (e.g. McCully, 1996). Much development activity is pursued without implementing environmental regulations because the social cost of these regulations in the short term is too high, resulting in long-term negative implications for the environment and overall quality of life for human populations (Lacher and Goldstein, 1997).

Not surprisingly, the maintenance of water quality in many tropical countries is primarily driven by human health concerns (PAHO, 1992). Environmental conditions are only considered as far as they may influence conditions for human health (Gladwell, 1993). Lack of safe drinking water is responsible for the high incidence of water-related tropical diseases (Gladwell and Bonell, 1990). Recent epidemics of cholera which occurred in Latin America (e.g. Peru, Ecuador, Colombia, Mexico, Guatemala, Panama and Brazil) in 1991 have been tied to rapidly expanding human populations and untreated municipal wastewater (PAHO, 1990; Witt and Reiff, 1991). In the Caribbean, less than 10% of the total domestic wastewater receives treatment before disposal. In West Africa there is generally no sewage treatment, with raw sewage often being discharged directly into streams and estuaries. In southern Asia, the main pollution is from three metropolitan areas in India: Bombay, Madras and Calcutta. An estimated 1.02×10^6 t of untreated sewage and industrial effluents are released into the Indian Ocean each year from these three cities alone (Linden, 1990).

The trajectory of conservation activities in developed temperate countries can similarly be traced back to a primary focus on human health which led to the development of protective legislation. In the United States, for example, the Clean Water Act of 1972 underpins current conservation activities. Over the last two decades the Clean Water Act has reduced annual discharges of toxic chemicals and raw sewage into US lakes and rivers by 453 000 t and 914×10^6 t respectively (Adler *et al.*, 1993).

In both the United States and Europe, emphasis on management of riverine systems to promote/accommodate commercial and sports fishing, recreational interests (e.g. river rafting) and hydropower development has now evolved towards a focus on issues of ecosystem integrity, native species and restoration (e.g. Boon *et al.*, 1992; Doppelt *et al.*, 1993; Naiman *et al.*, 1995; Karr *et al.*, this volume). The current focus on ecosystem integrity has led to widespread abandonment of exotic fish stocking practices (whereby sports fish are introduced into streams at the expense of native fish stocks), management of water releases from dams (to simulate natural flow regimes), and even to dam removal (see Karr *et al.*, this volume). These changes have been supported by a growing environmental ethic and awareness among aquatic scientists and the general populace (e.g. Pringle and Aumen, 1993; Dewberry and Pringle, 1994; Showers, this volume).

Protection, improvement and enforcement of existing legislation (and the creation of new legislation) must be a foundation of future conservation activities developed in temperate countries in order to ensure both short- and long-term safeguard of riverine ecosystems. The United States at present spends about 2.4% of its gross national product ($150 billion per year) on environmental protection, and significant gains have been made over the last 25 years. However, much more has to be done to protect riverine biointegrity. For example, legislation such as the Clean Water Act of 1972 has helped mitigate effects of point-source pollution into rivers and streams but has fallen short on the regulation of non-point-source pollution (e.g. Adler *et al.*, 1993). Over 30% of streams and rivers in the United States fail to meet existing chemical water quality standards and, when biological criteria are taken into account, almost 50% of water bodies nationwide are substandard (Adler *et al.*, 1993). It is estimated that over 680 t of pesticides end up in the Upper Mississippi River annually from agricultural run-off and, at its juncture with the Ohio River, the

Mississippi is carrying an estimated 100 t of phosphorus and nearly 80 t of nitrates each day.

While many tropical countries are busy struggling to implement at least some type of sewage treatment in urban areas (Gladwell, 1993), temperate countries such as the United States are attempting to upgrade existing sewage treatment facilities and to develop management and regulatory frameworks for rapidly evolving agricultural practices and technologies. For example, a current pressure affecting riverine systems in many parts of the United States is large-scale pig (hog) and poultry farming operations which are moving from serious non-point-sources of pollution to equally problematic point-source pollution, as waste disposal for such operations becomes centralized.

As mentioned earlier, it is difficult to generalize regarding the ecological status of streams in tropical versus temperate zones because of varying ecological and socio-economic conditions among countries in both regions. While this chapter has focused so far on characteristics of temperate-zone countries in North America and Europe, countries with transitional economies in Eastern Europe and the former Soviet Union (Khaiter *et al.*, this volume) share some interesting similarities with tropical developing countries: emerging riverine conservation strategies within these nations are also largely driven by human health concerns. In many countries, environmental degradation has progressed to the extent of actually limiting development, and the lack of development is correspondingly limiting environmental protection (Marek and Kassenberg, 1990). Many of these countries have faced regional degradation of freshwater resources on a scale greater than much of the Western World has experienced (Pringle *et al.*, 1993).

GLOBALIZATION

Tropical countries with developing and/or unstable economies are particularly vulnerable to negative aspects of globalization. Foreign currency earned through export is often needed to pay the interest costs on foreign debt and this situation has attracted foreign investment in industry and agribusiness which often operate in ways that are deleterious to the environment; such companies are often not restricted by environmental regulations that they would otherwise adhere to in temperate nations. Also, many technologies that are developed in temperate systems are often exported to tropical countries where they may have different, and often more harmful, effects than they had in temperate systems (e.g. because of the

unique ecological properties of tropical systems). Certain hydropower technologies provide a good case in point and are discussed in detail in a subsequent section of this chapter.

A growing problem is the globalization of trade in toxic wastes which is threatening to have serious negative effects on the quality of freshwater systems in tropical countries given inadequate disposal and storage. The total size of the toxic waste trade in Europe and around the world is rapidly increasing (Puckett, 1992) and an understanding of the interaction of both national and international politics makes it possible to track emerging problems in tropical countries. For example, as costs of waste disposal in developed countries increased in the 1980s, an initial market for toxics developed in Africa. However, when African countries began to seal their borders to the toxic trade, waste dealers in developed countries (largely in the temperate zone) turned to tropical countries in Latin America and the Caribbean (and also Eastern Europe). Southern Asia is emerging as the favourite target of waste trade (Leonard, 1993), as Latin America and the Caribbean have moved towards stricter regulations. Differences in ecotoxicological effects in tropical versus temperate aquatic systems are discussed later in this chapter.

On a positive note, international conferences held within the last decade provide guidance for tropical and temperate countries alike. For example, the United Nations Conference on Environment and Development (UNCED) produced 'Agenda 21', which included an entire chapter devoted to 'Protection of the quality and supply of freshwater resources: application of integrated approaches to the development, management and use of water resources' (Johnson, 1993). Also, the recently established 'Code of Conduct for Responsible Fisheries' includes guidelines on inland fisheries, fisheries management and aquaculture (Food and Agriculture Organization of the United Nations – FAO, 1995).

Important aspects of the hydrological cycle in tropical systems

Tropical aspects of the hydrological cycle are important considerations for riverine conservation since they enhance the vulnerability of riverine systems to human influences. In the humid tropics, water management and related conservation issues arise from the hydrological features of the region such as excess water, floods, and greater variation of flows than in temperate zones. These features contribute to the unique potential for hydropower generation, the increasing need for flood-control development, and the traditional emphasis on growing irrigated rice (Le Moigne and Kuffner, 1993). The vulnerability of riverine systems to these particular development pressures is discussed in more detail later.

Unique combinations of temperature and precipitation occur in the tropics which do not exist in temperate regions. Tropical climates are dominated by high inputs of solar radiation which result in rapid evaporation from warm surfaces in both terrestrial and aquatic environments. These aspects, combined with the erratic patterns and high intensities of rainfall, are major characteristics of the Intertropical Convergence Zone (or Intertropical Confluence) between the trade wind systems north and south of the Equator (Pereira, 1989). When large quantities of water vapour that have evaporated from warm tropical oceans condense into high intensity rainfall, they release latent heat energy which increases turbulence, contributing to major storms that are characteristic of tropical areas. The Intertropical Convergence Zone does not form a continuous belt, or move in a continuous or predictable manner; instead it fills and reforms to produce a tropical chain of major disturbances which cause highly variable rainfall regimes, making floods and droughts inescapable features of tropical latitudes (Pereira, 1989).

High temperatures and evaporation rates in tropical drylands (which are characterized by <600 mm of annual precipitation) reduce the stability of vegetation cover so that rangelands are more vulnerable to misuse. Temperatures up to 70°C have been recorded when bare soil is exposed, and organic matter is rapidly destroyed by oxidation under such conditions, resulting in a decrease in soil stability. Also, overgrazing can actively extend areas of desert.

The wet tropics are vulnerable because of the intensity and duration of rainfall. When forests are removed, severe erosion is often a consequence. For instance, from 20 to 200 t of soil ha^{-1} yr^{-1} are eroded from deforested slopes in Costa Rica (Hartshorn *et al.*, 1982). Also, large and intense annual rainfall volumes combined with high variability (i.e. periods during which little or no rain occurs) can result in high ratios of high to low flows in rivers and streams. Low flows can occur over extended periods and this often leads to serious problems in urban areas because of the corresponding low capacity for rivers at low flow to dilute and 'process' biodegradable wastes. High temperatures further reduce the potential for waste

degradation. The rainfall intensity regimes for which tropical urban stormwater control systems should be designed are poorly known. Rainfall intensities in many tropical urban areas are extremely high relative to temperate urban areas, typically exceeding the natural infiltration rates of soils so that even unpaved areas contribute substantially to run-off, flooding and high erosion (Bouvier, 1990). The impact of sediment deposited in streams and drains can be particularly severe as a result of highly erosive rainfall events. When sediments are deposited on flat stretches of rivers near a city, they can result in increases in flooding because of raised river-bed levels (Gladwell, 1993).

Critical habitats for conservation in tropical riverine systems

Many faunal components of tropical rivers are migratory and their movements are adapted to seasonally variable hydrological conditions. They are thus very vulnerable to human modifications of both the lateral (river channel–floodplains) and longitudinal (headwaters–mouth) connectivity of their habitat. A feature of tropical floodplain rivers is extensive lateral flooding which each year leads to temporary lacustrine conditions over vast areas. Many tropical fish taxa are adapted to live in both running water environments and lacustrine pools, depending on the season. Other taxa have adapted to migrate longitudinally along the river, covering great distances. This section focuses on two critical habitats for conservation in tropical rivers: (a) floodplain forests and (b) estuarine areas including mangrove forests.

FLOODPLAIN FORESTS/RIPARIAN VEGETATION

Tropical floodplains and riparian zones are critical riverine habitats which are vulnerable to stream regulation schemes, impoundment and conversion to other types of land use. While the ecology of *large* floodplains has received some attention (e.g. Welcomme, 1979; Junk, 1982), medium and small-sized inundateable areas adjacent to streams are also essential to the productivity of many rivers in the tropics and remain little explored (Golterman *et al.*, 1993).

In tropical rivers with extensive floodplains, Welcomme and Hagborg (1977) found that the most significant factor to influence the growth of fish in any single year was the area of flooded land, since this was an index of food availability during the prime growing season. During this period, fish can achieve 75% of their annual growth. The breeding and feeding cycles of many tropical fish species are closely tied to seasonal inundation of floodplains. The lakes that form in the floodplain serve as settling areas for alluvial materials, resulting in greater water transparency, and enhanced algal and zooplankton production. These environments are extremely important for young fish. In the Amazon, this explains why migratory fish move down clear-water and black-water tributaries to spawn near floodplain habitats (Goulding *et al.*, 1996). Lateral migrations of adult fish into flooded forest areas are often marked by intensive feeding upon allochthonous materials (Goulding, 1990; Goulding *et al.*, 1996) and many fish have special adaptations for feeding on fruits and seeds. Based on the abundance of seed- and fruit-eating fish species in the Amazon, Goulding makes the convincing argument that the flooded forests have been more significant as gene pools for *terra firma* (i.e. drier regions of the rain forest) than the *terra firma* itself (e.g. Alexander, 1994). The floodplains of tropical rivers are also critical habitat for many terrestrial and arboreal animals. They supply food (e.g. herbaceous vegetation, fruits, seeds, etc.) at times when it is less abundant in the uplands. In South America, for example, over half of the 200 mammal species in the Amazonian lowlands inhabit floodplain areas at some point during the year (Goulding *et al.*, 1996).

Logging of floodplain habitat is considered to be one of the major forces behind the decline of Amazonian fisheries (Goulding *et al.*, 1996). Although tropical floodplain habitats may constitute a relatively small proportion of a given watershed, their accessibility by logging operations has led to massive destruction, often with consequent negative effects on fish and other aquatic biota. Massive logging of floodplain forests also results in dwindling supplies of wood, which play an important role in maintaining the stability of the stream channel and in creating habitat heterogeneity for fish and other aquatic biota along the entire stream continuum – from small tributaries, to large rivers, estuaries and oceans. While the ecological importance of this wood (and the debris dams it creates) has been documented for temperate streams (e.g. Bilby and Likens, 1980; Gregory, 1992; Maser and Sedell, 1994), few studies have examined its role in tropical rivers. Goulding *et al.* (1996) note that there is a decrease in the wood material that is being carried by the Amazon. Now that much of the floodplain of the middle Amazon and its tributaries is already deforested,

logging operations are targeting the tidal forests of the estuary.

Management solutions should include establishment of floodplain parks and reserves where logging and other land-use activities are prohibited. Very few floodplain reserves exist in tropical areas. For example, currently there are no floodplain parks or reserves in the lower 2500 km of the Amazon River. The first reserve as one moves upstream is the Mamiraua Ecological Reserve (11 000 km^2) which was established in 1990. At least one-third of the reserve is flooded forest (Alexander, 1994). The most critical regions for conducting surveys of potential floodplain reserves are the middle and lower Amazon because of the massive deforestation that has already taken place there (Goulding *et al.*, 1996).

ESTUARINE AREAS

The ecological dynamics of tropical estuarine systems have important implications for river conservation since they are tied to the entire river network. Migratory biota (e.g. potamodromous catfish and amphidromous shrimp of the Amazon) link estuaries to inland waters. In the dry season (or in years with lower rainfall), saline waters penetrate upriver beneath the fresh water, enabling marine bottom-dwelling fish to move upstream (Lowe-McConnell, 1987). The fresh/salt interface moves upstream and downstream seasonally – over 200 km in the Amazon. In addition, water flowing from the mouths of tropical rivers often dilutes the sea for many kilometres.

The wise management and protection of tropical estuarine floodplains is critical for the stability and biointegrity of riverine systems upstream. This point is perhaps best illustrated by a consideration of the many taxa of tropical fish and crustaceans that migrate longitudinally along the river, spending a portion of their life cycle in the estuary. For example, most taxa of freshwater shrimps (Decapoda) migrate to the estuary as larvae and spend up to several months of their life in estuarine habitats associated with floodplain forests before they migrate back upstream to adult habitat as postlarvae (e.g. Hunte, 1978; Hobbs and Harte, 1982; Benstead *et al.*, 1999).

Tropical estuaries are vulnerable to exploitation because of their generally high fish productivity. For example, the lower Amazon and its estuary is the largest and most productive fishery in the entire Amazon drainage. Moreover, the health of this fishery is imperative for the survival of large and commercially important migratory fish taxa such as catfish which use estuarine areas as nursery habitats. Barthem and Goulding (1997) recently presented a migratory model for large catfish which illustrates the necessity for managing catfish harvests in the estuary. Catfish such as dourada (*Brachyplatystoma flavicans*) and piramutaba (*Brachyplatystoma vaillantii*) have evolved to use an immense area of the Amazon Basin stretching from the estuary to the base of the Andes. Since these catfish taxa are vulnerable to overharvesting in the estuary, it is recommended that the number of industrial boats allowed to exploit the estuary be reduced and that the mesh size used in fish nets be increased (Barthem and Goulding, 1997).

Mangrove forests are also critical habitats for tropical riverine conservation. Approximately 240×10^3 km^2 of mangrove vegetation cover the river deltas, lagoons and estuarine ecosystems of the tropics. These forests play important roles, including the maintenance of water quality and shoreline stability (through the control of nutrient and sediment delivery and distribution) and the provision of habitat, food and refugia for a multitude of both terrestrial and aquatic organisms of different life stages and trophic levels. Management of mangrove forest in south-east Asia and other areas of the tropics consists of selective logging practices which retain buffer strips to reduce erosion. The Matang Mangrove Forest Reserve in Malaysia is considered to be a good example of 'sustained yield' management of mangrove forest (FAO, 1985; Ong, 1995). The reserve operates on a 30-year cutting cycle; a specific density of trees is retained during harvesting as a seed source for regeneration. Buffer strips (3 m wide) are preserved alongside the river channel. Similar systems are operated in Thailand and Indonesia (Ong, 1995).

Vulnerability to specific development pressures

Development pressures discussed here include the toxic effects of agrochemicals, hydropower projects, and other hydrological alterations, such as water diversions and interbasin transfers. Such development pressures are by no means limited to tropical countries, but the unique combination of physical, ecological and socio-economic characteristics of tropical watersheds enhance their vulnerability to these pressures.

TOXIC EFFECTS OF AGROCHEMICALS

Riverine ecosystems in the wet tropics are extremely vulnerable to agrochemicals given the extensive use of pesticides, the behaviour of pollutants in tropical environments, the large amount of run-off originating from agricultural drainage systems, and the hydrodynamics of tropical aquatic ecosystems. Banana plantations can have particularly deleterious effects on rivers since pesticide use is intensive and river systems are often incorporated into the drainage system of the plantation in order to promote the run-off of excess water that would otherwise promote fungal growth (see Pringle *et al.*, this volume).

When viewing contamination of tropical aquatic ecosystems it is important not to make incorrect assumptions based on models derived from the temperate zone (Henriques *et al.*, 1997). Differences between tropical and temperate ecosystems affect the way water pollutants behave. While contaminant depletion (i.e. physical, chemical and biological degradation) rates can be higher in the tropics as a result of higher temperatures and sunlight (Sethunathan, 1989; Viswanathan and Krishna Murti, 1989), tropical aquatic ecosystems are characterized by higher rates of biological uptake, biological release and oxygen depletion, along with greater solubility of liquids and solids, greater biological impacts of nutrients and suspended solids, and lower toxicity thresholds (Sprague, 1985; Wolanski, 1992; Howe *et al.*, 1994).

Despite the high use of pesticides in tropical regions, regulatory infrastructure is often underdeveloped and/or not adequately enforced (Hilje *et al.*, 1987; Murray, 1994), relative to developed countries. This problem is exacerbated by the use of extremely toxic and environmentally deleterious agrochemicals in tropical developing countries, that are banned in developed countries (see Pringle *et al.*, this volume). For example, in developed countries, most organochlorines were restricted in the early 1970s. While the majority of these chemicals are now banned in Central America, they were banned or restricted one decade later than in developed countries (Wesseling and Castillo, 1992). Also, in many instances bans are ignored. For example, while organochlorine pesticides have been banned in Costa Rica since 1990, imports have continued (e.g. in 1993, 32 000 kg were brought into the country: Castillo *et al.*, 1997).

Ecotoxicology has focused almost exclusively on ecosystems in temperate zones (Lacher and Goldstein, 1997), and, consequently, little information exists for the tropics regarding environmental distribution, toxic effects on aquatic organisms, or general impacts on aquatic ecosystems (Castillo *et al.*, 1997). For example, published data in Central America relate mostly to pesticide residue levels (e.g. the more persistent organochlorines). More recent integrated studies have related exposure and effects (see review in Castillo *et al.*, 1997) and these studies are a starting point for future development of appropriate methods to assess environmental impacts in the tropics. Research is needed to improve our understanding of differences in the distribution, degradation, bioavailability and toxicity of pesticides between tropical and temperate environments and emphasis should be placed on the potential impacts of repeated and/or continual low-level exposures to a mixture of pesticides (Castillo *et al.*, 1997). Contemporary pesticides in aquatic systems must be identified, and those with high toxicity prioritized for future studies. Development activities in the tropics that merit immediate research are identified by Lacher and Goldstein (1997) and include large-scale agricultural activities, particularly bananas, pineapple and soybean farming (and also gold-mining with its associated heavy use of mercury).

HYDROELECTRIC PROJECTS

The potential for hydropower generation is very large in the humid tropics due to high river flows. It is particularly attractive in mountainous regions. A study conducted by the World Bank (1984) indicated that only 2% of the hydropower potential was developed in the humid tropical countries of Africa and Asia, and 7% in Latin America. The vulnerability of tropical streams to such development is an important concern given the many massive hydroelectric projects that are being proposed for tropical developing countries (Pringle *et al.*, this volume). Seventy-nine dams are either planned or exist in the Brazilian Amazon alone (Seva, 1990) and Brazil's state-owned power monopoly proposes to meet over half of Brazil's future electricity needs via hydroelectricity despite the drawbacks of the region's flat topography, wide floodplains, high river sediment loads, and abundant trees and wildlife.

Most of the impacts of river engineering are difficult or impossible to predict with certainty (McCully, 1996). This, combined with our limited understanding of how tropical rivers function, has contributed to great uncertainty regarding both short- and long-term environmental effects of hydroelectric projects in the tropics. For example, in temperate areas, organic matter decay within impounded reservoirs may proceed within a decade, while in the tropics it may take many decades or even centuries as a result of the large amount of plant biomass. Thorough clearing of vegetation in the submergence zone, before the

reservoir is filled, is particularly important in tropical areas (McCully, 1996), yet this rarely occurs. For example, large-scale flooding in Surinam submerged 1500 km^2 of rain forest (1% of the country) to create the Brokopondo Reservoir. Organic matter decomposition resulted in severe deoxygenation of the water and massive emissions of hydrogen sulphide. Workers at the dam wore masks for two years after the reservoir started to fill in 1964 and the cost of repairing acid-water damage to the dam's turbines totalled more than US$4 million (Van Der Heide, 1976). Studies conducted three years after the dam was created indicated that levels of oxygen in the river only began to recover about 110 km downstream of the dam, with severe effects on fish communities.

Legislation and/or legal requirements to clear vegetation from reservoir areas before they are submerged is particularly important in tropical areas. In many cases, these legal requirements exist but they are ignored. For example, despite a legal requirement to clear vegetation from all areas to be submerged by an impoundment, the Brazilian electricity utility Eletronorte cleared less than one-fifth of a 2250 km^2 area of rain forest inundated by the construction of the Tucuruí Dam (on the border between Argentina and Paraguay) and only 2% of the 3150 km^2 of forest inundated by the Balbina Dam, with disastrous environmental consequences. Consumption of oxygen by decomposing vegetation in the Yacyreta Reservoir is believed to have killed more than 120 000 fish found downstream after the first test of the dam's turbines in 1994 (International Rivers Network, 1994).

Tropical reservoirs are particularly prone to colonization by mats of floating aquatic macrophytes which shade lower waters, clog turbines, impede fishing boats and nets, and provide habitat for disease vectors such as mosquitoes and snails. For example, two years after the Brokopondo Reservoir began to fill, over half of the reservoir was covered with water hyacinth (*Eichhornia crassipes*) and the plant was partially brought under control by aerial spraying with 2,4-D which also poisoned many other plants and animals (Van Donselaar, 1989).

The complex life-history strategies and migratory behaviour of many aquatic taxa in tropical streams make them extremely vulnerable to stream fragmentation by dams. As discussed in a previous chapter (Pringle *et al.*, this volume), hydroelectric dams are considered potentially to be the most dangerous human activity to Amazonian fisheries in the near future (Bayley and Petrere, 1989; Goulding *et al.*, 1996). Dams not only disrupt upstream and downstream migrations along the longitudinal continuum, but they also disrupt the lateral migrations between the stream channel and floodplains through flow regulation. This can potentially affect biodiversity on an ecosystem and landscape level. Goulding *et al.* (1996) recommend river surveys from headwaters to mouth in order to develop hypothetical models of fish migrations and movements in and out of individual rivers within stream networks where dams are proposed. Such studies will indicate the necessity of fish ladders or other devices designed to accommodate the unique migratory movements of aquatic biota.

The lack of successful implementation of fish-pass devices associated with tropical dams is a compelling example of how temperate mitigation strategies are often ineffectively applied in tropical areas. Many tropical fishways have been based on the temperate salmon fish-pass model and are impassable for many native species. For example, the Yacyreta Dam on South America's Paraná River was fitted with fish elevators (at a cost of over US$30 million). These were designed to transport fish upriver, based on knowledge and experience with anadromous fish migrations on the Columbia River in North America (Treakle, 1992; World Bank, 1995). Little or no consideration was given to the fact that many of the fish species in the Paraná are potamodromous, migrating up and down the river several times during their life cycle.

WATER DIVERSIONS, IRRIGATION AND INTERBASIN TRANSFERS

While many streams that drain developed temperate and sub-tropical zones are being tapped for water well beyond the limits of maintaining a healthy continuous ecosystem (e.g. North America's Colorado River is barely a trickle by the time it reaches the sea), many large rivers in the tropics are just now experiencing this pressure. Rivers draining arid tropical regions are particularly vulnerable as human population levels increase and water needs spiral out of control.

There has been a rapid expansion of irrigation for rice production over the last three decades in many areas of the humid tropics (particularly in south-east Asia and Madagascar), as a result of the need to increase food production, the success of traditional rice production, and the development of modern agricultural techniques and improved rice varieties (Le Moigne and Kuffner, 1993). Rice cultivation systems are irrigation-intensive and the most common

irrigation method is the continuous flow of water from rice paddy to rice paddy which can lead to dewatering of streams. The area under irrigation in Indonesia (which is mainly devoted to rice) increased from about 4×10^6 ha in 1968 to almost 5.5×10^6 ha in 1980; in the Philippines the area increased from 740 000 ha to 1.2×10^6 ha; and in Thailand, from 1.8 to 2.6×10^6 ha in the same period (LeMoigne and Kuffner, 1993).

Rivers and streams draining tropical regions are particularly vulnerable to interbasin water transfers (IBTs) which have major implications for the diffusion of diverse faunal communities that were previously isolated – and could produce environmental changes with unforeseeable ramifications (see Davies *et al.*, this volume). Hazards include competition for resources, predation, and the spread of parasitic diseases among geographic isolates (O'Reilly-Sternberg, 1995; Davies *et al.*, this volume). Interbasin linkages in the Amazon Basin have been under consideration for at least two centuries (Humboldt and Bonpland, 1820–1822). A plan was even once proposed to link the Orinoco with the Amazon system and the Plata Basin (O'Reilly-Sternberg, 1995).

Problems of water scarcity in the arid tropics provide an impetus for broader strategic planning. For example, should countries have to meet certain standards before tapping international waters and adversely affecting neighbouring countries and riverine environments? As in temperate areas, alternatives to stream diversions include recycling a certain percentage of water consumed, implementing effective conservation and demand control measures, and full utilization of economically feasible domestic sources.

Priority research and management needs for tropical aquatic systems

As stressed throughout this chapter, there is a lack of basic information on the ecology of tropical aquatic systems on which to base ecologically sound water management and conservation strategies. Those few studies that address the management of tropical systems often do not consider the unique ecological characteristics of tropical ecosystems and/or the requirements of tropical aquatic biota (Pringle and Scatena, 1999a,b). Specific areas where there is a critical need for information on tropical rivers have been cited throughout this chapter and several key areas are discussed in more detail below.

There is a lack of fundamental knowledge concerning the biology of tropical aquatic organisms.

Very little is known regarding migration patterns, food supplies or breeding grounds for even commercially important fish species in tropical Asia (Dudgeon *et al.*, 1994) and the neotropics (Goulding *et al.*, 1996; Araujo-Lima and Goulding, 1997; Barthem and Goulding, 1997). Such information is important to the development of management strategies for fish, which should include regulation of commercial exploitation in vulnerable riverine ecosystems, such as estuaries, and protecting lateral and longitudinal connectivity within the stream network.

Another need is the development of monitoring techniques for tropical rivers. The biological and chemical indices of water quality that have been developed for temperate zones are often not appropriate for tropical systems (Pringle and Scatena, 1999a). One compelling example of this is provided by faecal coliforms which are used as standard indicators of recent faecal contamination in aquatic systems of the United States. Studies in tropical regions indicate that faecal coliforms are often high in 'pristine' streams (i.e. relatively undisturbed by human activity) and it appears that tropical ambient temperatures and humidity promote high coliform counts, including *Escherichia coli* (Carillo *et al.*, 1985).

Likewise, while aquatic insects have been used widely throughout North America and Europe in biomonitoring, in many tropical areas the taxonomy and ecology of aquatic insects is not well understood. Broad taxonomic-level biological indices, such as the Ephemeroptera/Plecoptera/Trichoptera Index, do not work for tropical rivers since entire orders (e.g. Plecoptera) are often absent from the fauna. Pringle and Ramírez (1998) propose that drift sampling be used as a standard complementary tool to benthic sampling in assessments of water quality and invertebrate community composition in tropical streams where invertebrate species with migratory life cycles are often major faunal components. For example, migratory freshwater shrimps are a major energetic link between tropical rivers and their estuaries and may serve as useful indicators (Pringle and Ramírez, 1998).

It is also necessary to identify sensitive organisms of ecological or economic value that can be used in toxicity testing. A literature survey by Castillo *et al.* (1997) reported that almost no studies involving the toxic effects of pesticides to aquatic organisms have been conducted in the tropics. The authors further suggest that freshwater organisms that seem good candidates because of their sensitivity, or because of their economic value, are the freshwater fish *Cichlasoma dovii*, the freshwater shrimp *Macrobrachium rosenbergii*, as well as species of Trichoptera and Ephemeroptera. Ultimately, regional water quality criteria must be developed for tropical systems along with acceptable risks and relevant models of evaluation (Castillo *et al.*, 1997).

The need for baseline information on effects of deforestation/logging in tropical aquatic systems, upon which to base management and policy decisions, is discussed in detail elsewhere (Pringle and Benstead, 2000). The few tropical studies that have related the effects of different logging practices to sediment yield have been conducted in relatively small catchments (Kasran, 1988; Douglas *et al.*, 1993; Yusop and Suki, 1994) and, therefore, it is difficult to extrapolate results to large drainage basins. While the conservation of aquatic biodiversity in timber-production forests is a central issue in conservation biology in temperate areas such as North America (e.g. Carlson *et al.*, 1990; Beschta, 1991; Naiman, 1992; Adams and Ringer, 1994), data on cumulative effects of logging on freshwater communities in the temperate zone are rare.

While some major concerns of hydropower projects in tropical regions were discussed previously in this chapter, other regional/global implications also merit serious investigation. In the last two decades it has become apparent that mercury contamination of fish can be a major problem in newly created reservoirs in the temperate zone (Hecky *et al.*, 1991; Rosenberg *et al.*, 1995; McCully, 1996), yet few studies have examined this phenomenon in tropical areas (but see Yingcharoen and Bodaly, 1993). It has also recently been shown that newly flooded reservoirs emit methane and become a net source of CO_2, contributing to the 'Greenhouse Effect' (McCully, 1996). Fearnside (1995) calculated the impact on global warming of Brazil's Balbina and Tucuruí Dams over the first 50 years of their existence by assessing the amount of forest that they flooded and the rate at which vegetation decayed. He calculated that Tucuruí had 60% as much impact on global warming as a coal-fired plant generating the same amount of electricity (and 50% more impact than a gas-fired power station). As stressed by Fearnside (1995), the scale of hydroelectric development contemplated for Amazonia makes this a potentially significant source of emissions of greenhouse gases in the future and this must be considered in cost–benefit analyses.

Finally, the river-basin concept has not been widely applied in water management in many tropical countries and national water quality monitoring programmes are often non-existent (e.g. Ongley,

1993). Integrated watershed management (i.e. management efforts that observe watershed boundaries and consider the entire river basin; see Hooper and Margerum, this volume) is an imperative for land-use planning, as are multi-scale management strategies that consider the longitudinal and lateral dimensions of aquatic habitats and the role of riparian and floodplain vegetation (Dudgeon, 1994).

Acknowledgements

I gratefully acknowledge National Science Foundation grant DEB-95-28434 for support in writing this chapter. Also, special thanks go to Phil Boon, Bryan Davies, Alonso Ramirez, William Lewis and an anonymous reviewer for their feedback on this chapter.

References

Adams, P.W. and Ringer, J.O. 1994. *The Effects of Timber Harvesting and Forest Roads on Water Quantity and Quality in the Pacific Northwest: Summary and Annotated Bibliography.* Special Publication of the Forest Engineering Department, Oregon State University, Corvallis.

Adler, R.W., Landman, J.C. and Cameron, D.C. 1993. *The Clean Water Act 20 Years Later.* Natural Resources Defense Council, Island Press, Washington, DC.

Alexander, B. 1994. People of the Amazon fight to save the flooded forest. *Science* 264:606–607.

Araujo-Lima, C. and Goulding, M. 1997. *So Fruitful a Fish: Ecology, Conservation and Aquaculture of the Amazon's Tambaqui.* Columbia University Press, New York.

Barthem, R. and Goulding, M. 1997. *The Catfish Connection: Ecology, Migration, and Conservation of Amazon Predators.* Columbia University Press, New York.

Bayley, P.B. and Petrere, M. 1989. Amazon fisheries: assessment methods, current status and management options. *Canadian Special Publications in Aquatic Sciences* 106:385–398.

Benstead, J.P., March, J.G., Pringle, C.M. and Scatena, F.N. 1999. Effects of a low-head dam and water abstraction on migratory tropical stream biota. *Ecological Applications* 9:656–668.

Beschta, R.L. 1991. Stream habitat management for fish in the northwestern United States: the role of riparian vegetation. *American Fisheries Society Symposium* 10:53–58.

Bilby, R.E. and Likens, G.E. 1980. Importance of organic debris dams in the structure and function of stream ecosystems. *Ecology* 61:1107–1113.

Boon, P.J. 1996. The conservation of fresh waters: temperate experience in a tropical context. In: Schiemer, F. and

Boland, K.T. (eds) *Perspectives in Tropical Limnology.* SPB Academic Publishing, Amsterdam, 333–334.

Boon, P.J., Calow, P. and Petts, G.E. (eds) 1992. *River Conservation and Management.* John Wiley, Chichester.

Bordeau, P., Haines, J.A., Klein, W. and Krishna Murti, C.R. (eds) 1989. *Ecotoxicology and Climate with Special Reference to Hot and Cold Climates.* John Wiley, New York.

Bouvier, C. 1990. Concerning experimental measurements of infiltration for runoff modelling of urban watersheds in Western Africa. *Hydrological Processes and Water Management in Urban Areas.* (Invited lectures and selected papers. Duisburg Conference, Urban 88, April 1988). IAHS Publication 198:43–49.

Carillo, M., Estrada, E. and Hazen, T.C. 1985. Survival and enumeration of the fecal indicators *Bifidobacterium adolescentis* and *Escherichia coli* in a tropical rainforest watershed. *Applied Environmental Microbiology* 50:468.

Carlson, J.Y., Andrus, C.W. and Froehlich, H.A. 1990. Woody debris, channel features, and macro-invertebrates of streams with logged and undisturbed riparian timber in northeastern Oregon, USA. *Canadian Journal of Fisheries and Aquatic Sciences* 47:1103–1111.

Castillo, L.E., De La Cruz, E. and Ruepert, C. 1997. Ecotoxicology and pesticides in tropical aquatic ecosystems of Central America. *Environmental Toxicology and Chemistry* 16:41–51.

Dewberry, T.C. and Pringle, C.M. 1994. Lotic science and conservation: moving toward common ground. *Journal of the North American Benthological Society* 13:399–404.

Doppelt, R., Scurlock, M., Frissell, C. and Karr, J. 1993. *Entering the Watershed.* Island Press, Washington, DC.

Douglas, I., Greer, T. and Bidin, K. 1993. Impacts of rainforest logging on river systems and communities in Malaysia and Kalimantan. *Global Ecology and Biogeography Letters* 3:245–252.

Dudgeon, D. 1994. The need for multi-scale approaches to the conservation and management of tropical inland waters. *Mitteilungen Internationalen Vereinigung für theoretische und angewandte Limnologie* 23:11–16.

Dudgeon, D., Arthington, A.H., Chang, W.Y.B., Davies, J., Humphrey, C.L., Pearson, R.G. and Lam, P.K. 1994. Conservation and management of tropical Asian and Australian inland waters: problems, solutions and prospects. *Mitteilungen Internationalen Vereinigung für theoretische und angewandte Limnologie* 24:369–386.

Food and Agriculture Organization of the United Nations (FAO) 1985. *Mangrove Management in Thailand, Malaysia and Indonesia.* FAO Environment Paper 4, Rome.

Food and Agriculture Organization of the United Nations (FAO) 1995. *Code of Conduct for Responsible Fisheries.* Rome.

Fearnside, P.M. 1995. Hydroelectric dams in the Brazilian Amazon as sources of 'greenhouse' gases. *Environmental Conservation* 22:7–19.

Gladwell, J.S. 1993. Urban water management problems in the humid tropics: Some technical and non-technical considerations. In: Bonell, M., Hufschmidt, M.M. and

Gladwell, J.S. (eds) *Hydrology and Water Management in the Humid Tropics: Hydrological Research Issues and Strategies for Water Management.* Cambridge University Press, Cambridge, 414–436.

Gladwell, J.S. and Bonell, M. 1990. An international programme for environmentally sound hydrological and water management strategies in the humid tropics. In: *Proceedings of the International Symposium on Tropical Hydrology and Fourth Caribbean Islands Water Resources Congress,* American Water Works Association Technical Publication Series TPS-90-2, 1–10.

Golterman, H.L., Burgis, M.J., Lemoalle, J. and Talling, J.F. 1993. Ecological characteristics of tropical waters: an outline. In: Bonell, M., Hufschmidt, M.M. and Gladwell, J.S. (eds) *Hydrology and Water Management in the Humid Tropics: Hydrological Research Issues and Strategies for Water Management.* Cambridge University Press, Cambridge, 367–394.

Goulding, M. 1990. *The Fishes and the Forest: Explorations in Amazonian Natural History.* The University of California Press, Berkeley.

Goulding, M., Smith, N.J.H. and Mahar, D.J. 1996. *Floods of Fortune: Ecology and Economy along the Amazon.* Columbia University Press, New York.

Gregory, K.J. 1992. Vegetation and river channel process interactions. In: Boon, P.J., Calow, P. and Petts, G.E. (eds) *River Conservation and Management.* John Wiley, Chichester, 255–270.

Hartshorn, G., Hartshorn, L., Atmella, A., Gomez, L.D., Mata, A., Morales, R. Ocampo, R., Pool, D., Quesada, C., Solera, C., Solorzaro, R., Stiles, G., Tosi, J.Jr., Umana, A., Villalobos, C. and Wells, R. 1982. *Costa Rica Country Environmental Profile: A Field Study.* Tropical Science Center, San Jose, Costa Rica.

Hecky, R.E., Ramsey, D.J., Bodaly, R.A. and Strange, N.E. 1991. Increased methylmercury contamination in fish in newly formed freshwater reservoirs. In: Suzuki, T., Imura, N. and Clarkson, T.W. (eds) *Advances in Mercury Toxicology.* Plenum, New York, 33–52.

Henriques, W., Jeffers, R.D., Lacher, T.E. and Kendall, R.J. 1997. Agrochemical use on banana plantations in Latin America: Perspectives on ecological risk. *Environmental Toxicology and Chemistry* 16:91–99.

Hilje, L., Castillo, L.E., Thrupp, L. and Wesseling, C. 1987. El uso de los plaguicidas en Costa Rica. Ed. Heliconia/UNED, San Jose, Costa Rica.

Hobbs, H.H. and Harte, C.W. 1982. The shrimp genus *Atya* (Decapoda: Atyidae). *Smithsonian Contributions to Zoology* No. 364.

Howe, G.E., Marking, L.L., Bills, T.D., Rach, J.J. and Mayer, F.L. 1994. Effects of water temperature and pH on toxicity of terufos, trichlorfon, 4-nitrophenol and 2,4-dinitrophenol to the amphipod *Gammarus pseudolimnaeus* and rainbow trout (*Oncorhynchus mykiss*). *Environmental Toxicology and Chemistry* 13:51–66.

Humboldt, A, and Bonpland, A. 1820–1822. *Voyage aux regions equinoxiales du Nouveau Continent fait en 1799,* 1800, 1801, 1802, 1803, et 1804, vols. 6 and 7 (1820), 8 (1822). Chez N. Maze, Libraire, Paris.

Hunte, W. 1978. The distribution of freshwater shrimps (Atyidae and Palaeomonidae) in Jamaica. *Zoological Journal of the Linnean Society* 64:135–150.

International Rivers Network (IRN) 1994. Yacyreta killing fish. *World Rivers Review,* Berkeley.

Johnson, S.P. 1993. *The Earth Summit: The United Nations Conference on Environment and Development (UNCED).* Chapter 18: Agenda 21: Protection of the quality and supply of freshwater resources: application of integrated approaches to the development, management and use of water resources, 334–360.

Junk, W.J. 1982. Amazonian floodplains: their ecology, present and potential use. *Revista Hydrobiologia Tropicales* 15:285–301.

Kasran, B. 1988. Effect of logging on sediment yield in a hill dipterocarp forest in peninsular Malaysia. *Journal of Tropical Forest Science* 1:56–66.

Lacher, T.E. and Goldstein, M.I. 1997. Annual review – tropical ecotoxicology: status and needs. *Environmental Toxicology and Chemistry* 26:100–111.

Le Moigne, G. and Kuffner, U. 1993. Water resource management issues in the humid tropics. In: Bonnell, M.M., Hufschmidt, M.M. and Gladwell, J.S. (eds) *Hydrology and Water Management in the Humid Tropics: Hydrological Research Issues and Strategies for Water Management.* Cambridge University Press, Cambridge, 556–568.

Leonard, A. 1993. South Asia: the new target of international waste traders. *Multinational Monitor,* December 1993, 21–24.

Linden, O. 1990. Human impact on tropical coast zones. In: *Nature and Resources,* UNESCO, 26:3–11.

Lowe-McConnell, R.H. 1987. *Ecological Studies in Tropical Fish Communities.* Cambridge University Press, New York.

Marek, M.J. and Kassenberg, A.T. 1990. The relationship between strategies of social development and environmental protection. In: Grodzinsk, W., Cowling, E.B. and Bandreymeyer, A.I. (eds) *Ecological Risks: Perspectives from Poland and the United States.* National Academy Press, Washington, DC, 41–59.

Maser, C. and Sedell, J.R. 1994. *From the Forest to the Sea: The Ecology of Wood in Streams, Rivers, Estuaries, and Oceans.* St Lucie Press, Delray Beach, Florida.

McCully, P. 1996. *Silenced Rivers: The Ecology and Politics of Large Dams.* Zed Books, Atlantic Highlands and New Jersey.

Meybeck, M., Chapman, D. and Helmer, R. 1989. *Global Freshwater Quality. A First Assessment.* Basil Blackwell, Oxford.

Mohd, Zakaria-Ismail 1994. Zoogeography and biodiversity of the freshwater fishes of Southeast Asia. *Hydrobiologia* 285:41–48.

Murray, D. 1994. *Cultivating Crisis: The Human Cost of*

Pesticides in Latin America. University of Texas Press, Austin.

Naiman, R.J. (ed.) 1992. *Watershed Management: Balancing Sustainability and Environmental Change*. Springer-Verlag, New York.

Naiman, R.J., Magnuson, J.J., McKnight, D.M. and Stanford, J.A. 1995. *The Freshwater Imperative: A Research Agenda*. Island Press, Washington, DC.

Ong, J.E. 1995. The ecology of mangrove conservation and management. *Hydrobiologia* 295:343–351.

Ongley, E.D. 1993. Water quality data programmes for developing land use and resource policies: Latin America. In: *Prevention of Water Pollution by Agriculture and Related Activities. Proceedings of the Food and Agriculture Organization (FAO) of the United Nations*, Santiago, Chile, 1992. Rome, Italy, 263–272.

O'Reilly-Sternberg, H. 1995. Waters and wetlands of Brazilian Amazonia. In: Nishizawa, T. and Uitto, J.I. (eds) *The Fragile Tropics of Latin America*. United Nations University Press, New York, 113–179.

Pan American Health Organization (PAHO) 1990. *The Situation of Drinking Water Supply and Sanitation in the American Region at the end of the Decade 1981–1990, and Prospects for the Future*. Volume 1. Washington, DC.

Pan American Health Organization (PAHO) 1992. Health and the environment. *Bulletin of PAHO* 26:370–378.

Pereira, H.C. 1989. *Policy and Practice in the Management of Tropical Watersheds*. Westview Press, San Francisco.

Postel, S.L. 1998. Water for food production: will there be enough in 2025? *BioScience* 48:629–637.

Postel, S.L., Daily, G.C. and Ehrlich, P.R. 1996. Human appropriation of renewable freshwater. *Science* 271:785–788.

Pringle, C.M. and Aumen, N.G. 1993. Current efforts in freshwater conservation. *Journal of the North American Benthological Society* 12:174–176.

Pringle, C.M. and Benstead, J.P. 2000. Effects of logging on tropical riverine ecosystems. In: Fimbel, R., Grajal, A. and Robinson, J. (eds) *Conserving Wildlife in Managed Tropical Forests*. Columbia University Press, New York (in press).

Pringle, C.M. and Ramírez, A. 1998. Use of both benthic and drift sampling techniques to assess tropical stream invertebrate communities along an altitudinal gradient, Costa Rica. *Freshwater Biology* 39:359–373.

Pringle, C.M. and Scatena, F.N. 1999a. Factors affecting aquatic ecosystem deterioration in Latin America and the Caribbean. In: Hatch, U. and Swisher, M.E. (eds) *Managed Ecosystems: The Mesoamerican experience*. Oxford University Press, 104–113.

Pringle, C.M. and Scatena, F.N. 1999b. Freshwater resource development: case studies from Puerto Rico and Costa Rica. In: Hatch, U. and Swisher, M.E. (eds) *Managed Ecosystems: The Mesoamerican Experience*. Oxford University Press, 114–121.

Pringle, C.M., Vellidis, G., Heliotis, F., Bandacu, D. and

Cristofor, S. 1993. Environmental problems of the Danube Delta. *American Scientist* 81:350–361.

Puckett, J. 1992. Dumping on our world neighbors: The international trade in hazardous wastes, and the case for an immediate ban on all hazardous waste exports from industrialized to less-industrialized countries. In: Ole Bergesen, H. *et al.* (eds) *Green Globe Yearbook 1992*. Fridtjof Nansen Institute, Norway, 93–106.

Rosenberg, D.M., Berkes, F., Bodaly, R.A., Hecky, R.E., Kelly, C.A. and Rudd, J.W.M. 1995. Large-scale impacts of hydroelectric development. *Environmental Review* 5:27–54.

Sethunathan, N. 1989. Biodegradation of pesticides in tropical rice ecosystems. In: Bordeau, P., Haines, J.A., Klein, W. and Krishna Murti, C.R. (eds) *Ecotoxicology and Climate*. SCOPE 38. John Wiley, Chichester, 247–264.

Seva, O. 1990. Works on the Great bend of the Xingu – a historic trauma? In: De O Santos, L.A. and De Andrade, L.M. (eds) *Hydroelectric Dams on Brazil's Xingu River and Indigenous Peoples*. Cultural survival report 30. Cultural Survival, Cambridge, Massachusetts, 19–35.

Sprague, J.B. 1985. Factors that modify toxicity. In: Rand, G.M. and Petrocelli, S.R. (eds) *Fundamentals of Aquatic Toxicology*. Taylor & Francis, Bristol, Pennsylvania, 124–163.

Treakle, K. 1992. *Briefing Paper No. 1: Yacyreta Hydroelectric Project II*. Bank Information Center, Washington, DC, August 1992.

Van Der Heide, S. 1976. Hydrobiology of the man-made Brokopondo Lake, Brokopondo Research Report, Suriname, Part II. *Natuurwetenschappelijke Studiekring Voor Suriname en de Nederlandse Antillen (NSFSNA)*, Utrecht.

Van Donselaar, J. 1989. The vegetation in the Brokopondo Lake Basin (Surinam) before, during and after the inundation, 1964–1972. *Brokopondo Research Report, Suriname, Part III, NSFSNA*, Utrecht.

Viswanathan, P.N. and Krishna Murti, C.R. 1989. Effects of temperature and humidity on ecotoxicology of chemicals. In: Bordeau, P., Haines, J.A., Klein, W. and Krishna Murti, C.R. (eds) *Ecotoxicology and Climate*. SCOPE 38. John Wiley, Chichester, 139–154.

Weitzman, S.H. and Weitzman, M. 1982. Biogeography and evolutionary diversification in neotropical freshwater fishes, with comments on the refuge theory. In: Prance, G.T. (ed.) *Biological Diversification in the Tropics*. Columbia University Press, New York, 403–422.

Welcomme, R.L. 1979. *Fisheries Ecology of Floodplain Rivers*. Longman, London.

Welcomme, R.L. and Hagborg, D. 1977. Towards a model of a floodplain fish population and its fishery. *Environmental Biology of Fishes* 2:7–24.

Wesseling, C. and Castillo, L.E. 1992. Plaguicidas en America Central: Algunas consideraçiones sobre las condiçiones de uso. *Proceedings, Primera Conferencia Centroamericana sobre Ecologia y Salud (ECOSAL I), San Salvador, El Salvador, 1–3 September*, 83–112.

Witt, V.M. and Reiff, F.M. 1991. Environmental health conditions and cholera vulnerability in Latin America and the Caribbean. *Journal of Public Health Policy* 12:450–463.

Wolanski, E. 1992. Hydrodynamics of tropical coastal marine systems. In: Hawker, D.W. and Connell, D.W. (eds) *Pollution in Tropical Aquatic Systems*. CRC Press, Boca Raton, Florida, 3–27.

World Bank 1984. *A Survey of the Future Role of Hydroelectric Power in 100 Developing Countries*. Washington, DC.

World Bank 1995. *Project Completion Report: Argentina Yacyreta Hydroelectric Project and Electric Power Sector Project*, 14 March 1995.

World Conservation Monitoring Centre (WCMC) 1992. *Global Biodiversity: Status of the Earth's Living Resources*. Chapman & Hall, New York.

World Resources Institute (WRI) 1994. *World Resources: 1994–1995*. Oxford University Press, New York.

Yingcharoen, D. and Bodaly, R.A. 1993. Elevated mercury levels in fish resulting from reservoir flooding in Thailand. *Asian Fisheries Science* 6:73–80.

Yusop, Z. and Suki, A. 1994. Effects of selective logging methods on suspended solids concentration and turbidity level in streamwater. *Journal of Tropical Forest Science* 7:199–219.

16

Special problems of urban river conservation: the encroaching megalopolis

K.E. Baer and C.M. Pringle

'We shape our buildings and afterwards our buildings shape our world.'

Winston Churchill (1979)

Introduction

The global landscape is in the midst of an urban transition on a scale that is greater than at any other time in history (Figure 16.1; World Resources Institute – WRI, 1996). Urbanization is playing an increasingly central role in degrading riverine ecosystems, affecting water quantity and quality for both human and non-human biota. As recently as 1990, most of the world's population still lived in rural areas; however, by the year 2025, it is estimated that 60% (or about 5 billion people) will be concentrated in cities (Young *et al.*, 1994).

Riverine ecosystems are extremely vulnerable to the effects of urban development. Many urban areas have been purposefully located near rivers for proximity to transportation, drinking water, irrigation, food supply and waste disposal. Urban degradation of river systems is an historic problem that can be traced back as far as the establishment of the first cities along the major river systems of the Tigris–Euphrates and Indus Rivers 7000 years ago (Meybeck *et al.*, 1989). Urban drinking water sources have repeatedly been degraded and then abandoned in the course of development (Leslie, 1987). For example, the Roman Tiber was used and then deserted for cleaner sources requiring the construction of aqueducts. Expanding urban populations, associated wastes and old or non-existent waste treatment systems have overtaxed the assimilative capacities of rivers. Urbanization threatens entire catchments; even international water bodies (e.g. the Caribbean, and the Baltic, Black and Mediterranean seas) have been affected, highlighting the critical need for immediate management and conservation efforts to protect water bodies from the impacts of urbanization. This chapter examines global issues surrounding urban river conservation with an emphasis on streams in the USA.

Urban areas and water resources

Urban areas occupy only 1% of the earth's surface (WRI, 1996), but they affect water resources well beyond the scope of a city's geographical boundaries. With a higher *per capita* use of natural resources in urban than in non-urban areas (WRI, 1996), urban areas clearly exert a disproportionate effect on the environment relative to non-urban areas.

McDonnell and Pickett (1990) describe urban areas functionally:

'Urbanization can be characterized as an increase in human habitation, coupled with increased per capita energy consumption and extensive modification of the landscape, creating a system that does not depend principally on local natural resources to persist.'

Mumford (1956) likened growing cities to parasitic organisms which consume increasing amounts of

Global Perspectives on River Conservation: Science, Policy and Practice.
Edited by P.J. Boon, B.R. Davies and G.E. Petts. © 2000 John Wiley & Sons Ltd.

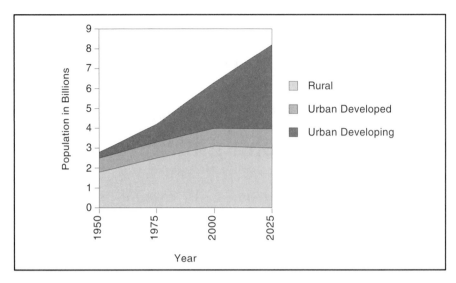

Figure 16.1   Urban population growth, 1950–2025. (From United Nations Population Division – UNPD, 1995)

resources, suggesting that 'the more highly a city seems developed as an independent entity, the more fatal are the consequences for the territory it dominates.'

Urban areas alter the landscape throught their effects on geomorphology, hydrology, energy flow and biogeography (Douglas, 1983), creating a highly altered landscape. Within this environment, streams and rivers face continuous disturbance from activities including, but not limited to, land clearance, construction, paving, lawn maintenance, industrial pollution, human and animal wastes and vehicle-related discharges (Schueler, 1991). Urbanization alters the hydrological cycle (Hengeveld and Vocht, 1982), affecting stream morphology, water quality, and habitat availability in aquatic systems (e.g. Dunne and Leopold, 1978; Klein, 1979; Schueler, 1991), among other effects qualitatively described in Table 16.1.

Regional trends of urbanization

Urban growth in developed regions during the 19th and 20th centuries was fuelled by an overall increase in population growth rates, technological advances, and access to new agricultural land which provided a plentiful food supply for city dwellers (Mumford, 1956). By 1995, more than 70% of the population in both Europe and North America lived in urban areas (United Nations Population Division – UNPD 1995). Today population shifts typically involve movement

Table 16.1   Major stream impacts caused by urbanization (modified from Schueler, 1991)

Changes in stream hydrology:
- Increase in magnitude and frequency of severe floods
- Increased frequency of erosive bankfull floods
- Increase in annual volume of surface run-off
- More rapid stream velocities
- Decrease in dry weather baseflow

Changes in stream morphology:
- Stream channel widening and downcutting
- Increased streambank erosion
- Shifting bars of coarse-grained sediment
- Elimination of pool/riffle structure
- Embedding of stream sediments
- Stream enclosure or channelization
- Stream crossings form fish barriers

Changes in stream water quality
- Massive sediment pulses
- Increased pollutant washoff
- Nutrient enrichment
- Bacterial contamination
- Increase in organic carbon loads
- Elevated levels of toxics, trace metals, hydrocarbons
- Increase in water temperature
- Trash/debris dams (jams)

Changes in habitat and ecology
- Shift from external to internal production
- Changes in diversity of aquatic insects
- Changes in diversity and abundance of fish
- Destruction of wetlands, riparian buffers and springs

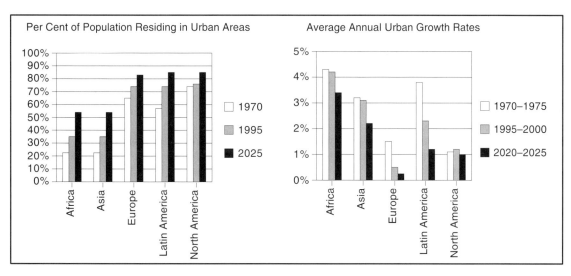

Figure 16.2 *Regional trends in urbanization, 1970–2025. (From UNPD, 1995)*

away from concentrated urban centres to sprawling metropolitan regions or to small and intermediate-sized cities (World Resources Institute, 1996).

The majority of urban growth will now occur in developing countries where 150 000 people join city populations daily (Figure 16.2; UNPD, 1995; WRI, 1996), responding to economic pressures and the promise of employment opportunities. Latin America and the Caribbean constitutes the most urbanized region in the developing world, with more than 70% of the region's population living in urban areas in 1995 (UNPD, 1995). Rapid urban growth continues especially in small and intermediate-sized cities (Gavidia, 1994). In contrast, human populations in Africa and Asia are approximately 30–35% urban (UNPD, 1995), but explosive urban growth is occurring at the rate of 4% per year. This trend is projected to continue for several decades and Asia and Africa are expected to be approximately 54% urban by the year 2075 (UNPD, 1995). Furthermore, most of this growth will take place in 'megacities' (cities with more than 8 million people); it is predicted that by the year 2015, 27 of the world's 33 megacities will be in developing countries (WRI, 1996).

Patterns of urban growth and their effects on aquatic systems vary from region to region and from city to city. In developed countries, the trend is towards suburbanization, 'sprawl', or conurbation, in which cities grow outward and together, losing their distinctiveness and altering greater areas of land; in the USA, the area classified as urban increased from 21 to 26×10^6 ha (or from 23 to 28% of the country's area)

between 1982 and 1992 (WRI, 1996). In developing regions, however, growth fringing cities is dominated by poor communities. This is in sharp contrast to many North American cities where downtown urban centres are left empty and economically depauperate as affluent commuters return to the suburbs. Each growth pattern is associated with its own set of environmental issues.

In developing countries, such fringe growth often occurs in environmentally fragile areas (WRI, 1996); meanwhile, in developed countries, inner cities often become more polluted as deteriorating infrastructure and services cannot be maintained by a dwindling tax base, while suburban landscapes are saturated with lawn and pool care chemicals which are non-point sources of pollution in urban waterways.

Urban socio-political effects on river conservation

Urban development profoundly affects the social environment as well as the physical environment. Mumford (1956) described the transition from a village to a city as a process in which people become increasingly alienated from, and disinterested in, the local natural environment which fosters the 'illusion of a complete independence from nature' (Lee, 1993). While cities undoubtedly provide opportunities in terms of education, employment and culture that contribute to population increases in these areas, such advantages are gained only with environmental tradeoffs. Douglas (1983) wrote that as 'cities have grown, these environmental problems have grown so severe that the liberation the cities provide appears to be reduced by the restrictions imposed by the urban environment'. However, such environmental problems can also be the impetus to bring environmentally concerned groups together to improve local environmental conditions, often simultaneously strengthening local communities (Douglas, 1983).

Water supply and use patterns are also directly affected by urbanization. Politically, power often resides in urban areas which increasingly gain control of water from rural and agricultural users (Postel, 1992). Worster (1985) writes: 'control over water has again and again provided an effective means of consolidating power within human groups', supporting the concept of a 'hydraulic society'. Although the control of water may not be the dominant force shaping a civilization's power structure, there is no doubt that access to, and control over, water resources bestows great power (Cosgrove, 1990) and it is often a major source of political tension (e.g. Reisner, 1986; Leitman *et al.*, 1993; Wishart *et al.*, this volume). Recognizing this, in the USA, minority groups are increasingly demanding a voice at the water policy table (e.g. *Race, Poverty and the Environment*, 1992), and this is an important first step in protecting urban rivers. Social and political stability, as well as human and ecosystem health, depend on an adequate and equitable water supply (Postel, 1992), emphasizing the need for aquatic conservation in burgeoning urban areas for the sake of both human and non-human biota.

In the USA, increased awareness regarding urban water issues has come from the academic community (e.g. Naiman *et al.*, 1995), government agencies (e.g. National Urban Runoff Program of the US Environmental Protection Agency (US EPA)), non-governmental environmental organizations (e.g. American Rivers Urban Rivers Program – ARURP), and from grassroots groups (e.g. Coalition to Restore Urban Waters – CRUW). Just as the Greek word for city, 'polis', means a state or society characterized by a sense of community, successful urban river conservation must be based on many levels of community involvement (see Showers, this volume).

Threats to river systems caused by urbanization

Urbanization poses a myriad threats to waterways and it is difficult to single out any one disturbance because of their synergistic and continuous nature. Examples include: (a) changes in flood hydrology (frequency, magnitude); (b) changes in channel geomorphology (deposition, erosion); (c) pollution from sewage, industrial effluents, and run-off from impervious surfaces via stormwater; (d) thermal stress; and (e) invasion by exotic species.

Threats to local streams and rivers vary throughout the urbanization process. Site clearing, construction and existing urban development all affect local streams in different ways. Vegetation clearance, for instance, may cause increased water temperatures, a reduction in organic carbon inputs and increased erosion and sedimentation; once construction is complete, however, sediment yields may decline and stream hydrology and channel morphology will continue to evolve, responding to land-use change in the watershed. Water quality in urban areas is also highly variable, responding to seasonal use of pesticides, leaking sewer lines and periods of increased water abstractions. While all of the interacting effects of urbanization on streams and rivers are not understood, it is clear that a number of disturbances combine cumulatively to change and degrade waterways.

The location of a given urban area within a drainage network is also a major factor determining the rate and magnitude of urban effects throughout the river ecosystem. An urban area located in the upper part of a basin, for example, may amplify peak flows at the basin mouth, whereas an urban area sited closer to the mouth of the basin may experience two flood peaks – the rapid urban one and a second draining a less-developed upper catchment (Rhoads, 1995). In addition, urban areas can affect ecosystem processes occurring both upstream and downstream. Although downstream communities often experience the brunt of upstream urban environments (Soto, 1996), it is becoming increasingly clear that urbanization can set off a cascade of effects upstream (Figure 16.3) that are often overlooked (Pringle, 1997). Such effects range from genetic and species level changes, such as reduced genetic flow and variation in isolated upstream populations, to ecosystem-level changes (e.g. nutrient cycling) that can occur in headwater systems as a result of downstream urbanization (Pringle, 1997). This is of particular concern, given that many large urban areas originate in coastal regions near the mouths of large rivers and subsequently encroach inland, potentially affecting the entire drainage upstream.

EFFECTS ON HYDROLOGY AND GEOMORPHOLOGY

Hydrologically, urban aquatic ecosystems and their associated biota are subject to discontinuous or 'flashy'

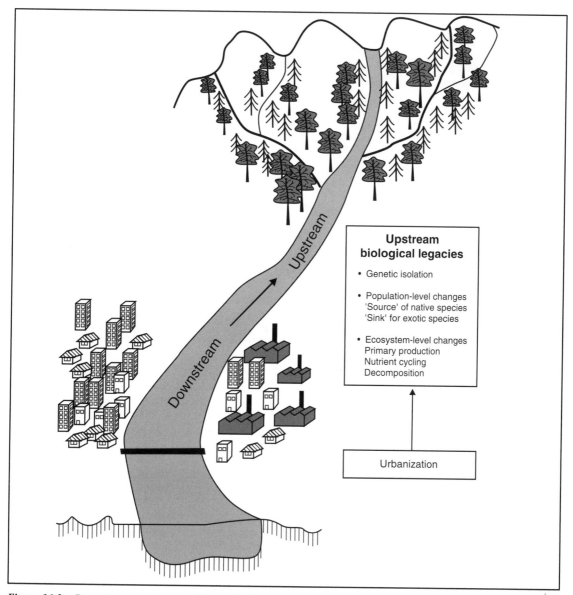

Figure 16.3 *Downstream–upstream effects of urbanization on the biology of riverine ecosystems (modified from Pringle, 1997)*

discharge regimes caused by the increased area of impervious surfaces (Dunne and Leopold, 1978; Hengeveld and Vocht, 1982; Ferguson and Suckling, 1990; Arnold and Gibbons, 1996). Elevated peak flows during storms flush toxics, trace metals, nutrients, and sediments from urban areas into riverine ecosystems (Mikalsen, 1993).

Urban streams undergo a number of transformations relating to hydrological and geomorphological changes. The overriding philosophy in urban areas is to drain water off streets and parking areas as quickly as possible (historically surface run-off has been viewed as a 'common-enemy' (Noland, 1981)). For this reason, many first- and second-order streams are buried or placed in culverts (Dunne and Leopold, 1978), and the systematic 'decapitation' of hills and the burial of valleys has eliminated many of the natural channels and

Figure 16.4   *Urban storm run-off flowing into a storm drain in Athens, Georgia (Photograph by Katherine Baer)*

tributaries that cross the landscape (Mount, 1995). Remaining overland flow in urban areas is diverted into drains through a system of culverts and channels (Figures 16.4 and 16.5).

This enhancement of 'hydraulic efficiency' and decrease in 'hydraulic roughness' alters important relationships between channel transport and storage of water and sediment (Graf, 1975; Dunne and Leopold, 1978; Riley, 1998). The combination of increased hydraulic efficiency, lower infiltration capacity from increased impervious surfaces, and higher total run-off results in changes in the hydrograph of urban catchments (Figure 16.6), leading to a significant alteration in flood volume, frequency and timing (Hirsch *et al.*, 1990; Rhoads, 1995). Peak discharge tends to increase in direct relation to the proportion of the catchment that is covered by impervious surfaces and is served by storm sewers (Mount, 1995). Time-to-peak discharge decreases in urban streams, reflecting the loss of storage capacity and increased water velocities (Leopold, 1991).

Morphologically, urban streams respond to increased sediment loads caused during basin development by aggrading, sometimes accumulating significant areas of floodplain (Graf, 1975; Rhoads, 1995), which may cause channels to become smaller

(Riley, 1998). Following completion of urban construction and corresponding decreases in sediment yield (Wolman, 1967; Mount, 1995; Rhoads, 1995), 'hungry' or sediment-free water combines with increased power of run-off to erode, incise and enlarge the channel (Arnold *et al.*, 1982; Hirsch *et al.*, 1990; Rhoads, 1995; Rosgen, 1996; Riley, 1998).

In extreme cases of channel enlargement, a stream's ability to transport sediment is decreased. Sediment will again begin to aggrade (Rhoads, 1995), and in some cases may even evolve towards a braided channel (Arnold *et al.*, 1982), and what Riley (1998) terms an 'urban quasi-equilibrium channel'. In addition, the size of channel sediments often decreases, degrading in-stream habitat (Schueler, 1991).

Further, and more directly affecting stream hydrology and geomorphology, is the channelization (Figure 16.7) and levee construction that frequently accompanies urbanization, disrupting the equilibrium of river systems (Brookes, 1988). Channel straightening increases slope resulting in increased stream power which often leads to bed and bank scour as the river attempts to re-establish its natural pattern (Mount, 1995). Levees deprive a river of access to its floodplain; peak discharges are raised and lag times shortened, thus increasing the overall hazard of flooding in the

Figure 16.5 *Urban storm drain during rain in Atlanta, Georgia (Photograph by Katherine Baer)*

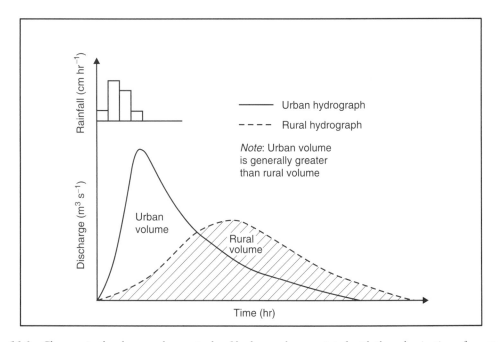

Figure 16.6 *Changes in the shape and magnitude of hydrographs associated with the urbanization of a watershed. (Reproduced with permission from Jeffrey F. Mount, California Rivers and Streams: the conflict between fluvian process and land use. Copyright © 1995 The Regents of the University of California)*

Figure 16.7   *A channelized, concrete, urban 'stream' in Huntsville, Alabama (Photograph by Katherine Baer)*

basin (Mount, 1995). Increased river levels induce bed scour compounding the impact of straightening. Channelization can have dramatic effects on river systems both downstream and upstream. Upstream effects, such as headward erosion or knickpoint migration, are often ignored (Hartfield, 1993) but can rejuvenate extensive networks (Smith and Patrick, 1991; Patrick *et al.*, 1993; Mount, 1995). The consequences of this erosion are becoming increasingly evident in coastal-plain streams throughout the south-eastern USA which suffer severe bank erosion, channel widening, depth reduction and the consequent elimination or reduction of biotic communities and their habitats. Biological consequences of headcutting include decreased diversity, reduced productivity and the extirpation of faunal groups from individual stream systems (Hartfield, 1993; Patrick *et al.*, 1993).

EFFECTS ON WATER QUALITY

Water quality in urban areas is affected by point and non-point-source pollution. Major point sources of urban pollution include municipal sewage, industrial outfalls from intensive chemical industries such as tanneries, metal plating operations, pulp mills and refineries, and air emissions from power plants and industry. Non-point sources include sediment, pollutant-laden storm run-off from impervious surfaces, and baseflow from aquifers that have been contaminated with sewage or industrial chemicals (WRI, 1996).

Urban sewage is the primary threat to local urban waters. Sewage-borne pathogens such as cholera bacteria, hepatitis viruses, salmonellae and shigellas pose significant risks to human health (WRI, 1996), particularly in developing countries where 90% or more of the sewage is released untreated into the nearest water body (Bartone, 1990; Pringle *et al.*, this volume). Aquatic biota are also severely affected by the volume of urban sewage and major declines in fish catches associated with urban sewage have been documented (WRI, 1996; see also Gopal *et al.*, this volume).

Even in some developed countries, only a portion of the sewage receives conventional treatment (WRI, 1996). Many major cities and towns in central and eastern Europe lack primary sewage treatment facilities for domestic and industrial wastes and Budapest, for example, empties an estimated three-quarters of its sewage untreated into the Danube (Pringle *et al.*, 1993).

Moreover, cities with conventional sewage treatment systems are not necessarily preserving or improving water quality. Cities experiencing rapid growth may not have maintained these sewage treatment facilities and related infrastructure in working order. In the city of Atlanta, Georgia, urban streams are contaminated with levels of faecal coliform bacteria up to 50 times above acceptable legal limits (Seabrook, 1997a). Atlanta's problems stem from an ageing sewer system with leaking pipes, overloaded and underfunded wastewater treatment plants, and delays in building treatment facilities at three of the city's combined sewage overflow (CSO) sites (Seabrook, 1997b). Sewer pipes are so old and overburdened that even a light shower can cause sewage to back-up into local manholes causing 'faecal fountains'.

In countries such as the USA, where urban sewage treatment has improved over the last 20 years, non-point-source pollution such as stormwater run-off continues to threaten the integrity of aquatic systems (Adler *et al.*, 1993). Urban streams integrate by-products of urban lifestyles, reflecting trends such as pesticide use for lawn care and swimming pool maintenance (United States Geological Survey – USGS, 1994). Urban run-off can exceed the output from industrial sources and sewage treatment plants for some pollutants. For example, a single year's run-off from the Washington, DC, metropolitan area was found to carry 3.8 million to 19 million litres of oil, 180 t of zinc, 29 t of copper and 10 t of lead (Environmental Defense Fund – EDF, 1992). Because much of urban run-off is generated as non-point-source pollution, it is also difficult to control (Adler, 1994).

For instance, sediment can be a major pollutant in the early phases of urban development. Sediment delivery in areas newly cleared for urban construction has approached the equivalent of many decades of natural erosion (Wolman and Schick, 1967). Sediment yields have been recorded that are 10–100 times higher than rates for natural areas (Mount, 1995; Rhoads, 1995). Improved site management using bunding and interceptor traps to reduce run-off is often incorporated into building controls, minimizing the problem where such 'best practice' is applied. Furthermore, once urbanization is complete, erosion rates often fall back to near normal or even less than normal because of sewering of run-off and the stabilization of the land surface (Wolman, 1967; Mount, 1995; Rhoads, 1995). However, urban sediment often carries with it contaminants. For instance, sediment in urban rivers draining western

Lake Michigan has higher concentrations of metals including cadmium, copper, mercury, nickel, lead and zinc than from other sites draining non-urban catchments (Peters *et al.*, 1998). Elsewhere, concentrations of metals in bed sediments have been found to increase in direct proportion to the amount of industrial and transportation land use in the catchments (Frick *et al.*, 1998). Organic compounds such as polycyclic aromatic hydrocarbons and phthalates have also been found in highest concentrations in sediment in streams draining urban and suburban catchments (Frick *et al.*, 1998; Peters *et al.*, 1998).

Another common water quality problem in urban streams is 'shock pollution'. This phenomenon occurs when highly polluted water, which is retained in urban basins during dry weather run-off, is suddenly displaced by the beginning of storm run-off (Mount, 1995). Pollutants which build up on the surfaces of urban areas include lead chlorides and nitrates derived from vehicle exhaust, street litter, animal faeces and fertilizers – all of which combine to create a chemical and biological oxygen demand within run-off (Schueler, 1991). The polluted water that accumulates during dry weather is often anaerobic with high organic nitrogen and phosphorus content. When it is suddenly discharged into a stream or river it can cause massive fish kills (Mount, 1995).

Changes in stream thermal regime are also linked to several aspects of urbanization. Destruction of riparian zones for construction, landscaping and other bank alterations may decrease shading and increase water temperature (Klein, 1979). The widening of streams associated with urbanization also affects stream temperature by enhancing the degree of heat lost or gained, leading to greater temperature fluctuations, rendering water temperatures warmer in the summer and colder in the winter (Klein, 1979; Galli, 1991). Finally, run-off will increase in temperature as it passes over warmed surfaces such as roads, thereby delivering heated water to receiving streams (Schueler, 1991).

EFFECTS ON WATER QUANTITY

Many countries face urban water supply problems (Smil, 1993; WRI, 1996; Pringle and Scatena, 1999a). While local water shortages are a major concern in the world's large megacities, water quantity problems are increasingly troublesome in many smaller urban areas as well (Pringle and Scatena, 1999b; Pringle *et al.*, this volume).

Although municipal water use apparently accounts for less than one-tenth of the world's overall water use (WRI, 1996), urbanization increases the *per capita* demand for domestic water as well as increasing industrial demand for water, thus causing water shortages. As cities expand, water demands can quickly surpass local water supplies, a problem exacerbated by poor urban water management. In many cities in developing countries, leaky pipes and illegal connections waste between 20 and 50% of public water supplies (Lee, 1993). Developed countries often face similar problems as a result of ageing or poorly managed infrastructure; an estimated 25% of water supplies in the United Kingdom leaks from pipes to groundwater (WRI, 1996).

Over-exploitation of urban water sources may force cities to search for adequate drinking water supplies well beyond the city boundaries. In Mexico City, for example, overuse of groundwater has caused subsidence; large areas of downtown Mexico City have sunk approximately 9 m since the early 1900s (Ezcurra and Mazari-Hiriart, 1996). The city now obtains water from a river system 127 km away (Postel, 1992). In the USA, water allocation in the arid west has long been an issue of contention. Cities such as Los Angeles, Phoenix and Las Vegas thrive at the cost of regulated rivers and massive diversion schemes (Reisner, 1986). Even in the humid, south-eastern USA, proposed expansion of municipal water withdrawal has required a federal compact to allocate water between the states of Georgia, Alabama and Florida in a situation locally referred to as the 'tri-state water war' (Upper Chattahoochee Riverkeeper, 1997a). Traditionally, concentrated urban pollution forces cities to obtain their water supplies from widely disparate geographic locations upstream from pollution sources. In Shanghai, for example, water pollution caused the city to move the water intake 40 km upstream at a cost of US$300 million (WRI, 1996).

EFFECTS ON AQUATIC BIOTA

Changes in stream habitat, production, water quality, and loss or degradation of riparian zones and floodplains associated with urbanization all contribute to affect instream communities. Urban areas along streams can be thought of as 'selective environmental filters' (Poff, 1997), affecting biotic integrity and connectivity along stream corridors.

Taxonomic surveys of urban stream biota reflect degraded conditions, but it is difficult to evaluate the ultimate causes of community change given the many different sources of degradation (Pitt, 1995). Nonetheless, stream fauna can serve as effective bioindicators in urban areas. In several studies where chemical water quality data for streams draining urban land resulted in a healthy prognosis, aquatic insects indicated otherwise (e.g. Whiting and Clifford, 1983; Garie and McIntosh, 1986).

Urban streams seem to be less functionally diverse, reflecting the dominance of depositional habitats and food resources limited to fine particulate organic matter (FPOM) (Pedersen and Perkins, 1986). A decrease in the commonly used Ephemeroptera–Plecoptera–Trichoptera (EPT) index (e.g. Plafkin *et al.*, 1989) has been found below many urban areas (e.g. Whiting and Clifford, 1983; Garie and McIntosh, 1986), and this decrease represents a decline in species (mostly Plecoptera) using coarse particulate organic matter (CPOM) and in taxa requiring clean substrate. CPOM in urban streams can be limited by riparian zone destruction, thereby decreasing allochthonous CPOM input, and by abrasive hydraulic processes that can speed the transformation of CPOM to FPOM, reduce downstream transport time, and bury available CPOM in sediment (Paul, 1998).

Urban streams generally show a decrease in the proportion of shredders (species shredding material as a food source) and an increase in the proportion of a few collector–gatherer taxa (species collecting and gathering material as a food source) (Crawford and Lenat, 1989). Lower food quality may also contribute to changes in community composition (Pedersen and Perkins, 1986). These functional changes in streams draining urban areas have implications for ecosystem functioning since organic matter processing may be significantly altered. It is a challenge to evaluate the effect of run-off on stream biota because the diversity of contaminants contained in the run-off are likely to interact with in-stream processes (Borchardt and Statzner, 1990). Similar changes in fish communities in urban streams have also been recorded (e.g. Weaver and Garman, 1994; May *et al.*, 1997; Dennehy *et al.*, 1998).

Stream reaches above or below urban areas are also vulnerable to invasion by exotic species that are common in degraded urban areas. Urban areas can potentially act as 'source' populations of exotic species. For example, the red shiner, *Cyprinella lutrensis*, is a cyprinid native to the Mississippi River drainage. In 1978, small numbers of *C. lutrensis* were observed in the Apalachicola–Chattahoochee–Flint River system in Georgia, presumably introduced as discarded 'bait bucket' fish (DeVivo, 1995). The species has since become established in urban streams in the Atlanta area. It has become dominant or co-dominant at the expense of native species, often comprising up to 90% of the fish population in urban streams (DeVivo, 1995). Its success is apparently due to its ability to tolerate high turbidity, sedimentation, and flashy hydrographs and, perhaps, to exclude native fish through competition and hybridization with native congenerics. Reproducing populations of *C. lutrensis* will most likely remain intact as long as in-stream habitat remains degraded (DeVivo, 1995, 1996).

Conservation and policy: managing the street–river connection

The policy structure guiding the management of urban water resources is as varied as the threats posed to these waters. Internationally, there is no overarching framework guiding urban water management. International environmental principles are occasionally agreed upon, but are rarely binding or fully effective, and urban water management is not considered a global priority compared with global warming or population growth (e.g. Stevens, 1997).

POLICY AND MANAGEMENT IN DEVELOPING COUNTRIES

In developing countries, urban water policy is closely linked to environmental health and social issues. The 'brown agenda' refers to the synergistic issues of pollution, environmental hazards and poverty that are tied to broad indicators such as productivity, disease rate and ecological health (Bartone *et al.*, 1994). Brown agenda issues specifically note and address the lack of a potable water supply, sanitation services and wastewater treatment in developing countries (WRI, 1996). Provision of clean water and prevention of waterborne disease are integral parts of any urban environmental management programme (Bartone *et al.*, 1994). The WRI (1996) identified three components crucial for successful urban water management: (a) adoption of appropriate technology and standards; (b) involvement of local communities; and (c) improvement of the operation and maintenance of existing facilities.

Appropriate technology will differ based on local custom and geographical constraints and it is clear that there is a strong need for community involvement to help define the best-suited technology. Although

funding for infrastructure in developing countries may seem cost-prohibitive to local governments, changes in operation and maintenance procedures can provide needed cost recovery. For instance, because water supply is a public health issue, many governments subsidize the cost of water, leaving little money to expand existing systems; unfortunately, most of the subsidy often goes to the wealthier portion of a given society, leaving poorer members without access to clean water (WRI, 1996). Reducing illicit connections to water systems and improving maintenance may help further to provide water to more people without increasing costs (WRI, 1996).

In Karachi, Pakistan, the Orangi Pilot Project (OPP) has been successful at bringing water and sanitation to the Orangi settlement. The OPP was designed to involve community members and tailored to address local needs. While a traditional sewage system requires hooking all residential units to the main trunk line, the condominial system used in Karachi gathers an entire block's waste and disposes of cumulative wastes at one point (WRI, 1996). Residents provided labour and selected neighbourhood managers who operate the block system (Bartone *et al.*, 1994). This system was built at one-tenth of the cost of a conventional sewage system and has provided a basis from which to start other community-based environmental programmes in Karachi (WRI, 1996). Several organizations including the Urban Management Programme of the United Nations Centre for Human Settlements (UNCHS), the Metropolitan Environmental Improvement Programme of the United Nations Development Programme and the World Bank, and the Sustainable Cities Programme of the UNCHS all work to strengthen and demonstrate sustainable environmental planning strategies among cities in developing countries (Bartone *et al.*, 1994).

POLICY AND MANAGEMENT IN THE USA

In the USA, the Clean Water Act (CWA) of 1972 does not provide urban waters with any overall special protection, although the Act does include sections that address typical urban problems such as combined sewer overflows and stormwater run-off (Adler *et al.*, 1993). Although treatment technology in the USA is advanced, many municipalities continue to pollute downstream waters with excess phosphorus and raw sewage (e.g. Helton, 1997). At present, combined sewers serve approximately 43 million people in over 1000 municipalities throughout the USA (Nonpoint Source News–Notes, 1996), and urban run-off is attributed as a leading cause of water quality impairment nationwide (United States Environmental Protection Agency – US EPA, 1992). While the CWA has improved point-source regulation, it has not been successful in protecting many urban creeks from problems associated with sedimentation, channelization and diffuse run-off (Adler *et al.*, 1993).

For example, USA policy regarding the control and management of stormwater illustrates the difficulties of regulating non-point-source pollution in densely populated areas. Although the CWA was passed in 1972, final regulations concerning urban stormwater control were not published until November 1990, almost 20 years after the passage of the CWA (Federal Register, 1990). Maintaining that there were too many outfalls to regulate, the US EPA (the agency responsible for implementing the CWA) initially exempted many stormwater outfalls from permit requirements (Federal Register, 1973). However, after recognizing that urban stormwater run-off was a significant problem, the US Congress required reduction of pollutants to the 'maximum extent possible' (MEP) through management control techniques (Federal Register, 1990). The new regulations required a comprehensive stormwater management programme focusing on 'at-source' controls such as detention ponds, infiltration basins and education efforts throughout the catchment as opposed to end-of-the-pipe controls, which treat effluent from a single generator such as a wastewater treatment plant or an industry. Without specified permit limitations (or required at-source controls), however, municipal stormwater management remains elusive; without accountability there is little incentive to maintain pre-development discharge or to improve the quality of urban run-off (Baer, 1996).

Specific federal legislation to address urban rivers has been introduced, although without ultimate success. The Waterways Restoration Act of 1995, for example, was introduced in the US Congress to target urban stream restoration. The Bill would have amended the Watershed and Flood Protection Act (also known as the Natural Resources Conservation Service Small Watershed Program) to allow funds from this traditionally agriculturally based programme to be used in a wider variety of circumstances. The new focus would have allowed funding for urban, community-based projects with the following objectives: (a) flood damage reduction; (b) erosion control; (c) stormwater management; and (d) water quality enhancement, with funding priorities for low-income, economically depressed areas. The Bill, however, did not make it

through the committee process, the first step of the congressional process (Jones, 1996).

On the state and city level within the USA, urban river conservation varies. Some counties or cities will fund urban stream protection projects, but a state rarely has a specific urban streams policy. The state of California is unique for its Urban Streams Restoration Act passed in 1984. The Act emphasizes using non-structural techniques to restore the 'ecological viability of creek environments located in predominantly urban areas' (California Water Code §7048). This Act has encouraged stream restoration by providing funds and technical assistance to community groups using stream restoration to alleviate erosion and flooding problems (Mcdonald, 1995).

Many cities in the USA have purchased land surrounding their water supply reservoirs to protect these supplies from encroaching development (Leslie, 1987). As water becomes the factor limiting urban growth and water treatment becomes more costly, the need to preserve a permanent water supply catchment is more acute. For example, to avoid high costs of a new filtration system, New York City recently made an agreement with upstate communities to help protect the 5180 km^2 catchment. The city committed close to US$300 million to buy buffer land and an additional US$1.2 billion to upgrade or build wastewater treatment plants on tributary streams and to fix septic systems (Revkin, 1996).

The important role of non-governmental organizations

Besides governmental laws and programmes, there are a large number of non-profit and/or grassroots organizations that focus either entirely or in part on the health of urban streams. Foremost among these organizations in the USA is the Coalition to Restore Urban Waters (CRUW), which was founded in 1993 in Berkeley, California (Mcdonald, 1995). CRUW acts as an umbrella organization for a growing number of groups such as the California Urban Creeks Council and Friends of the Chicago River, allowing communication and exchange of ideas (Nixon, 1995). Since that time, the coalition has grown to include close to 400 member organizations such as community groups, state and local agencies, schools, conservation corps, and individuals (J. Vincentz, Isaac Walton League, personal communication). CRUW also sponsors the 'Friends of Trashed Rivers' annual conference and publishes a quarterly newsletter.

Community-based urban stream protection projects have special urgency because these efforts often attempt to revive disenfranchised neighbourhood groups. The unique thread uniting CRUW member groups is that stream restoration in inner cities is an effort to restore both community and ecology (Mcdonald, 1995). In this sense, urban stream restoration overlaps with the environmental justice movement (e.g. *Race, Poverty and the Environment: A Newsletter for Social and Environmental Justice*, 1992) which seeks to empower economically disadvantaged communities who rarely have a voice in environmental decisions that may adversely affect their local area and personal health (see also Showers, this volume). For example, the California Urban Creeks Council was founded by offering a community-acceptable 'alternative' restoration project to avoid an Army Corps of Engineers channelization through an African–American community in North Richmond, California (Nixon, 1995). The group has since tackled many other projects such as the revitalization of urban creeks throughout the Bay Area as well as the rest of the state (Mcdonald, 1995).

Other more broadly focused environmental groups may also target urban rivers as part of their programme. For example, the organization American Rivers sponsors an urban rivers symposium which highlights and awards groups throughout the nation that are involved in urban restoration and education efforts (American Rivers, 1996a).

The community-based nature of stream protection efforts in cities often leads to local projects that vary widely in their scope and goals. The academic definition of the term 'restoration' (e.g. National Research Council –NRC, 1992) may differ from a community definition that encompasses social, economic and aesthetic concerns such as the need for safe local parks or a visually pleasing, clean stream. Project achievements vary from habitat improvement to stream monitoring and from bio-engineered erosion control to passage of river protection legislation and innovative education programmes such as 'Get Hooked on Fishing not on Drugs' (American Rivers, 1996b).

To connect urban residents to their local resources, an initial step for local groups is to identify their nearest stream and to delineate the watershed in which they live; this can be more challenging than it sounds because many urban streams have been buried in culverts and a system of storm drains, leaving people to identify their 'sewershed' rather than 'watershed' (*Volunteer Monitor*, 1995). For example, 'Friends of the Los Angeles River' was founded when a small community group cut through a fence surrounding a channelized, cement-lined rivulet (a tributary of the Los Angeles River), thus rediscovering a local resource (Nixon, 1995). Another activity that draws attention to urban waterways is stream 'daylighting' whereby culverts are unearthed and the creek is 'brought to light' (Mcdonald, 1996).

Recommendations for conservation and management

The conservation of urban stream systems depends on a multifaceted approach that encompasses planning, education and community involvement. Regional planning efforts that maintain riparian corridors and protect floodplains will both protect urban streams and highlight their importance to city residents. Preservation of natural landscape features (e.g. wetlands, floodplains and tree cover) results in increased filtration and absorption of stormwater run-off. It is always easier to maintain landscape features than to mimic them with man-made structures. Natural features can also enhance property values and provide needed corridors for wildlife in developed areas (Tourbier, 1994).

In Portland, Oregon, the Urban Growth Boundary (UGB) was established to guide growth within the Portland area and to limit sprawl outside of the UGB, forcing citizens to address issues of housing and environmental quality within the city (Kasowski, 1994). Minneapolis, Minnesota, has ringed its lakes with trail systems and connected lakes with trails along riparian zones to make greenway corridors throughout the urban landscape. Successful steps towards restoring urban rivers in Chicago, Illinois, are documented in Karr *et al.* (this volume).

Local guidelines for development can also help to reduce run-off and enhance infiltration. For example, local limits can be set for the amount of impervious surface allowed within a catchment (Arnold and Gibbons, 1996). Minimizing connections between adjacent expanses of impervious surfaces can also help to reduce run-off and allow for greater infiltration (Greenfield and LeCouteur, 1994). Impervious surfaces such as parking lots can be replaced with permeable pavements to allow infiltration (Ferguson, 1994). After a comprehensive study of the relationships between impervious surfaces and water quality, the City of Olympia, Washington, recommended a 20% reduction in impervious surface cover (City of Olympia, 1995).

Zoning and local ordinances can be used positively to protect and enhance urban stream environments. Zoning laws that encourage cluster developments and preserve open space and wetlands as functioning systems will help protect urban streams. Unfortunately, zoning ordinances in the USA have often headed in the opposite direction – towards increased urban sprawl; activities such as housing and shopping are often segregated from each other in zoning plans, thus encouraging strip mall development and a car-dependent, impervious surface-demanding society (Kuntsler, 1996).

Environmentally sustainable development practices can be encouraged through economic incentives; linking lowered property tax assessments to cluster development is one example (Atlanta Regional Commission, 1993). In addition, municipalities can provide an economic incentive to reduce the total amount of impervious surface. Some towns have adopted charges based on the total area of impervious surface. The town of Spring Grove, Pennsylvania, implemented a stormwater facilities fee to pay for the construction and maintenance costs of a stormwater control structure. All new construction in the town is assessed a fee of US$0.15 per square foot of impervious surface. The fees are placed in escrow for future use (Greenfield and LeCouteur, 1994). Fees could further be used to finance a stormwater utility responsible for stormwater management (Lindsey, 1990). Impervious surfaces could also be reduced through an impervious surface banking programme; like wetlands mitigation or air quality credits, a developer could replace an existing paved area with a permeable surface in order to pave a new lot (Baer, 1996).

Many new developments are using natural landscape features as a design component to address typical urban water problems. For example, an innovative type of stormwater control is being piloted at a residential development in the state of Maryland, where each of 199 lots will have a rain garden at the low point of the site. The shallow, landscaped gardens combine grasses, shrubs, and trees to manage stormwater run-off and to maximize pollutant removal through physical, biological and chemical processes. The gardens are designed to make the development environmentally functional by preserving and enhancing the site's natural filtration characteristics (Nonpoint Source News–Notes, 1995).

Human behavioural changes are integral for urban stream conservation to be successful. Active water conservation by urban residents will reduce the amount of water consumed in urban areas. In Boston, Massachusetts, for example, a Long Range Water Supply Program effectively reduced the average daily water demand from 4442 m^3 to 3407 m^3. This programme used educational outreach, plumbing retrofits, pollution prevention, and detection of leaks to achieve such dramatic results (WRI, 1996).

Environmental education is a vital component in most urban conservation projects (e.g. American Rivers, 1996b). Volunteer stream monitoring is a popular activity designed to increase awareness and to involve citizens in protecting local stream health. In urban areas, some groups have adapted monitoring techniques to specific issues such as stormwater run-off. In Texas, the Urban Watch Group monitors stormwater outfalls with a kit designed to test for all of the mandated EPA-regulated parameters (Drinkwin, 1995). Another stormwater-related urban education activity is 'storm-drain stencilling'. Many urban dwellers consider storm drains as rubbish collectors for used car oil, yard wastes, and other debris. To educate the public and to establish the 'street–river connection', storm drain stencilling is a simple idea that has been used throughout the USA by community groups that often work with local Public Works Departments. Stencils range from messages such as 'Don't dump here, drains to creek' to attractive sketches of turtles and other riverine organisms which encourage people to think about and recognize catchment drainage patterns (Hunter, 1995).

Urban landscapes are engineered to have water bypass riparian zones and other natural landscape features, making any point in the watershed effectively right next to the nearest stream. Efforts must thus focus on reducing run-off quantity and enhancing run-off quality at the source, allowing the entire landscape to function as a 'riparian zone' (Baer, 1996). Education for urban stream conservation thus depends on helping urban residents to see themselves as part of a functioning catchment. Projects such as the Proctor Creek Watershed Initiative teach citizens to find and then learn about their local streams and catchments through hands-on monitoring and exploration activities (Upper Chattahoochee Riverkeeper, 1997b).

Urban areas, by definition, are places where people dominate the ecosystem. High population density and resource use in urban areas make streams and rivers vulnerable to extreme degradation, visible in many cities. However, it is this same resource, the community, which can be harnessed to mitigate and correct such situations. Whether working to improve water quality through a sanitation project in urban Pakistan (WRI, 1996) or restoring habitat in a US stream (Mcdonald, 1995), community involvement is critical to the successful conservation of urban streams and rivers. As human populations expand, so will our cities. An awareness of the location and importance of water in our urban homes may help restore rivers and streams – and possibly our sense of community as well.

Acknowledgements

National Science Foundation grant DEB-95-28434 (to the University of Georgia) partially supported the writing of this chapter. We thank Michael J. Paul and Jon Benstead for their constructive feedback on the manuscript.

References

Adler, R. 1994. Reauthorizing the Clean Water Act: looking to tangible values. *Water Resources Bulletin* 30:799–807.

Adler, R.W., Landman, J.C. and Cameron, D.M. 1993. *The Clean Water Act 20 Years Later*. Natural Resources Defense Council, Island Press, Washington, DC.

American Rivers 1996a. *Newsletter No. 1*, Spring. Washington, DC.

American Rivers 1996b. *Urban Lifelines: A Casebook for Successful Urban River Projects*. March. Washington, DC.

Arnold, C.L. and Gibbons, C.J. 1996. Impervious surface coverage: the emergence of a key environmental indicator. *Journal of the American Planning Association* 2:243–258.

Arnold, C.L., Boison, P.J. and Patton, P.C. 1982. Sawmill Brook: an example of rapid geomorphic change related to urbanization. *Journal of Geology* 90:155–166.

Atlanta Regional Commission 1993. *Municipal Storm Water Management Manual: A Guide for Developing a Comprehensive Municipal Stormwater Management Program and Complying with NPDES Permit Application Rules*. Atlanta, Georgia.

Baer, K.E. 1996. When it rains it drains: stormwater management in metropolitan Atlanta. Unpublished Master's Thesis, University of Georgia, Athens, Georgia.

Bartone, C. 1990. Water quality and urbanization in Latin America. *Water International* 15:3.

Bartone, C., Bernstein, J., Leitmann, J. and Eigen, J. 1994. *Toward Environmental Strategies for Cities: Policy Considerations for Urban Environmental Mangement in Developing Countries*. Urban Management Programme, The World Bank. Washington, DC.

Borchardt, D. and Statzner, B. 1990. Ecological impact of urban stormwater runoff studied in experimental flumes: population loss by drift and availability of refugial space. *Aquatic Sciences* 52:299–314.

Brookes, A. 1988. *Channelized Rivers: Perspectives for Environmental Management*. John Wiley, Chichester.

California Water Code §7048. *West's Annotated California Codes Water Code Sections 7000 to 12569*. West Publishing Co., St Paul, Minnesota.

City of Olympia 1995. *Impervious Surface Reduction Study: Final Report*. City of Olympia Public Works Department, Olympia, Washington.

Cosgrove, D. 1990. An elemental division: water control and engineered landscape. In: Cosgrove, D. and Petts, G.E. (eds) *Water, Engineering and Landscape – Water Control and Landscape Transformation in the Modern Period*. Bellhaven Press, London, 1–11.

Crawford, J.K. and Lenat, D.R. 1989. *Effects of Land Use on the Water Quality and Biota of Three Streams in the Piedmont Province of North Carolina*. US Geological Survey, Water Resources Investigations Report 89-4007, North Carolina.

Dennehy, K.F., Litke, D.W., Tate, C.M., Qi, S.L., McMahon, P.B., Bruce, B.W., Kimbrough, R.A. and Heiny, J.S. 1998. Water quality in the South Platte River Basin, Colorado, Nebraska, and Wyoming, 1992–1995. *US Geological Survey Circular* 1167, Reston, Virginia.

DeVivo, J.C. 1995. Impact of introduced red shiners, *Cyprinella lutrensis*, on stream fishes near Atlanta, Georgia. In: Hatcher, K. (ed.) *Proceedings of the 1995 Georgia Water Resources Conference*, Carl Vinson School of Government, University of Georgia, Athens, Georgia, 95–98.

DeVivo, J.C. 1996. Fish assemblages as indicators of water quality within the Apalachicola–Chattahoochee–Flint (ACF) River Basin. Unpublished Master's Thesis, University of Georgia, Athens, Georgia.

Douglas, I. 1983. *The Urban Environment*. Edward Arnold, London.

Drinkwin, J. 1995. Urban watch: a new approach to monitoring urban nonpoint source pollution. *The Volunteer Monitor* 7:4–5.

Dunne, T. and Leopold, L.B. 1978. *Water in Environmental Planning*. W.H. Freeman & Co., New York.

Environmental Defense Fund (EDF) 1992. *How Wet is a Wetland: The Impacts of the Proposed Revisions to the Federal Wetlands Manual*. Environmental Defense Fund, Washington, DC.

Ezcurra, E. and Mazari-Hiriart, M. 1996. Are mega-cities viable? A cautionary tale from Mexico City. *Environment* January/February:6–35.

Federal Register 1973. *Environmental Protection Agency National Pollution Discharge Elimination System* 38:1362–1370.

Federal Register 1990. *Environmental Protection Agency National Pollution Discharge Elimination System Permit Application Regulations for Storm Water Discharges* 55(222):47990–48075.

Ferguson, B.K. 1994. *Stormwater Infiltration*. Lewis Publishers, Boca Raton, Florida.

Ferguson, B.K. and Suckling, P.W. 1990. Changing rainfall–runoff relationships in the urbanizing Peachtree Creek

Watershed, Atlanta, Georgia. *Water Resources Bulletin* 26:313–322.

Frick, E.A., Hippe, D.J., Buell, B.R., Couch, C.A., Hopkins, E.H., Wangsness, D.J. and Garrett, J.W. 1998. Water quality in the Apalachicola–Chattahoochee–Flint River Basin, Georgia, Alabama, and Florida, 1992–1995. *US Geological Survey Circular* 1164, Reston, Virginia.

Galli, F.J. 1991. *Thermal Impacts Associated with Urbanization and Stormwater Management Best Management Practices.* Metropolitan Washington Council of Governments/Maryland Department of Environment, Washington, DC.

Garie, H.L. and McIntosh, A. 1986. Distribution of benthic macroinvertebrates in a stream exposed to urban runoff. *Water Resources Bulletin* 22:447–455.

Gavidia, J. 1994. Housing and land in large cities of Latin America. In: *Enhancing the Management of Metropolitan Living Environments in Latin America.* United Nations Centre for Regional Development, Nagoya, Japan.

Graf, W.L. 1975. The impact of suburbanization on fluvial geomorphology. *Water Resources Research* 11:690–692.

Greenfield, J. and LeCouteur, B.M. 1994. *Chesapeake Bay Community Action Guide: A Step-by-step Guide to Improving the Environment in Your Neighborhood.* Metropolitan Washington Council of Governments, Washington, DC.

Hartfield, P. 1993. Headcuts and their effect on freshwater mussels. Conservation and management of freshwater mussels. In: *Proceedings of a Conference on Conservation and Management of Freshwater Mussels.* Upper Mississippi River Conservation Committee, Rock Island, Illinois, 131–141.

Helton, C. 1997. Report finds pattern of sewage spills. *The Atlanta Journal/The Atlanta Constitution*, 31 March.

Hengeveld, H. and Vocht, C.P. 1982. *Role of Water in Urban Ecology: Developments in Management and Urban Planning.* Elsevier Scientific Publishing, New York.

Hirsch, R.M., Walker, J.F., Day, J.C. and Kallio, R. 1990. The influence of man on hydrologic systems. In: Wolman, M.G. and Riggs, H.C. (eds) *Surface Water Hydrology.* The Geological Society of America, Boulder, Colorado, 329–359.

Hunter, R. 1995. Storm drain stenciling: the street–river connection. *The Volunteer Monitor* 7:8–10.

Jones, E. 1996. *Charting a New Course for Waterway Restoration in the 104th Congress.* Coalition for Restoring Urban Waters (CRUW) Newsletter, Winter, Berkeley, California.

Kasowski, K. 1994. Portland's urban growth boundary. *The Urban Ecologist*, Spring.

Klein, R.D. 1979. Urbanization and stream quality impairment. *Water Resources Bulletin* 15:948–963.

Kuntsler, J.H. 1996. Home from nowhere. *The Atlantic Monthly* 278:43.

Lee, Y.F. 1993. Urban water supply and sanitation in developing countries. In: Nickum, J.E. and Easter, K. (eds) *Metropolitan Water Use Conflicts in Asia and the Pacific.* Westview Press, Boulder, Colorado, 19–35.

Leitman, S., Bethea, S. and Carr, C. 1993. Participation by environmental NGOs in management decisions in the Apalachicola–Chattahoochee–Flint River Basin. In: Hatcher, K.J. (ed) *Proceedings of the 1993 Georgia Water Resources Conference*, Athens, Georgia, 18–22.

Leopold, L.B. 1991. Lag times for small drainage basins. *Catena* 18:157–171.

Leslie, T.C. 1987. *Abandoned Water Supplies of Metro Atlanta: Past, Present and Future.* Presented at the annual meeting of the Georgia Water and Pollution Control Association, 19 August 1987.

Lindsey, G. 1990. Charges for urban runoff: issues in implementation. *Water Resources Bulletin* 26:117–125.

May, C.W., Horner, R.R., Karr, J.R., Mar, B.W. and Welch, E.B. 1997. Effects of urbanization on small streams in the Puget Sound Lowland Ecoregion. *Watershed Protection Techniques* 2:483–494.

Mcdonald, M. 1995. A combination on behalf of restoration: the coalition to restore urban waters. *Restoration and Management Notes* 13:98–103.

Mcdonald, M. 1996. Daylighting Blackberry Creek: Unearthing an urban stream. *Coalition to Restore Urban Waters (CRUW) Newsletter*, Spring, Berkeley, California.

McDonnell, M.K. and Pickett, S.T.A. 1990. Ecosystem structure and function along urban–rural gradients: an unexploited opportunity for ecology. *Ecology* 71:1232–1237.

Meybeck, M., Chapman, D. and Helmer, R. (eds) 1989. *Global Freshwater Quality: A First Assessment.* United Nations Environment Programme and World Health Organization, Geneva.

Mikalsen, T. 1993. Managing the quality of urban streams in Georgia. In: Hatcher, K.J. (ed.) *Proceedings of the 1993 Georgia Water Resources Conference.* Athens, Georgia, 284–292.

Mount, J.F. 1995. *California Rivers and Streams: The Conflict Between Fluvial Processes and Land Use.* University of California Press, Berkeley, California.

Mumford, L. 1956. The natural history of urbanization. In: Thomas, W.E. (ed.) *Man's Role in Changing the Face of the Earth.* The University of Chicago Press, Chicago, Illinois, 382–398.

Naiman, R.J., Magnuson, J.J., McKnight, D.M. and Stanford, J.A. (eds) 1995. *The Freshwater Imperative: A Research Agenda.* Island Press, Washington, DC.

National Research Council (NRC) 1992. *Restoration of Aquatic Ecosytems.* National Academy Press, Washington, DC.

Nixon, W. 1995. Trashed urban rivers and the people who love them. *The Amicus Journal* Fall:25–28.

Noland, G.A. (ed.) 1981. Floodwater. In: *Cal Jur 3d.* 3:119–126. Bancroft-Whitney Co., San Francisco.

Nonpoint Source News–Notes 1995. *Maryland Developer Grows 'Rain Gardens' to Control Residential Run-off.* Terrene Institute, Alexandria, Virginia 42:5–7.

Nonpoint Source News–Notes 1996. *Combined Sewer*

Overflow Technology Based Control Due in January. Terrene Institute, Alexandria, Virginia 45:1.

Patrick, D.M., Ross, S.T. and Hartfield, P.D. 1993. Fluvial geomorphic considerations in the management and stewardship of fluvial ecosystems. In: *Proceedings of a Symposium on Riparian Ecosystems in the Humid U.S.: Functions, Values, and Management.* National Association of Conservation Districts, Washington, DC, 90–99.

Paul, M.J. 1998. Stream ecosystem function along a land-use gradient. Unpublished Doctoral dissertation, University of Georgia, Athens, Georgia.

Pedersen E.R. and Perkins, M.A. 1986. The use of benthic invertebrate data for evaluating impacts of urban runoff. *Hydrobiologia* 139:13–22.

Peters, C.A., Robertson, D.M., Saad, D.A., Sullivan, D.J., Scudder, B.C., Fitzpatrick, F.A., Richards, K.D., Stewart, J.S., Fitzgerald, S.A. and Lenz, B.N. 1998. Water quality in the Western Lake Michigan Drainages, Wisconsin and Michigan, 1992–1995. *US Geological Survey Circular* 1156, Reston, Virginia.

Pitt, R.E. 1995. Biological effects of urban runoff discharges. In: Herricks, E.E. (ed.) *Stormwater Runoff and Receiving Systems, Impact Monitoring and Assessment.* CRC Lewis Publishers, Boca Raton, Florida, 127–162.

Plafkin, J.T., Barbour, M.T., Porter, K.D., Gross, S.K. and Hughes, R.M. 1989. *Rapid Bioassessment Protocols for Use in Streams and Rivers.* Benthic macroinvertebrates and fish. EPA 444/4-89/001. Office of Water Regulations and Standards, US EPA, Washington, DC.

Poff, N.L. 1997. Landscape filters and species traits: towards mechanistic understanding and prediction in stream ecology. *Journal of the North American Benthological Society* 16:391–409.

Postel, S. 1992. *The Last Oasis: Facing Water Scarcity.* Worldwatch Environmental Alert Series. W.W. Norton & Company, New York.

Pringle, C.M. 1997. Exploring how disturbance is transmitted upstream: going against the flow. *Journal of the North American Benthological Society* 16:425–438.

Pringle, C.M. and Scatena, F.S. 1999a. Factors affecting aquatic ecosystem deterioration in Latin America and the Caribbean. In: Hatch, U. and Swisher, M.E. (eds) *Managed Ecosystems: The Mesoamerican Experience.* Oxford University Press, 104–113.

Pringle, C.M. and Scatena, F.S. 1999b. Freshwater resource development: case studies from Puerto Rico and Costa Rica. In: Hatch, U. and Swisher, M.W. (eds) *Managed Ecosystems: The Mesoamerican Experience.* Oxford University Press, 114–121.

Pringle, C.M., Vellidis, G., Heliotis, F., Bandacu, D. and Crisofor, S. 1993. Environmental problems of the Danube Delta. *American Scientist* 81:350–361.

Race, Poverty and the Environment: A Newsletter for Social and Environmental Justice 1992. Volume III, No. 2. California Rural Legal Assistance Foundation and the Earth Island Institute Urban Habitat Program, San Francisco, California.

Reisner, M. 1986. *Cadillac Desert: The American West and its Disappearing Water.* Viking Penguin Inc., New York.

Revkin, A.C. 1996. U.S. praises plan to protect New York City's watershed. *New York Times,* 11 September:B12.

Rhoads, B.L. 1995. Stream power: a unifying concept for urban fluvial geomorphology. In: Herricks, E.E. (ed.) *Stormwater Runoff and Receiving Systems, Impact Monitoring and Assessment.* CRC Lewis Publishers, Boca Raton, Florida, 65–75.

Riley, A.L. 1998. *Restoring Streams in Cities.* Island Press, Washington, DC.

Rosgen, D.L. 1996. *Applied River Morphology.* Wildland Hydrology, Pagosa Springs, Colorado.

Schueler, T.R. 1991. *Mitigating the Adverse Influence of Urbanization on Streams: A Comprehensive Strategy for Local Governments.* Watershed restoration sourcebook. Anacostia Restoration Team, College Park, Maryland.

Seabrook, C. 1997a. EPA slams metro streams. *The Atlanta Journal/The Atlanta Constitution.* 22 March, C10.

Seabrook, C. 1997b. River in peril: how Atlanta's sewers threaten the Chattahoochee. *The Atlanta Journal/The Atlanta Constitution.* 29 June, C6–C7.

Smil, V. 1993. *Global Ecology: Environmental Change and Social Flexibility.* Routledge, London.

Smith, L.M. and Patrick, D.M. 1991. Erosion, sedimentation, and fluvial systems. In: Kiersch, G.A. (ed.) *The Heritage of Engineering Geology: The First Hundred Years,* Volume 3. Geological Society of America, Boulder, Colorado, 169–181.

Soto, L. 1996. Downstream currents. *The Atlanta Journal/The Atlanta Constitution.* 22 February, B4.

Stevens, W.K. 1997. Five years after Environmental Summit in Rio, little progress. *The New York Times,* 17 June, B14.

Tourbier, J.T. 1994. Open space through stormwater management. Helping to structure growth on the urban fringe. *Journal of Soil and Water Conservation* 49:14–21.

United Nations Population Division (UNPD) 1995. *World Urbanization Prospects: The 1994 Revision.* United Nations, New York.

United States Environmental Protection Agency (USEPA) 1992. *The National Water Quality Inventory 1992 Report to Congress.* Washington, DC.

United States Geological Survey (USGS) 1994. *Do the Pesticides I Use Contaminate the Rivers Everyone Uses?* US Geological Survey National Water Quality Assessment Program, Atlanta, Georgia.

Upper Chattahoochee Riverkeeper 1997a. *Riverchat Winter Issue.* Atlanta, Georgia.

Upper Chattahoochee Riverkeeper 1997b. *Riverchat Spring Issue.* Atlanta, Georgia.

Volunteer Monitor (San Francisco) 1995. Monitoring urban watersheds. 7(2): Fall.

Weaver, L.A. and Garman, G.C. 1994. Urbanization of a watershed and historical change in a stream fish assemblage. *Transactions of the American Fisheries Society* 123:162–172.

Whiting, E.R. and Clifford, H.F. 1983. Invertebrates and urban runoff in a small northern stream, Edmonton, Alberta, Canada. *Hydrobiologia* 102:73–80.

Wolman, M.G. 1967. A cycle of erosion and sedimentation in urban river channels. *Geografiska Annaler* 49A:385–395.

Wolman, M.G. and Schick, A.P. 1967. Effects of construction on fluvial sediment: urban and suburban areas of Maryland. *Water Resources Research* 3:451–562.

World Resources Institute (WRI) 1996. *A Guide to the Global Environment: The Urban Environment, 1996–1997*. Oxford University Press, New York.

Worster, D. 1985. *Rivers of Empire: Water, Aridity, and the Growth of the American West*. Pantheon Books, New York.

Young, G.J., Dooge, J.C.I. and Rodda, J.C. 1994. *Global Water Resource Issues*. Cambridge University Press, Cambridge.

C Key Constraints

17

River size as a factor in conservation

E.H. Stanley and A.J. Boulton

Introduction

Other chapters in this book explore how physical, ecological, political and socio-economic factors influence river conservation efforts around the world. Directly or indirectly, each of these factors relates to the size of the river. In turn, definitions and perceptions of the 'size' of a river vary widely, and it is essential to effective legislation, for example, that these be expressed as unambiguously as possible.

In a physical sense, river size can be defined by catchment area, discharge, channel capacity and other geomorphological features (Richards, 1982; Newson, 1994). In a cultural sense, perceptions of river 'size' may vary, reflecting the experiences of the observer and the social values attached to the goods and services (*sensu* Meyer, 1997) provided by the river. Large rivers should attract the attention of conservationists because they are more threatened by exploitation and river regulation than smaller streams. However, large rivers are also more difficult to conserve because of their sheer size, the fact that they drain a variety of land uses and often different political states, and the potentially greater number of threatening processes they may face. Consequently, the more manageable nature of smaller streams means that these latter systems often receive disproportionately greater attention to their conservation and restoration.

This chapter reviews the definitions of river size, ecological scale and spatial hierarchy in a conservation context. Most studies recognize the importance of a catchment-scale approach but divide river systems into 'sub-catchments' or some intermediate spatial scale for assessment, management and conservation purposes. We explore the success of several of these strategies based on river size, seeking generalities that may guide future conservation efforts. We challenge water managers to consider explicitly the implications of river size in conservation of entire catchments, recognizing that selection of equal-sized areas for management along a river course (e.g. regions based on municipal boundaries) is unlikely to prove a useful strategy for *ecological* conservation even though this may be expedient politically.

Descriptions of river size

Running waters range in size from ephemeral rills that carry run-off for hours to days after rain up to massive permanent rivers whose catchments may span several countries and whose discharge alters ocean salinity for

Global Perspectives on River Conservation: Science, Policy and Practice.
Edited by P.J. Boon, B.R. Davies and G.E. Petts. © 2000 John Wiley & Sons Ltd.

several kilometres from the mouth of the estuary. Several approaches have been used to quantify river size, and each system comes with its own limitations. These restrictions are an inevitable product of the frame of reference adopted by each particular system.

Geomorphological descriptors of stream size are either based on channel development within the drainage (e.g. Strahler, 1957; Scheidegger, 1965; Shreve, 1966) or on local channel cross-sectional features of width, depth and water velocity (i.e. hydraulic geometry; Newson, 1994). At the catchment scale, the most commonly used approach to express river size is based on 'stream order' and refers to the sum of tributaries on a map of a given scale (usually 1:24 000 or 1:25 000). In the Strahler (1957) method, all small, exterior channels are considered 'first-order' streams. When two first-order streams merge, they form a second-order stream and so on (Figure 17.1a). However, several limitations of this method have been identified. The most significant difficulty involves the inconsistent relationship between changes in stream order and changes in discharge. When a large number of minor order tributaries join a larger order channel (see left-hand branch of Figure 17.1a), discharge increases, but this increase is not reflected by an increase in stream order. To overcome this inconsistency, Shreve (1966) suggested that stream order be calculated as the sum of the tributaries (Figure 17.1b). This system gives a classification that better matches discharge patterns, but is cumbersome to compute for large streams. Other restrictions of the stream order system include a lack of agreement regarding the treatment of ephemeral and intermittent channels or the scale of map to use when ordering stream channels, and differences in the size of channels sharing the same order from one region to another (Hughes and Omernik, 1981).

Recent advancements in digitization and remote sensing techniques enable us to measure drainage area readily. Drainage area is a reliable measure of stream size within similar climatic regions because there is usually a strong region-specific relationship between drainage area and discharge. However, comparisons of river size by means of drainage area between, for example, temperate and arid regions may be misleading owing to changes in the rainfall–run-off relationship (Graf, 1988). A 500 km^2 basin in a desert may be drained by an intermittent channel, whereas a broad, permanent river is likely to export water from a similar sized basin in temperate or humid areas (cf. Puckridge *et al.*, 1998). Hughes and Omernik (1983) recommend the combined use of drainage area and

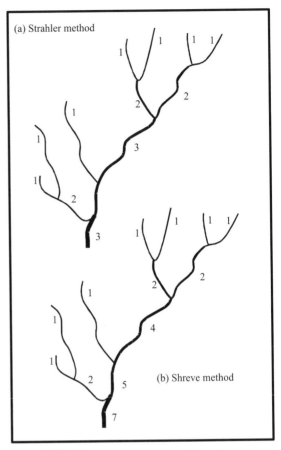

Figure 17.1 *Two ways to measure river size: (a) Strahler (1957) and (b) Shreve (1966). Note the different methods yield different values for the larger channels*

mean annual discharge per unit area (i.e. unit discharge, m^3 s^{-1} km^{-2}) to describe stream size because these two metrics relate catchment and stream characteristics of biological significance. They also provide a comparable assessment of stream size regardless of map scale, stream permanence, or presence of other water bodies in the channel. While unit discharge and catchment area offer many advantages for expressing stream size, considerable error still can arise when assessing small arid catchments, basins where groundwater and topographic divides differ, and streams with limited gauge records (Hughes and Omernik, 1983).

Although ecologists and hydrologists often have contrasting views of the same reach of stream (Gordon *et al.*, 1992), most ecological papers describing stream size tend to adopt the approaches outlined above

(Bisson and Montgomery, 1996). Stream order, channel width and mean discharge are commonly used, although their limitations for ecological prediction, especially in dryland regions, are recognized (Walker *et al.*, 1995). Usually, site descriptions are at the scale of habitat or reach (see later) and, while this limited framework is relevant for local ecological factors, it ignores upstream and downstream environments that are also potentially relevant (Allan, 1995). Furthermore, most ecologists tend to think of streams as linear systems, but, in reality, they are complex, branching entities. We have little appreciation of how ecological processes may be influenced by different branching patterns or spatial configuration of channels (Fisher, 1997). Clearly, the frame of reference adopted when describing stream and river size affects the assessment. The first step in defining system size for river conservation is to recognize these biases and restrictions.

Ecological significance of river size

Physical, chemical and biological processes vary with river size, and shifts in ecological patterns and processes associated with increasing channel size have been incorporated into the River Continuum Concept (RCC; Vannote *et al.*, 1980). Fundamentally, this concept identifies important differences in ecological processes such as energy flow, organic matter breakdown, and community structure in stream and river channels of different sizes along a longitudinal continuum. Differences in the magnitudes and rates of many of these variables are governed by differences in discharge, channel width, channel depth, and other size-related features. While the application of specific RCC tenets has geographic restrictions (Winterbourn *et al.*, 1981; Brussock *et al.*, 1985), this model highlights the importance of changing channel size in dictating a large suite of processes within the river, and between the river and its surrounding landscape.

If this observation seems trite, uncritical extrapolation of the results of ecological studies on one river size to another is alarmingly common, and dangerously misleading. For example, a 1993 literature review on riparian processes designed to summarize scientific data for policy makers in the Environment Protection Authority in New South Wales, Australia, lumped conclusions from upland streams and lowland rivers together with little consideration of river size. Although this document was ostensibly an 'in-house' publication of limited peer review and circulation, it

exerted a disproportionately great impact on river management. However keenly such generalizations may be sought, serious mistakes are made when river size and issues of scale are ignored. Scientists and water managers should work together to identify spatial and temporal restrictions on such generalizations; this is another of the potential friction points in the turbulent boundary between the two cultures (Cullen, 1986).

Scale and spatial hierarchies of the river environment

Scale has been defined as the period of time or space over which information signals are integrated or smoothed to give a message (Allen and Starr, 1982). Without exploring these authors' approach in detail, we can see intuitively that scale influences our perception of the available information and, hence, our predictive capacity. Failure to consider the effects of scale has frequently confounded explanations and generalizations in ecology (Jackson, 1991; May, 1994) and scale issues have also played a key role in conservation biology, especially in terrestrial ecosystems (e.g. reserve size, migratory patterns). Many authors bemoan the mismatch between most scientific research (small scale) and the appropriate scale for conservation (large scale) (e.g. May, 1994; Porritt, 1994; Carpenter, 1996). For rivers, many threatening processes such as salinization, non-point-source pollution, sedimentation, interbasin transfers and water abstraction, occur over large spatial extents (see Davies *et al.* 1992; Boon, 1996). These impacts must be assessed and tackled at an equivalent large scale. Thus, we are left with a need to specify the appropriate spatial scale for successful conservation of rivers, and to encourage effective scientific research at the appropriate scale.

Spatial scale is also relevant for practical reasons. Conservation or restoration of aquatic ecosystems must adopt a large enough area to minimize negative effects of boundary conditions, just as in terrestrial reserves. Project managers must be able to exert influence over areas where threatening processes occur, the area to be conserved must be large enough so that samples for monitoring can be collected, and the project must be of an affordable size (National Research Council – NRC, 1992). Furthermore, the 'spatial hierarchy' of the habitat must be understood so that its conservation and management can be considered in context – conserving a single stream reach without regard for upstream or downstream

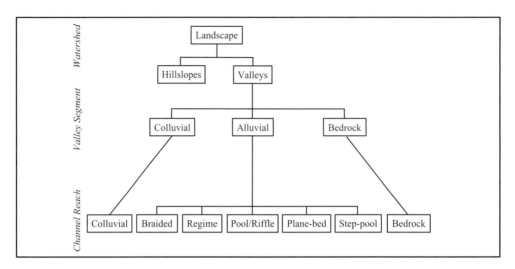

Figure 17.2 *Hierarchical subdivision of catchments ('watersheds') into valley segments and channel reaches. The reaches are arranged along a gradient according to whether or not their substrate is controlled by sediment supply from adjacent hillslopes (supply limited) or by fluvial transport from upstream (transport limited). (After Montgomery and Buffington, 1993)*

environments will restrict or, worse still, prohibit the success of management efforts (Muhar, 1996).

In this sense, the hierarchical frameworks of stream structure, such as that proposed by Frissell *et al.* (1986) and Montgomery and Buffington (1993) can be used to help guide conservation strategies. These frameworks identify: (a) structures that are being targeted for conservation; and (b) the physical and ecological processes that may affect these structures, and therefore affect the success of conservation efforts. One type of hierarchy, proposed by Montgomery and Buffington (1993), subdivides catchments (they use the term 'watersheds') into hill slopes and valley segments (Figure 17.2) before further dividing the alluvial channels into smaller habitat units. One distinctive feature of this classification is the inclusion of a 'gradient' of sediment supply. In transport-limited segments, the amount of sediment present is controlled by the frequency of floods that move sediment into and out of local channel areas. At the other extreme, sediments in supply-limited reaches are controlled by the volume of sediment entering the stream (Montgomery and Buffington, 1993). This classification is relevant to conservation because of the importance of sediment supply as both an important ecological factor as well as a threatening process (e.g. channel erosion, siltation: Wood and Armitage, 1997).

The most commonly used classification is that described by Frissell *et al.* (1986), which spans the

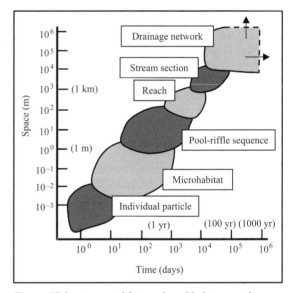

Figure 17.3 *A spatial hierarchy of habitats within a river superimposed on the temporal scale over which physical change occurs (modified from Frissell et al., 1986)*

spatial scale of drainage network (catchment) to microhabitat (Figure 17.3). Its value lies in representing the hierarchical relationship among the structural units, defining these clearly, and allowing the physical environment of a stream to be classified in a systematic fashion by identifying the homogeneous units that comprise these heterogeneous systems. Their scheme divides a drainage basin into a series of channel *segments* or stream sections. Different segments can have different attributes (e.g. a narrow first-order channel segment versus a broader third-order segment) but all of these units are of a similar size and are affected by similar physical processes. Segments, in turn, are composed of a series of *reaches* that begin and end at points where the channel slope or valley floor morphology changes sharply. Individual reaches can be broken down into *pools* and *riffles* (Figure 17.3), distinguished in terms of differences in depth, flow velocity and substrate. Finally, within an individual pool or riffle, distinct *microhabitats* (e.g. detrital patches, gravel accumulations, algal mats, etc.) are recognizable.

Because of the correlation between the spatial and temporal scales over which physical change occurs in rivers (Figure 17.3), the Frissell *et al.* (1986) description of stream structure is more than a spatial classification based on size. At the individual particle level (a spatial scale scale of millimetres to centimetres), processes that move and redistribute particles usually operate at a time-scale of days to weeks. On the other hand, physical changes in stream sections and drainage networks occur over centuries or longer (Figure 17.3). Such a spatial and temporal hierarchy has major applications for conservation. Responses to restoration or rehabilitation efforts at the pool–riffle or reach scale may not be evident for decades (Figure 17.3), and this approximate space–time correlation is useful to show managers and landholders so that they do not expect instant results. In this way, river size is related intimately to time-scale, and changes as a result of management or human activities may take long periods to occur or reverse.

For any given spatial scale, distinct sub-units of similar size, and formed by processes occurring at a similar time-scale, are identifiable. While the specific identity of stream structure at a given scale may vary regionally, hierarchical classifications of streams and rivers have proved invaluable in studies of fish habitat analysis (Hawkins *et al.*, 1993), stream hydrology (Stanley *et al.*, 1997), nutrient cycling (Fisher *et al.*, 1998), riparian zones (Gregory *et al.*, 1991) and river restoration (Muhar, 1996). For river conservation, these classification frameworks provide a context for targets of conservation. Is the target of conservation a discrete physical unit of the stream hierarchy, or are several patch types being included? For example, if conservation is motivated by a goal of maintaining a single species of non-migratory fish, then efforts may

focus on the preservation of relatively small spatial-scale units or habitats that are required by the targeted taxon (e.g. riffles for darter species). However, if the goal is to improve the richness of the entire fish community, then the spatial scale of the study will need to be expanded to include riffles and pools in both low- and high-gradient reaches. That is, the spatial scale of conservation efforts must be enlarged. Such a framework also illustrates the folly of reductionism – focusing on too small a scale for species conservation while ignoring the broader-scale linkages that may indirectly govern species persistence.

River classification strategies

For agencies responsible for the management and conservation of riverine (and other) resources, some systematic means of identifying targets of conservation or land evaluation is essential. Agencies must reconcile diverse issues such as site prioritization, feasibility, and regional policies appropriate both to ecological attributes and societal concerns for a particular area. As discussed above, issues of river size work their way into this equation as well; policies aimed at conserving small streams may be inappropriate for larger river systems, hence classification schemes should be able to distinguish between different-sized systems. Resource management and conservation efforts need to be targeted at appropriate spatial scales, and be based on appropriate physical units. In short, the inherent complexity of drainage basins must be reconciled in some systematic fashion so that there is a reasonable match between the target of conservation and management efforts and the efforts themselves.

While within-basin classifications like those described above are relatively easy to apply, classification of multiple catchments is more challenging. Finding common ground, for example in hydrological variables (Puckridge *et al.*, 1998), is a challenge. River classification systems can be broadly divided into ones aimed at describing basic physical and/or ecological patterns and those that have been developed as frameworks for environmental assessment (O'Keeffe and Uys, this volume). Continental rivers have been classified using flow regime (Haines *et al.*, 1988; Poff and Ward, 1989; Puckridge *et al.*, 1998), catchments or catchment units (Seaber *et al.*, 1987), physiographic location (Omernik, 1995), pollution status (Wright *et al.*, 1989), and various geomorphological attributes (e.g. Schumm, 1977; Brussock *et al.*, 1985; Rosgen, 1985).

In the United States, resource management agencies typically use a catchment-oriented framework of hydrologic units (Seaber *et al.*, 1987) or ecoregions (*sensu* Bailey 1995; Omernik 1995). The former system is a four-tiered hierarchical classification in which major drainage systems or regions are broken down into small basins or reaches. Twenty-one water resource regions have been defined in the United States, each of which represents either a major drainage system (e.g. the Missouri River drainage), part of a major drainage system (e.g. the Upper Mississippi River drainage), or the combined drainage areas of several rivers in a coastal region (e.g. the Texas Gulf coast region). Each of the 21 water resource areas is composed of several smaller catchments or river reaches and their associated tributaries referred to as sub-regions. The final two levels of the hierarchical classification include accounting units and, finally, cataloguing units (Seaber *et al.*, 1987). Management and research are matched to the appropriate spatial scale, and the hydrologic units that make up that particular scale. The hydrologic unit system is widely used as a framework for water quality assessment and regulation because of the well-established relationship between water quality and drainage basin attributes (Omernik and Bailey, 1997). In Australia, a similar classification exists. The continent is divided into 12 drainage divisions, composed of 245 river basins (Smith, 1998). Although these are used for reporting run-off and rainfall, they are of limited use for groundwater conservation because of the poor match between surface and groundwater 'catchments'. They are also usually subdivided further for management and assessment, depending on the specific aim (see later).

Most ecological studies of rivers have focused on the scientifically tractable scales of habitat, reach, or segment whereas the catchment scale is considered more suitable for understanding the relationship between 'the stream and its valley' (Hynes, 1975) and for total catchment management (Hooper and Margerum, this volume). Practical obstacles to large-scale studies are being overcome. Rapid technological advances in geographic information systems (GIS) and image processing (IP) provide the means to analyse spatial data at the catchment scale (Johnson and Gage, 1997) and should prove invaluable tools for monitoring river conservation and restoration efforts at appropriately broad scales. Although these large spatial-scale approaches are within our grasp, we must recognize that suitable *time* scales must also be adopted and long-term data sets are essential to supplement these larger spatial scales (Allan and

Johnson, 1997). Assessment of the success of conservation efforts in rivers will reflect the variable chosen. In turn, this is influenced by the scale at which the data are collected and the study design (Allan and Johnson, 1997). We conclude that the level of the 'sub-catchment' represents a compromise between tractability, available scientific knowledge, and an intermediate level in the hierarchy.

The sub-catchment approach

The sub-catchment approach has been used in New South Wales, Australia, to classify the State's rivers into three broad categories of degradation or 'stress' (reviewed in Boulton *et al.*, 2000). In some river systems such as the Macleay River that drains the New England Tablelands to the eastern coast, the upper sub-catchments of the headwaters are classed as severely 'stressed' because they receive run-off from grazing lands that have been heavily fertilized and planted with pasture (Reid *et al.*, 1997). Further down its course, the river cascades several hundred metres over the edge of the escarpment set in a National Park (gazetted somewhat by default as the terrain is steep and rocky, and unsuited for agriculture). Water quality improves and sub-catchments are rated as moderately stressed. As the river emerges onto the coastal plain where agricultural land use predominates again, some sub-catchments are classed as severely stressed whereas others, draining less settled regions or coastal reserves, are considered 'unstressed'. In this case, the sub-catchment approach has the resolution to distinguish between consecutive sections along a river, and to assist managers to identify sections for river rehabilitation or conservation.

More commonly, the sub-catchments comprising the upper reaches of most New South Wales rivers are classified as less stressed than their counterparts downstream. As the inputs of various land uses on the floodplains of the middle reaches enter and accumulate in the river, downstream sub-catchments receive a rating of moderate or highly stressed. Thus, a catchment may be compared with an adjacent one by assessing the status of the sub-catchments although this should not be done on a simply additive basis. Such an approach also underpins a strategy of conservation of unstressed sections and, where feasible, rehabilitation or restoration of stressed sub-catchments (Boulton *et al.*, 2000).

The sub-catchment approach has also proved useful in other river conservation and assessment programmes. SERCON (System for Evaluating Rivers for Conservation) is used to assess the conservation value of rivers in the United Kingdom (Boon *et al.*, 1997; O'Keeffe and Uys, this volume). This approach entails gathering field survey data on physical features of the river corridor, supplementing these data with physical, chemical and biological data from relevant sources, and finally translating the collated data into scores that can be used for river evaluation (for details, see Boon *et al.*, 1998). Rivers are evaluated in discrete sections rather than entire watercourses – these sections (Evaluated Catchment Sections, ECSs) are from 10 to 30 km, and match the same spatial scale as many of the sub-catchment assessments described for the New South Wales study. Other examples exist in the United States (e.g. 50–200 km^2 sub-catchments used to reduce the impacts of forestry on lotic ecosystems, Collins and Pess, 1997a,b), France (Cohen *et al.*, 1998), and elsewhere.

The ecoregion approach

Perhaps the sub-catchment approach is too reductionist under some circumstances and may result in an unwieldy number of management areas. As mentioned earlier, an alternative spatial framework for conservation and resource management is the use of ecoregions. A physiographically large and diverse area (e.g. the continental United States) can be divided into regions within which climate, geology, soils, vegetation and other factors are relatively homogeneous (Bailey, 1995; Omernik, 1995). Because an ecoregion comprises several similar ecosystems, management and conservation issues and strategies across the ecoregion should also be similar. The same does not always hold true for hydrologic units of comparable spatial extent because their catchments typically span several diverse environments. Hence, the aquatic ecoregional framework is most appropriate for the selection of regional reference or benchmark sites (Hughes *et al.*, 1986; Omernik and Bailey, 1997) or to establish regional recovery criteria for degraded stream systems (Hughes *et al.*, 1990).

From an aquatic perspective, the ecoregion approach has received criticism because of inconsistent relationships between ecoregions and specific aquatic biological variables. One recent example is the demonstration that categories of river size may be better predictors of fish community structure than ecoregion identity (Newell and Magnuson, 2000). Ecoregions may be useful in the design of research

and management strategies for a suite of environmental resources integrating aquatic and terrestrial components, but we question the utility of this approach at a finer scale. Our major concern is that because small streams and large rivers co-occurring in a given ecoregion will probably differ ecologically because of their size, they should not be lumped together uncritically.

How does size affect the success of river conservation programmes?

As far as we know, there has never been an explicit attempt to assess the effect of river size on the success of river conservation programmes. Boon (1996) proposed six key areas that contribute to the success of freshwater conservation programmes and tabulated some specific examples within these areas. Using this as a template, we suggest a number of ways that river size influences these key areas (Table 17.1). The significance of river size to virtually all aspects of scientific research on running waters is obvious – even taxonomic descriptions involving brief habitat descriptions must distinguish between small streams and large rivers at the coarsest scale. Conservation efforts must focus on applying our knowledge of how compartments of river systems (e.g. floodplains, riparian zones, the hyporheic zone) function and interact at a range of scales. We recognize that the rates, magnitudes and relevance of these compartments vary with river size (see Boulton *et al.*, 1998, for examples in the hyporheic zone), and this would apply to conservation efforts as well.

Although the political aspects in Table 17.1 refer largely to the existence of a political and administrative infrastructure, the scale at which this infrastructure functions best is dictated partly by river size. Catchment and sub-catchment physical scales (depending on river size) may be matched by state and council boundaries and management agencies. However, in the same way that sub-catchments must be viewed within the spatial hierarchy of the catchment, political subgroups must cooperate in a framework that works at catchment or ecoregional levels – and this is less common. Even for large rivers that span only state borders within a country, there can be social and political friction (e.g. Kingsford *et al.*, 1998), confounding efforts to develop resources such as cotton farming with a realistic catchment-wide assessment of the availability of water and the natural flow regime.

Table 17.1   *The relevance of river size to some key areas that contribute to the success of freshwater conservation programmes. Relevance is represented as moderate (*), high (**), and essential (***). Key areas are those proposed by Boon (1996)*

Key area	Specific examples	Relevance
Scientific	Systematic inventory of habitats and species	***
	Techniques for classifying/evaluating rivers	***
	Studies of processes underpinning effective conservation (e.g. understanding riparian zones)	***
	Taxonomic descriptions (habitat preferences)	*
Political	Existence of statutory conservation bodies (focus)	*
	Existence of environmental management agencies	*
	An overall political will to include resource development as part of an overall strategy for conservation (catchment scale)	**
Legislative and regulatory	Statutory mechanisms for habitat protection	***
	Regulatory bodies for Total Catchment Management	***
	Legal framework for minimizing environmental damage (e.g. pollution licences)	**
	Appropriate procedures for environmental impact assessment (including explicit statement of spatial scale)	***
	Policies regulating water abstraction (amounts and timing)	**
Technical	Application of latest methods for combating environmental damage (e.g. nutrient removal)	**
	Availability and reliability of content of technical manuals for reducing environmental impacts (e.g. for forestry, road construction)	**
	GIS, IP and remote sensing applications for integrating smaller scales into catchment–regional scale maps	***
Educational	Inclusion of conservation principles in school curricula and continued adult education programmes	***
	Explicit description of the significance of river size in educating the public about conserving rivers	***
Ethical	Willingness to include conservation principles in all aspects of resource management	*
	Recognition of the global significance of conservation and preservation of species and biodiversity	**

Related to the key area of political issues is that of legislative and regulatory mechanisms. River size must be considered when drafting the scope of most of these regulations. Habitat and species protection depends on selection of the appropriate spatial scale. Similarly, Total Catchment Management (or Integrated Catchment Management) and procedures for effective environmental impact assessment (Harvey, 1998) explicitly rely on the choice of study area, whereas pollution control must consider cumulative impacts along a river course because these sub-catchments are interrelated laterally, longitudinally and often vertically by common groundwater pathways.

Technical areas involve the application of scientific information to field and management situations. As in the example above, these manuals are often 'grey literature' that has sometimes been subjected to only limited, if any, scientific review. Ignorance of the significance of river size to the conclusions and generalizations in the manuals can have serious repercussions for effective management strategies in

the field. River size is clearly relevant to technologies that seek to integrate spatial scales (e.g. GIS and IP; Johnson and Gage, 1997), and for sampling strategies to assess water quality or the effectiveness of techniques for combating river pollution (Table 17.1). It is easier to detect and remedy the effects of a point source in a small stream than the diffuse inputs in a large river (Allan, 1995).

Education at all levels from school to continued training of professionals in the water industry will always be essential. In this context, an understanding of the significance of river size is relevant. Similarly, ethical considerations must be based on sound scientific knowledge and recognition of global issues of biodiversity conservation and responsibility. For example, conservation of migratory waterbirds entails maintenance of all the links in the chain of resting places along their flight paths that may span the Northern and Southern Hemispheres. Ethical issues underpinned the development of the Ramsar convention that now provides large-scale legislative protection of a number of wetlands worldwide.

Conclusions

The complex physical structure of streams and rivers, regardless of their size, presents a plethora of challenges to researchers and managers alike. For the purposes of conservation and restoration, many of us, often subconsciously, make distinctions between small, intermediate and large systems. Small streams and their drainage basins may be considered as discrete units, and these are the popular choice for research programmes by forest hydrologists, biogeochemists and resource managers (Likens *et al.*, 1977; Hornbeck and Swank, 1992). While we have learned a great deal about the ecology of small, upland catchments (probably due to their tractability), our understanding of large, lowland rivers is embarrassingly thin. At present, most approaches to river conservation in lowland rivers are dominated by studies that focus only on channel and floodplain habitats. We struggle with the enormity of large catchments so one of our first reactions is to narrow our focus to the aquatic environment. Fortunately, growing accessibility to technology in remote sensing and quantitative landscape methods are helping overcome these difficulties. Recent estimates of catchment nutrient budgets for the largest US hydrologic units (Smith *et al.*, 1997) signify steps in the right direction – now we need to assist conservation efforts similarly.

Larger rivers are harder to conserve than smaller ones because they drain diverse catchments, often contain unique, basin-specific biota, and are exposed to a range of human impacts and multiple and often conflicting uses. Their courses and catchments may span several political states, leading to conflicts in water management and conservation practices. Consequently, it is reasonable to say that the distinctiveness of a river increases as a function of its size. The acquisition of unique ecological characteristics restricts the ability to use other large rivers as references for conservation and restoration. Even if we could, there are few unaltered large rivers that can be used as reference points. Finally, reliable long-term and broad-scale data are lacking for most large rivers, and we recognize the perils of extrapolation of results from studies at the finer scale of the habitat, reach or segment. All these are size-related issues that must be considered in river conservation.

We conclude that the intermediate spatial unit of the sub-catchment may be the most appropriate size for conservation and management as long as its location in the drainage hierarchy is acknowledged. Tractable and relatively homogeneous, the sub-catchment scale has proved a useful starting point for classification of the conservation status of entire river basins. Land use tends to be less variable than at the basin scale, and practical management is more feasible. Indeed, examination of sub-catchments or watershed units appears to be a reasonable compromise between whole catchment (total drainage) and ecoregional approaches to resource management. However, we must always realize that our perception of conservation success is strongly influenced by the choice of variable, the scale of its measurement, and the context in which it is presented. Size *is* important – physically, conceptually and culturally.

Acknowledgements

We thank Drs Phil Boon, Pierre Marmonier and Chris Peterson for discussions and insights into this topic. Trish Piper assisted with preparation of this chapter. Comments from two anonymous reviews improved early drafts of this manuscript.

References

Allan, J.D. 1995. *Stream Ecology: Structure and Function of Running Waters*. Chapman & Hall, London.

Allan, J.D. and Johnson, L.B. 1997. Catchment scale analysis of aquatic ecosystems. *Freshwater Biology* 37:107–111.

Allen, T.F.H. and Starr, T.B. 1982. *Hierarchy: Perspectives for Ecological Complexity*. University of Chicago Press, Chicago.

Bailey, R.G. 1995. *Ecosystem Geography*. Springer-Verlag, New York.

Bisson, P.A. and Montgomery, D.R. 1996. Valley segments, stream reaches, and channel units. In: Hauer, F.R. and Lamberti, G.A. (eds) *Methods in Stream Ecology*. Academic Press, San Diego, 23–52.

Boon, P.J. 1996. The conservation of fresh waters: temperate experience in a tropical context. In: Schiemer, F. and Boland, K.T. (eds) *Perspectives in Tropical Limnology*. SPB Academic Publishing, Amsterdam, 333–344.

Boon, P.J., Holmes, N.T.H., Maitland, P.S., Rowell, T.A. and Davies, J. 1997. A system for evaluating rivers for conservation ('SERCON'): development, structure and function. In: Boon, P.J. and Howell, D.L. (eds) *Freshwater Quality: Defining the Indefinable?* The Stationery Office, Edinburgh, 299–326.

Boon, P.J., Wilkinson, J. and Martin, J. 1998. The application of SERCON (System for Evaluating Rivers for Conservation) to a selection of rivers in Britain. *Aquatic*

Conservation: Marine and Freshwater Ecosystems 8:597–616.

Boulton, A.J., Findlay, S., Marmonier, P., Stanley, E.H. and Valett, H.M. 1998. The functional significance of the hyporheic zone in streams and rivers. *Annual Reviews of Ecology and Systematics* 29:59–81.

Boulton, A.J., Boon, P.J., Muhar, S. and Gislason, G.M. 2000. Making river conservation work: integrating science, legislative policy, and public attitudes. *Verhandlungen der Internationalen Vereinigung für theoretische und angewandte Limnologie* (in press).

Brussock, P.P., Brown, A.V. and Dixon, J.C. 1985. Channel form and stream ecosystem models. *Water Resources Bulletin* 21:859–866.

Carpenter, S.R. 1996. Microcosm experiments have limited relevance for community and ecosystem ecology. *Ecology* 77:677–680.

Cohen, P., Andrianmahefa, H. and Wasson, J. 1998. Towards a regionalization of aquatic habitat: distribution of mesohabitats at the scale of a large river basin. *Regulated Rivers: Research & Management* 14:394–404.

Collins, B.D. and Pess, G.R. 1997a. Evaluation of forest practices from Washington's Watershed Analysis Program. *Journal of the American Water Resources Association* 33:969–996.

Collins, B.D. and Pess, G.R. 1997b. Critique of Washington's Watershed Analysis Program. *Journal of the American Water Resources Association* 33:997–1010.

Cullen, P. 1986. The turbulent boundary between water science and water management. *Freshwater Biology* 24:201–209.

Davies, B.R., Thoms, M.C. and Meador, M. 1992. The ecological impact of inter-basin water transfers and their threats to river basin integrity and conservation. *Aquatic Conservation: Marine and Freshwater Ecosystems* 2:325–349.

Fisher, S.G. 1997. Creativity, idea generation, and the functional morphology of streams. *Journal of the North American Benthological Society* 16:305–318.

Fisher, S.G., Grimm, N.B., Marti, E. and Gomez, R. 1998. Hierarchy, spatial configuration, and nutrient cycling in a desert stream. *Australian Journal of Ecology* 23:41–52.

Frissell, C.A., Liss, W.J., Warren, C.E. and Hurley, M.D. 1986. A hierarchical framework for stream habitat classification: viewing streams in a watershed context. *Environmental Management* 10:199–214.

Gordon, N.D., McMahon, T.A. and Finlayson, B.L. 1992. *Stream Hydrology: An Introduction for Ecologists.* John Wiley, Chichester.

Graf, W.L. 1988. *Fluvial Processes in Dryland Rivers.* Springer-Verlag, Berlin.

Gregory, S.V., Swanson, F.J., McKee, W.A. and Cummins, K.W. 1991. An ecosystem perspective of riparian zones. *BioScience* 41:540–552.

Haines, A.T., Finlayson, B.L. and McMahon, T.A. 1988. A global classification of river regimes. *Applied Geography* 8:255–272.

Harvey, N. 1998. *Environmental Impact Assessment – Procedures, Practice, and Prospects in Australia.* Oxford University Press, Melbourne.

Hawkins, C.P., Kershner, J.L., Bisson, P.A., Bryant, M.D., Decker, L.M., Gregory, S.V., McCullough, D.A., Overton, C.K., Reeves, G.H., Steedman, R.J. and Young, M.K. 1993. A hierarchical approach to classifying stream habitat features. *Fisheries* 18:3–12.

Hornbeck, J.W. and Swank, W.T. 1992. Watershed ecosystem analysis as a basis for multiple use management of eastern forests. *Ecological Applications* 2:238–247.

Hughes, R.M. and Omernik, J.M. 1981. Use and misuse of the terms, watershed and stream order. In: Krumholz, L. (ed.) *Warmwater Stream Symposium*, Southern Division of the American Fisheries Society, Bethesda, Maryland, 320–326.

Hughes, R.M. and Omernik, J.M. 1983. An alternative for characterizing stream size. In: Fontaine, T.D. and Bartell, S.M. (eds) *Dynamics of Lotic Ecosystems.* Ann Arbor Science, Ann Arbor, Michigan, 87–102.

Hughes, R.M., Larson, D. and Omernik, J.M. 1986. Regional reference sites: a method for assessing stream potentials. *Environmental Management* 10:629–635.

Hughes, R.M., Whittier, T., Rohm, C. and Larson, D. 1990. A regional framework for establishing recovery criteria. *Environmental Management* 14:673–683.

Hynes, H.B.N. 1975. The stream and its valley. *Verhandlungen der Internationalen Vereinigung für theoretische und angewandte Limnologie* 19:1–15.

Jackson, J.B.C. 1991. Adaptation and diversity of reef corals. *BioScience* 41:475–482.

Johnson, L.B. and Gage, S.H. 1997. Landscape approaches to analysis of aquatic ecosystems. *Freshwater Biology* 37:113–132.

Kingsford, R.T., Boulton, A.J. and Puckridge, J.T. 1998. Challenges in managing dryland rivers crossing political boundaries: lessons from Cooper Creek and the Paroo River, central Australia. *Aquatic Conservation: Marine and Freshwater Ecosystems* 8:361–378.

Likens, G.E., Bormann, F.H., Eaton, J.S. and Johnson, N.M. 1977. *Biogeochemistry of a Forested Ecosystem.* Springer-Verlag, New York.

May, R.M. 1994. The effects of spatial scale on ecological questions and answers. In: Edwards, P.J., May, R.M. and Webb, N.R. (eds) *Large-scale Ecology and Conservation Biology.* Blackwell Scientific, London, 1–17.

Meyer, J.L. 1997. Stream health – incorporating the human dimension to advance stream ecology. *Journal of the North American Benthological Society* 16:439–447.

Montgomery, D.R. and Buffington, J.M. 1993. *Channel Classification, Prediction of Channel Response, and Assessment of Channel Condition.* Washington State Timber, Fish, and Wildlife Agreement. Report TFW-SH10-93-002, Department of Natural Resources, Olympia, Washington.

Muhar, S. 1996. Habitat improvement of Austrian rivers with regard to different scales. *Regulated Rivers: Research & Management* 12:471–483.

National Research Council (NRC) 1992. *Restoration of Aquatic Systems: Science, Technology, and Public Policy*. National Academy Press, Washington, DC.

Newell, P.R. and Magnuson, J.J. 2000. The importance of ecoregion versus drainage area on fish distribution in the St. Croix River and its Wisconsin tributaries. *Environmental Biology of Fishes* (in press).

Newson, M. 1994. *Hydrology and the River Environment*. Clarendon Press, Oxford.

Omernik, J.M. 1995. Ecoregions: a spatial framework for environmental management. In: Davis, W. and Simon, T. (eds) *Biological Assessment and Criteria: Tools for Water Resource Planning and Decision Making*. Lewis Publishers, Boca Raton, Florida, 49–62.

Omernik, J.M. and Bailey, R.G. 1997. Distinguishing between watersheds and ecoregions. *Journal of the American Water Resources Association* 33:935–950.

Poff, N.L. and Ward, J.V. 1989. Implications of streamflow variability and predictability for lotic community structure: a regional analysis of streamflow patterns. *Canadian Journal of Fisheries and Aquatic Sciences* 46:1805–1818.

Porritt, J. 1994. Translating ecological science into practical policy. In: Edwards, P.J., May, R.M. and Webb, N.R. (eds) *Large-Scale Ecology and Conservation Biology*. Blackwell Scientific, London, 345–354.

Puckridge, J.T., Sheldon, F., Walker, K.F. and Boulton, A.J. 1998. Flow variability and the ecology of large rivers. *Marine and Freshwater Research* 49:55–72.

Reid, N., Boulton, A., Nott, R. and Chilcott, C. 1997. Ecological sustainability of grazed landscapes on the Northern Tablelands of New South Wales, Australia. In: Klomp, N. and Lunt, I. (eds) *Frontiers in Ecology: Building the Links*. Elsevier, Oxford, 117–130.

Richards, K. 1982. *Rivers: Form and Process in Alluvial Channels*. Methuen, London.

Rosgen, D.L. 1985. A stream classification system. In: *Riparian Ecosystems and their Management, Interagency North American Riparian Conference*. General Technical Report ROM-120, Rocky Mountain Forest and Range Experimental Station, US Department of Agriculture Forest Service, Fort Collins, Colorado, 91–95.

Scheidegger, A.E. 1965. The algebra of stream order numbers. *US Geological Survey Professional Paper* 525B:187–189.

Schumm, S.L. 1977. *The Fluvial System*. John Wiley, New York.

Seaber, P.R., Kapinos, F.P. and Knapp, G.L. 1987. *Hydrologic Units Maps*. US Geological Survey Water Supply Paper 2294, US Department of the Interior, Geological Survey, Denver, Colorado.

Shreve, R.L. 1966. Statistical law of stream numbers. *Journal of Geology* 74:17–37.

Smith, D.I. 1998. *Water in Australia: Resources and Management*. Oxford University Press, Melbourne.

Smith, R.A., Scharz, G.E. and Alexander, R.B. 1997. Regional interpretation of water-quality monitoring data. *Water Resources Research* 33:2781–2798.

Stanley, E.H., Fisher, S.G. and Grimm, N.B. 1997. Ecosystem expansion and contraction in streams. *BioScience* 47:427–435.

Strahler, A.N. 1957. Quantitative analysis of watershed geomorphology. *Transactions of the American Geophysical Union* 38:913–920.

Vannote, R., Minshall, G.W., Cummins, K.W., Sedell, J.R. and Cushing, C.E. 1980. The river continuum concept. *Canadian Journal of Fisheries and Aquatic Sciences* 37:130–137.

Walker, K.F., Sheldon, F. and Puckridge, J.T. 1995. A perspective on dryland river ecosystems. *Regulated Rivers: Research & Management* 11:85–104.

Winterbourn, M.J., Rounick, J.S. and Cowie, B. 1981. Are New Zealand streams really different? *New Zealand Journal of Marine and Freshwater Research* 15:321–328.

Wood, P.J. and Armitage, P.D. 1997. Biological effects of fine sediment in the lotic environment. *Environmental Management* 21:203–217.

Wright, J.F., Armitage, P.D., Furse, M.T. and Moss, D. 1989. Prediction of invertebrate communities using stream measurements. *Regulated Rivers: Research & Management* 4:147–155.

18

Problems and constraints in managing rivers with variable flow regimes

A.J. Boulton, F. Sheldon, M.C. Thoms and E.H. Stanley

Introduction

Many contemporary ecosystem theories developed to explain how rivers function originated from research on temperate, perennial streams (Williams, 1988). Not surprisingly, recommendations for river management and restoration (e.g. Petts and Calow, 1996) and water policies and legislation (e.g. Johnson, 1993) share similar origins. However, uncritical extrapolation of theories developed in permanent lotic ecosystems to intermittent and ephemeral streams can prove perilous and even misleading (Boulton and Suter, 1986; Williams, 1987, 1996). For example, extremes of flooding and drying (variable flows) largely structure stream assemblages and regulate ecosystem processes in most intermittent streams (e.g. Feminella, 1996; Miller and Golladay, 1996; Stanley *et al.*, 1997). Conversely, biological interactions such as predation by fish that often are excluded by a temporary flow regime may be more important in perennial streams. Flooding occurs in most perennial streams as well (Resh *et al.*, 1988) but drying is rare except during severe drought when the fauna is devastated by desiccation (e.g. Hynes, 1958).

Rivers and streams naturally vary in flow although we must specify temporal scale when we use the term 'variable'. In this chapter, we are discussing whole river systems and adopting a multi-year temporal scale appropriate to the definition of flow regime (see below). Thus, the most highly variable flow regimes usually occur in intermittent and ephemeral rivers, especially those in dryland areas. Here, the coefficients of variation of annual flows are, on average, 467%

greater than those from humid and temperate regions (Davies *et al.*, 1994). This hydrological variability seems to be associated with increased habitat and food web complexity (Thoms and Sheldon, 1997). Although it is likely that the persistence of many species in dryland rivers relies on maintenance of intermittency (Walker *et al.*, 1995; Boulton, 1999), there are few scientific data to support this hypothesis because we often lack information about the ecology of the river before regulation. Such data are a fundamental requirement for managing these types of rivers and raises the question: How should we manage rivers with variable flow regimes when we have so little information?

Historically, water management practices in arid and semi-arid zones have been driven by a single priority – human demand for water (Biswas, 1996; Postel, 1996). Demand for water in some dryland areas such as Mexico has escalated as surface waters have become polluted or dried through excessive pumping of groundwater, leading to reservoir construction and other regulatory structures (Grimm *et al.*, 1997). River regulation and, more recently, interbasin water transfers (IBTs) (Davies *et al.*, 1992, 1994) are imposed most intensively upon rivers with highly variable flow regimes (including natural intermittency) to sustain human agriculture and navigation (Walker *et al.*, 1997; Kingsford *et al.*, 1998). The issue is made more complex by a Western human perception that a 'healthy' river flows all year round; many of the more ambitious river regulation projects have had technological and intellectual input from experts living in well-watered regions. This is despite evidence

Global Perspectives on River Conservation: Science, Policy and Practice.
Edited by P.J. Boon, B.R. Davies and G.E. Petts. © 2000 John Wiley & Sons Ltd.

that indigenous farmers in arid areas may have been highly efficient at harvesting water using extremely simple technology not involving river regulation (Lavee *et al.*, 1997).

There are now numerous examples of the disastrous ecological consequences of poor management of rivers with variable flow regimes (e.g. Davies *et al.*, 1992; Lyons *et al.*, 1992; Walker *et al.*, 1997). We have learned from these that effective water management must extend beyond simply engineering solutions to guarantee a reliable water supply as we try to attain the slippery goal of sustainable development of our water resources. This chapter reviews the ecological importance of flow variability and the problems caused by removing this variability through river regulation. Constraints and solutions to managing these rivers with variable flow regimes are outlined.

Flow regimes and hydrological variation

As flow mediates many ecological processes in rivers and streams, measurements of hydrological *variation* that are ecologically relevant (reviews in Poff, 1996; Richter *et al.*, 1996, 1997; Puckridge *et al.*, 1998) are essential 'trend indicators' (Walker *et al.*, 1996) for their successful management. Historically, streams have been classified based on the predictability and duration of their flow (e.g. Usinger, 1956; Clifford, 1966; Bayly and Williams, 1973; Abell 1984; Williams, 1987; Comin and Williams, 1994) although some of these authors admit the difficulty of imposing categories on a continuum. One generally accepted scheme distinguishes four main categories of streams:

- ephemeral streams – flow briefly (<1 month) with irregular timing and usually only after unpredictable rain has fallen;
- intermittent or temporary streams – flow for longer periods (>1–3 months), regularly have an annual dry period coinciding with prolonged dry weather;
- semi-permanent streams – flow most of the year but cease flow during dry weather (<3 months), drying to pools. During wetter years, flow may continue year round;
- permanent streams – perennial flow. Only cease flow during rare extreme droughts.

More recently, there has been a growing awareness of the importance of both flow variability and the multivariate aspects of rivers' flow regimes, and these have had articulate advocates for the use of this

information to formulate management practices (Walker *et al.*, 1995; Richter *et al.*, 1996, 1997). Comparisons of rivers based on hydrological characteristics of their flow regimes (derived from long-term (>15 yr) data sets) have proved successful in generating testable predictions about their biota and dominant ecological processes (e.g. Poff and Ward, 1989; Poff, 1996). Further, we can now identify potential effects of river regulation upon some of these hydrological features in rivers with variable flow regimes (Walker *et al.*, 1995; Puckridge *et al.*, 1998). For example, alteration of stage amplitude caused by river regulation in the River Murray, Australia, has resulted in truncated flows in spring, contributing to declines in stocks of native cod and other indigenous fish species in the last few decades (Rowland, 1989).

For perennial temperate rivers, short-term hydrographs (<10 years) may suffice to represent their overall hydrology. However, in ephemeral and intermittent systems, a much longer time frame is required to encompass the variable hydrological behaviour adequately (Walker *et al.*, 1995; Puckridge *et al.*, 1998). Managing rivers must involve several time-scales because short-term solutions may have adverse long-term effects. Walker *et al.* (1995) recognize three scales of hydrological behaviour:

- the *flood pulse* (an increase followed by a decrease in river stage, or discharge)
- the *flow history* (the sequence of pulses before any point in time)
- the *flow regime* (the long-term, statistical generalization of flow behaviour)

For all rivers, each flood pulse has a complex character with unique patterns of stage, amplitude, flood timing, flood duration, rate of flood rise and fall, and flood frequency (Figure 18.1). The hydrological features of this flood pulse have ecological ramifications for rivers with variable flow regimes (Table 18.1). Each individual flood is unique and may differ in ways that have diverse biological consequences. Furthermore, the flow history is obviously important and influences the effects of subsequent flood pulses.

In a flow regime analysis of 20-year hydrographs for 52 large rivers, Puckridge *et al.* (1998) identified a number of independent measures of hydrological variability, each with biological significance. These include variability in flood pulse timing, variability in flood pulse duration, variability in flow magnitude,

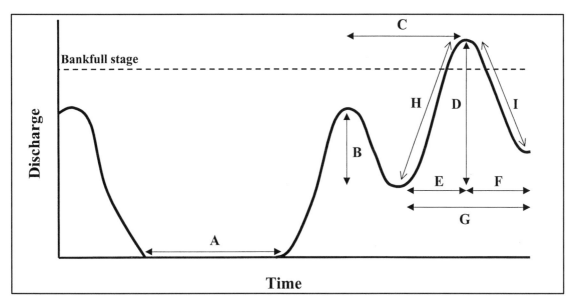

Figure 18.1   *Aspects of the flood pulse that may be ecologically important (Table 18.1): A = duration of zero flow (and often, discontinuity of channel pools); B = amplitude of falling limb ('drawdown'); C = interval since last flood peak; D = amplitude of rising limb; E = duration of rising limb; F = duration of falling limb; G = interval since last flood minimum; H = slope of rising limb; I = slope of falling limb. Modified from Walker* et al. *(1995)*

Table 18.1   *Some ecological ramifications of the hydrological features of the flood pulse. Letters in parentheses refer to Figure 18.1. See text for more details and references. Alterations to these aspects of the flood pulse by the management of flow variability will have repercussions on these ecological processes*

Feature	Ecological consequences
Duration of zero flow (A)	Deteriorating conditions in remaining pools, declines in aquatic species richness, isolation and drying of channel and floodplain wetlands, survival of resting eggs and seeds, establishment of terrestrial floodplain vegetation.
Amplitude of falling limb ('drawdown') (B)	Extent of drawdown affects degree of isolation of floodplain wetlands from main channel, can influence fish recruitment with repercussions on invertebrates and waterbirds, and affects extent of inundation of littoral habitats.
Interval since last flood peak (C)	Coupling of timing with seasonal cycles of spawning cues may affect recruitment of fish and other biota with seasonal life cycles, saturation and slump of river banks, establishment of floodplain vegetation.
Amplitude of rising limb (D)	Related to stage amplitude and size of flood; large floods usually enable extensive breeding/recruitment of most river and floodplain species, inundation of large areas of channel and floodplain, releases of nutrients and hatching of resting stages, germination of floodplain vegetation.
Duration of rising and falling limbs (E and F), interval since last flood minimum (G)	These influence the length of time that the floodplain is inundated and hence the time for recolonization of floodplain wetlands, fish, invertebrate and plant growth on the floodplain, successional changes in biota. Cumulative effects and survival and growth since previous flood relate to interval since last flood minimum.
Slope of rising and falling limbs (H and I)	Rates of change of the flood pulse influence types of species favoured by the flood (e.g. steep rising limb may dislodge biota typical of standing water habitats; steep falling limbs may strand taxa), affect survival and recruitment of species, influence flow rates and erosion/deposition of sediments.

variability in the rates of rise and fall of a flood pulse, and variability in flooding frequency. Multivariate analysis of these aspects of variability from the hydrographs separated groups of 'tropical' and 'dryland' rivers. The large dryland rivers in central Australia (Cooper Creek and the Diamantina River) with markedly variable flow regimes are the most distinctive. These systems have long-term hydrographs characterized by periods of extreme flooding followed by extensive periods of low or no flow, and contrast sharply with the flow regimes of tropical rivers such as the Mekong (Figure 18.2).

Timing of flow is crucial and intimately related to the geography of the catchment. For example, intermittent streams in the Mojave Desert, USA, flow during spring, fed by upper basin snowmelt and seasonal rains. In contrast, similar systems in the Chihuahuan Desert (USA-Mexico) flow in response to summer convective storms typical of this desert (Grimm *et al.*, 1997). On a longer time-scale, flow frequency and magnitude of many south-western US streams and rivers depend on El Niño/Southern Oscillation climate patterns (Molles and Dahm, 1990). A relatively small decrease of 10% in mean annual precipitation may reduce surface run-off in these streams by 30–50%, having a substantial impact on their aquatic communities (Dahm and Molles, 1992). Management of these rivers with variable flow regimes must account for factors associated with climate change although it is difficult to separate these effects from those of other human activities, especially in dryland areas where variability is inherent (Grimm *et al.*, 1997). For example, increased aridity through global warming may be superimposed on human water extraction leading to extended periods of drying with their concomitant impacts (e.g. Stanley *et al.*, 1994, 1997). These climate changes may also mean that water supplies in some arid regions such as the Great Basin, USA, will not support present human demand (Flaschka *et al.*, 1987).

Although we focus on hydrological variation primarily in dryland areas because that is where problems and constraints are most obvious, many of the issues discussed here apply equally validly to more temperate regions such as the United Kingdom. For example, chalk streams contain a diverse and productive biota (Wright, 1992). Recent periods of drought (1988–1992) exacerbated by groundwater abstraction led to concerns about recovery when normal flows resumed. Interestingly, although the impacts of drought were severe on stream invertebrates, recovery was rapid (Wood and Petts, 1994). Castella *et al.* (1995) suggest that the naturally

variable flows in upland streams in these temperate regions means that, within limits, the invertebrate communities are capable of tolerating extreme low flows and may rapidly recolonize from the many streams nearby. Species diversity in naturally intermittent streams in the area is also high (Wright *et al.*, 1984) and this may be a key factor in recolonization pathways.

Comparing perennial and intermittent streams

It is useful to address the differences in physical, chemical and biological features between perennial and intermittent streams that give rise to particular problems in conservation management (Table 18.2). As we discuss later, many of these 'problems' relate to Western perceptions of how a 'true river' should be. Many of the ecological features that typify a stream with a variable flow regime are not predictable by some of the conventional, deterministic models of river ecosystems (e.g. the River Continuum Concept, Vannote *et al.*, 1980) and require modifications (Walker *et al.*, 1995).

Amplitudes in physical and chemical conditions, particularly in drying pools, far exceed those in permanent streams (Table 18.2; Williams and Hynes, 1977; Boulton and Suter, 1986; Williams, 1987, 1996). As rivers dry, conductivity tends to rise through evapo-concentration. Water temperatures also rise (>30°C) and dissolved oxygen saturation falls (Boulton and Lake, 1990). In some receding pools, leaf leachate concentration increases and pH may fall to as low as 4.5, further exacerbating conditions for the aquatic biota. Survival of these harsh conditions includes a range of physiological and behavioural responses (Berra *et al.*, 1989; Boulton, 1989; Boulton *et al.*, 1992; Stanley *et al.*, 1994; Feminella, 1996; Williams, 1996). The concentrating effect may also enhance biological interactions in the drying pools (Boulton and Suter, 1986; Stanley *et al.*, 1994). These range from intensifying competitive interactions for space and moisture (e.g. the snail *Physella*: Stanley *et al.*, 1994) to heavy predation by terrestrial and aquatic invertebrates and vertebrates (e.g. Hynes, 1975; Gray, 1980; Abell, 1984).

The numbers of species of water plants, invertebrates and fish are generally lower in intermittent streams compared with nearby permanent streams of the same size and geomorphology (Table 18.2; Wright *et al.* 1984; Boulton and Suter 1986; Williams, 1987, 1996). For water plants that are usually submerged or floating, periodic drying poses a serious limitation

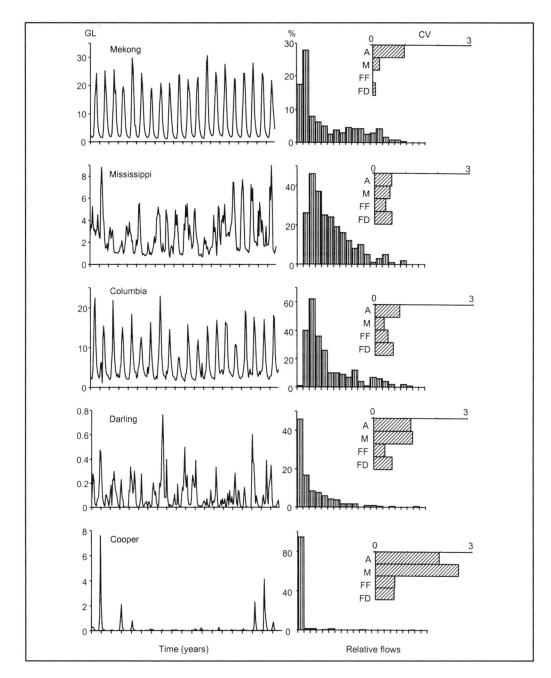

Figure 18.2 *Examples of variable flow regimes from a tropical river (Mekong; 1925–1944), a warm-temperate, semi-arid continental river (Mississippi; 1928–1947), a warm-temperate, semi-arid river (Columbia; 1880–1899), a warm-temperate, semi-arid river (Darling; 1924–1943), and a warm-temperate, arid river (Cooper Creek; 1973–1992). Extended 15–25 year hydrographs and corresponding relative flow–frequency distributions are given. The coefficient of variation (CV), a relative measure of the extent of the variable flows, is shown for annual flows (A), monthly flows (M), flood frequency (FF) and flood duration (FD). Discharge data are in gigalitres (GL); data are from United Nations Educational, Scientific and Cultural Organization – UNESCO (1971). Modified from Walker et al. (1995) and Puckridge et al. (1998)*

Table 18.2   *Differences between intermittent and perennial streams and rivers that may influence conservation management practices (see text for full details). These differences are generalizations and some exceptions exist, emphasizing the need for case-specific management*

Feature	Perennial streams	Intermittent streams
Variability of flow at a range of scales	Lower	Higher
Duration of zero or low flows	Lower	Higher
Amplitude of physical and chemical conditions	Lower	Higher
Extremes of environmental conditions	Lower	Greater
Species richness of submerged or floating aquatic plants	Higher	Lower
Species richness of aquatic invertebrates	Higher	Lower
Diversity, species richness and body size of fish	Higher	Lower
Potential importance of the hyporheic zone	Lower	Higher
Pulsing of deposition and transport of material (e.g. leaf litter, sediments)	Lower	Higher
Importance of biological interactions for most of the time	Higher	Lower
Spectrum of habitats over a long time period	Lower	Higher
Perceived value to humans	Higher	Lower
Scientific knowledge	Higher	Lower

unless they can produce desiccation-resistant propagules (e.g. seeds and tubers). Emergent plants are more tolerant of fluctuations in water level but are adversely affected by rapid changes in water level or channel instability (Fox, 1996) – features that typify many intermittent streams. River regulation by weirs along the River Murray has altered the extent of fluctuation of the water level, changing the composition of the littoral fringe of emergent macrophytes and favouring establishment of tolerant and widespread dominant species (Walker *et al.*, 1994).

Invertebrates that either lack desiccation-tolerant stages or are poor recolonists will be eliminated from intermittent streams when they dry (Boulton, 1989; Stanley *et al.*, 1994). Permanent streams will contain both species that are opportunistic and found in nearby intermittent streams as well as species with long-lived aquatic stages (>1 yr, e.g. some megalopterans and stoneflies) or limited powers of dispersal (e.g. some crayfish and caddisflies). Similarly, most species of riverine fish cannot tolerate drying or the harsh physical and chemical conditions in receding pools, and are restricted to permanent streams (Larimore *et al.*, 1959; Meffe and Minckley, 1987; Matthews *et al.*, 1988; Closs and Lake, 1996). However, some rely on intermittent channels for spawning (Erman and Hawthorne, 1976).

The relative magnitude of ecosystem components may differ between intermittent and permanent streams. Compared with permanent rivers, subsurface flow in the hyporheic zone of many gravel and sand-bed intermittent streams represents a substantial proportion of the total discharge (Valett *et al.*, 1990). Exchanges of water between the surface and subsurface zones influence ecosystem processes such as algal productivity, respiration, and nutrient cycling (reviews by Findlay, 1995; Jones and Holmes, 1996; Brunke and Gonser, 1997). Drying may sever these hydrological linkages, changing a range of ecosystem processes (Stanley and Valett, 1992). The usual balance between upwelling (movement of hyporheic water to the surface) and downwelling (surface water infiltrating into the hyporheic zone) tips almost completely towards the latter flux during drying. Microbial respiration continues while sediments remain moist or saturated, consuming available carbon and oxygen, and potentially shifting hyporheic metabolism towards anaerobic processes. This has profound effects on nitrogen transformations, phosphorus availability, and the potential for the hyporheic zone to serve as a refuge for surface-dwelling organisms (Boulton and Stanley, 1995; Stanley and Boulton, 1995) but is a phenomenon not generally known by most river managers.

The ecological significance of high flow variability

It may appear that variable flows and intermittency have largely negative effects, adversely affecting water quality during the drying phase and limiting the diversity of water plants, invertebrates and fish. Yet, the significance of the comparison is not that 'permanent is better' but that river systems with highly variable flow regimes are *different* and call for a different approach to their management. Efforts to reduce this flow variability in order to increase biodiversity or to 'restore' the river system to one that better fits a Western perception of a 'healthy' river will prove disastrous.

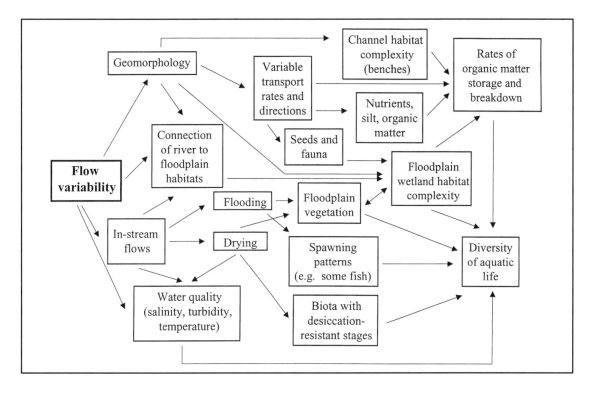

Figure 18.3   *The importance of flow variability to biota and ecosystem processes in rivers. Note the complexity of interactions among variables; many others have been omitted for clarity. After Thoms and Sheldon (1997)*

Flow variability is important in a number of ways (Figure 18.3). Geomorphologically, variable flows maintain the complexity of the in-stream environment in some semi-arid rivers. Thoms and Sheldon (1997), using historic flow and channel survey data that pre-date significant water resource development in the Barwon–Darling River, Australia, showed that the cross-sectional morphology of the unregulated river was complex and characterized by a series of flat surfaces or 'benches'. These benches provided aquatic habitats during high flow events and enabled the accumulation and temporary storage of organic matter. The more variable the flow regime (especially in terms of flood flows), the greater the number of benches present (Thoms and Sheldon, 1997; Table 18.3).

Ecologically, flow variability underpins rates of most ecosystem processes and transport of organisms, nutrients, organic carbon and other materials within rivers and on their floodplain. Many lowland rivers are intimately linked to their floodplains (Walker *et al.*, 1995) and the variability of flow creates a diversity of habitats over time. During high flows, water spreads across the floodplain inundating lakes and billabongs. Drainage patterns are not always consistent and, depending upon the flow history, geomorphology of the floodplain, and human activity (e.g. water abstraction, levee construction), floodplain lakes may lie dry for years or decades. Reflooding at these different frequencies leads to a diverse biota, exemplified by the different taxa that emerge from desiccation-resistant stages in floodplain sediments from areas differing in inundation frequency (Boulton and Lloyd, 1992; Jenkins and Boulton, 1998). Similarly, assemblages of aquatic invertebrates in three central Australian rivers with highly variable flow regimes (Diamantina, Cooper and lower Darling) were structured by the flooding frequency of the various habitats rather than by the microhabitat features that influenced invertebrate assemblage structure in the less-variable River Murray (Sheldon, unpublished data).

Thus, there is compelling evidence that the variable flows of rivers promote a diversity of physical and chemical conditions, and these in turn lead to habitat patchiness and increased biodiversity. Variable flows

Table 18.3 *Flow variability measures (Colwell indices) and the number of in-channel benches associated with two flow conditions (natural or current) at sites along the Barwon–Darling River system*

	Mungindi		Collarenebri		Wilcannia	
	Natural	Current	Natural	Current	Natural	Current
Predictability	0.25	0.47	0.27	0.38	0.35	0.47
Consistency	0.30	0.42	0.23	0.34	0.28	0.41
Contingency	0.05	0.06	0.03	0.04	0.05	0.07
No. of benches	4	3	7	4	6	5

have repercussions for the survival of wetland ecosystems on the floodplain as well, and management recommendations must span in-stream and off-stream issues. As river regulation is essential for human uses of water resources, we must identify effective compromises that enable us to use the water and also release enough for the environment to sustain natural river ecosystem processes. To do this, we must have a better understanding of which *specific* features of flow variability are crucial, and this is a future research goal.

Removing variability – impacts of regulation

One of the main aims of flow regulation is to provide a reliable and constant water supply. By definition, this entails preventing intermittency or artificially creating it below a dam. Water must be harnessed from high flows and released during dry weather, meaning that most flood peaks are dampened or removed completely. Regulation in arid and semi-arid areas often involves interbasin transfers (Davies *et al.*, 1992, 1994, this volume), water storage, and groundwater abstraction (Crabb, 1997). These practices can alter groundwater recharge patterns, leading to the cessation of permanent flow in some areas (Stromberg *et al.*, 1996; Grimm *et al.*, 1997; Mount, 1997). Not surprisingly, pressure to regulate river flow is greatest in arid and semi-arid zones where human populations are increasing, and the limiting resource is water (Agnew and Anderson, 1992; Davies *et al.*, 1994). Worse still, there is a serious mismatch between the 'boom-and-bust' ecology of rivers with variable flow regimes and the sophistry of the short-term economic gains from regulation for irrigation (Walker *et al.*, 1997).

In the Barwon–Darling River, water diversions are now equivalent to approximately 60% of the long-term flow and, during individual flood pulses, water abstractions can remove up to 70% of the daily flows (Thoms, 2000). Extensive water resource development

in this river since 1965 has significantly altered flow variability at many sites. Analyses of hydrologic predictability, constancy and contingency (Colwell, 1974) for a 40-year period using daily simulated flows for 'natural' and 'current' conditions indicate that flows have become more predictable and constant in the Barwon–Darling following water resource development (Table 18.3). Flows are now maintained at constant levels for long periods so that water can be abstracted. Since 1965, the low-flow channel has enlarged with the virtual removal of many of the in-channel benches, and consequent habitat decline. These changes associated with flow regulation in the river resemble those documented by Thoms and Walker (1993) in the lower River Murray.

In the United States, the most famous example of geomorphological changes associated with dampening of flows is the Colorado River as it passes through the Grand Canyon. Following the completion of Glen Canyon Dam in 1963, the annual cycle of sediment scour and deposition was profoundly altered as size and frequency of floods and sediment loads below the dam were significantly reduced. Over the last 35 years, sand-bar size has declined, vegetation has encroached on the channel, and sediment and debris inputs from tributaries have created large deltaic fans that have constricted river flow and filled in backwater areas used for spawning by native fish (Stevens *et al.*, 1995; Collier *et al.*, 1997). This suite of alterations inspired managers to stage the river's first 'controlled flood' in March 1996. Some 900×10^6 m^3 of water were released over seven days to scour out large woody debris and establish vegetated terraces, and to allow the re-establishment of gravel bars and beaches more typical of pre-regulation geomorphology. The success of this artificial flood is still being monitored and this information will be valuable for future management options.

The impacts of damping flow variability are widespread at a range of scales. Unfortunately, empirical evidence of these impacts at large scales for many environmental indicators is lacking because of the intractability of sampling entire river ecosystems (including their floodplains) over long periods when the flow regime is extremely variable (Walker *et al.*, 1995). However, management issues need to be addressed at this catchment scale if appropriate judgements and legislation concerning environmental flow allocations and other palliative measures are to be developed. Although there are cogent arguments for assessing the impacts of regulation as a massive large-scale experiment (cf. Walters, 1990), we are hampered by inadequate pre-regulation data and the nagging feeling that reversibility of the impact is not an option. We seek a compromise between maintenance of intermittency as an extreme of the allocation of environmental flows to mimic a natural regime and the demands placed by a thirsty human population enjoying the benefits of irrigation and making 'the desert bloom' (Plummer, 1994). The reverse – production of intermittent conditions downstream of a dam on a formerly perennial stream – has only lately provoked substantial concern and serious consideration of environmental flow allocations (see later).

In the face of these apparently mutually exclusive uses of this resource, are there management practices that might at least minimize the impacts of flow regulation in rivers with highly variable flow regimes? How can ecologists help draft legislation that will protect intermittent rivers from excessive regulation but without their recommendations being rendered irrelevant by political, social and economic factors? It is widely recognized that the existence of scientific data does not alone lead to effective management (see Karr *et al.*, this volume; Schofield *et al.*, this volume) but there is a need for ecologists to suggest possible solutions to management issues facing these particular rivers. We acknowledge that a generic solution is unlikely (see earlier) but identify several common themes that might assist policy-makers in drafting legislation that preserves the crucial flow variability that maintains these ecosystems.

Issues in the management and conservation of rivers with variable flows

MISMATCH BETWEEN REALITY AND PUBLIC PERCEPTION/PROFESSIONAL TRAINING

This chapter began with the general viewpoint that ecological theory, and consequently management and legislation associated with river systems, are founded on empirical evidence derived from perennial rivers. Public and scientific perception is often an unrecognized player here – many Westerners tend to view a dry river-bed with concern because it does not match their perceptions of what a river should be.

In the south-western United States, perception and use of water is a double-edged sword. On one hand is the belief that a healthy river is one that supports sustained flow and that an attractive environment is lush and green. On the other is the belief that water should be used to the fullest extent and that any not saved for irrigation or other human uses is being wasted (Reisner, 1986). The re-authorization of the Clean Water Act has highlighted this conflict in the United States (Maurer, 1995). This legislation sets guidelines for state maintenance of water quality standards for point-source discharges. Because many river systems in the west are intermittent, effluent releases are likely to represent a large percentage (and in some cases, all) of the flow below the release point. In such situations, meeting water quality standards in an 'effluent-dominated' system requires that the effluent itself be at or near established clean water standards – a far more expensive demand than for discharges that empty into large perennial systems and thus are diluted to accepted levels relatively quickly.

Inevitably, dischargers have been inclined to remove effluent releases into ephemeral or intermittent channels when faced with the high cost of compliance. However, because of the perception that a flowing river is better than a dry one, state regulatory agencies have been willing to compromise. This has led to the unusual situation of water quality standards being relaxed for such effluent-dominated systems because of the belief that it is better to have sustained (albeit low quality) water than a predominantly dry channel. Similar examples occur in the Santa Cruz River, Arizona (Grimm *et al.*, 1997), and some tributaries of the River Murray Basin flowing from country towns (Crabb, 1997) where sewage effluent may comprise the main flow during the dry season.

The logic of effluent release is based on two assumptions (Maurer, 1995): first, that ephemeral streams do not support viable aquatic communities, and, second, that effluent-dependent systems provide 'net ecological benefits' such as habitat restoration and increased species diversity by maintaining permanent flow. Clearly, there are problems with this logic. In some regions, temporary streams and rivers have quite diverse assemblages and considerable faunal overlap with adjacent perennial sites (e.g. Boulton and Suter,

1986). Alternatively, temporary streams may support biota that are 'temporary stream specialists' (Williams and Hynes, 1977; Williams, 1987; Boulton and Lake, 1992a) or that use these sites for special purposes such as spawning (Erman and Hawthorne, 1976). Further, the poorly diluted or undiluted effluent inevitably has deleterious effects upon the biota of these systems and in the downstream receiving waters, regardless of the tolerance of the organisms to intermittency (e.g. Bromley and Por, 1975).

MISMATCH BETWEEN ACCEPTED WATER QUALITY CRITERIA AND NATURAL CONDITIONS IN INTERMITTENT RIVERS

Historically, water quality criteria have been based on chemical and physical characteristics (as well as some microbiological variables, e.g. *Escherischia coli* counts) but increasingly, biological variables are becoming popular because of the perceived advantages of biomonitors (see Bishop *et al.*, 1995; Chessman, 1995). Furthermore, there is increasing recognition of the potential value of ecosystem measures as indicators of the 'health' of a system (e.g. Davies, 1994). However, the point was made earlier that at certain stages of the flow regime, water quality of intermittent streams naturally deteriorates and the diversity of intolerant biota declines. Unless this is understood, uncritical application to intermittent rivers of water quality criteria and biological indicator species used for assessing the health of permanent rivers will prove misleading.

For example, we reviewed the indicator families for an index that has been proposed for rapid bioassessment of river health in Australia using invertebrates identified to the family level (Chessman, 1995). Of the 15 families (out of a total of 111) that were considered most pollution-insensitive (i.e. scored the maximum of 10 – see Table 1 in Chessman, 1995), only three (Helicopsychidae, Leptophlebiidae and Ptilodactylidae) have been commonly recorded from relatively pristine temperate intermittent streams (Boulton and Lake, 1992b); the other 12 families are typical of high-altitude, fast-flowing cool mountain streams and are generally stenothermal (e.g. Ameletopsidae, Coloburiscidae). When arid-zone intermittent streams are included in the comparison, only the Leptophlebiidae (a widespread mayfly family that is especially diverse in Australia) remains. Conversely, there are 15 'tolerant' families that scored low (i.e. <4), and all but one (Erpobdellidae)

are common in intermittent streams in both arid and temperate zones (Boulton, 1988; Boulton and Lake, 1992b; Boulton and Williams, 1996; Boulton unpublished data). While this in no way diminishes the general value of Chessman's (1995) index, it does illustrate that intermittency apparently exerts a similar influence to pollution in permanent streams (as detected by this single measure), and that uncritical application of the index to intermittent rivers could be misleading. Sampling during naturally inimical periods (e.g. drying) would exacerbate this misconception.

MAKING MANAGEMENT DECISIONS WITHOUT EMPIRICAL DATA

If ecologists' perceptions of the water quality of intermittent rivers are problematic, they are nothing compared with the problems facing water managers. It is likely that this is the first paper suggesting that special policies that include periods of intermittent flow be drafted for regulated rivers that were historically intermittent. Such policies should be based on recognition of the flow regime and flow history (*sensu* Puckridge *et al.*, 1998) rather than simply the hydrograph. Unfortunately, long-term data on flow regimes for many intermittent rivers may not exist. Policies also should be based on integrated and adaptive flow management ideals (Banens *et al.*, 1994; Thoms *et al.*, 1996) rather than being devised to favour a single process or group of the biota (e.g. only fish recruitment or waterbird nesting).

The need to manage flows in intermittent rivers is demonstrated clearly by examining conditions in the Murray–Darling Basin, Australia. Much of the basin is classified as semi-arid and many of the rivers were naturally intermittent. Water resource development has meant that some 80% of the total flow from the combined river systems in the basin is now diverted for off-stream use and the major rivers are subject to 'drought' flows in 60% of years compared with 5% under natural flows (Cullen *et al.*, 1996). In some rivers, the frequency of intermittency has increased whereas in others the maintenance of minimum flows through weir and dam construction means that the natural drying phase of the flow regime has been eliminated (Thoms *et al.*, 1996). In most of the basin's rivers, the magnitude, frequency, seasonality, duration and variation of flows have been modified with serious consequences for the ecology of the river (Walker *et al.*, 1995). These changes are implicated in the declining numbers of native fish species and the provision of ideal conditions for toxic blue–green algal blooms (Cullen *et al.*, 1996; Crabb, 1997).

Although empirical data for the above are limited, the Australian state governments acknowledged the degradation and established a 'Cap' aimed to ensure that there would be no net growth in diversions of water from the rivers of the Murray–Darling Basin (Cox and Baxter, 1996). The level of the 'Cap' is the volume of water that would have been diverted under 1993/94 levels of development. This was a landmark decision, made all the more impressive by the fact that several Australian states are drained by the River Murray. Each state had to develop and revise its own water policies to meet the Cap. These included the Victorian 'Bulk Entitlement' process, establishment of a set of 'River Flow Objectives' for each river system in New South Wales, initiation of 'Water Allocation Management Plans' (WAMP) for strategic catchments in Queensland, and the establishment of water needs for key ecosystems in the Australian Capital Territory. At present, individual state policies do not recognize the need for variable flow conditions or the importance of a drying phase for many of the rivers within the basin that were intermittent before regulation. However, there is general agreement among the states that a balance needs to be struck between consumptive and in-stream uses of water within the basin (Cox and Baxter, 1996).

Since 1993, the Council of Australian Governments (COAG) has introduced a series of new administrative water reforms. It proposed that all consumptive and non-consumptive water entitlements are to be allocated and managed through various planning systems and based on a catchment perspective of the resource. Major components of the proposed reforms include:

- market-trading of water rights as a basis of water allocation;
- separation of purchaser, provider and regulator functions to remove the disadvantage of the private sector tendering for the provision of services;
- government management services to become open to market competition;
- the removal of government 'technical expertise' and the use of external consultant expertise.

Concerns exist with some of the reforms, especially those that relate to the control of water resources to market and administrative controls. Nevertheless, for the first time in water resource development in Australia, water is to be allocated for environmental purposes and this is to be achieved using the best

available scientific information, no matter how scant. However, the 'Cap' has proved politically highly contentious and several ecologists argue that it freezes diversions at a level far beyond sustainable limits.

ENVIRONMENTAL FLOW ALLOCATIONS FOR INTERMITTENT RIVERS

Environmental flow allocations are becoming accepted as a valid approach to returning water to the environment but attention must be paid to the quality as well as quantity of water. A variety of approaches exists for assessing allocations (Gordon *et al.*, 1992; Arthington and Pusey, 1993; Karim *et al.*, 1995) but most require massive amounts of data whereas managers require solutions immediately. This has led to the formation of 'scientific panels' that try to determine how best to use the present flows (within tight economic and social constraints) while still maintaining a natural flow regime (cf. Swales and Harris, 1995; Richter *et al.*, 1996, 1997; Thoms *et al.*, 1996). Unfortunately, there are few scientists experienced in the ecology of intermittent rivers and their advice may reflect their experience with permanent rivers, potentially with disastrous results. It takes some courage to recommend a period of drought!

In Australia, 'scientific panels' are being used increasingly to provide relatively rapid recommendations for flow management. These follow a contextual framework, developed by Thoms *et al.* (1996), whereby recommendations are developed from an ecosystem perspective. There are three driving principles in the framework: ecosystem health, philosophical issues on river ecosystem functioning, and water management issues. Up to 1997, there have been seven scientific panels conducted on various rivers within the Murray–Darling Basin providing flow recommendations with the best available scientific information. Panel recommendations include reinstating near-natural periods of low flow and/or cessation of flow within the flow regime of regulated intermittent rivers. For example, the scientific panel assessing environmental flow requirements for the Barwon–Darling River recognized that drought was an ecological phenomenon as significant to the river ecosystem as flooding (Thoms *et al.*, 1996) and recommended that the natural frequency, duration and intensity of drought conditions be maintained. For other large intermittent rivers like the Barwon–Darling, it may be more beneficial to focus management of environmental flows around the natural frequency and duration of drought-like conditions. As these unregulated rivers naturally have low discharge, it is the low-flow band that is most often impacted by development. However, one of the problems with managing environmental flows based on low-flow conditions appears to be the limitations of many models in accurately modelling discharges within the low-flow band (Thoms *et al.*, 1996), along with inadequate amounts of hydrological data (Walker *et al.*, 1995).

There are some concerns that environmental releases of water may be ineffectual in some intermittent and ephemeral river systems. In 1990, the Oanob Dam was completed across the Oanob River, an ephemeral river in Namibia (Jacobson *et al.*, 1995). By 1993, enough water had accumulated to allow an 'artificial flood' without jeopardizing water supply to nearby human settlements. Jacobson *et al.* (1995) suggest these releases may be ineffectual for several reasons. Scaling these releases to natural floods is not feasible. The artificial flood released in 1993 had a peak discharge of 15 m^3 s^{-1} and totalled <5 × 10^6 m^3. It travelled less than 30 km downstream. Average annual flood discharges in the river before regulation exceeded 150 m^3 s^{-1} and inundated much larger areas of floodplain. A second key factor is the different character of these floods. Natural floods carry heavy loads of silt, organic matter and nutrients that are deposited on the floodplain and through the riparian forest downstream, enriching the area. The artificial flood carried almost no organic material, and only low amounts of silt and nutrients (Jacobson *et al.*, 1995). In short, water alone may not be enough.

CHANGES IN PERCEPTIONS TO A LIMITING RESOURCE

One prevailing issue in many parts of the world is the 'use it or lose it' approach to water policy. For example, in many western states in the United States, water rights are lost if water use decreases (Plummer, 1994). Development and use of western US water resources has a complex but well-documented history (Worster, 1985; Reisner, 1986) driven by priorities completely unrelated to aquatic ecosystem health. Enormous government investments in large-scale projects such as dams and interbasin transfers have produced abundant and inexpensive water to locales that have historically been unfamiliar with such wealth. Consequently, *per capita* water use in arid states such as Arizona and New Mexico is almost twice the national average (Maddock and Hines, 1995).

Nonetheless, many major urban centres have begun to develop and implement water management strategies that emphasize both demand management (reducing water usage) along with the more traditional supply management (maximizing the water supply). Demand management strategies include buy-back plans from irrigators, re-use of treated wastewater, household conservation, and inverted water rate charges to discourage wasteful water use. In contrast, supply management strategies rely on increased use of groundwater and interbasin transfers. In the end, the persistent philosophy and regulatory structure which treat water as an inexpensive commodity for human use has led both to the perception of rivers as conduits that deliver this commodity and to the devastation of the ecological integrity of major south-western US rivers. There are marked parallels with the history of water resource exploitation in the Murray–Darling Basin in Australia (Crabb, 1997) and we suspect that our limited examples are not unique.

A NEED FOR SPECIALIST POLICIES

Given the problems described above, do we need a specialist policy to regulate releases (and holdings) of flow in regulated intermittent rivers? Formulating such a policy based on our present knowledge of intermittent ecosystems may seem premature, especially given that we cannot easily extrapolate from experiences gained in perennial rivers. However, the demand is urgent and we must learn from our management practices. Thus, we suggest that such a policy is needed and that it should address:

- the importance of the dry phase (of variable duration and timing) to intermittent rivers;
- the importance of irregularity, gauging the variability on the pre-regulated flow regime (if such data are adequate);
- the necessity to assess 1 and 2 based on flow regime not hydrograph;
- the need for integrated flow management that does not allocate flows based on a few, readily identified water users (e.g. fish, waterbirds) but takes the whole system into account;
- the relationship between water quality and quantity, recognizing that cues to using the floodplain may rely on subtle changes in water temperature, etc. and that the water of 'artificial floods' may differ from natural flood-water in important ecological characteristics (e.g. sediment; particulate organics);

- maintenance of variability of flows to promote diversity of habitat types over large time and spatial scales;
- explicit recognition of public perception of intermittency as a 'problem', and educational programmes to remedy this concern.

Prognosis

Ephemeral and intermittent streams exemplify the extreme of rivers with variable flow regimes, and are globally widespread. The formulation of policies and legislation for ephemeral systems must take into account that intermittent streams and rivers usually occur in regions where the competition for water is high and it is often the environmental needs of the systems that are neglected. Regulation to meet human demands means that the natural variability in flooding and drying is modified either by removing water from the system and increasing the frequency of drying or by rendering the system permanent for water supply, thus removing the all-important dry phase. Combined with this interference is a Western perception that dry river-beds and associated droughts are natural disasters and should be prevented.

The severe environmental degradation apparent in many rivers with variable flow regimes worldwide (e.g. the USA, Namibia, Australia) appears to have generated a new and more dynamic approach to managing these rivers. There is growing recognition that successful management must be based on the natural flow regime, that the dry phase is as significant as flooding, and that this must be incorporated into policies for water resource management. Management of intermittent systems must be proactive and the natural flow regime (including flow characteristics such as flow magnitude, antecedent conditions and seasonality) must be analysed to assess environmental flow requirements. Moreover, each flood must be considered on its own merit. Technology may allow for provision of individual floods but the limitations of planned water releases (e.g. scaling, water quality) must be recognized. However, each release constitutes a 'large-scale experiment' and, despite problems of replication and detecting long-term effects, we should focus on using these events to aid adaptive management of intermittent river systems. On a policy side, this approach to water resource management must be incorporated into the licence agreements of water users, and efforts should be made to educate stakeholders about the value of maintaining the variable flow regimes that underpin the ecology of these rivers.

Acknowledgements

We are grateful to Bryan Davies for initially suggesting this topic, and challenging us to integrate ecological theory with management issues facing streams with variable flow regimes. We also appreciate the support and assistance from the editors during the long gestation of this work. For insightful comments, additional information, and constructive thoughts on earlier drafts, we thank Phil Boon, Bryan Davies, Avital Gasith, Jim Puckridge, Vince Resh and Keith Walker.

References

Abell, D.L. 1984. Benthic invertebrates of some California intermittent streams. In: Jain, S. and Moyle, P. (eds) *Vernal Pools and Intermittent Streams.* University of California, Davis Institute of Ecology Publication 28:46–60.

Agnew, C. and Anderson, E. 1992. *Water Resources in the Arid Realm.* Routledge, London.

Arthington, A.H. and Pusey, B.J. 1993. In-stream flow management in Australia: methods, deficiencies and future directions. *Australian Biologist* 6:52–60.

Banens, R.J., Blackmore, D.J., Lawrence, B.W., Shafron, M.C. and Sharley, A.J. 1994. A program for the sustainable management of the rivers of the Murray–Darling basin. In: Uys, M.C. (ed.) *Classification of Rivers and Environmental Health Indicators.* Proceedings of a joint South African/ Australian workshop, Cape Town, South Africa. Water Research Commission Report No. TT 63/94, 287–300.

Bayly, I.A.E. and Williams, W.D. 1973. *Inland Waters and their Ecology.* Longman, Melbourne.

Berra, T.M., Sever, D.M. and Allen, G.R. 1989. Gross and histological morphology of the swimbladder and lack of accessory respiratory structures in *Lepidogalaxias salamandroides*, an aestivating fish from Western Australia. *Copeia* 4:850–856.

Bishop, K.A., Pidgeon, R.W.J. and Walden, D.J. 1995. Studies on fish movement dynamics in a tropical floodplain river: prerequisites for a procedure to monitor the impacts of mining. *Australian Journal of Ecology* 20:81–107.

Biswas, S.P. 1996. Global water scarcity: issues and implications with special reference to India. *Verhandlungen der Internationalen Vereinigung für theoretische und angewandte Limnologie* 26:115–121.

Boulton, A.J. 1988. Composition and dynamics of macroinvertebrate communities in two intermittent streams. Unpublished PhD Thesis, Monash University, Victoria.

Boulton, A.J. 1989. Over-summering refuges of aquatic macroinvertebrates in two intermittent streams in central Victoria. *Transactions of the Royal Society of South Australia* 113:23–34.

Boulton, A.J. 1999. Why variable flows are needed for invertebrates of semi-arid rivers. In: Kingsford, R.T. (ed.) *A Free-flowing River: The Ecology of the Paroo River*. NSW National Parks and Wildlife Service, Hurstville, 113–128.

Boulton, A.J. and Lake, P.S. 1990. The ecology of two intermittent streams in Victoria, Australia. I. Multivariate analyses of physicochemical features. *Freshwater Biology* 24:123–141.

Boulton, A.J. and Lake, P.S. 1992a. The ecology of two intermittent streams in Victoria Australia. II. Comparisons of faunal composition between habitats, rivers and years. *Freshwater Biology* 27:99–121.

Boulton, A.J. and Lake, P.S. 1992b. The macroinvertebrate assemblages in pools and riffles in two intermittent streams (Werribee and Lerderderg Rivers southern central Victoria). *Occasional Papers of the Museum of Victoria* 5:55–71.

Boulton, A.J. and Lloyd, L.N. 1992. Flooding frequency and invertebrate emergence from dry floodplain sediments of the River Murray Australia. *Regulated Rivers: Research & Management* 7:137–151.

Boulton, A.J. and Stanley, E.H. 1995. Hyporheic processes during flooding and drying in a Sonoran Desert stream. II. Faunal dynamics. *Archiv für Hydrobiologie* 134:27–52.

Boulton, A.J. and Suter, P.J. 1986. Ecology of temporary streams – an Australian perspective. In: De Deckker, P. and Williams, W.D. (eds) *Limnology in Australia*. CSIRO/Dr W. Junk Melbourne/Dordrecht, 313–327.

Boulton, A.J. and Williams, W.D. 1996. Aquatic biota. In: Twidale, C.R., Tyler, M.J. and Davies, M. (eds) *The Natural History of the Flinders Ranges*. Royal Society of South Australia, Adelaide, 102–112.

Boulton, A.J., Stanley, E.H., Fisher, S.G. and Lake, P.S. 1992. Over-summering strategies of macroinvertebrates in intermittent streams in Australia and Arizona. In: Robarts, R.D. and Bothwell, M.L. (eds) *Aquatic Ecosystems in Semi-arid Regions: Implications for Resource Management*. NHRI Symposium Series 7, Environment Canada, Saskatoon, 227–237.

Bromley, H.J. and Por, F.R. 1975. The metazoan fauna of a sewage-carrying wadi Nahal Soreq (Judean Hills, Israel). *Freshwater Biology* 5:121–133.

Brunke, M. and Gonser, T. 1997. The ecological significance of exchange processes between rivers and groundwater. *Freshwater Biology* 37:1–33.

Castella, E., Bickerton, M., Armitage, P.D. and Petts, G.E. 1995. The effects of water abstractions on invertebrate communities in U.K. streams. *Hydrobiologia* 308:167–182.

Chessman, B.C. 1995. Rapid assessment of rivers using macroinvertebrates: a procedure based on habitat-specific sampling family level identification and a biotic index. *Australian Journal of Ecology* 20:122–129.

Clifford, H.F. 1966. The ecology of invertebrates in an intermittent stream. *Investigations of Indiana Lakes and Streams* 7:57–98.

Closs, G.P. and Lake, P.S. 1996. Drought, differential mortality and the coexistence of a native and an introduced fish species in a south east Australian intermittent stream. *Environmental Biology of Fishes* 47:17–26.

Collier, M.P., Webb, R.H. and Andrews, E.D. 1997. Experimental flooding in the Grand Canyon. *Scientific American* 213:82–89.

Colwell, R.K. 1974. Predictability, constancy and contingency of periodic phenomena. *Ecology* 55:1148–1153.

Comin, F.A. and Williams, W.D. 1994. Parched continents: our common future? In: Margalef, R. (ed.) *Limnology Now: A Paradigm of Planetary Problems*. Elsevier Science, 473–527.

Cox, W. and Baxter, P. 1996. *Setting the Cap*. Report of the Independent Audit Group, Murray–Darling Basin Ministerial Council, Canberra.

Crabb, P. 1997. *Murray–Darling Basin Resources*. Murray–Darling Basin Commission, Canberra.

Cullen, P., Humphries, P., Thoms, M.C., Harris, J. and Young, W. 1996. Environmental flow allocations – an ecological perspective. In: *Managing Australia's Inland Waters: Roles for Science and Technology*. Prime Minister's Science and Engineering Forum, Canberra, 54–71.

Dahm, C.N. and Molles, M.R. 1992. Streams in semi-arid regions as sensitive indicators of global climate change. In: Firth, P. and Fisher, S. (eds) *Global Climate Changes and Freshwater Ecosystems*. Springer-Verlag, New York, 250–260.

Davies, B.R., Thoms, M.C. and Meador, M. 1992. The ecological impact of inter-basin water transfers and their threats to river basin integrity and conservation. *Aquatic Conservation: Marine and Freshwater Ecosystems* 2:325–349.

Davies, B.R., Thoms, M.C., Walker, K.F., O'Keeffe, J.H. and Gore, J.A. 1994. Dryland rivers: their ecology, conservation and management. In: Calow, P. and Petts, G.E. (eds) *The Rivers Handbook*. Vol. 2, Blackwell Scientific, Oxford, 484–512.

Davies, P.M. 1994. Ecosystem processes: A direct assessment of river health. In: Uys, M.C. (ed.) *Classification of Rivers and Environmental Health Indicators*. Proceedings of a Joint South African/Australian Workshop, Cape Town. South Africa Water Research Commission Report No. TT 63/94, 119–128.

Erman, D.C. and Hawthorne, V.M. 1976. The quantitative importance of an intermittent stream in the spawning of rainbow trout. *Transactions of the American Fisheries Society* 6:675–681.

Feminella, J.W. 1996. Comparison of benthic macroinvertebrate assemblages in small streams along a gradient of flow permanence. *Journal of the North American Benthological Society* 15:651–669.

Findlay, S. 1995. Importance of surface–subsurface exchange in stream ecosystems: the hyporheic zone. *Limnology and Oceanography* 40:159–164.

Flaschka, I.M., Stockton, C.W. and Boggess, W.R. 1987. Climatic variation and surface water resources in the Great Basin region. *Water Resources Bulletin* 23:47–57.

Fox, A.M. 1996. Macrophytes. In: Petts, G. and Calow, P. (eds) *River Biota – Diversity and Dynamics*. Blackwell Science, Oxford, 27–44.

Gordon, N.D., McMahon, T.A. and Finlayson, B.L. 1992. *Stream Hydrology – An Introduction for Stream Ecologists*. John Wiley, Chichester.

Gray, L.J. 1980. Recolonization pathways and community development of desert stream macroinvertebrates. Unpublished PhD Thesis, Arizona State University, Tempe, Arizona.

Grimm, N.B., Chacon, A., Dahm, C.N., Hostetler, S.W., Lind, O.T., Starkweather, P.L. and Wurtsbaugh, W.W. 1997. Sensitivity of aquatic ecosystems to climatic and anthropogenic changes: the Basin and Range, American Southwest and Mexico. *Hydrological Processes* 11:1023–1041.

Hynes, H.B.N. 1958. The effect of drought on the fauna of a small mountain stream in Wales. *Verhandlungen der Internationalen Vereinigung für theoretische und angewandte Limnologie* 13:826–833.

Hynes, J.D. 1975. Annual cycles of macro-invertebrates of a river in southern Ghana. *Freshwater Biology* 5:71–83.

Jacobson, P.J., Jacobson, K.N. and Seely, M.K. 1995. *Ephemeral Rivers and their Catchments: Sustaining People and Development in Western Namibia*. Desert Research Foundation of Namibia, Windhoek.

Jenkins, K.M. and Boulton, A.J. 1998. Community dynamics of invertebrates emerging from reflooded lake sediments: flood pulse and aeolian influences. *International Journal of Ecological and Environmental Science* 24:179–192.

Johnson, M. 1993. The water industry overseas – lessons for Australia. In: Johnson, M. and Rix, S. (eds) *Water in Australia*. Pluto Press, New South Wales, Australia, 138–165.

Jones, J.B. and Holmes, R.M. 1996. Surface–subsurface interactions in stream ecosystems. *Trends in Ecology and Evolution* 11:239–242.

Karim, K., Gubbels, M.E. and Goulter, I.C. 1995. Review of determination of instream flow requirements with special application to Australia. *Water Resources Bulletin* 31:1063–1077.

Kingsford, R.T., Boulton, A.J. and Puckridge, J.T. 1998. Challenges in managing dryland rivers crossing political boundaries: lessons from Cooper Creek and the Paroo River, central Australia. *Aquatic Conservation: Marine and Freshwater Ecosystems* 8:361–378.

Larimore, R.W., Childers, W.F. and Heckrotte, C. 1959. Destruction and re-establishment of stream fish and invertebrates affected by drought. *Transactions of the American Fisheries Society* 88:261–285.

Lavee, H., Poeson, J. and Yair, A. 1997. Evidence of high efficiency water-harvesting by ancient farmers in the Negev Desert, Israel. *Journal of Arid Environments* 35:341–348.

Lyons, J.K., Pucherelli, M.J. and Clark, R.C. 1992. Sediment transport and channel characteristics of a sand-bed portion of the Green River below Flaming Gorge Dam, Utah, USA. *Regulated Rivers: Research & Management* 7:219–232.

Maddock, T.S. and Hines, W.G. 1995. Meeting future public water supply needs: a southwest perspective. *Water Resources Bulletin* 31:317–329.

Matthews, W.M., Cashner, R.C. and Gelwick, F.P. 1988. Stability and persistence of fish faunas and assemblages in three midwestern streams. *Copeia* 1988:945–955.

Maurer, E.F. 1995. *Neglected Places: Water Quality Standards and Ephemeral Systems*. US Environmental Protection Agency, Office of Policy Analysis, Washington, DC.

Meffe, G.K. and Minckley, W.L. 1987. Persistence and stability of fish and invertebrate assemblages in a repeatedly disturbed Sonoran Desert stream. *American Midland Naturalist* 117:177–191.

Miller, A.M. and Golladay, S.W. 1996. Effects of spates and drying on macroinvertebrate assemblages of an intermittent and a perennial prairie stream. *Journal of the North American Benthological Society* 15:670–689.

Molles, M.C. and Dahm, C.N. 1990. A perspective on El Niño and La Niña: global implications for stream ecology. *Journal of the North American Benthological Society* 9:68–76.

Mount, T.J. 1997. Geology and groundwater characteristics – Geological controls on saline groundwater inflow to the Darling River, Bourke, NSW. In: Thoms, M.C., Gordon, A. and Tatnell, W. (eds) *Researching the Barwon–Darling*. CRC for Freshwater Ecology, Canberra, 10–33.

Petts, G. and Calow, P. 1996. *River Restoration*. Blackwell Science, Oxford.

Plummer, J.L. 1994. Western water resources: the desert is blooming, but will it continue? *Water Resources Bulletin* 30:595–603.

Poff, N.L. 1996. A hydrogeography of unregulated streams in the United States and an examination of scale-dependence in some hydrological descriptors. *Freshwater Biology* 36:71–91.

Poff, N.L. and Ward, J.V. 1989. Implications of streamflow variability and predictability for lotic community structure: a regional analysis of streamflow patterns. *Canadian Journal of Fisheries and Aquatic Sciences* 46:1805–1818.

Postel, S. 1996. *Dividing the Waters: Food, Security, Ecosystem Health and the New Politics of Scarcity*. Worldwatch Paper 132, Worldwatch Institute, Washington, DC.

Puckridge, J.T., Sheldon, F., Walker, K.F. and Boulton, A.J. 1998. Flow variability and the ecology of large rivers. *Marine and Freshwater Research* 49:55–72.

Reisner, M. 1986. *Cadillac Desert*. Penguin Books, New York.

Resh, V.H., Brown, A.V., Covich, A.P., Gurtz, M.E., Li, H.W., Minshall, G.W., Reice, S.R., Sheldon, A.L., Wallace, J.B. and Wissmar, R. 1988. The role of disturbance in stream ecology. *Journal of the North American Benthological Society* 8:433–455.

Richter, B.D., Baumgartner, J.V., Powell, J. and Braun, D.P.

1996. A method for assessing hydrologic alteration within ecosystems. *Conservation Biology* 10:1163–1174.

Richter, B.D., Baumgartner, J.V., Wigington, R. and Braun, D.P. 1997. How much water does a river need? *Freshwater Biology* 37:231–249.

Rowland, S.J. 1989. Aspects of the history and fishery of the Murray Cod *Maccullochella peeli* (Mitchell) (Percicthyidae). *Proceedings of the Linnean Society of New South Wales* 111:201–213.

Stanley, E.H. and Boulton, A.J. 1995. Hyporheic processes during flooding and drying in a Sonoran Desert stream. I. Hydrologic and chemical dynamics. *Archiv für Hydrobiologie* 134:1–26

Stanley, E.H. and Valett, H.M. 1992. Interactions between drying and the hyporheic zone in a desert stream. In: Firth, P. and Fisher, S.G. (eds) *Global Climate Change and Freshwater Ecosystems*. Springer-Verlag, New York, 234–249.

Stanley, E.H., Buschman, D.L., Boulton, A.J., Grimm, N.B. and Fisher, S.G. 1994. Invertebrate resistance and resilience to intermittency in a desert stream. *American Midland Naturalist* 131:288–300.

Stanley, E.H., Fisher, S.G. and Grimm, N.B. 1997. Ecosystem expansion and contraction in streams. *BioScience* 47:427–436.

Stevens, L.E., Schmidt, J.C., Ayers, T. and Brown, B.T. 1995. Flow regulation, geomorphology and Colorado River marsh development in the Grand Canyon Arizona. *Ecological Applications* 5:1025–1039.

Stromberg, J.C., Tiller, R. and Richter, B. 1996. Effects of groundwater decline on riparian vegetation of semiarid regions: The San Pedro, Arizona. *Ecological Applications* 6:113–131.

Swales, S. and Harris, J.H. 1995. The expert panel assessment method (EPAM): A new tool for determining environmental flows in regulated rivers. In: Harper, D.M. and Ferguson, A.J.D. (eds) *The Ecological Basis for River Management*. John Wiley, New York, 125–134.

Thoms, M.C. 2000. Bank erosion in a semi arid river: extent and possible causes following a flood in the Barwon–Darling River. *Australian Journal of Soil and Water Conservation* 38: (in press).

Thoms, M.C. and Sheldon, F. 1997. River channel complexity and ecosystem processes: the Barwon–Darling River. In: Klomp, N. and Lunt, I. (eds) *Frontiers in Ecology*. Elsevier Science, Oxford, 193–206.

Thoms, M.C. and Walker, K.F. 1993. Channel changes associated with two adjacent weirs on a regulated lowland alluvial river. *Regulated Rivers: Research & Management* 8:103–119.

Thoms, M.C., Sheldon, F., Roberts, J., Harris, J. and Hillman, T.J. 1996. *Scientific Panel Assessment of Environmental Flows for the Barwon–Darling River*. Report to the NSW Department of Land and Water Conservation, Sydney.

United Nations Educational, Scientific and Cultural Organization (UNESCO) 1971. *Discharge of Selected Rivers of the World*. Volume 2. UNESCO, Paris.

Usinger, R.L. 1956. Introduction to aquatic entomology. In: Usinger, R.L. (ed.) *Aquatic Insects of California*. University of California Press, Berkeley, 3–49.

Valett, H.M., Fisher, S.G. and Stanley, E.H. 1990. Physical and chemical characteristics of the hyporheic zone of a Sonoran Desert stream ecosystem. *Journal of the North American Benthological Society* 9:201–215.

Vannote, R.L., Minshall, G.W., Cummins, K.W., Sedell, J.R. and Cushing, C.E. 1980. The river continuum concept. *Canadian Journal of Fisheries and Aquatic Sciences* 37:130–137.

Walker, J., Alexander, D., Irons, C., Jones, B., Penridge, H. and Rapport, D. 1996. Catchment health indicators: an overview. In: Walker, J. and Reuter, D.J. (eds) *Indicators of Catchment Health: A Technical Perspective*. CSIRO, Melbourne, 3–18.

Walker, K.F., Boulton, A.J., Thoms, M.C. and Sheldon, F. 1994. Effects of water-level changes induced by weirs on the distribution of littoral plants along the River Murray, South Australia. *Australian Journal of Marine and Freshwater Research* 45:1421–1438.

Walker, K.F., Sheldon, F. and Puckridge, J.T. 1995. A perspective on dryland river ecosystems. *Regulated Rivers: Research & Management* 11:85–104.

Walker, K.F., Puckridge, J.T. and Blanch, S.J. 1997. Irrigation development on Cooper Creek central Australia – prospects for a regulated economy in a boom-and-bust ecology. *Aquatic Conservation: Marine and Freshwater Ecosystems* 7:63–73.

Walters, C. 1990. Large scale management experiments and learning by doing. *Ecology* 71:2060–2068.

Williams, D.D. 1987. *The Ecology of Temporary Waters*. Timber Press, Portland, Oregon.

Williams, D.D. 1996. Environmental constraints in temporary fresh waters and their consequences for the insect fauna. *Journal of the North American Benthological Society* 15:634–650.

Williams, D.D. and Hynes, H.B.N. 1977. The ecology of temporary streams. II. General remarks on temporary streams. *Internationalen Revue der gesamten Hydrobiologie* 62:53–61.

Williams, W.D. 1988. Limnological imbalances: an antipodean viewpoint. *Freshwater Biology* 20:407–420.

Wood, P.J. and Petts, G.E. 1994. Low flows and recovery of macroinvertebrates in a small regulated chalk stream. *Regulated Rivers: Research & Management* 9:303–316.

Worster, D. 1985. *Rivers of Empire: Water, Aridity, and the Growth of the American West*. Pantheon Books, New York.

Wright, J.F. 1992. Spatial and temporal occurrence of invertebrates in a chalk stream, Berkshire, England. *Hydrobiologia* 248:11–30.

Wright, J.F., Hiley, P.D., Cooling, D.A., Cameron, A.C., Wigham, M.E. and Berrie, A.D. 1984. The invertebrate fauna of a chalk stream in Berkshire, England, and the effect of intermittent flow. *Archiv für Hydrobiologie* 99:179–199.

19

A biogeographical approach to interbasin water transfers: implications for river conservation

B.R. Davies, C.D. Snaddon, M.J. Wishart, M.C. Thoms and M. Meador

Introduction

Water availability has placed important constraints on the location and size of early human settlements. However, rapid increases in technology and engineering over the past century or more, have reduced this dependence. Urban sprawls and agricultural developments now flourish in environments that, without major engineering and technological advances, could not possibly support them. Many examples spring to mind. In South Africa, Gauteng Province (essentially the cities of Johannesburg and Pretoria, the industrial heartland of the country) is entirely supported by an array of interbasin water transfers (IBTs) (Van Niekerk *et al.*, 1996; Basson, 1997; Davies and Day, 1998). The desert cities of Phoenix (Reisner, 1986) and Las Vegas in the United Sates and Adelaide in Australia (Davies *et al.*, 1992) all rely heavily on the importation of water from distant regions. Modern industrialized agriculture is also completely dependent upon irrigation technologies that now rely largely on IBT schemes (e.g. McCully, 1996; Wackernagel and Rees, 1996).

Not only have IBTs extended the limits of development of human society beyond the sustainability of the immediate environment, but their construction and use have serious implications for the river systems that are manipulated. Although impoundment is frequently a part of many IBTs, river regulation by IBTs leads to more complex problems than those caused by simple storage of water (Day, 1985; Snaddon, 1998). While the effects of river regulation by impoundment have been well documented (e.g. Ward and Stanford, 1979, 1983a,b; Lillehammer and Saltveit, 1984; Petts, 1984; Craig and Kemper, 1987; Petts *et al.*, 1989), the ecological effects of IBTs are relatively unknown. In this chapter there is a focus on the implications of IBTs for river conservation in terms of the breakdown of biotic integrity that such schemes impose. The definition of an IBT adopted in this chapter follows that of Davies *et al.* (1992), *viz*: 'the transfer of water from one geographically distinct river catchment or basin to another, or from one river reach to another'. Further, the term biotic integrity is used here as defined by Karr (1991), as 'the ability to support and maintain a balanced, integrated, adaptive community of organisms having a species composition, diversity and functional organization comparable to that of the natural habitat of the region'.

This chapter addresses the biotic effects of IBTs at differing scales, from biogeographical regions and subregions to basin scales, and finally to that of specific river reaches. Temporal considerations associated with the operation of IBTs are also considered. These include, for instance, the evolutionary implications resulting from the breakdown in catchment integrity, long-term changes in discharge patterns, disruption of seasonal cycles and local short-term effects and recovery.

Global Perspectives on River Conservation: Science, Policy and Practice.
Edited by P.J. Boon, B.R. Davies and G.E. Petts. © 2000 John Wiley & Sons Ltd.

Both donor and recipient river reaches often undergo substantial and relatively permanent geomorphological changes as a result of major changes to the natural hydrograph, in terms of the magnitude and timing of required flows. In the case of donor systems, these changes are associated with a loss of water, while in the receiving river, the effects are associated with increases in discharge, although these may be highly variable. Indeed, Davies *et al.* (1992) have emphasized that there are a wide variety of discharge types associated with IBTs: pulsed, continuous, aseasonal, and so on. Furthermore, these effects are influenced by the nature of the transfer route – tunnels/pipelines/canals (Figure 19.1(b)). Coupled to alterations in flow are a suite of water quality changes and other physical effects (Figure 19.1(a)). These include changes in major ion and nutrient concentrations, altered thermal characteristics and pH, and transport and retention characteristics, all of which are important determinants of riverine community structure (Snaddon *et al.*, 2000). A comprehensive list of the physical, chemical and biological effects of IBTs has been compiled by Snaddon and Davies (1997) and Snaddon *et al.* (2000).

A recent review of IBTs on a global basis (Snaddon *et al.*, 2000) has demonstrated a variety of sizes, operational criteria and effects of such schemes. It has become clear from this study that manipulation of river ecosystems through IBTs is occurring on a global scale. River conservation, and the conservation of riverine biodiversity, are therefore seriously compromised through the unresearched transfer and mixing of previously isolated biotas, communities and populations. The chapter will address this concern by using information from three regions: continental Australia, South Africa, and the continental United States of America.

Patterns of scale

The effects of IBTs are expressed across a wide range of spatial and temporal scales; rivers are dynamic systems that operate across all these scales. For example, the basin operates as a biogeographical entity with great importance for speciation processes. At the other end of the scale, individual populations have adapted to microhabitat processes and forces, with a gradation between the two extremes (e.g. Statzner *et al.*, 1988). A similar range of temporal scales exists, from the geological time required for the formation of large river basins (tens or even hundreds of millions of

years), to microscales in milliseconds, in terms of the bacterial processes involved in the decomposition of a leaf pack (e.g. Minshall, 1988). In almost all cases, IBTs transfer water between different river reaches, often between different basins, and sometimes between ecoregions.

SPATIAL PATTERNS

Regional and sub-regional scales

One of the primary motivations for this chapter is the fact that river basins are not only biogeographical entities by definition, but, very frequently, they exhibit similar attributes that can be used to delineate ecological regions. The definition of such regions has been attempted using different approaches. In South Africa, for example, the regional approach has been adapted and applied to rivers, where the delineation of biotic sub-regions (e.g. King, 1995), more recently named 'bioregions' (Eekhout, 1996), has been based on the distribution patterns of riparian vegetation, fish and riverine macroinvertebrates. The definition of ecoregions is based upon the philosophy that physical processes drive ecological processes at a landscape scale, which in turn are responsible for driving observed patterns in biological productivity, and associated patterns of biodiversity (Brown *et al.*, 1996). It can be assumed, therefore, that catchments can be grouped geographically into ecoregions, each characterized by distinct indigenous species assemblages (Thackway and Cresswell, 1995; Brown *et al.*, 1996).

Eighteen bioregions are recognized in South Africa (Figure 19.2; Eekhout, 1996). Current, and some proposed, IBT schemes in southern Africa have been superimposed on a map of these bioregions (Figure 19.2), in order to illustrate the frequency at which bioregional boundaries have been crossed. A further refinement of the spatial patterns of biodiversity in South Africa has focused on the definition of sub-regions within each bioregion (Eekhout, 1996). These encompass similar river zones (as defined by Rowntree and Wadeson, 1998) within the same bioregion: for example, fynbos lowland rivers. The 'Fynbos Biome' is the smallest and most diverse plant community in the world (in terms of species per unit area) and is confined to the Western Cape of South Africa. In most cases, as suggested by Davies *et al.* (1992), donor and recipient systems are in different river zones, within the same or in separate catchments. River zones have distinctive faunas, and so the transfer of water and organisms between sub-regions could have significant effects on

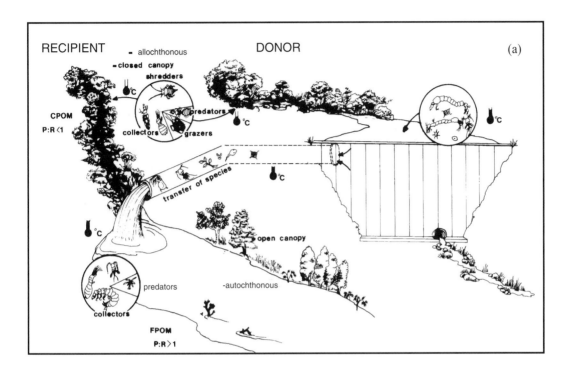

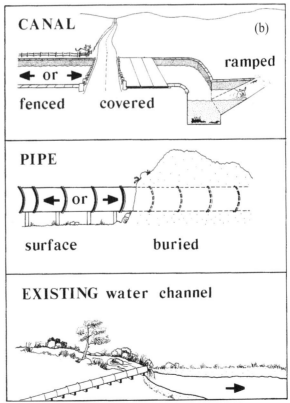

Figure 19.1 *(a) A summary of some of the biological effects of an IBT from an impoundment on the middle reaches of a donor river, to the upper reaches of a recipient. Thermometer symbols represent changes in temperature, which is cooler in the headwaters, and warmer in the impoundment and below the IBT outlet. P:R is the production to respiration ratio, and the invertebrates drawn in the circles represent the functional feeding groups that are expected to dominate in these systems. CPOM = coarse particulate organic matter; FPOM = fine particulate organic matter. (b) There are various water transfer routes utilized in IBTs. These include canals, which can be open or covered, with fences and ramps, to provide an exit for trapped animals; pipes which are buried or laid on the surface; and existing water channels which are often used to transport the diverted water to and from storage reservoirs. The direction of the water through canals and pipes can, in some cases, be reversed*

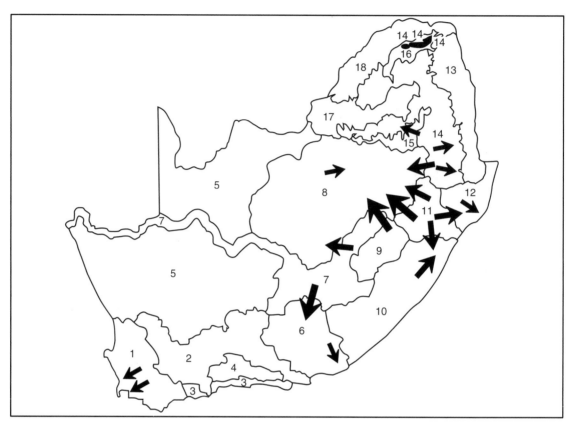

Figure 19.2   *Eighteen aquatic bioregions of South Africa with IBTs overlaid as arrows. (Modified from Eekhout, 1996.) 1, Fynbos; 2, Alkaline interior; 3, Southern coastal; 4, Southern inland; 5, Arid interior; 6, Drought corridor; 7, Orange; 8, Vaal; 9, Montane; 10, Eastern seaboard; 11, Tugela; 12, St Lucia complex; 13, Lowveld; 14, Northern uplands; 15, Highveld source; 16, Northern plateau; 17, Bushveld basin; 18, Limpopo*

the nature of recipient rivers and their biotic community structure.

In Australia, a more detailed study of bioregions has delineated 80 such regions. This was based on a suite of more than 100 variables, including climate, lithology/geology, soils, landform, vegetation, flora and fauna, land use and other special features (Thackway and Cresswell, 1995). The characterization procedure used a hierarchical information structure, integrating a complex array of data to produce a series of regional maps, the coarsest layer being at a scale of 1:15 000 000 (Figure 19.3). Smaller regions tend to be located in close proximity to the coast and have comparatively high annual rainfalls compared with the larger drier inland regions. Associated with these climatic variations are marked differences in the landform and soil characteristics of coastal and inland regions – a result of the geomorphological history of Australia. As a consequence, there are distinct differences in the

floral and faunal attributes and sustainable land-use activities of each region (Heathcote, 1988). Transfers of large volumes of water in Australia generally involve the export of water from basins in the wetter coastal fringes to drier inland basins. In so doing, biogeographical barriers are traversed. There are 16 IBTs in Australia that cross biogeographic regions (Table 19.1; Figure 19.3). The majority, 62.5%, either involve the transfer of water into or out of the Murray–Darling Basin, or within Tasmania.

The use of biotic distributions for describing ecological regions contrasts with the more common practice of generating spatial patterns based on physical and chemical variables. For instance, in South Africa, statistical analyses of large inorganic chemical databases have been used to divide the country's rivers into water quality management regions (Day and King, 1995; Day *et al.*, 1998). In the USA, major river drainages are used to define *hydrological*

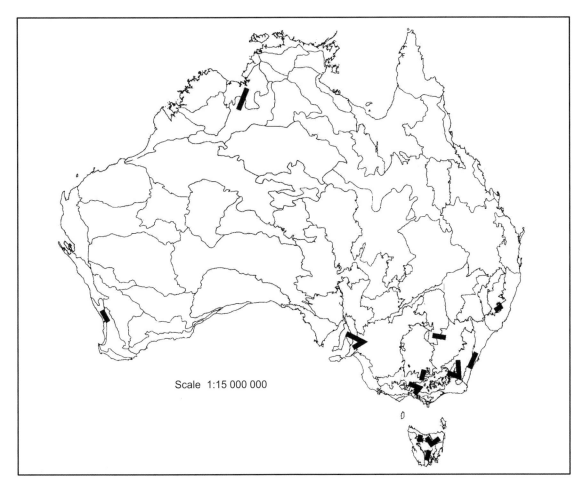

Figure 19.3: *Australian biogeographical regions with IBT overlays (bars)*

Table 19.1 *Australian interbasin transfers (IBTs) that traverse bioregional boundaries*

Donor region	Recipient region
Swan Coastal Plain	Jarrah Forest
Victoria Bonaparte	Ord–Victoria Plains
Murray–Darling Depression	Lofty Block
	Eyre and Yorke Blocks
South-eastern Highlands	Sydney Basin
	South-east Coastal Plain
	Victoria Volcanic Plain
Brigalow Belt South	Darling Riverine Plain
New England Tableland	NSW North Coast
Riverina	Victorian Midlands
Australian Alps	South-eastern Highlands
	NW south-western Slopes
Central Highlands	Tasmanian Midlands
	Woolnorth
Tasmanian Midlands	Ben Lomond
	Woolnorth

regions. The United States Water Resources Council – USWRC (1978) has established 21 water resources regions, 18 in the conterminous USA (Figure 19.4) and one each for Alaska, Hawaii and the Caribbean. Each region represents a natural basin, or hydrological area, that contains the drainage of a major river, or the combined drainages of two or more rivers. The majority of major IBT schemes have been constructed in the western USA, where annual renewable water supplies are lower, and consumption of available supply is higher than in the east (Foxworthy and Moody, 1986). Petsch (1985) has reported a total of 111 IBTs between water resource sub-regions in the western USA.

The delineation of bioregions or other ecologically meaningful regions provides an overview of the quality and ecological state of a resource (Omernik and Bailey, 1991), and the linking of such regions through IBTs

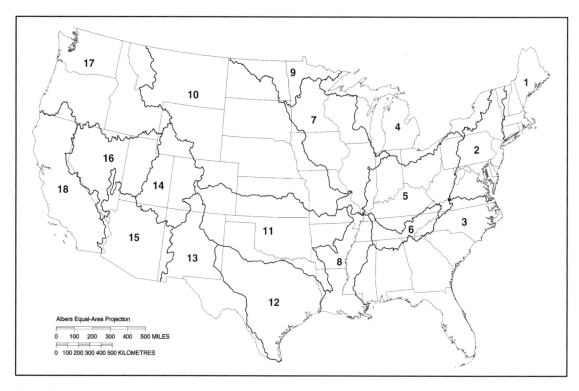

Figure 19.4   *Hydrological regions of the continental United States of America (see Table 19.2)*

should be considered with caution. While it is clear that regional boundaries are currently traversed by IBT routes, there is insufficient evidence that this has, or will have, effects on donor and recipient systems at such a scale.

Basin scale

Defined by topography, river basins or catchments represent natural 'islands' in which the biota is essentially restricted by its ability to disperse. As such, catchments provide natural isolating mechanisms through which local selective pressures, along with processes of genetic drift, can provide the driving force for micro-evolution. Genetic variability in a population is important for adaptive flexibility and future evolution. In this context, the transfer of water between catchments in which populations express high levels of differentiation runs the risk of reduction of the overall genetic complement (and hence, fitness) of a species, or its variability, which could subsequently impede its ability to evolve or to respond to environmental fluctuations. Such transfers and genetic mixing of separate populations can also result

in the loss of the genetic distinction of local endemic forms.

The breakdown of the natural biogeographical integrity of river basins caused by IBTs has been raised as an issue of importance (e.g. Davies *et al.*, 1992). Loss of species through altered flow regimes in the donor and recipient systems, as well as mixing of gene pools of donor populations with those of the recipient, must have important long-term implications for the evolution of species in both systems. In addition, the risks associated with the transfer of non-indigenous species (i.e. not occurring naturally in a country or catchment) from one system to another, as well as competitors, disease vectors, invasive species and parasites, are significant, although they are virtually unresearched worldwide. In this context, there appear to be no technologies available to prevent biological transfers of any sort. Indeed, where installation of what could be termed 'safety devices' (e.g. pressure valves, screens) has been part of a transfer scheme (e.g. the Orange River Project – ORP – in South Africa), they appear to have even failed to prevent the transfer of fish (Cambray and Jubb, 1977a,b).

Differing degrees of genetic differentiation have been observed in various aquatic invertebrate populations (McArthur *et al.*, 1992; Robinson *et al.*, 1992; Their, 1994; Hughes *et al.*, 1995; Thomas *et al.*, 1997). In many instances, high levels of genetic differentiation have been observed between populations both within, and between, geographically separate catchment areas. For instance, Hughes *et al.* (1996) demonstrated high levels of genetic differentiation between populations of the freshwater shrimp, *Caradina zebra* (Decapoda, Atyidae). Genetic differences were demonstrated between sub-catchments of the Tully River and between the Tully and the Herbert Rivers in Australia. At present, plans for a substantial transfer of water between the Tully and the Herbert systems threaten the genetic integrity of the highly differentiated populations of *C. zebra* (Hughes *et al.*, 1996). In northern California, fixed allelic differences between populations of the caddisfly, *Helicopsyche borealis*, from catchments separated by distances of between 7 and 11 km, actually suggest a number of separate reproductively isolated species, rather than genetic variants of a single species (Jackson and Resh, 1992).

Several other recent investigations give rise to concern for the increasing propensity towards the redistribution of water for human use. Populations of *Barbus callensis* collected in different streams, rivers and reservoirs in the north-western parts of Tunisia show ecological and genetic isolation from other Tunisian populations, with greater affinities to Algerian populations (Kraiem, 1993; Berrebi *et al.*, 1995). This population differentiation was only partly correlated with the ecological characteristics of the rivers they inhabit and there was no indication of any genetic cline, but rather a discontinuity between populations in the north-westernmost catchments and other Tunisian populations. This differentiation probably has a palaeo-ecological origin, related not only to adaptation to ecological conditions, but also to the difficulties in colonizing the catchments. Molecular investigations support suggestions that there is some sort of historical impediment to the migration of other fish species between Morocco and the western Algerian highlands, and those river basins to the east, in eastern Algeria and Tunisia (Machordom and Doadrio, 1993).

Along with these genetic changes at the population level there are a suite of other processes acting over a range of temporal scales. As well as the obvious evolutionary implications of introducing (and mixing) individuals of the same species from previously isolated populations, there are more immediate effects through the introduction of 'new' species, including pests and disease vectors. The Garrison Diversion Project (GDP) in the USA provides a compelling case of the potential problems caused by biota introduced through a water transfer. The project was designed to divert water from the Missouri River in North Dakota to irrigation areas, with return flows to Lake Winnipeg, Canada. At least nine fish species native to the Missouri River, but not to Manitoba, Canada, could be expected to invade the new environments made available by the GDP (Oetting, 1977). They are the shovelnose sturgeon *Scaphirhynchus platorynchus*, paddlefish *Polyodon spathula*, shortnose gar *Lepisosteus platostomus*, gizzard shad *Dorosoma cepedianum*, rainbow smelt *Osmerus mordax*, river carpsucker *Carpiodes carpio*, smallmouth buffalo *Ictiobus bubalus*, Utah chub *Gila atraria*, and the endangered pallid sturgeon *Scaphirhynchus albus*. Populations of three species in particular – the Utah chub, gizzard shad and rainbow smelt – could be expected to increase in lakes Manitoba and Winnipeg, to the detriment of walleye *Stizostedion vitreum*, sauger *S. canadense*, lake whitefish *Coregonus clupeaformis*, and other commercial and sports fish species. Walleye, sauger and whitefish populations may disappear when non-indigenous populations become established, possibly bringing about the total collapse of the commercial fisheries in the two lakes (Keys, 1984). Biologists proposed the addition of a fish screen and a sand filtration system to inhibit the transfer of biota. However, the Canadian Government has remained unconvinced that such additions will provide a sufficient guarantee against the introduction of non-native biota. The issue of introduced biota remains unresolved, and the future of the GDP remains unclear.

Although more-or-less anecdotal, there are cases in southern Africa where genetic transfers have occurred and 'new' species have been introduced. For example, four new fish species have been transferred to the Great Fish River in the Eastern Cape from the Orange River through the ORP. These include the small-mouth yellow fish *Barbus aeneus*, Orange River mud fish *Labeo capensis*, the catfish *Clarias gariepinus*, and the rock barbel *Gephyroglanis sclateri* (Cambray and Jubb, 1977a,b). These introductions occurred despite the installation of pepper-pot valves and screens at the draw-off from the donor reservoir. In addition, populations of three fish species common to both systems have also been transferred from the Orange to the Great Fish: the mud mullet *Labeo umbratus*, the mirror carp *Cyprinus carpio* and the chubby-head minnow *Barbus anoplus*.

Reach scales

It has been suggested that the links between riverine geomorphology and ecological functioning are at their strongest at the level of the river reach (Rowntree and Wadeson, 1998). Channel morphology provides a physical template which determines habitat conditions. Within the hierarchical classification of the physical features of a river, proposed by Rowntree and Wadeson (1998), the reach is defined as 'a length of channel within which the local constraints on channel form are uniform, which has a characteristic channel pattern . . . and degree of incision and within which a characteristic assemblage of channel morphologies occur'. It has been shown that IBTs have significant effects on discharge and, thus, on channel form (Snaddon *et al.*, 2000) and, as such, local biotic changes at the level of the reach are likely to be of similar significance.

Very few studies have collected data on biotic community changes due to water transfers. However, recent South African studies (Snaddon, 1998; Snaddon and Davies, 1998) have clearly demonstrated marked effects on riverine communities in recipient river reaches. A three-year study on the Riviersonderend–Berg–Eerste River Government Water Scheme (RBEGS), which transfers water from the Theewaterskloof impoundment on the Riviersonderend system into the Berg and Eerste Rivers in the Western Cape Province of South Africa, highlighted important invertebrate community changes. The transfer into the Berg River occurs only during summer months, in order to augment flow in the Berg to meet irrigation demands. The benthic invertebrate communities in the Berg River were sampled over a number of seasons between 1994 and 1997, in order to determine the responses of these communities to the transfer. Community changes below the IBT outlet appeared to be fairly similar to those occurring below any impoundment, and were probably associated with the alterations in flow and water chemistry.

Towards the end of the summer irrigation season, the invertebrate communities below the IBT were markedly different from those above it. The downstream communities had significantly greater numbers of individuals, and lower overall richness and numbers of taxa (Figure 19.5). One of the taxa that responded to the IBT was a collector–filterer hydropsychid trichopteran which showed a significant increase in abundance below the IBT. This is believed to be a response to the release into the Berg River of

large quantities of zooplankton (effectively a major increase in coarse particulate organic matter – CPOM) through the transfer tunnel from the donor impoundment: zooplankton can form a large proportion of the hydropsychid diet (e.g. Chutter 1968, 1971). It was further noted that the species of hydropsychid that was expected to occur in the study reaches of the Berg River, *Cheumatopsyche afra*, was absent from below the IBT in summer, while the species characteristic of river reaches further downstream, *Cheumatopsyche thomasseti*, was recorded in large numbers below the outlet. *C. thomasseti* prefers a neutral pH, and fairly high temperatures and total dissolved solids concentrations, and is tolerant of a wide range of conditions, while *C. afra* is a more sensitive species (Harrison and Elsworth, 1958; Scott, 1983, 1985).

Biotope scales

The term 'biotope' has become recognized as the biotic and abiotic environment of an aquatic community or assemblage of species (J.M. King, University of Cape Town, personal communication), while the 'hydraulic biotope' refers only to the flow- and substratum-related aspects of that environment (e.g. Rowntree and Wadeson, 1998).

Water transfer schemes can create new biotopes or habitats suitable for the establishment of new populations, or that allow the extension of the range of extant populations. In many cases, the creation of habitat will favour pest or pioneer species. For example, the creation of habitat suitable for the spread of disease-vector snails (schistosomiasis) was a major concern during the construction of the Jonglei Canal in the Sudan (Brown *et al.*, 1984). It was suspected that the snails would colonize the margins of the main canal and draw-off canals. In addition, the Jonglei Canal interrupts the east–west drainage across the floodplain between the IBT intake and release points; this will lead to an increase in permanent and semi-permanent wetlands on the eastern side of the canal. Aquatic snails would benefit from this available habitat.

Another example can be drawn from the ORP in South Africa, where an increase in pest simuliid (Diptera) species was recorded in the recipient, the Great Fish River. A shift in dominance of species of Simuliidae was noted in the recipient reaches, where the pest species, *Simulium chutteri*, increased to the detriment of the original benign populations of *Simulium adersi* and *Simulium nigritarse*, which were

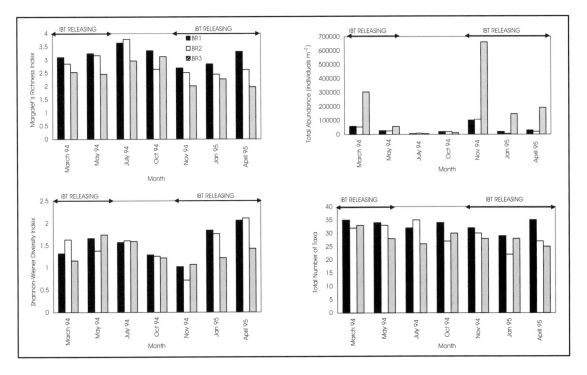

Figure 19.5 *Bar graphs showing changes in benthic macroinvertebrate structure in the Berg River, Western Cape, South Africa, as the result of an IBT. BR1 (solid bars) and BR2 (open bars) are sites above the IBT, while BR3 (cross-hatched) is below the outlet from the IBT (after Snaddon, 1998)*

previously co-dominant. *Simulium chutteri* now causes severe damage to livestock in the lower reaches of the river: the feeding activities of swarms of adult females cause stock damage and disturbance during spring (see also Scott *et al.*, 1972; O'Keeffe, 1982, 1985). All of these shifts in the invertebrate fauna can be directly attributed to the changes in flow regime caused by the transfer, particularly the loss of flow variability, and the shift from a seasonal to a perennial river. This has led to an increase in the total area of available erosional habitats, which are favoured, in particular, by simuliid larvae.

TEMPORAL PATTERNS

Many IBTs cease for some part of the year, particularly during non-irrigation seasons or when emergency water supplies are not required. Thus, if not impounded, the recipient rivers involved have the opportunity to recover to a certain extent before transfer recommences. Recovery will depend on several factors such as the robustness of the systems (Hildrew and Giller, 1994), the period during which the transfer

ceases, and the timing of releases (Snaddon, 1998). The robustness of the system will depend on its proximity to sources of species, and the ability of species to re-colonize the affected river reaches.

Evidence of temporal recovery downstream of an IBT can be drawn from the South African transfer scheme referred to above, the RBEGS. During winter, when water transfer ceased, the communities below the IBT recovered to resemble the unaffected communities recorded above the IBT (Figure 19.6). This temporal recovery reached its maximum towards the end of the winter, before the IBT tunnel discharged the first summer release (Snaddon, 1998). The IBT is situated on the upper reaches of the Berg River, and the river reaches above the tunnel are relatively pristine, with a fairly high species richness. Thus, it is likely that drift and other natural dispersal mechanisms lead to rapid recovery of the system. The implications of this for conservation are clear: if reaches of river above and below IBT release points are in relatively good condition, and if the release patterns of the IBT are seasonal, then good recovery is ensured during those seasons when releases do not occur (Snaddon, 1998).

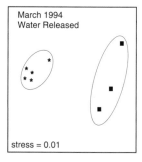
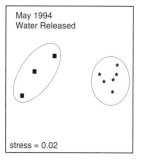
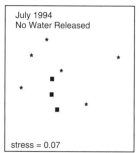
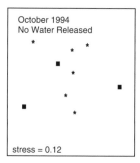
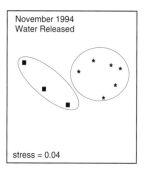
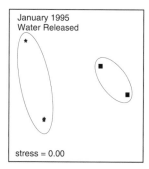

Figure 19.6   *Multi-dimensional scaling (MDS) ordination plots of invertebrate community data collected from the Berg River, South Africa, illustrating the recovery of the invertebrate community during winter months. During summer, above-IBT (stars), and below-IBT (squares) samples form distinct clusters, while in winter, the samples were not significantly different from each other. Ellipses delineate macroinvertebrate communities that had compatible Bray–Curtis similarity indices (after Snaddon, 1998)*

At the other end of the timescale – millennia and more, rather than seasonal – the implications of IBTs for biological communities in Australian rivers are immense, given the antiquity of the catchments and biogeographical regions of that continent. Australian ecosystems are different from those in the Northern Hemisphere in terms of their structure and functioning (e.g. Lake *et al.*, 1994), owing to their long evolution. Continental Australia dates back to pre-Tertiary times (65 Ma). In comparison, glaciation in parts of Britain, Europe and the USA during the Quaternary (2 Ma) provided a new starting point in landscape evolution: events of this magnitude occurred during the Permian in Australia. Hence, Australian landscapes have been affected by events over the last 250 Ma compared to 25 000 years in most parts of the Northern Hemisphere. Australia's river basins and biogeographical regions are also very old. For example, Ollier and Pain (1994) suggest that the present drainage divide between the inland Murray–Darling Basin and the coastal catchments resulted from rifting during the late Cretaceous–early Tertiary (80–60 Ma). This also

produced a new continental margin, which essentially set the gross character of the coastal catchments. Further upwarping of the eastern margins of Australia, to form a series of other divides, provided an impetus for a flow reversal in the Barwon–Darling system, from a north-east to a south-west-flowing drainage system (Thoms, 1997). Continued up-warping at the northern extremities of the Murray–Darling Basin resulted in a relatively new development of the drainage network of its western rivers, such as the Paroo, Warrego, Culgoa and Maranoa systems. Whilst the main tributaries of the Murray–Darling are approximately 65 Ma old, these systems are probably the youngest drainage networks in the Barwon–Darling catchment.

Implications for conservation and management of stream ecosystems

The International Union for the Conservation of Nature (IUCN) definition of conservation specifies three objectives (International Union for the

Table 19.2 *Volume of water transfers ($\times 10^6$ m^3) in 1982 within and among hydrological regions of the conterminous United States and numbers of introduced fish species. Water transfer data are taken from Petsch (1985) and Mooty and Jeffcoat (1986). Non-indigenous fish species data were provided by the Florida Caribbean Science Center, United States Geological Survey (written communication)*

Hydrological number and region	Volume of water transfers			Number of non-native fish species
	Within region	Into region	Out of region	
1 New England	9.5	0.7	0	79
2 Mid-Atlantic	5 104.9	45.7	1 174.1	109
3 South Atlantic–Gulf	164.5	31.8	35.9	197
4 Great Lakes	1.9	1 150.1	2 767.7	69
5 Ohio	35.7	1.2	1.8	102
6 Tennessee	0.5	1.8	5.3	70
7 Upper Mississippi	2.6	2 759.5	4.5	49
8 Lower Mississippi	0	84.3	30.2	34
9 Souris–Red–Rainy	0	0	0	18
10 Missouri Basin	2 597.4	542.4	15.4	96
11 Arkansas–White–Red	51.9	281.8	7.1	87
12 Texas–Gulf	34.7	0	51.2	83
13 Rio Grande	0	163.4	1.6	89
14 Upper Colorado	178.7	15.4	1 012.6	52
15 Lower Colorado	905.0	0	5 019.2	98
16 Great Basin	0.9	97.2	6.2	75
17 Pacific Northwest	20.0	28.8	1.7	83
18 California	5 479.1	4 699.3	28.8	111
Average	810.9	550.2	564.6	84

Conservation of Nature/United Nations Environmental Programme/World Wide Fund for Nature – IUCN/UNEP/WWF, 1991), to:

- maintain essential ecological processes and life support systems
- ensure that the utilization of species and ecosystems is sustainable
- preserve genetic diversity

Topographic isolation and environmental fluctuations together drive speciation events within river basins. The breakdown of natural biogeographical barriers through the transport, or transfer, of water between basins compromises the genetic integrity of aquatic populations. There has been limited attention given to such considerations in the development, construction and operation of IBTs worldwide, despite increases in the redistribution of water and the establishment of complex distribution and delivery networks. Environmental impact assessments typically focus on shorter temporal processes affecting community structure and the occurrence and recovery of specific taxa. Although such assessments are relatively easy and affordable, they do not provide adequate protection for river systems and processes. In

the Tunisian example discussed earlier, the transfer of north-western fish populations to other streams and reservoirs within the country would result in genetic mixing of once distinct populations. The irony for conservation is that while it is now evident that IBTs are responsible for genetic mixing of this nature, researchers are not even in a position to estimate what was there in the first place. Clearly a great deal of research is needed to clarify these concerns.

Degradation of river systems and alteration of riverine biota is a *global* issue. The introduction of non-indigenous species is one of the most pervasive and damaging human impacts on the rivers in the USA (Drake *et al.*, 1989) (Table 19.2). At least 4542 non-indigenous species, including several thousand plant and insect species and several hundred non-native vertebrate, mollusc, fish and plant-pathogen species, have established free-living populations (United States Congress Office of Technology Assessment – USCOTA, 1993). Approximately 15% of these non-indigenous species have caused severe harm, affecting agriculture, industry, human health and the natural environment. Hybridization and (particularly with fish) interbreeding with hatchery-reared counterparts of different genetic make-up is an especially critical issue. Hybridization was found to be a factor in 38% of the recorded extinctions of North American fish species (Allan and Flecker, 1993). Thus, the breakdown of natural barriers by water transfer is not the only means of introducing non-native fish species in the USA. In fact, 44% of fish species introductions have been carried out intentionally by stocking, as a matter of agency policy, while less than 2% of fish species introductions result from water transfers. This, however, is not to gainsay the importance of IBTs, either now or in the future, simply because of the growing human pressures for water supply, as well as the large scale of many extant and proposed IBTs. Furthermore, these data are for fish, and the lack of similar statistics for the introductions through IBTs of other non-indigenous aquatic organisms, including disease vectors (waterborne and water-based viruses, bacteria, helminths and crustaceans, etc.), aquatic plants, phytoplankton, periphyton, zooplankton and aquatic invertebrates, is of grave concern. The extent of such introductions has not been quantified, and this can only be attempted for proposed IBTs, and not for extant schemes where the damage has already been done. There is, however, an urgent need to assess extant schemes, as it is possible that potential transfers have not yet taken place and might still be avoided.

It is clear from the evidence presented here that IBTs have had, and are having, profound effects on medium- to long-term speciation processes and, hence, upon the track of evolution itself. Whilst considerable controversy at present surrounds the release of genetically modified organisms into the environment (*vide* the Greenpeace action at Downing Street, London, in February, 1999, when 4 t of genetically engineered soya bean products were dumped in order to bring public and political attention to the potential problems of genetically modified organisms), water redistribution systems in the form of IBTs have effectively been 'engineering' new or hybrid populations and communities in recipient rivers. In addition, they have almost certainly reduced genetic diversity. Accordingly, the implications for species and system conservation are significant and disturbing, and we urge a rethink of present design, operation and management of IBTs worldwide.

For instance, the vast majority of extant and proposed IBTs are the direct result of the response of water managers and politicians to perceived problems of water supply for rapidly expanding urban areas. More often than not, curbing water demand is not part of the planning process; nor, for that matter, is the problem of losses of water in urban areas through leakage (e.g. Davies and Day, 1998). It should be a general principle that occupants of catchments with perceived water deficits should, through appropriate legislation, show that all possible water-demand management measures have been implemented *before* any IBT from a donor catchment is even considered, let alone approved and constructed. In terms of the main thrust of this chapter – biotic integrity of river basins – it is imperative for river conservation that the promoters of future IBTs secure adequate research for the development of inventories of species and their uniqueness to basins, before transfers are allowed to take place. Furthermore, research has shown (e.g. Snaddon *et al.*, 2000) that the operational characteristics of many IBTs leave much to be desired and that impacts may be lessened, and the recovery of receiving reaches may be affected, by more carefully planned and ecologically sensitive release patterns.

References

Allan, J.D. and Flecker, A.S. 1993. Biodiversity conservation in running waters. *Bioscience* 43:32–43.

Basson, M.S. 1997. *Overview of Water Resources Availability and Utilisation in South Africa*. Booklet prepared for the Department of Water Affairs and Forestry, Pretoria.

Berrebi, P., Kraiem, M.M., Doadrio, I., El-Gharbi, S. and Cattaneo-Berrebi, G. 1995. Ecological and genetic differentiation of *Barbus callensis* populations in Tunisia. *Journal of Fish Biology* 47:850–864.

Brown, C.A., Eekhout, S. and King, J.M. 1996. *National Biomonitoring Programme for Riverine Ecosystems: Proceedings of Spatial Framework Workshop*. National Biomonitoring Programme Report Series, 2, Institute for Water Quality Studies, Department of Water Affairs and Forestry, Pretoria.

Brown, D.S., Fison, T., Southgate, V.R. and Wright, C.A. 1984. Aquatic snails of the Jonglei Region, southern Sudan, and transmission of trematode parasites. *Hydrobiologia* 110:247–271.

Cambray, J.A. and Jubb, R.A. 1977a. Dispersal of fishes *via* the Orange-Fish Tunnel, South Africa. *Journal of the Limnological Society of Southern Africa* 3:33–35.

Cambray, J.A. and Jubb, R.A. 1977b. The Orange-Fish Tunnel. *Piscator* 99:4–6.

Chutter, F.M. 1968. On the ecology of fauna of stones-in-the-current in a South African river supporting a very large *Simulium* (Diptera) population. *Journal of Applied Ecology* 5:531–561.

Chutter, F.M. 1971. Hydrobiological studies in the catchment of the Vaal Dam, South Africa. Part 2. The effects of stream contamination of the fauna of stones-in-current and marginal vegetation biotopes. *Internationale Revue der gesamten Hydrobiologie* 56:227–240.

Craig, J.F. and Kemper, J.V. (eds) 1987. *Regulated Streams: Advances in Ecology*. Plenum Press, New York.

Davies, B.R. and Day, J.A. 1998. *Vanishing Waters*. UCT Press and Juta Press, Cape Town.

Davies, B.R., Thoms, M.C. and Meador, M. 1992. The ecological impacts of inter-basin water transfers, and their threats to river basin integrity and conservation. *Aquatic Conservation: Marine and Freshwater Ecosystems* 2:325–349.

Day, J.C. 1985. *Canadian Interbasin Diversions*. Inquiry on Federal Water Policy, Research Paper 6, Natural Resources Management Program, Simon Frazer University, Burnaby, British Columbia.

Day, J.A. and King, J.M. 1995. Geographical patterns, and their origins, in the dominance of major ions in South African rivers. *South African Journal of Science* 91:299–306.

Day, J.A., Dallas, H.F. and Wackernagel, A. 1998. Delineation of management regions for South African rivers based on water chemistry. *Aquatic Ecosystem Health and Management* 1:183–197.

Drake, J.A., Mooney, H.A., Di Castri, F., Groves, R.H., Kruger, K.J., Rejmanek, M. and Williamson, M. 1989. *Biological Invasions: A Global Perspective*. John Wiley, New York.

Eekhout, S. 1996. Biogeographic regions for South African rivers. In: Brown, C.A., Eekhout, S. and King, J.M. (eds) *National Biomonitoring Programme for Riverine Ecosystems: Proceedings of a Spatial Framework Workshop*. National Biomonitoring Programme Report Series No. 2, Institute for Water Quality Studies, Department of Water Affairs and Forestry, Pretoria, 19–26.

Foxworthy, B.L. and Moody, D.W. 1986. National perspective on surface-water resources. In: Moody, D.W., Chase, E.B. and Aronson, D.A. (eds) *US Geological Survey, National Water Summary 1985, Hydrologic Events and Surface-Water Resources*, Water-Supply Paper 2300, 51–68.

Harrison, A.D. and Elsworth, J.F. 1958. Hydrobiological studies on the Great Berg River, Western Cape Province, Part 1. General description, chemical studies and main features of the flora and fauna. *Transactions of the Royal Society of South Africa* 35:125–226.

Heathcote, R.L. 1988. *The Australian Experience: Essays in Australian Land Settlement and Resource Management*. Longman Cheshire, Melbourne.

Hildrew, A.G. and Giller, P.S. 1994. Patchiness, species interactions and disturbance in the stream benthos. In: Giller, P.S., Hildrew, A.G. and Raffaelli, D.G. (eds) *Aquatic Ecology. Scale, Pattern and Process*. The 34th Symposium of the British Ecological Society with the American Society of Limnology and Oceanography, University College, Cork. Blackwell Science, Oxford, 21–62.

Hughes, J.M., Bunn, S.E., Kingston, D.M. and Hurwood, D.A. 1995. Genetic differentiation and dispersal among populations of *Paratya australiensis* (Atyidae) in rainforest streams in southeast Queensland, Australia. *Journal of the North American Benthological Society* 14:158–173.

Hughes, J.M., Bunn, S.E., Hurwood, D.A., Choy, S. and Pearson, R.G. 1996. Genetic differentiation among populations of *Caridina zebra* (Decapoda: Atyidae) in tropical rainforest streams, northern Australia. *Freshwater Biology* 36:289–296.

International Union for Conservation of Nature and Natural Resources/United Nations Environment Programme/World-Wide Fund for Nature (IUCN/UNEP/WWF) 1991. *Caring for the Earth: A Strategy for Sustainable Living*. The World Conservation Union, Gland.

Jackson, J.K. and Resh, V.H. 1992. Variation in genetic structure among populations of the caddisfly *Helicopsyche borealis* from three streams in northern California, U.S.A. *Freshwater Biology* 27:29–42.

Karr, J.R. 1991. Biological integrity: a long-neglected aspect of water resource management. *Ecological Applications* 1:66–84.

Keys, D.L. 1984. National environmental policy, foreign policy, and the Garrison Diversion Unit. *The Environmental Professional* 6:223–234.

King, J.M. 1995. *Report Back on Trip to Australia*. Report to Council for Scientific and Industrial Research and the Institute for Water Quality Studies, Pretoria.

Kraiem, M.M. 1993. Variability analysis in barbel populations (*Barbus callensis*) of Tunisia. In: Poncin, P., Berrebi, P., Philippart, J.C. and Ruwet, J.C. (eds) *Biology of the European, African and Asiatic Barbus. Proceedings of the International Round Table Barbus II.* 6–8 July 1993, Volume 13, 159–162.

Lake, P.S., Barmuta, L.S., Boulton, A.J., Campbell, I.C. and St Clair, R.M. 1994. Australian streams and Northern Hemisphere stream ecology: comparisons and problems. *Proceedings of Ecological Societies of Australia* 14:61–81.

Lillehammer, A. and Saltveit, S.J. (eds) 1984. *Regulated Rivers*. Universitetsforlaget AS, Oslo.

Machordom, A. and Doadrio, I. 1993. Phylogeny and taxonomy of North African barbels. In: Poncin, P., Berrebi, P., Philippart, J.C. and Ruwet, J.C. (eds) *Biology of the European, African and Asiatic Barbus. Proceedings of the International Round Table Barbus II.* 6–8 July 1993, Volume 13, 218.

McArthur, J.V., Leff, L.G. and Smith, M.H. 1992. Genetic diviersity of bacteria along a stream continuum. *Journal of the North American Benthological Society* 11:269–277.

McCully, P. 1996. *Silenced Rivers: The Ecology and Politics of Large Dams*. ZED Books, London.

Minshall, G.W. 1988. Stream ecosystem theory: A global perspective. *Journal of the North American Benthological Society* 7:263–288.

Mooty, W.S. and Jeffcoat, H.H. 1986. *Inventory of Interbasin Transfers of Water in the Eastern United States*. US Geological Survey Open-File Report, 86-148, Tuscaloosa.

Oetting, R.E. 1977. How the Garrison Diversion Project

affects Canada. *Canadian Geographic Journal*, October/ November: 38–45.

O'Keeffe, J.H. 1982. The regulation of simuliid populations in the Great Fish River. *Institute of Freshwater Studies, Annual Report*, 15, Rhodes University, Grahamstown, 1–9.

O'Keeffe, J.H. 1985. The blackfly problem in the Great Fish River. *The Naturalist* 29:3–10.

Ollier, C.D. and Pain, C.F. 1994. Landscape evolution and tectonics in south-eastern Australia. *AGSO Journal of Australian Geology and Geophysics* 15:335–345.

Omernik, J.M. and Bailey, R.G. 1991. Distinguishing between watersheds and ecoregions. *Journal of the American Water Resources Association* 33:935–949.

Petsch, H.E. 1985. *Inventory of Interbasin Transfers of Water in the Western Conterminous United States*. US Geological Survey Open-File Report, 85-166, Lakewood.

Petts, G.E. 1984. *Impounded Rivers*. John Wiley, Chichester.

Petts, G.E., Armitage, P. and Gustard, A. (eds) 1989. Fourth International Symposium on Regulated Streams. *Regulated Rivers: Research & Management* 3.

Reisner, M. 1986. *Cadillac Desert*. Viking Penguin, New York.

Robinson, C.T., Reed, L.M. and Minshall, G.W. 1992. Influence of flow regime on life history, production and genetic structure of *Baetis tricaudatus* (Ephemeroptera) and *Hesperoperla pacifica* (Plecoptera). *Journal of the North American Benthological Society* 11:278–289.

Rowntree, K.M. and Wadeson, R.A. 1998. *A Hierarchical Geomorphological Model for the Classification of Selected South African Rivers*. Report to the Water Research Commission, Pretoria.

Scott, K.M.F. 1983. Hydropsychidae (Trichoptera) of Southern Africa with keys. *Annals of the Cape Provincial Museums (Natural History)* 14:299–422.

Scott, K.M.F. 1985. Order Trichoptera (caddis flies). In: Scholtz, C.H. and Holm, E. (eds) *Insects of Southern Africa*. Butterworths, Durban, 327–340.

Scott, K.M.F, Allanson, B.R. and Chutter, F.M. 1972. *Orange River Project*. Working Group for Orange River Project, Hydrobiology of the Fish and Sundays Rivers, Research Report of the CSIR, National Institute for Water Research, Pretoria, 306.

Snaddon, C.D. 1998. Some of the ecological effects of a small inter-basin water transfer on the receiving reaches of the Upper Berg River, Western Cape. Unpublished MSc Thesis, University of Cape Town, South Africa.

Snaddon, C.D. and Davies, B.R. 1997. *An Analysis of the Effects of Inter-Basin Water Transfers in Relation to the New Water Law*. Report prepared for the Department of Water Affairs and Forestry, January 1997.

Snaddon, C.D. and Davies, B.R. 1998. A preliminary assessment of the effects of a small inter-basin water transfer, the Riviersonderend–Berg River Transfer Scheme, Western Cape, South Africa, on discharge and invertebrate community structure. *Regulated Rivers: Research & Management* 14:421–441.

Snaddon, C.D., Davies, B.R. and Wishart, M. 2000. *A Global Overview of Inter-Basin Water Transfer Schemes, with an Appraisal of their Ecological, Socio-Economic and Socio-Political Implications, and Recommendations for their Management* (in press). Report to the Water Research Commission, Pretoria.

Statzner, B., Gore, J.A. and Resh, V.H. 1988. Hydraulic stream ecology: observed patterns and potential application. *Journal of the North American Benthological Society* 7:307–360.

Thackway, R. and Cresswell, I.D. 1995. *An Interim Biogeographic Regionalisation for Australia: A Framework for Establishing the Natural System of Reserves*. Australian Nature Conservation Agency, Canberra.

Their, E. 1994. Allozyme variation among natural populations of *Holopedium gibberum* (Crustacea; Cladocera). *Freshwater Biology* 31:87–96.

Thomas, P.E., Blinn, D.W. and Keim, P. 1997. Genetic and behavioural divergence among desert spring amphipod populations. *Freshwater Biology* 38:137–143.

Thoms, M.C. 1997. The physical character of the Barwon–Darling river system. In: Thoms, M.C., Gordon, A. and Tatnell, W. (eds) *Researching the Barwon–Darling*. Co-operative Centre for Freshwater Ecology, Canberra, 34–43.

United States Congress Office of Technology Assessment (USCOTA) 1993. *Harmful Non-Indigenous Species in the United States: Summary*. US Government Printing Office, Report Number OTA-F-566, Washington, DC.

United States Water Resources Council (USWRC) 1978. *The Nation's Water Resources 1975–2000, Volume 2: Water Quantity, Quality, and Related Land Considerations*. Washington, DC.

Van Niekerk, P.H., Van Rooyen, J.A., Stoffberg, F.A. and Basson, M.S. 1996. *Water for South Africa into the 21st Century*. Paper presented at the IAHR Biennial Congress, Sun City, South Africa, 5–7 August, 1996.

Wackernagel, M. and Rees, W. 1996. *Our Ecological Footprint: Reducing Human Impact on the Earth*. New Society Publishers, Gabriola Island, British Columbia.

Ward, J.V. and Stanford, J.A. (eds) 1979. *The Ecology of Regulated Streams*. Plenum Press, New York.

Ward, J.V. and Stanford, J.A. 1983a. The serial discontinuity concept of lotic ecosystems. In: Fontaine, T.D. and Bartell, S.M. (eds) *Dynamics of Lotic Ecosystems*. Ann Arbor Science Publishers, Michigan, 29–42.

Ward, J.V. and Stanford, J.A. 1983b. The intermediate disturbance hypothesis: An explanation for biotic diversity patterns in lotic ecosystems. In: Fontaine, T.D. and Bartell, S.M. (eds) *Dynamics of Lotic Ecosystems*. Ann Arbor Science Publishers, Michigan, 347–356.

D   Conservation in Practice

20

The role of classification in the conservation of rivers

J.H. O'Keeffe and M. Uys

Introduction

'What's the use of their having names', the gnat said, 'if they won't answer to them?'

'No use to *them*', said Alice, 'but it's useful to the people that name them, I suppose.'

This quotation from Lewis Carroll illustrates a few of the fundamental limitations of classification systems, and in particular emphasizes the artificial nature of classification – a process which is intended to help us to simplify, organize and understand complex and variable groups so that we can work with them more easily, but is in no way an inherent property of the things being classified. We would like to start by offering the following definition of classification, so that it is clear from the start what we mean when we refer to classification:

Classification is an artificial process in which people attempt to divide a group of objects/systems/ideas into discrete groups according to a set of criteria, decided on by the classifier.

We do this in order to make the group conceptually manageable – e.g. we split the group of living things into different Phyla according to our perceptions of the origins of their morphology.

Much debate has been engendered about the usefulness of classification as a tool. For example, it has been pointed out that *every* river is different in many ways from every other, and it is therefore fruitless to attempt to classify them, since the classifier will inevitably end up with as many groupings as there are rivers.

This, however, is a myopic point of view. The point of a classification system may not be to identify all the distinctive features of a group – it is far more likely to be aimed at clarifying the similarities between subgroups, and the use of hierarchical sets of criteria provides a way of defining similarities (or differences) at different levels of resolution for different purposes.

In view of the emphasis on variability and heterogeneity in ecological river research at present (Palmer and Poff, 1997), it may seem anachronistic to be concentrating on classification, which is essentially

Global Perspectives on River Conservation: Science, Policy and Practice.
Edited by P.J. Boon, B.R. Davies and G.E. Petts. © 2000 John Wiley & Sons Ltd.

an attempt to group like with like. However, heterogeneity and classification are not mutually exclusive, and classification systems are often used to demonstrate the different types of groups or their variability. Poff and Ward (1989) used a regional analysis of stream flow patterns to suggest a classification of flow types, such as 'intermittent flashy' and 'mesic groundwater'. Such a classification is used to describe discrete groupings of similar flow regimes for what is in fact a continuum, and a multivariate continuum at that.

TWO KINDS OF CLASSIFICATION: TYPING AND EVALUATION

Typing is the term which we shall apply to the process of grouping like with like – to identify more or less homogeneous groups according to one or more criteria. The homogeneity is based simply on the similarity of the criterion or criteria chosen, and is objective, with no implication of assessment or assigning values. Strahler's (1957) stream order classification is an example of such a system using only a single criterion, while the river zonation scheme of Illies (1961) is an example of a multi-criteria system using physical variables such as temperature, water velocity, substrate and altitude to define the zones. Statistical techniques such as cluster analysis, multiple discriminant scaling and principal component analysis are all objective attempts to group units on the basis of shared or unshared characteristics.

Evaluation is the term which we shall apply to the process of assigning a value (usually some composite index) which encapsulates the status of the subject in respect of the attribute of interest. The exercise of evaluation usually incorporates typing, but implies additional value judgements, which must usually be subjective. An example of such a system is SERCON (Boon *et al.*, 1997), a broad-based technique for evaluating rivers with respect to criteria such as naturalness, representativeness, rarity and species richness, as well as more objective physical descriptors. Scores and weightings are used to provide index values (from 0 to 100) and quality bands (from A to E) which indicate the status and importance of a river for conservation.

The proposed classification system for water resources for the new South African Water Act (Republic of South Africa, 1998; Palmer *et al.*, this volume), is another example of an evaluation system, but in this case based on less detailed information. The present condition of a water resource may be assigned to

Table 20.1 *Present state classes for rivers, based on ecological integrity and ecosystem health, defined for rivers under the new South African Water Law (Republic of South Africa, 1998)*

Class	Description
A	• Unmodified, natural • The reserve (that quantity and quality of water calculated as required to maintain aquatic ecosystem functioning) has not been decreased • The resource capability has not been exploited
B	• Largely natural with few modifications • The reserve has been decreased to a small extent • A small change in natural habitats and biota may have taken place but the ecosystem functions are essentially unchanged
C	• Moderately modified • The reserve has been decreased to a moderate extent • A change of natural habitat and biota have occurred, but the basic ecosystem functions are still predominantly unchanged
D	• Largely modified • The reserve has been decreased to a large extent • Large changes in natural habitat, biota and basic ecosystem functions have occurred
E	• Seriously modified • The reserve has been seriously decreased and depletion regularly exceeds the amount of water required to maintain ecosystem functioning • The loss of natural habitat, biota and basic ecosystem functions is extensive
F	• Critically modified • The reserve has been critically decreased and there is never enough water to maintain ecosystem functioning • Modifications have reached a critical level and the resource has been modified completely with an almost total loss of natural habitat and biota. In the worst instances the basic ecosystem functions have been destroyed and the changes are irreversible

a class from A to F, in which A is virtually natural and unmodified, and F is critically modified (Table 20.1).

The resource is then assigned an 'Ecological Management Class' which encapsulates the objectives for future management as classes A, B, C or D (classes E and F are considered to be beyond the long-term sustainability of a natural resource, and the objective must therefore be to improve its condition).

The roles of typing and evaluation

The basic purpose of typing is to group like with like. This is the traditional aim of classification. The purpose of typing systems is primarily for the users to place their stream of interest into a context from which other attributes of the stream can generally be inferred. In the case of widely accepted methods such as those already referred to, the context is usually clear and unequivocal, and it is the extent to which inferences can be drawn that is the major cause of controversy. For example, although there is general agreement as to what constitutes a third-order stream, there appears to be little agreement about the extent of the ecological similarities between different third-order streams.

The well-known hierarchical framework for stream habitat classification of Frissell *et al.* (1986) is another example of a typing exercise. The authors describe the system as 'a general approach for classifying stream systems in the context of the watersheds that surround them . . . entailing an organized view of spatial and temporal variation among and within stream systems'. They list the uses of the classification as:

• establishment of monitoring stations
• determination of local impacts of land-use practices
• generalization from site-specific data
• assessment of basin-wide cumulative impacts of human activities on streams and their biota

These are the normal objectives of a general typing system. The central theme of this kind of classification is the organization of a large body of information into discrete packages which can be more easily assimilated and understood. The system is used to extrapolate knowledge gained in specific studies to similar rivers, without the need to repeat the studies in each case.

The problem inherent in typing is that the boundaries between groups are seldom discrete, and must therefore be to some extent arbitrary. The degree of similarity between rivers also varies, so that the user of the typing system must judge the extent to which conclusions based on one river can be extrapolated to another. In addition, where the typing domain is as broad as 'all streams', Frissell *et al.* (1986) point out two further problems. First, different variables may be important in different locations, and different processes control the form and development of landscapes, watersheds and streams. Second, the controlling or constraining variables change with the time-frame in which the system is viewed (i.e. over geological time the slope of a stream is a changing dependent variable, but over a few years or decades it is relatively constant). To overcome these problems, Frissell *et al.* (1986) adopted a spatial and temporal hierarchical approach.

The typical role of an evaluation system is to facilitate the ranking or assessment of the present status of a river relative to that of other similar rivers, or relative to the state of least impact. This allows the user to distinguish degraded sites from healthy ones, identify the cause of the impact, then monitor the response of the system to remedial action. Evaluation, which may be preceded by a typing exercise, attempts to assign a value – usually in the form of an index – which encapsulates the status of the subject (e.g. a river reach) in terms of the aspect of interest. Whereas a typing system can be general, within the constraints of the criteria used to define similarity, an evaluation system has to be specific to the parameters which are to be assessed, and tailor-made methods are required for each field of application. For example, a system designed to assess the biotic diversity of a river will not equally well assess its water quality status, unless the two are closely related.

O'Keeffe (1997) identified a number of conservation features for which evaluation methods have been developed (Table 20.2). There is clearly considerable overlap between some of these terms, and there are a host of others covering, for example, landscape evaluation, river corridor assessment, and recreational potential. All such methods have in common the aim to quantify how good or bad the state of a river is in relation to some perceived desirable state. Few of these methods have escaped criticism, mainly because they rely on value systems that are ultimately subjective, and different 'experts' often hold different views on the priorities.

Evaluation is an essential prerequisite to the drawing up of conservation objectives and management plans, and for establishing the priority uses of any river. In the USA, the Water Pollution Control Act Amendments, Public Law 92-500, charges the authorities with 'evaluating, restoring and maintaining the chemical, physical and biological integrity of the nation's waters' (Polls, 1994). In South Africa, the new Water Act requires the Minister of Water Affairs to 'prescribe a system for classifying water resources' and the Director-General of the Department of Water Affairs and Forestry to 'use the classification system . . . to determine the class and resource quality objectives of all or part of water resources considered to be significant' (Republic of South Africa, 1998).

Entrenchment of these terms and concepts in national legislation is a measure of their importance in protecting natural resources, but there is a danger that the subjectivity and imprecision inherent in

Table 20.2   *Terms used to describe the environmental condition and value of rivers (from O'Keeffe, 1997)*

Conservation status	A measure of the extent to which a river has been modified from its natural state.
Conservation importance	A measure of the value of a river for conservation, including natural, socio-economic and cultural aspects. Natural aspects include the rarity or uniqueness of the river, its biodiversity, levels of endemism, geographical position and size, special features, and fragility. Socio-economic aspects include the number of people directly dependent for water supplies, sanitation, subsistence, or recreation, and the economic value of agriculture and industry dependent on the river. Cultural aspects include the historical and archaeological value of the river, its importance in rituals and rites of passage, the use of riparian plants for building or traditional medicines, and the intrinsic and aesthetic value of the river for those who live within or visit the catchment.
Conservation values	Generally include both intrinsic values and non-material utility values, including scientific (ecological and cultural), aesthetic, social and historic values.
Conservation significance	Can only be established in the context of a regional, state, or national evaluation of the resource. For example, state significance: the area contains the only, or otherwise significant, population(s) of a species in the state.
Conservation potential	The ecological potential of a stream and its sensitivity to natural and human disturbance. Kleynhans (1994) points out that ecological potential is undefined, but can be interpreted as referring to the current potential of the riverine habitat to support aquatic biological communities. Implicit in this is the potential for the improvement of existing habitat conditions in order that they can support a more diverse biota.
Natural value	The survival of all indigenous species of flora and fauna, both rare and commonplace, in their natural communities and habitats, and the preservation of representative samples of all classes of natural ecosystems and landscapes which give a region its own recognizable character.
Ecological health	The condition when a system's inherent potential is realized, its condition is stable, its capacity for self-repair, when perturbed, is preserved, and minimal external support for management is needed.
Ecological integrity	A living system exhibits integrity if, when subjected to disturbance, it sustains an organizing, self-correcting capability to recover toward an end-state that is 'normal' or 'good' for that system. End states other than pristine or naturally whole may be taken to be 'normal and good'.
Biological integrity	The ability to support and maintain a balanced, integrated, adaptive community of organisms having a species composition, diversity and functional organization comparable to that of natural habitats of the region.
Habitat integrity	The maintenance of a balanced, integrated composition of physical, chemical and habitat characteristics on a temporal and spatial scale that are comparable to the characteristics of natural habitats of the region.

evaluation techniques will hamper the implementation of such measures and leave them open to challenge in constitutional courts. As Polls (1994) points out, it is difficult for a public authority to know whether or not it is maintaining the biological integrity of water resources when the definition contains terms that are very difficult to quantify such as 'balanced, integrated, adaptive community' and 'functional organization' (see Table 20.2). Even 'natural' conditions are extremely difficult to pin down in systems as variable and stochastic as rivers: Should the 'natural' baseline be taken as that before any human interference, or before industrial human interference? How many years does a baseline dataset need to cover before the range of natural conditions is adequately described?

The terms and techniques described in Table 20.2 cover a range of subjectivity:

- conservation status is generally a measurable concept if the definition of the baseline natural state can be agreed;

- measures of integrity are similarly based on the difference between natural and present conditions;
- conservation potential, importance and values, and significance are all subjective assessments to a greater or lesser extent, but methods have been developed to ensure consistency of evaluation as far as possible;
- river health requires the greatest degree of value judgement, and can only be defined in terms of some general consensus.

Classification exercises for specific uses

Traditional river classification systems are well illustrated by the general river typing exercise of Frissell *et al.* (1986), that of wetlands by Cowardin *et al.* (1979), or the ecoregion approaches of Omernik (1987) and Lotspeich (1980), and of Harrison (1959) for South Africa. These systems are designed for multiple uses and are not generally considered for specific applications.

Classifications based on evaluation systems have traditionally had a narrower focus. Models for assessing conservation status, natural value and habitat integrity have been developed for many regions. These include that of Collier (1993) for New Zealand; O'Keeffe *et al.* (1987) and Kleynhans (1996) for South Africa; MacMillan and Kunert (1990) for Victoria (Australia); and SERCON (System for Evaluating Rivers for Conservation) for the UK (Boon *et al.*, 1997). These models all comprise a scoring system for different components (usually weighted) resulting in an overall score (or scores) reflecting the condition of the river under investigation.

A wide range of biomonitoring systems have been developed to assess water quality. These include methods devised by the US Environmental Protection Agency (Plafkin *et al.*, 1989; Karr *et al.*, this volume), and RIVPACS (River InVertebrate Prediction And Classification System) developed by the Institute of Freshwater Ecology in the UK (Wright *et al.*, 1984, 1997). The latter system uses a limited suite of environmental variables to predict the fauna at any river site in the UK in the absence of stress.

Indices that compare the observed fauna with the predicted fauna are then used for site appraisal (Wright *et al.*, 1997). At an International RIVPACS Conference held in Oxford in September 1997 (Wright *et al.*, 2000) there seemed to be a general consensus that these multivariate methods have a greater predictive capability than the more traditional multimetric approaches.

The development of classification systems has been driven more by a desire to organize information than to solve particular problems. It is probably more useful to allow the problem to drive the development of the classification, and this has been the trend more recently. Figure 20.1 represents a hierarchy of many of the classification methods that have been developed for different uses. The advantage of 'customized' classifications catering for particular purposes is to some extent offset by the additional effort of developing and applying each system. However, the enormous number and diversity of classifications currently available for rivers indicates that there has never been a shortage of people willing to engage in this task.

Different approaches to classification

Equipped with the reason for the classification (or the question to be answered), and knowing whether the

system involves typing or evaluation (or both), the approach to formulation needs to be considered. Brierley (1994) comments that the approaches to river classification should to be open-ended so that the extraordinary can be accommodated along with the ordinary. He comments that there is 'no magic number of models that defines the spectrum of morphological complexity demonstrated by rivers'. Norris (1994) suggests that a general guide for determining the effectiveness of a classification is to consider its ability to partition variance – that is, to account satisfactorily for the natural geographical variation in the ecological features of undisturbed aquatic ecosystems.

The approach adopted in this chapter will include a consideration of:

- the philosophy behind the classification
- the foundation of the classification
- the structure and the scale/extent of the classification
- the type of data to be used to develop the system ('top-down', 'bottom-up', or a combination)
- the status of rivers to be included in developing the classification (natural or present conditions)
- the river attributes to be included in the classification (structural, functional and/or compositional)
- the format of the classification (static, dynamic and/or interactive)

THE PHILOSOPHY BEHIND THE CLASSIFICATION

Conventionally, rivers are typed by recognizing geographical, geological, climatic or biotic boundaries between them, i.e. on the basis of the outer limits of features which characterize them (e.g. Hart and Campbell, 1994). The alternative is to differentiate between river types on the basis of their core characteristics or inner limits, such that the resulting groupings are units viewed along a continuum rather than as discrete entities (Uys and O'Keeffe, 1997). Both approaches have advantages and disadvantages. As Brierley (1994) comments:

'Individual systems fit in particular positions along the continuum of stream power and grain-size trends. Forcing each situation into an end-member river style, however, may severely constrain the suitability of management options, as the sense of system dynamic may be fundamentally

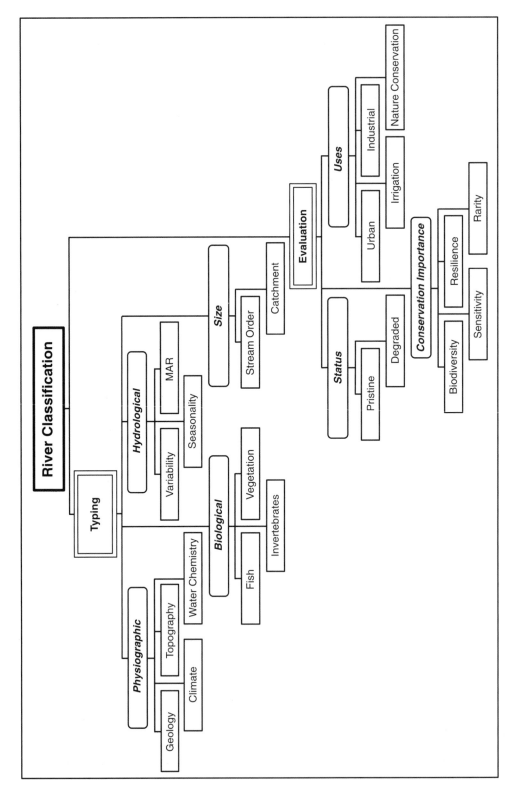

Figure 20.1   *A hierarchy of different kinds of classification that have been used to describe different kinds and states of rivers. The list is by no means complete, but provides a cross-section of the subjects and scales used for classification*

flawed. Conversely, interpretation of a river as intermediary in character between two or more other styles may be no more than describing the system as a meaningless mean.'

THE FOUNDATION OF THE CLASSIFICATION

The foundation, or starting point of the classification is determined by the problem to be addressed, and the end-user requirements. A spatial classification which groups rivers on the basis of a number of shared catchment characteristics, for example, may provide information for land-use management or assessment of conservation status. Grouping rivers on the basis of system response to time-linked variables (e.g. seasons) may be used to manage the timing and intensity of resource use. A classification for electing biological monitoring sites may focus on biotic distributions. Conventionally, classifications comprise two or more basic parameters, and multivariate procedures are used to distinguish groupings.

STRUCTURE AND EXTENT OF THE CLASSIFICATION

Most contemporary classification systems are hierarchical in nature – that is, they incorporate a number of different scales, nested hierarchically such that a system at one level encompasses its sub-systems at lower levels, one within the next. The geographical, spatial and temporal hierarchy such as that developed by Frissel *et al.* (1986) accommodates the variety of different processes and influences operating on systems of dissimilar size or slope either within or between different geographical regions. Naiman *et al.* (1992) took the approach that 'ideally, a classification scheme should be based on a hierarchical ranking of linkages between the geologic and climatic settings, the stream habitat features and the biota'.

THE TYPE OF DATA TO BE USED

The data used to develop a classification system can be selected from the scale at which top-level independent ('driving') physical variables such as climate, topography and geology operate, down to the scale of lower-level dependent ('response') variables including geomorphology, soils, vegetation, and stream fauna. A number of systems use a combination of top-down and bottom-up variables.

The classifications of Omernik (1986) for the United States, or Briggs *et al.* (1990) for New Zealand initially follow a top-down approach, using independent physical variables to define the similarities between rivers. The resulting classifications are verified by means of a bottom-up approach, referring to faunal similarities. The RIVPACS predictive model for rapid biological assessment of water quality and the state of aquatic ecosystems (Wright *et al.*, 1984, 1997) follows a similarly dual approach to generate site-specific predictions of the fauna to be expected in the absence of environmental stress, for comparison with the observed fauna. River typing is performed using 'least affected' sites, and a bottom-up approach based on aquatic invertebrates. This process generates a number of river groups, each representing a different reference condition. The higher-level environmental variables of a test site are used in establishing a match with those of a relevant reference group, to generate an observed (test) versus expected (reference) ratio, and enabling the assessment of the test site.

For the bottom-up classification of South African rivers developed by Eekhout *et al.* (1997), distribution records of fauna and flora were the primary basis of the boundaries between ecoregions. These boundaries were then modified on the basis of specialist input relating to driving physical variables.

Both top-down and bottom-up approaches have advantages and disadvantages. The top-down approach can use easily available data such as geological maps, but relies on the assumption that the distribution of the biota is intimately linked to these higher-level variables. The bottom-up approach uses the biotic information directly, but such data are usually far from comprehensive (King *et al.*, 1992) and are very labour-intensive to collect (Pennak, 1971). Undoubtedly, the best option is to use both higher- and lower-level data, using the biotic distributions to check on the boundaries suggested by changes in the physical variables.

THE STATUS OF RIVERS TO BE INCLUDED IN THE CLASSIFICATION

The decision on whether to base the classification solely on 'least affected' or 'assumed natural' rivers, or to include modified rivers, depends on the function of the classification. If a set of reference groups are required for the purposes of evaluation, then the typing exercise will initially focus only on 'least affected' sites. This is the approach adopted by systems such as RIVPACS.

In the case of a typing exercise which is not aimed at evaluation, the inclusion or exclusion of modified sites is a fairly complex choice. To include artificial effects with natural variables will result in arbitrary groupings which reflect neither natural affinities nor classes of degradation. For example, the rivers of the south-western Cape of South Africa exhibit naturally low pH values (down to pH 4) owing to the underlying highly leached sandstones. Acidity in many of the Transvaal rivers is similarly low, but this is due to the effects of mine-dump run-off. To group such rivers together would be meaningless in relation to the processes involved, and would provide little predictive capacity about other aspects of the rivers.

On the other hand, few rivers are not modified in some ways, and an exercise designed to type river systems for reasons other than evaluation would, at best, generate a partial picture if modified river systems were excluded from the analysis. Their inclusion is best facilitated by selecting a general, top-down approach. Independent physical variables are more reliable than the biota, which may have been modified in ways that are difficult to detect. Using the argument that classification should be based on natural variables rather than measures of disturbance (Lotspeich, 1980), Lotspeich and Platts (1982) follow a top-down typing exercise including catchment characteristics. This accommodates the authors' opinion that the entire system needs to be viewed as the final integrated product of its catchment in order to type it satisfactorily. The constraint on this type of system is that the resulting classification at this scale of resolution is likely to be general in nature and lacking in specific applicability.

RIVER ATTRIBUTES TO CONSIDER IN CLASSIFICATION

A major consideration in the development of classification systems is which river attributes to include. Some guidance is provided by Noss (1990) whose nested hierarchy illustrates the relationship of the three primary attributes of ecosystem diversity (structural, functional and compositional) to the full range of ecological systems (genes to landscapes) (see Figure 20.2).

Most river classification systems use structural and compositional rather than functional characteristics as their criteria for similarity, because data on physical structures (e.g. geomorphology) and the river biota are more accessible and easier to handle than functional information such as flow patterns and sediment loads.

There is an awareness, however, of the importance of including process-related characteristics. Naiman *et al.* (1992) call for the integration of structural and functional characteristics as a fundamental attribute of an enduring classification system, and Noss (1990) points out that functional information is much more likely to reveal meaningful patterns and to provide predictive capability.

Classification systems based on physical variables (structural attributes) such as hydrology and geomorphology have most often been used in the environmental field on the assumption that these variables define the types and abundance of habitats available, which in turn regulate the number, types and abundances of species that will survive in a river. The acceptance of these relationships implies a basic assumption that biotic components (compositional attributes) and species interactions (functional characteristics) are not of primary significance in ultimately determining species composition. The suggestion by Pennak (1971) that primarily physical features of rivers should be used in classification endorses the assumption that these attributes create the habitat conditions that will provide the templet (*sensu* Southwood, 1977) for the biota. These assumptions may to some extent be justifiable in the rivers of arid regions, where conditions are harsh, stochastic and highly variable, and where physical rather than biotic processes seem to limit the appearance and survival of species. There are, however, a number of types of biotic interaction which are likely to affect the species composition even of arid rivers: in particular, introduced aliens, parasitism and disease, and the evolution of endemic species in isolated rivers.

Certain evaluation methods make use of a wide range of components: SERCON (Boon *et al.*, 1997) is based on many of the nature conservation criteria defined by Ratcliffe (1977), including naturalness, representativeness, rarity and species richness, but also including physical diversity, impacts and a section for special features. The system can be used either with reference to a printed manual or on computer, and appears to be highly versatile. Within the seven major criteria, SERCON evaluates 35 structural and compositional attributes, such as fluvial features, fish, invertebrates, Red Data Book species, wintering birds on floodplains, and impacts such as acidification, water abstraction and recreational pressures. Such a system is clearly specific to the purposes of nature conservation assessment, and to a particular geographical region, although the attributes could be modified to suit other areas.

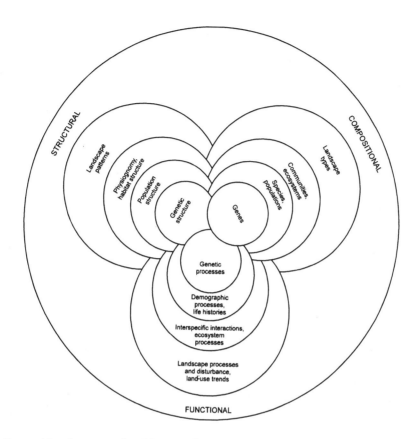

Figure 20.2 *Compositional, structural and functional attributes of biodiversity at four levels of organization. From Noss (1990). Reproduced by permission of Blackwell Science, Inc*

THE FORMAT OF THE CLASSIFICATION

The majority of river typing systems are static in the sense that they provide a set of discrete groups or types of rivers, and are not equipped for modification or extension; and non-interactive in the sense that the user is unable to communicate with the system. The modern challenge is to move towards simple, problem-driven classifications which are both dynamic and interactive in nature. To be effective in serving the needs of conservation, typing systems should be simple, practical, innovative, accessible and useful.

There is great scope for creativity and originality here. The global trend towards computerization, and the rapid growth in interaction and cooperation between people and machines, offers avenues to data access, online analysis, real-time communication, and the use of the Internet for information sharing and decision support. The interface between research and management is facilitated by the application of one or

a combination of decision support technologies (tools). These techniques can be divided into a number of categories. The application of one or more of these is known as a Management Support System (MSS):

Decision Support Systems (DSS): DSS have no universal definition, but are conceptually interactive computer-based systems, which help decision-makers utilize data and models to solve unstructured problems.

Group Support Systems (GSS) and Group Decision Support Systems: Systems designed to improve the work of groups with the aid of information technology.

Expert Systems (ES): These are typically decision-making or problem-solving computer packages that can achieve a level of performance equivalent or superior to that of a human expert in some specialized and usually narrow problem area. Expertise is transferred by the specialist to the

Table 20.3   *Examples of tools at different levels of complexity to aid in classification and decision-making (Turban and Aronson, 1998)*

Phase	Description	Examples of tools
Early	Compute, crunch numbers, summarize, organize	Calculators, early computer programs, statistical models, simple management science models
Intermediate	Find, organize and display decision-relevant information	Database management systems, Management Information Systems, filing systems, management science models
Current	Perform decision-relevant computations on decision-relevant information; organize and display the results; query-based and user-friendly approach, what-if analysis	Financial models, spreadsheets, trend exploration, operations research models, decision support systems
	Interact with decision-makers to facilitate formulation and execution of the intellectual steps in the process of decision-making	Expert systems, executive information systems
Just beginning	Complex and fuzzy decision situations, expanding to collaborative decision-making and machine learning	Second-generation expert systems, group DSS, neural computing

computer, and this knowledge is then stored so that others can use it.

Artificial Neural Networks (ANN): These systems cater for situations in which there is little or no available data, information or knowledge, and people are making decisions based on partial, incomplete or inaccurate information. This is a common situation in river ecology, where many of the systems included in analysis may be unknown or only partially known, or simply unpredictable and variable by nature. In these circumstances, decision-makers transfer their learning from previous experiences to similar new situations to formulate a course of action. ANNs (neural computing) introduce the concept of the computer learning. This emerging technology uses a pattern recognition approach, which gives it the capability to work despite missing data.

Hybrid Support Systems: This is the concept of integrating a number of decision support technologies (and constructing new ones), in an attempt to provide a successful solution to the conservation or management problem. Many complex problems require several MSS. The problem-solver then has access to a number of tools, which can either be used independently such that each tool solves one aspect of the problem, or loosely or tightly integrated as one hybrid system.

Second-generation expert systems, group DSS, and neural computing will facilitate the incorporation of complex decisions and the inclusion of fuzzy logic (to accommodate for uncertainty), collaborative decision-making and machine learning (see Table 20.3).

The ability to transfer data and information electronically *via* the Internet, and the capacity to design multi-scale, multi-media, interactive software in an easily accessible format (such as that provided by CD-ROM or web-site) provides unlimited opportunities for collaboration between institutions within countries and internationally, in formulating new conceptual designs for river classification (but see Wishart *et al.*, this volume, Part II).

Discussion

We would define the main aims of river conservation as the maintenance or rehabilitation of as much of the natural compositional, structural and functional biodiversity as is essential for and compatible with the agreed objectives for the system. This may not be a universally agreed definition, but it acknowledges the value of riverine resources for the sustenance and development of human populations (see Gardiner and Perala-Gardiner, this volume). Conservation of rivers as natural undisturbed systems is rarely possible, and it is therefore necessary for the stakeholders in a catchment to agree to a set of objectives for any river before it can be managed for conservation (see Showers, this volume). In cases of very high conservation importance the objectives may be governed by the need to protect the natural attributes of the system unmodified. For urban rivers (Baer and Pringle, this volume), or others which are already substantially disturbed and exploited, the objectives may deal more with the maintenance of functions such as natural water purification or bank stability, which are still operational and necessary for sustainable development, rather than unrealistic ambitions to re-establish natural species richness. This perspective is particularly important in less developed and water-poor countries such as South Africa (Davies and Wishart, this volume; Wishart *et al.*, this volume, Part II).

Given some set of conservation objectives (which may not necessarily align with the above), conservation agencies must then assess the variability, abundance, status and importance of the rivers of their region, and determine how these have changed and are changing, both as a result of natural phenomena and human interference. Any attempt to manage rivers without such assessments will result in *ad hoc* decisions, the effects of which cannot be estimated.

The role of classification in river conservation is therefore to type and to evaluate in order to provide a coherent framework for management decisions. Specifically, classification systems should be useful for at least one of the following:

- establishing groups of similar rivers
- choosing representative rivers or river sites
- identifying unique rivers or sites
- describing the present status of rivers
- evaluating trends over time
- designing monitoring systems
- extrapolating knowledge from well-researched rivers to those about which less is known
- setting conservation objectives

Although classification systems should contribute to these aims, it is a mistake to imagine that such systems are more than organizational frameworks to assist decision-making. The definition of boundaries between groups, the choice of sites, the interpretation of status, the evaluation of trends, the decision as to how far knowledge can be extrapolated, and the choice of objectives all require specific expertise in which the use of classification should improve the reliability and consistency of the results.

Naiman *et al.* (1992) suggested that there is a consensus emerging on the fundamental attributes for an enduring classification system. They suggested that it should:

- encompass broad spatial and temporal scales
- integrate structural and functional characteristics under various disturbance regimes
- convey information about the underlying mechanisms controlling in-stream features

- accomplish these features at low cost and at a high level of uniform understanding among resource managers

They concluded that no existing classification system adequately meets all these criteria. Perhaps the answer to this is that classification is always a trade-off between resolution and general applicability. There are generally applicable systems (such as that of Frissell *et al.*, 1986), but these lack the resolution and predictive capabilities of the purpose-built systems such as RIVPACS. There appears to be a trend towards developing targeted classification systems for specific purposes, which can operate within the broad framework of a general system such as that of Frissell *et al.* (1986).

One specialized application of classification demonstrates the advantages of specifically targeted systems. Partly fuelled by the controversy over elephant culling in the Kruger National Park, South African National Parks has recently reviewed and redesigned its entire wildlife monitoring and management policy. One of the results of this review has been the development of a process based on an 'Objectives Hierarchy' culminating in 'Thresholds of Potential Concern (TPCs)' (Rogers and Bestbier, 1997) to represent end-points at which some management action must be taken. The Objectives Hierarchy converts high-level objectives (e.g. visions and mission statements) which set the culture of an organization but are too general to be directly implemented, through a hierarchy of key elements, objectives and sub-objectives, into TPCs which can be audited and measured (see examples in Table 20.4). One advantage is that each of the TPCs may appear obscure when viewed in isolation, but its context and purpose can be recognized by backtracking through the hierarchy. If a threshold is reached, the management action may be to adopt immediate remedies, or to institute further investigations, or simply to review the suitability of the TPC. In each case, the aim is to sound a preliminary alarm (raise a red flag) before the trend has reached crisis proportions (hence Threshold of *Potential* Concern).

The system is still experimental, but has already resulted in researchers and managers designing integrated monitoring and management systems, in which the information generated feeds directly into the decision-making process, in turn feeding back into modifications of the monitoring design, a process illustrated in Figure 20.3.

Perhaps the greatest hurdle to be overcome is that of unrealistic expectations by potential users, and particularly the managers. To return to the theme of

Table 20.4 *Examples from an Objectives Hierarchy leading to the identification of Thresholds of Potential Concern, for the rivers of the Kruger National Park, South Africa. Adapted from Rogers and Bestbier (1997). The examples are drawn from a hierarchical tree which contains many other branches, concerning all aspects of the Park, which are not shown here*

Vision	To maintain biodiversity in all its natural facets and fluxes and to provide human benefits in keeping with the mission of South African National Parks in a manner which detracts as little as possible from the wilderness qualities of the Kruger National Park (KNP)
Key Element	Biodiversity – Ecosystem Management
Objective	To manage the KNP as part of the Lowveld Savanna and river catchments in such a manner as to conserve and restore its varied structure, function and composition over time and space
Sub-objective (SO)	Aquatic systems – To maintain the intrinsic biodiversity (hydrological, geomorphic and biotic) of the aquatic ecosystems as an integral component of the landscape, and where necessary restore or simulate natural structure, function and composition
Aquatic systems SO	Riverine – To maintain, and whenever necessary, restore natural river ecosystem health and biodiversity, particularly through promoting Integrated Catchment Management
Riverine Objectives	Diversity – To ensure the intrinsic attributes and role of each river as part of the landscape diversity, in such a way as to allow the natural fluctuation over space and time in structure, function and composition
Riverine Diversity Objective	Monitoring – To determine whether biodiversity changes are approaching or exceeding stated Thresholds of Potential Concern (TPCs), and to determine the role of human influence in this. • Catalogue riverine biodiversity • Identify long-term fluctuations and scales • Establish TPCs • Integrate desired future conditions and monitoring programmes • Design and integrate a practical monitoring programme
Example TPCs for the Sabie River	Channel Types – Directional loss of bedrock influence and water surface area (due to increased sedimentation and/or reduced flow) Riparian Vegetation – Loss of vigorous recruitment of *Breonadia salicina* (indicator tree species) in anastomosing sections Fish – 33% drop in Fish Community Integrity Index value and/or a decrease or loss of certain indicator groups or species

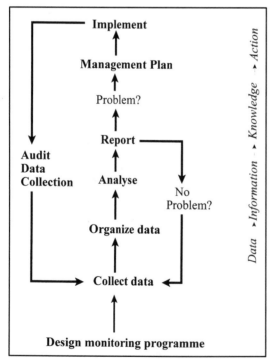

Figure 20.3 *Converting monitoring into management. This schematic illustrates the steps and feedback loops by which data collected in a monitoring programme should be used to influence management, and how management actions should serve to direct and refine the type of monitoring undertaken*

the 'Introduction', classification is an artificial and limited tool which can provide more or less objective groupings. The use that is made of these groupings, the predictions and extrapolations arising from them, and the values that are assigned to them are additional to the classification exercise, not part of it, and will still depend on the accuracy and reliability of the available expertise and information. Classification is ultimately an exercise in data organization; it can provide invaluable help when making judgements and decisions, but it should not be confused with those subjective processes.

References

Boon, P.J., Holmes, N.T.H., Maitland, P.S., Rowell, T.A. and Davies, J. 1997. A system for evaluating rivers for conservation (SERCON): development, structure and function. In: Boon, P.J. and Howell, D.L. (eds) *Freshwater Quality: Defining the Indefinable?* The Stationery Office, Edinburgh, 299–326.

Brierley, G.J. 1994. River reach analysis as a geomorphic tool for river classification. In: Uys, M.D. (ed.) *Classification of Rivers and Environmental Health Indicators.* Proceedings of a Joint South African/Australian Workshop. SA Water Research Commission Report No. TT 63/94, 111–117.

Briggs, B.J.F., Duncan, M.J., Jowett, I.G., Quinn, J.M., Heckey, C.W., Davies-Colley, R.J. and Close, M.E. 1990. Ecological characterization, classification, and modelling of New Zealand rivers: an introduction and synthesis. *New Zealand Journal of Marine and Freshwater Research* 24:277–304.

Collier, K. 1993. *Towards a Protocol for Assessing the Value of New Zealand Rivers.* Science and Research Series 58. National Institute of Water and Atmospheric Research Ltd, New Zealand.

Cowardin, L.M., Carter, V., Golet, F.C. and Laroe, E.T. 1979. Classification of wetlands and deepwater habitats of the United States. FWS/OBS – 79/31, Fish and Wildlife Service, US Department of the Interior, Washington, DC.

Eekhout, S., King, J.M. and Wackernagel, A. 1997. *Classification of South African Rivers.* Volume 1. SA Department of Environmental Affairs and Tourism, Pretoria.

Frissell, C.A., Liss, W.J., Warren, C.E. and Hurley, M.D. 1986. A hierarchical framework for stream habitat classifications. Viewing streams in a waterland context. *Environmental Management* 10:199–214.

Harrison, A.D. 1959. *General Statement on South African Hydrobiological Regions.* Report No. 1, Project 6.8 H, NIWR, CSIR, Pretoria.

Hart, B.T. and Campbell, I.C. 1994. Assessment of river 'health' in Australia. In: Uys, M.C. (ed.) *Classification of Rivers and Environmental Health Indicatiors.* Proceedings of

a Joint South African/Australian Workshop. SA Water Research Commission Report No. TT 63/94, 177–189.

Illies, J. 1961. Versuch einer allgemeinen biozonotischen Gliederung der Fliessgewasser. *Internationale Revue der Gesamten Hydrobiologie* 46:205–213.

King, J.M., De Moor, F.C. and Chutter, F.M. 1992. Alternative ways of classifying rivers in Southern Africa. In: Boon, P.J., Calow, P. and Petts, G.E. (eds) *River Conservation and Management.* John Wiley, Chichester, 213–228.

Kleyhans, C.J. 1994. The assessment of conservation status for rivers. In: Uys, M. (ed.) *Classification of Rivers and Environmental Health Indicators.* Proceedings of a Joint South African/Australian Workshop. SA Water Research Commission Report No. TT 63/94, 255–270.

Kleynhans, C.J. 1996. A qualitative procedure for the assessment of the habitat integrity status of the Luvuvhu River (Limpopo system) South Africa. *Journal of Aquatic Ecosystem Health* 5:1–14.

Lotspeich, F.B. 1980. Watersheds as the basic ecosystem. This conceptual framework provides a basis for a natural classification system. *Water Resources Bulletin* 16:581–586.

Lotspeich, F.B. and Platts, W.S. 1982. An integrated land–aquatic classification system. *North American Journal of Fisheries Management* 2:138–149.

MacMillan, L. and Kunert, C. 1990. *Conservation Value and Status of Victorian Rivers. Part 1. Methodology: Classification, Nature Conservation Evaluation and Strategies for Protection.* Faculty of Environmental Design and Construction Research, Royal Melbourne Institute of Technology, Melbourne.

Naiman, R.J., Lonzarich, D.G., Beechie, T.J. and Ralph, S.C. 1992. General principles of classification and the assessment of conservation potential in rivers. In: Boon, P.J., Calow, P. and Petts, G.E. (eds) *River Conservation and Management.* John Wiley, Chichester, 93–123.

Norris, R. 1994. Rapid biological assessment, natural variability, and selecting reference sites. In: Uys, M. (ed.) *Classification of Rivers and Environmental Health Indicators.* Proceedings of a Joint South African/ Australian Workshop. SA Water Research Commission Report No. TT 63/94, 129–166.

Noss, R.F. 1990. Indicators for monitoring biodiversity: a hierarchical approach. *Conservation Biology* 4:355–364.

O'Keeffe, J.H. 1997. Methods of assessing conservation status for natural fresh waters in the southern hemisphere. In: Boon, P.J. and Howell, D.L. (eds) *Freshwater Quality: Defining the Indefinable?* The Stationery Office, Edinburgh, 369–386.

O'Keeffe, J.H., Danilewitz, D.B. and Bradshaw, J.A. 1987. An expert system approach to the assessment of the conservation status of rivers. *Biological Conservation* 40:69–84.

Omernik, J.M. 1986. Ecoregions of the conterminous United States. *Annals of the Association of American Geographers* 77:118–125.

Palmer, M.A. and Poff, N.L. 1997. The influence of

environmental heterogeneity on patterns and processes in streams. *Journal of the North American Benthological Society* 16:169–173.

Pennak, R.W. 1971. Toward a classification of lotic habitats. *Hydrobiologia* 38:321–334.

Plafkin, J.L., Barbour, M.T., Porter, K.D., Gross, S.K. and Hughes, R.M. 1989. *Rapid Bioassessment Protocols for Use in Streams and Rivers: Benthic Macroinvertebrates and Fish.* US Environmental Protection Agency Report No. EPA/440/4-89-001. Assessment and Watershed Division, Washington, DC.

Poff, N.L. and Ward, J.V. 1989. Implications of streamflow variability and predictability for lotic community structure. A regional analysis of streamflow patterns. *Canadian Journal of Fisheries and Aquatic Sciences* 46:1805–1817.

Polls, I. 1994. How people in the regulated community view biological integrity. *Journal of the North American Benthological Society* 13:598–604.

Ratcliffe, D.A. (ed.) 1977. *A Nature Conservation Review.* Cambridge University Press, Cambridge.

Republic of South Africa 1998. *National Water Act. No. 36 of 1998.* Published by the South African Government Gazette, Pretoria.

Rogers, K. and Bestbier, R. 1997. *Development of a Protocol for the Definition of the Desired State of Riverine Systems in South Africa.* SA Department of Environment Affairs, Pretoria.

Southwood, T.R.E. 1977. Habitat, the templet for ecological strategies? *Journal of Animal Ecology* 46:337–365.

Strahler, A.N. 1957. Quantitative analysis of watershed geomorphology. *Transactions of the American Geophysical Union* 38:913–920.

Turban, E. and Aronson, J.E. 1998. *Decision Support Systems and Intelligent Systems.* Prentice-Hall, New Jersey.

Uys, A.C. and O'Keeffe, J.H. 1997. Simple words and fuzzy zones: early directions for temporary river research in South Africa. *Environmental Management* 21:517–531.

Wright, J.F., Moss, D., Armitage, P.D. and Furse, M.T. 1984. A preliminary classification of running-water sites in Great Britain based on macro-invertebrate species and the prediction of community type using environmental data. *Freshwater Biology* 14:221–256.

Wright, J.F., Moss, D., Clarke, R.T. and Furse, M.T. 1997. Biological assessment of river quality using the new version of RIVPACS (RIVPACS III). In: Boon, P.J. and Howell, D.L. (eds) *Water Quality: Defining the Indefinable?* The Stationery Office, Edinburgh, 102–108.

Wright, J.F., Sutcliffe, D.W. and Furse. M.T. 2000. *Assessing the Biological Quality of Fresh Waters: RIVPACS and Similar Techniques.* Special Publication of the Freshwater Biological Association, Ambleside (in press).

21

Popular participation in river conservation

K.B. Showers

Introduction

The 1992 United Nations Conference on Environment and Development (UNCED), often referred to as 'The Earth Summit', focused attention on the importance of popular participation in environmental issues. The Summit's resulting documents, *The Rio Declaration* and *Agenda 21: Programme of Action for Sustainable Development*, recognized networks of formal, informal and grassroots movements active in environmental conservation as important sources of action at the local level (UNCED, 1993). *Agenda 21* called for governments to seek new techniques for popular participation in planning, decision-making and implementation of projects.

Members of the public have participated in river conservation since the late 19th century either in association with government-sponsored programmes or in formal or informal citizen groups. In some instances partnerships evolved between citizen groups and government agencies, while in others citizen groups opposed government decisions that endangered river quality or function. Since the Earth Summit, some nations and international agencies have increased efforts to identify and work with citizen groups concerned with river conservation.

This chapter provides a general discussion of different conceptualizations of conservation and reviews categories of river conservation activities. Examples from around the world demonstrate citizen involvement, the significance of international computer networks to river conservationists, and the relationship between environmental rights and human rights.

Attitudes to rivers and river conservation

The definitions of a river, river conservation and acceptable uses of river environments depend upon how nature and ecosystem components are valued. Lemons and Saboski (1994) suggest that theories of the value of nature can be broadly grouped into three categories:

- anthropocentrism – where all value in non-human nature is instrumental value and dependent upon contributions to some human values;
- inherentism – where all value in non-human nature is dependent on human consciousness, but some of this value does not derive from human values;
- intrinsicalism – where some value in nature is independent of human values and human consciousness.

Anthropocentrism allows the commodification of all aspects of nature, assigning monetary value in terms of utility to human society. Components of an ecosystem that are recognized as useful are referred to as 'natural resources'. Assessment of the worth of natural resources depends upon a knowledge of their biology and ecology in order to link them to human benefits. Those elements of the natural world that do not have measurable conventional values are considered to be 'non-resources' (Lemons and Saboski, 1994). This approach has led to a description of the natural world as an accumulation of separate objects, each with an economic value.

Considering function to be as important as natural resources, Costanza *et al.* (1997) found a way to value

Global Perspectives on River Conservation: Science, Policy and Practice.
Edited by P.J. Boon, B.R. Davies and G.E. Petts. © 2000 John Wiley & Sons Ltd.

ecosystem function in economic terms. Arguing that the earth is a 'very efficient least-cost provider of human life-support systems', the researchers identified 17 categories of 'ecosystem services' to which monetary value could be attached (Costanza *et al.*, 1997). The category 'lakes/rivers' was valued for water regulation, water supply, waste treatment, food production and recreation providing a 'total global flow value' (per hectare services × area of biome) of $1700 × 10^9$ yr^{-1} (Costanza *et al.*, 1997). Conservation of 'lakes/rivers' would mean ensuring the perpetuation of these 'services'.

Value systems based upon inherentism or intrinsicalism do not express value solely in monetary terms. Instead, they stress other species' rights to existence as well as the importance of inanimate elements of the landscape. Systems based upon inherentism and intrinsicalism are less dependent upon detailed information about ecosystem components and function because commodification of every aspect is not required for valuation. In riverine ecosystems concern would be not only for the survival of all species associated with the river, but also for river processes. Human uses of a river would be assessed in terms of their impacts on other species and on the river's integrity.

Arguments about conservation stem from these very different philosophies of the value of non-human nature. Boon (1994) summed up the divergence when arguing that conservation has become a 'clearly circumscribed activity' rather than 'an attitude that permeates the way society and environment interact'.

'Conservation as an activity' reduces conservation to one of several competing uses of a river, to be selected only if the benefits of conservation to human society economically outweigh the lack of conservation. Economic and political interests may well see a greater cost in conservation and greater benefit in non-conserving uses than do members of the public. In urban areas, river management decisions are often based on how much a river can be used without affecting its 'use value' (Box and Walker, 1994). One consequence of defining conservation as simply a competing use is that conservation becomes contested and politicized.

'Conservation as a way of life' sets human use of ecosystem components in the context of the long-term existence and functioning of an ecosystem. With this attitude, human use is moderated by consideration of its impact upon all aspects of an ecosystem, so that possible long-term destruction will modify plans with short-term political or economic gains. This approach also supports notions of common property resources and public space, whose maintenance and use benefit communities. Conservation as an activity and conservation as a way of life, therefore, are not synonymous and will not result in the same definitions of problems or of solutions.

Despite a society's predication on any of these philosophies of value, individuals within them may embrace alternative value systems. National and international debates about economic growth, modernization and progress reflect clashes among these value systems. Industrial societies are dominated by an anthropocentric view of the natural world. They have what Oyadomari (1989) calls an 'economy-first' value system that requires rupturing the bonds between human society and nature so that elements of the environment can be commodified and viewed as raw materials for the production process (Oyadomari, 1989; Maltby, 1991; Asopa, 1993; Pretty and Pimbert, 1995). However, not all members of such societies embrace this value system.

In the United Kingdom (UK), researchers at Middlesex University's Flood Hazard Research Centre implemented a series of studies of public attitudes towards river water quality in the mid 1980s. Initially begun as a study for the Department of the Environment to evaluate the benefits of sewerage schemes, the project developed into a study of the value of river corridors (Green *et al.*, 1988; Green and Tunstall, 1993). The research revealed that people supported spending money on pollution reduction for reasons not quantifiable by an anthropocentric neo-classical economic system (Green and Tunstall, 1993). The people interviewed felt a moral obligation to protect and conserve, they wanted to leave a pleasant environment for future generations, and they wanted to ensure public health (Green *et al.*, 1988).

Many non-industrial societies are guided by perspectives of inherentism or intrinsicalism. Their way of life depends upon the continued function of an intact ecosystem. Religion, taboos, folklore, local environmental knowledge and traditions stress the importance of different elements of the ecosystem, its function, and human respect for it. This has not prevented societies from over-exploiting specific ecosystem components, but it has provided a non-commodified valuation system that placed humanity in a relationship with other species, which limited consumption.

However, as these societies become stressed by population pressures on finite ecosystems or overwhelmed by (integrated into) larger economies, the underlying values which can function to conserve nature are disrupted. Ecological destruction frequently coincides with the collapse of these cultural values, especially when elements of the landscape are given increased monetary value (Oyadomari, 1989; Coordinating Body for Indigenous People's Organizations of the Amazon Basin – CBIPOAB, 1990; Grove, 1992; Pearce, 1992; Asopa, 1993; Osemeobo, 1994).

Restoration of the broken bonds between human beings and the natural world requires an active awareness of the need for an attitude of conservation. Modern conservationists recognize industrial society's separation of society from ecosystem function (what Maltby calls 'uncoupling') and question its economy-first value system (see, for example, Webb and Smyth, 1984; Maltby, 1991). Many advocate trying to live with a conservation attitude by consuming as little as possible, creating minimal amounts of waste, and working to guide their societies as a whole in those directions in order to reintegrate them with their ecosystem. Others pursue the industrial notion of separation to its conclusion: the total exclusion of human beings from nature. River conservation for them is the abolition of all human interactions with riverine ecosystems.

Those seeking to foster conservation as an attitude in daily life may not agree with those who would enforce the separation of human beings from nature, but both may find common concern with members of societies having perspectives of inherentism or intrinsicalism. Conservation as an activity can unite non-industrial, industrial and post-industrial people who share an attitude of conservation.

Types of river conservation activities

While the destructiveness of hydraulic societies was written about in the Middle East in the fourth millennium BC, the origins of modern conservationism can be traced directly to observations of ecological destruction associated with the 18th century expansion of European commercial interests in the tropics (Grove, 1995). Some of the earliest experiments in forest conservation, pollution control and fisheries protection were implemented by French reformers on the island of Mauritius between 1768 and 1810 (Grove, 1995). Until the mid 19th century colonial governments were influenced by growing scientific communities to enact nature protection policies and legislation at home and in colonial territories (Grove, 1992, 1995). Popularly based European nature conservation organizations grew out of these traditions (Grove, 1992). The river conservation movement in the United States of America (USA) was launched by the desire to protect California's Tuolumne River from damming at the turn of the 19th century (Palmer, 1986).

As long as conservation groups sought to protect and preserve regions not considered to have potential for economic activity, the conflict among differing philosophies of natural value was not always apparent. However, measures to protect or restore rivers directly affected by industrial production, and plans for infrastructure development, challenged the dominant regional and state political economies. This was as true in the past as it is today. Both proponents and opponents of river conservation activities can include local government officials, business leaders, and water officials – as well as ordinary citizens. Conservation is, therefore, a political act.

Citizen concern about river integrity has been fundamental to the development of modern environmental law and the concept of the environmental rights both of human beings and of nature. The 17-year struggle (1963–1980) to protect the Hudson River in New York State from a pumped water electrical generation plant at Storm King Mountain marked the beginning of the modern era of US environmental law (Cronin and Kennedy, 1997). For the first time the legal system recognized the right of citizens to challenge polluters and developers on behalf of nature, and the first environmental impact statement (EIS) was ordered by a court. The logic requiring the Storm King EIS was codified in the 1969 National Environmental Policy Act (NEPA) (Cronin and Kennedy, 1997). Environmental impact assessment is, therefore, a direct result of popular participation in river protection.

CONSERVATION OR POLITICAL OPPOSITION?

In industrial societies, Oyadomari (1989) and Fleischer (1993) both argue, it is a mistake to confuse opposition to environmental degradation with a wider concern for conservation. Because of the linkages between economic development and politics, environmental destruction can be seen to be symbolic of the state, and opposition to the state can, therefore, be expressed through opposition to an environmentally disruptive project. Such was the case with both the construction of the Cahora Bassa dam on the Zambezi River in

Mozambique and the plan to build the Gabcikovo Barrage on the Danube River in Hungary.

Soon after the plans were made public, the Cahora Bassa Dam became a symbol of the struggle between Portuguese colonialism and African nationalism (*Barclays National Review*, 1973; Schreyogg and Steinmann, 1989). Supporters of the African nationalist cause mounted an international anti-dam campaign in Europe and North America. The international activists were successful in pressuring companies in – and the governments of – Sweden, West Germany, Italy, UK, USA and Canada to withdraw from funding and participation in the project (Schreyogg and Steinmann, 1989).

Although the campaign failed to halt dam construction, the now widely used tactics of disinvestment, corporate shareholder lobbying, and political pressure were pioneered in this campaign (Schreyogg and Steinmann, 1989). International objections to the dam were on purely political grounds, however. Environmental concerns were only voiced by a few biologists in semi-popular publications (Davies, 1975a,b; Tinley, 1975).

Hungary's Middle Danube Hydroelectric Dam similarly became the symbol of an unpopular political regime (Fleischer, 1993; McCully, 1996). Since it was impossible to express dissatisfaction with the government, popular opposition to the government coalesced in opposition to the dam. What began as the public expression of one biologist's concern for environmental impact grew into a social movement, called the Danube Circle, with international support (Fleischer, 1993). After four years of protest, in the autumn of 1989 the Hungarian Government suspended construction. This coincided with the collapse of East European communist governments. Two years later the government announced the cancellation of the intergovernmental agreements for the dam project (Fleischer, 1993).

Fleischer (1993) cautions against interpreting the social movement against the dam as an expression of commitment to conservation. Hungarian society is in transition from a post-feudal economy to one of a 19th century style of (unregulated) capitalism. Although the government took the Slovak Republic to the World Court to dispute the diversion of the Danube River, it proposed two smaller dams on the river (Fodor, 1997; Hungarian Ministry for Foreign Affairs, 1997). Post-communist Hungary has very active citizen-based environmental organizations which continue the work to protect the Danube and its inland sea delta, the Szigetkoz (Fodor, 1997).

In the past, when industrial demand for 'natural resources' affected non-European nations, local residents responded with protests and social movements. These have long been interpreted as simply being political resistance to colonialism. However, new scholarship concerned less with the state and more with the attitudes and interests of ordinary people has contributed to a literature of popular resistance to environmental destruction. Less has been written about rivers than about other aspects of watersheds and watershed management (see, for example, MacKenzie, 1988; De La Court, 1990; Martinez-Alier, 1991; Showers and Malahleha, 1992; Asopa, 1993; Wilson, 1995; Showers, 1996).

There is no question that people in both non-industrial and industrialized societies have objected to interventions which have altered fundamental river properties. What is unclear is whether this opinion and its attendant action are manifestations of river conservation. On one level such a distinction is academic; the end result is activity to preserve river function. For organizers, strategists and planners of conservation agendas, however, the distinction is crucial.

CONSERVATION ACTIONS

Modern river conservation activities can range from protecting a natural or near natural stream to restoring a degraded stream ecosystem (Boon, 1992). This can include limiting access to or use of a river, its floodplain or watershed; monitoring water quality and mitigating against the consequences of use; clearing out rubbish and pollutants; removing flood control structures or dams; replanting riparian areas and reintroducing aquatic species. Depending upon the attitude of government (local, regional, national) to a particular stream, popular participation in conservation activities can be implemented in cooperation with government programmes or in opposition to commercial or government activities. Around the world river protection, clean-up and restoration have been initiated both by individual, concerned citizens and by government bodies.

The form of action taken by conservationists has, to a large extent, been shaped by the political culture of the nation in which they live. River conservationists with access to decision-making structures have been able to work with them to organize local clean-up and restoration campaigns, establish water quality monitoring networks and support enactment of a range of legislation limiting activities affecting river ecosystems at municipal, regional and national levels. Such activities were either implemented in cooperation with existing government structures or by newly created special-purpose statutory entities. Conservationists without access to decision-making bodies have had to resort to public oppositional activities. Citizen groups working with, or in spite of, governments have often concluded that conservation education must be a component of their projects.

Considerable activity has been associated with the rehabilitation of streams degraded by over-use or pollution. Consensus can usually be built around pollution detectable by the senses or clearly associated with public health problems. The primary areas of negotiation in these instances have concerned obtaining required funds and the creation of enabling or protective legislation. *Agenda 21* recommended that the 'polluter pays' principle, currently operating in France, Germany and the Netherlands, be applied so that entities responsible for water pollution are made financially responsible for its clean-up (UNCED, 1993; Howell and Mackay, 1997).

The restoration of heavily engineered and altered rivers to create more natural ecosystems was a relatively new area of activity in the 1990s. Very little was known about the science and economics of removing structures and restoring stream channels. Stimulated by the Earth Summit's call for the preservation of biodiversity, the European Union granted funds to establish river restoration demonstration projects in Denmark and the UK (Holmes and Nielsen, 1998). Out of this grew the River Restoration Centre in the UK and the European Centre for River Restoration based in Denmark (Appendix I). These agencies provide information, networking and models for partnerships to restore river and floodplain ecosystems (River Restoration Project, 1994; Holmes and Nielsen, 1998).

In the USA, advocacy of dam removal as a component of river restoration has united indigenous people, fishers, environmental scientists, recreational river users, NGOs and government agencies (Shuman, 1995; Loomis, 1996). Dam removal became a serious option in December 1994. The Federal Energy Regulatory Commission (FERC), which licenses hydroelectric dams, announced that it had the authority to order dam removal at the owner's expense – instead of relicensing it – if this were in the public interest (Reynolds, 1995). Since the relicensing procedure includes a public comment period, opponents were empowered to present public interest arguments for dam removal.

CONFLICT AND CONSERVATION

The greatest conflicts in river conservation have been about uses of streams that would completely disrupt or change their function. Expanding industrial societies' anthropocentrism often views rivers solely as sources of water for irrigation or distant cities, and for hydroelectric power. Water reaching the sea at a river's mouth is referred to as 'spillage'. Engineering projects could harness the rivers and prevent this 'waste'.

Those who value rivers from perspectives of inherentism or intrinsicalism do not agree that engineering interventions disrupting a river's ecological functioning are essential for a region or a nation's development. There are alternative means of solving development requirements. The fundamentally different approaches to the valuation of nature make building consensus about river use difficult. All too often the gap between the two positions has been filled with antagonistic rhetoric, and economic and political power have been used in attempts to end discussion or debate.

Local residents opposed to river engineering projects in the late 20th century were called backward, primitive, or in need of education and development by authorities promoting the schemes (Adams, 1992; Coetzee, 1995; Pretty and Pimbert, 1995; Harring, 1996). International supporters of anti-engineering campaigns were called 'militant conservationists', 'sinister manipulators' 'environmental extremists', 'green terrorists', and accused of seeking a playground or paradise after having destroyed their own countries, or wanting to ensure that the poor remain poor (Coetzee, 1995; Scudder, 1995; Harring, 1996; *The Namibian*, 1997a,b; Inambao, 1997). Proponents of engineering projects have grouped project opponents with preservationists who would exclude all human use of nature and referred to both as 'environmental fundamentalists' (Scudder, 1995).

POPULAR PARTICIPATION

Consistent with the approach advocated in *Agenda 21*, the most successful river conservation campaigns have been those with a strong local base. Conservation as an activity can unite or polarize local communities. Conservation goals can be damaged when non-local people impose their attitudes and solutions rather than supporting local concerns and initiatives. This can happen in both non-industrial and industrial/post-industrial societies (Reed, 1984; Anger, 1989;

CBIPOAB, 1990; Kottak and Costa, 1993; Pretty and Pimbert, 1995).

The concept of participation is contentious, however, since people understand the term differently. Participation can mean access to power and control. Pretty and Pimbert (1995) present a typology of participation. Four of the seven listed participatory processes – passive participation, participation in giving information, participation by consultation and participation for material incentives – have no lasting effect in people's lives or on conservation.

Pretty and Pimbert argue that only three forms of participation – functional participation, interactive participation and self-mobilization – can produce long-term economic and environmental success. Functional participation is when people form groups to meet predetermined objectives related to a project. This type of participation usually occurs after major decisions have been taken. The participating groups are usually dependent upon external initiators or facilitators. Interactive participation occurs when people participate in joint analysis leading to action plans and the formation of new local groups. These groups have control over local decision-making. Self-mobilization results when people take initiatives independent of external institutions to change systems. These groups organize themselves and plan collective action.

Examples of popular participation in river conservation

RIVER PROTECTION

Detailed discussion of past and current government sponsored river protection activities is beyond the scope of this chapter. The importance of popular participation and a trend towards its inclusion in government river regulation and management are illustrated by examples from the UK and USA. Citizen involvement in the initiation of citizen/government partnerships in the absence of river protection legislation is represented by an example from Brazil.

Absence of participation: Wild and Scenic Rivers Act, USA

In the USA, the 1968 Wild and Scenic Rivers Act was enacted by the Federal Government to prevent further hydroelectric dam construction and other engineering projects on designated rivers or stretches of rivers throughout the country. The Act authorized a range of Federal Government actions and restrictions, but provided little opportunity for popular participation (Reed, 1984).

Implementation of the Act was resisted in many locations (Palmer, 1993). Lack of consultation allowed concerns about losing control over ways of life or local economies to flourish. Reed (1984) describes how misunderstandings and distortions led the residents of the St Joe River watershed, northern Idaho, to oppose inclusion of their river in the Wild and Scenic Rivers Act.

This anti-classification action was not necessarily 'anti-conservation'. In the state of Idaho the Federal Government controls more than 60% of the land, and property owners resist federal regulation of privately held land. The same people who opposed the Wild and Scenic Rivers Act endorsed local zoning controls to protect the river (Reed, 1984). The dispute had to do with local participation and control, not simply with attitudes about conservation.

Participation by consultation: Sites of Special Scientific Interest, UK

The traditional government approach of regulation and enforcement was modified in the British designation of Sites of Special Scientific Interest (SSSI) to include participation by consultation with affected landowners and users. The statutory conservation agencies (English Nature, Countryside Council for Wales, and Scottish Natural Heritage) can designate a section of a river or an entire river an SSSI if there are areas of flora, fauna, or physiographic features of special interest (Boon, 1991, 1995). The conservation agencies can negotiate with landowners along the river's banks to arrive at management agreements 'in harmony with nature'.

Efforts are being made to expand the use of these management agreements to encourage positive conservation practices rather than merely preventing damaging activities. The process of notification and consultation is discussed in Boon (1991, 1995) and Howell and Mackay (1997) and described as applied to the protection of a river in the English Midlands (River Blythe) in Box and Walker (1994).

Self-mobilizing and interactive participation: the Massachusetts Watershed Initiative, USA

Self-mobilizing participation by citizen groups led to interactive participation with the state to design and

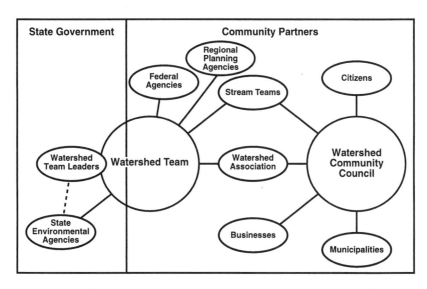

State Government

Community Partners

Figure 21.1 The structure of the Massachusetts Watershed Initiative (after EoEA, 1998)

implement policies and programmes for river restoration and protection in Massachusetts, USA. In 1992 grassroots river action groups from across the state formed the Massachusetts Watershed Coalition (MWC). The Coalition approached the Massachusetts Executive Office of Environmental Affairs (EoEA) with a proposal to restore Massachusetts' rivers to fishable and swimmable quality and protect their function by developing public/private partnerships on a watershed basis (MWC, undated; EoEA, 1998).

A dialogue between the Coalition and the state resulted in the Massachusetts Watershed Initiative (Watershed Initiative Steering Committee – WISC, 1995; River Network, 1998). The Initiative's purpose was to 'shift from top-down, federal- and state-driven environmental management to bottom-up, locally focused environmental management' (EoEA, 1998). A main component of the Initiative is EoEA's reorientation of environmental agencies to serve watershed-based decision-making (EoEA, 1998).

Implementation began with the highly successful Neponset River Watershed Pilot Project in December 1994. From this experience a Watershed Management Methodology – which is both a *structure* and a *process* – was formulated. Features considered to be essential for the Methodology's success are (EoEA, 1998):

- the co-leadership roles of the state, watershed associations or other citizen groups, the business community and municipalities in implementing the watershed approach;

- bottom-up resource assessment, planning and involvement of all interests;
- the sub-watershed focus of problem identification and Watershed Action Plan development;
- the goal of targeted allocation of limited dollars to watershed priorities, according to where the most environmental protection for available funds can be achieved.

The Watershed Management Methodology (WMM) structure (see Figure 21.1) calls for the creation of three new entities – Watershed Teams, Watershed Community Councils, and Stream Teams – while strengthening the institutional capacity of the existing citizen groups, called Watershed Associations.

Watershed Teams, convened by EoEA, include representatives of state and federal environmental agencies, citizen groups and interested individuals. These teams are responsible for the implementation of watershed management activities. They are equally accountable to the Secretary of Environmental Affairs and to the community (EoEA, 1998).

In contrast to the Watershed Teams, Watershed Community Councils (WCCs) are convened by a citizen group and represent all of the interests in the watershed – including municipalities, business, landowners, citizen groups, recreational users and representatives of state and federal agencies with programmes in the watershed (EoEA, 1998). The Watershed Teams and WCCs have the same direction and the mutual goal of each other's success. Their other goals and tasks may or may not be the same.

Linking the Watershed Teams and WCCs are the citizen-based Watershed Associations and the newly created sub-watershed Stream Teams. Stream Teams consist of five to ten people from the business community, municipal government structures and interested citizens who work together on a regular basis to assess the quality, identify problems and to recommend solutions for the stretch of a stream flowing through their community (EoEA, 1998).

This *structure* was created to ensure the *process* of the WMM. A Five Year/Phase Planning Cycle, with each year building on previous years, consists of: Year 1: Outreach and education (of the Watershed Team and the community); Year 2: Research (data collection, including water quality); Year 3: Assessment (interpretation of results); Year 4: Planning and implementation (including permitting, compliance and enforcement) and Year 5: Evaluation. This cycle will be implemented by a series of Annual Work Plans and a Five Year Watershed Action Plan (EoEA, 1998).

Both the Annual and Five Year Action plans are written primarily by the state-led Watershed Team. However, to ensure full public participation and contribution, the Watershed Action Plan development and approval process is: (a) to be preceded by public notice and personal invitation to all interests for participation; (b) to include a public comment period; (c) to balance strategic short-term and long-term actions (for broader consensus from the community); and (d) to include a sub-watershed focus (EoEA, 1998).

The state-wide implementation of the Massachusetts Watershed Initiative began in 1997. For a detailed description of the Watershed Initiative see EoEA (1998); for discussion of the cultural and political aspects of retooling a state government toward community-based and integrated environmental decision-making see Jewell *et al.* (1998).

Functional participation in the absence of legislation: Japurá and Solimões Rivers, Brazil

Governments that do not initiate river protection legislation can be persuaded to enter into partnerships with citizen groups to protect riverine ecosystems. A Brazilian wildlife biologist worked with a functional participation model to protect the floodplain at the junction of the Japurá and Solimões Rivers in north-west Brazil in the early 1990s. This area is covered by the little understood *várzea* ecosystem – rain forest that

is flooded by at least 10 m of water for more than seven months each year (Alexander, 1994). The biologist negotiated with the government to set aside 11 000 km^2 as the Mamirauá Ecological Station, and solicited funds from international donors to support the station and ecological research. Local river people, called *caboclos*, were allowed to continue to reside in the area, but agreed upon certain use restrictions. They were trained to serve as game wardens and permanent caretakers. In the first years of existence these collaborative arrangements reportedly resulted in employment for local people, a decline in illegal fishing and logging, and increased research about a little-known riverine ecosystem (Alexander, 1994).

MONITORING RIVER QUALITY

Monitoring water quality is an important, although time-consuming, component of river protection and restoration. Volunteer monitoring programmes can be both an educational tool and a government resource, since salaries are not required to collect the data. Regular data collection provides school-based programmes with an 'outdoor classroom', instilling future citizens with a knowledge of aquatic systems and, in community-based programmes, fosters an awareness of the consequences of human activities for river systems.

Educators in Denmark have linked school curricula with stream monitoring. The Biology Teachers Association of Denmark's project called 'Blue Stream' collects data for a database of all participating observation points (Iversen *et al.*, this volume).

The Global Rivers Environmental Network (GREEN) encourages and supports schools and communities to monitor local river quality by regular water sampling. GREEN provides watershed education resources, maintains web-sites on the Internet, and works closely with business, government, community and educational organizations. GREEN-Europe and Oz GREEN (Australia) coordinate activities in their respective regions (Appendix I).

Community-based stream monitoring programmes exist in many nations. There are non-school voluntary stream monitoring programmes in Costa Rica (Pringle *et al.*, this volume). In Australia more than 200 000 people in over 200 monitoring groups and networks were involved with environmental observations in the late 1990s, including stream and river monitoring. To facilitate networking, the Australian Conservation

Foundation publishes an annotated directory describing these groups (Appendix I).

In the USA, the non-governmental organization River Network was established to help people organize river protection at the local, state and regional level, and to acquire and conserve river lands. The River Network publishes information about different aspects of grassroots river and watershed protection, fundraising and non-profit management, river conservation jobs, and a River and Watershed Conservation Directory. It also maintains a web-site (Appendix I). Kentucky Water Watch maintains a web-site listing national and international volunteer monitoring websites with useful information (Appendix I).

Citizen groups in the USA can affiliate with stream monitoring programmes run by state environmental conservation departments. In the late 1990s approximately 24 000 volunteers monitored more than 985 streams, including some that would not have been monitored otherwise. The volunteer groups were financed by city, state and federal programmes, charitable foundations, universities and research centres, and/or private corporations. Many volunteers raised money locally to purchase equipment or host training sessions. The US Environmental Protection Agency (EPA) facilitates citizen participation by supporting a semi-annual newsletter (*The Volunteer Monitor*) and by maintaining a web-site on the Internet containing a directory of voluntary monitoring groups, information, and links to other water quality resources and institutions (Appendix I).

PREVENTING DISRUPTION OF RIVER FUNCTIONING

Rivers and wetlands have been channelized, dammed, diverted and straightened for centuries. However, in the late 20th century citizen groups emerged around the world to oppose further interventions in riverine ecosystems. Examples from France and Botswana show how government opinion and action can change in response to citizens' campaigns for river protection.

Rhine–Rhône Canal, France

In 1961 a US$9 billion plan was put forward to link the Rhine and Rhône Rivers, permitting year-round navigation between Rotterdam at the mouth of the Rhine River in the North Sea, and Marseilles at the Rhône's delta in the Mediterranean Sea ('Le canal Rhin–Rhône', undated). Called the Rhine–Rhône Canal, the project entailed channelization and

straightening of the Rhône, the Saône and the Doubs rivers in France. Initially ignored, the plan was revived in 1978 by a newly elected conservative government.

As work began on the Rhône, opposition grew in the valleys of the Saône and Doubs because 3700 ha of agricultural land would have to be acquired for the project. This opposition grew into a 16-year campaign (1981–1997) against the project. First local anti-canal committees were formed (Comités de Liaison Anti-Canal – CLAC). Then, to increase their strength, the groups united to form the Coalition Saône–Doubs vivant-Sundgau vivant-WWF. By 1997 the coalition included fishers, consumers, ecology politicians and scientific societies as well as almost 200 federations and nature protection associations from Alsace and the watersheds of the Saône, Doubs and Rhône ('Le canal Rhin–Rhône', undated). The coalition instituted litigation and held rallies. It also engaged in tactics to slow down or block the project, such as buying land required for the engineering works to ensure lengthy expropriation processes.

When the Prime Minister of France attended a local public inquiry in the summer of 1996, four-fifths of the people opposed the canal ('Le canal Rhin–Rhône', undated). During the June 1997 national election campaign, politicians from the Green, Socialist and Communist parties ran on a coalition platform that included a promise to cancel the project. Seven Green Party members were elected, the canal was cancelled and the 1000 ha of acquired land was either returned or sold (European Rivers Network, 1997a).

Okavango Delta, Botswana

The system of government in Botswana is predicated upon consultation between the government and the people at a local public meeting called a *kgotla*. Failure to consult those who will be affected by a decision is considered to be a serious affront (Rothert, 1997).

The Botswana Department of Water Affairs did not consult with northern Botswana's Okavango Delta area residents in 1985 when it began to plan the Southern Okavango Integrated Water Development Project (SOIWDP). The project would involve dredging the main waterway in the lower Delta and constructing three reservoirs. (Burman, 1996).

When earthmoving equipment arrived unannounced in November 1990 to implement the SOIWDP, local people demanded a *kgotla*. They were concerned about dredging. An area of the Delta subjected to minor dredging 30 years before had failed to recover (Pottinger, 1997). A *kgotla* in December ended within 15 minutes because the residents demanded that more senior officials speak. A January 1991 *kgotla* attended by the Minister of Mineral Resources and Water Affairs lasted seven hours as 700 villagers expressed their concerns and asked that the project be stopped (Burman, 1996).

The Tshomorelo Okavango Conservation Trust (TOCT) was formed to campaign against the project and contacted the South African-based Okavango Wildlife Society and Greenpeace International for support (Burman, 1996; Thomas, 1996). The government invited Greenpeace to make a study tour of the Delta (Thomas, 1996). Their report recommended the project be stopped (Thomas, 1996). As opposition mounted the government suspended the project.

In 1991 the government asked the International Union for the Conservation of Nature and Natural Resources (IUCN) to make an independent evaluation of the project (Scudder, 1995). The IUCN report concurred with that of Greenpeace and suggested alternative non-engineering strategies (Scudder, 1995; for report see IUCN, 1992, 1993). The government did not formally accept or discuss the IUCN report (Scudder, 1995). However, in May 1992 the project was shelved. In 1996 Botswana designated the Okavango Delta System as its first Wetland of International Significance, making it the world's largest Ramsar Site (Burman, 1996; Kabii, 1997).

RIVER REHABILITATION AND RESTORATION

Members of the public concerned about aesthetics, water quality, public health, fish stocks or river integrity have initiated river rehabilitation and restoration programmes. The first steps in river rehabilitation are usually physical clean-up of the stream-bed and banks and pollution abatement. In some instances a group's goal is to go beyond rehabilitation to restoration of a stream's full ecological function. This can require the removal of engineering works such as culverts, lining materials, erosion control measures, or dams. Many citizen groups link education and the revival of non-anthropocentric values of rivers to their physical clean-up and restoration activities. Examples from Kobe (Japan), Kingfisher (Canada) and Berkeley (USA) suggest the range of local and community efforts.

Toga River, Japan

The 7 km Toga River is a typical Japanese urban river flowing from the Rokko Mountains to its outlet in the Seto Inland Sea at the Port of Kobe (Hashimoto,

1991). The Kobe section of the river is lined with concrete. In the 1970s citizens became concerned about the river's water quality and society's loss of traditional values about water. The Group to Protect the Toga River was formed in 1977 to clean and protect one section of the river. From this a city-wide campaign developed that had an educational component to revive traditional values of respect for water and rivers.

Shuswap River, Canada

In 1981 the salmon sport fishery on the Shuswap River in British Columbia was closed. Residents of the town of Kingfisher decided to establish a community hatchery in an attempt to rectify the situation (Harvey, 1995). With a small government grant to pay for equipment, 15 full-time and 100 part-time volunteers from the town staffed the facility. The sport fishery was reopened in 1988.

In the process of restocking the salmon fishery, the residents realized the importance of environmental education. In January 1994 an education centre for school children built next to the hatchery opened. The following summer an international exchange programme began with students from a Japanese fishing village (Harvey, 1995).

Strawberry Creek, USA

Strawberry Creek flows from the hills of Berkeley, California, across the campus of the University of California at Berkeley, through a tunnel under the city to its mouth in the San Francisco Bay. Since the mid 1880s the stream's habitat has been degraded by deforestation, stream-bank engineering, channel diversion, and sewage and chemical contamination (Charbonneau and Resh, 1992). A 1987 water quality study commissioned by the University resulted in the formation of the Chancellor's Advisory Committee on Strawberry Creek Environmental Quality to oversee creek restoration.

The Committee included not only university faculty and staff but also representatives from the city of Berkeley and a federal government research laboratory abutting the stream. The Committee's goals were to restore the ecological integrity of the creek, promote its teaching and research values, and provide an innovative example of urban creek restoration. Priorities were reducing pollution and accelerated stream-bank erosion. Within two years of the initial study's inception, water quality had so improved that restocking of fish could begin. Increased public awareness resulting from on- and off-campus environmental education programmes was demonstrated by an upsurge in pollution reporting (Charbonneau and Resh, 1992).

Obstacles to participation in river conservation

The preceding examples come from nations where there is some public access to decision-making bodies and where citizens can take initiative. When political structures are not responsive to citizen opinion, activists organize popularly based campaigns and use economic tactics to exert pressure. Those opposing dams and water diversions throughout the world have used such strategies.

Discussion of campaigns against dams is beyond the scope of this chapter. However, there is a considerable literature about the struggles against the James Bay Project in Canada and the many dams proposed for the Narmada and other rivers in India. Anti-dam tactics and campaigns in the USA, Australia (Tasmania), Brazil, Thailand and India have been reviewed by McCully (1996).

Authoritarian governments in some nations make environmental discussion difficult, if not impossible (Weinberg, 1991). McCully (1996) has argued that large, externally funded hydroelectric schemes are closely associated with dictatorships. Where democratic processes prevail, popular opposition to river engineering projects affects their implementation.

Amplifying local voice: international networking

When multinational interests are involved in river degradation or engineering projects, local opposition campaigns have not usually been sufficient. There have always been international connections among conservationists, but with the growth of computer networks and their use by non-government organizations (NGOs) and social movements in the 1980s, local groups and marginalized people have been able to publicize their concerns, soliciting and receiving national and international support.

In the absence of international regulation and regulatory mechanisms, environmental activists linked by computer networks have played an extremely important role by applying pressure on key institutions that have some regulatory powers (Frederick 1993; Young, 1993; Murphy 1994; Preston, 1994; Dale, 1996; Montavalli and Landskroner, 1996). Computer networks connect activists throughout the world so that pressure can be put on head offices, subsidiaries and foreign governments, as well as local governments and businesses (Lane, 1990; O'Brien, 1992; Sallin, 1994). River conservation activists have been highly successful in employing these tactics in their struggles against industrial river pollution and large-scale hydroelectric dams. Many NGOs use the Association for Progressive Communication (APC) computer network which is distinct from, but can connect to, the Internet (Appendix I).

RIVER CONSERVATION IN CYBERSPACE

The European Rivers Network (ERN) provides information in English, French and German about rivers and campaigns in Europe and the rest of the world at a web-site on the Internet called RiverNet (Appendix I). They also publish 'Riverfax', a bulletin sent by fax 2–4 times per year in English, French, Russian and German to those without access to computer communications. Editions in Spanish and Turkish were being planned in 2000, and other languages could be made available on request.

The International Rivers Network (IRN) provides information at a web-site in English, Spanish, Portuguese and German about rivers and campaigns in the USA and the rest of the world (Appendix I).

Other organizations concerned about rivers from around the world are listed in RiverLinks, a web page accessible from both the ERN and IRN web sites.

RIVER CONSERVATION, ENVIRONMENTAL RIGHTS AND HUMAN RIGHTS

Within the context of multinational opposition action, strong local organization and direction is crucial. In the late 1980s and 1990s conservationists realized that by joining forces with human rights organizations to support the land claims and environmental rights of indigenous people, the environment could be protected in many parts of the world that had yet to be exploited on an industrial scale. *Agenda 21* recognized that guaranteeing indigenous rights was an important aspect of environmental protection.

The close connection between indigenous rights and environmental protection, and the ability of international networking to give voice to cultural minorities, is illustrated below by campaigns to stop pollution of the OK Tedi River in Papua New Guinea and a series of dams on the BíoBío River in Chile.

OK Tedi River, Papua New Guinea

The OK Tedi River in western Papua New Guinea was severely polluted by copper and gold mines managed by BHP, a corporation with multinational investment and Australia's largest company (Anderson, 1994). Objections to this pollution by members of 200 clans resident downstream of the mines were ignored. In May 1994 the residents launched a lawsuit for A\$4 billion – the largest civil claim ever filed in Australia (Anderson, 1994). An international support campaign was initiated to persuade the company to clean-up its operations, restore the river and compensate the victims of industrial pollution. The company was not asked to stop mining, just to stop polluting.

The German environmental group Stanberg Institute lobbied German companies to disinvest from BHP, resulting in a 20% withdrawal of holdings in the OK Tedi mines. According to the chairman of BHP, Coles Myer, the international campaign caused 'global attention [to be] focused on the mining industry as a whole, using OK Tedi as an example of how wicked and horrible mining companies are and how much damage they cause to the environment' (Myer and Clarke, 1996). The campaign was 'part of the ground swell of popular NGO-led opinion about the mining industry in the Third World' (Myer and Clarke, 1996).

On 14 June 1996 Jerry Ellis, the Head of BHP Minerals, announced an out of court settlement of the OK Tedi dispute: A\$110 million in compensation, A\$7 million in legal fees, and promise of a tailings disposal system for mine wastes. The principal lesson learned, Ellis said, was 'the absolute paramount importance of having relations right with communities in which you operate' (Myer and Clarke, 1996).

BioBío River, Chile

During the 1980s the Empresa Nacional de Electricidad, SA (ENDESA), Chile's largest private company, formulated plans for a series of six hydroelectric dams on the free-flowing BíoBío River. The Pehuenche Indians, whose land would be inundated, opposed the dams. Their objections were ignored, and the first dam was built.

Chile began to develop democratic structures in the early 1990s as it emerged from years of military dictatorship. In October 1993 the Chilean Indigenous Peoples Law came into effect, guaranteeing autonomy for indigenous people over their lands and the right to

refuse all offers for resettlement (BíoBío Update #3, 1996). This was followed in 1994 by the Chilean Environmental Law, which required public participation in environmental impact assessments (BíoBío Fact Sheet, undated). These two new laws were used to fight against further damming of the BíoBío.

The Pehuenche Indians worked with national environmental and human rights groups under the Grupo de Acción por el BíoBío (GABB) in Santiago de Chile. Together they launched local campaigns and legal action against ENDESA and sought international support. GABB issued press releases, fact sheets and updates that were faxed and sent by electronic mail nationally and internationally. Representatives of indigenous peoples from North and South America held a meeting in the Upper BíoBío, issued a Friendship and Cooperation Agreement and planned supportive demonstrations. (BíoBío Update #7, 1996).

Local and international activists lobbied donors in the Nordic nations and the World Bank's International Finance Corporation (which financed the first dam), as well as the Chilean Government, to stop the project and support the rights of the Pehuenche Indians (BíoBío Fact Sheet, undated). Calls for specific action were posted on computer networks. During the 120-day citizen comment period for the environmental impact statement, a call for scientists to help 'form a unified, multidisciplinary team that will present a well-founded opinion about the EIA' was posted on the computer networks (BíoBío Update #5, 1996). GABB was able to submit a professional and critical report. In August the Chilean National Commission on the Environment declared the EIA to be unsatisfactory (European Rivers Network, 1997b).

The fight to stop the Ralco Dam was seen as a crucial test for Chilean democracy. The new laws guaranteeing public participation and citizens rights were used against the wealth and power of a major corporation (BíoBío Fact Sheet, undated). In 1997 the Goldman Environmental Prize for grassroots environmentalism was awarded to Juan Pablo Orrego, the leader of GABB, for making the campaign to protect one of the world's last free-flowing rivers the defining environmental issue of post-Pinochet Chile (European Rivers Network, 1997a,b). In 1998 he was awarded the Right Livelihood Award for his work (Orrego, 1998).

However, the struggle continued. A 1998 peaceful occupation of roads and the construction site resulted in a delay of several months until some compensation claims were resolved. A preventive ruling delivered on 8 September 1999 again halted construction, but was reversed by an appeal court decision. The environmental and human rights campaign against the Ralco Dam has been ensnared in the larger debate within Chilean society between supporters of the former Pinochet government and its opponents, who control different parts of the government and judicial system (BioBío Update #17, 1999).

Discussion

Different systems for valuing nature underlie all environmental discussion, whether acknowledged or not. Dominant economic forces (capitalist and socialist) in the 19th and 20th centuries were informed by anthropocentrism and saw the 'conquest of nature' as a sign of civilization and a demonstration of power. In the late 20th century some governments passed legislation limiting the uses of rivers or protecting specific stretches of rivers. Citizens have used these regulations to protect, rehabilitate or restore streams.

People concerned about their rivers have developed educational programmes in schools and community institutions. The long-term goal of these programmes is to teach children and adults about the properties of a stream and its function. Some educational efforts have had as a goal the reassertion of traditional value systems of inherentism or intrinsicalism.

River protection is not always supported by governments or laws. In some places citizens have had to organize themselves into groups and coalitions to put pressure upon economic entities and government agencies to achieve river protection. At times protest against an unpopular government or policy has overshadowed environmental protest. Environmental campaigners have welcomed mass popular support, but have not lost sight of environmental issues.

The rivers of old industrial nations have long been used and abused. Concerns in these nations are removal of pollutants or the rehabilitation and restoration of river ecosystems. In newly industrializing or non-industrial nations, the protection of river function is still a major concern. As dams and diversions are proposed by industry or central government, rural ways of life and the very existence of minority cultural groups are threatened. These rural and minority voices have been strengthened through the international networking capabilities of non-governmental organizations, particularly through computer networks. River activists have joined with cultural survival groups in struggles to preserve river function. The struggles for

human rights, indigenous rights, environmental rights and river protection have become inseparable.

It is not only indigenous or traditional cultures that have environmental rights. Citizens of urban industrial areas have insisted that rivers are public resources to be used and enjoyed by all, rather than monopolized by a few economic interests. In making these claims, they assert inherent and intrinsic values as important components of their lives and sense of well-being. All over the world basic human rights are a major component of environmental protection.

Acknowledgements

I would like to thank Louise Fortmann, Brian Martin Murphy and Rhonda Janke for comments and suggestions, Ed Himlan, Alice Rojko and Manuel King for artwork.

References

Adams, W.M. 1992. *Wasting the Rain: Rivers, People and Planning in Africa.* University of Minnesota Press, Minneapolis, Minnesota.

Alexander, B. 1994. People of the Amazon fight to save the flooded forest. *Science* 265:606–607.

Anderson, I. 1994. Villagers sue mine over 'ruined river'. *New Scientist* 142:4.

Anger, D. 1989. 'No queremos el refugio': conservation and community in Costa Rica. *Alternatives (Canada)* 16:18–22.

Asopa, S.K. 1993. Environmental conservation through the people's movements in India. *Environmental Education and Information* 12:297–306.

Barclays National Review 1973. The story of Cabora Bassa. March: 10–13.

BioBío undated and 1996. International Rivers Network, Berkeley CA. Fact Sheet (undated), Update #3 (31 January), Update #5 (2 April), Update #7 (5 December), Update #17 (7 October). http://www.irn.org/irn/programs/biobio/

Boon, P.J. 1991. The role of Sites of Special Scientific Interest (SSSIs) in the conservation of British rivers. *Freshwater Forum* 1:95–108.

Boon, P.J. 1992. Essential elements in the case for river conservation. In: Boon, P.J., Calow, P. and Petts, G.E. (eds) *River Conservation and Management.* John Wiley, Chichester, 11–33.

Boon, P.J. 1994. Nature conservation. In: Maitland, P.S., Boon, P.J. and McLusky, D.S. 1994. *The Fresh Waters of Scotland: A National Resource of International Significance.* John Wiley, Chichester, 555–576.

Boon, P.J. 1995. The relevance of ecology to the statutory protection of British rivers. In: Harper, D.M. and

Ferguson, A.J.D. 1995. *The Ecological Basis for River Management*. John Wiley, Chichester, 239–250.

Box, J.D. and Walker, G.J. 1994. Conservation of the Blythe, a high quality river in a major urban area of England. *Aquatic Conservation: Marine and Freshwater Ecosystems* 4:75–85.

Burman, E. 1996. Botswana NGOs question. *Ecodecision* 22:60–62.

Charbonneau, R. and Resh, V.H. 1992. Strawberry Creek on the University of California, Berkeley campus: a case history of urban stream restoration. *Aquatic Conservation: Marine and Freshwater Ecosystems* 2:293–307.

Coetzee, C. 1995. They'll have to shoot all Himbas first. *Tempo*, Windhoek, Namibia, 12 February 1995.

Coordinating Body for Indigenous People's Organizations of the Amazon Basin (CBIPOAB) 1990. It's our rain forest. *Mother Jones* 15:47.

Costanza, R., D'Arge, R., De Groot, R., Farber, S., Grasso, M., Hannon, B., Limburg, K., Naeem, S., O'Neill, R.V., Paruelo, J., Raskin, R.G., Sutton, P. and Van Den Belt, M. 1997. The value of the world's ecosystem services and natural capital. *Nature (London)* 387:253–259.

Cronin, J. and Kennedy, R.F. 1997. *The Riverkeepers: Two Activists Fight to Reclaim our Environment as a Basic Human Right*. Scribner, New York.

Dale S. 1996. *McLuhan's Children: The Greenpeace Message and the Media*. Between the Lines, Toronto.

Davies, B. 1975a. What's happening to the Zambezi? *African Wildlife* 29:18–21.

Davies, B. 1975b. They pulled the plug out of the Lower Zambezi. *African Wildlife* 29:26–28.

De La Court, T. 1990. *Beyond Brundtland: Green Development in the 1990s*. Zed Press, London.

European Rivers Network 1997a. Rhine–Rhone Canal Cancelled. Press release, Le Puy, France.

European Rivers Network 1997b. Goldman Prize 97 for Chilean Bío Bío River activist. 4 April 1997, Le Puy, France.

Executive Office of Environmental Affairs (EoEA) 1998. *The Massachusetts Watershed Initiative and its Implementation, Status Report, June 1998*. Massachusetts Executive Office of Environmental Affairs, Boston.

Fleischer, T. 1993. Jaws on the Danube: water management, regime change and the movement against the Middle Danube Hydroelectric Dam. *International Journal of Urban and Regional Research* 17:429–443.

Fodor, A. 1997. Two more dams for the Danube? Press Release 25 November 1997, European Rivers Network. http://www.rivernet.org/press_e2htm#Appeal

Frederick, H. 1993. Computer networks and the emergence of global civil society. In: Harasim, L.M. (ed.) *Global Networks: Computers and International Communication*. MIT Press, Cambridge, Massachusetts.

Green, C.H. and Tunstall, S.M. 1993. *The Ecological and Recreational Value of River Corridors: An Economic Perspective*. Flood Hazard Research Centre, Middlesex University, Middlesex.

Green, C.H., Tunstall, S.M. and House, M.A. 1988.

Evaluating the Benefits of River Water Quality Improvements. Flood Hazard Research Centre, Middlesex Polytechnic, Middlesex.

Grove, R.H. 1992. Origins of western environmentalism. *Scientific American* 267:42–47.

Grove, R.H. 1995. *Green Imperialism: Colonial Expansion, Tropical Island Edens and the Origins of Environmentalism, 1600–1860*. Cambridge University Press, Cambridge.

Harring, S.L. 1996. Namibian villagers reject dam proposal: but government is determined. *Development Dialogue* 5:13, August–September, Harare, Zimbabwe.

Harvey, A. 1995. Kingfisher, BC: a community of environmental ambassadors. *Harrowsmith Country Life* 20:9.

Hashimoto, Y. 1991. Citizen's movements to protect the environment of rivers flowing into the Seto Inland Sea – an example of a citizen's movement along the Toga River. *Marine Pollution Bulletin* 23:621–622.

Holmes, N.T.H. and Nielsen, M.B. 1998. Restoration of the Rivers Brede, Cole and Skerne: A joint Danish and British EU-LIFE Demonstration Project, I – Setting up and delivery of the project. In: Hansen, H.O., Boon, P.J., Madsen, B.L. and Iverson, T.M. (eds) *River Restoration: The Physical Dimension, Special Issue, Aquatic Conservation: Marine and Freshwater Ecosystems* 8:185–196.

Howell, D.L. and Mackay, D.W. 1997. Protecting freshwater quality through legislation: enforcement, inducement or agreement. In: Boon, P.J. and Howell, D.L. (eds) *Freshwater Quality: Defining the Indefinable?* The Stationery Office, Edinburgh, 457–481.

Hungarian Ministry for Foreign Affairs. 1997. Gabcikovo-Nagymaros case of Hungary and Slovakia before the World Court in the Hague. 3 March 1997. http://www.meh.hu/KUM/defhu.htm

Inambao, C. 1997. Epupa gets green light, Nujoma slams 'sinister manipulators'. *The Namibian*, Windhoek, Namibia, 19 August 1997.

International Union for Conservation of Nature and Natural Resources (IUCN) 1992. *The IUCN Review of the Southern Okavango Integrated Water Development Project: Final Report*. October, IUCN, Gland.

International Union for Conservation of Nature and Natural Resources (IUCN) 1993. *The IUCN Review of the Southern Okavango Integrated Water Development Project*. The IUCN Wetlands Programme, IUCN, Gland.

Jewell, P., Gildesgame, M. and VanDusen, M. 1998. *The Massachusetts Watershed Initiative: Opportunities and Challenges in Reshaping Government*. Massachusetts Executive Office of Environmental Affairs, Boston.

Kabii, T. 1997. Okavango – the world's largest Ramsar site. *Ramsar Archives*, 24 February 1997. http://www.iucn.org/themes/ramsar/w.n.okavango.htm

Kottak, C.P. and Costa, A.C.G. 1993. Ecological awareness, environmentalist action, and international conservation strategy. *Human Organization* 52:335–343.

Lane, G. 1990. *Communications for Progress*. Catholic Institute for International Relations, London.

Le canal Rhin–Rhône (suite): le canal de l'absurde, undated. http://www.rivernet.org/rhinrhon/rhrho2_f.htm

Lemons, J. and Saboski, E. 1994. The scientific and ethical implications of Agenda 21: biodiversity. In: Brown, N.J. and Quiblier, P. (eds) *Ethics and Agenda 21: Moral Implications of a Global Consensus*. United Nations Environment Programme, United Nations Publications, New York.

Loomis, J.B. 1996. Measuring the economic benefits of removing dams and restoring the Elwha River: results of a contingent valuation survey. *Water Resources Research* 32:441–447.

Maltby, E. 1991. Wetland management goals: wise use and conservation. *Landscape and Urban Planning* 20:9–18.

MacKenzie, J.M. 1988. *The Empire of Nature: Hunting, Conservation and British Imperialism*. Manchester University Press, Manchester.

Martinez-Alier, J. 1991. Ecology and the poor: a neglected dimension of Latin American history. *Journal of Latin American Studies* 23:621–639.

Massachusetts Watershed Coalition (MWC), undated. *MWC Program Summary*. Leominster, Massachusetts.

McCully, P. 1996. *Silenced Rivers: The Ecology and Politics of Large Dams*. Zed Press, London.

Montavalli, J. and Landskroner, R. 1996. Cruising the Green Net. *E – The Environmental Magazine* 7:34.

Murphy, B.M. 1994. Broadcasting crises through new channels in the Post-New World Information order era: alternative news agencies and the computer networks of non-governmental organizations. *Journal of International Communication* 1:88–111.

Myer, C. and Clarke, N. 1996. BHP, OK Tedi and the building industry. Australian Broadcast Corporation, Radio National Transcripts, The Business Report, 14 June 1996.

O'Brien, R. 1992. APC computer networks: global networking for change. *Canadian Journal of Information Science* 17:16–24.

Orrego, J.P. 1998. *Acceptance Speech, Right Livelihood Award*, 9 December 1998, Stockholm, Sweden. European Rivers Network, Le Puy, France.

Osemeobo, G.J. 1994. The role of folklore in environmental conservation: evidence from Edo State, Nigeria. *International Journal of Sustainable Development and World Ecology* 1:48–55.

Oyadomari, M. 1989. The rise and fall of the nature conservation movement in Japan in relation to some cultural values. *Environmental Magazine* 13:22–33.

Palmer, T. 1986. *Endangered Rivers and the Conservation Movement*. University of California Press, Berkeley.

Palmer, T. 1993. *The Wild and Scenic Rivers of America*. Island Press, Washington, DC.

Pearce, F. 1992. First aid for the Amazon. *New Scientist* 133:42–47.

Pottinger, L. 1997. Namibian pipeline project heats up. *World Rivers Review*, February 1997.

Preston, S. 1994. Electronic global networking and the 1992 Rio Summit and beyond. *Swords Into Ploughshares* 3(2), Spring 1994. spsis@american.edu

Pretty, J.N. and Pimbert, M.P. 1995. Beyond conservation ideology and wilderness myth. *Natural Resources Forum* 19:5–14.

Reed, S.W. 1984. The scenic St. Joe: a study of resistance to federal river protection. *Northwest Environmental Journal* 1:171–185.

Reynolds, P. 1995. Barrage of fire. *International Water Power and Dam Construction* 47:52–53.

River Network 1998. *Massachusetts Background Report, Four Corners Watershed Innovators Initiative*, 5 October 1998. River Network, Portland, Oregon.

River Restoration Project 1994. *Institutional Aspects of River Restoration in the UK, Part One. Summary*, March 1994. River Restoration Project, Huntingdon.

Rothert, S. 1997. Okavango chiefs rally to oppose Namibian pipeline project. *World Rivers Review*, August 1997.

Sallin, S. 1994. The Association for Progressive Communications: a co-operative effort to meet the information needs of non-governmental organizations. Harvard-CIESIN Project on Global Environmental Change Information Policy, Cambridge, USA.

Schreyogg, G. and Steinmann, H. 1989. Corporate morality called into question: the case of Cabora Bassa. *Journal of Business Ethics* 8:677–685.

Scudder, T. 1995. The big dam controversy and environmental fundamentalism: musings of an anthropologist. *Development Anthropology* 13:8–18.

Showers, K.B. 1996. Soil erosion in the Kingdom of Lesotho and development of historical environmental impact assessment. *Ecological Applications* 6:653–664.

Showers, K.B. and Malahleha, G.M. 1992. Oral evidence in historical environmental impact assessment: soil conservation in the 1930s and 1940s. *Journal of Southern African Studies* 18:276–296.

Shuman, J.R. 1995. Environmental considerations for assessing dam removal alternatives for river restoration. *Regulated Rivers: Research & Management* 11:249–261.

The Namibian 1997a. Mines ministry loses cool over Himba trip. Windhoek, Namibia, 18 June 1997.

The Namibian 1997b. Ministry's approach is embarrassing. Windhoek, Namibia, 20 June 1997.

Thomas, A. 1996. NGO advocacy, democracy and policy development: some examples relating to environmental policies in Zimbabwe and Botswana. In: Potter, D. (ed.) *NGOs and Environmental Policies in Asia and Africa*. Frank Cass, London, 38–65.

Tinley, K. 1975. Marromeu: wrecked by the big dams. *African Wildlife* 29:22–25.

United Nations Conference on Environment and Development (UNCED) 1993. *Agenda 21: Programme of Action for Sustainable Development*. United Nations Department of Public Information, United Nations, New York City.

Watershed Initiative Steering Committee (WISC) 1995. *The*

Massachusetts Watershed Approach and its Implementation: Status Report, October 1995. The Watershed Initiative Steering Committee, Massachusetts Executive Office of Environmental Affairs, Boston.

Webb, L.J. and Smyth, D.M. 1984. Ecological guidelines and traditional empiricism in rural development acknowledgements. *The Environmentalist* 4:99–105, Suppl.7.

Weinberg, B. 1991. *War on the Land: Ecology and Politics in Central America*. Zed Press, London.

Wilson, K.B. 1995. 'Water used to be scattered in the landscape': local understandings of soil erosion and land use planning in southern Zimbabwe. *Environment and History* 1:281–296.

Young, J.E. 1993. *Global Network: Computers in a Sustainable Society*. Paper 115, World Watch Institute, Washington, DC.

Appendix I: Contact details for organizations cited in the text

Association for Progressive Communications (APC)
http://www.apc.org
APC Secretariat, e-mail information: apc-info@apc.org

Australian Conservation Foundation
http://www.acfonline.org.au
e-mail: reception@acfonline.org.au

European Centre for River Restoration
e-mail: ECRR@dmu.dk

European Rivers Network
http://www.rivernet.org
e-mail: ern@rivernet.org

Global Rivers Environmental Education Network (GREEN)
http://www.earthforce.org/green/
e-mail: green@earthforce.org

Oz GREEN (Australia) has programmes in India, Nepal, and Papua New Guinea as well as Australia
http://www.ozgreen.org.au
e-mail: lennox@ozgreen.org.au

International Rivers Network (IRN)
http://www.irn.org

Kentucky Water Watch Volunteer Monitoring Groups On-Line
http://www.state.ky.us/nrepc/water/vm.htm

Massachusetts Watershed Coalition
e-mail: mwc@ma.ultranet.com

The River Network
http://www.teleport.com/~rivernet/

National Office e-mail: info@rivernetwork.org

Eastern Office e-mail: rivernet2@aol.com

Northern Rockies Office e-mail: montanazac@aol.com

River Restoration Centre
http://www.qest.demon.co.uk/rrc/rrc.htm
e-mail: rrc@cranfield.ac.uk

US Environmental Protection Agency (EPA) Volunteer Monitoring
http://www.epa.gov/owow/monitoring/vol.html

The Volunteer Monitor, National Newsletter of Volunteer Water Quality Monitoring
http://www.epa.gov/owow/monitoring/volunteer/vm_index.html

22

The role of legislation in river conservation

C.G. Palmer, B. Peckham and F. Soltau

In memoriam

After preparing the penultimate draft of this chapter, Brian Peckham died very suddenly from lung cancer. Brian was not just amazing in the way he coped with being a quadriplegic – though he was – he was an extraordinary person in himself. His great gift to everyone who knew him was just that, knowing him. We experienced his formidable intelligence, wicked twinkle and infectious guffaw; his warm hugs and quick riposte; his integrity; his clear, robust faith; his curiosity and questioning, his optimism and his pragmatism.

'He was a verray parfit gentil knight . . .'
(Chaucer, *Canterbury Tales – General Prologue*)

Introduction

Ultimately, it is in legal structures that society expresses its values, and environmental conservation is one of the emerging values of the late 20th century. This chapter explores the role of legislation in securing and defining the manner in which river ecosystems can be, and are, conserved.

As a primary practical mode of expressing and enforcing human values, law can play a major role in river conservation. There are, however, key factors which determine the success of the three main stages in legislation – policy development, legal drafting and implementation – especially if the ultimate goal is the sustainable, integrated management of freshwater ecosystem health (as is consistent with 'Agenda 21' resulting from the 1992 United Nations Conference on Environment and Development, in Rio de Janeiro).

These key factors include a good ecological understanding of river functioning (Boon, 1995; Palmer, 1999; Moss, 2000), together with adequate monitoring, referencing and classification of riverine ecosystems (Howell and Mackay, 1997; Pollard and Huxham, 1998) and, most importantly, an integrated approach which takes account of terrestrial–aquatic links and accepts the catchment as the unit of legislation and management (Newson, 1992; Department of Water Affairs and Forestry – DWAF, 1996a; Pollard and Huxham, 1998; Moss, 2000). It is also vital to understand the social and economic values that interact with environmental values in decision-making, as legislation can only succeed if its requirements are practicable, and reflect values accepted by – or at least not actively opposed by – society (Hahlo and Kahn, 1968). The political and moral will of the legislators plays a leading role in determining legal content (Hart 1984). Finally, there needs to be the capacity to implement the law. Frequently there are limitations to policing – places may be inaccessible, and, particularly in developing countries, skills, personnel and financing may be limited. The use of inducement and agreement needs to be actively explored as adjuncts to enforcement in implementation (Howell and Mackay, 1997). The goal of legislation in promoting the 'constructive interplay of sound ecological science and societal values which ensures ecological integrity and allows a sustainable yield of ecosystem goods and services' (Pollard and Huxham, 1998) requires the multidisciplinary cooperation of legal specialists, water resource managers and users, land-use specialists and aquatic scientists (Caponera, 1992).

However, the complexity both of riverine ecosystems

Global Perspectives on River Conservation: Science, Policy and Practice.
Edited by P.J. Boon, B.R. Davies and G.E. Petts. © 2000 John Wiley & Sons Ltd.

and society makes each key factor dependent on other factors. For example, a good understanding of river function is not enough: 'we as limnologists may have failed to ensure that our knowledge has been effectively applied in the drafting of protective legislation' (Moss, 2000). Ecological information needs to be communicated and ecological concepts require accessible, effective advocacy. Integrated catchment management is the ultimate challenge (Hooper and Margerum, this volume), requiring among other things, working links between all levels of government (local, regional and national) and between the usually separate spheres of agriculture, water administration and conservation. There is also a need to move from the narrow assessment of chemical water quality to the broad assessment of resource quality, which includes biotic and physical integrity, and habitat assessment (Howell and Mackay, 1997).

In many ways rivers force an integrated approach: they traverse landscapes, reflecting both ecoregional gradients and anthropogenic change, provoking a recognition of the limitations of 'fortress' conservation with the focus on special sites (Boon, 1995; Moss, 2000) and encouraging integration. It is not surprising that integrated catchment management initiatives come more frequently from the aquatic than the terrestrial perspective (e.g. the European Water Framework Directive (Pollard and Huxham, 1998)).

The opportunity to revise completely the legal basis for the administration of water resources is rare. We have therefore used the South African experience of a comprehensive water law review to illustrate the pitfalls and successes of a legislation development process, and to provide an insight into the role which legislation can play in river conservation. We include a comparative perspective, using examples from the USA, Australia and Finland as representatives of countries with a wide range of climatic regimes (legislation from all four countries is contained in Tables 22.1–3, grouped according to whether its impact is broadly on quantity, quality or river conservation respectively).

The South African water law review

Since the first democratic elections in 1994, legal reform in South Africa has been aimed at the removal of the *apartheid* system of government. Part of this legal reform was an initiative to review comprehensively the (now repealed) Water Act (No. 54 of 1956). Although the primary driving force behind this review was the political and social goal of equitable access to water resources by all South Africans, it also provided the opportunity to include a sound legal basis for river – and other aquatic ecosystem – conservation.

The process of water law review in South Africa was a remarkable one, with widespread public participation and consultation (some 2000 pages of comment being received on the review process and the need for reform alone). New legal provisions changing the way in which water will be allocated and managed were tested against both the South African Constitution, and the legal systems in other countries.

One of the first products of the water law review process in South Africa was the development and acceptance of a set of principles on which the new law would be based (DWAF, 1996b). Key principles were developed to guide the process of water resource management on a sustainable basis, and protection of aquatic ecosystems (DWAF, 1996a). The principles formed the basis of the subsequent *White Paper on a National Water Policy for South Africa* (DWAF, 1997a) which in turn became the policy basis for the drafting of the law, the National Water Act (No. 36 of 1998).

The key Principle from a river conservation perspective reads as follows: 'The quantity, quality and reliability of water required to maintain the ecological functions on which humans depend shall be reserved so that the human use of water does not individually or cumulatively compromise the long term sustainability of aquatic and associated ecosystems.'

The process of communicating the ecological basis for this principle (detailed in Palmer, 1999) depended upon the understanding by lawyers and water resource managers of certain basic concepts: that the shape of a river channel from source, through the estuary, to the sea, is a product of geomorphological and hydrological processes; that water quantity (specifically the amount of rainfall that becomes run-off) together with geology and topography, results in a particular channel form (Gordon *et al.*, 1992); that the combination of discharge, velocity, slope and substrate results in the availability of a range of physical and hydraulic habitats (Statzner *et al.*, 1988; Newson, 1996; Rowntree, 1996), and that it is these habitats that become the abiotic template of the ecological niche and determine aspects of biotic diversity; and finally, that as the quantity of water in a river is altered by human activities (either artificially increased or decreased, or the seasonality of flow changed) the nature of aquatic habitats, and therefore the biota, is altered.

Similarly, hydrology and geology combine to produce a natural pattern of water quality, which is

Tables 22.1–22.3 Explanatory note to comparative tables of legislation. A sample of water legislation from four jurisdictions – Australia, Finland, the United States and South Africa – was chosen and grouped under three headings: water quantity and water supply, water quality, and conservation legislation and related provisions. Such a division is a rough one, and certain laws contain provisions relevant to more than one heading. The tables give an indication first, of what laws (if any) a sample country has enacted in a particular field, and second how such legislation compares with that of other countries. Interesting to note is the extent to which general water management laws incorporate conservation provisions and what 'mix' of legislation has been chosen

Table 22.1 *Water quantity and water supply*

Legislation	Provisions
1. Australia	
1.1 Commonwealth enactments: Snowy Mountains Hydro-Electric Agreements Act; National Water Resources (Financial Assistance Act) 1978; Environmental Protection (Impact of Proposals) Act 1975; River Murray Waters Agreement Act 1917 (and amendments)	Environmental legislation is not within jurisdiction of Commonwealth Parliament. Only interstate rivers are partly covered by Commonwealth (federal) legislation under the federal trade and commerce jurisdiction. Acts deal with: federal financial involvement in water resource management, soil conservation and nature conservation; as well as environmental impacts of activities of federal agencies (Environmental Protection (Impact of Proposals) Act 1975); regulation of the operation and maintenance of waterworks and water entitlements from shared rivers.
1.2 Individual states' Water Acts include: Water Act 1912 (New South Wales); Water Act 1989 (Victoria); Water Resources Act 1989 (Queensland); Water Resources Act 1990 (South Australia)	Most of these Acts are characterized by placing under Crown control the power over the allocation and utilization of water resources. Thus although some characteristics of riparian rights may survive within particular Acts, there is no general recognition of riparian-based water rights. Water policies are imposed by means of a system of licence or permit by the state. Permits to utilize water or undertake other projects in relation to rivers are not dependent on occupation or ownership of adjacent riparian areas. Licensing procedures – involving applications and official valuation of intended activities – apply in respect of all activities affecting watercourses and water resources, and cover issues such as abstraction, irrigation, drainage, flood control, etc. Environmental protection is largely incidental, as the legislation is aimed primarily at water utilization issues. The tendency to move away from riparianity has been strengthened by the introduction of transferable water abstraction entitlements.
2. Finland	
Water Act 1961	Water, whether in lakes, rivers or other water resources, is privately owned. However, the state, through a system of Water Courts, exerts control over the use of water. Permits are required for activities such as water supply, hydropower, log-floating and water-flow regulation. Permit conditions to protect natural flow, fishery, recreational or traffic access are prescribed where necessary. Permits will not be issued if the activity causes significant damaging changes in the ecosystem.
3. United States	
3.1 Federal: Water Resources Planning Act 42 USC 1962	This Act created the Water Resources Council and directed the adoption of particular principles in water resource management. The work of the Council has led to the adoption of certain principles and standards for planning water and related land resources which utilize cost–benefit analysis as one of the evaluation tools, but requires the consideration of such factors as environmental quality, social well-being, and regional economic development.
3.2 States: A multitude of legislative enactments, in addition to common law rules apply	In the drier Western states (e.g. Texas) the prior water appropriation regime predominates. Appropriators have the right to a quantity of water used beneficially, to the exclusion of other less senior appropriators and the ecological needs of the environment. In the eastern states the riparian system, which entitles riparian owners to a 'reasonable' share of the water, was adopted and still exists. There are 'mixed' systems such as in California which neither adhere entirely to the prior appropriation nor to the riparian system. There is a growth in permitting and concern for the maintenance of in-stream flows.
4. South Africa	
National Water Act 36 of 1998	This piece of legislation deals comprehensively with the management of water resources. As regards access to water it replaces the riparian system with a system of administrative allocations. The management of water resources is largely devolved onto catchment management agencies. These agencies must when deciding on allocations take into account, amongst other things, factors relevant to the protection and conservation of rivers. Water allocation subject to the 'Reserve', which is that quantity and quality of water necessary to satisfy basic human needs, and to protect aquatic ecosystems to ensure sustainable use of that particular water resource.
Water Services Act 108 of 1997	Sets out framework for ensuring that the constitutional (s 27(1)) right to adequate water and sanitation services is realized.

Table 22.2   *Water quality legislation*

Legislation	Provisions
1. Australia	
1.1 Commonwealth enactments	Federal structure and extensive state regulation means that there is no general Commonwealth legislation dealing with water quality (but see River Murray Waters Agreement Act 1917, as amended, which does deal with water quality). National Water Quality Management Strategy is intended as a basis for future Australian water quality framework legislation.
1.2 States: various state enactments	General water legislation also contains provisions regarding water pollution, creating offences in respect of polluting activities. Water Act 1912 (New South Wales) bans pollution of water by any harmful substance described in the Act and creates a criminal sanction. Specific, sectoral anti-pollution legislation also exists, such as the New South Wales Clean Water Act 1970 and regulations, which classify waters according to existing and future uses and establish acceptable loads of pollutants in areas where different covering capacities exist. In other states integrated pollution legislation is contained in a single item of legislation – for instance the Environment Protection Act 1970 (Victoria). The integrated approach in Victoria is interesting in that all pollution emissions must be in compliance with state environmental protection policy on the particular category of activity. There is a general ban on pollution but discharges to waters must comply with the above state environment protection policy criteria in respect of aquatic reserves, parks and forests, estuaries, coastal waters, and general surface waters, each of which has its own separate water protection policy.
2. Finland	
Water Act 1961	Control of water pollution not integrated, regulated by means of a permit system. General prohibition on granting permits if the project has major detrimental effects on the natural surroundings or to aquatic ecology. Principle of best available technology (BAT) is applied. Non-point-source pollution, for instance caused by agriculture, remains a problem but the European Union Directive concerning the discharge of nitrates from agriculture (No. 91/676/EEC) is to be implemented in legislation.
3. United States	
3.1 Federal: The Clean Water Act 33 USC 1362 (7)	Aims at *restoring and maintaining the chemical, physical and biological integrity of the nation's waters,* elimination of all pollutant discharges into navigable waters. Creates a system of permitting, which applies to all proposed activities involving wetlands or other water bodies, with permits required under section 404 regulations where impacts considered substantial.[a] Apart from taking into account environmental effects such as those on wetlands, fish and wildlife, water quality[b], s 404 allows exemptions e.g. ongoing agricultural or forestry activities, maintenance of structures, farm ponds, irrigation and drainage channels, road construction and maintenance in mining areas and sedimentation control activities. S 304 of the Act permits individual control strategies in difficult situations for control of pollution going into rivers. These apply water quality standards, determined in some states by regional water quality control boards – in many cases water quality control plans for particular basins have been developed. These often contain water quality objectives and in particular the concept of beneficial use being unduly affected by discharges from non-point sources is identified.
3.2 Individual states	Many versions of water pollution, environmental and river statutes exist.
4. South Africa	
National Water Act 36 of 1998	Section 21 lists a broad range of 'water uses' which can only be exercised with a licence, under exemption or subject to a general authorization. Discharging waste or disposing of waste in a manner which may have a detrimental impact on a water resource will usually require a licence, general authorization or exemption. Prior to granting a licence or authorization the responsible authority must consider a range of factors – amongst others, the catchment management strategy applicable to that water resource and the quality of the water required for the Reserve (see above: Water quantity and water supply). Licences may be issued subject to conditions, covering, for instance, permissible chemical content of waste water.
Conservation of Agricultural Resources Act 43 of 1983	Has incidental significance for water quality protection, provides for control of non-point-source pollution by run-off of agricultural chemicals, soil erosion, etc.
Minerals Act 50 of 1991	The Act provides for regulations to protect water sources and prevent water pollution.

[a] National Environmental Policy Act EIA regulations may also apply, and statutes e.g. the Fish and Wildlife Act, Migratory Marine Game Fish Act, the Fish and Wildlife Coordination Act, the National Historic Preservation Act, the Interstate Land Sales Full Disclosure Act and the Endangered Species Act.
[b] Subject also to state water pollution control laws.

Table 22.3   *Conservation legislation and related provisions*

Legislation	Provisions
1. Australia	
1.1 Commonwealth enactments	No general competence on the part of the Commonwealth to deal with nature conservation, save where the impacts of federal agencies are concerned (Environmental Protection (Impacts of Proposals) Act 1974) or by agreement with the states. Statutes such as the National Parks and Wildlife Conservation Act 1975 and the Endangered Species Protection Act 1992 (providing for listing of endangered species and the protection of ecological communities) may be of relevance to river conservation.
1.2 States Rivers and Foreshores Improvement Act 1948 (NSW); River Improvement Act 1958 (Victoria); River Improvement Trust Act 1940 (Queensland); Waterways Conservation Act 1976 (Western Australia); etc.	Various enactments impose government controls on the management of rivers and catchments. Typically these control activities which will affect watercourses: removal of timber or vegetation from banks, bed or shores of rivers and tidal waters; activities altering the natural course of rivers; erosion of river banks and beds including riparian areas; siltation; artificial alteration of river courses; inflow of sea water into river courses, etc. In some states, general catchment management schemes are in existence. Catchment Management Act 1989 (NSW) provides for the establishment of catchment management committees, etc. to consider the impact of all natural resources exploitation on the general well-being of a catchment.
Environmental Planning and Assessment Act 1979 (NSW)	Environmental impact assessment requirements are set in respect of particular types of activities affecting catchments, riparian areas and rivers generally.
Water Supply Authorities Act 1987 (NSW)	Contains provisions relating to the promotion of fishing to water resource use; necessity of integrated catchment management and land-use planning; community needs; conservation of natural resources; prevention and control of pollution; etc.
2. Finland	
Water Act 1961	Contains regulations relating to the conservation and protection of certain biotopes, e.g. wetlands, groves, etc. Activities affecting protected biotopes require permits. Act allows the prescription of regulations in respect of natural flow, fisheries, recreation, etc.
Forestry Act 1996	Provides limitation on right to drain woodlands or wetlands for forestry purposes; protection of certain biotopes; degradation of land.
Wilderness Act 1991	Limits land uses in particular areas in keeping with, for example, Forestry Act and the Water Act.
Building Act 1958	Subjects development activities to permit controls, has implications for water use planning, planning for natural resource protection and management. Provides for controls on land-use planning, building of towns and attendant construction, as well as the protection of soil and vegetation. Enables the identification of protected areas on the basis of conservation value. The general landscape is protected against detrimental effects of land-use decisions. Enables the compulsory expropriation of particular areas, e.g. in connection with the establishment of plan reserves and can be utilized in respect of river bank areas, open water space, etc. Makes possible the creation of shore plans and the regulation of settlement on shorelines, islands and other water areas. In regulating such settlements, discharges, water supply, conservation values, etc. are taken into account.
Environmental Impact Assessment Act 1994	Formulated to bring Finnish Law in line with the EU Directive on EIAs of 1985 as well as the 1991 EEC Convention on Trans-boundary Impact Assessment. Applies to major projects only.
Nature Conservation Act 1997	Contains various provisions obliging landowners to ensure preservation of sound environmental values. Provisions aimed at protecting the environment and its quality against degradation. Provisions exist for protective management, conservation, restoration of damaged aquatic environments, etc. Contains provisions to maintain biological diversity, promote natural beauty and preservation of the landscape and the sustainable utilization of natural resources. Provides for creation of nature conservation areas on state-owned lands and permits the compulsory or voluntary acquisition of land for conservation purposes.
3. United States	
3.1 Federal Clean Water Act 33 USC 1362 (7)	The permitting process for discharges takes into account environmental effects on wetlands, fish and other wildlife.

continues overleaf

Table 22.3   *(Continued)*

Legislation	Provisions
National Environmental Policy Act 1969 (NEPA)	Provides for compulsory EIA in certain circumstances. EPA guidelines in respect of mitigating adverse environmental impacts of human activities become effective herein, and EIA assessment regulations exist – involve detailed reviews: e.g. public interest, potential environmental benefits, economic factors, mitigation of adverse environmental impacts.
Other federal legislation	Fish and Wildlife Coordination Act 1958 (16 USC 661) and Endangered Species Act.
3.2 States	Large number of general conservation and river conservation enactments.
4. South Africa	
National Water Act 36 of 1998	The following are defined as 'water uses' and require a licence or authorization: impeding or diverting the flow of water, altering the bed, banks, course or characteristics of a watercourse. Provides that commercial afforestation and the cultivation of declared crops may be regulated as stream flow reduction activities. Declaration of controlled activities, one example being irrigation of land with waste or water containing waste, which is subject to authorization in terms of the Act.
Lake Areas Development Act 39 of 1975	Permits the state to acquire control of private land situated within proclaimed lake areas, as well as state land in those areas with a view to protecting resources and to control, manage and develop state land in those areas. Two areas have been proclaimed lake areas, i.e. Knysna National Lake Area and Wilderness National Lake Area. The Act provides for the issuing of regulations regarding *inter alia* the use of the relevant areas.
Conservation of Agricultural Resources Act 43 of 1983	Provides *inter alia* for prevention of erosion and interference with water sources by protecting vegetation, also by permitting action to be taken to eradicate invasive plants, weeds, etc. and in this regard also complements the Mountain Catchment Areas Act 63 of 1970 by being applicable over mountain catchment areas not subject to that Act. Has incidental significance for conservation purposes by protecting wetland vegetation, etc.
National Forests Act 84 of 1998	Provides for protection and management of forests. Contains principles to be considered and applied when making decisions affecting forests, namely that biological diversity, ecosystems and habitats, and other natural resources, especially water and soil, should be conserved.
Mountain Catchment Areas Act 63 of 1970 (MCA)	Provides structures and regulates the conservation, use, management and control of land in designated mountain catchment areas with a view to protecting catchment areas from degradation, and to conserving and protecting the quality and quantity of water collected in such catchments. Permits the Minister to declare certain areas mountain catchments; to give directions with reference to the manner in which land is used and managed, to prevent soil erosion and protect the natural environment as well as removal of certain categories of vegetation, etc. These areas may be proclaimed over private land or public land.
Regulations in terms of section 21 of the Environment Conservation Act 73 of 1989	The regulations identify activities for which an EIA process must be initiated. A number of activities are relevant to river conservation: construction of canals, water transfer schemes between catchments, dams and weirs affecting river flow, and the reclamation of inland water, including wetlands.

the sum total of the chemical elements dissolved and suspended in the water (Dallas and Day, 1993). Biota evolve under particular conditions, and can only tolerate water quality changes within particular ranges. Therefore, even if the amount and pattern of water is adequate for ecological functioning, water quality is another critical factor on which biological integrity depends.

Early drafts of the Principle included reference to physical changes (such as those imposed by channel engineering) but multidisciplinary discussion led to the somewhat oversimplified focus on water quality and quantity, recognizing the primary human impacts of altering flow patterns by impoundment, abstraction and interbasin transfers; and of water

quality impacts from effluent disposal and diffuse pollution. All these activities threaten biological integrity and ecosystem functioning. The law therefore needed to provide for river protection, both for the intrinsic value of the natural environment and because humans use and depend on natural river functions, particularly the self-cleansing capacity of rivers (Ashton *et al.*, 1995).

The Principle acknowledges that if the quality of the water, and the discharge and patterns of water flow, are altered too greatly, river ecosystem structure and functioning can deteriorate and even collapse, to the detriment of human dependence on the system. It seeks to reserve as a right, particular natural patterns of water quality and quantity for the river itself, so as to ensure the long-term functioning of the system. The heart of a legal basis of river conservation may thus be the 'right' of the river to its own water – both its pattern of flow and the quality of the water (although it must be noted that no legal system currently recognizes a legal right on the part of a river, or anything other than humans *per se*, as bearing legal rights).

Thus water quantity and quality – two concepts with which lawyers and resource managers were familiar – were used to encompass and convey the links between natural and altered physical, chemical and biotic patterns. Of particular importance is the link between flow changes and habitat availability, and in the integration of water quantity *and* quality in river management (Howell, 1994), for example, notes a lack of such integration in Scotland). We therefore go on to discuss legislative aspects of quantity and quality, from South African and comparative perspectives, before evaluating the South African legislation development process.

Water quantity

Although methods for the quantification of the in-stream flow requirements of rivers are internationally well established (Estes and Osborne, 1986; King and Tharme, 1994; Tharme, 1996), the question 'How much water does a river need?' continues to be raised (Richter *et al.*, 1997). The 'rights' of rivers and other watercourses to their own water can, in theory, be realized. In certain jurisdictions, such as the western states of the USA (MacDonnell *et al.*, 1989), the concept of the river's 'right' to, and need of, its own water is well established. The foremost problem is a legal one – rights in water have been appropriated or allocated to private individuals over a long period without regard to the needs of rivers. Such rights are frequently classified as property rights in the legal system concerned, and therefore the reduction of the entitlement of individuals, in order to accommodate the requirements of rivers means infringing property rights. Compensation is frequently required in such circumstances (see Table 22.3, Finland). As is so often the case in the environmental and conservation field, legislative action is constrained by private property rights and interests.

In water-scarce regions the problem is exacerbated because the protection of the environmental integrity of rivers effectively becomes a water allocation issue, since water must be 'reserved' (DWAF, 1997a) for conservation purposes. At this point it is useful to consider the legal basis for the administration of water allocation. Three main categories of human activity exert impacts on natural patterns of water flow in rivers: abstraction, impoundment and interbasin transfers.

Legal control of abstraction is generally achieved by 'prior appropriation' or by an administrative system of permits – in the former system the right to abstract water is often based on a 'first come, first served' in perpetuity approach and is applied, for example, in the State of Colorado, USA. Alternatively water may be allocated under the 'riparian principle' whereby the ownership of land adjacent to a river confers rights of use, as was the case in South Africa under the 1956 Water Act. Under the permit system the right to use water is granted by permit, for specified periods, as in South Africa now, and in some Australian states. In some places hybrid systems have evolved combining two or more of the above – in California, for example, where the principles of an earlier riparian approach have remained in a later prior appropriation system (Getches, 1984; Caponera, 1992). The building of dams is often subject to requirements for Environmental Impact Assessments (EIAs) – thus the National Environmental Protection Act and the Clean Water Act, *inter alia*, in the USA; the Water Act and the Nature Conservation Act in Finland; and the Environment Conservation Act (No. 73 of 1989) in South Africa, all impose such controls (Table 22.3). Impoundment construction and operation can also be constrained by the in-stream flow requirements of the river. Interbasin transfers can similarly be controlled both in construction and operation by EIA requirements and in-stream flow requirements. EIA requirements are often also influenced by water quality control needs, as lowering of flows by damming, interbasin transfers, or other forms of abstraction may increase the relative pollution load of a river, thus bringing into effect pollution-focused EIA requirements (Getches, 1984).

Legislation mandating EIAs serves as an example of how river conservation can be furthered outside the ambit of traditional 'conservation' statutes such as the US Wild and Scenic Rivers Act (Table 23.3, US). EIAs coupled with permit requirements (Table 22.1) frequently offer means for river protection couched in other terminology. An illustration of the way in which this process would operate in the USA is provided by the Court of Appeal for the Second Circuit judgement in a case involving opposition to the granting of a licence to a hydroelectric power company for the operation of a pumped storage plant in the Hudson

River Valley. In its judgement the following was stated by the Court:

> 'The . . . Project is to be located in an area of unique beauty and major historical significance. The highlands and gorge of the Hudson offer one of the finest pieces of river scenery in the world. . . . Petitioners' contention that the Commission must take these factors into consideration in evaluating the . . . Project is justified by the history of the Federal Power Act. (2d Cir 1965).'

The benefits of EIA for conservation should not be overstated. Nevertheless, the possibilities afforded scientists through the medium of EIA is considerable – e.g. Section 102 of NEPA requires all federal government agencies to:

> include in every recommendation or report on proposals for legislation, and other major federal actions significantly affecting the quality of the human environment, a detailed statement by the responsible official on (among others) . . . the environmental impact of the proposed action; . . . any adverse environmental effects which cannot be avoided, should the proposal be implemented; . . . alternatives to the proposed action; . . . the relationship between local short term uses of man's environment and the maintenance and enhancement of long term productivity; and . . . any irreversible and irretrievable commitment of resources which would be involved in the proposed action should it be implemented.

The open texture of the section's wording left much scope for the courts to shape the practical content of the concept, and they obliged in the large volume of litigation which followed, crafting 'the statute into a meaningful tool of environmental protection'. Finally, US policy reform is often guided by the views of the courts when reviewing licence or permit applications.

COMPARATIVE PERSPECTIVE (TABLE 22.1)

In-stream flow protection has generally advanced most in arid areas, with the western USA as a good example. MacDonnell *et al.* (1989) report on a comprehensive examination of the laws and programmes in these states which provide legal protection for in-stream flows. One of the legal concepts used to enact such

protection is that of the public trust (Sax, 1970), which holds that the state, as trustee for the people, retains some residual right in natural resources. The concept of the public trust has been used by courts and legislatures to outflank private property rights in water. The concept supports the argument that it is in the national interest to protect stream flow (MacDonnell and Rice, 1989). Another is the right to in-stream flows to establish or maintain fisheries, as recognized in Wyoming, and which has led to modifications of the state's original prior appropriation approach to water rights (Squillace, 1989).

Interestingly, in-stream flow protection has also been linked to the water rights of native North American Indians; for example, the Supreme Court of Wyoming has held that Indian tribes occupying the Wind River Indian reservation were entitled to a reserved water right on the Big Horn River (Stanton, 1990). This award was quantified entirely on the basis of an agricultural purpose for the reservation, whereas Stanton (1990) notes that since the treaty specifically preserved hunting rights, and since the tribes subsisted on indigenous wildlife at the time the treaty was signed, and would continue to do so, the court should have awarded additional reserved water rights in an amount sufficient to maintain wildlife populations at historic levels on the reservation. Although water allocation to Indian reservations may be a vehicle for in-stream flow protection in the arid western states, Boomgaarden (1990) warns of the excessive investment in litigation of the reserved rights quantification issue and suggests that resource protection would be better served by negotiation than litigation – a comment which reflects the necessity for societal acceptance of in-stream flow protection as a necessary adjunct to appropriate legal instruments. Proposed solutions generally revolve around balancing legitimate property interests (Moore, 1989), rather than recognition of environmental values *per se*.

Endangered species protection has also been promoted as a vehicle for water resource protection in the American West, where 'implementation of the Endangered Species Act may prove to be the litmus test of the extent to which existing water use patterns . . . can accommodate contemporary environmental values' (Moore *et al.*, 1996). Congress's earlier intervention to remove some of the effectiveness of the Act in this regard is enlightening. In response to the courts' acceptance of the view that the Act prohibited the construction of the $100 million Tellico Dam, which would have destroyed the habitat of the snail darter (a small fish, listed as an endangered species), the Act was amended to enable the authorities to override it in certain circumstances (Tennessee Valley Authority *v* Hill 437 US 153 1978; Schoenbaum and Rosenberg, 1991).

Howell and Mackay (1997) discuss the role of inducement in effective legislation. A related issue is the heatedly debated role of water markets. The concept is that where water becomes scarce, and has historically been used in relatively inefficient ways (e.g. the aerial irrigation of low-value crops), greater profitability and efficiency result from the economic incentives of a market sytem where rights to water use can be bought. Chile has been held up as the model of a successful water rights market (Bauer, 1995; Hearne, 1995), but little attention was paid to the environmental consequences of this trading. In a consideration of water allocation problems during the severe Californian drought of 1992, Howitt (1994) and Israel and Lund (1995) drew attention to the fact that instream requirements are frequently undervalued and need to be more carefully taken into account. Legal impediments to decreasing the volumes available to irrigation and other established users are manifold and complicated, but could be partially overcome by temporary transfers (Gould, 1989), a view also relevant to the protection of environmental requirements. The Australian experience sounds the warning that water markets can operate to the detriment of adequate protection of water quantity for the environment, and that water allocations for instream flow requirements should be explicitly excluded from the market arena (Sturgess and Wright, 1993; Stringer, 1995; Langford, 1996).

INTERNATIONAL LAW

Rivers frequently cross or form political borders, and transboundary issues are common. In a coherent review of current international trends Korhonen (1996) concludes 'International endeavours to protect ecosystems and species have paralleled the inadequate and fragmentary approaches characterising watershed management on national levels'. This is quite clear from the attempts of the USA and Mexico to agree on a collaborative approach to the protection of the Rio Grande, where critical water quality issues were compounded by water scarcity in an arid region (Bowman, 1996). In Europe, a good example is provided by the 1992 Helsinki Convention on the Protection and Use of Transboundary Watercourses and International Lakes, which has been joined by most European countries. Among its aims are the

prevention of unacceptable transboundary environmental impacts due to pollution, and the protection of the ecological well-being of such waters, and the adoption of the *polluter pays*, and the *precautionary* principles. The concept of *sustainability* is central to its provisions. It provides for EIA as a primary instrument in achieving these ends (Birnie and Boyle, 1995).

Birnie and Boyle (1995) discuss the 1992 Convention on Biodiversity, noting it is currently the only global treaty dealing with resource protection at the landscape scale appropriate to river ecosystems. Article 1 of the Convention requires 'the conservation of ecosystems and natural habitats and the maintenance and recovery of viable populations of species in their natural surroundings'. More specifically Article 8 obliges parties to establish a system of protected areas and to develop guidelines for their establishment, to promote environmentally sound and sustainable development in areas adjacent to protected areas, and to rehabilitate and restore degraded ecosystems and promote the recovery of threatened species. Lastly parties must 'prevent the introduction of, or control or eradicate those alien species which threaten ecosystems, habitats or species'. Szekely (1992) and Szekely *et al.* (1996) provide an example of a multilateral treaty for the protection of the natural resources of North America which provides protection at the scale necessary for river protection.

The efficacy of provisions contained in international conventions depends, in most countries, on the implementation of those provisions in domestic legislation. Implementation, whether legislative or also administrative, frequently lags behind ratification. In this respect legislation emanating from the European Union has an advantage – certain laws are directly binding, while directives must be accurately translated into the law of the various member states within a certain time limit. Thus the provisions of the forthcoming Water Framework Directive will be enacted in domestic legislation in all the EU member states. This legislation will introduce catchment management in all the EU member states and shifts the emphasis from chemical to biological quality goals (Pollard and Huxham, 1998).

Customary international law contains a number of principles relevant in this context, as well as that of protecting the quality of water in international rivers. The *principle of good faith* in international relations, and *the principle of good neighbourliness* between states, both long-standing principles of international law, have acquired increasing significance in connection with the transboundary environmental impacts of activities in one state upon neighbouring states.

These imply, in practice, that states carrying out, or permitting, activities which carry a significant risk of causing appreciable harm – including environmental harm – in another state are obliged to give timeous notice to the state concerned. Lang *et al.* (1991) note that 'the basic idea . . . is to prevent the commission of unlawful transboundary interferences with the natural resources or the environment of other states'. One implication of this is that EIAs may be required on the part of the state where the activity is being carried out, to determine potential transboundary impacts of the activity, and to permit timeous steps to be taken to prevent harm therefrom.

SOUTH AFRICA

Under the old Water Act (1956), water allocation was based largely on the riparian principle. Thus land ownership was linked to the right to use water. In this way access to water was inextricably part of the social inequalities which were the consequence of the policy of apartheid. In the National Water Act there is a move to an administrative system of authorizing water use (Table 22.1). The advantages of this are both political and practical, as the link between land ownership and water access is broken (land issues will take considerable time to resolve), and an administrative system provides greater opportunities for control and greater flexibility in adjusting water use to suit changing economic, social and environmental circumstances. However, it can have the disadvantage of bureaucratic inefficiency. Pigram (1993) points out the advantages of the administrative allocation of water in Australia, where, as in South Africa (DWAF, 1997b), both the natural availability of water and demand for water, varies widely across the continent. Under an administrative system of water allocation, practical and inexpensive ways must be devised to quantify and to allocate the necessary quantities and patterns of flow for river conservation – the in-stream flow requirements (IFR) of the river.

River conservation has often been associated with specially protected areas, such as the Sites of Special Scientific Interest in the United Kingdom (Howell, 1994; Boon, 1995) or rivers designated under the Wild and Scenic Rivers Act in the USA (Table 22.3). Useful as such provisions are, general water management statutes have implications for river conservation – even within protected areas. For instance, under the old South African Water Act (1956), the availability of water for environmental purposes was generally subordinated to the rights of other categories of users: the powers of the National Parks Board to demand river flows for environmental purposes within National Parks was specifically subordinated to rights of abstraction upstream, outside the parks. The new law provides for the Ecological Reserve – that quantity (volumes and flow patterns in time and space) of water and the quality of the integrated resource (described in the next section), which will ensure the long-term, sustainable use of water resources. The only water which will be allocated by right in South Africa will be the water for the Ecological Reserve, and water for basic human needs (the UN-designated volume for health and hygiene) – all other water allocation for use will be by permit. Methods to quantify rapidly the Ecological Reserve on a national basis are being developed, and will be based on the South African IFR method, the Building Block Methodology (BBM) (King and Louw, 1998). Where an impoundment is being planned, the IFR allocation can be accompanied by operating rules for the impoundment which are linked to environmental cues such as rainfall events. In effect, the BBM is used to build a modified flow regime which resembles the natural regime, including key environmental cues such as the first flushing flows of the season.

Water quality

Rivers have long been the receptacles of human waste, and in well-watered regions water quality is frequently the prime environmental issue, as pollution becomes a problem before scarcity of water. However, it is in arid regions where competition for water has forced a recognition of the connection between water quality and quantity, since in such regions reduced flows exacerbate pollution problems as there is little capacity for dilution (Shafron *et al.*, 1990; Ashton *et al.*, 1995). Thus, quality issues have tended to overshadow others in legislation in almost all jurisdictions.

Pollution control in rivers has received legal attention since the time of the Romans (Haslam, 1990). Modern pollution control legislation is comprehensive and varied, encompassing such concepts as integrated pollution control, where standards are set which are consistent for air, land and water; the polluter pays principle, where the economic costs of pollution are borne by the polluter rather than other water users; and the precautionary principle which puts the onus on the polluter to prove that the discharge will not have a seriously deleterious effect. Control can be effected both by the setting of objectives for environmental integrity, and the imposition of controls – with in-stream limits to the concentrations of various pollutants. Some systems rely on the setting of quality standards for waters receiving a discharge of a particular substance and the fixing of varying discharge limits to achieve the quality standards, while other systems set uniform discharge limits for particular subtances based on the best available technology criterion (Howell and Mackay, 1997). Most administrative systems are based on permits (Table 22.2). Some of the complicating factors include the synergistic, additive and antagonistic interaction of chemicals in solution which can nullify the efficacy of limits set for the concentrations of individual water quality variables, and the problems associated with sprawling urban environments (particularly in developing countries) where inadequate sanitation results in diffuse-source pollution (Weeks, 1982; Pegram *et al.*, 1996).

Howell and Mackay (1997) state that British regulators have been fairly successful in utilizing enforcement (or the threat thereof) combined with education and negotiation to achieve less polluted waters. However, the usefulness of the enforcement approach depends on resources and skills being made available. This is typically a concern in developing countries. There is evidence that dedicated agencies (such as the Environmental Protection Agency (EPA) in the USA and the Environment Agency in England and Wales) which are empowered to prosecute offenders are more effective in securing enforcement. The high cost of this command-and-control approach to pollution has awakened interest in the use of agreements between regulators and the industries concerned or particular plants, the setting of targets, tax incentives or inducements (subsidies). Experience by the authorities responsible for Sites of Special Scientific Interest in the UK has shown that where objectively quantifiable standards are not readily available, and private rights in property must be accommodated, the use of inducements, as opposed to enforcement, can be very useful in improving water quality (Howell and Mackay, 1997).

Pollution control legislation also confronts its limits in regard to diffuse pollution of water resources. Such pollution commonly arises in four ways: atmospheric acid deposition, accidental spillages resulting in unauthorized discharges (e.g. the Sandoz incident on the Rhine), land-use activities such as fertilizer application in agriculture, and historical pollution from abandoned waste dumps. The first and second

instances, especially, highlight the importance of integrated pollution control for river conservation. Diffuse pollution caused by agricultural practices is by its nature difficult to police. In Europe there is the Nitrates Directive (91/676/EEC) (Table 22.2, Finland), but the difficulty with regulation outside defined protection areas means that the legislation will have to focus on the establishment of codes of good practice, guidelines and financial support (Howell and Mackay, 1997). It is ironic that subsidies are to be granted by the EU to set aside land in order to reduce agricultural surpluses – land that could be used for the establishment of buffer zones and re-establishment of waterside habitats (Howell, 1994) – when it was subsidies that contributed in the first place to the current problem of diffuse pollution from agriculture.

A general problem relating to the application and enforcement of legislation is that most legal systems make it quite difficult for persons or groups to litigate in order to have laws enforced. Thus individuals and also (particularly) interest groups are often powerless in cases where the state neglects to act against polluters or does not enforce its conservation statutes. In legal terms the question is one of standing – judges have usually been extremely reluctant to recognize the right of individuals and groups to bring such actions through the courts. Standing can be varied by legislation, by recognizing the right of groups, such as conservation bodies, to bring the breach of statutes and regulations before the courts and to obtain relief. Against this background the South African National Environmental Management Act (107 of 1998) strikes a bold blow. Section 31(1) of the Act provides that:

> Any person or group of persons may seek appropriate relief in respect of any breach or threatened breach of any provision of this Act . . . or any other statutory provision concerned with the protection of the environment or the use of natural resources – (a) in that person's or group of person's own interest; . . . (d) in the public interest; and (e) in the interest of protecting the environment.

Most legal systems contain, apart from the administrative regulations and associated criminal sanctions, various civil controls based on, *inter alia*, the possibility of civil litigation against polluters who infringe the rights of others (McLoughlin and Bellinger, 1993). These are generally rooted in utilitarian values such as property or personal health considerations, rather than aesthetic or environmental ones. Howell and Mackay (1997) raise the question of the role of inducements such as tax incentives for zero discharge, and agreement as more effective means of water quality control than enforcement.

COMPARATIVE PERSPECTIVE (TABLE 22.2)

In the United States, the promulgation of the Clean Water Act USA 1251–1376 of 1972 coordinated pollution regulation throughout the USA and redressed the effects of the somewhat less stringent legislation of some individual states. The goal set out in the Act was to eliminate discharge of pollutants by 1985 and to reach the stage of swimmable and fishable water by 1993, and finally *to restore and maintain the chemical, physical and biological integrity of the nation's waters.* The Act includes ambient water quality standards and effluent standards, and is implemented by the EPA which is empowered to identify streams and reaches which require more stringent control to achieve adequate in-stream water quality conditions. The EPA standard is based on the concept of total maximum daily loads and available capacity which is divided among users. Permit requirements include application of best available technology economically achievable (Getches, 1984).

In a comparative study of international water law by the Finnish Environment Institute (FEI, 1996) water legislation of a range of countries was considered:

- *Australia*: Water pollution control in Australia is mainly based on independent state enactments outside the general water legislation (e.g. Brunton, 1994), and includes such mechanisms as quality standards, a licence system and monitoring. The polluter pays principle is applied (Table 22.2).
- *European Union*: A 1992 European Union Directive for Integrated Pollution Prevention and Control affects all member countries.
- *Finland*: Finland has a clearly defined set of environmental statutes and integrated water legislation which includes a specific pollution control Act. Like other Scandinavian countries, Finland has a system which gives ecological values protection in the face of industrial interests. Pollution permits are subject to the precautionary principle, but there are still problems with diffuse/ non-point-source pollution from the agricultural sector (Table 22.2).
- *Sweden*: Sweden's pollution prevention legislation requires integrated pollution control and is administered *via* a renewable permit system.

- *France*: France is interesting in that water resource management is carried out on the basis of catchments or basins, which facilitates an integrated approach. Although pollution permits are used, they are subject to a pre-control system which regulates pre-treatment facilities in treatment works. Once the treatment capacity is appropriate, permits are subject to risk assessment. The objective of regulations is to satisfy the needs of public health, agriculture and industry, as well as for the protection of the aquatic environment.
- *Israel*: Israel, an arid country, has a general prohibition on pollution and then exemptions *via* permits.

SOUTH AFRICA

A recent review of South African water resources (DWAF, 1997b) shows South Africa as an arid country with unevenly distributed water resources, many of them far from the site of industrial development. It estimates that 'the country's conventional water resources will be fully utilised before 2030'. In the face of a critically limited resource the Department of Water Affairs and Forestry (DWAF) has had a rapidly evolving water quality management policy, moving from general and special standards to receiving-water quality objectives during the 1980s (Van Der Merwe and Grobler, 1990) where the aim was to maintain water resources in a state 'fit for use' by water resource users. Initially the environment was viewed as a user, but this concept put environmental protection in dangerous competition with economically beneficial users. Much sounder is the current policy (DWAF, 1997a) which recognizes that all users depend on the resource and therefore resource protection is essential before optimal resource utilization can be effected.

Because of the lack of water for dilution purposes the prevention of pollution in rivers has a wide and integrated focus. The first step was an acknowledgement of the 'silent services' provided by naturally functioning ecosystems – especially rivers. In terms of water quality the key service is the self-cleansing capacity of rivers where organic pollution such as sewage is processed biologically and water quality is improved. This resulted in a recognition of the economic advantages of maintaining a functioning riverine ecosystem.

The concept of water quality was replaced by that of 'resource quality' in order to recognize that the 'silent services' are not dependent simply on the physical and chemical characteristics of the water itself, but on the health of all components of the riverine ecosystem. Resource quality includes the water and its dissolved constituents (the traditional focus of water quality management – Boon and Howell, 1997), the sediments, the adjacent riparian land and all biota. 'It is the healthy functioning of the whole ecosystem which gives the ecosystem its capacity to recover from . . . human use' (DWAF, 1997a).

Two strategies were adopted and are reflected in the new water law, the National Water Act of 1998. The first is a system of resource-based objectives that define the state of health the river should sustain. The second is a system of source-based controls which includes a permit system aimed at reducing or eliminating the production of potential pollutants which could harm river ecosystems. Like Israel, arid South Africa has opted for a general prohibition of pollution with exemptions, some general, others specific, and administered *via* a permit system.

In a process similar to that described for the United Kingdom (Howell and Mackay, 1997), the development of resource objectives depended on the identification of reference conditions, and various classification systems. As early as 1994, South Africa and Australia compared progress in river classification and the development of environmental health indices (Uys, 1994). The proposed classification system is based on the recognition that rivers can function at various levels of health and biotic integrity, and that South Africa is a developing country where the self-cleansing capacity of rivers must be used for the processing of appropriate waste (Ashton and Van Vliet, 1997; Ashton *et al.*, 1995). Therefore rivers and river reaches will be classified according to a system of biotic health and integrity (Roux *et al.*, 1994, 1996). A classification system can be used to provide management objectives which allow differing degrees of resource protection – and differing degrees of the risk of environmental degradation. For example a resource may be classified in the classes A, B, C or D in each of five categories: water quantity, water quality, in-stream habitat, riparian habitat and biota. The class of river then has implications for the appropriate Ecological Reserve as regards water quantity. For example:

Class A: the river has a natural variability and disturbance regime; allow no modification;

Class B: there are slight modifications to the natural conditions; use in-stream flow requirements (IFR) methods to provide for quantities of water that allow only slight risk to intolerant organisms;

Class C: the river is considerably modified; set IFR requirements which allow moderate risk only to intolerant biota;

Class D: the river is severely impacted, but essential natural processes are still functional; set IFR requirements that may result in high risk of the loss of tolerant biota.

This system clearly integrates water quantity, through the setting of appropriate IFR allocations, and water quality. Environmental water quality guidelines (DWAF, 1996c), derived from local and international tolerance databases, provide information on the link between in-stream water quality and the response of the biota. From a river conservation perspective the other critical initiative is the implementation of a national biomonitoring programme which has taken shape after collaboration internationally and includes a review of biomonitoring programmes in the United Kingdom, Australia and Canada (Uys *et al.*, 1996).

Although South Africa is some way from the sophisticated requirements of integrated pollution control, the context of resource quality management will be the river catchment, with strong emphasis on land-use planning and integrated catchment management (DWAF, 1996c).

Conclusions

The South African water law review was successful from several points of view. A wide range of people were involved in the process, from developing the basic principles to drafting the law, with ecological input from representatives of the Southern African Society of Aquatic Scientists. The basis for implementation by agreement (Howell and Mackay, 1997) was laid during an extensive consultative process (despite continuing under-representation from rural communities). Controversy, particularly with the agricultural, forestry, mining and industrial sectors, focused on the change in water allocation from the riparian to the permit system. At no stage were the environmental provisions for integrated catchment management, the setting of integrated resource quality objectives, and the quantification of the Ecological Reserve, challenged or opposed by any user-sector or political party. Indeed, the remarkable Water Conservation Campaign used the climate of legal review to launch a national programme of clearing alien vegetation from catchments, together with increasing awareness of household water conservation. The campaign combined economic incentives in water tariffing with job creation to produce a high-profile programme which raised awareness of the role of catchment processes, and increased stream flow in previously infested catchments. The problem of limited implementation capacity was approached by drafting an 'enabling' Act which clearly defines the scope and intention of the legislation, but has left regulations to be developed with time and increasing capacity. Although the old legislation has been repealed, old regulations will apply as long as they are not in conflict with the new law, and until they are replaced by new regulations.

Some of the serious pitfalls that loom include the lack of capacity at a local government level (the municipality) which is responsible for sewage disposal. There is the long tradition of civil resistance by non-payment for services to be overcome, and there is inadequate capacity to develop or to maintain adequate sewage treatment works. This, combined with a rapidly growing population, will threaten ecosystem health.

There is an urgent need for communication and education at all levels of society about the concept of the Ecological Reserve and how it can best be implemented. Although the quantification of ecological water quantity requirements is well advanced with the application of the BBM (King and Louw, 1998), a similar method to define ecological water quality requirements was not available. Drawing on the international ecotoxicological literature, and a water quality modelling approach, methods to quantify biotic water quality requirements are well under way. At present they still suffer from the limitation of 'compromise ecotoxicology . . . that emphasises the reaction of a few tolerant species in the laboratory and not the natural community in the habitat' (Moss, 2000), but there are research initiatives in the use of indigenous riverine fauna in ecotoxicology (Palmer *et al.*, 2000).

Perhaps most heartening is the response of the responsible government department to the new legislation. There is an intense focus on successful implementation. Stressed catchments have been identified, and the first formal integrated determination of the Ecological Reserve is under way. It remains to be seen whether there will be the political will to change water allocations and discharge permits in the catchment so as to effect increases in ecosystem health. Equally heartening is the strong emphasis on the catchment, but, as pointed out by both

Moss (2000) and Pollard and Huxham (1998), lack of integration with other authorities, notably agriculture, could prove a stumbling block.

Worldwide, rivers are under threat from human use. From the perspective of the biologist or conservationist, involved in understanding the functioning of river ecosystems and chronicling the distribution, and often the demise, of riverine biota, it may seem that adequate protective legislation is the answer to dwindling and deteriorating rivers. Undoubtedly clear, modern legislation, based on the best ecological understanding of the ecosystem, can be part of the answer. However, it is equally, if not more important, to win the hearts and minds of the general population. There is a relatively small part of the global population that is sufficiently well fed, clothed and housed, to pay attention to matters of environmental conservation. And yet it is often poor rural populations that depend totally on run-of-river flow who are most affected by the consequences of environmentally poor river development, as the flow of their river reduces and becomes uncertain, and water quality deteriorates. Waterborne disease is a major health problem in all developing countries.

As South Africa emerged from a repressive political regime, one of the rallying cries was the call to 'give voice to the voiceless'. Of all things the environment is voiceless. River conservation on every continent will depend strongly on the advocacy of environmental scientists who will enunciate environmental requirements. It is not enough to publish learned papers, important though these are. Scientists need to emerge from their discipline and engage in the business of making information accessible to a wide range of audiences. Sectors of society which may sometimes be perceived as threats to conservation or as unimportant, such as the rural poor or inhabitants of informal settlements, must also be brought into the fold. Conservation is often regarded as a luxury which they cannot afford, yet it is their dwellings which are often situated in unsuitable places where they are prone to flooding, and it is they who succumb to waterborne diseases. In South Africa and elsewhere in the world a nascent environmental justice movement is linking poverty with environmental issues (Glazewski, 2000). The challenge for scientists is to cooperate with such movements and involve people from affected communities in decision-making.

Information gleaned by scientists should be aimed at the general population who are the political 'consumers' and ultimately have the capacity to sway political activities; at government departments and water resource managers who create policy and implement it; and at the law-makers who draft the legislation. It is also important to recognize that the step prior to legislation is policy development, and it is probably in this area that professional scientists can make their best contributions. The United Nations, through the Food and Agriculture Organization (FAO), has put considerable effort into making the mechanisms of sound policy development available to developing countries (FAOUN, 1995a,b,c). The challenge to communicate the importance of understanding how rivers function, and how healthy, functioning rivers can benefit society, lies firmly in the court of the environmental scientist.

Acknowledgements

The following people have contributed significantly to the development of these ideas: the South African Water Law Review team – particularly Bill Rowlston, Henk Van Vliet, Heather MacKay, Alison Howman, Johan Wessels and Hennie Schoeman; the Finnish Environment Institute team: Marianne Lindström and Kari Kuusiniemi; and those centrally involved in the development of in-stream flow requirement methodologies: Jackie King, Jay O'Keeffe, Rebecca Tharme, Delana Louw, Denis Hughes and Andrew Bath. Carolyn Palmer was supported by the Water Research Commission while working on the South African Water Law Review. We are grateful to Phil Boon, an external reviewer and Bill Rowlston for their constructive editorial input.

References

Ashton, P.J. and Van Vliet, H.R. 1997. South African approaches to river water quality protection. In: Laener, A. and Dunette, D. (eds) *River Quality: Dynamics and Restoration.* CRC Press, Lewis Publishers, New York, 403–411.

Ashton, P.J., Van Vliet, H.R. and MacKay, H.M. 1995. The use of environmental capacity concepts in water quality management: a South African perspective. *New World Water 1995,* 45–48.

Bauer, C. 1995. Against the current? Privatization, markets, and the state in water rights: Chile 1979–1993. PhD Dissertation, University of California, Berkeley.

Birnie, P.W. and Boyle, A. 1995. *Basic Documents on International Law and the Environment.* Clarendon Press, Oxford.

Boomgaarden, L.J. 1990. Water Law – quantification of

Federal Reserved Indian Water Rights – 'Practicably Irrigable Acreage' under fire: the search for a better legal standard. *Land and Water Law Review* LWLRDC 25:417–434.

Boon, P.J. 1995. The relevance of ecology to the statutory protection of British rivers. In: Harper, D.M. and Ferguson, A.J.D. (eds) *The Ecological Basis for River Management.* John Wiley, Chichester, 239–250.

Boon, P.J. and Howell, D.L. (eds) 1997. *Freshwater Quality: Defining the Indefinable?* The Stationery Office, Edinburgh.

Bowman, J.A. 1996. The Rio Grande – a confluence of waters, nations and cultures. In: Wessels, J.J. and Schoeman, H.N. (eds) *Water Rights – Determination and Apportionment. Research Notes for the Water Law Review* 5:36–46. Department of Water Affairs and Forestry, Pretoria.

Brunton, N. 1994. Water pollution law in New South Wales and Victoria: current status and future trends. *Environmental and Planning Law Journal* 11:39–70.

Caponera, D.A. 1992. *Principles of Water Law and Administration.* Balkema, Rotterdam.

Dallas, H.F. and Day, J.A. 1993. *The Effect of Water Quality Variables on Riverine Ecosystems: A Review.* WRC Special Report, Project No. 351, Pretoria.

Department of Water Affairs and Forestry (DWAF) 1996a. *The Philosophy and Practice of Integrated Catchment Management: Implications for Water Resources Management in South Africa.* WRC Report TT 81/96, Pretoria.

Department of Water Affairs and Forestry (DWAF) 1996b. *Water Law Principles: Discussion Document.* Pretoria.

Department of Water Affairs and Forestry (DWAF) 1996c. *South African Water Quality Guidelines. Volume 7. Aquatic Ecosystems.* Pretoria.

Department of Water Affairs and Forestry (DWAF) 1997a. *White Paper on a National Water Policy for South Africa.* Pretoria.

Department of Water Affairs and Forestry (DWAF) 1997b. *Overview of Water Resources Availability and Utilisation in South Africa.* Pretoria.

Estes, C.C. and Osborne, J.F. 1986. Review and analysis of methods for quantifying in-stream flow requirements. *Water Resources Bulletin* 22:389–398.

Finnish Environment Institute (FEI) 1996. *South African Water Law Review – A Comparative Study by the Finnish Environment Institute.* DWAF Water Law Review, Pretoria.

Food and Agriculture Organization of the United Nations (FAOUN) 1995a. *Reforming Water Resources Policy – A Guide to Methods, Processes and Practices.* FAO Irrigation and Drainage Paper 52, Rome.

Food and Agriculture Organization of the United Nations (FAOUN) 1995b. *Methodology for Water Policy Review and Reform.* Water Reports 6. Rome.

Food and Agriculture Organization of the United Nations (FAOUN) 1995c. *Water Sector Policy Review and Strategy Formulation – A General Framework.* FAO Land and Water Bulletin 3, Rome.

Getches, D. 1984. *Water Law*. West Publishing, St Paul, Minnesota.

Glazewski, J. 2000. *Envronmental Justice and the New South African Democratic Legal Order*. *Acta Juridica* 1999. Juta, Cape Town (in press).

Gordon, N.D., McMahon, T.A. and Finlayson, B.L. 1992. *Stream Hydrology – An Introduction for Ecologists*. John Wiley, New York.

Gould, G.A. 1989. Transfer of Water Rights. *Natural Resources Journal* 29:457–477.

Hahlo, H.R. and Kahn, E. 1968. *The South African Legal System and its Background*. Juta, Cape Town.

Hart, H.L.A. 1984. *The Concept of Law*. Oxford University Press, Oxford.

Haslam, S.M. 1990. *River Pollution: An Ecological Perspective*. Bellhaven Press, London.

Hearne, R. 1995. The market allocation of natural resources: Transactions of water use rights in Chile. PhD Dissertation, University of Minnesota.

Howell, D.L. 1994. Role of environmental agencies. In: Maitland, P.S., Boon, P.J. and McLusky, D.S. (eds) *The Fresh Waters of Scotland: A National Resource of International Significance*. John Wiley, Chichester, 577–611.

Howell, D.L. and Mackay, D.W. 1997. Protecting freshwater quality through legislation: enforcement, inducement or agreement? In: Boon, P.J. and Howell, D.L. (eds) *Freshwater Quality: Defining the Indefinable?* The Stationery Office, Edinburgh, 457–481.

Howitt, R.E. 1994. Empirical analysis of water market institutions: the 1991 California water market. *Resource and Energy Economics* 16:357–371.

Israel, M. and Lund, J.R. 1995. Recent California water transfers: implications for water management. *Natural Resources Journal* 35:1–32.

King, J.M. and Louw, D. 1998. Instream flow assessments for regulated rivers in South Africa using the Building Block Methodology. *Aquatic Ecosystem Health and Management* 1:109–124.

King, J.M. and Tharme, R.E. 1994. *Assessment of the Instream Flow Incremental Methodology and Initial Development of Alternative Instream Flow Methodologies for South Africa*. Report No. 295/1/94, Water Research Commission, Pretoria.

Korhonen, I.M. 1996. Riverine ecosystems in international law. *Natural Resources Journal* 36:481–520.

Lang, W., Nuehold, H. and Zamenek, K. 1991. *Environmental Protection and International Law*. Graham and Trotman, London.

Langford, J. 1996. An Australian approach to the sustainable use of water. In: Wessels, J.J. and Schoeman, H.N. (eds) *Water Rights – Determination and Apportionment. Research Notes for the Water Law Review* 5:36–46. Department of Water Affairs and Forestry, Pretoria.

MacDonnell, L.J. and Rice, T.A. 1989. National interests in in-stream flows. In: MacDonnell, L.J., Rice, T.A. and Shupe, S.J. (eds) *Instream Flow Protection in the West*. Island Press, Covelo, California, 69–86.

MacDonnell, L.J., Rice, T.A. and Shupe, S.J. (eds) 1989. *Instream Flow Protection in the West*. Island Press, Covelo, California.

McLoughlin, J. and Bellinger, E.G. 1993. *Environmental Pollution Control*. Graham and Trotman, London.

Moore, M.R. 1989. Native American water rights: efficiency and fairness. *Natural Resources Law Journal* 29:763–791.

Moore, M.R., Mulville, A. and Weinberg, M. 1996. Water allocation in the American West: endangered fish versus irrigated agriculture. *Natural Resources Journal* 36:319–357.

Moss, B. 2000. Conservation – of freshwaters or of the *status quo*? *Verhandlungen der Internationalen Vereinigung für theoretische und angewandte Limnologie* (in press).

Newson, M.D. 1992. *Land, Water and Development: River Basins and their Sustainable Management*. Routledge, London.

Newson, M.D. 1996. *An Assessment of the Current and Potential Role of Fluvial Geomorphology in Support of Sustainable River Management Practice in South Africa*. Report KV83/96, Water Research Commission, Pretoria.

Palmer, C.G. 1999. The application of ecological research in the development of a new water law in South Africa. *Journal of the North American Benthological Society* 18:132–142.

Pegram, G.C., Quibell, G. and Görgens, A.H.M. 1996. *Non-Point Sources in South Africa – A Situation Assessment*. Workshop Starter Document. Report K5/665/0/1, Water Research Commission, Pretoria.

Pigram, J.J. 1993. Property rights and water markets in Australia: an evolutionary process towards institutional reform. *Water Resources Research* 29:1313–1319.

Pollard, P. and Huxham, M. 1998. The European Water Framework Directive: a new era in the management of aquatic ecosystem health? *Aquatic Conservation: Marine and Freshwater Ecosystems* 8:773–792.

Richter, B., Baumgartner, J.V., Wigington, R. and Braun, D. 1997. How much water does a river need? *Freshwater Biology* 37:231–249.

Roux, D.J., Thirion, C., Smidt, M. and Everett, M.J. 1994. A procedure for assessing biotic integrity in rivers – application to three river systems flowing through the Kruger National Park, South Africa. IWQS Report No N 0000/00REQ/0894. Department of Water Affairs and Forestry, Pretoria.

Roux, D.J., Jooste, S.J.H. and MacKay, H.M. 1996. Substance-specific water quality criteria for the protection of South African freshwater ecosystems: methods for derivation and initial results for some inorganic substances. *South African Journal of Science* 92:198–206.

Rowntree, K.M. (ed.) 1996. *The Hydraulics of Physical Biotopes – Terminology, Inventory and Calibration*. Report KV84/96, Water Research Commission, Pretoria.

Sax, J.L. 1970. The public trust doctrine in natural resource law: effective judicial intervention. *Michigan Law Review* 68:471–566.

Schoenbaum, T.J. and Rosenberg, R.H. 1991. *Environmental Policy Law*. Foundation Press, Westbury, New York.

Shafron, M., Croomer, R. and Rolls, J. 1990. Water quality. In: MacKay, N. and Eastburn, D. (eds) *The Murray*. Murray–Darling Basin Commission, Canberra, 147–165.

Squillace, M. 1989. A critical look at Wyoming water law. *Wyoming University Land and Water Law Review* 24:307–346.

Stanton, D.M. 1990. Is there a reserved water right for wildlife on the Wind River Indian Reservation? A critical analysis of the Big Horn River General Adjudication. *South Dakota Law Review* 35:326–340.

Statzner, B., Gore, J.A. and Resh, V.H. 1988. Hydraulic stream ecology: observed patterns and potential application. *Journal of the North American Benthological Society* 7:307–360.

Stringer, D. 1995. Water markets and trading developments in Victoria. *Water*, March/April:11–14.

Sturgess, G.L. and Wright, M. 1993. *Water Rights in Rural New South Wales – The Evolution of a Property Rights System*. Centre for Independent Studies Ltd, Policy Monograph 26. St Leonards, NSW.

Szekely, A. 1992. Establishing a region for ecological cooperation in North America. *Natural Resources Journal* 32:563–581.

Szekely, A., Beesley, A. and Utton, A.E. 1996. Ciuxmala model draft treaty for the protection of the environment and the natural resources of North America. *Natural Resources Journal* 36:591–633.

Tharme, R. 1996. *Review of International Methodologies for the Quantification of the Instream Flow Requirements of Rivers*. Discussion Document, Department of Water Affairs and Forestry, Water Law Review, Pretoria.

Uys, M.C. (ed.) 1994. *Classification of Rivers and Environmental Health Indicators – a Joint South African/ Australian Workshop*. Report No. TT63/94 Water Research Commission, Pretoria.

Uys, M.C., Goetsch, P.-A. and O'Keeffe, J.H. 1996. *National Biomonitoring Programme for Riverine Ecosystems: Ecological Indicators, a Review and Recommendations*. NBP Report Series No. 4. IWQS, Department of Water Affairs and Forestry, Pretoria.

Van Der Merwe, W. and Grobler, D.C. 1990. Water quality management in the RSA: Preparing for the future. *Water SA* 16:49–54.

Weeks, C.R. 1982. Pollution in urban runoff. In: Hart, B.T. (ed.) *Water Quality Management: Monitoring Programs and Diffuse Runoff*. Water Studies Centre, Chrisholm Institute of Technology, Melbourne.

2d Cir (1965) Scenic Hudson Preservation Society vs Federal Power Commission, 354 F.2d 608, 610.

23

River restoration in developed economies

G.E. Petts, R. Sparks and I. Campbell

Introduction

'The geometrical precision of an aqueduct signifies the engineer's vision of water flow, a bounded channel form that has become the common conception of how even a natural river should appear (Cosgrove, 1990).'

Tamed rivers were once a symbol of advanced civilizations. The 'training' of river channels for navigation and flood control, 'reclamation' of floodplains for productive farmland, 'conservation' of water by using large storage reservoirs to prevent 'wasting' run-off to the sea, the use of high dams to 'harvest' the energy from flowing water, and the large-scale transfer of water to make deserts bloom, came to symbolize social advancement and technological prowess (Scarpino, 1985; Reisner, 1986; Cosgrove and Petts, 1990; White, 1995). In all long-settled regions of the world subject to large-scale administration, the modernist vision of development has involved mega-projects of land and water management designed to extract wealth from rural areas for dispersal through cities. The 1990s have witnessed a major change in social philosophy, with global recognition of the need for *sustainable* environmental management and the protection of biodiversity; in developed countries, attention has turned to environmental restoration. This chapter discusses the restoration of river corridors in developed economies by reference to experience in the UK, USA and Australia.

In the UK, the creation of the modern cultural landscape began quite suddenly with the onset of the Neolithic period about 6000 BP (Rackham, 1997). The natural vegetation cover is forest with only small areas

of moorland and grassland (Rackham, 1997). Lime, hazel, oak and elm would dominate southern areas, with birch and pine in the north, and alder around pools and fens. Most of the uplands had been deforested by 4000 BP and floodplain woods were cleared by about 2500 BP. By the Roman period (AD 40–410) much of the lowlands had experienced a long history of cultivation. More than 5000 water-mills with weirs to control river levels are recorded in the Domesday Book of 1086. In 1665 an Act was passed 'for making divers rivers navigable' – most rivers were channelized by 1880. The long history of change in the UK (Table 23.1) contrasts with the more recent environmental changes in the USA (Table 23.2) and Australia (Table 23.3) that are largely confined to the past 200 years of European settlement. Australia's population of 16 million is small in relation to the area of the continent, but the aridity of the country and the concentration of the population in the higher rainfall areas, along the east coast and south-west, has resulted in a disproportionately large impact on Australian rivers.

The context

THE LEGACY OF ENVIRONMENTAL IMPACTS

From the early 19th century, the 'taming' of rivers was led by the popular pioneering vision of the human struggle to tame nature, and entrepreneurs motivated by the desire for economic growth. In 1853, Charles Ellet wrote:

'The banks of the Ohio and Mississippi, now broken by the current and lined with fallen

Global Perspectives on River Conservation: Science, Policy and Practice.
Edited by P.J. Boon, B.R. Davies and G.E. Petts. © 2000 John Wiley & Sons Ltd.

Table 23.1 *Human impacts on rivers in the UK (modified after Petts et al., 1989)*

Date (years BP)	Impact
5000–1000	Clearance of the uplands
2000–1000	Deforestation of the lowlands and loss of floodplain wetlands
2000–350	Drainage schemes and channelization for navigation and water power (mills)
700–	Overfishing
350–100	Extensive land drainage Channelization, dredging Extensive floodplain reclamation Pollution of small streams and, from 1750, larger rivers; mainly faecal pollution
250–	Introduction of exotic species (especially after 1850). Problem species today include: giant hogweed (*Heracleum mantegazzianum*), Japanese knotweed (*Fallopia japonica*) and Himalayan balsam (*Impatiens glandulifera*)
100–	Era of large schemes; storage reservoirs and interbasin transfers Pollution increasingly involving industrial as well as domestic wastes; many large rivers (e.g. lower Thames) 'dead' by 1965
30–	Post 1965 period of 'clean-up' and restoration; 1988–92 drought in south-east England drew attention to problem of over-abstraction, especially in groundwater-dominated catchments

Table 23.2 *Human impacts on rivers in North America (based upon Poff et al., 1997)*

Date (years BP)	Impact
Pre-350	Tillage, irrigation and deforestation may have had local effects. Fires set to maintain prairies and savannas in many catchments
350–170	Fur trade decimates beavers in northern USA and Canada. Beaver dams once moderated flows of headwater streams all over North America
250–190	Smaller rivers are dammed for water power in populated areas, blocking migrations of fish. Local pollution from cities and towns
150–70	Ploughing of prairies and drainage of wetlands for agriculture. Increase in sediment loads and flow fluctuations Channelization of streams, leveeing of floodplains Rising pollution from growing population centres
80–30	Era of massive public works: large dams, canals, levees, interbasin transfers (of water and organisms) Shift to industrialized, row-crop agriculture increases flow extremes and non-point-source pollution (sediment, pesticides, excessive nutrients)
30–	Social concern for the environment prompts effective federal and state legislation to control point-source pollution, protect wild and scenic rivers, stop the loss of wetlands, and assess the environmental impacts of development projects. Water resource development agencies acquire new authorities to mitigate past environmental damage. Non-point-source pollution and invasive aquatic species begin to be addressed in legislation

trees, . . . may yet, in the course of a very few years, be cultivated and adorned down to the water's edge . . . the grass will hereafter grow luxuriantly along the caving banks; all material fluctuations of the waters will be prevented, and the level of the river's surface will become nearly stationary. Grounds, which are now frequently inundated and valueless, will be tilled and subdued. . . . The channels will become stationary. . . . The Ohio first, and ultimately the Missouri and Mississippi, will be made to flow forever with a constant, deep, and limpid stream.'

In the USA, a philosophy of 'wise utilization' of natural resources for regional and national economic development came to underpin government policy in the Great Depression of the 1930s when public-works projects generated much-needed employment:

'(I)n the interests of the national welfare there must be national control of all running waters of the United States, from the desert trickle that might make an acre or two productive to the

Table 23.3 *Human impacts on rivers in Australia over the past 200 years, since European colonization*

Area(s)	Impacts
Eastern Australia south of Townsville, Murray–Darling Basin and Tasmania	Catchment clearance for agriculture and urban development. Damming for domestic water supply, irrigation use, and hydropower
Northern Australia and more arid regions	Stock grazing, causing particular problems in riparian areas, and vegetation burning
General	Introduction of exotic species of invertebrates (e.g. the snail *Potamopyrgus antipodarum*), fish (the European trout *Salmo trutta*, the carp *Cyprinus carpio* and the mosquito fish *Gambusia holbrooki*), and riparian plants (e.g. willows, *Salix* spp)

rushing flood waters of the Mississippi.' (US National Resources Planning Board 1934, cited in National Research Council – NRC, 1992).

There followed 30 years of construction of integrated systems of projects within large river basins (e.g. the Tennessee, Columbia, Colorado, and Mississippi–Missouri Rivers), consisting of new large, multipurpose storage reservoirs and the expansion and improvement of inland navigation systems. Thus, natural order was replaced by human order (Cosgrove, 1990). The need to conserve flora and fauna, let alone entire ecosystems, was not simply viewed as a constraint on progress, as it is in many developing economies today; it was irrelevant.

In reality, the legacy of 'progress' over the past 200 years, i.e. since the Industrial Revolution, has been pollution, increased flooding, prolonged droughts, habitat destruction and species extinctions. Today, salinization and eutrophication are common signals of catchment disturbance in semi-arid and temperate catchments respectively (see e.g. Pringle, this volume); regulated rivers, degraded habitats and loss of biodiversity are the common inheritance. In 1864, George Perkins Marsh – the 'mighty prophet of modern conservation' (Lowenthal, 1965) – provided a comprehensive and graphic exposé, not only of the damage which humans had done to the Earth, but also of the potential long-term consequences. For more than a century, important academic symposia, such as *Man's Role in Changing the Face of the Earth* (Thomas, 1956), continued to highlight the impacts of humans on natural ecosystems. However, it was not until the 1960s that a new scientific method and political recognition came together to initiate widespread changes in approaches to the environment.

From a scientific perspective, the 1960s witnessed the beginning of the 'quantitative revolution'. For the first time, detailed analyses of environmental impacts were produced. Previously, these had been inhibited by a paucity of field data. However, full utilization of the information yielded by this 'revolution' was constrained by the isolationist and reductionist philosophies that dominated scientific research for two decades. The major outcome of this phase of scientific endeavour was to demonstrate that virtually all the world's rivers had been significantly altered by human impacts well before the modern era of scientific investigation (Oglesby *et al.*, 1972; Petts, 1984; Davies and Walker, 1986).

From a political perspective, it was the dramatic impact of Rachel Carson's book, *Silent Spring* (1963) that aroused public concern about contamination of

air and water and the threat to wildlife and human health, which provided the catalyst for change in policy and practice. Then, in the 1980s, Bruce Brown's (1982) *Mountain in the Clouds* and Jeremy Purseglove's (1988) *Taming the Flood*, in their very different ways, presented to a wide audience vivid accounts of rivers in crisis. The Age of River Restoration had begun.

THE RISE OF ENVIRONMENTAL AWARENESS

During the past two decades river management has come to address the 'integrity triad' (Sparks, 1995) involving the coordinated management of chemical, physical and biological processes, although Brookes (1996) noted that even today, there have been few, if any, studies which effectively integrate morphological, water quality and biological data. Its development evolved around three actions. First, 'environmental management' focused almost exclusively on pollution control. Second, flow protection measures were introduced at planned dam sites, and habitat refuges – 'protected areas' – were established for fish and wildlife. Third, 'nature-like' channel designs were introduced. Its roots in Europe are found in the mid 19th century but until the mid 1960s these actions were of a small-scale and local nature.

A conservation movement grew out of the concerns of fishermen and hunters for the loss and degradation of habitat for fish and wildlife, and reduced opportunities for outdoor recreation. By 1890 in the UK, these groups had succeeded in securing minimum flows to protect migratory fisheries below dams (Petts, 1988). In the USA, early achievements were the creation of a national fish and wildlife refuge system, including refuges along major rivers that serve as migratory corridors and nesting areas for birds (Scarpino, 1985). Terms such as 'rehabilitation' and 'enhancement' have been in general use in federal legislation and interagency agreements dating back to 1937 (NRC, 1992).

In Australia, considerable concern has been expressed about river management since the first European settlement. Settlement at Sydney Cove commenced in 1788 and the first source of water was the 'Tank' stream. Within the first five years of settlement, regulations were passed forbidding the felling of trees within 50 feet (15.24 m) of the stream, and fences and intercepting drains were constructed on both banks to protect it. This does not appear to have been successful, and in 1802 Governor King informed the Colonial Office that he was taking vigorous action against polluters (Powell, 1976):

'If any person whatever is detected throwing any filth into the stream of fresh Water, cleaning Fish, Washing, erecting pig-styes near it, or taking water up but at the tanks, on conviction before a Magistrate, their Houses will be taken down and forfeit £A 35 for each Offence to the Orphan Fund.'

King was also concerned about uncontrolled clearing of riparian vegetation along the Hawkesbury River north of Sydney, which he thought contributed to erosion and flooding. In 1803 he directed (Powell, 1976):

'From the improvident method taken by the first Settlers on the side of the Hawkesbury and creeks cutting down timber and cultivating the banks, many acres of ground have been removed, lands inundated, houses, stacks of wheat, and stock washed away by former floods, which might have been prevented in some measure if the trees and other native plants had been suffered to remain, and instead of cutting any down to have planted others to bind the soil of the banks closer, and render them less liable to be carried away by every inconsiderable flood.
. . . it is hereby directed that no settler or other person to whom ground is granted or leased on the sides of any river or creek where timber is now growing, do on any account cut down or destroy, by barking or otherwise, any tree or shrub growing within two rods [*c.* 5 m] of the edge of the bank except for an opening one rod wide to have access to the water.'

Today, the 'Tank' stream is an underground drain!

Pollution control

In the UK, pollution of the aquatic environment was regarded as an environmental problem as early as the 13th century (Sweeting, 1994). Chronic conditions were reached in the 1840s, and all commercial fishing in the Thames through London had ceased by 1850. In England, 1858 became known as the 'Year of the Great Stink' because the condition of the London section of the River Thames was so bad. The magazine *Punch* in that year (volume XV, p. 16) characterized the Thames as one vast foul, stinking gutter. Problems of gross contamination by untreated sewage, high organic suspended solids and high ammonia levels, with high biochemical oxygen demand (BOD), were compounded

by the accumulation of industrial and domestic refuse and, increasingly, by effluent from heavy industry. The solutions were advanced by the resolution of a series of Royal Commissions on Pollution. The Pollution of Rivers Act (1876) was enforced by county councils, established by the 1888 Local Government Act (Sheail, 1988).

The foundation for controls on the polluting load discharged by industrial and sewage effluents was laid by the 1912 Royal Commission on Sewage Disposal. Flow and quality of the discharge to sewers were controlled by the Public Health Acts (1937 and 1947) and control was given to the Water Authorities under such legislation as the 1951 and 1961 Rivers (Prevention of Pollution) Acts. Improvements in river water quality were evident by 1970 and ammonia and BOD levels had been reduced considerably by 1985. The Control of Pollution Act was passed in 1974 but significant sections relating to water quality were not implemented until 1985. The passage of the Bill coincided with the reorganization of the water industry, following the 1973 Water Act, when 10, catchment-based, Water Authorities in England and Wales were given responsibility for the control of the whole water cycle including water supply, effluent treatment, land drainage, pollution control, and recreation and fisheries.

Similarly, in the USA, during the 1960s Congress enacted a series of Clean Air and Clean Water Acts and created the Federal Water Pollution Control Administration. In 1977 the US Clean Water Act moved beyond control of pollution to consider '*restoration* of the physical, chemical and biological integrity of the nation's waters'. However, despite a much broader mission and a new name, the Environmental Protection Agency (EPA) continued to concentrate on the chemical part of the integrity triad. In Australia, concern for river pollution became a political issue only in the 1960s and 1970s (Senate Select Committee – SCC, 1970) but grew to encompass concerns for the effects of salinity, impacts of land use and river regulation. The occurrence of a toxic cyanobacterial bloom over 1000 km of the Darling River in 1993 – shortly before a national election – focused the attention of politicians like never before on the need to manage and restore rivers, and provided a major stimulant to the establishment of a National River Health Programme in 1993.

Channel form

Simultaneously with the advancement of effective pollution control measures, developments were contributing to a re-examination of the traditional management of rivers. Not the least important of these was the growing concern over the increasing cost of flood damage, despite engineering efforts to control flooding, and the documentation of the long-term environmental effects of dams and channelization schemes. In western Europe, there is more than 20 years of experience of 'designing with nature' (Brookes, 1988). Early efforts to restore stretches of small streams were made by enthusiasts in the 1960s and 1970s, notably in the German state of Baden–Württemberg (Larsen, 1994). The first guidelines on 'nature-related river training and maintenance' date from around 1980, and the emphasis on 'renaturalization' has been strengthened by laws in most German states. In Baden–Württemberg, a recent modification of the water law requires that a 'nature-like' condition must be strived for; the conservation and restoration of the ecological functioning of a watercourse has to be given priority.

Similarly, in Denmark, the 1982 Watercourse Act provided an instrument to safeguard the physical environment of streams by focusing on ecologically acceptable channel-maintenance practices, incorporating special provisions for stream restoration and the potential for financial support for such projects (Nielsen, 1996; Iversen *et al.*, this volume). Elsewhere in Europe, Environmental Assessment legislation implemented following the European Commission Directive of 1985 makes provision for certain types of projects to be assessed prior to an application for planning consent and, in the UK, this has resulted in environmental issues receiving greater attention in the decision-making process. The *New Rivers and Wildlife Handbook* (Ward *et al.*, 1994) provides a comprehensive review of a range of rehabilitation projects carried out in the UK over the past 20 years. However, progress has been slow with most countries focusing on water quality and public-health-related issues, and fisheries (usually fish stocking: see e.g. Petts and Calow, 1996); and in most countries the length of channelized river restored with 'nature-like' features remains less than 1% (Brookes and Shields, 1996).

Flow regime

Traditionally, river flow patterns, which are critical to the maintenance and restoration of ecological integrity, were of concern only in relation to dilution of effluent, assimilation of organic wastes, and flushing of solid waste materials. However, 20 years of research during the 1970s and 1980s firmly established the importance of: (a) the flood pulse; (b) channel-forming discharges; (c) the interaction of flow and channel form; and (d) exchanges between surface waters and groundwaters in sustaining the biological diversity and productivity of river corridors (Petts and Amoros, 1996). However, the incorporation of 'environmental' needs into flow allocation remains a key issue for water-resource and river-basin managers (Petts and Maddock, 1994).

In England and Wales this issue has been addressed by applying a Minimum Acceptable Flow (MAF) concept, introduced in the Water Resources Act (1963). However, no statutory MAFs have been set. The main disadvantage of the statutory MAF procedure was perceived to be its inflexibility; being too prescriptive and encouraging abstraction to a maximum limit where considerable uncertainly exists in defining that limit (National Rivers Authority – NRA, 1996). Nevertheless, the MAF 'concept' has become embedded in abstraction licensing practice and there has been widespread use of flow controls as conditions attached to abstraction licences (Table 23.4). The conditions have been set to protect downstream interests giving regard to the principle of 'first come, first served', i.e. the historic sequence in which licences were granted with more recent licences having a higher Minimum Residual Flow (MRF) (Table 23.4) than older ones (a process known as 'stacking'). Such 'pseudo-mafs' have been used in about 1500 cases to protect water quality, fisheries, and downstream abstractors, and to prevent saline intrusion. A major problem is that the 'pseudo-mafs' reflect historic uses rather than present-day needs and rarely give due regard to the now widely recognized needs of river ecosystems. This latter issue is currently being advanced in discussion of Environmentally Acceptable Flow Regimes (NRA, 1996; Petts, 1996; Petts *et al.*, 1999).

The USA has two separate legal systems for regulating water flow. East of the Mississippi River, rainfall is generally plentiful and water shortages were historically not as much a problem as degradation of water quality, especially in urban areas. The legal principle was 'reasonable use', initially by riparian landowners and then through allocation systems to supply remote users (Lamb and Doerksen, 1990), as long as that use did not degrade or diminish the water available to downstream users (Ausness, 1983). Most eastern states have some provision to reserve streamflows in time of shortage, but the statutes vary widely in terms of effectiveness and enforcement (Lamb and Doerksen, 1990). In the arid lands to the west of the

Table 23.4   *Terms used in abstraction licences to protect flows for downstream users including fisheries and conservation interests. (Based upon experience in England and Wales; National Rivers Authority – NRA, 1996)*

The 'Prescribed Flow' (PF) is a generic term for any flow 'prescribed' in a statute which must not be diminished by abstraction.

Until it becomes 'prescribed' by statute, the flow which must not be diminished by abstraction is often termed 'Minimum Residual Flow (MRF)'.

The PF/MRF may comprise a single, simple control, or comprise a complex sequence of controls to meet specific conditions:

Hands-off Flow (HOF): the flow below which an abstraction must cease.
Maintained Flow (MF): the flow on a regulated river that shall be maintained by groundwater pumping, reservoir releases or interbasin transfer.

HOF and MF may vary during the year to meet seasonal needs and special provisions may be made at specified times (e.g. to prohibit abstraction for a time-limited period during summer spates to protect fish migration, or to provide time-limited reservoir releases to meet fisheries needs).

Also, they are both subject to variation by Drought Orders made by the Secretary of State upon application by the Environment Agency or a water undertaker, under powers conferred by Act of Parliament, to meet deficiencies in the supply of water due to exceptional shortages of rain. They can authorize abstraction from an unlicensed source, override the conditions pertaining to an abstraction licence or limit the amount of water which may be taken from a source.

Mississippi, the 'appropriative doctrine' applies. The first user of the water establishes a right to continued use, for beneficial purposes (drinking water for people or livestock, irrigation, industry, etc.). When scarce, those who established their appropriative rights last must stop using water until the needs of those with antecendent rights are satisfied. The Federal Government has reserved water rights in the west to fulfil the purpose of substantial federal holdings (national forests, parks, fish and wildlife refuges, wild and scenic rivers; Lamb and Doerksen, 1990). Current changes in water allocation relate to: (a) the assertion of prior water rights by native American nations in basins where water is already over-allocated, and the purchasing of water rights from agricultural users by (b) the expansion of cities in the west, and (c) the assertion of conservation and environmental groups to maintain in-stream flows for aquatic organisms.

Some conservation groups argue that all the water that flows in rivers and that lies on the land is used

ecologically, and is important for biodiversity; in their view, an excess of water does not exist. In contrast, the water supply industry has promoted the view that water is a fundamental resource which, like other essentials, must be exploited with a clear view of the needs of the future and a clear understanding that the water supply needs of people are paramount. Perhaps a more pragmatic approach is one that views natural ecosystems as an integral part of the resource and, as such, are bona fide users of water. This view requires that some legitimate constraint be applied to other water demands.

The importance of flow regime is well illustrated in Australia where the variation of discharge is exceptionally high (McMahon *et al.*, 1992) and where the stream biota is well adapted to this variability. Whether or not the high level of flow variation is necessary for the maintenance of Australian rivers is uncertain but it is distinctly possible. The role of spates as a disturbance or resetting mechanism in Australian streams has received some attention (e.g. Lake *et al.*, 1985) but the results are inconclusive, and probably at too fine a scale to allow extrapolation to the importance of flow variation to Australian rivers as a whole. Closs and Lake (1996) have demonstrated that the high frequency of low-flow periods plays an important role in allowing the native fish *Galaxias olidus* successfully to compete with the exotic brown trout (*Salmo trutta*) in the Lerderderg River, Victoria. They suggest that the maintenance of a constant flow at artificially high levels may disadvantage native fauna, compared with exotic species. Similarly, in the arid south-west USA, thunderstorms produce scouring floods that wash introduced species downstream, while native species have behavioural adaptations that enable them to persist (Meffe, 1984; Minkley and Deacon, 1991). Poff and Ward (1989) and Poff *et al.* (1997) describe regional variations in flow regimes across North America and their importance for the locally adapted organisms.

THE THEORETICAL CONTEXT

Today, the focus for developed economies is on quality of life and on investment in the environment, expressed by symbols of social advancement: restored landscapes, conserved species and protected wilderness areas. Since J.A. Gore published *The Restoration of Rivers and Streams* in 1985, considerable progress has been made in advancing the fundamental science necessary to underpin sustainable river management (e.g. Gore and Petts,

Table 23.5 *Terms relating to the management of river ecosystems. The definitions are based upon Boon (1992); NRC, (1992); Brookes and Shields (1996) and Rhoads and Herricks (1996)*

Preservation
The maintenance of functions and characteristics of an ecosystem in their desired states . . . not requiring rehabilitation.

Limitation
Usually applied to higher quality rivers; actions to limit catchment development, often involving land-use and conservation planning.

Mitigation
Actions to avoid, to reduce, or to compensate for the effects of environmental damage.

Restoration
The return of an ecosystem to a close approximation of its condition prior to disturbance. Both the structure and the functions of the ecosystem are restored.

Rehabilitation
Partial return to a pre-disturbance condition, usually linked to fish or wildlife habitat. To return degraded habitats to a pre-existing condition (e.g. dredging backwaters that have filled with sediment; forming riffles, or changing the planform of channelized reaches; or planting riparian buffer strips).

Enhancement
Usually applies to any improvement in environmental quality but is often used when making conditions optimal for a highly valued species, such as game and sport fish.

Creation
Bringing into being an ecosystem that previously did not exist on the site.

Naturalization
The creation of morphological and ecological configurations that are compatible with the magnitudes and rates of fluvial processes driven by the contemporary catchment ecosystem.

Dereliction
When a river has become severely degraded, the most pragmatic management option in the short- and medium-terms may be to accept the *status quo*. Resources may be better directed to higher quality sites within which projects may have a fair chance of success.

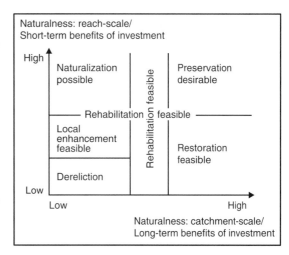

Figure 23.1 *Alternatives for ecologically sensitive river management according to the magnitude of environmental changes at both the catchment (flow, sediment and water quality regimes reflecting land use) and the local (reach) scale. 'Naturalness' refers to fluvial processes: at the catchment scale these include water flow, sediment load and water quality; at the reach scale, naturalness refers to channel dynamics and to channel–floodplain connectivity. Severely degraded systems are located to the bottom left, with pristine systems to the upper right of the diagram. Definitions of terms are given in Table 23.5*

1989; Boon *et al.*, 1992; Harper and Ferguson, 1995; Petts and Calow, 1996). Such works cover the range of processes involved in advanced plans for sustainable river management, including pollution control, channel maintenance and flow regulation, as well as controls on the human use of rivers and controls on biota.

In 1990 the US National Research Council convened a committee to examine the emerging science of restoration ecology in relation to aquatic ecosystems. The report of the committee was published as a book in 1992 that described 'the status and functions of surface water ecosystems; the effectiveness of on-going restoration efforts; the technology associated with those efforts; and the research, policy, and institutional reorganisation required to begin a national strategy for aquatic ecosystem restoration' (NRC, 1992).

Concepts are embodied in terms, and the terms relating to the restoration of ecosystems are based on the degree of human alteration, the attributes or services that are to be recovered, and the degree of human action required (Table 23.5; Figure 23.1). Preservation is distinct from restoration and creation in that the desired functions and characteristics are presumed to exist. However, this does not imply that the ecosystem is pristine and has not been subject to changes over the years, but that it is performing in an 'acceptable' and 'desirable' manner. Preservation does, however, imply a need for management in order to maintain the functions and characteristics, including weed and pest control in the system. Preservation has been mistakenly linked to mitigation under the assumption that a preserved system (e.g. a wetland) at one location offsets or mitigates the loss of another. In this context, it must be noted that preservation cannot compensate for losses at other locations; instead, *the result is a net loss* (NRC, 1992).

According to the NRC (1992):

'The goal (of restoration) is to emulate a natural, functioning, self-regulating system that is integrated with the ecological landscape in which it occurs. . . . Merely recreating the form without the functions, or the functions in an artificial configuration bearing little resemblance to a natural resource, does not constitute restoration.'

The words 'emulate' and 'approximate' (see below and in Table 23.5) are important:

'It is axiomatic that no restoration can ever be perfect; it is impossible to replicate the biogeochemical and climatic sequence of events over geological time that led to the creation and placement of even one particle of soil, much less to exactly reproduce an entire ecosystem. Therefore, all restorations are exercises in approximation and in the reconstruction of naturalistic rather than natural assemblages of plants and animals with their physical environment (NRC, 1992).'

However, in many cases, river management cannot address the return of an ecosystem to a pristine condition and the objective of restoring one which closely resembles unstressed surrounding areas, as proposed by Gore (1985), may be a realistic target in only a few cases. Disturbance has affected catchments as well as river corridors directly. Thus, practical river management seeks to establish a self-regulating system that is integrated with the contemporary – or projected – catchment in which it occurs. Restoration aims at least to approximate ecosystem structure and functioning prior to human disturbance, and contrasts with the terms 'creation', 'enhancement' and 'rehabilitation' that imply putting a landscape to a new or altered use to serve a particular human purpose. In the USA, the terms rehabilitation and enhancement are often used together, as in the Habitat Rehabilitation and Enhancement Program on the Upper Mississippi River System (US Water Resources Development Act of 1986).

Rhoads and Herricks (1996) introduced the term 'naturalization' as an environmental quality objective for rivers characterized by long-term human interference. Such situations commonly arise because the catchment has been intensively disturbed and/or transformed and its restoration is impractical. The characteristics of developed societies – urban areas and intensive agriculture – must be seen as permanent features of the catchment over the time-scale of river management and planning. Naturalization is broadly consistent with aims of restoration in that the process seeks to produce functionally diverse, self-regulating ecological systems. However, they differ in their explicit recognition of the permanence and importance of human influences upon the driving set of fluvial processes which may bear little resemblance to those processes operative within an unmodified catchment in a given ecoregion.

Restoration may be seen as a long-term goal, but in many cases it will require major investment and catchment-scale management of land and water use (Figure 23.1). Naturalization may be the best practicable short-term target, realizing the benefits of investment over a time-scale that may be appreciated by the society that provides the resources. However, naturalization measures should not constrain long-term restoration planning. Rather, it should be seen as a stepping-stone, in both practical and educational ways, to achieving restoration objectives in the longer term.

Current policy and practice

Both policy and practice in river management vary substantially among countries and often among administrative regions/areas within countries. Most countries have some form of hierarchical political/management structure but different practices and policies also reflect real physical differences between areas. This is well illustrated by Australia where in the Murray–Darling Basin the major management issue is water quality, particularly in relation to salinity and factors contributing to cyanobacterial blooms. In southern coastal streams around Brisbane, Melbourne and Sydney two issues dominate: abstraction for water supply and agricultural impacts of catchment clearing, removal of riparian vegetation and nutrient and contaminant run-off. In Tasmania, the impact of hydroelectric power development has been the overwhelming river conservation issue, highlighted by the Franklin Dam blockade in 1982.

The Australian Government plays a role by coordinating state agency activities and providing funds to state and local government agencies and non-government organizations. It has a direct role only in the management of a limited number of areas, chiefly those that are listed under Ramsar or World Heritage agreements and thus subject to international treaty. State government agencies are responsible for managing water-supply infrastructure for domestic and agricultural supply and hydroelectric power generation. They are also responsible for setting water quality guidelines, monitoring and managing water quality, and allocating environmental flows. At the level of small catchments and river reaches, management responsibilities and activities are carried out by community groups, such as Landcare groups, Shire Councils and local river-improvement trusts. There are also other management organizations, of which the most significant is the Murray–Darling Basin Commission. This organization comprises government ministers from the four states through which the Murray–Darling River system extends (New South Wales, Queensland, South Australia and Victoria) as well as the national government, and has responsibility for environmental management of the Murray–Darling system. It does not have the power to override the states, but rather operates in a cooperative consensual mode.

Examples of restoration in practice

'Environmentally sensitive' river management currently involves three components. The first is data collection for monitoring and assessment, a good example of which is the 'Long Term Resource Monitoring Program (LTRMP) of the upper Mississippi River System'. The LTRMP is a cooperative effort between the federal agency charged with maintaining navigation and providing flood protection (US Army Corps of Engineers – USACE); the federal agency responsible for gathering hydrological and biological data on the nation's inland waters (US Geological Survey – USGS); the US Fish and Wildlife Service (which maintains the system of wildlife refuges along the upper Mississippi); and the natural resource agencies from the five states along the river (USACE, 1997a,b; USGS, 1998). The purpose of the LTRMP is to assess trends in plant and animal populations, sediment and water quality, and the morphology of the river and its backwaters, determine the causes of the trends, and make recommendations for improved management.

The second component of environmentally sensitive river management is programmes of research (e.g.

Special Issues of *Regulated Rivers: Research & Management*; 12 (4–5), 1996: *Remedial Strategies in Regulated Rivers*; and *Aquatic Conservation: Marine and Freshwater Ecosystems* 9(1), 1999: *Restoring Aquatic Ecosystems*). The third is restoration demonstrations (see e.g. Special Issue of *Aquatic Conservation: Marine and Freshwater Ecosystems* 8(1), 1998 on *River Restoration: The Physical Dimension*). Although the integrated management of large rivers and their basins has been the focus of several symposia (e.g. Special Issue of *Regulated Rivers: Research & Management* 11(1), 1995: *Sustaining the Ecological Integrity of Large Floodplain Rivers*), for the most part, rehabilitation schemes have focused on small streams or short reaches of larger rivers (e.g. Brookes and Shields, 1996; De Waal *et al.*, 1998).

EUROPEAN PERSPECTIVE

In Europe, the need to advance knowledge of channel restoration led to a Demonstration Project partly funded by the European Community Life Fund (Brookes, 1996; Holmes and Nielsen, 1998). The project was started in 1993 and aims (i) to demonstrate the benefits of channel restoration and (ii) to motivate and to train all those involved with river management. Three rivers were selected for restoration, two in lowland England and one in Denmark, with one site of 3–12 km on each. Independently, a number of pilot projects, typically on reaches of 2 km or less, are being undertaken elsewhere and in Baden–Württemberg plans are being formulated to 'renaturalize' 160 km of the River Danube (Larsen, 1994).

Frameworks have also been established for the management of international river basins. A good example of the latter is the Rhine Action Plan (Van Dijk *et al.*, 1995). In 1963 'Europe's largest sewer' was the subject of an international agreement, the Convention of the International Commission for the Protection of the Rhine (ICPR), involving five states bordering the Rhine and, since 1976, the European Union. Effective pollution control led to the recovery of biological communities from the early 1980s. Following the major Sandoz incident in which nearly 30 000 kg of toxic chemicals were flushed into the river at Basle, a Rhine Action Programme for ecological rehabilitation was agreed in 1987. The specific ecological goal is the return of Atlantic Salmon (*Salmo salar*). The first focus was on water quality, the second, habitat improvement. The latter aims (a) to encourage the return of migratory fish, to allow free fish migration and to restore spawning and nursery grounds, and (b) to restore all connected habitats to allow the development of self-regulating biocoenoses with intact food chains. The Rhine Action Programme is not only the first programme for a major transboundary river to include clear ecological objectives; it will also serve as an important demonstration project.

USA PERSPECTIVE

In the USA in 1994 there were 238 stream and river monitoring programmes staffed by volunteers in 45 of 50 states and in the District of Columbia (United States Environmental Protection Agency – USEPA, 1994). Since then, the number of programmes has more than doubled, according to an updated USEPA report that is currently in press. One example of how citizen interest has sparked government action is provided by the Illinois River Basin. Virtually every major tributary and headwater in the Basin has a 'Friends of the . . . River' group. Requests from these groups for technical assistance led to the formation of an Illinois River Basin Strategy Team, a group of leaders in business, agriculture, and conservation, chaired by the Lt Governor. A directory of on-going model projects and model approaches to river restoration was published in July 1995; most of the model projects had a strong volunteer component and many were initiated by local residents. During 1996, 150 technical experts and citizens from the basin participated in a year-long process that culminated in the Integrated Management Plan for the Illinois River Watershed (Kustra 1997a,b). The vision for the Illinois River adopted by the Team is noteworthy, because a healthy riverine ecosystem comes first, and socio-economic activities must be *compatible* with that objective: 'A naturally diverse and productive Illinois River Valley that is sustained by natural ecological processes and managed to provide for compatible social and economic activities.'

In April 1998, the Illinois River Coordinating Council (IRCC) was formed to institutionalize the Plan and to ensure that implementation continued beyond the terms of office of the current Governor and Lt Governor. The US Department of Agriculture and the state of Illinois have jointly committed US$459 million over the next 15 years to reduce soil erosion and sedimentation, to improve water quality, and to enhance fish and wildlife habitat in the basin. The programme includes incentives to landowners to stop farming their most erodible land

and to restore wetlands and riparian zones (see also Showers, this volume).

AUSTRALIAN PERSPECTIVE

In Australia, the National River Health Programme, originally established by the Australian Government in 1993, uses a standard rapid bio-assessment technique (based mainly on macroinvertebrate assemblages) at over 1200 sites nationwide, (i) to provide an initial objective assessment of the state of Australian rivers, and (ii) to commence a continuing monitoring programme that can feed into the national state of the environment assessment programme. Hitherto, there was no database adequate to allow any broad nationwide assessment of the state of Australian rivers. A programme of research and restoration demonstrations of stream riparian systems has been established by the Land and Water Resources Research and Development Corporation and a number of projects have aimed to restore in-stream habitat. For example, on the Broken River in northern Victoria, the local river management board together with the State Department of Natural Resources have embarked on a project to restore the river as habitat for trout cod (*Maccullochella macquariensis*), a native fish considered endangered in Victoria. Works carried out as part of the programme include bank stabilization, removal of exotic willows, placement of large woody debris – a process known as re-snagging – and fish stocking. Such programmes often include a monitoring component to demonstrate effectiveness (Tennant *et al.*, 1996).

But by far the most extensive and most significant programme for stream restoration in Australia is that carried out under the Landcare Programme. This is a community-based programme first established in Victoria in 1986 and later extended nationally. As the name suggests, the programme is primarily aimed at reducing land degradation, but land degradation is virtually inseparable from stream degradation. In 1995–96 the Australian Government provided A$77.6 million to Landcare groups. These groups usually involve 20–30 local people working in an area between 4×10^3 ha and 14×10^6 ha with total annual funding for any individual group of about A$15 000. A major activity of Landcare groups has been the replanting of riparian vegetation along streams, with the Federal Government money being used to buy trees and fencing materials, and the group members providing the labour (Campbell, 1994; Campbell *et al.*, 1998). In some cases the aim of tree planting has been to reduce soil erosion and to restabilize stream banks, but in other cases tree planting is seen as part of a long-term strategy to reduce salinization problems (e.g. Masterson, 1996). Unfortunately, most of the Landcare restoration projects are being undertaken with little or no monitoring, so that the success of these projects will be difficult to assess.

Major unresolved issues

There are four key questions that underpin much of current scientific research on degraded river systems:

- Is it feasible to preserve or restore functioning large river ecosystems?
- How effective is local, reach-scale 'habitat' management, compared with catchment-scale management of the key hydrological processes?
- What is the target for river management?
- What do people want?

The major issue for river conservation is how to move from management of water quality and fish and wildlife habitat to the broader and harder-to-define goal of ecosystem restoration. Spatial scale is important because large rivers cannot be wholly protected within parks and reserves that include most of the catchment, whereas small streams can. Examples include the upper Yellowstone River in Yellowstone National Park in Wyoming and the Hoh River in Olympic National Park in the state of Washington. Less than 2% of the combined length of all rivers in the USA qualifies for protection under the National Wild and Scenic Rivers Act, and the Act focuses on preserving relatively undisturbed channel *segments*, rather than on maintaining and restoring stream and river ecosystems (Doppelt *et al.*, 1993). Some would argue that restoration is not even a viable policy option for large rivers.

Rivers are products of their catchments and some large rivers and their catchments may be so altered that preservation is no longer an option: restoration is unfeasible, and in most cases only rehabilitation for selected uses and limited naturalization is feasible. However, large floodplain rivers are particularly dependent upon lateral connectivity between the channel(s) and adjacent floodplains, at least during the annual flood. Where this remains in a relatively natural condition, preservation and restoration of substantial – but not all – segments of rivers are still

viable options, as is the restoration of many processes, functions and services. The NRC (1992) identified reaches of the Mississippi that retain at least half the predisturbance floodplain and a seasonal flood pulse. The flood regime is the master variable that drives river–floodplain ecosystems, so these reaches are much less altered than many other large rivers in the developed world, where water regimes are controlled by humans and decoupled from the precipitation pattern. A large aquatic system where restoration has progressed sufficiently to indicate at least partial success is the Kissimmee River–Lake Okeechobee–Everglades ecosystem in central and south Florida (NRC, 1992; Toth, 1995). Although the proportion of semi-natural rivers and streams left in the USA may be small in relation to total river kilometres, even a fraction of 5.1×10^6 km can represent substantial riverine corridors!

Restoration of urban streams remains problematical (see Baer and Pringle, this volume). Water quality problems and channelization appear to limit options. Even where no industrial or domestic discharges are released into the streams, urban run-off is often characterized by high levels of metals, especially zinc, from galvanized roofing materials, and high levels of hydrocarbons and lead from fuels and lubricants used by motor vehicles (e.g. see Weeks, 1982). Urban streams also have high nutrient and pesticide levels, partly from public and domestic gardens, and they frequently have high bacterial levels resulting from leaking sewerage systems and from domestic pets. For some Australian urban streams, easements have been maintained which may contain remnant riparian vegetation, or at least sufficient space in which riparian vegetation can be re-established. There are a number of community projects in most Australian state capital cities carrying out replanting and restoration of urban riparian vegetation. In many cases urban streams have been contained within concrete surface or underground drains and, in some cases, the easement has been used for highway construction. In such areas the restoration of the river corridor is not possible. However, the extent to which physical habitat naturalization can achieve ecological benefits without changes to the fundamental processes in severely disturbed urban rivers remains to be evaluated. The enhancement of floodplain areas along urban rivers may require the isolation of the river from the floodplain and backwaters to prevent further accumulation of contaminated sediments; degraded areas must be isolated, dredged or dried out (but problems may arise from the mobilization of heavy metals on oxidation), then protected from the river as much as possible. In some places, rivers may need to be protected from contaminated sites on the floodplain.

In small catchments, and as part of a long-term policy, the re-establishment of a more natural water and sediment regime is possible, through upland catchment treatment, detention basins in urban areas, de-channelization of selected tributary reaches, restoration of selected wetlands and riparian zones, modification of dams and their operating procedures, and reconnection of more of the floodplain to the rivers.

With regard to targets, the national perspective varies between countries and is probably dependent upon the degree, type and timing of landscape alteration. In Europe, where catchments, streams and rivers were extensively altered centuries ago, it may not be feasible to restore substantial lengths of rivers and, besides, the managed and naturalized systems have their own values defined in historical, aesthetic, recreational and conservation terms. Indeed, a popular landscape in England remains the rolling green pastures, punctuated by clumps of trees, bordering broad streams or ribbon lakes – fundamental elements of the landscapes created by the 18th century landscape 'improver' Lancelot ('Capability') Brown (Burke, 1971). Such landscapes are often perceived to be 'open' and 'safe', providing individual and family 'space' within a densely populated country, whereas wet woodland corridors – the natural landscape – are often seen as 'dark, damp and dangerous' (Purseglove, 1988; Petts, 1990).

In the USA, many writers and conservation organizations both formed and reflected public belief that the wilderness experience had shaped the character of the nation and was important in the development of individuals in the modern era (Nash, 1967; Scarpino, 1985). Intensive settlement and river alterations have occurred recently enough, particularly in the American West, that substantial declines in natural resources have been observed by people who are still living. White (1995) noted the importance of recent memory and social value in regard to the declines of the salmon runs: 'Salmon symbolize nature in the Pacific Northwest; the experience of taking them has become a quintessential Northwest experience. Salmon are not just fish on the Columbia [River]; they are tokens of a way of life.' Thus, the Army Corps of Engineers and the Environmental Protection Agency (1990) signed a memorandum of agreement that defined restoration as 'measures taken to return the existing fish and wildlife

habitat resources to a *modern historic condition*'. This is defined as the pre-dam condition – 1880s to 1920s.

The 'modern historic condition' may be a realistic reference standard in the USA but in the more intensively developed areas of the world a 'post-modern' condition might be more appropriate. Human impacts disequilibrated large river systems by altering water and sediment regimes. Dams created reservoirs, pools and marshes that enhanced boating, fishing and hunting, as well as controlled floods and supplied water, electricity and year-round navigation. These rivers are now approaching a new 'equilibrium' condition; a new functioning and self-regulating ecosystem is being created by natural processes working within the constraints and limits imposed by human actions. An appropriate target for management would be to guide and to facilitate this adjustment process, reducing the human constraints and limits, to allow a new system to develop as naturally as practicable. This pragmatic perspective recognizes the multiple uses required of river systems increasingly through the new millennium when demands for reliable water supply for domestic, industrial and agricultural uses, navigation, hydropower production, and flood control, compete with recreation, amenity, and the preservation and conservation of biodiversity. However, the goal must be to preserve and restore functioning ecosystems to the maximum extent possible, to allow ecosystems to take care of the species. This will require acceptance of the annual variation in abundances of plant and animal populations – of good and bad fishing years – and explicit recognition of the critical role of disturbance regimes in maintaining and creating biodiversity as well as the relaxation of some human constraints on natural variation.

The way ahead

In many countries this millennium will witness a change in the policy and practice of natural resource management, including the management of rivers. The old paradigm was sustained yield. This is well exemplified by the US natural resource agencies, such as the National Marine Fisheries Service (commercial fisheries), Fish and Wildlife Service, Soil Conservation Service, Forest Service, and the Bureau of Reclamation which built dams to store and supply water in the arid West. The paradigm for the Corps of Engineers was to build the infrastructure for national economic development. The need to protect the environment was viewed as a constraint on development, yield and use. While no new paradigm has been fully accepted, a shift is occurring and the emerging paradigm will be based on two principles: ecosystem management and collaborative decision-making (Cortner and Moote, 1994). Under the new paradigm, maintenance and restoration of ecosystem structure and functioning will be included in agency mission statements.

The paradigm shift will also involve a move away from decision-making by technical experts towards procedures that promote consensus building among stakeholders, with technical expertise and analysis of alternatives provided by the agencies. Kemmis (1990) describes how the current US system of required public hearings on public-works projects often leads to stalemate. No one goes to 'hear'; rather, the stakeholders go to influence the agency decision-makers by putting forward the strongest possible case for their position and against the position of other stakeholders. The outcome is often less than that which the stakeholders could have achieved by talking to each other. Nevertheless, public participation is critical for the success of any river restoration project. Without the support of riparian landowners it is not possible to achieve significant ecological improvements. Public participation is also a key component of public education. Public participation in waterways monitoring has been widely used in Australia through programmes such as 'Streamwatch', which have proved extraordinarily effective as a public education strategy promoting community 'ownership' of streams (Australian Nature Conservation Agency – ANCA, 1995). Similarly, in the USA there are programmes (e.g. within the Illinois River Basin Integrated Management Plan, see above) which use local rivers and streams as a theme to integrate elementary and secondary school curricula in the sciences and humanities.

The technical, social, political and institutional barriers to managing complex regional ecosystems, and the role of simulation models to inform stakeholders of the consequences of management alternatives, have been described (Holling, 1978; Walters, 1986; Lee 1993; Gunderson *et al.*, 1995). These authors argue that uncertainties inherent in management of ecosystems requires an adaptive management approach, where management constantly improves and adjusts, based on improved understanding of how the ecosystem works and on constantly updated information on the status and response of the ecosystem obtained by monitoring critical indicators. Management should include experimentation: some actions may be undertaken in

order to learn more about the managed system, as well as to achieve more traditional objectives.

'Naturalization' appears to be the best practicable option: it is defined (Brookes and Shields, 1996, and illustrated by Rhoads and Herricks, 1996) as the process that determines the morphological and ecological configurations which are compatible with contemporary magnitudes and rates of fluvial processes. However, it is also clear that considerable research gaps remain to be filled before analytical approaches can be developed to determine the configurations which produce naturally functioning ecological systems. The transferability of knowledge and experience of river restoration must be approached with caution as the significance of 'ecological distinctiveness' is not understood. In Australia, for example, the distinctiveness of the freshwater biota has been well documented (e.g. Williams and Campbell, 1987), while several authors have commented on the apparently striking evolutionary convergences between many Australian species and species apparently filling similar niches elsewhere (e.g. Campbell, 1990).

In areas with transitional and developing economies, economic growth, social advancement and political security demand the wise use of natural resources. Nevertheless, in many of these areas, rapid development is seen as necessary for survival (e.g. see Wishart *et al.*, this volume, Part II). In some areas the continuing degradation of the environment is threatening the long-term welfare of societies, but the immediate need for water, food, and fuelwood, to sustain life for another day or week, overrides any long-term plan. In other areas, rapid development is fuelled by the unsustainable exploitation of natural resources and large-scale, technological solutions to resource management and hazard control. The term 'pharaonic work' has been used in Brazil to refer to massive civil works that mortgage an entire society to benefit the elite (the technocrats, industrial and construction firms, and politicians; e.g. McCully, 1996) and to enhance national prestige, but which bring little long-term economic return to the society as a whole (Veja, 1987, cited in Fearnside, 1989).

Even in areas of water scarcity with rapid population growth and grinding poverty, sustainability philosophies must be encouraged. One step may be for developed economies to ensure the availability of trained personnel of sufficient quality and in sufficient numbers to advance education and public awareness, and to encourage the formation of volunteer and stakeholder groups. They could do more to describe the long-term effects (both beneficial and negative) of their own large development projects, so that developing nations do not repeat the same mistakes (Sparks, 1992). Given the historical experience and analytical and predictive capabilities that have accrued in the developed nations, there is little excuse now for failing to consider and to evaluate consequences of projects anywhere in the world, including projects that are supposed to mitigate past damage.

The vast majority of the people who are physically or economically displaced by unsustainable developments lack political power, and the best hope for more rational, socially equitable, and sustainable development comes from the lending institutions (e.g. the World Bank). These institutions are bankrolled by developed nations which should ensure that issues of sustainability and social equity are factored into lending decisions. Grassroots political movements worldwide are also beginning to make their cases directly to these lending institutions, in addition to attempting to influence their own governments. In any case, future development polices and practices in all countries must be underpinned by sustainable ecosystem management based upon collaborative decision-making.

Acknowledgements

We are pleased to acknowledge editorial support by Ruth Sparks and valuable comments by Bryan Davies. RS gratefully acknowledges support from the US National Science Foundation/Environmental Protection Agency Partnership for Environmental Research and from the cooperative federal–state Long Term Resource Monitoring Program on the Upper Mississippi River System.

References

Ausness, R.C. 1983. Water right legislation in the East – a program for reform. *College of William and Mary Law Review* 24:547–590.
Australian Nature Conservation Agency (ANCA) 1995. *Waterwatch, a National Snapshot.* National Waterwatch Facilitator. Canberra.
Boon, P.J. 1992. Essential elements in the case for river conservation. In: Boon, P.J., Calow, P. and Petts, G.E. (eds) *River Conservation and Management.* John Wiley, Chichester, 11–33.
Boon, P.J., Calow, P. and Petts, G.E. (eds) 1992. *River Conservation and Management.* John Wiley, Chichester.
Brookes, A. 1988. *Channelized Rivers.* John Wiley, Chichester.

Brookes, A. 1996. River restoration experience in Northern Europe. In: Brookes, A. and Shields, F.D. (eds) *River Channel Restoration.* John Wiley, Chichester, 233–268.
Brookes, A. and Shields, F.D. (eds) 1996. *River Channel Restoration,* John Wiley, Chichester.
Brown, B. 1982. *Mountain in the Clouds: A Search for the Wild Salmon.* Touchstone, New York.
Burke, G. 1971. *Towns in the Making.* Edward Arnold, London.
Campbell, I.C. 1990. The Australian mayfly fauna: composition, distribution and convergence. In: Campbell, I.C. (ed.) *Mayflies and Stoneflies, Life Histories and Biology.* Kluwer Academic Press, Dordrecht, 149–153.
Campbell, C.A. 1994. *Landcare – Communities Shaping the Land and the Future.* Allen & Unwin, Sydney.
Campbell, I.C., Boon, P.J., Madsen, B.L. and Cummins, K.W. 1998. Objectives and approaches in lotic and riparian restoration. *Verhandlungen Internationalen Vereinigung für theoretische und angewandte Limnologie* 26:1295–1302.
Carson, R. 1963. *Silent Spring.* Hamish Hamilton, London.
Closs, G. and Lake, P.S. 1996. Drought, differential mortality and the co-existence of a native and an introduced fish species in a south east Australian intermittent stream. *Environmental Biology of Fish* 47:17–26.
Cortner, H.J. and Moote, M.A. 1994. Trends and issues in land and water resources management: setting the agenda for change. *Environmental Management* 18:167–173.
Cosgrove, D. 1990. An elemental division: water control and engineered landscape. In: Cosgrove, D. and Petts, G.E. (eds) *Water, Engineering and Landscape.* Belhaven, London, 1–11.
Cosgrove, D. and Petts, G.E. (eds) 1990. *Water, Engineering and Landscape.* Belhaven, London.
Davies, B.R. and Walker, K.F. (eds) 1986. *The Ecology of River Systems.* Dr W. Junk, Dordrecht.
De Waal, L.C., Large, A.R.G. and Wade, P.M. (eds) 1998. *Rehabilitation of Rivers: Principles and Implementation.* John Wiley, Chichester.
Doppelt, B., Scurlock, M., Frissell, C. and Karr, J. 1993. *Entering the Watershed: A New Approach to Save America's River Ecosystems.* The Pacific Rivers Council, Island Press, Washington, DC.
Ellet, C. 1853. *The Mississippi and Ohio Rivers.* Lippincott, Grambo and Co., Philadelphia.
Fearnside, P.M. 1989. Brazil's Balbina Dam: environment versus the legacy of the pharaohs in Amazonia. *Environmental Management* 13:401–423.
Gore, J.A. (ed.) 1985. *The Restoration of Rivers and Streams.* Butterworth Publishers, Boston.
Gore, J.A. and Petts, G.E. (eds) 1989. *Alternatives in Regulated River Management.* CRC Press, Boca Raton, Florida.
Gunderson, L.H., Holling, C.S. and Light, S.S. (eds) 1995. *Barriers and Bridges to the Renewal of Ecosystems and Institutions.* Columbia University Press, New York.
Harper, D.M. and Ferguson, A.J.D. (eds) 1995. *The Ecological Basis for River Management.* John Wiley, Chichester.

Holling, C.S. (ed.) 1978. *Adaptive Environmental Assessment and Management*. John Wiley, Chichester.

Holmes, N.T.H. and Nielsen, M.B. 1998. Restoration of the rivers Brede, Cole and Skerne: a joint Danish and British EU-LIFE demonstration project, I – setting up and delivery of the project. *Aquatic Conservation: Marine and Freshwater Ecosystems* 8:185–196.

Kemmis, D. 1990. *Community and Politics of Place*. University of Oklahoma Press, Norman Publishing Division of the University.

Kustra, B. 1997a. *Integrated Management Plan for the Illinois River Watershed*. Office of the Lt Governor, State of Illinois, Springfield.

Kustra, B. 1997b. *Integrated Management Plan for the Illinois River Watershed: Technical Report*. Office of the Lt Governor, State of Illinois, Springfield.

Lake, P.S, Barmuta, L.A., Boulton, A., Campbell, I.C. and StClair, R.D. 1985. Australian streams and Northern Hemisphere stream ecology: comparisons and problems. In: Dobson, J.R. and Westoby, M. (eds) *Are Australian Ecosystems Different? Proceedings of the Ecological Society of Australia* 14:61–82.

Lamb, B.L. and Doerksen, H.R. 1990. Instream water use in the United States – water laws and methods for determining flow requirements. In: *National Water Summary 1987 – Hydrologic Events and Water Supply and Use*. US Geological Survey, Water Supply Paper 2350, US Geological Survey, Reston, Virginia.

Larsen, P. 1994. Restoration of river corridors: German experiences. In: Calow, P. and Petts, G.E. (eds) *The Rivers Handbook*. Blackwell Scientific, Oxford, 419–440.

Lee, K.N. 1993. Rebuilding confidence: salmon, science and law in the Columbia Basin. *Environmental Law* 21:745–805.

Lowenthal, D. (ed.) 1965. *Man and Nature*. Balknap Press of Harvard University Press, Cambridge, Massachusetts.

Masterson, S. 1996. *Demonstration and Evaluation of Riparian Restoration in the Blackwood Catchment*. Land and Water Resources Research and Development Corporation (LWRRDC) Milestone Report 1 – BCW 1, Australia.

McCully, P. 1996. *Silenced Rivers. The Ecology and Politics of Large Dams*. Zed Books, London.

McMahon, T.A., Finlayson, B.L., Haines, A.T. and Srikanthan, R. 1992. *Global Runoff. Continental Comparisons of Annual Flows and Peak Discharges*. Catena Press, Cremlingen.

Meffe, R.H. 1984. Effects of abiotic disturbance on coexistence of predator and prey fish species. *Ecology* 65:1525–1534.

Minkley, W.L. and Deacon, J.E. (eds) 1991. *Battle Against Extinction: Native Fish Management in the American West*. University of Arizona Press, Tucson.

Nash, R. 1967. *Wilderness and the American Mind*. Yale University Press, New Haven.

National Rivers Authority (NRA) 1996. *Determination of Minimum Flows*. Bristol, R & D Note 449.

National Research Council (NRC) 1992. *Restoration of Aquatic Ecosystems: Science, Technology and Public Policy*. National Academy Press, Washington, DC.

Nielsen, M.B. 1996. Lowland stream restoration in Denmark. In: Brookes, A. and Shields, F.D. (eds) *River Channel Restoration*. John Wiley, Chichester, 269–292.

Oglesby, R.T., Carson, C.A. and McCann, J.A. (eds) 1972. *River Ecology and Man*. Academic Press, New York.

Petts, G.E. 1984. *Impounded Rivers*. John Wiley, Chichester.

Petts, G.E. 1988. Regulated rivers in the UK. *Regulated Rivers: Research & Management* 2:201–220.

Petts, G.E. 1990. The role of ecotones in aquatic landscape management. In: Naiman, R. (ed.) *The Roles of Ecotones in Aquatic Landscapes*. Cambridge University Press, Cambridge, 227–261.

Petts, G.E. 1996. Allocating water to meet in-river needs. *Regulated Rivers: Research & Management* 12:353–365.

Petts, G.E. and Amoros, C. (eds) 1996. *Fluvial Hydrosystems*. Chapman & Hall, London.

Petts, G.E. and Calow, P. (eds) 1996. *River Restoration*. Vol. 2. Blackwell Scientific, Oxford.

Petts, G.E. and Maddock, I. 1994. Flow allocation for in-river needs. In: Calow, P. and Petts, G.E. (eds) *The Rivers Handbook*. Blackwell Scientific, Oxford, 289–307.

Petts, G.E., Moller, H. and Roux, A.L. (eds), 1989. *Historical Change of Large Alluvial Rivers: Western Europe*. John Wiley, Chichester.

Petts, G.E., Bickerton, M.A., Crawford, C., Lerner, D.N. and Evans, D. 1999. Flow management to sustain groundwater-dominated ecosystems. *Hydrological Processes* 13:497–513.

Poff, N.L. and Ward, J.V. 1989. Implications of streamflow variability and predictability for lotic community structure: a regional analysis of streamflow patterns. *Canadian Journal of Fisheries and Aquatic Sciences* 46:1805–1818.

Poff, N.L., Allan, J.D., Bain, M.B., Karr, J.R., Prestegaard, K.L., Richter, B.D., Sparks, R.E. and Stromberg, J.C. 1997. The natural flow regime. *BioScience* 47:769–784.

Powell, J.M. 1976. *Environmental Management in Australia 1788–1914. Guardians, Improvers and Profit – An Introductory Survey*. Oxford University Press, Melbourne.

Purseglove, J. 1988. *Taming the Flood. A History and Natural History of Rivers and Wetlands*. Oxford University Press, New York.

Rackham, O. 1997. *The History of the Countryside*. Phoenix, London.

Reisner, M. 1986. *Cadillac Desert: the American West and its Disappearing Water*. Penguin Books, New York.

Rhoads, B.L. and Herricks, E.E. 1996. Naturalization of headwater streams in Illinois: challenges and possibilities. In: Brookes, A. and Shields, F.D. (eds) *River Channel Restoration*. John Wiley, Chichester, 331–368.

Scarpino, P.V. 1985. *Great River: An Environmental History of the Upper Mississippi, 1890–1950*. University of Missouri Press, Columbia.

Senate Select Committee (SSC) 1970. *Report of the Senate Select Committee on Water Pollution*. Australian Government Publishing Service, Canberra.

Sheail, J. 1988. River Regulation in the UK. *Regulated Rivers: Research & Management* 2:221–232.

Sparks, R.E. 1992. Risks of altering the hydrologic regime of large rivers. In: Cairns, J., Niederlehner, B.R. and Orvos, D.R. (eds) *Predicting Ecosystem Risk*. Princeton Scientific, Princeton, New Jersey, 119–152.

Sparks, R.E. 1995. Need for ecosystem management of large rivers and their floodplains. *BioScience* 45:168–182.

Sweeting, R.A. 1994. River pollution. In: Calow, P. and Petts, G.E. (eds) *The Rivers Handbook*. Blackwell Scientific, Oxford, 23–32.

Tennant, W., Gooley, G. and Douglas, J. 1996. *Evaluation of Operational Management Works and Aquatic Habitat Restoration in the Mid-Broken River Catchment*. Progress Report November 1996, Department of Natural Resources and Environment, Melbourne.

Thomas, W.L. (ed.) 1956. *Man's Role in Changing the Face of the Earth*. University of Chicago Press, Chicago.

Toth, L.A. 1995. Principles and guidelines for restoration of river/floodplain ecosystems – Kissimmee River, Florida. In: Cairns, J. (ed.) *Rehabilitating Damaged Ecosystems*, 2nd edition. Lewis/CRC, Boca Raton, Florida, 49–73.

United States Army Corps of Engineers (USACE) 1997a. *An Evaluation of the Upper Mississippi River System Environmental Management Program*. Report to Congress. Rock Island District, Rock Island, Illinois.

United States Army Corps of Engineers (USACE) 1997b. *An Evaluation of the Upper Mississippi River System Environmental Management Program. Appendix A: The Long Term Resource Monitoring Program*. Rock Island District, Rock Island, Illinois.

United States Environmental Protection Agency (USEPA) 1994. *National Directory of Volunteer Environmental Monitoring Programs*, 4th edition. EPA 841-B-94-001. USEPA Office of Water Research, Washington, DC.

United States Geological Survey (USGS) 1998. *Ecological Status and Trends of the Upper Mississippi River System*. USGS Biological Resources Division, Environmental Management Technical Center, Onalaska, Wisconsin.

Van Dijk, G.M., Marteijn, E.C.L. and Schilte-Wulwer-Leidig, A. 1995. Ecological rehabilitation of the River Rhine: plans, progress and perspectives. *Regulated Rivers: Research & Management* 11:377–388.

Walters, C.J. 1986. *Adaptive Management of Renewable Resources*. Macmillan, New York.

Ward, D., Holmes, N. and José, P. 1994. *The New Rivers and Wildlife Handbook*. Royal Society for the Protection of Birds, National Rivers Authority, and Royal Society for Nature Conservation, Sandy, Bedfordshire.

Weeks, C.R. 1982. Pollution in urban runoff. In: Hart, B.T. (ed.) *Water Quality Management – Monitoring Programs and Diffuse Runoff*. Water Studies Centre, Chisholm Institute of Technology, Melbourne.

White, R. 1995. *The Organic Machine: The Remaking of the Columbia River*. Hill and Wang, a Division of Farrar, Dtraus and Giroux, New York.

Williams, W.D. and Campbell, I.C. 1987. Major components and distribution of the freshwater fauna. In: Walton, D.W. (ed.) *The Fauna of Australia*, Volume 1, Australian Biological Resources Survey (ABRS), Canberra, 156–183.

24

Integrated watershed management for river conservation: perspectives from experiences in Australia and the United States

B.P. Hooper and R.D. Margerum

Introduction

A river valley is an integrated system, in which there are strong ecological relationships between the rivers and land systems of the valley. Rivers act as hydrological conduits receiving water from precipitation, infiltration and groundwater movement, transferring water across the landscape to watershed outlets, such as another river, lakes, estuaries or oceans. The ecological health of rivers reflects the ecological health of the land systems in the basin, and the impacts of land-management practices on riverine ecological processes.

Linking water and land-resource management is an attractive and sound approach, especially if done on a watershed basis (Burton, 1988a). A watershed is characterized by a clearly defined boundary, determined by its geomorphological, geological and hydrological history. The nature of hydrological linkages suggests that a river basin is a natural unit of management for river conservation. However, river basin boundaries may not reflect groundwater or ecological boundaries, prevailing social networks or administrative boundaries (e.g. Davies and Walker, 1986). This implies that other management units may be more appropriate, such as bioregions or regions defined by social networks and information flows. However, while, a regional approach may be appropriate at times, watersheds and river basins provide the most useful management area for river conservation,

because a watershed divide is the natural boundary between watershed runoff and stream flow.

We suggest that integrated watershed management (IWM) provides a model for river conservation. The increased recognition of ecosystem complexity and heightened awareness of competing uses means that natural resource management can no longer be based on single-issue management planning. An integrated approach using a full range of stakeholders not only recognizes the ecological complexity of river systems, but also acknowledges the political and institutional difficulties of planning and policy implementation (see Showers, this volume).

The use of stakeholders who view management of rivers in a watershed context is radically different from past government practices and, in some locations, studies demonstrate a reluctance to adopt the new paradigm of participatory management. The challenge remains to develop robust institutional arrangements based on participatory decision-making and stakeholder involvement, that can withstand the vagaries of government-funding cutbacks, political interference and changing public perceptions of the efficacy of IWM.

This chapter analyses IWM as a planning process, defining it, describing selected examples of its use in Australia and the United States, and discussing achievements in, and impediments to, implementing IWM.

Global Perspectives on River Conservation: Science, Policy and Practice.
Edited by P.J. Boon, B.R. Davies and G.E. Petts. © 2000 John Wiley & Sons Ltd.

Integrated watershed management

HISTORICAL PERSPECTIVE

Water resource managers both in Australia and the United States long ago recognized the need for regional approaches to land and water management. In Australia, the first efforts emerged from the growing cities of Sydney and Melbourne, which were concerned about protecting their drinking water supplies (Pigram, 1986; Burton, 1988b). As early as the 1880s, the US Government sponsored regional basin planning efforts on the Mississippi system to address flood control and navigation, and regional studies of the irrigation potential of western lands (Holmes, 1972). Resource managers promoted concepts such as 'integrated' and 'unified' river-basin management (North *et al.*, 1981). However, the focus was on the integration of economically productive uses of water such as irrigation, flood control, and to a lesser extent recreation. Projects such as the Snowy River Scheme in Australia and the Hoover Dam in the United States were viewed as engineering marvels that demonstrated the ability of humans to 'tame' the wilderness and demonstrated to the world the potential of the young nations. Opposition to these projects was limited, and the public battle between John Muir and Gifford Pinchot over the proposed Hetch Hetchy Reservoir in Yosemite National Park was an unusual challenge to the development model (Wilkinson, 1992).

In the early part of the 1900s, several severe droughts both in Australia and the United States led to dramatic losses of topsoil in agricultural lands. Increased awareness of soil erosion and the impacts it could have on farm productivity resulted in extensive efforts to promote conservation practices among farmers. Agricultural extension specialists in both countries promoted soil conservation techniques, resulting in some of the first efforts to address comprehensively land management on a regional scale (Holmes, 1972; Burton, 1988b).

Increased concern for the environment in the 1960s and 1970s marked a new era in water-resource management. The emergence of ecology as a discipline stressed a broader vision for managing bioregions and ecosystems (e.g. Odum, 1969; Risser, 1985). Ecologists highlighted the interconnectedness of ecosystems (Risser, 1985) and the problems that emerged when they were viewed in isolation (Odum, 1977, 1986). Both in the United States and Australia there were dramatic changes in environmental legislation, with new policies to reduce water pollution. The point-source controls and extensive investment in sewage treatment resulted in dramatic improvements in many watersheds. However, in the United States, efforts to control non-point-source pollution under Section 208 of the Clean Water Act had a much more limited effect. Researchers suggest a number of reasons for this, including the inability of planners to develop consensus on water quality goals (Edgmon, 1980) and 'the naiveté of those who failed to grasp the intergovernmental, interagency, and conceptual complexity of the program' (Breithaupt, 1980).

More recently, states, local government and citizens in Australia and the United States have begun to refocus their efforts on the watershed and the need to integrate not only point- and non-point-source pollution, but to integrate the range of land- and water-management activities that occur within a watershed. Learning from past problems, these approaches are using more collaborative efforts that include a range of government and non-government stakeholders.

DEFINITION

The terms 'integration' and 'watershed approach' describe many activities, but are often used differently by different people. Table 24.1 lists several watershed-based approaches and their components.

Our definition of IWM includes all of the activities listed in Table 24.1 to varying degrees, depending on the goals and objectives identified by the participating stakeholders. Several terms have been used to describe an integrated approach to natural resources management, including Total Catchment Management, Integrated Catchment Management, Integrated Environmental Management, and Ecosystem Management. While no one definition captures the concept, it can be summarized as one involving the coordinated management of land and water resources within a region (river valley or bioregion), with the objectives of conserving or rehabilitating the resource and the environment, maintaining biodiversity, minimizing land degradation, and achieving specified and agreed land- and water-management and social objectives (Hooper, 1997a).

The literature on integrated approaches reveals several common themes regarding its substance and process (Margerum, 1997; Figure 24.1). IWM is holistic, in that it addresses the range of activities in a given watershed or environmental system (Mitchell and Hollick, 1993). Gilbert (1988) points out that this requires managers to take into account the character of the ecosystem and the jurisdictional setting. The approach is also interconnective, meaning that it

Table 24.1 *Typology of watershed approaches*

Watershed terminology	Types of activities and approaches
Watershed education and information	• School water-watch programmes • Watershed information systems • Information materials and brochures
Watershed monitoring and planning	• Ambient monitoring • Watershed studies and assessments • Local and regional management planning
Watershed permitting	• Discharge permits by watershed • Watershed modelling • Compliance monitoring
Watershed-based soil conservation	• Farmer information and education • Farm planning • Restoration and rehabilitation • Incentives and demonstration projects
Watershed-based stormwater control	• Land-use planning • Stormwater management • Plan assessment and review • Site design
Integrated Watershed Management	• Goal-based approach to management • Strategic regional-scale river basin management (often across competing jurisdictions) • Combines several or all of the above approaches

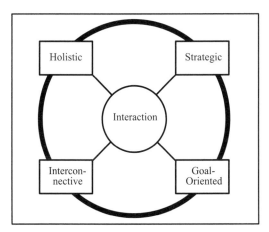

Figure 24.1 *Substance and process of integrated watershed management (Margerum, 1997)*

addresses biophysical relationships and socio-economic relationships. Mitchell and Hollick (1993) suggest that integrated management approaches should be directed to 'weighing concerns about enhancing economic development, protecting the integrity of natural systems, and satisfying social norms and values.'

For the collaborative planning process to be effective, participants must use a goal-oriented approach (Johnson and Agee, 1988; Grumbine, 1991; Mitchell and Hollick, 1993). This entails a proactive approach that envisages desired states rather than reacting to existing conditions (Jarobe, 1986). Furthermore, when

participants develop common goals, they help encourage coordination by working towards endpoints (Bührs, 1991). Stakeholders must also produce an approach to implementation that is strategic. Participants will start with a comprehensive view of the system, but they must 'narrow the focus to key aspects of systems and concentrate efforts on the tasks that achieve system goals' (Margerum and Born, 1995). Similarly, Lang (1986) emphasizes that participants must use a strategic approach that focuses planning early, biases planning towards implementation, regards planning as a management tool, and makes planning a continuous process.

Carrying out an integrated approach requires interaction across the full range of stakeholders and public involvement, not only because it is important to successful implementation, but also because stakeholders and the public are essential to building a holistic, interconnective, goal-oriented and strategic approach (Mitchell and Hollick, 1993; Margerum and Born, 1995; Showers, this volume). This interactive and collaborative approach produces a more inclusive view of management than could be created by any one organization or individual.

The implications of integrated watershed management are two-fold. First, an integrated approach is not a quick fix, but a long-term process (Margerum, 1996). It takes time for stakeholders, organizations and individuals, to come together, to reach consensus, and to begin implementing a strategy.

Second, an integrated approach requires planned change. Government organizations must change policies, allocate new resources, and change priorities; individuals must change their practices and actions. Without change, the process will simply be a process for information sharing (Walther, 1987).

Integrated watershed management in practice – selected experiences in the United States and Australia

THE AUSTRALIAN EXPERIENCE

In Australia, IWM emerged in the mid 1980s and still today many government agencies, community groups and individuals are struggling to implement the integrated approach. The term Integrated Catchment Management (ICM) is more commonly used to describe IWM in Australia and will be used in this section. A notable feature of Australian ICM is the strong emphasis placed on bottom-up/participatory decision-making using Federal Government initiatives to stimulate local action in the states and territories of Australia.

ICM is not federally funded, nor is it a national policy enshrined in legislation. It is a national policy and has emerged as an accepted philosophy by practitioners in many levels of government and by resource management organizations and by individuals, and it is strongly supported by both national and state governments. At the national level, ICM is the approach used in the National Water Quality Management Strategy, and the Murray–Darling Basin Commission, and is also an implementation procedure for the ecologically sustainable development of water resources (Hooper, 1997b). The management of natural resources at the national level is the jurisdiction of several government agencies, but the primary initiative lies within the agricultural resource and environmental management agencies Agriculture, Forests and Fisheries Australia, and Environment Australia. In 1997, these departments commenced a national programme, the National Heritage Trust (NHT), built around the concepts of ecologically sustainable development and ICM, including national programmes to encourage local initiatives in watershed management (Environment Australia and Department of Primary Industries and Energy, 1997). In these programmes, the Commonwealth Government's role is policy initiatives, with state governments and local communities initiating programmes of action for

watershed management. The NHT built on earlier achievements in natural resources management in the National Soil Conservation Programme (1983–1992), the Federal Water Resources Assistance Programme (1984–1992), the Decade of Landcare Plan (1989+) and the National Landcare Programme (1993+). The National Landcare Programme funds group action at the local, farm scale, and also was the first national programme to endorse ICM.

The first state government initiative in Australian ICM was the Total Catchment Management Policy (1986) of the New South Wales Government, following a review outlining the philosophy and approach to implement ICM in that state (Burton, 1985). Similar ICM legislation or policies followed in New South Wales (1989), Queensland (1990), Victoria (1994 and 1997) and Western Australia (1998). Tasmania, South Australia and the Northern Territory have also developed initiatives to integrate natural resources management, but a watershed focus is less apparent.

Other notable events in the recent history of similar approaches in IWM include:

- the establishment in 1986 of the Murray–Darling Basin Commission (MDBC – a four-state commission for the integrated management of natural resources for one-seventh of the Australian continent);
- the adoption in 1987 of a state-wide Salt Action Programme by Victoria;
- the introduction in 1997 of Catchment Management Authorities in Victoria;
- commonwealth Government programmes to establish water research centres, and cooperative research centres with a water focus, which research and promote ICM (such as the Centre for Water Policy Research, Armidale);
- commonwealth Government river-health initiatives which promote river conservation within an ICM context;
- national workshops on ICM in 1988 and 1997 (Australian Water Resources Council – AWRC, 1988; Laut and Taplin, 1989) which reviewed ICM achievements.

EXPERIENCE IN THE UNITED STATES

In the United States, IWM is being strongly encouraged by the US Environmental Protection Agency (US EPA): 'It is now generally recognized that the critical environmental issues facing society are so intertwined that a comprehensive, ecosystem-based approach is required. It is also recognized that solving environmental problems will depend in many cases on local governments and local citizens' (US EPA, 1994b). Federal agencies such as the Forest Service, the Army Corps of Engineers, and the Bureau of Reclamation are also promoting watershed-based approaches. Throughout the country, hundreds of community-based watershed organizations have been formed, some of which are now being supported by local, state and federal funding. Development and implementation of IWM are largely up to the states, which are using several different approaches. A number of states have initiated integrated approaches, including Connecticut, Florida, New Jersey and North Carolina (US EPA 1992, 1993, 1994a,b; Wilson, 1994). The experiences of Oregon and Wisconsin illustrate two approaches.

Oregon

A range of people and programmes have helped initiate watershed management efforts in Oregon, including local governments, the Oregon Department of Water Resources, the Oregon State Extension and non-governmental organizations. The structure and membership of these efforts varies, but they generally include state agencies, local governments, non-governmental organizations and citizens (Horton *et al.*, 1996; Natural Resources Law Centre, 1996). For example, federal and state agencies, local government, first nations' peoples, interest groups, and individuals are involved in an effort to coordinate management of the Grand Ronde River in north-eastern Oregon. In particular, the group is focused on recent declines in salmon and steelhead populations, which have been linked to decline in riparian vegetation, and sedimentation from mining, logging, grazing and wildfires (Natural Resources Law Centre, 1996).

In 1993, the Oregon Legislature passed the Watershed Health Program that encourages the formation of local watershed councils and provides funding for watershed projects. In response, new associations have formed in Coos Bay, Middle and Upper Rogue rivers, and the Umatilla River. Thus far, most of these efforts have focused on building consensus and on-the-ground restoration efforts. Funding for these efforts is provided by the state's Watershed Health Program, the Governor's Watershed Enhancement Board, and through funding supplied by participating state and federal agencies (Natural Resources Law Centre, 1996) .

Wisconsin

For over a decade, the state of Wisconsin has supported a 'Priority Watershed Program,' which addresses land and water management on a sub-watershed basis using a cooperative approach between state government, the state extension service, county personnel, local farmers, community residents, and university researchers (Born and Sonzogni, 1995; Wisconsin Department of Natural Resources, 1995). These efforts have focused primarily on non-point-source pollution from agricultural and urban areas, providing funding to individual landowners and local governments to share the costs of controlling run-off. The efforts are led by a stakeholder group composed of state and local government representatives, farmers and other concerned citizens. Coordination and implementation is carried out by county land conservation agents, who are partially funded by the state. The agents work closely with staff from the Department of Natural Resources, and often with staff from the state extension service (Wisconsin Department of Natural Resources, 1995). The state is now expanding this model by simultaneous consideration of non-point and point-source pollution through stakeholder committees. In particular, the state is interested in identifying the most cost-effective approaches to achieving watershed goals, which may include point-source polluters funding non-point-source pollution control (in lieu of additional investment in point-source treatment).

Assessment of practice in Australia and the United States

INTRODUCTION

A national survey of the effectiveness of watershed management in Australia revealed that the philosophy and products of integrated approaches are well understood (AACM and Centre for Water Policy Research, 1995). Additional research of ICM committees in Queensland and New South Wales found that many watershed committees have made considerable progress in identifying issues and developing strategies (Margerum, 1996). However, significant process problems remain in implementing integrated approaches (Centre for Water Policy Research, 1995; Hooper, 1995a,b, 1996; Margerum, 1996). The overall shortcoming of IWM in Australia and the United States is the lack of clear processes to plan, implement and evaluate IWM.

ACHIEVEMENTS AND STRENGTHS

Endorsement of the IWM philosophy

Several decades of experience in IWM in Australia and the United States confirm that the philosophy of IWM is commonly accepted and forms the basis of substantial government action. Many communities and their governments in both countries have shared information, resolved conflicts, reached consensus and identified the products which they seek from IWM activities. The acceptance of this approach has progressed beyond the rhetoric of integration. Stakeholders have redefined the role of government in natural resources management, and have enlisted the power and capability of local government, local community organizations and individuals to deliver improved outcomes. The focus of these efforts has been the arrest of land degradation and water quality improvement, using watershed management committees as the principal agents for IWM implementation.

Effectiveness of local community institutions

Local institutions have proved to be effective in watershed management. The most effective IWM process is one where local land and water management programmes have been developed by local communities who have cut through hierarchical communication barriers found in and between government agencies. IWM is more effective when a full array of major local-resource users, such as agriculture, mining, forestry and tourism are included.

WEAKNESS

Tension between bottom-up consultation and top-down policy

Many policies that govern management activities in watersheds do not consider interrelationships with other policies and ecosystem functioning. Watershed stakeholders are struggling with how bottom-up consultation and community participation is linked with top-down flows of policy and government investment. This is particularly apparent where local management actions are complicated by numerous environmental management programmes emanating from many different government agencies.

Lack of integration of economic development with ecological management

Interconnections between ecological and economic processes are often poorly understood, and IWM is rarely linked with regional economic development programmes. Addressing these issues may require:

- economic instruments, such as market mechanisms, to address sustainability questions;
- processes for considering farm business profitability in watershed planning;
- a clear definition of property rights in relation to implementing land management practices;
- Decision Support Systems to help evaluate watershed management policies and programmes. Geographic Information Systems (GIS) can also be used to incorporate regional and local economic assessments, identify the drivers of landscape and land-use changes, and assess the impacts of alternative policies.

Institutional barriers to effective coordination

There are institutional barriers at the national, state, local government and community levels that impede the implementation of IWM activities. These barriers operate both between, and within, government agencies. Few IWM efforts have been able to foster regular or systematic horizontal linkages to coordinate actions among watershed managers. Coordination is often left to *ad hoc* processes that overload the time and resources of watershed committees and coordinators.

Lack of evaluation processes

There are few, if any, formal processes for evaluating the outcomes of IWM, including a lack of economic analysis of watershed management policies and programmes. Economic analyses can be used to identify priorities and to allocate costs and benefits between the public and private sectors. This allows identification of co-financing, or cost-sharing arrangements, with private sector beneficiaries of IWM.

Failure to assign priorities

Watershed management strategies often focus on symptoms rather than causes, and on remedial actions rather than preventative measures. Watershed planning efforts that have set priorities and identified demonstrable products have increased their success in implementation. Priority ranking can identify the most critical remedial and preventative actions and focus the day-to-day activities of coordinators and watershed committees.

FUTURE DIRECTIONS

Watershed management committees throughout Australia and the United States have limited resources, very few staff, and little or no authority. Their role is to foster better environmental outcomes by improving private and public management practices. Therefore, improving IWM will require support and change by all levels of government. Chief among these changes are the needs for vertical integration of natural resource management policies and programmes, and horizontal coordination of watershed management actions.

Vertical integration

Vertical integration refers to the need to resolve the top-down objectives and organization of government agencies and programmes and the bottom-up issues and objectives identified by watershed management committees.

Addressing these conflicting objectives requires stakeholders to recognize conflicts, and the participating organizations and committees to develop processes for resolving them. A review of Australian activities recommended a strategy that would define a process for public sector investment in IWM, use a contract system to fund IWM and to account for public-sector investment, and that would develop better communication linkages between organizations. The strategy would also include a Regional Natural Resource Management Plan developed through community consultation and stakeholder groups. The plan would identify sources of funding, produce cost-sharing arrangements and identify funding priorities. A Natural Resource Coordinating Council in each state or territory would set state and regional priorities, and help to coordinate natural resource management (AACM and Centre for Water Policy Research, 1995)

Horizontal coordination

A review of watershed management in New South Wales and Queensland, found that implementation of

IWM also requires greater and coordinated integration of decision-making at the watershed level (Margerum, 1996). Plans and strategies provide a general framework for coordinated action, but they cannot anticipate changing conditions or adjust to new information and conflicts. IWM requires adaptive approaches that permit continuous information exchange and conflict resolution. The problem has been that adaptive coordination is generally left to *ad hoc* approaches that are often reactive and usually time-consuming for coordinators.

One of the keys to adaptive management is institutional mechanisms and structural changes that encourage greater information sharing and coordination among decision-makers. This is not just coordination among stakeholders on a watershed committee, but among the array of people who make decisions that affect the environment on an everyday basis. To achieve this, stakeholders must identify critical issues or objectives that require adaptive management, identify the decision-makers associated with the issue or objective, and establish new fora for information sharing, conflict resolution and coordination. For example, state agencies, local government and a port authority on the Trinity Inlet near Cairns, Australia, have developed a joint process for reviewing development permits. The organizations share data, technical staff jointly review development applications, and the organizations make a joint recommendation (Margerum, 1996). Similarly, on the upper Wisconsin River, in the United States, state and federal agencies, dam operators, local governments and interest groups have all been participating in the review of dam relicensing permits (Margerum, 1995).

INDICATORS OF SUCCESS

IWM is explicitly appealing. It suggests that if a more comprehensive set of resource issues and values are examined and used in watershed decision-making, then more effective resource management can be achieved. However, while this approach has stimulated innovation in Australian and American watershed management, there is a need to develop indicators of success and progress. Some work has already been completed by Synnott (1991), Syme *et al.* (1993) and Eigeland and Hooper (2000). Such indicators can help governments to assess whether or not their investment in IWM is successful. Indicators will also help successful watershed management committees to support their cause. Three sets of indicators are:

- a process to assess the effectiveness of IWM decision-making
- multi-dimensional indicators of changes to ecosystem health
- indicators of social and economic gains

Conclusions

Australian and US IWM is progressing, and its participative, stakeholder-driven approach is showing signs of success. We maintain that watershed management organizations must integrate their plans and actions into the decision-making structures and processes of local, state and national management organizations. These organizations must decide whether they support IWM as an organizing principle and practical management tool for natural resource management, or whether IWM will ultimately become only a forum for citizen consultation and information exchange.

Despite its shortcomings, IWM has demonstrated valuable benefits to natural resource management in watersheds. We maintain that this approach can also readily be expanded to address river conservation, provided that natural resources and environmental planning and management are linked to the watershed in which the river is located. The lessons learned from the Australian and US experiences can be applied to other countries and jurisdictions in which river conservation is being practised.

References

AACM and Centre for Water Policy Research 1995. *Enhancing the Effectiveness of Catchment Management Planning. Final Report.* AACM International, Adelaide.

Australian Water Resources Council (AWRC) 1988. *Working Papers for the National Workshop on Integrated Catchment Management.* Melbourne. Australian Water Resources Council, Canberra.

Born, S.M. and Sonzogni, W. 1995. Towards integrated environmental management: strengthening the conceptualization. *Environmental Management* 19:167–183.

Breithaupt, J. 1980. 208 and land use planning. In: Lamb, B.L. (ed.) *Water Quality Administration: A Focus on Section 208.* Ann Arbor Science, Ann Arbor, Michigan.

Bührs, T. 1991. Strategies for environmental policy coordination: the New Zealand experience. *Political Science* 43:1–29.

Burton, J.R. 1985. *Development and Implementation of Total Catchment Management Policy in New South Wales. A Background Paper.* The University of New England, Armidale.

Burton, J.R. 1988a. The environmental rationale for integrated catchment management. *Proceedings, National Workshop on Integrated Catchment Management.* Australian Water Resources Council, Conference Series No. 16, Appendix 5. Victorian Government Printing Office. Melbourne, 1–19.

Burton, J.R. 1988b. Catchment management in Australia. *Civil Engineering Transactions CE* 30:145–152.

Centre for Water Policy Research 1995. *A Review of the Hawkesbury–Nepean Catchment Management Trust.* Report to the Minister of Land and Water Conservation. CWPR, University of New England, Armidale.

Davies, B.R. and Walker, K.F. 1986. *The Ecology of River Systems. Monographiae Biologicae,* 60. Dr W. Junk, Dordrecht.

Edgmon, T.D. 1980. Is 208 planning a technical process? In: Lamb, B.L. (ed.) *Water Quality Administration: A Focus on Section 208.* Ann Arbor Science, Ann Arbor, Michigan.

Eigeland, N. and Hooper, B.P. 2000. Indicators and resource use. *Water* (in review).

Environment Australia and Department of Primary Industries and Energy 1997. *Natural Heritage Trust. A Better Environment for Australia in the 21st Century.* http://www.nht.gov.au/

Gilbert, V.C. 1988. Cooperation in ecosystem management. In: Agee, K. and Johnson, D.R. (eds) *Ecosystem Management for Parks and Wilderness.* University of Washington Press, Seattle, Washington.

Grumbine, R.E. 1991. Cooperation or conflict? Interagency relationships and the future of biodiversity for U.S. parks and forests. *Environmental Management* 15:27–37.

Holmes, B.H. 1972. *A History of Federal Water Resources Programs, 1800–1960.* Economic Research Service, US Department of Agriculture.

Hooper, B.P. 1995a. *A Scoping Paper on Water Quality Management in Malpas Dam and its Catchment. A Report to Armidale City Council.* Centre for Water Policy Research, University of New England, Armidale.

Hooper, B.P. 1995b. *Best Management Practice for Integrated Land and Water Management. Final Report on Project #M334 to the Murray–Darling Basin Commission.* Centre for Water Policy Research, University of New England, Armidale.

Hooper, B.P. 1996. *Review of the Dawson Valley Development Association. Project Report.* Centre for Water Policy Research, University of New England, Armidale.

Hooper, B.P. 1997a. Improving watershed management in Australia using an innovative integrated resources management in Australia. *ASCE Journal of Professional Issues and Practice* 123:57–61.

Hooper, B.P. 1997b. The Australian experience in sustainable water management. In: Mitchell, B. and Shrubsole, D. (eds)

Practising Sustainable Water Management: Canadian and International Experiences. Canadian Water Resources Association, Cambridge, Ontario, 236–259.

Horton, R.L., Duncan, D.J. and Prevost, M. 1996. *Citizen Directed Watershed Management: The Oregon Experience.* Proceedings of Watershed '96 Conference, http://www.epa.gov/watershed/OWOW/watershed/Proceed/horton.html

Jarobe, K.P. 1986. The structural approach to planning and policy making. In: Dluhy, M.J. and Chen, K. (eds) *Interdisciplinary Planning: A Perspective for the Future.* Center for Urban Policy Research, New Brunswick, New Jersey, 43–59.

Johnson, D.R. and Agee, J.K. 1988. Introduction to ecosystem management. In: Agee, J.K. and Johnson, D.R. (eds) *Ecosystem Management for Parks and Wilderness.* University of Washington Press, Seattle, Washington, 1–12.

Lang, R. 1986. Introduction. In: Lang, R. (ed.) *Integrated Approaches to Resource Planning and Management.* The Banff Centre, Calgary, Alberta, 1–9.

Laut, P. and Taplin, B.J. 1989. *Catchment Management in Australia in the 1980s.* CSIRO Division of Natural Resources. Divisional Report, 89/3. Canberra.

Margerum, R.D. 1995. Examining the practice of integrated environmental management: towards a conceptual model. PhD Dissertation, University of Wisconsin-Madison.

Margerum, R.D. 1996. *Integrated Environmental Management: A Framework for Practice.* Discussion Paper No. 6. Centre for Water Policy Research. University of New England, Armidale.

Margerum, R.D. 1997. Integrated approaches to environmental planning and management. *Journal of Planning Literature* 11:459–475.

Margerum, R.D. and Born, S.M. 1995. Integrated environmental management: moving from theory to practice. *Journal of Environmental Planning and Management* 38:371–391.

Mitchell, B. and Hollick, M. 1993. Integrated catchment management in Western Australia: transition from concept to implementation. *Environmental Management* 17:735–743.

Natural Resources Law Centre 1996. *The Watershed Source Book: Watershed-Based Solutions to Natural Resource Problems.* University of Colorado, Boulder.

North, R.M., Dworsky, L.B. and Allee, D.J. 1981. Editors' introductory notes and summary. In: Allee, D.J., Dworsky, L.B. and North, R.M. (eds) *Unified River Basin Management – Stage II.* American Water Resources Association, Minneapolis, 1–15.

Odum, E.P. 1969. The strategy of ecosystem development. *Science* 164:262–270.

Odum, E.P. 1977. The emergence of ecology as a new integrative discipline. *Science* 195:1289–1283.

Odum, E.P. 1986. Introductory review: perspective of ecosystem theory and application. In: Polunin, N. (ed.) *Ecosystem Theory and Application.* John Wiley, New York, 1–13.

Pigram, J.J.J. 1986. *Issues in the Management of Australia's Water Resources.* Longman Chesire, Melbourne.

Risser, P.G. 1985. Toward a holistic management perspective. *Bioscience* 35:414–418.

Syme, G., Butterworth, J.E. and Nancarrow, B.E. 1993. *National Whole Catchment Management. A Review and an Analysis of Processes.* CSIRO Division of Water Resources. Consultancy Report No. 93/30, Canberra.

Synnott, M. 1991. A review of ICM in Australia. *Land and Water Research News,* Issue No. 8. pp. 42–47.

US Environmental Protection Agency (USEPA) 1992. *The Watershed Protection Approach: Annual Report 1991.* Washington, DC.

US Environmental Protection Agency (USEPA) 1993. *Geographic Targeting: Selected State Examples.* Office of Water, Washington, DC.

US Environmental Protection Agency (USEPA) 1994a. *National Water Quality Inventory: 1992 Report to Congress.* Office of Water, Washington, DC.

US Environmental Protection Agency (USEPA) 1994b. *The Watershed Protection Approach: Statewide Basin Management.* Office of Wetlands, Oceans and Watersheds, Washington, DC.

Walther, P. 1987. Against idealistic beliefs in the problem-solving capacities of integrated resource management. *Environmental Management* 11:439–446.

Wilkinson, C.F. 1992. Crossing the next meridian: land, water and the future of the West. Island Press, Washington, DC.

Wilson, R.R. 1994. An integrated river management model: the Connecticut river management program. *Journal of Environmental Management* 41:337–348.

Wisconsin Department of Natural Resources 1995. *1994 Wisconsin Water Quality Report to Congress.* Madison, Wisconsin.

RIVERS INDEX

SUBJECT INDEX

TAXONOMIC INDEX